Lecture Notes in Mathematics

Edited by A. Dold and B. Eckmann

763

Algebraic Topology Aarhus 1978

Proceedings of a Symposium held at Aarhus, Denmark, August 7 – 12, 1978

UNIVERSITY OF AARHUS
50TH ANNIVERSARY
11 SEPTEMBER 1978

Edited by
J. L. Dupont and I. H. Madsen

Springer-Verlag
Berlin Heidelberg New York 1979

Editors

Johan Louis Dupont
Ib Henning Madsen
Matematisk Institut
Aarhus University
8000 Aarhus C
Denmark

AMS Subject Classifications (1970): 55-02, 57-02

ISBN 3-540-09721-X Springer-Verlag Berlin Heidelberg New York
ISBN 0-387-09721-X Springer-Verlag New York Heidelberg Berlin

Printing and binding: Beltz Offsetdruck, Hemsbach/Bergstr.
2141/3140-543210

P R E F A C E

In August 1978 a Symposium on Algebraic Topology was held in Aarhus. The Symposium was part of the scientific activity in connection with the 50th anniversary of Aarhus University, and was supported by the Danish Science Foundation and Aarhus University's Research Fund.

The meeting was structured as a series of plenary talks together with special sessions on Homotopy Theory, Characteristic Classes and Bordism, Algebraic K- and L-Theory, Transformation Groups and Geometry of Manifolds.

These Proceedings contain manuscripts submitted by the invited speakers and are for the most part accounts of the talks given.

We would like to thank everyone who contributed to the success of the Symposium.

Aarhus, June 1979 Johan Dupont, Ib Madsen

TABLE OF CONTENTS

TRANSFORMATION GROUPS:

GEOMETRY OF MANIFOLDS:

Decompositions of Loop Spaces and Applications to Exponents

by

F.R. Cohen, J.C. Moore, and J.A. Neisendorfer*

We shall discuss product decompositions of loop spaces and their application to exponents of homotopy groups. The fundamental definition is: A simply connected space X has exponent p^k at the prime p if p^k annihilates the p-primary component of $\Pi_q(X)$ for all q.

It is not at all obvious that there exist nontrivial simply connected finite complexes which have exponents. But, in the 1950's, I.M. James (for p=2 [9]) and H. Toda (for p odd [14]) showed that the odd dimensional sphere S^{2n+1} has exponent p^{2n} at p. Based on computational evidence, M. Barratt then conjectured that S^{2n+1} has the lesser exponent p^n at any odd prime p.

Barratt also asked the following question [2]. Given a double suspension for which the identity map has finite order, does it have an exponent? Barratt obtained bounds on the exponent of the homotopy groups. However, these bounds grew as the dimension of the homotopy group got larger. The existence of an exponent for the simplest example, a 2-connected mod p^r Moore space, remained open.

In the 1960's, B. Gray [7] showed that Barratt's exponent conjecture for S^{2n+1} was best possible. For p odd, he constructed infinitely many elements of order p^n in $\Pi_*(S^{2n+1})$. These elements stabilized mod p to elements in the image of J.

Recently, the thesis of P. Selick [12] gave a proof of the first nontrivial case of Barratt's conjecture. The sphere S^3 has exponent p at an odd prime p.

Studying the exponent problem for Moore spaces led us to new methods, valid for primes greater than 3. We began by considering Samelson products in the mod p homotopy Bockstein spectral sequence of the loops on a 2-connected mod p^r Moore space. With this machinery, we detected infinitely many elements of order p^{r+1} in the homotopy groups. Further investigation of this higher torsion gave two

* The authors were supported in part by the National Science Foundation.

theorems. First, a 2-connected mod p^r Moore space has exponent at most p^{2r+1}. Second, S^{2n+1} has exponent p^n at any prime p greater than 3.

Somewhat stronger and more geometric results are true. Suppose that S^{2n+1} has been localized at p greater than 3 and let $(\Omega^{2n+1} S^{2n+1})_o$ denote the component of the basepoint in the 2n+1 fold loop space. Then the p^n-th power map of $(\Omega^{2n+1} S^{2n+1})_o$ is null homotopic. Similarly, the p^{2n+1}-st power map of the 4-fold loop space of a 2-connected mod p^r Moore space is null homotopic. In fact, all exponent theorems we know reduce to null homotopic power maps in some iterated loop space.

We get our exponent theorems as consequences of product decompositions of loop spaces. These product decompositions are related to an old theorem of Serre [13] and also to the Hilton-Milnor theorem [10]. Recall Serre's theorem: if all spaces are localized away from 2, then ΩS^{2n+2} decomposes up to homotopy type into the product $S^{2n+1} \times \Omega S^{4n+3}$.

Serre proves his theorem as follows. The integral Pontrjagin ring of ΩS^{2n+2} is a tensor algebra on a single generator of degree 2n+1. We may interpret a tensor algebra as a universal enveloping algebra of a free lie algebra. Serre's decomposition comes from a decomposition of this universal enveloping algebra into a tensor product. He constructs a geometric realization of this algebraic decomposition by using Samelson products to construct maps.

Serre's method for decomposing loop spaces is our method also with two differences. First, his example was simple enough that he had no need to use the language of universal enveloping algebras. Second, we work with mod p^r homology and use Samelson products in mod p^r homotopy. This means that there is a Bockstein structure which must be compatible with our decompositions.

It is by way of this Bockstein structure that higher torsion in homotopy influences the construction of product decompositions for loop spaces. One such product decomposition gives a map $\Omega^2 S^{2n+1} \to S^{2n-1}$ of spaces localized at p greater than 3. This map has degree p on the bottom cell of $\Omega^2 S^{2n+1}$ and it gives a factorization of the double loop of the degree p map on $\Omega^2 S^{2n+1}$. From this factorization, the exponent theorem for spheres follows.

1. How to prove a decomposition theorem

We shall prove a product decomposition theorem for the loops on a mod p^r Moore space where the top cell is even and p is an odd prime. The Bockstein structure is an important part of the proof. Hence, this example shows how differential Lie algebras enter the proof of a product decomposition. But higher torsion plays no role so that higher order Bocksteins do not enter. Higher order Bocksteins play a role in some of the decompositions given in the next section. If we ignore Bocksteins, the technique below can be used to prove such classical theorems as Serre's decomposition of ΩS^{2n+2} localized away from 2 and the Hilton-Milnor theorem.

We will not give all details below. Those statements which are not proved here are treated in our paper [3].

Let p be a prime and let p^r: $S^m \to S^m$ denote the degree p^r map. The Moore space $S^{m-1} \cup_{p^r} e^m$, denoted $P^m(p^r)$, is the cofibre of the map p^r: $S^{m-1} \to S^{m-1}$.

Let $S^m\{p^r\}$ denote the homotopy theoretic fibre of the map p^r: $S^m \to S^m$.

Theorem. If p is an odd prime with $n > 0$, then there is a homotopy equivalence

$$S^{2n+1}\{p^r\} \times \Omega \bigvee_{k=o}^{\infty} P^{4n+2kn+3}(p^r) \to \Omega P^{2n+2}(p^r).$$

Proof: Since $P^{2n+2}(p^r)$ is the suspension $\sum P^{2n+1}(p^r)$, the Bott-Samelson theorem gives that $H_*(\Omega P^{2n+2}(p^r); Z/p^rZ)$ is the tensor algebra $T(u,v)$ with generators u and v of degrees 2n and 2n+1, respectively. The Bockstein β related to the short exact coefficient sequence $Z/p^rZ \to Z/p^{2r}Z \to Z/p^rZ$ satisfies $\beta v = u$. This Bockstein gives $T(u,v)$ the structure of a differential algebra.

The tensor algebra $T(u,v)$ is the universal enveloping algebra UL where $L=L(u,v)$, the free differential Lie algebra generated by u and v.

There is a natural map of differential Lie algebras $L \to A$ where $A = \langle u,v \rangle$, the abelian differential Lie algebra generated by u and v.

The kernel of $L \to A$ is the commutator sub-Lie algebra [L,L]. Subalgebras of free Lie algebras are free; hence [L,L] is free. It is generated by countably many generators, $y_k = \mathrm{ad}^k(u)\,[v,v]$ and $z_k = \mathrm{ad}^k(u)\,[u,v]$ with $k \geq 0$. That $\beta\, y_k = 2z_k$ follows from the derivation property of the Bockstein.

Given any short exact sequence $L_1 \to L_2 \to L_3$ of Lie algebras which are free modules, UL_2 and $UL_1 \otimes UL_3$ are isomorphic UL_1 modules. Hence, UL is isomorphic to $U[L,L] \otimes UA$ as a differential $U[L,L]$ module. This is the algebraic version of our decomposition theorem. The proof will be completed by giving maps $S^{2n+1}\{p^r\} \to \Omega P^{2n+2}(p^r)$ and $\Omega \bigvee_{k=o}^{\infty} P^{4n+2kn+3}(p^r) \to \Omega P^{2n+2}(p^r)$ such that the induced maps in mod p^r homology are isomorphisms onto U A and U[L,L], respectively. For, the loop multiplication of $\Omega P^{2n+2}(p^r)$ allows us to multiply these maps to get a map $S^{2n+1}\{p^r\} \times \Omega \bigvee_{k=o}^{\infty} P^{4n+2kn+3}(p^r) \to \Omega P^{2n+2}(p^r)$. The Künneth theorem implies that this map is a mod p^r homology isomorphism. Since these spaces are (already) localized at p, it is a homotopy equivalence.

Extending the fibration sequence $S^{2n+1}\{p^r\} \to S^{2n+1} \to S^{2n+1}$ to the left gives a principal fibration sequence $\Omega S^{2n+1} \to S^{2n+1}\{p^r\} \to S^{2n+1}$. The Serre spectral sequence of the latter shows that $H_*(S^{2n+1}\{p^r\}; Z/p^rZ)$ and $H_*(\Omega S^{2n+1}; Z/p^rZ) \otimes H_*(S^{2n+1}; Z/p^rZ)$ are isomorphic $H_*(\Omega S^{2n+1}; Z/p^rZ)$ modules. The Bockstein maps degree 2n+1 isomorphically onto degree 2n.

Since the composition $S^{2n+1} \xrightarrow{p^r} S^{2n+1} \to P^{2n+2}(p^r)$ is null, there is a homotopy commutative diagram in which the rows are fibration sequences up to homotopy.

$$\begin{array}{ccccccc}
\Omega S^{2n+1} & \to & S^{2n+1}\{p^r\} & \to & S^{2n+1} & \xrightarrow{p^r} & S^{2n+1} \\
\downarrow & & \downarrow & & \downarrow & & \downarrow \\
\Omega P^{2n+2}(p^r) & \xrightarrow{=} & \Omega P^{2n+2} & \to & * & \to & P^{2n+2}(p^r)
\end{array}$$

(* denotes a point).

The map $S^{2n+1}\{p^r\} \to \Omega P^{2n+2}(p^r)$ is equivariant with respect to the left actions of ΩS^{2n+1} on $S^{2n+1}\{p^r\}$ and $\Omega P^{2n+2}(p^r)$. It is now easy to see that the map induces an isomorphism of $H_*(S^{2n+1}\{p^r\}; Z/p^rZ)$ onto UA.

To construct the second map, it is convenient to introduce the James construction J(X)[8]. Let X be a (pointed) space. Then J(X) is the free monoid generated by the points of X subject to the single relation that the basepoint is the unit. There is an injection $X \to J(X)$ which is continuous in the natural topology on J(X). Given a topological monoid G and a map $X \to G$, there is a unique extension to a multiplicative map $J(X) \to G$. The loop space made up of loops of arbitrary length is a topological monoid. The suspension Σ: $X \to \Omega\Sigma X$ induces a multiplicative homotopy equivalence from $J(X)$ to $\Omega\Sigma X$ for connected X. Up to homotopy, $J(X)$ and $\Omega\Sigma X$ are interchangeable.

We also need some knowledge of mod p^r homotopy groups. If X is a space, $\Pi_m(X; Z/p^rZ)$ is the set of homotopy classes of maps from $P(p^r)$ to X.

If p is greater than 3 and ΩX is simply connected, then $\Pi_*(\Omega X; Z/p^rZ)$ is a differential Lie algebra. The differential is the Bockstein and the Lie bracket is the Samelson product. The Hurewicz map ϕ is a morphism of differential Lie algebras [11].

If p equals 3, the Jacobi identity for Samelson products fails. Since we do not use it, this proof is valid if p equals 3.

Let μ and ν be homotopy classes in $\Pi_*(\Omega P^{2n+2}(p^r); Z/p^rZ)$ such that $\phi(\mu) = u$ and $\phi(\nu) = v$. The iterated Samelson product $ad^k(\mu)\ [\nu,\nu]$ defines a map $P^{4n+2kn+2}(p^r) \to \Omega P^{2n+2}(p^r)$. Add these maps to get a map $\bigvee_{k=o}^{\infty} P^{4n+2kn+2}(p^r) \to \Omega P^{4n+2kn+2}(p^r)$. Extend to the James construction to get a map $\Omega \bigvee_{k=o}^{\infty} P^{4n+2kn+3}(p^r) \to \Omega P^{2n+2}(p^r)$. The induced map in mod p^r homology is an isomorphism onto $U[L,L]$ since $[L,L]$ is generated by $\phi(ad^k(\mu)[\nu,\nu])$ and their Bocksteins.

2. Decompositions of localized loop spaces

Many decomposition theorems hold only if we localize at a prime. Accordingly, we assume that all spaces are localized at a prime p in the remainder of this paper. Some spaces are localized at the start, for example, mod p^r Moore spaces or fibres of degree p^r maps, but, for example, we write S^m to denote the result of localizing the m-sphere at p.

All decompositions theorem below are decompositions of localized homotopy types. The decomposition theorems for the loop spaces of spheres, Moore spaces, and certain related spaces provide a perspective on parts of classical homotopy theory. Complete details are given in our papers [3,4,5,6].

Localized at an odd prime, even dimensional homotopy theory is expressible in terms of odd dimensional homotopy theory. To be precise, we list a result of Serre [13], a related result on the fibres of degree p^r maps, and the decomposition theorem proved in the preceding section.

Theorem 1. At an odd prime p,

a) ΩS^{2n+2} decomposes into $S^{2n+1} \times \Omega S^{4n+3}$

b) if $n > 0$, $\Omega S^{2n+2}\{p^r\}$ decomposes into $S^{2n+1}\{p^r\} \times \Omega S^{4n+3}\{p^{2r}\}$

c) if $n > 0$, $\Omega P^{2n+2}(p^r)$ decomposes into $S^{2n+1}\{p^r\} \times \Omega \bigvee_{k=0}^{\infty} P^{4n+2kn+3}(p^r)$.

(An important technical point is: In all the loop space decompositions of this section, the decomposition is given by multiplying maps of the factors.)

The summands in the bouquet of 1 c all have odd dimensional top cells. The Hilton-Milnor theorem, i.e. $\Omega\Sigma(X \vee Y)$ decomposes into the (weak) product $\Omega\Sigma X \times \prod_{i=0}^{\infty} \Omega\Sigma(X^{\wedge i} \wedge Y)$, applies to this bouquet. Furthermore, for p odd, a smash of Moore spaces $P^n(p^r) \wedge P^m(p^r)$ decomposes into the bouquet $P^{n+m}(p^r) \vee P^{n+m-1}(p^r)$. Infinite iteration of this and of 1c expresses $\Omega P^{2n+2}(p^r)$ as an infinite product of fibres $S^{2k+1}\{p^r\}$ and of loop spaces $\Omega P^{2\ell+1}(p^r)$.

Trying to understand something about the homotopy theory of mod p^r Moore spaces with odd dimensional top cells led us to consider the homotopy theoretic fibre $F^{2n+1}\{p^r\}$ of the pinch map $P^{2n+1}(p^r) \to S^{2n+1}$ which pinches the bottom cell of the Moore space to a point. It fits into the following homotopy commutative diagram in which the rows and columns are fibration sequences up to homotopy.

$$\begin{array}{ccccccc}
\Omega S^{2n+1}\{p^r\} & \to & E^{2n+1}\{p^r\} & \to & P^{2n+1}(p^r) & \to & S^{2n+1}\{p^r\} \\
\downarrow & & \downarrow & & \downarrow \| & & \downarrow \\
\Omega S^{2n+1} & \longrightarrow & F^{2n+1}\{p^r\} & \to & P^{2n+1}(p^r) & \to & S^{2n+1} \\
\downarrow \Omega p^r & & \downarrow & & \downarrow & & \downarrow p^r \\
\Omega S^{2n+1} & \underset{=}{\longrightarrow} & \Omega S^{2n+1} & \longrightarrow & * & \to & S^{2n+1}
\end{array}$$

In this diagram, * denotes a point and the space $E^{2n+1}\{p^r\}$ is the homotopy theoretic fibre of two maps.

Let C(n) be the homotopy theoretic fibre of the double suspension $\Sigma^2: S^{2n-1} \to \Omega^2 S^{2n+1}$. If $n > 1$, then for p greater than 3 or for p=3 and $r > 1$, we show in [3, 4,5] that there exists a quite complicated bouquet of Moore spaces, $P(n) = \bigvee_{\alpha} P^{n_\alpha}(p^r)$, and a space $T^{2n+1}(p^r)$ so that the following theorem is true.

Theorem 2. a) $\Omega F^{2n+1}\{p^r\}$ decomposes into $S^{2n-1} \times \prod_{k=1}^{\infty} S^{2p^k n-1}\{p^{r+1}\} \times P(n)$.

b) $\Omega P^{2n+1}(p^r)$ decomposes into $T^{2n+1}(p^r) \times P(n)$.

c) if p is greater than 3, $\Omega E^{2n+1}\{p^r\}$ decomposes into

$$C(n) \times \prod_{k=1}^{\infty} S^{2p^k n-1}\{p^{r+1}\} \times P(n).$$

The natural loop map induces an injection $\Pi_{2p^k n-2}(S^{2p^k n-1}\{p^{r+1}\})$, which is $Z/p^{r+1}Z$, into the homotopy of $\Omega P^{2n+1}(p^r)$. Higher torsion of order p^{r+1} is detected and we should expect higher order Bocksteins to enter into the proof of Theorem 2. Except for complications of this sort, the proof proceeds in the same manner as the proof given in the preceding section. The proof of 2a may be outlined as follows. The mod p homology of $\Omega P^{2n+1}\{p^r\}$ is a universal enveloping algebra of a free Lie algebra. Compute it and the mod p homology Bockstein spectral sequence. Then produce a decomposition of the universal enveloping algebra which is compatible with the Bockstein spectral sequence. Finally, give a geometric realization of the algebraic decomposition by using Samelson products. Details are in [3].

The argument in the paragraph following Theorem 1 combines with 2b to show that $\Omega P^m(p^r)$ decomposes into an infinite product of the fibres $S^{2k+1}\{p^r\}$ and the spaces $T^{2n+1}(p^r)$ provided that $m \geq 4$ and either p is greater than 3 or p is 3 and r is greater than 1. These factors are indecomposable.

Hence, to understand the homotopy theory of these Moore spaces (modulo the combinatorics of counting the factors), we must understand the factors. Both types of factors relate to spheres.

By definition, there is a fibration sequence $S^{2n+1}\{p^r\} \to S^{2n+1} \longrightarrow S^{2n+1}$. The resulting exact homotopy sequence makes plausible the fact that the homotopy groups of $S^{2n+1}\{p^r\}$ are the mod p^r homotopy groups of S^{2n+1}, i.e. $\Pi_*(S^{2n+1}\{p^r\}) = \Pi_{*+1}(S^{2n+1}; Z/p^rZ)$ if p is odd.

The spaces $T^{2n+1}(p^r)$ are not quite so close to spheres. But, with the same hypotheses as in Theorem 2, we have:

<u>Theorem 3.</u> There are fibration sequences:

a) $S^{2n-1} \times \prod_{k=1}^{\infty} S^{2p^k n-1}\{p^{r+1}\} \to T^{2n+1}(p^r) \to \Omega S^{2n+1}$

b) if p is greater than 3,

$$C(n) \times \prod_{k=1}^{\infty} S^{2p^k n-1}\{p^{r+1}\} \to T^{2n+1}(p^r) \to \Omega S^{2n+1}\{p^r\}.$$

Moore spaces with top dimensions 2 and 3 were not included in the previous discussion. The space $P^2(p^r)$ is not simply connected, its fundamental group is Z/p^rZ and an explicit construction shows that its universal cover has the homotopy type of a bouquet of p^r-1 copies of S^2.

The space $P^3(p^r)$ is less well understood, but we can prove:

Theorem 4. If p is an odd prime, then there is a space X such that $\Omega P^3(p)$ decomposes into $\Omega P^{2p+1}(p) \times X$.

Theorem 4 is reminiscent of Selick's decomposition [12]:

Theorem 5. If p is an odd prime, then there is a space Y such that $\Omega^2 S^{2p+1}\{p\}$ decomposes into $\Omega^2 S^3 < 3 > \times Y$, where $S^3 < 3 >$ is the 3-connected cover of S^3.

For p greater than 3, Selick has pointed out that Y is C(p). His argument is based on a fibration sequence $D(n) \to \Omega^2 S^{2n+1}\{p\} \to C(n)$ derived from Theorem 8.

Quite a few loop spaces decompose into products. In fact, the following is true.

Theorem 6. Let X be a space localized at p for which the mod p homology is nontrivial. Then $\Omega\Sigma^2 X$ decomposes into a product of nontrivial spaces unless p=2 and X is a sphere or p is odd and X is an odd dimensional sphere.

To give a fairly complete list of the methods applicable to exponent theory, we need only add the fibrations of James (at 2) [9] and of Toda (at an odd prime) [14]:

Theorem 7. There are fibration sequences localized at a prime p:

a) if p=2, $S^n \to \Omega S^{n+1} \to \Omega S^{2n+1}$

b) if p is odd, $J_{p-1}(S^{2n}) \to \Omega S^{2n+1} \to \Omega S^{2pn+1}$ and $S^{2n-1} \to \Omega J_{p-1}(S^{2n}) \to \Omega S^{2pn-1}$,

where $J_{p-1}(S^{2n})$ denotes the filtration of the James construction consisting of products of length less than p.

3. Factoring degree p maps

We use decompositions of loop spaces to demonstrate the existence of a two-sided homotopy inverse for the double suspension Σ^2: $S^{2n-1} \to \Omega^2 S^{2n+1}$, up to a multiple of a prime p. This result implies the exponent theorem for spheres.

Theorem 8. Localized at p greater than 3 with $n > 1$, there exists a map $\Pi: \Omega^2S^{2n+1} \to S^{2n-1}$ such that the composition $\Pi\Sigma^2$ is homotopic to $p: S^{2n-1} \to S^{2n-1}$ and the composition $\Sigma^2\Pi$ is homotopic to $\Omega^2p: \Omega^2S^{2n+1} \to \Omega^2S^{2n+1}$.

Proof: Recall the decomposition of $\Omega F^{2n+1}\{p\}$ in Theorem 2. Projection on the first factor of this decomposition is a map $\Pi_1: \Omega F^{2n+1}\{p\} \to S^{2n-1}$.

Loop the diagram of fibrations which precedes Theorem 2. It remains a diagram of fibrations. In this new diagram, there is a map $\partial: \Omega^2S^{2n+1} \to \Omega F^{2n+1}\{p\}$. Let Π be the composition $\Pi_1\partial$.

Since Σ^2 represents an element in $\Pi_{2n-1}(\Omega^2S^{2n+1})$, the exact homotopy sequence of the fibration sequence $\Omega F^{2n+1}\{p\} \to \Omega P^{2n+1}(p) \to \Omega S^{2n+1}$ shows that $\Pi\Sigma^2$ is p.

In the decomposition of $\Omega F^{2n+1}\{p\}$, all factors except S^{2n-1} are present in the decomposition of $\Omega E^{2n+1}\{p\}$, which is the homotopy theoretic fibre of $t: \Omega F^{2n+1}\{p\} \to \Omega^2S^{2n+1}$. It follows that t factors as $\Sigma^2\Pi_1$. But $\Omega^2p = t\partial = \Sigma^2\Pi_1\partial = \Sigma^2\Pi$.

4. Applications to exponents

Localized at an odd prime p, S^{2n+1} is an H-space [1]. The degree p map $p: S^{2n+1} \to S^{2n+1}$ may be defined as the p-th power map. Hence, p induces multiplication by p on homotopy groups. Also note that, if X is an H-space, the two multiplications on ΩX are homotopic.

Hence, Theorem 8 implies:

Corollary 9. If n is greater than 1 and p is greater than 3, then the kernel and cokernel of the double suspension $\Sigma^2: \Pi_*(S^{2n-1}) \to \Pi_*(\Omega^2S^{2n+1})$ are annihilated by p.

Selick has shown that S^3 has exponent p at p if p is odd [12]. Hence, Corollary 9 implies:

Corollary 10. If p is greater than 3, then S^{2n+1} has exponent p^n at p.

We need the following theorem from [3].

Theorem 11. If p is odd, then $\Omega^3S^{2n+1}\{p^r\}$ has a null homotopic p^r-th power map.

Hence, Selick's decomposition (Theorem 5 in this paper) implies that $\Omega^3S^3 < 3 >$ has a null homotopic p-th power map. Since $\Omega^3S^3 < 3 > = (\Omega^3S^3)_0$, the component of the basepoint in Ω^3S^3, a bit of work with Theorem 8 gives the following stronger forms of 9 and 10.

Theorem 12. If p is greater than 3, then the fibre C(n) of Σ^2 is an H-space with a null homotopic p-th power map.

Theorem 13. If p is greater than 3, $(\Omega^{2n+1} S^{2n+1})_o$ has a null homotopic p^n-th power map.

The James and Toda fibrations at the end of section 2 can be used to show that $(\Omega^{2n+1} S^{2n+1})_o$ has a null homotopic p^{2n}-th power map for any prime p.

Consider Moore spaces $P^m(p^r)$ with $m \geq 4$. A computation shows that no power map induces 0 in the mod p reduced homology of $\Omega P^m(p^r)$. Hence, $\Omega P^m(p^r)$ has no null homotopic power maps. However:

Theorem 14. Let $m \geq 4$. If p is greater than 3 or p=3 and $r > 1$, then $\Omega^4 P^m(p^r)$ has a null homotopic p^{2r+1}-st power map.

We do not believe that Theorem 14 is best possible. We suspect that $\Omega^2 P^m(p^r)$ has a null homotopic p^{r+1}-st power map.

Proof of Theorem 14: In the second paragraph which follows Theorem 2, we show that $\Omega P^m(p^r)$ decomposes into an infinite product of spaces $S^{2k+1}\{p^r\}$ and $T^{2\ell+1}(p^r)$. Theorem 11 says that $\Omega^3 S^{2k+1}\{p^r\}$ has a null homotopic p^r-th power map, so it suffices to show that $\Omega^3 T^{2\ell+1}(p^r)$ has a null homotopic p^{2r+1}-st power map. This is a consequence of Theorems 3,11, and 12. If $p > 3$, then 3b implies that $\Omega^3 T^{2\ell+1}(p^r)$ is the total space of a fibration where the fibre and base have null homotopic p^{r+1}-st and p^r-th power maps, respectively, Hence, it has a null homotopic p^{2r+1}-st power map. If p=3 and $r > 1$, a little more work with 3a gives the same result.

Corollary 15. Let $m \geq 4$. If $p > 3$ or p=3 and $r > 1$, then $p^{2r+1} \Pi_*(P^m(p^r)) = 0$.

5. Open problems

This section gives a list of some open problems.

(a) Does S^{2n+1} have exponent 3^n at 3? Selick proves this if n=1 and we suspect that it is true in general.

(b) Barratt and Mahowald conjecture that S^{2n+1} has exponent $2^{\phi(2n+1)}$ at 2 where $\phi(j)$ is the number of integers x, $0 < x \leq j$, such that $x \equiv 0,1,2$, or 4 mod 8.

(c) Barratt and we conjecture that $P^m(p^r)$ has exponent p^{r+1} if p is odd and $m \geq 4$.

(d) More generally, let p be any prime. Barratt conjectures that a double suspension (of a connected space) with an identity map of order p^r has exponent p^{r+1}.

(e) Even more generally, we conjecture that any simply connected finite complex with totally finite dimensional rational homotopy groups has an exponent at all primes.

(f) Is the fibre $C(n)$ of the double suspension Σ^2: $S^{2n-1} \to \Omega^2 S^{2n+1}$ a loop space? Let $D(n)$ be the fibre of the map Π: $\Omega^2 S^{2n+1} \to S^{2n-1}$ in Theorem 8. We conjecture that, if p is greater than 3, then $\Omega D(pn)$ is $C(n)$. We have verified this for n=1.

(g) Suppose p is an odd prime. Does there exist a loop space or even an H-space such that the Samelson product makes the mod p homotopy into a free Lie algebra on at least two generators?

(h) Does there exist a fibration sequence $S^{2n-1} \to B(n) \to \Omega S^{2n+1}$, localized at an odd prime p, such that the connecting homomorphism $\Pi_{2n}(\Omega S^{2n+1}) \to \Pi_{2n-1}(S^{2n-1})$ is multiplication by p? If n=p, Toda shows that one exists with $B(p) = \Omega S^3 < 3 >$.

(i) For any simply connected finite complex with nontrivial mod p homology, we conjecture that there is always an element of infinite height in the mod p Pontrjagin ring of the loop space.

Northern Illinois University and Temple University

Princeton University

Fordham University

References

1. J.F. Adams, The sphere, considered as an H-space mod p, Quart J. Math. Oxford Ser. (2), 12 (1961), 52-60.
2. M.G. Barratt, Spaces of finite characteristic, Quart, J. Math. Oxford Ser. (2), 11 (1960), 124-136.
3. F.R. Cohen, J.C. Moore, and J.A. Neisendorfer, Torsion in homotopy groups, to appear in Ann. of Math.
4. __________, The double suspension and exponents of the homotopy groups of spheres, to appear.
5. __________, Moore spaces have exponents, to appear.
6. __________, James-Hopf invariants and homology, to appear.
7. B. Gray, On the sphere of orgin of infinite families in the homotopy groups of spheres, Topology, 8 (1969), 219-232.
8. I.M. James, Reduced product spaces, Ann. of Math., 62 (1955), 170-197.
9. I.M. James, On the suspension sequence, Ann. of Math., 65 (1957), 74-107.
10. J.W. Milnor, On the construction FK, in Algebraic Topology - a student's guide by J.F. Adams, Cambridge Univ. Press, 1972.
11. J.A. Neisendorfer, Unstable homotopy theory modulo an odd prime, to appear.
12. P.S. Selick, Odd primary torsion in $\Pi_k(S^3)$, to appear in Topology.
13. J-P. Serre, Groupes d'homotopie et classes de groupes abeliens, Ann. of Math., 58 (1953), 258-294.
14. H. Toda, On the double suspension E^2, J. Inst. Polytech. Osaka City Univ. Ser. A, 7 (1956), 103-145.

ON THE BI-STABLE J-HOMOMORPHISM

K.Knapp

Introduction.

The best understood part of the stable homotopy groups of spheres is the image of the stable J-homomorphism

$$J: \pi_n(SO) \longrightarrow \pi_n^s(S^0)$$

As computed in [1] $J(\pi_n(SO))$ is a cyclic group of order r,where r is 2 if $n\equiv 0$ or 1 mod 8 (n>0), r=denominator of $B_k/4k$ (B_k is the k-th Bernoulli-number) if n=4k-1, and zero otherwise. The map J factors through an even more stable J-homomorphism

$$J': \pi_n^s(SO) \longrightarrow \pi_n^s(S^0)$$

called the bistable J-homomorphism. By applying the Hopf construction to the evaluation map $SO(m)\times S^{m-1} \longrightarrow S^{m-1}$ we get a stable map which induces J' in stable homotopy.

One reason for interest in J' is given by its geometrical interpretation if one identifies stable homotopy $\pi_n^s(X)$ with reduced framed bordism $\tilde{\Omega}_n^{fr}(X)$. Whereas im(J) consists of the elements in Ω_*^{fr} represented by the standard spheres and their various framings, im(J') consists of the elements given by twisting the framings of all framed manifolds which bound a framed manifold. To be more precise, after identifying $\pi_n^s(SO)$ with $\tilde{\Omega}_n^{fr}(SO)$ the map J' can be described as follows: An element in $\tilde{\Omega}_n^{fr}(SO)$ is given by a triple $[M,\Phi,f]$ where $f:M\rightarrow SO$ is a map and Φ a framing of M such that $[M,\Phi]=0$. The map f defines an automorphism $\tilde{f}$ of the trivial bundle $M\times \mathbb{R}^n$ (n large) and we can twist the framing Φ by composing Φ with $\tilde{f}$ to get a new framing Φ^f. The image of $[M,\Phi,f]$ under J' is then $[M,\Phi^f]$.

The stable J-homomorphism is induced in homotopy by a map

$J: SO \longrightarrow Q = \Omega^\infty S^\infty$. If J were an infinite loop map with respect to the the usual infinite loop space structures on SO and Q, this would imply $im(J') = im(J)$. But clearly the image of J' is much larger than $im(J)$. Therefore the difference between $im(J)$ and $im(J')$ is some sort of measure of the deviation of J from an infinite loop map. So one is interested in knowing $im(J')$.

G.W.Whitehead conjectured that J' is onto (in strictly positive dimensions) and this conjecture was supported by the fact that the 2-primary part of $\pi_*^s(S^0)$ is in $im(J')$ (this is a corollary of the Kahn-Priddy Theorem).

In terms of framed bordism, the statement that J' is onto would mean that for every dimension $n>0$,there exists a framed manifold (M,Φ) which bounds and gives all other elements in this dimension by twisting the framing Φ .

The purpose of this note is to prove that J' is not onto. Since complete proofs will appear in an other context, the following is only a summary.

§1. J-homomorphism and S^1-transfer

The main tools in working with J' are the e-invariant on $\pi_*^s(P_\infty\mathbb{C})$ and the S^1-transfer. Given an element $[M,\phi,f]$ in $\tilde{\Omega}_n^{fr}(P_\infty\mathbb{C}) \cong \pi_n^s(P_\infty\mathbb{C}^+)$, by pulling the universal S^1-bundle back via the map f,we get an induced S^1-bundle $(\tilde{M},\pi,M)$ over M. We can lift the framing ϕ of M to get a framing $\pi^*\phi$ on $\tilde{M}$. Then $[M,\phi,f] \longmapsto [\tilde{M},\pi^*\phi]$ defines the S^1-transfer:

$$t: \pi_n^s(P_\infty\mathbb{C}^+) \longrightarrow \pi_{n+1}^s(S^0)$$

<u>Remark</u>: If we identify $\Omega_n^{fr}(P_\infty\mathbb{C})$ with the bordism group of equivariantly framed free S^1-manifolds, then t is simply the forgetful map.

The connection with the bistable J-homomorphism is as follows: In [2] Becker and Schultz proved

<u>Theorem:</u> $$Q(S^1 \wedge P_\infty \mathbb{C}^+) \simeq F_{S^1}$$

where F_{S^1} is the limit of the spaces of S^1-equivariant self maps of spheres with free S^1-action. Because unitary maps are S^1-equivariant, we have a forgetful map $U \longrightarrow F_{S^1} \simeq \Omega^\infty S^\infty(S^1 \wedge P_\infty \mathbb{C}^+)'$. Let θ be the adjoint of this map. The stable map

$$\theta : U \longrightarrow S^1 \wedge P_\infty \mathbb{C}^+$$

fits into the commuting diagram

$$\begin{array}{ccc} \pi_n^s(S^1 \wedge P_\infty \mathbb{C}^+) & \xrightarrow{\ t\ } & \pi_n^s(S^0) \\ \uparrow \theta & \nearrow J'_U & \\ \pi_n^s(U) & & \end{array}$$

where J'_U is the complex analogue of J'. In looking at the odd-primary part only, it is clear that we can work equally well with J'_U instead of J'. So $\operatorname{im}(J')$ is contained in $\operatorname{im}(t)$.

<u>Remark:</u> Because there is a map $\omega: S^1 \wedge P_\infty \mathbb{C}^+ \longrightarrow U$ such that $J'_U \omega_* = t$, we actually have $\operatorname{im}(J') = \operatorname{im}(t)$.

The transfer t has an important property, namely t raises the degree of the filtration F^i associated with the Adams spectral sequence for BP by at least 1 ([4]). So if $x \in F^2 \pi_n^s(S^0)$ is in the image of t, it must come from $F^1 \pi_{n-1}^s(P_\infty \mathbb{C}^+)$. But filtration 1 problems can be treated by means of the e-invariant.

§2. <u>The e-invariant on $\pi_*^s(P_\infty \mathbb{C}^+)$</u>

There are several possible definitions of the e-invariant. We define e as the functional Hurewicz map using the following commuting diagram of Bockstein sequences

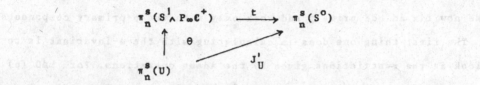

$$\begin{array}{ccccccccc}
\longrightarrow \Omega^{fr}_{2n}(X;Q) & \longrightarrow & \Omega^{fr}_{2n}(X;Q/\mathbb{Z}) & \xrightarrow{\beta} & \Omega^{fr}_{2n-1}(X) & \xrightarrow{q} \\
\downarrow & & \downarrow h & & \downarrow h_{\mathbb{Z}} & \\
K_o(X) \rightarrow K_o(X;Q) & \xrightarrow{r} & K_o(X;Q/\mathbb{Z}) & \longrightarrow & K_1(X) & \longrightarrow
\end{array}$$

For x in the kernels of q and $h_{\mathbb{Z}}$ we set

$$e(x) := r^{-1} \circ h \circ \beta^{-1}(x) \quad \in \frac{K_o(X;Q)}{K_o(X) + H_{2n}(X;Q)}$$

We now fix an odd prime p and work only with the p-primary components.

The first thing one does in calculating with the e-invariant is to look at the restrictions given by the Adams operations. For $k \not\equiv 0$ (p) we have a stable Adams operation ψ^k_n in K-homology $K_o(X;Q/\mathbb{Z}_{(p)})$. We normalize ψ^k_n in such a way, that $\psi^k_n = \text{id}$ on $\tilde{K}_o(S^{2n})_{(p)}$; then it is clear that $h(\pi^s_{2n}(X;Q/\mathbb{Z}_{(p)})) \subset \ker(\psi^k_n - 1)$. Let X be a space with torsion-free homology, then

$$\text{im}(e) \subset \ker(\psi^k_n - 1)/H_{2n}(X;Q) \subset K_o(X;Q/\mathbb{Z}_{(p)})/H_{2n}(X;Q)$$

Our principal tool for calculating these upper bounds for im(e) is the algebraic K-theory introduced by Quillen. Seymour [5] has constructed a multiplicative cohomology theory $Ad^n_k(X)_R$ (k a prime, $1/k \in R$, R a torsionfree commutative ring) which fits into the exact sequence

$$Ad^n_k(X)_R \xrightarrow{k} K^n(X;R) \xrightarrow{\psi^k_n - 1} K^n(X;R) \xrightarrow{j} Ad^{n+1}_k(X)_R$$

$Ad^o_k(X)_R$ is based on vector bundles invariant under the Adams operation ψ^k. We choose a prime k generating $(\mathbb{Z}/p^2\mathbb{Z})^*$, set $R = \mathbb{Z}_{(p)}$ and denote $Ad^n_k(X)_R$ simply by $Ad^n(X)$.

To the cohomology theory Ad there corresponds a homology theory Ad_*

and connected versions denoted by A^* and A_*. The groups A^* are isomorphic to the higher algebraic K-groups for the finite field $\mathbb{F}_k$ constructed by Quillen.

For spaces like $P_\infty\mathbb{C}$ it is then easy to see that the upper bound $\ker(\psi_n^k-1)/H_{2n}(P_\infty\mathbb{C};\mathbb{Q})$ is isomorphic to $A_{2n-1}(P_\infty\mathbb{C})$. The interpretation of $\ker(\psi_n^k-1)/H_{2n}(P_\infty\mathbb{C};\mathbb{Q})$ as a homology group allows one to determine its order fairly easily:

<u>Theorem 1</u>. Let p be an odd prime. Then the order of $\tilde{A}_{2n-1}(P_\infty\mathbb{C})$ is given by

$$\nu_p(|\tilde{A}_{2n-1}(P_\infty\mathbb{C})|) = \sum_{i=1}^{\left[\frac{n-1}{p-1}\right]} (1+\nu_p(i)) - \nu_p(n!)$$

where $\nu_p(x)$ denotes the power of p in the number x and $[x]$ the largest integer r with $r \leq x$.

We know $K^0(P_\infty\mathbb{C}) = Z[[x]]$ where $x+1 = H$ = Hopf bundle and $K_0(P_\infty\mathbb{C})$ = free Z-module generated by elements b_j dual to the powers of x, that is $\langle x^i, b_j\rangle = \delta_{ij}$. It seems to be difficult to find those linear combinations of the b_i's which are in $\ker(\psi_n^k-1)$, even though $\psi_n^k(b_j)$ is easy to calculate. But there is a simple way of constucting elements in $A_{2n-1}(P_\infty\mathbb{C})$:

Let $R(\mathbb{Z}/p^r)$ be the representation ring of the group $\mathbb{Z}/p^r$. We have

$$R(\mathbb{Z}/p^r) = \mathbb{Z}[\lambda]/(\lambda^{p^r}-1)$$

Then there is an isomorphism [3]:

$$K_0(B\mathbb{Z}/p^r;\mathbb{Q}/\mathbb{Z})_{(p)} \cong R(\mathbb{Z}/p^r) \otimes (\mathbb{Q}/\mathbb{Z})_{(p)}$$

Now representations are more closely related to the ψ^k and this allows one to compute the elements in $\ker(\psi_n^k-1)$ for $K_0(B\mathbb{Z}/p^r;\mathbb{Q}/\mathbb{Z}_{(p)})$. One can write down these elements explicitly:

$$\Big(\sum_{\substack{s=1\\ s\not\equiv 0(p)}}^{p^{r-i}} s^n \cdot \lambda^{sp^i}\Big)/p^{r-i} \in \ker(\psi_n^k - 1)$$

(This computes $Ad_{2n-1}(B\mathbb{Z}/p^r)$.)

Via the canonical map $\pi: B\mathbb{Z}/p^r \longrightarrow P_\infty\mathbb{C}$ we get elements in $A_{2n-1}(P_\infty\mathbb{C})$ and formulas for them.

Proposition 2. If n is fixed and r large, then

$$\pi_* : A_{2n-1}(B\mathbb{Z}/p^r) \longrightarrow A_{2n-1}(P_\infty\mathbb{C})$$

is surjective.

Example: $p=3$, $n=5$.

$$x = \pi_*(\lambda^3 + 2^5 \cdot \lambda^6)/3) = (b_3 - b_4 + b_5)/3 \in A_9(P_\infty\mathbb{C})$$

This element is in the image of e and its inverse image in $\pi_9^s(P_\infty\mathbb{C})_{(3)}$ is mapped under t to $\beta_1 \in \pi_{10}^s(S^o)_{(3)}$, the first element in the odd-primary cokernel of J.

This method of describing elements in $A_*(P_\infty\mathbb{C})$ gives also:

Theorem 3. Let p be an odd prime. Then the number of cyclic summands in $\tilde{A}_{2n-1}(P_\infty\mathbb{C})$ is given by

$$\left[\frac{\log((n+1)/(s+1))}{\log p}\right]$$

where $n=t(p-1)+s$ and $0 < s \leq p-1$.

The next task is to find the image of the e-invariant in this group:

Problem: What is im(e) in $A_{2n-1}(P_\infty\mathbb{C})$?

This is a difficult question and I don't know the answer in general. There is a simple method to find elements in the cokernel of

$$e : \pi_{2n-1}^s(P_\infty\mathbb{C}^+) \longrightarrow A_{2n-1}(P_\infty\mathbb{C}).$$

We look at spaces or spectra X where $\pi_*^s(X)$ is known. For example we can take the spectra $\underline{MU}$ or $\underline{BP}$. We have

$$\pi^s_{2n-1}(\underline{MU}) = \Omega^U_{2n-1} \text{ (complex bordism in odd dimension) } = 0,$$

but $A_{2n-1}(\underline{MU})$ is not zero. We can map $P_\infty\mathbb{C}$ into $\underline{MU}$ and if an element $x \in A_{2n-1}(P_\infty\mathbb{C})$ is mapped onto a nonzero element in $A_*(\underline{MU})$ it can never come from stable homotopy.

Example: $p = 5$.

$$A_{277}(P_\infty\mathbb{C}) = \mathbb{Z}/_{5^7} + \mathbb{Z}/_5$$

$$\text{im}(e) = \mathbb{Z}/_{5^6} + \mathbb{Z}/_5$$

We map $P_\infty\mathbb{C}$ as follows:

$$P_\infty\mathbb{C} \longrightarrow P_\infty\mathbb{C}/P_{118}\mathbb{C} \cong P_\infty\mathbb{C}^{119H} \xrightarrow{\ g\ } S^{2\cdot 119}{}_\wedge \underline{MU}$$

where g is induced by the classifying map of $119\cdot H$. Then it turns out that the image of the generator of $\mathbb{Z}/5^7$ is nonzero. This example will be important later. The next step is to get control over the transfer in filtration 1.

§3 The S^1-transfer on filtration 1

The transfer $t: \pi^s_n(P_\infty\mathbb{C}^+) \longrightarrow \pi^s_{n+1}(S^0)$ can be represented by a stable map

$$\tau : S^1{}_\wedge P_\infty\mathbb{C}^+ \longrightarrow S^0$$

and we can form its cofibre sequence. It turns out that the cofibre is a well known space, namely the Thomspace of the bundle $-H$ over $P_\infty\mathbb{C}$.

Theorem 4.

$$S^0 \longrightarrow P_\infty\mathbb{C}^{\mathbb{C}-H} \longrightarrow P_\infty\mathbb{C}^{\mathbb{C}} \xrightarrow{\ \tau\ } S^1$$

is a cofibre sequence.

We map this sequence into the sequence of spectra $S^o \longrightarrow \underset{\sim}{MU} \longrightarrow \overline{\underset{\sim}{MU}} = \underset{\sim}{MU}/S^o$ to get the commuting diagram

$$\begin{array}{ccccccc}
S^o & \longrightarrow & P_\infty \mathbb{C}^{\mathbb{C}-H} & \longrightarrow & P_\infty \mathbb{C}^{\mathbb{C}} & \xrightarrow{\tau} & S^1 \\
\| & & \downarrow \bar{f} & & \downarrow f & & \| \\
S^o & \longrightarrow & \underset{\sim}{MU} & \longrightarrow & \overline{\underset{\sim}{MU}} & \xrightarrow{\partial} & S^1
\end{array}$$

where $\bar{f}$ is induced by the classifying map of $-H$.

Now the boundary map

$$\partial : \pi^s_{2n-1}(\overline{\underset{\sim}{MU}}) \longrightarrow \pi^s_{2n-2}(S^o)$$

is an isomorphism, so we have described t by f_*. The point is now that ∂ also raises filtration by 1 in the Adams spectral sequence, that is to say elements in $F^2\pi^s_*(S^o)$ must come from $F^1\pi^s_*(\overline{\underset{\sim}{MU}})$ under ∂. Because ∂ is an isomorphism we have reduced the problem of computing

$$t : F^1\ \pi^s_*(P_\infty \mathbb{C}^+) \longrightarrow F^2\ \pi^s_*(S^o)$$

to a problem in filtration 1. But the behaviour of f in filtration 1 can be detected using the K-theory e-invariant (This is true only for torsion free spaces; for spaces with torsion in homology one must use the MU or BP e-invariant). Thus

<u>Theorem 5</u>. Let $x \in \pi^s_{2n-1}(P_\infty \mathbb{C})_{(p)}$. If $e(f_*(x)) \in A_{2n+1}(\overline{\underset{\sim}{MU}})$ is zero, then $t(x) \in F^3$.

For every prime $p>3$, there exists a family of elements $\beta_i \in \pi^s_*(S^o)_{(p)}$ having filtration 2 ([6]). We now apply Theorem 5 to the case of β_{p+1}:

$n = (p+1)(p^2-1)-p$

$$A_{2n-1}(P_\infty\mathbb{C}) = \mathbb{Z}/p^{p+2} + \mathbb{Z}/p$$

$$\mathrm{im}(e) = \mathbb{Z}/p^{p+1} + \mathbb{Z}/p$$

(we only need to know that the generator of $\mathbb{Z}/p^{p+2}$ is not in im(e)) as in the example for $p=5$ above. One finds $f_*(\mathrm{im}(e)) = 0$ in $A_{2n+1}(\widetilde{MU})$, but $\pi^s_{2n}(S^0)_{(p)} \cong \mathbb{Z}/p$, generated by β_{p+1} which is in F^2_{BP}, and $t(\pi^s_{2n-1}(P_\infty\mathbb{C}^+)) \subset F^3 = \{0\}$. Thus

<u>Theorem 6</u>. Let p be an odd prime, then β_{p+1} is not in the image of the bistable J-homomorphism.

If we look back to the definition of J' , we see that no manifold representing β_{p+1} can be reframed to bound.

<u>Remarks</u>:

1. $\beta_1,\ldots\ldots,\beta_p$ are in the image of J'.

2. For $p=3$ there exist only parts of the β-family and it is known that β_4 does not exist.

3. For $r\geq 2$, $p\geq 5$ the transfer maps

$$t\colon \pi^s_n(B\mathbb{Z}/p^r) \longrightarrow \pi^s_n(S^0)_{(p)} \qquad (n>0)$$

(which are onto for $r=1$ by the Kahn-Priddy theorem) also fail to be onto.

References

[1] J.F.Adams: On the groups J(X)-II, Topology 3(1965) 137-171

[2] J.C.Becker,R.E.Schultz: Equivariant function spaces and stable homotopy theory,Comment.Math.Helv.44(1974) 1-34

[3] K.Knapp: On the K-homology of classifying spaces, Math.Ann. 233(1978) 103-124

[4] K.Knapp: Rank and Adams filtration of a Lie group, Topology 17 (1978) 41-52

[5] R.M.Seymour: Vector bundles invariant under the Adams operations, Quart.J.Math.Oxford(2) 25(1974) 395-414

[6] L.Smith: On realizing complex bordism modules.Applications to the homotopy of spheres, Amer.J.Math. 92(1970) 793-856

SOME HOMOTOPY CLASSES GENERATED BY η_j

MARK MAHOWALD

In [4] an infinite family of homotopy classes $\{\eta_j\}$ were described. They represent a generator of a summand in $\pi^S_{2^j}(S^0)$. The standard name in the Adams spectral sequence for these classes is $h_1 h_j$. Their order and their composition properties are not known. Ravenel has pointed out to the author that they should belong to a "v_2-periodic" family. This note does not quite prove this but does strongly support this. We exhibit some particular compositions with η_j. Before we state the principle result we will give some notation. Let $\mu_k \in \pi_{8k+1}(S^0)$ be the element described in [1]. This class has Adams filtration $4k + 1$ and May-Tangora name $P^k h_1$. The class $\eta\mu_k$ is also non-zero for each k. Its name is $P^k h_1^2$. Let $\rho_k \in \pi_{8k+7}(S^0)$ be the generator of the image of J. The classes $\eta\rho_k$ and $\eta^2\rho_k$ are also non-zero. These classes have names $P^k c_0$ and $P^k h_1 c_0$. (The name for ρ_k is too complicated to give, see [5].)

<u>Theorem 1</u>: For all $k < 2^{j-4}$ the compositions $\mu_k\eta_j$, $\eta\mu_k\eta_j$, $\rho_k\eta_j$ and $\eta\rho_k\eta_j$ are essential. Their names in the Adams spectral sequence are $P^k h_1^2 h_j$, $P^k h_1^3 h_j$, $P^k c_0 h_j$ and $P^k c_0 h_1 h_j$.

The first two of these compositions are closely related to the classes $\beta_{2^j/4\ell+i}$, $i = 2,3$ as given by Ravenel [10], §7. The other, it would seem, correspond to $\beta_{2^{j+1}/4\ell+i}$, $i = 2,3$. These classes are not discussed in [10] but our proof seems to suggest this relationship.

If $4\eta_j = 0$ and $k < 2^{j-4}$, then one can show that $\{P^k h_1 h_j\}$ and $\{h_j a_k\}$ are surviving cycles in the Adams spectral sequence where a_k generates $\mathrm{Ext}_A^{4k,12k-1}(Z_2,Z_2)$. For each j, there are twice as many classes in the family $(P^k h_j c_0)$ also. It seems that their complete description requires information about the Kervaire invariant.

This theorem has an interesting implication in the homotopy of $P_n^k = RP^k/RP^{n-1}$ which we will describe but will not prove here. First we give some notation. We suppose N is an integer and $n = \frac{N}{2}$. Let $j: S^{2^N-1} \to P_{2^N-n}$ be a degree 1 map given by the vector field solution: Let $\epsilon(N) = [(\frac{n}{4} + 1)/4]$.

<u>Theorem 1.2</u>. The map j induces a monomorphism on the image of the J-homomorphism (at the prime 2) through dimension $8\epsilon(N) - 3$.

Since $\epsilon(N) \to \infty$ as $N \to \infty$ this gives immediately

Corollary 1.3. The map $S^{-1} \to P_{-\infty}^{\infty}$ induces a monomorphism on the image of the J-homomorphism.

For the definition of $P_{-\infty}^{\infty}$ and for remarks concerning the possible interest this corollary might have see [9].

In §2 we will recall the definition of η_j and prove some properties about them which we need. Some of these are also interesting in themselves, for example, Proposition 2.2.

In §3 some rather elementary homotopy observations are described. There is very little new in this section, but the results are not as familiar as they might be and are interesting.

In §4 some spectra which we will use are described.

Finally, in §5 the proof of the main theorem is given.

§2 Some Properties of $\{\eta_j\}$

In this section we wish to recall some properties of $\{\eta_j\}$ defined in [4]. In that paper a collection of elements was defined by the following construction. Let $S^9 \to B^2O$ be a generator. This gives a map $f: \Omega^2S^9 \to O$ and if $J: O \to \Omega^\infty S^\infty S^0$ is the J-homomorphism then $J\circ f$ can be considered as a spectruum map $J\circ f: \Omega^2S^0 \to S^0$. In [4] it is shown that for every $i \geq 3$ there is a stable map $g_i: S^{2^i} \to \Omega^2S^9$ so that Jfg_i is in the coset $\{h_ih_j\}$ in the Adams spectral sequence. The arguments there also prove that if we take $S^j \to B^2O$, $j = 5,3,2$ we get $f_j: \Omega^2S^j \to O$ and maps $g_{i,j}$ exist for $2^j \geq i - 1$ so that $Jf_jg_{i,j}$ is in the coset $\{h_1h_i\}$. Note that this coset is trivial if $i = 2$. Let $\eta_{i,j}$ be the homotopy class of $Jf_jg_{i,j}$. The following follows from definition.

<u>Proposition 2.1</u> $\eta_{i,j}^2 = 2\eta_{i+1,j}$ and $2\eta_{i+1,j} \neq 0$ if $i + 1 \geq 6$.

The following is possibly surprising.

<u>Proposition 2.2.</u> The element $\eta_{i,j}$ is not necessarily equal to $\eta_{i,j'}$ if $j \neq j'$. (They both represent h_1h_i of course.)

<u>Proof.</u> We will sketch the proof that $\eta_{5,9} - \eta_{5,5} = \{q\}$ where $q \in \mathrm{Ext}_A^{6,38}(Z_2,Z_2)$ as labeled by May and Tangora [6]. The point is that $\{q\} \in \langle \kappa^2, {}_{2\iota,\,\nu}^{\eta,2\iota,\eta} \rangle$ [3]. If we use Ω^2S^5 then at the fourth level we have a four cell complex $\Sigma^9(RP^6/RP^2) \to S^0$. The map is trivial over the 13 skeleton and this gives $\Sigma^9(RP^6/RP^4) \to S^0$ and the composite $S^{14} \to \Sigma^9(RP^6/RP^4) \to S^0$ is $\sigma\sigma + \kappa$ in Toda's notation [8]. On the other hand, the 2 level part of $\Omega^2S^9 \to S^0$ is just $\Sigma^9(RP^6/RP^4) \to S^0$ with 14 skeleton representing just $\sigma\sigma$. Now the rest of the argument should be clear.

This result indicates very clearly that elements of higher filtration could make various arguments about η_j's complicated. Of course, if there were the Kervaire invariant elements of order 2, θ_j, then $\eta'_{j+1} \in \langle \theta_j, 2\iota, \eta \rangle$. This family would have more tractable properties. Some results here could be improved when (if?) the Kervaire invariant conjecture is settled. The results we will discuss in this paper will refer to $\eta_{j,5}$ and for the balance of this note $\eta_j = \eta_{j,5}$.

§3. Some elementary homotopy theory

If $\alpha \in \pi_j(S^0)$, we let $M_\alpha = S^0 \cup_\alpha e^{j+1}$. Let $Y = M_\eta \wedge M_{2\iota}$.

Proposition 3.1. There is a map $v_1 \colon \Sigma^2 Y \to Y$ such that all iterates of v_1, $\Sigma^{2n} Y \xrightarrow{v_1} \Sigma^{2n-2} Y \xrightarrow{v_1} \cdots \longrightarrow \Sigma^4 Y \xrightarrow{v_1} \Sigma^2 Y \xrightarrow{v_1} Y$ are essential.

Proof. Let $\overline{A}_1$ be a space such that the sub algebra A_1 of A, the Steenrod algebra, generated by Sq^1 and Sq^2, acts freely with one generator. Such a space is easily constructed. It is easy to see that $\pi_j(\overline{A}_1) = 0$ if $j \neq 0, \neq 3$ and $j < 5$. Hence there is a map $Y \to \overline{A}_1$ of degree 1 on the zero cell. It is easily verified by squaring operations, that $Y \to \overline{A}_1 \to \Sigma^3 Y$ is a cofiber sequence. The connecting homomorphism is the map v_1. The composite $S^2 \to \Sigma^2 Y \to Y \to bu \wedge M_\iota$ is a generator and thus Bott periodicity shows that all iterates are essential.

Proposition 3.2 A map $f \colon \Sigma^j Y \to bo$ is essential, $j > -6$, if and only if the composite $\Sigma^{j+2} Y \xrightarrow{v_1} \Sigma^j Y \xrightarrow{f} bo$ is essential.

Proof. Suppose the composition is not essential. Then f could be extended to a map $\tilde{f}: \Sigma^j\bar{A}_1 \to bo$. Let Z be the Spanier Whitehead dual of $\bar{A}_1$. In particular suppose there are maps $S^6 \xrightarrow{a} \bar{A}_1 \wedge Z \xrightarrow{b} S^6$ so that $S^{6+j} \xrightarrow{a} \Sigma^j\bar{A}_1 \wedge Z \xrightarrow{\tilde{f}\wedge id} bo \wedge Z$ is essential if and only if $\tilde{f}$ is. The complex Z is also free over A_1. Hence $bo \wedge Z \simeq K(Z_2,0)$. Hence $\tilde{f}$ is essential only if $6 + j = 0$. The other way, of course, is immediate.

Proposition 3.3. Let $bo \xrightarrow{r} bu$ be the usual map. A map $f: \Sigma^j Y \to bo$ is essential if and only if the composite rf is essential.

Proof. Let $S^3 \xrightarrow{a} Y \wedge Y \xrightarrow{b} S^3$ be the duality maps. Note that Y is self dual. The map f is essential if and only if $S^{3+j} \to \Sigma^j Y \wedge Y \to bo \wedge Y$ is essential. But $bo \wedge Y = bu \wedge M_{2^i}$. The map $r: bo \to bu$ gives $bo \wedge Y = bu \wedge M \to bu \wedge Y = bu \wedge M \vee \Sigma^2 bu \wedge M$ as inclusion in the first factor. Thus $S^{j+3} \xrightarrow{f\wedge id\circ a} bo \wedge Y \xrightarrow{r\wedge id} bu \wedge Y$ is essential if and only if $(f \wedge id)\circ a$ is essential.

Consider the following sequence of maps

3.4 $$bo \leftarrow S^0 \xleftarrow[\eta]{} \Sigma M_{2^i} \xleftarrow[\eta\wedge\eta]{} \Sigma^2 M_{2^i} \wedge M_{2^i} \xleftarrow{\mu} \Sigma^3 Y \xleftarrow{v_1} \Sigma^5 Y \leftarrow \cdots$$

All the maps are obvious except possibly μ. It is easily seen that $Y \subset M \wedge M \wedge M$ and μ is the composite $\Sigma Y \to \Sigma M \wedge M \wedge M \xrightarrow{id\wedge\eta} M \wedge M$. Clearly the composition of the first three maps is essential.

Proposition 3.5 The composite of 3.4 $\Sigma^3 Y \to bo$ is essential.

Proof. We really wish to show that $\Sigma^3 M_{2^i} \to S^0 \to bo$ is essential. There is a space J which is the fiber of a map $bo \to \Sigma^4 b$ spin. We have

$$\begin{array}{ccc} & & \Sigma^3 b\ \mathrm{spin} \\ & & \downarrow g \\ \Sigma^3 M_{2\iota} & \xrightarrow[\eta^3]{} S^0 \to & J \\ & & \downarrow k \\ & & bo \end{array}$$

The map g cannot factor η^3 since $\pi_3\ (\Sigma^3 b\ \mathrm{spin}) \approx Z$. Hence $k\eta^3$ is essential. This is what we need.

Corollary 3.6 All the composites of 3.4 are essential.

§4. Some results about particular ring spectra.

To prove Theorem 1 we will use a ring spectrum discussed in [5] Chapter 6. Let $S^5 \xrightarrow{f} B^2O$ be a generator. Let $\Omega f: \Omega S^5 \to BO$ be the loop map. Let X_5 be the Thom spectrum of Ωf. In [5] it is proved.

Proposition 4.1 (a) X_5 is a ring spectrum with a unit.

(b) $X_5 \wedge X_5 = \Omega S^5_+ \wedge X_5 = \bigvee_{i\geq 0} \Sigma^{4i} X_5$

(c) There is a map $\ell_5: X_5 \to bo$ which is the spin bundle orientation. This map induces an epimorphism in homotopy.

In addition we need this old result. (For a proof see [5] §6.2).

Proposition 4.2 (Brayton Gray and M. G. Barratt.) The space $X_5 \wedge Y$ is also a ring spectra. If $X_2 = X_5 \wedge Y$ then $X_2 \wedge X_2 = \sum_{i\geq 0} X_2$. The space $X_5 \wedge M_\eta = X_3$ is also a ring spectrum with $X_3 \wedge X_3 = \bigvee_{i\geq 0} \Sigma^{2i} X_3$.

The following is less well known.

Proposition 4.3 There is a map $g: X_2 \to BP \wedge M_{2\iota}$ and $g_*(\pi_*(X_2))$ includes $Z_2(v_1, v_2)$.

Proof. There is a map $g': X_3 \to BP$ of degree 1 in dimension zero. Indeed, $\pi_{2k+1}(BP) = 0$ and all the cohomology of X_3 is in even dimensions. Since $(X_3)^5 = (BP)^5$, gv_1 is in the image of g'_*, hence, all its powers. In $\pi_5(X_5)$ there is a Z_2 which is represented by $S^5 \xrightarrow[\eta^{\#}]{} M_v \to X_5$. Let $h: \Sigma^5 M \to X_5$ be an extension. Since v_2 is the composite $S^6 \xrightarrow{2\iota^{\#}} S^0 u_v e^4 u_\eta e^6 \to BP$ we have the following commutative diagram

$$\begin{array}{ccccc} S^6 & \xrightarrow{v_2} & BP & & \\ \uparrow & & \uparrow g' & & \\ \Sigma^5 M^6 & \xrightarrow{h} & X^5 & \xrightarrow{i} & X_3 \end{array}$$

Since $\pi_4(X_5) = Z$ generated by a class of Adams filtration 3 h is also non-trivial in X_3. By duality $i \circ h$ is equivalent to a map $S^6 \to X_3 \wedge M = X_2$ and since the composite $\Sigma^5 M \to BP$ is essential. This is what we wish to show since again all the iterates will be non-zero.

The following result is a May-Tangora differential. It also represents part of the formula of Ravenel [7]. Using 4.2 we have the composite $X_2 \xrightarrow{i \wedge S^0} X_2 \wedge X_2 = \bigvee_{i \geq 0} \Sigma^i X_2 \to \Sigma^{2^j} X_2$ which we will call h_j.

Proposition 4.4 If $h_j: X_2 \to \Sigma^{2^j} X_2$ as above then $h_{j*} v_2^{2^j - 2} = v_1^{2^j - 2}$.

Proof. Consider the short exact sequence

$$Y = (X_2)^3 \to (X_2)^7 \xrightarrow{\rho} \Sigma^4 Y$$

The generator of $\pi_4(\Sigma^4 Y)$ under the boundary homomorphism hits the class of Adams filtration 1 in $\pi_3(Y)$. In $\pi_5(Y)$ there is no class of

Adams filtration 2 or higher so $\partial_*(v_1) = 0$. The class $v_2 \epsilon \pi_6((X_2)^7)$ satisfies $\rho_* v_2 = v_1$. This is the proposition if $j = 2$. From the construction of X_2 we get a commutative diagram

$$\begin{array}{ccc} \Sigma X_2^{7k} & \xrightarrow{\rho} & \Sigma^{4k}(X_2^{4k})^{3k} \\ \uparrow & & \uparrow \\ \Lambda^k[(X_2)^7] & \xrightarrow{\rho^k} & \Lambda^k(\Sigma^4 Y) \end{array}$$

of $4k = 2^j$ then $\rho = h_j$ and the proposition follows.

We need to calculate some more homotopy of X_5. A class $\alpha \epsilon \pi_*(M)$ is bo primary if it is essential in $M \wedge bo$.

Proposition 4.5 Let $\alpha \epsilon \pi_j(M)$ be a bo primary class. Then $S^{j+5} \xrightarrow{\alpha} \Sigma^5 M \xrightarrow{h} X_5$ is essential where h is described in 4.3.

Proof. Let $X_5 \to bo$ be the k-theory orientation. Then $X_5 \to bo \to \Sigma^4 b\,\mathrm{spin} \to \Sigma^6 M \wedge bo$ is null homotopic since we have a commutative diagram

$$\begin{array}{ccccc} X_5 & \to & bo & & \\ \downarrow & & \downarrow & \searrow & \\ \Sigma^4 bo & \to & \Sigma^4 b\,\mathrm{spin} & \to & \Sigma^6 M \wedge bo \end{array}$$

Thus $X_5 \to bo$ lifts to $X_5 \to \overline{bo}$ which is the fiber of $bo \to \Sigma^6 M \wedge bo$. The map $h\colon \Sigma^5 M \to X_5 \to \overline{bo}$ is essential and this completes the proof.

We have a commutative diagram

$$\begin{array}{ccc} X_5 & \to & \overline{bo} \\ \uparrow \bar{h} & & \uparrow \tilde{h} \\ \Sigma^5 X_5 & \to & \Sigma^5 bo \end{array}$$

where $\bar{h}$ extends the essential map $S^5 \to X_5$ and $\tilde{h}$ is the composite

$\Sigma^5 bo \to \Sigma^5 M \wedge bo \to \overline{bo}$. Let $X_5 \cup C\Sigma^5 X_6 \xrightarrow{k} \overline{bo} \cup C\Sigma^5 bo$ be the cofiber map. The following is clear from the above.

Proposition 4.6 The induced map in homotopy, k_*, is an epimorphism.

Proposition 4.7 If ρ_{k-1} generates the image of J in 8k-1 stem then the composite $S^{8k} \xrightarrow[\rho_{k-1}]{} S^1 \xrightarrow[\eta]{} X_5 \to \overline{bo} \cup C\Sigma^5 bo$ is essential.

Proof. Consider the following diagram

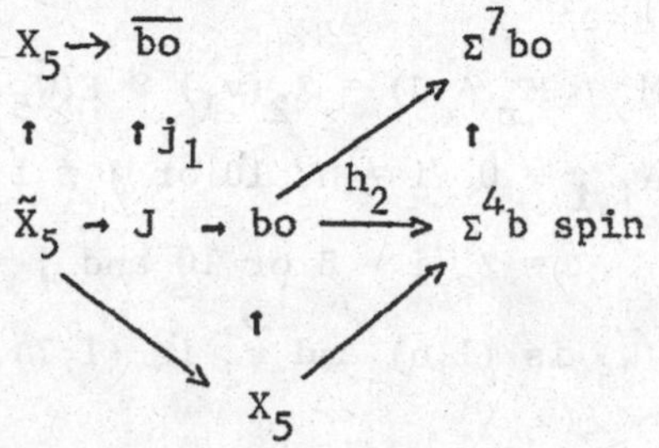

where $\tilde{X}_5$ is the fiber of ρ and the other spaces have been defined. The class in $\pi_{8k}(J)$ maps nontrivially under j_{1*}. But this class represents $\rho_k \eta$ and this is what we wished to show.

We also need to calculate the v_1-periodic homotopy of X_i. By [5] this is equivalent to calculating $\pi_*(X_i \wedge J)$ where J is the fiber of the map $bo \to \Sigma^4 b$ spin which detects Sq^4. First we have

Proposition 4.8 Let ν generate π_3. Then $\pi_*(M_\nu \wedge J)$ is the homology of a chain complex $0 \to \pi_* bo \oplus \pi_*(\Sigma^5 M_{2\iota} \wedge bo) \to \pi_*(\Sigma^7 b\text{ spin}) \to 0$.

Proof. Consider the sequence

$$M_\nu \wedge J \to M_\nu \wedge bo \to M_\nu \wedge \Sigma^4 b\text{ spin}.$$

This is equivalent to

$$M_\nu \wedge J \to bo \vee \Sigma^4 \xrightarrow{f} \Sigma^4(b\ spin) \vee \Sigma^8(b\ spin).$$

The composite $\Sigma^4 bo \to bo \vee \Sigma^4 bo \xrightarrow{f} \Sigma^4(b\ spin) \vee \Sigma^8(b\ spin) \to \Sigma^4 b\ spin$ has degree 1 in dimension 4 and such a map has fiber $\Sigma^5 M_{2\iota} \wedge bo$.

From this we have immediately

Proposition 4.9 $\pi_*(X_2 \wedge J) = Z_2(a_8) \otimes \pi_*(M_\nu \wedge M_\eta \wedge M_{2\iota} \wedge J)$. Using this we can restate Proposition 4.8 as:

Proposition 4.10 $\pi_*(M_\nu \wedge M_\eta \wedge J) = Z_2(v_1) \otimes E(\tilde{W}_5, \tilde{V}_2) \oplus A$ where A, as a graded group, is $A_{j,i} = 0 \quad i \neq 8,\ 10$ or $j \neq 1$

$= Z_2 \quad i = 8$ or 10 and $j = 1$ and

where the bidegree of $\tilde{W}_5$ is $(1,6)$ and $\tilde{v}_2$ is $(1,7)$.

Proposition 4.11 The Hurewicz image of $\pi_*(X_2)$ in $\pi_*(X_2 \wedge J)$ includes $Z_2(v_1, av_1^2) \otimes E(\tilde{W}_5, \tilde{v}_2)$.

Proof. From 4.4 we see that av_1^2 is in the image and is v_2^2. By direct calculation we get $\tilde{W}_5$ and $\tilde{v}_2$ also. Hence all the products are also in the image.

Label the classes described by 4.10 so that $v_2' \to \tilde{v}_2$, and $W_5 \to \tilde{W}$. Then if $BP\langle 2\rangle$ is the spectrum of Baas [2], we have immediately

Proposition 4.12 The map $\pi_*(X_2) \to \pi_*(BP\langle 2\rangle \wedge M_{2i}) = Z_2(v_1, v_2)$ maps $Z_2(v_1, v_2') \subset \pi_*(X_2)$ isomorphically to $Z_2(v_1, v_2)$.

Consider the following diagram

$$\begin{array}{ccccc} \Sigma^{2^j-1}(X_2)^{2^j-1} & \xrightarrow{\ell} & (X_2)^{2^j-1} & \xrightarrow{m} & (X_2)^{2^{j+1}-1} \\ j_1\downarrow & & j_2\downarrow & & j_3\downarrow \\ \Sigma^{2^j-1}(X_2)^{2^j-1}\wedge J & \xrightarrow{\ell\wedge 1} & (X_2)^{2^j-1}\wedge J & \xrightarrow{m\wedge 1} & (X_2)^{2^{j+1}-1}\wedge J \end{array}$$

where $(X_2)^k$ refers to the k-skeleton of X_2. Proposition 4.10 asserts that $\ell \wedge 1$ induces the zero map in homotopy. Also $\partial_* a^{2^{j-3}} = 1$ where ∂_* is the boundary homomorphism in the bottom row and a is the class a_8 of 4.10. Proposition 4.10 implies that if $a^{2^{j-3}} v_1^{2^{j-2}}$ is in the image of j_{3*}. Proposition 4.12 implies that if $a^{2^{j-3}} v_1^k \in \operatorname{im} j_{3*}$ then $k \geq 2^{j-2}$. Indeed, let $\tilde{v}_2^{2^{j-2}}$ be such that $j_{3*}\tilde{v}_2^{2^{j-2}} = a^{2^{j-3}} v_1^k$. Then for $p\colon X_2 \to BP\langle 2\rangle$, $p_*\tilde{v}_2^{2^{j-2}} = v_2^{2^{j-2}}$ and in $BP\langle 2\rangle$ this class is not divisible by v_1. This proves the first part of

Proposition 4.12 a) If there is a class $b \in \pi_*((X_2)^{2^{j+1}-1})$ such that $\partial_* b = v_1^k$ then $k \geq 2^{j-2}$.

b) If there is a class $C \in \pi_*((X_2)^{2^{j+1}-1})$ such that $\partial_* C = \tilde{v}_2 v_1^k$ then $k \geq 2^{j-2} + 1$.

The second part follows in an almost identical manner.

§5. Proof of Theorem 1

We are now ready to prove Theorem 1. Let S_1 be the fiber of the map $S^0 \to X_5$.

Proposition 5.1 The map $\eta_j\colon S^{2^j} \to S^0$ lifts to a map $\eta_j\colon S^{2^j} \to S_1$.

Proof. There is an obvious map $\Sigma\Omega^2 S^5 \xrightarrow{a} \Omega S^5$ and the Thom complex of $\Omega f \circ a$ is $S^0 \cup_f C\Omega^2 S^5$ and hence we have a commutative diagram

$$\begin{array}{ccccc} S_1 & \to & S^0 & \to & X_5 \\ \uparrow & & & & \\ \Omega^2S^5 & \to & S^0 & \to & S^0 \cup_F C\Omega^2 S^5 \end{array}$$

The definition of η_j gives the lifting.

Proposition 5.2 In $\pi_*(S_1)$, $\mu_k\bar{\eta}_j$, $\mu_k\eta\bar{\eta}_j$, $\rho_k\bar{\eta}_j$ and $\eta\rho_k\bar{\eta}_j$ are essential for all k.

Proof. Since $X_5 \wedge X_5 = \bigvee_{i\geq 0} \Sigma^{4i}X_5$ there is a map of $S_1 \to \Sigma^{2^j-1}X$ and $S^{2^j} \xrightarrow[\bar{\eta}_j]{} S_1\Sigma^{2^j-1}X_5 \to \Sigma^{2^j-1}\overline{bo}$ is just $S^{2^j} \to S^{2^j-1} \xrightarrow[\iota_0]{} \Sigma^{2^j-1}\overline{bo}$. The definition of μ_k and Proposition 4.7 imply $\mu_k\bar{\eta}_j$ and $\rho_k\bar{\eta}_j$ are essential. The composite $\Sigma^{2^j+8k+2}M \xrightarrow{\eta} S^{2^j+8k+1} \xrightarrow[\mu_k]{} S^{2^j} \to S_1 \to \Sigma^{2^j-1}\overline{bo}$ is essential by 3.5. If the map to S_1 on the 2^j+8k+2 cell were trivial then there would have to be an infinite summand in S_1 mapping to a generator. This does not happen for filtration reasons. Thus $\eta\mu_k\bar{\eta}_j$ is essential in S_1. The argument is the same for $\eta\rho_k\bar{\eta}_j$.

Now what remains to the proof of Theorem 1 is to show that the map $S_1 \to S^0$ does not have too many of these classes in its kernel. Note that the results of [5] assert that these classes cannot all be nonzero in S^0 since there are only a finite number of v_1-periodic generators.

Consider the sequence $\Sigma^{-1}X_5 \xrightarrow{j} S_1 \xrightarrow{\rho} S^0$. If $\mu_k\eta_j = 0$ then there is a g: $S^{2^j+8k+1} \to \Sigma^{-1}X_5$ so that $i \circ g = \mu_k\bar{\eta}_j$.

Now consider the sequence $\Sigma^{2^j-1}(X_5)^{2^j-1} \xrightarrow{\bar{\ell}} (X_5)^{2^j-1} \xrightarrow{\bar{m}} (X_5)^{2^{j+1}-1}$. The composite $\bar{m}g$ is $\mu_k\eta$. Let $M \xrightarrow{\bar{\eta}} S^0$ be the composite

$M \xrightarrow{\rho} S^1 \xrightarrow[\eta]{} S^0$ where ρ has degree 1. Then $g\tilde{\eta}$ may be essential but if $k: X_5 \to X_2$ then $kg\tilde{\eta}$ is trivial. This gives

$$\begin{array}{ccccc} \Sigma^{8k+1+2^j}(S^0 \cup_{\tilde{\eta}} CM) & & & & \\ \downarrow g\# & & & & \\ \Sigma^{-1}X_5 \cup C\Sigma^{2^j+8k+1}M & \xrightarrow{\ell} & \Sigma^{2^j-1}X_5 & \to & \Sigma^{-1}X_5 \\ \downarrow k & & \downarrow k' & & \\ \Sigma^{-1}X_2 & \xrightarrow{\ell'} & \Sigma^{2^j-1}X_2 & & \end{array}$$

By 3.3 $k'\ell g\#$ is essential. Thus $kg\#$ is essential. If $k < 2^{j-4}$ then the composite

$$\Sigma^{2^j+2^{j-1}-3}Y \xrightarrow{(2^{j-2}-4k-2)v_1} \Sigma^{8k+1+2^j}S^0 \cup_{\tilde{\eta}} CM$$

$$\to \Sigma^{-1}X_2 \to \Sigma^{2^j-1}X_2 \to \Sigma^{2^j-1}bu \wedge M$$

is $v_1^{2^{j-2}}$. Proposition 4.13 implies $k \geq 2^{j-4}$ for this to happen.

Next suppose $\pi\mu_k\eta_j = 0$. If this happens then in the sequence $\Sigma^{2^j-1}(X_5)^{2^{j-1}} \xrightarrow{\bar{\ell}} (X_5)^{2^j-1} \xrightarrow{\bar{m}} X_5^{2-1}$ the infinite cyclic class in $\pi_{2^j-1+8l+4}(\Sigma^{-1}X^{2^j-1})$ maps under J to a class divided by 2. Thus there is a map of $\Sigma^{2^j+8k+2}M \xrightarrow{g} \Sigma^{-1}X_5$ so that the composite $S^{2^j+8k+2} \to \Sigma^{2^j+8k+2}M \xrightarrow{g} \Sigma^{-1}X_5 \to S_1$ is $\pi\mu_k\bar{\eta}_j$. The map g can be extended to $\Sigma^{2^j+8k+2}Y \xrightarrow{\tilde{g}} \Sigma^{-1}X_5 \cup C\Sigma^{2^j+8k+3}M$ to give a commutative diagram

$$\begin{array}{ccc} \Sigma^{2^j+8k+2} & & \\ \downarrow \tilde{g} & & \\ \Sigma^{-1}X_5 \cup C\Sigma^{2^j+8k+3} & \xrightarrow{\ell'} & \Sigma^{2^j-1}X_5 \\ \downarrow k & & \downarrow k' \\ \Sigma^{-1}X_2 & \longrightarrow & \Sigma^{2^j-1}X_2 \end{array}$$

and $k'\ell'\tilde{g}$ is essential. The argument is finished as above.

The argument concerning $\rho_k\eta_j$ is similar. We use 4.13.b and note that by 4.7 the composite $S^{8k-1} \to \Sigma^{2^j-1}X_5$ can be extended to a map $\Sigma^{8k-2+2^j}Y \to \Sigma^{2^j-1}X_5 \to \Sigma^{-1}BP\langle 2\rangle$ which is essential and represents $v_2v_1^{4k-2}$. Thus if the composite $\Sigma^{8k-2}Y \to \Sigma^{2^j-1}X_5 \to (X_5)^{2^j-1}$ is zero then $\Sigma^{2^j+2^{j-1}+2}Y \xrightarrow{v_1^q} \Sigma^{8k-2+2^j}Y \to \Sigma^{-1}X_5 \to \Sigma^{-1}BP\langle 2\rangle$, where $q = 2^{j-2}-4k+2$, represents $v^{2^{j-2}+1}$ and since this class is not divisible by v_1, we again are finished. $\eta\rho_k\eta_j$ is handled similarly.

REFERENCES

[1] J. F. Adams, On the groups J(X)-IV, Topology 5 (1966), 21-71.

[2] N. A. Bass, Bordism theories with singularities, Proc. of Aarhus Conference (1970).

[3] M. G. Barratt, M. E. Mahowald and M. C. Tangora, Some differentials in the Adams spectral sequence II. Topology 9 (1970), 309-316.

[4] M. E. Mahowald, A new infinite family in ${}_2\pi_*^s$, Topology 6 (1977), 249-256.

[5] M. E. Mahowald and A. Unell, Lectures on Bott periodicity in stable and unstable homotopy at the prime 2, submitted to Springer Lecture Notes Series.

[6] M. E. Mahowald and M. Tangora, Some differentials in the Adams spectral sequence, I, Topology 6 (1967), 349-369.

[7] D. Ravenel, The structure of BP_*BP modulo an invariant prime ideal, Topology 15 (1977), 149-153.

[8] H. Toda, Composition methods in homotopy groups of spheres, Princeton University Press, 1962.

[9] J. F. Adams, Operations of the Nth kind in K-theory and what we don't know about RP^∞. Sym. on Algebraic Topology, Oxford Cambridge Press (1974), 1-5.

[10] D. Ravenel, A novice's guide to the Adams Novikov spectral sequence, Springer Lecture Notes # 658, p. 404-475.

APPLICATIONS AND GENERALIZATIONS OF THE APPROXIMATION THEOREM

by J. P. May

In its basic form, the approximation theorem referred to provides simple combinatorial models for spaces $\Omega^n\Sigma^n X$, where X is a connected based space. The first such result was given by James [26], who showed that $\Omega\Sigma X$ is equivalent to the James construction MX. The unpublished preprint form of Dyer and Lashof's paper [25] gave an approximation to $QX = \varinjlim \Omega^n\Sigma^n X$, and Milgram [41] gave a cellular model for $\Omega^n\Sigma^n X$ for all finite n.

Starting from Boardman and Vogt's spaces $\mathcal{C}_{n,j}$ of j-tuples of little n-cubes [5], Dold and Thom's treatment of the infinite symmetric product NX in terms of quasifibrations [24], and the category theorists' comparison between finitary algebraic theories and monads (as for example in Beck [4]), I gave a new approximation $C_n X$ to $\Omega^n\Sigma^n X$ in [36]. This model has proven most useful for practical calculational purposes when $n > 1$, and it is its applications and generalizations that I wish to discuss here. This will be a survey of work by various people, and I would like to mention that I have also given a survey of other recent developments in iterated loop space theory in [39], updating my 1976 summary [38].

The first section will give background, mention miscellaneous relevant work, and discuss generalizations, notably Caruso and Waner's recent homotopical approximation to $\Omega^n\Sigma^n X$ for general non-connected spaces X [9,11]. The second and third sections will outline the two main lines of applications. Both are based on certain stable splittings of $C_n X$, due originally to Snaith [47]. One line, initiated by Mahowald [33] and with other major contributors Brown and Peterson [7,8] and Ralph Cohen [21], is primarily concerned with a detailed analysis of the pieces in the resulting splitting of $\Omega^2 S^q$ and leads to new infinite families of elements in the stable stems. The other line, primarily due to Fred Cohen, Taylor, and myself [16-19] but also contributed to by Caruso [10] and Koschorke and Sanderson [30], is based on a detailed analysis of the splitting maps and their homotopical implications and leads to an unstable form of the Kahn-Priddy theorem, among various other things. These lines, and thus sections 2 and 3, are essentially independent of each other.

§1. Background and generalizations

The construction of the approximating spaces is naively simple. Suppose given a collection of Σ_j-spaces $\mathcal{C}_j$ with suitable degeneracy operators $\sigma_i : \mathcal{C}_{j-1} \to \mathcal{C}_j$, $1 \leq i \leq j$. (Here Σ_j is the $j^{\underline{th}}$ symmetric group.) Given a space X with (nondegenerate) basepoint $*$, construct a space $CX = \coprod \mathcal{C}_j \times_{\Sigma_j} X^j/(\approx)$, where the equivalence relation is generated by

$$(c, x_1, \ldots, x_j) \approx (c\sigma_i, x_1, \ldots, x_{i-1}, x_{i+1}, \ldots, x_j) \quad \text{if } x_i = * .$$

See [16, §§1, 2] for details, examples, naturality properties, etc. If each $\mathcal{C}_j = \Sigma_j$, the resulting space is MX. If each $\mathcal{C}_j$ is a point, the resulting space is NX.

We shall largely be concerned with examples $C(Y,X)$ obtained from the configuration spaces $\mathcal{C}_j(Y) = F(Y,j)$ of j-tuples of distinct points of Y, and we let $B(Y,j)$ be the orbit (braid) space $F(Y,j)/\Sigma_j$. $B(R^2,j) = K(B_j, 1)$, where B_j is Artin's group of j-stranded braids. Following Koschorke and Sanderson [30], we think of $C(Y,X)$ as the space of pairs (L, λ), where L is a finite (unordered) subset of Y and $\lambda : L \to X$ is a function. Here we impose the equivalence relation generated by

$$(L, \lambda) \approx (L - \{y\}, \lambda \,|\, L - \{y\}) \quad \text{if } \lambda(y) = * .$$

The crucial example $C_n X$ may be described similarly, with the sets L taken to have affine embeddings $I^n \to I^n$ with disjoint interiors as their elements; see [36, §4]. We think of the interior of I^n as R^n, by abuse, and obtain a natural homotopy equivalence

$$g : C_n X \to C(R^n, X)$$

by restriction of little n-cubes to their center points. (The proof that g is an equivalence in [36, 4.8] is not quite right; it is corrected in [14, p. 485] and also in [30].)

Define $\alpha_n : C_n X \to \Omega^n \Sigma^n X$ by letting $\alpha_n(L, \lambda) : S^n \to X \wedge S^n$ be specified on points $s \in S^n = I^n/\partial I^n$ by

$$\alpha_n(L,\lambda)(s) = \begin{cases} \lambda(c) \wedge c^{-1}(s) & \text{if } s \in \operatorname{Im} c \text{ for } c \in L \\ * & \text{if } s \notin \operatorname{Im} c \text{ for } c \in L . \end{cases}$$

The approximation theorem [36, 6.1] asserts that α_n is a weak equivalence if X is connected. Segal [45] and Cohen [14] later proved that α_n is a group completion in the general, non-connected, case. This means that $\pi_0 \Omega^n \Sigma^n X$ is the universal group associated to the monoid $\pi_0 C_n X$ (which is easy [36, 8.14]) and that $H_*\Omega^n\Sigma^n X$ is obtained from the Pontryagin ring $H_* C_n X$ by localizing at its submonoid $\pi_0 C_n X$. Of course, the same conclusions hold for $\alpha_n g^{-1}: C(R^n, X) \to \Omega^n\Sigma^n X$.

In the case $X = S^0$, McDuff [40] gave another proof of the group completion property, viewing it as a special case of a general homological relationship between $C(M, S^0)$ and the space of sections of the tangent sphere bundle of M for suitable manifolds M. She also gave a construction $C^{\pm}(M, S^0)$ of pairs of finite sets in M, with points thought of as positive and negative particles with suitable annihilation properties. While this construction is of some interest and yields a homotopical approximation to the space of sections of a bundle, the bundle in question is not the tangent sphere bundle of M. In particular, $C^{\pm}(R^n, S^0)$ fails to be a homotopical approximation to $\Omega^n S^n$.

Various other people have tried to obtain homotopical (rather than merely homological) approximations to $\Omega^n\Sigma^n X$ for general non-connected spaces X. The problem is quite delicate. It is very easy to give intuitive arguments for plausible candidates but very hard to pin down correct details. Such an approximation has recently been obtained by Caruso and Waner [11]. They construct a model $\widetilde{C}_n X$ for $\Omega^n\Sigma^n X$ by use of partial little cubes $c' \times c'': [a,b] \times I^{n-1} \to I^n$, where $c': [a,b] \to I$ is a linear map (increasing or decreasing) and $c'': I^{n-1} \to I^{n-1}$ is an affine little (n-1)-cube. These partial little cubes are required to appear in closed configurations, which may be thought of as piecewise linear maps $I^n \to I^n$ given by a piecewise linear path in the first coordinate and a single affine embedding in the remaining coordinates. The labels in X of the component partial little cubes of a closed configuration are all required to be the same.

Caruso [9] has generalized both this result and the original approximation theorem by proving that $\Omega^n C(Y, \Sigma^n X)$ is a group completion of $C(Y \times R^n, X)$ and obtaining a homotopical approximation $\widetilde{C}_n(Y, X)$ to $\Omega^n C(Y, \Sigma^n X)$. The case when Y is a point reduces to the earlier approximations, and the use of general Y allows a quick inductive reduction of the entire result to the case n = 1. Further generalizations, based on a combination of the ideas of Caruso and McDuff, are in the works.

Related to these ideas are Cohen and Taylor's extensive calculations of $H_* C(M, X)$ for certain manifolds M [20]. Their arguments work best when $M = N \times R^n$ for some positive n. In particular, they give complete information on the rational homology of $C(M, X)$ for such M, these calculations having direct implications for Gelfand-Fuks cohomology.

There is also an equivariant generalization of the approximation theorem, the present status of which is discussed in [39, §5].

However, we shall mainly concentrate on the specific approximations

$$C(R^n, X) \xleftarrow{g} C_n X \xrightarrow{\alpha_n} \Omega^n\Sigma^n X$$

and their applications in the rest of this paper.

Perhaps I should first explain just why these approximations are so useful a tool. One reason is that $C_n X$ has internal structure faithfully reflecting that of iterated loop spaces. Much of this is captured by the assertion that C_n is a monad and α_n is a map of monads [36, 5.2]. Less cryptically, the three displayed spaces all have actions by the little cubes operad $\mathcal{C}_n$ and g and α_n are both $\mathcal{C}_n$-maps [16, 6.2 and 36, 5.2]. In particular, they are H-maps, but this is only a fragment of the full structure preserved. Further, there are smash and composition products

$$C_m X \wedge C_n Y \to C_{m+n}(X \wedge Y) \quad \text{and} \quad C_n X \times C_n S^0 \to C_n X$$

which are carried by the α_n to the standard smash and composition products for iterated loop spaces [36, §8]. There is a similar and compatible smash product on the $C(R^n, X)$, but there is no precise point set level (as opposed to homotopical) composition product on the $C(R^n, X)$.

By specialization to $X = S^0$, $\alpha_n : C_n S^0 \to \Omega^n S^n$ is a map of topological monoids. As Cohen has recently observed [13], this leads to approximations for the localizations of the classifying spaces BSF(n) for S^n-fibrations (with section). Indeed, $C_n S^0 = \coprod_{q\geq 0} \mathcal{C}_{n,q}/\Sigma_q$ and $\Omega^n S^n = \coprod_{q\in Z} \Omega^n_q S^n$, where $\Omega^n_q S^n$ consists of the maps of degree q. If p is any prime, then the map of classifying spaces

$$B\alpha_n : B(\coprod_{i\geq 0} \mathcal{C}_{n,p^i}/\Sigma_{p^i}) \to B(\coprod_{i\geq 0} \Omega^n_{p^i} S^n)$$

is an equivalence, and the universal cover of the target is the localization of BSF(n) away from p. When $n = \infty$, this result is due to Tornehave and is a special case of a general phenomenon [37, VII §5].

A key reason for the usefulness of the approximation theorem is that spaces of the general form CX come with an evident natural filtration. The successive (and equivalent) quotients of $C(R^n, X)$ and $C_n X$ are

$$D_q(R^n, X) = F(R^n, q)^+ \wedge_{\Sigma_q} X^{[q]} \quad \text{and} \quad D_{n,q} X = \mathcal{C}^+_{n,q} \wedge_{\Sigma_q} X^{[q]},$$

where $X^{[q]}$ denotes the q-fold smash power of X. Just as the simplicial version of the James construction admits the splitting $\Sigma MX \simeq \bigvee_q \Sigma X^{[q]}$ found by Milnor [42], so Kahn in 1972 proved that Barratt's simplicial model [2] for QX splits stably as the wedge of its filtration quotients; Kahn has just recently published a proof [27], a different argument having been given by Barratt and Eccles [3]. When word of Kahn's splitting reached Cambridge, where I was lecturing on the approximation theorem, Snaith [47] worked out a corresponding stable splitting

$$\Sigma^{\infty} C_n X \simeq \bigvee_{q \geq 1} \Sigma^{\infty} D_{n,q} X ,$$

where Σ^{∞} is the stabilization functor from spaces to spectra (denoted Q_{∞} in all my earlier papers). New proofs of such splittings by Cohen, Taylor, and myself [16] are the starting point of the work discussed in section 3. Incidentally, by work of Kirley [29], these splittings for $n \geq 2$ cannot be realized after any finite number of suspensions (see also [16, 5.10]).

There are two points of view on these splittings. One can either ignore how they were obtained and concentrate on analyzing the pieces or one can concentrate on the splitting maps and see what kind of extra information they yield. These two viewpoints are taken respectively in the following two sections.

The crucial reason for the usefulness of the approximation theorem is that we have very good homological understanding of the filtered spaces $C_n X$. Historical background and complete calculations of $H_* QX$ and $H_* C_{\infty} X$ (including $X = S^0$, when the latter is $\sum_{q \geq 0} H_* B\Sigma_q$) are in [14, I]. Cohen [14, III] has given complete calculations of $H_* \Omega^n \Sigma^n X$ and $H_* C_n X$. (Some minor corrections are in [12, App] and also in [7].) Here "complete" means as Hopf algebras over the Steenrod algebra, with full information on all relevant homology operations. Since these operations are nicely related to the geometric filtration, complete calculations of all $H_* D_{n,q} X$ drop out. Here homology is understood to be taken mod p for some prime p, but we also give complete information on the Bockstein spectral sequences of all spaces in sight.

A major drawback to these calculations is that they give inductive formulae for the Steenrod operations, but not a global picture. One wants to know $H^* D_{n,q} X$ as a module over the Steenrod algebra A. The solution to this dualization problem for $n = 2$ and X a sphere is basic to the work of the next section, and we shall also say what little is known when $n > 2$. Before turning to this, however, I should mention the related work of Wellington [51]. He has solved the analogous dualization problem for the algebra structure, giving a precise global description of $H^* \Omega^n \Sigma^n X$ for all connected X (or all X if $p = 2$; corrections of [51] are needed when $p > 2$.) He has also studied the problem of determining the A-annihilated primitive elements in $H_*\Omega^n\Sigma^n X$. More is said about this in [39, §4] (but the description there should have been restricted to connected X).

§2. The spaces $D_q(R^n, S^r)$ and the Brown-Gitler spectra

The work discussed in this section began with and was inspired by Mahowald's brilliant paper [33]. I shall reverse historical order by first discussing the work of Brown and Peterson [7,8] and Ralph Cohen [21] on the structure of the spaces $D_q(R^n,S^r) \simeq D_{n,q}S^r$ and then briefly explaining the use of this analysis for the detection of elements of the stable stems. I am very grateful to Cohen for lucid explanations of some of this material. (In case anyone has not yet noticed, it is to be emphasized that there are two different Cohens at work in this area.)

Let $\zeta_{n,q}$ be the q-plane bundle

$$F(R^n, q) \times_{\Sigma_q} (R^1)^q \to B(R^n, q).$$

With Thom spaces of vector bundles defined by one-point compactification of fibres followed by identification of all points at infinity, it is obvious that the Thom space of $\zeta_{n,q}$ is precisely $D_q(R^n, S^1)$. Replacing R^1 by R^r in this construction, the resulting bundle is the r-fold Whitney sum of $\zeta_{n,q}$ with itself and the resulting Thom space is $D_q(R^n, S^r)$. Here $D_1(R^n, S^r) \simeq S^r$ and $D_q(R^n, S^r)$ is (rq-1)-connected. Let $j_{n,q}$ denote the order of $\zeta_{n,q}$ (or better, of its associated S^q-fibration). We have the following evident periodicity (see e.g. [37, III §1]).

__Lemma 2.1.__ $D_q(R^n, S^{r+j_{n,q}})$ is equivalent to $\Sigma^{qj_{n,q}} D_q(R^n, S^r)$.

Thus the first problem in analyzing the spaces $D_q(R^n, S^r)$ is to determine the numbers $j_{n,q}$. The following lemma summarizes what is presently known. Let $\nu_p(j)$ denote the p-order of j (the exponent of p in j).

__Theorem 2.2.__ (i) $j_{2,q} = 2$ for all $q \geq 2$.

(ii) $j_{n,2} = 2^{\phi(n-1)}$, where $\phi(n-1)$ is the vector fields number (namely the number of $i \equiv 0, 1, 2, 4 \mod 8$ with $0 < i < n$).

(iii) $j_{n,q}$ divides $j_{n,q+1}$ (and, trivially, $j_{n,q}$ divides $j_{n+1,q}$).

(iv) For an odd prime p, $\nu_p(j_{n,q}) = 0$ for $q < p$ and $\nu_p(j_{n,p}) = [n-1/2]$.

(v) For any prime p, $\nu_p(j_{n,q}) = \nu_p(j_{n,p^i})$ if $p^i \le q < p^{i+1}$.

(vi) $j_{4,4} = 12$.

None of these is very hard. Part (i) was first proven by Cohen, Mahowald, and Milgram by use of various results of mine in infinite loop space theory, but Brown later found the trivial trivialization of $2\zeta_{2,q} = \zeta_{2,2q}$ displayed in [15]. For (ii), $B(R^n, 2) \simeq RP^{n-1}$ and $\zeta_{n,2}$ is the canonical bundle $\eta \oplus \varepsilon$. Parts (iii)-(v) are in Yang [52] and were also proven by Kuhn. Part (vi) is an unpublished result of F. Cohen. It remains to determine the numbers $\nu_p(j_{n,p^i})$ for $i \ge 2$ and $n \ge 3$, and this is an interesting and apparently difficult problem.

In connection with this, the only general work on the K-theory of spaces $\Omega^n\Sigma^n X$ for $1 < n < \infty$ that I am aware of is the computation by Saitoti [43] and Snaith [48] of $K_*(\Omega^2\Sigma^3 X; Z_2)$ for X a finite torsion-free CW-complex. However, there is work in progress by Kuhn.

As would be expected from the lemma, much more is known about the spaces $D_q(R^n, S^r)$ when $n = 2$ than when $n > 2$. Before restricting to $n = 2$, however, we summarize the results of Brown and Peterson [8] in the general case. Their main result gives the following splitting of certain of the $D_q(R^n, S^r)$. It is proven by using (ii) of Theorem 2.2, the Thom construction, and various structure maps from [36] to write down explicit splitting maps and then using F. Cohen's calculations in [14, III and IV] to check that they do indeed produce a splitting. We adopt the convention that $D_0(R^n, X) = S^0$.

<u>Theorem 2.3</u>. Let $t \ge 1$ (except that $t > 1$ if $n = 2, 4,$ or 8). Then $D_q(R^n, S^{2^{\phi(n-1)}t - n})$ is homotopy equivalent to

$$\bigvee_{i=0}^{[q/2]} D_{q-2i}(R^{n-1}, S^{2^{\phi(n-1)}t-n}) \wedge D_i(R^n, S^{2^{\phi(n-1)+1}t-n-1}).$$

As usual, $[m]$ denotes the greatest integer $\le m$. Brown and Peterson [8] also observe that the Nishida relations imply the following 2-primary cohomological periodicity.

<u>Lemma 2.4</u>. Let $\psi(n-1)$ be defined by $2^{\psi(n-1)-1} < n \le 2^{\psi(n-1)}$. Then

$$\tilde{H}^* D_q(R^n, S^{2^{\psi(n-1)}+r}) \cong \Sigma^{q2^{\psi(n-1)}} \tilde{H}^* D_q(R^n, S^r).$$

Since one can add a multiple $2^{\psi(n-1)}s$ to any $2^{\psi(n-1)}t-n$ so as to reach a number $2^{\phi(n-1)}u-n$, the previous lemma and theorem have the following consequence.

Corollary 2.5. As a module over the mod 2 Steenrod algebra, $\widetilde{H}^* D_q(R^n, S^{2^{\psi(n-1)}t-n})$ is isomorphic to

$$\sum_{i=0}^{[q/2]} \widetilde{H}^* D_{q-2i}(R^{n-1}, S^{2^{\psi(n-1)}t-n}) \otimes \widetilde{H}^* D_i(R^n, S^{2^{\psi(n-1)}t-n-1}) .$$

Brown and Peterson [8] note one further cohomological splitting.

Proposition 2.6. As a module over the mod 2 Steenrod algebra,

$$\widetilde{H}^* D_q(R^n, S^{2^{\psi(n-1)}t}) \cong \sum_{i=0}^{[q/2]} \Sigma^{(q-2i)2^{\psi(n-1)}t} K_i^*$$

where $K_0^* = Z_2$ and K_i^*, $i > 0$, is dual to the sub A-module of $\widetilde{H}_* D_{2i}(R^n, S^{2^{\psi(n-1)}t})$ spanned by all monomials in the $Q^I(\iota)$ not divisible by ι (where ι is the fundamental class of $S^{2^{\psi(n-1)}t}$ and the suspensions are realized by multiplication with ι^{q-2i}).

These last two cohomological splittings may or may not be realizable geometrically. Brown and Peterson conjecture that they exhaust the 2-primary possibilities in the sense that $\widetilde{H}^* D_q(R^n, S^r)$ has no non-trivial direct summands as an A-module unless $r \equiv 0$ or $r \equiv -n \mod 2^{\psi(n-1)}$. They point out explicitly at the end of [8] that, at least when $n = 3$, there can be finer splittings than those displayed when the specified congruences are satisfied. The analysis is not yet complete and there remains much work to be done. In particular, virtually nothing is known about the explicit global A-module structure of the indecomposable summands when $n \geq 3$.

The splittings of Cohen, Taylor, and myself in [17] (see Theorem 3.7 below) together with Lemma 2.1 and an easy homological inspection (compare [17, 3.3]) imply the following analog of Theorem 2.3.

Theorem 2.7. $D_q(R^n, S^{j_{n,q}t})$ is stably equivalent to

$$\Sigma^{j_{n,q}tq}(S^0 \vee \bigvee_{i=2}^{[q/2]} B(R^n, 2i)/B(R^n, 2i-1)) .$$

Specializing now to the case $n = 2$, note that Lemma 2.1 and Theorem 2.2(i) imply that

$$(1) \quad D_q(R^2, S^{2r+1}) \simeq \Sigma^{2qr} D_q(R^2, S^1) \quad \text{and} \quad D_q(R^2, S^{2r}) \simeq \Sigma^{2qr} D_q(R^2, S^0) .$$

Here $D_q(R^2, S^0) = B(R^2, q)^+$. (The disjoint basepoint was omitted in [15, p. 226].) Since $\phi(1) = 1$ and $D_q(R^1, S^t) \simeq S^{tq}$, Theorem 2.3 implies that

$$D_q(R^2, S^{2t}) \simeq \bigvee_{i=0}^{[q/2]} \Sigma^{2t(q-2i)} D_i(R^2, S^{4t+1}), \quad t \geq 1 .$$

Setting $t = 1$ and combining with (1), we find the splitting

$$(2) \qquad \Sigma^{2q} D_q(R^2, S^0) \simeq S^{2q} \vee (\bigvee_{i=1}^{[q/2]} \Sigma^{2q} D_i(R^2, S^1)) .$$

This splitting is also immediate from Theorem 2.7 and [17, 3.3]. Its original proof is in Brown and Peterson [7].

Clearly, then, analysis of the stable homotopy type of $\Omega^2 S^{r+2} \overset{s}{\simeq} \bigvee_{q \geq 1} D_q(R^2, S^r)$ reduces to analysis of the stable homotopy type of the spaces $D_q(R^2, S^1)$. We therefore abbreviate

$$X_q = D_q(R^2, S^1)$$

in what follows. We fix a prime p and localize all spaces and spectra at p. The results to follow are due to Mahowald [33], Brown and Gitler [6], and Brown and Peterson [7] at $p = 2$ and to R. Cohen [21] at $p > 2$.

The starting point of the analysis of the X_q is the determination of their mod p cohomologies. Let χ be the conjugation in the mod p Steenrod algebra A. Define

$$M(q) = A/A\{\chi(\beta^{\varepsilon} P^i) \mid i > q \text{ and } \varepsilon = 0 \text{ or } 1\}.$$

If $p = 2$, we let $P^i = Sq^i$ and suppress the Bockstein. Davis' result [23] that

$$\chi P^{p(i+1)} = (-1)^{i+1} P^{p^i} P^{p^{i-1}} \cdots P^p P^1 , \quad p(i+1) = 1 + p + \cdots + p^i ,$$

greases some of the computations. The following result is not too hard to prove by direct inductive calculation from F. Cohen's results on $H_* X_q$ [14, III].

Mahowald's original argument when $p = 2$ [33] is somewhat different. Abbreviate $M([q/p]) = M[q/p]$.

Theorem 2.8. $\tilde{H}^*(X_q; Z_2) \cong \Sigma^q M[q/2]$ and there is a stable 2-primary cofibration sequence

$$\Sigma X_{2q-1} \xrightarrow{f} X_{2q} \xrightarrow{g} \Sigma^{2q} X_q$$

which on mod 2 cohomology realizes the $2q^{th}$ suspension of the short exact sequence

$$0 \longleftarrow M(q-1) \xleftarrow{\beta} M(q) \xleftarrow{\alpha} \Sigma^q M[q/2] \longleftarrow 0,$$

where α is the A-map specified by $\alpha(\Sigma^q 1) = \chi(Sq^q)$ and β is the natural A-map.

Here ΣX_{2q-1} may be replaced by its 2-local equivalent $\Sigma^2 X_{2q-2}$. The key to the cofibration is the construction of g which, as Milgram pointed out, is an easy exercise in the use of the classical James maps. The rest follows by use of the geometric and homological properties of the spaces $C_2 S^r$ in [36] and [14,III]; see [15, Thm. 2] or [33, 5.5]. The analog at odd primes is more complicated but proceeds along similar lines [21]. Let M^r denote the Moore space $S^r \cup_p e^{r+1}$.

Theorem 2.9. Localize all spaces at $p > 2$. Then the following conclusions hold.

(i) X_q is contractible unless $q \equiv 0$ or $q \equiv 1 \mod p$.

(ii) $X_1 \simeq S^1$ and $X_{pq+1} \simeq \Sigma X_{pq}$ if $q > 0$.

(iii) $X_{p^2 q} \wedge M^{2r(p-1)} \simeq X_{p^2 q+pr}$ if $q > 0$ and $1 \leq r \leq p-1$.

(iv) $\tilde{H}^*(X_{pq}; Z_p) \cong \Sigma^{2q(p-1)} A/A\{\chi(\beta^{\varepsilon} P^i) \mid pi+\varepsilon > 0\}$; in particular, $\tilde{H}^*(X_{pq}; Z_p) \cong \Sigma^{2q(p-1)} M[q/p]$ if $q \not\equiv 0 \mod p$.

(v) There is a stable map

$$g: X_{p^{i+2}+p} \to \Sigma^{2p^{i+1}(p-1)} X_{p^{i+1}+p}$$

which on mod p cohomology realizes $\Sigma^{2(p^{i+1}+1)(p-1)}\alpha$, where $\alpha: \Sigma^{2p^i(p-1)} M(p^{i-1}) \to M(p^i)$ is the A-map specified by $\alpha(\Sigma^{2p^i(p-1)} 1) = \chi(P^{p^i})$.

In order to obtain homotopical information from these cohomological calculations, one wants to determine k-invariants. Now Brown and Gitler [6] have displayed certain spectra with the same cohomology as the X_q when $p = 2$, and R. Cohen [21] has generalized their constructions to odd primes. Combining results, one obtains the following theorem.

Theorem 2.10. There exist finite p-local CW-spectra $B(q)$ with the following properties.

(i) $H^*(B(q); Z_p)$ is isomorphic to the A-module $M(q)$.

(ii) If $i: B(q) \to K(Z_p, 0)$ represents the generator, then the induced map $i_*: B(q)_r(X) \to H_r(X; Z_p)$ is an epimorphism for all spaces X if either

$$p = 2 \text{ and } r \leq 2q+1 \quad \text{or} \quad p > 2 \text{ and } r \leq 2p(q+1) - 1.$$

(iii) If M is a compact smooth n-manifold embedded in R^{n+j} with normal bundle ν and Thom space $T\nu$, then $i_*: B(q)^r(T\nu) \to H^r(T\nu; Z_p)$ is an epimorphism if either

$$p = 2 \text{ and } n+j-r \leq 2q+1 \quad \text{or} \quad p > 2 \text{ and } n+j-r \leq 2p(q+1) - 1.$$

(iv) $\pi_k B(q)$ is a known Z_p-vector space (a certain quotient of the Λ-algebra in degree k) if either

$$p = 2 \text{ and } k < 2q \quad \text{or} \quad p > 2 \text{ and } k < 2p(q+1) - 2.$$

Here (3) follows from (2) by Alexander duality, since M is an n+j S-dual of $T\nu$. Particularly for the case $p > 2$, a little generalization of this situation is appropriate. Let T be an m S-dual of some finite CW-complex and let $\nu \in H^s(T; Z_p)$. Say that (T, ν) is adapted to $M(q)$ if

(a) $p = 2$ and $m-s \leq 2q+1$ or $p > 2$ and $m-s \leq 2p(q+1) - 1$; and

(b) the kernel of $\bar{\nu}: A \to H^*(T; Z_p)$, $\bar{\nu}(a) = a\nu$, is $A\{\chi(\beta^\epsilon P^i) \mid i > q\}$.

Brown and Peterson [7] at $p = 2$ and R. Cohen [21] at $p > 2$ proved the following characterization of the spectra $B(q)$.

Theorem 2.11. Let E be a spectrum which is trivial at all primes other than p and satisfies $H^*(E; Z_p) \cong \Sigma^s M(q)$. Suppose that for some (T, ν) adapted to $M(q)$ there exists a map $\tilde{\nu}: \Sigma^\infty T \to E$ such that $\tilde{\nu}^*$ carries the generator of $\Sigma^s M(q)$ to ν. Then E is equivalent to $\Sigma^s B(q)$.

They used this characterization to prove the following basic theorem on the structure of the X_q. Recall that Σ^{∞} denotes the stabilization functor from spaces to spectra.

Theorem 2.12. At $p = 2$, $\Sigma^{\infty}X_q$ is equivalent to $\Sigma^q B[q/2]$. At $p > 2$, $\Sigma^{\infty}X_{pq}$ is equivalent to $\Sigma^{2q(p-1)}B[q/p]$ if $q \not\equiv 0 \mod p$.

Because of Theorem 9(iii), it suffices to consider $q \equiv 2 \mod p$ when $p > 2$. In both cases, one inductively constructs (T_q, ν_q) adapted to $M[q/p]$, with ν_q of degree q or $2q(p-1)$, together with maps $\tilde{\nu}_q$ from $\Sigma^{\infty}T_q$ to $\Sigma^{\infty}X_q$ or $\Sigma^{\infty}X_{pq}$ which realize ν_q. Modulo certain constructions and calculations based on the geometric and homological properties of the C_2S^r in [36] and [14, III], the construction of the (T_q, ν_q) and $\tilde{\nu}_q$ reduces to the verification of the following result.

Theorem 2.13. For each $i \geq 0$, there is a compact smooth closed manifold of dimension 2^i if $p = 2$ or $p^{i+2} + p - 2$ if $p > 2$ with stable normal bundle ν_i and a bundle map

$$f_i : \nu_i \to \zeta_{2,2^i} \quad \text{if } p = 2 \quad \text{or} \quad f_i : \nu_i \to \zeta_{2,p^{i+2}+p} \quad \text{if } p > 2$$

such that $w_{2^i-1}(\nu_i) \neq 0$ if $p = 2$ or $(Tf_i)^* P^{p(i+1)}u_i) \neq 0$ if $p > 2$, where $p(i+1) = 1 + p + \dots + p^i$ and u_i generate $H^*(T\zeta_{2,p^{i+2}+p})$ as an A-module.

When $p > 2$, the cohomological condition looks like but isn't a statement about Wu classes (since the relevant bundles $\zeta_{n,q}$ are not orientable).

Unfortunately, the proof of this last theorem is not very pleasant, being based on a detailed study of the Adams spectral sequance for the suspension spectrum of $T(\zeta_{2,q}) \wedge K(Z_p, 1)$ for the relevant q. This analysis is based on ideas of Mahowald [33]; the maps g displayed in Theorems 2.8 and 2.9 provide the key geometric input.

Several people have considered the problem of directly constructing explicit manifolds as claimed in the theorem above. However, as far as I know, not even the requisite four-dimensional manifold has been concretely identified. The maps f_i are called Mahowald orientations, or braid orientations. On the classifying space level, one is asking for a lift of $\nu_i : M_i \to BO$ to the appropriate braid space $B(R^2, q) = K(B_q, 1)$.

F. Cohen [12] has used results of [36 and 37] to prove that there is a commutative diagram

$$\begin{array}{ccc} K(B_\infty, 1) & \xrightarrow{i} & K(\Sigma_\infty, 1) \\ \bar{\alpha}\downarrow & & \downarrow j \\ \Omega^2 S^3 & \xrightarrow{\bar{\eta}} & BO \end{array},$$

where i is induced by the natural homomorphism $B_\infty \to \Sigma_\infty$, j is induced by the regular representation $\Sigma^\infty \to O$, $\bar{\alpha}$ is a homology isomorphism, and $\bar{\eta}$ is the second loop map induced from the non-trivial map $S^1 \to BO$. (This implies, and was the original proof of, (i) of Theorem 2.2.) Mahowald [33] proved that the Thom spectrum $M\bar{\eta}$ of $\bar{\eta}$ is $K(Z_2, 0)$, and it follows that the Thom spectrum of ji is also $K(Z_2, 0)$. (See Lewis [31] for a thorough study of the Thom spectra of maps.) In turn, this implies that any element of $H_*(X; Z_2)$ for any space X is represented by a braid orientable manifold. This fact says that such manifolds must abound, but they are still very hard to find directly. F. Cohen [12] shows that solv-manifolds and certain Bieberbach manifolds admit braid orientations.

In view of all this, it is very natural to ask precisely what it means geometrically for an n-manifold M to admit a braid orientation. Sanderson [44] answers this question by associating to a braid orientation of M a canonical immersion $N \times I \to M$, $\dim N = n-1$, factored through an embedding of $N \times I$ in $M \times R^2$ and showing how the normal bundle of M can be recovered from the immersion. I shall say no more about this here since Sanderson's paper appears in these proceedings.

Incidentally, if $\bar{\eta}_n$ is the restriction of $\bar{\eta}$ to $F_nC_2S^1 \subset C_2S^1 \simeq \Omega^2S^3$, then Mahowald [33] also proved that $H^*(M\bar{\eta}_n; Z_2) \cong M(n)$. R. Cohen [22] has just recently proven that $M\bar{\eta}_n$ is in fact equivalent to $B(n)$.

Returning to our main line of development, we sketch how the results above lead to infinite families of elements in the ring π_*^s of stable homotopy groups of spheres. The essential point is the connection of the spaces X_q to Ω^2S^{r+2} on the one hand and the relative ease of analyzing the homotopy groups $\pi_*B(q)$ on the other. Mahowald's work actually preceded that summarized above. In particular, he conjectured Theorem 2.12 (its implicit assumption being an error caught by Adams in a preliminary version of [33]). However, given the results above, Mahowald's arguments go as follows.

Let $b: S^8 \to BO$ generate $\pi_8BO = Z$ and let $\bar{b}: \Omega S^9 \to BO$ be the unique loop map which restricts to b on S^8. Let $\tilde{b}: \Sigma\Omega^2S^9 \to BO$ be the adjoint of $\Omega\bar{b}$, namely the composite of $\bar{b}$ with the evaluation map $\Sigma\Omega^2S^9 \to \Omega S^9$. It is easy to check that $\tilde{b}^*(w_{2i}) \neq 0$ for $i \geq 3$ [33, p. 250]. It is well-known (see [14, II. 5.15] and [31, 2.3.4]) that the Thom spectrum $M\tilde{b}$ is the cofibre of the map $\phi: \Sigma^\infty \Omega^2 S^9 \to \Sigma^\infty S^0 = S$ of spectra adjoint to the composite map of spaces

$$\Omega S^9 \xrightarrow{\Omega \bar{b}} \Omega_0 BO \simeq SO \xrightarrow{j} SF = Q_1 S^0 \xrightarrow{*[-1]} Q_0 S^0 \subset QS^0$$

(where the subscripts indicate the relevant components). Of course, $Sq^{2^i} \mu \neq 0$ for $i \geq 3$, where μ is the Thom class of $M\tilde{b}$. Now

$$\Sigma^\infty \Omega^2 S^9 \simeq \bigvee_{q \geq 1} \Sigma^\infty D_q(R^2, S^7) \simeq \bigvee_{q \geq 1} \Sigma^\infty(\Sigma^{6q} X_q),$$

hence ϕ is a sum of maps $\phi_q: \Sigma^\infty(\Sigma^{6q} X_q) \to S$. Let

$$Y_i = \Sigma^{2^{i-2}+2^{i-1}} X_{2^{i-3}} \quad \text{and} \quad f_i = \phi_{2^{i-3}}: \Sigma^\infty Y_i \to S, \quad i > 3.$$

Mahowald's main result [33, Thm. 2] reads as follows. Let Cf denote the cofibre of a map f (of spaces or spectra).

<u>Theorem 2.14.</u> The spaces Y_i and stable maps f_i have the following properties.

(i) Y_i has dimension $2^i - 1$, $H^{2^i-1}(Y_i; Z_2) = Z_2$, and Y_i is $(2^i - 2^{i-3} - 1)$-connected.

(ii) $Sq^{2^i} \mu \neq 0$, where μ generates $H^0(Cf_i; Z_2)$.

(iii) There is a (stable) map $g_i: S^{2^i} \to Y_i$ whose composite with the projection $Y_i \to Y_i/Y_i^{2^i-2} \simeq S^{2^i-1}$ is essential.

Here (i) is immediate from the filtration and homology of C_2S^1 and (ii) is immediate from the paragraph above. Part (iii) requires a little more work since $\pi_{2^i} Y_i = \pi_{2^i-3} B(2^{i-4})$, and this is the first group beyond the range of Theorem 2.10 (iv) and the first in which 4-torsion can occur. (The preprint version of [7] gave a range one higher, this being an error caught by Mahowald.) Nevertheless, the explicit construction of the B(q) makes detection of the required g_i via the Adams spectral sequence quite easy.

Standard arguments show that $f_i \circ \Sigma^\infty g_i: \Sigma^\infty S^{2^i} \to S$ projects to $h_1 h_i$ in $E_2^{2, 2^i+2}$ of the mod 2 Adams spectral sequence. Thus the $h_1 h_i$ are permanent cycles and represent non-zero elements η_i in π_*^s. This was the starting point of Mahowald's Aarhus talk, in which he noted that one could start the argument above with $b: S^4 \to BO$ instead of $b: S^8 \to BO$ and so obtain a different family of

η_i's. He asserted that $2\eta_{i+1} = \eta_i^2$ for both families, this being zero if $i = 4$ and non-zero if $i \geq 5$ (for both families) but the reader is referred to Mahowald's contribution [35] to these proceedings for further information.

R. Cohen [21] gives the following analogous development at odd primes. Let $b: S^{2p-2} \to BSO$ generate $\pi_{2p-2}BSO = Z$. The composite $Bj \circ b: S^{2p-2} \to BSF$ determines a loop map $\bar{b}: \Omega S^{2p-1} \to BSF$ and $\Omega\bar{b}$ has adjoint $\tilde{b}: \Sigma\Omega^2 S^{2p-1} \to BSF$. Here $\tilde{b}^*(w_{p^i}) \neq 0$ for $i \geq 1$, where w_k is the kth mod p Wu class. The Thom spectrum $M\tilde{b}$ is the cofibre of the adjoint $\phi: \Sigma^\infty \Omega^2 S^{2p-1} \to S$ of the composite

$$\Omega^2 S^{2p-1} \xrightarrow{\Omega\bar{b}_p} \Omega BSF \simeq SF = Q_1 S^0 \xrightarrow{*[-1]} Q_0 S^0 \subset QS^0.$$

Of course, $P^{p^i}\mu \neq 0$ for $i \geq 1$, where μ is the Thom class of $M\tilde{b}$. Now

$$\Sigma^\infty \Omega^2 S^{2p-1} \simeq \bigvee_{q \geq 1} \Sigma^\infty D_q(R^2, S^{2p-3}) \simeq \bigvee_{q \geq 1} \Sigma^\infty(\Sigma^{(2p-4)q} X_q),$$

hence ϕ is a sum of maps $\phi_q: \Sigma^\infty(\Sigma^{(2p-4)q} X_q) \to S$. Let

$$Y_i = \Sigma^{(2p-4)p^i} X_{p^i} \quad \text{and} \quad f_i = \phi_{p^i}: \Sigma^\infty Y_i \to S, \; i > 1.$$

<u>Theorem 2.15.</u> The spaces Y_i and stable maps f_i have the following properties (where cohomology is taken mod p).

(i) Y_i has dimension $2(p-1)p^i - 1$, $H^{2(p-1)p^i-1}(Y_i)$ and $H^{2(p-1)p^i-2}(Y_i)$ are both Z_p with respective generators y and x such that $\beta(x) = y$, and Y_i is $(2(p-1)p^i - 2p^{i-1} - 1)$-connected.

(ii) $P^{p^i}\mu = \hat{x}$ and $\Gamma_{i-1}\mu = \hat{y}$ (up to non-zero constants), where μ generates $H^0(Cf_i)$, $\hat{x}$ and $\hat{y}$ are the images in $H^*(Cf_i)$ of Σx and Σy in $H^*(\Sigma Y_i)$, and Γ_i is the secondary cohomology operation associated to the Adem relation for $P^{(p-1)p^i}P^{p^i}$; the second equation holds modulo zero indeterminacy.

(iii) There is a (stable) map $g_i: S^{2(p^i+1)(p-1)-3} \to Y_i$ such that $P^1\bar{x} \neq 0$ in $H^*(Cg_i)$, where $\bar{x}$ pulls back to $x \in H^*(Y_i)$.

Here (i) is again immediate from the filtration and homology of C_2S^1, the first part of (ii) is immediate from the paragraph above, and the second part of (ii) is a direct consequence in view of Liulevicius' factorization of P^{p^i} via secondary cohomology operations [32]. For (iii), Theorems 2.9(iii) and 2.12 give that

$$X_{p^i} \wedge M^{2(p-1)} \simeq X_{p^i+p} \overset{s}{\simeq} \Sigma^{2(p-1)(p^{i-1}+1)} B(p^{i-2}).$$

This time the desired element is just within the range of Theorem 2.10(iv), which gives a homotopy class

$$\mu_{p(i-1)} \in \pi_q B(p^{i-2}), \quad q = 2p(p^{i-2}+1) - 4 .$$

The required map g_i is obtained by suspension of the composite of $\Sigma^{2(p-1)(p^{i-1}+1)}\mu_{p(i-1)}$ with the projection $X_{p^i+p} \to \Sigma^{2p-1}X_{p^i}$ induced by the projection $M^{2(p-1)} \to S^{2p-1}$.

Standard arguments now show that $f_i \circ \Sigma^\infty g_i : \Sigma^\infty S^{2(p-1)(p^i+1)-3} \to S$ projects to $h_0\lambda_{i-1}$ in $E_2^{3,2(p-1)(p^i+1)}$ of the mod p Adams spectral sequence. Here $\lambda_i \in E_2^{2,2(p-1)p^{i+1}}$ corresponds to Γ_i and is also denoted b_1^i (or $b_{1,i}$ or b_i); up to sign, it is the p-fold symmetric Massey product $\langle h_i, \ldots, h_i \rangle$. Thus the $h_0\lambda_i$ are permanent cycles which represent non-zero elements ζ_i of order p in π_*^s. Of course, ζ_i extends to $\overline{\zeta}_i : \Sigma^{2(p-1)(p^{i+1}+1)-2} M^{-1} \to S$, where M^{-1} is the mod p Moore spectrum with bottom cell in dimension -1. If $\pi: M^{-1} \to S$ is the projection onto the top cell, then Cohen shows further that $\overline{\zeta}_i$ is represented by $\pi^*(h_0 h_{i+1})$ in E_2 of the mod p Adams spectral sequence converging to the stable cohomotopy of M^{-1}, but the method fails to detect the $h_0 h_{i+1}$ themselves.

§3. Splitting theorems; James maps and Segal maps

The material to be discussed here is simpler, but in an earlier stage of development, than that discussed in the previous section. It is potentially at least as rich, and should lead to a later generation of concrete homotopical applications. The calculations of the previous section presupposed from iterated loop space theory only the geometric properties of the approximation $C(R^n, X) \simeq \Omega^n\Sigma^n X$, the existence of the stable splitting $C(R^n, X) \overset{s}{\simeq} \bigvee_{q\geq 1} D_q(R^n, X)$, and understanding of the homologies of $C(R^n, X)$ and the pieces $D_q(R^n, X)$. The analogous information for first loop spaces would be the approximation $MX \simeq \Omega\Sigma X$, the splitting $\Sigma MX \simeq \bigvee_{q\geq 1} \Sigma X^{[q]}$, and the homologies of MX and the $X^{[q]}$. The latter information is utterly trivial, and the James approximation acquires much of its force from homological calculation of the James maps $j_q : MX \to MX^{[q]} \simeq \Omega\Sigma X^{[q]}$ whose adjoint James-Hopf maps $h_q : \Sigma MX \to \Sigma X^{[q]}$ yield the splitting. For example, it was just such homological information which led to the homological understanding of the key maps g of Theorems 2.8 and 2.9.

The deepest part of the theory to follow (and the part in most rudimentary form) will in principle lead to complete information on the homological behavior of the James maps $j_q: C(R^n, X) \to QD_q(R^n, X)$ whose adjoint stable James-Hopf maps $h^s_q: \Sigma^\infty C(R^n, X) \to \Sigma^\infty D_q(R^n, X)$ yield the stable splitting. However, while the geometry leading to such computations is more or less understood, we have not yet begun the actual calculations. Thus the present state of the theory is analogous to the status of the original approximation theorem after the work of [36] but before that of [14].

Before proceeding further, I should say that virtually everything discussed in this section is joint work of Cohen, Taylor, and myself [16-19] and also Caruso [10], the only exception being the closely related work of Koschorke and Sanderson [30].

I shall first explain the various splitting theorems of [16 and 17] and then discuss the multiplicative properties of the James maps and certain analogous maps, the definition of which is based on ideas of Segal [46]. We shall see that an unstable version of the Kahn-Priddy theorem follows directly from these properties, and we shall obtain a result on the 2-primary exponent of the homotopy group groups of spheres as an obvious corollary. Another fairly immediate application is a simple proof of Mahowald's theorem [34, 6.2.8] on how to represent $K(Z,0)$ as a Thom spectrum. Nevertheless, I am sure that the most interesting applications belong to the future.

Return to the general context established in section one. A collection of Σ_j-spaces $\mathcal{C}_j$ with degeneracy operators is denoted $\mathcal{C}$ and called a coefficient system. A collection $\underline{X} = \{X_q\}$ of based spaces with all the formal properties that would be present if X_q were the qth power of a based space X is called a Π-space. Given $\mathcal{C}$ and $\underline{X}$, there results a filtered based space $C\underline{X}$. See [16,§§1,2] for details of this generalization of the construction CX of section one. X_q might be $P \wedge X^q$ for based spaces P and X, and this example leads to useful "parametrized" splitting theorems. However, the reader may prefer to think of X_q as X^q.

The splitting theorems of [16] all fit into a single general framework which we now sketch. Let $\mathcal{C}$ and $\mathcal{C}'$ be coefficient systems and let q be given. Let

$$D_q(\mathcal{C}, \underline{X}) = F_q C\underline{X}/F_{q-1}C\underline{X} = \mathcal{C}^+_q \wedge_{\Sigma_q} X_{[q]} ,$$

where $X_{[q]}$ is the quotient of X_q by the generalized fat wedge present in X_q for a

Π-space $\underline{X}$. These spaces for $\mathcal{C}'$ will be irrelevant, and we abbreviate $D_q(\mathcal{C},\underline{X}) = D_q\underline{X}$.

A James system $\mathcal{C} \to \mathcal{C}'$ is a collection of maps $\mathcal{C}_r \to \mathcal{C}'_{(q,r-q)}$ such that certain simple diagrams commute [16, 4.1]. A James system induces a James map

$$j_q : C\underline{X} \to C'D_q\underline{X}$$

for any Π-space $\underline{X}$ [16, 4.2]. In practice, $C'X$ is an H-space for spaces X (but not for general Π-spaces). If we are given James systems $\mathcal{C} \to \mathcal{C}'$ for $1 \leq q \leq r$, then we can define $k_r : C\underline{X} \to C'(\bigvee_{q=1}^{r} D_q\underline{X})$ to be the sum over q of the composites of the j_q with the evident inclusions $C'D_q\underline{X} \to C'(\bigvee_{q=1}^{r} D_q\underline{X})$. We can also restrict k_r to the finite filtrations of $C\underline{X}$. Any chosen basepoint of $\mathcal{C}'_1$ gives rise to a natural map $\eta : X \to C'X$ for spaces X. We can thus write down the following diagram:

(*)

$$\begin{array}{ccccc}
F_{r-1}C\underline{X} & \xrightarrow{\iota} & F_rC\underline{X} & \xrightarrow{\pi} & D_r\underline{X} \\
\downarrow{\scriptstyle k_{r-1}} & & \downarrow{\scriptstyle k_r} & & \downarrow{\scriptstyle \eta} \\
C'(\bigvee_{q=1}^{r-1} D_q\underline{X}) & \xrightarrow{C'\iota} & C'(\bigvee_{q=1}^{r} D_q\underline{X}) & \xrightarrow{C'\pi} & C'D_r\underline{X}
\end{array}$$

Here ι and π are used generically for evident cofibrations and quotient maps. In practice, the left square always commutes on the nose (which allows passage to limits over r when we have James systems for all q) and the right square at least commutes up to homotopy.

We are interested in homotopical splittings, but we digress momentarily to discuss homological splittings. The infinite symmetric product NX comes from the coefficient system $\mathcal{N}$ with each $\mathcal{N}_j$ a point. The unique maps $\mathcal{C}_r \to \mathcal{N}_{(q,r-q)}$ give a James system for each q. Here (*) commutes. Applying N to (*) and using the natural transformation $NN \to N$, we obtain the commutative diagram

$$\begin{array}{ccccc}
NF_{r-1}C\underline{X} & \xrightarrow{N\iota} & NF_rC\underline{X} & \xrightarrow{N\pi} & ND_r\underline{X} \\
\downarrow & & \downarrow & & \| \\
N(\bigvee_{q=1}^{r-1} D_q\underline{X}) & \xrightarrow{N\iota} & N(\bigvee_{q=1}^{r} D_q\underline{X}) & \xrightarrow{N\pi} & ND_r\underline{X}
\end{array}$$

Using the fact that N converts cofibrations to quasifibrations and the relationship between N and integral homology [24], and playing games with parameter spaces P, we derive the following general homological splitting theorem [16, 4.10].

Theorem 3.1. For all coefficient systems $\mathcal{C}$, Π-spaces $\underline{X}$, Abelian groups G, and $r \geq 1$ (including $r = \infty$), $\tilde{H}_*(F_r C\underline{X}; G)$ is isomorphic to $\sum_{q=1}^{r} \tilde{H}_*(D_q\underline{X}; G)$. These isomorphisms are natural in $\mathcal{C}$, $\underline{X}$, and G and commute with Bockstein operations.

The case $\mathcal{C} = \mathcal{N}$ recovers Steenrod's homological splitting [49] of the reduced symmetric powers $F_r NX$, and this example shows that we could not hope for a stable homotopical splitting without some restriction on $\mathcal{C}$.

For the homotopical splittings, $\mathcal{C}$ will always be Σ-free in the sense that Σ_j acts freely on $\mathcal{C}_j$ for each j. Returning to the general context, assume given a natural map $\beta_t : C'X \to \Omega^t\Sigma^t X$ for spaces X and some $t \geq 1$. The key example is $C'X = C(R^t, X)$ and $\beta_t = \alpha_t g^{-1}$ as in section one. Composing the diagram (*) with β_t and taking adjoints, we obtain a homotopy commutative diagram

$$\begin{array}{ccccccc}
\Sigma^{t-1}D_r\underline{X} & \xrightarrow{\delta} & \Sigma^t F_{r-1}C\underline{X} & \xrightarrow{\iota} & \Sigma^t F_r C\underline{X} & \xrightarrow{\pi} & \Sigma^t D_r\underline{X} \\
 & & \downarrow{\scriptstyle \tilde{k}_{r-1}} & & \downarrow{\scriptstyle \tilde{k}_r} & & \| \\
 & & \bigvee_{q=1}^{r-1} \Sigma^t D_q\underline{X} & \overset{\iota}{\dashleftarrow\dashrightarrow} & \bigvee_{q=1}^{r} \Sigma^t D_q\underline{X} & \xrightarrow{\pi} & \Sigma^t D_r\underline{X}
\end{array}$$

Assuming inductively that $\tilde{k}_{r-1}$ is an equivalence, a trivial diagram chase implies that $\delta \simeq 0$ in the top cofibration sequence. This implies that $\tilde{k}_r$ is an equivalence. The same sort of argument works when $t = \infty$.

With $\mathcal{C} = \mathcal{C}' = \mathcal{M}$, where $\mathcal{M}_j = \Sigma_j$, this recovers and generalizes Milnor's splitting [42] of ΣMX [16, 3.7].

Theorem 3.2. For all Π-spaces $\underline{X}$ and $r \geq 1$ (including $r = \infty$), $\Sigma F_r M\underline{X}$ is naturally equivalent to $\bigvee_{q=1}^{r} \Sigma X_{[q]}$. The equivalence is given by sums over q of restrictions of James-Hopf maps $h_q : \Sigma MX \to \Sigma X_{[q]}$.

Note that no connectivity hypothesis is needed; we use a map $\beta_1 : MX \to \Omega\Sigma X$ but do not require it to be an equivalence. It is an immediate consequence that there is a natural weak equivalence $\Sigma\Omega\Sigma X \simeq \bigvee_{q \geq 1} \Sigma X^{[q]}$ for all connected based spaces X.

In [16, §5], we introduce "separated" coefficient systems. For such $\mathcal{C}$, if $\mathcal{B}_q = \mathcal{C}_q/\Sigma_q$ and $\mathcal{C}^{(q)}$ is the coefficient system given by the configuration spaces $F(\mathcal{B}_q, j)$, there are tautological James systems $\mathcal{C} \to \mathcal{C}^{(q)}$ for each $q \geq 1$. When $\mathcal{C} = \mathcal{C}(Y)$ is itself the configuration space coefficient system of a space Y, we shall write down the resulting James maps explicitly below. If $\mathcal{B}_q$ embeds in R^t for $q \leq r$, then $\mathcal{C}^{(q)}$ maps to $\mathcal{C}(R^t)$ and we can apply the theory above with $\mathcal{C}' = \mathcal{C}(R^t)$. The product of any coefficient system with a separated system is separated, and by use of tricks involving both pairs of projections

$$\mathcal{C} \longleftarrow \mathcal{C} \times \mathcal{C}(R^\infty) \longrightarrow \mathcal{C}(R^\infty) \quad \text{and} \quad \mathcal{C}^{(q)} \longleftarrow \mathcal{C}^{(q)} \times \mathcal{C}(R^\infty) \longrightarrow \mathcal{C}(R^\infty)$$

we prove the following general splitting theorem [16, 8.2]. The second pair of projections plays a critical role in the naturality, leads to a uniqueness assertion for the stable James-Hopf maps, and allows one to avoid any choices of embeddings. This precision is crucial to the deeper theory discussed at the end of the section. On the other hand, as discussed in [16, §5], use of embeddings gives a good hold on the destabilization properties of the James-Hopf maps.

Theorem 3.3. For all Σ-free coefficient systems $\mathcal{C}$, Π-spaces $\underline{X}$, and $r \geq 1$ (including $r = \infty$), $\Sigma^\infty F_r C\underline{X}$ is equivalent to $\bigvee_{q=1}^{r} \Sigma^\infty D_q \underline{X}$, naturally in $\mathcal{C}$ and $\underline{X}$. The equivalence is given by restrictions of stable James-Hopf maps $h_q^s : \Sigma^\infty C\underline{X} \to \Sigma^\infty D_q \underline{X}$.

Specializing either to $\mathcal{C} = \mathcal{C}(R^n)$ or $\mathcal{C} = \mathcal{C}_n$, compatibly in view of $g: \mathcal{C}_n \to \mathcal{C}(R^n)$, we obtain the following immediate consequence [16, 8.4].

Corollary 3.4. For all connected based spaces X, there is a natural equivalence in the stable category between $\Sigma^\infty \Omega^n \Sigma^n X$ and $\bigvee_{q \geq 1} \Sigma^\infty D_q(R^n, X)$, $n \geq 1$ or $n = \infty$, and these equivalences are compatible as n varies.

Such equivalences were first obtained by Snaith [48], but our proof has a number of advantages (discussed in [16, §7]). In particular, it is not clear that Snaith's splitting maps $\Sigma^t F_r C_n X \to \Sigma^t D_{n,q} X$ can be extended over all of $\Sigma^t C_n X$; that is, they are not given by globally defined James-Hopf maps.

We shall come back to these splittings shortly, but I want first to explain the further splittings obtained in [17], which partially remove the restriction to connected spaces in the corollary above.

In [17, §1], we introduce the notion of a "directed" coefficient system. The details are rather delicate and the range of examples is peculiar; $\mathcal{M}$ and $\mathcal{N}$ are directed but the $\mathcal{C}_n$ are not; $\mathcal{C}(Y)$ is directed if Y is an open manifold but is not directed if Y is a compact ANR. When $\mathcal{C}$ is directed, there are inclusions

$$\zeta_r : \mathcal{C}_r \times_{\Sigma_r} X_r \to \mathcal{C}_{r+1} \times_{\Sigma_{r+1}} X_{r+1}$$

for a Π-space $\underline{X}$, and we define $\overline{C}\underline{X}$ to be the resulting colimit. If $\mathcal{C} \to \mathcal{C}'$ is a James system, there result James maps

$$\overline{j}_q : \overline{C}\underline{X} \to C'\overline{D}_q\underline{X},$$

where $\overline{D}_q\underline{X}$ is a certain space equivalent to the cofibre of ζ_{q-1}. Just as before, we define $\overline{k}_r : \overline{C}\underline{X} \to C'(\bigvee_{q=1}^{r} \overline{D}_q\underline{X})$ by summing the $\overline{j}_q$ for $q \leq r$ and write down the diagram

$$\begin{array}{ccccc}
\mathcal{C}_{r-1} \times_{\Sigma_{r-1}} X_{r-1} & \xrightarrow{\iota} & \mathcal{C}_r \times_{\Sigma_r} X_r & \xrightarrow{\pi} & \overline{D}_r\underline{X} \\
\downarrow \overline{k}_{r-1} & & \downarrow \overline{k}_r & & \downarrow \eta \\
C'(\bigvee_{q=1}^{r-1} \overline{D}_q\underline{X}) & \xrightarrow{C'\iota} & C'(\bigvee_{q=1}^{r} \overline{D}_q\underline{X}) & \xrightarrow{C'\pi} & C'\overline{D}_r\underline{X}
\end{array}$$

From here, the derivation of splitting theorems is precisely the same as in the discussion above, and we obtain the following theorems [17, §2].

Theorem 3.5. For all directed coefficient systems $\mathcal{C}$, Π-spaces $\underline{X}$, Abelian groups G, and $r \geq 1$, there are isomorphisms

$$\tilde{H}_*(\mathcal{C}_r \times_{\Sigma_r} X_r; G) \cong \sum_{q=1}^{r} \tilde{H}_*(D_q\underline{X}; G) \quad \text{and} \quad \tilde{H}_*(\overline{C}\underline{X}; G) \cong \sum_{q \geq 1} \tilde{H}_*(D_q\underline{X}; G)$$

which are natural in $\mathcal{C}$, $\underline{X}$, and G and commute with Bocksteins.

With $\mathcal{C} = \mathcal{N}$, this recovers Steenrod's isomorphisms [49]

$$\tilde{H}_*(X^r/\Sigma_r; G) \cong \sum_{q=1}^{r} \tilde{H}_*(X^q/\Sigma_q, X^{q-1}/\Sigma_{q-1}; G)$$

for the unreduced symmetric powers of a space.

Theorem 3.6. For all Π-spaces $\underline{X}$ and $r \geq 1$, there are natural equivalences

$$\Sigma X_r \simeq \bigvee_{q=1}^{r} \Sigma(X_q/\mathrm{Im}\, X_{q-1}) \quad \text{and} \quad \Sigma \overline{M}\underline{X} \simeq \bigvee_{q \geq 1} \Sigma(X_q/\mathrm{Im}\, X_{q-1}).$$

For spaces X, $\overline{M}X$ is the weak infinite product of countably many copies of X and the successive quotients are X^q/X^{q-1}, X^{q-1} being embedded as points with last coordinate the basepoint.

Theorem 3.7. For all Σ-free directed coefficient systems $\mathcal{C}$ and all Π-spaces $\underline{X}$, there are natural equivalences

$$\Sigma^{\infty}(\mathcal{C}_r \times_{\Sigma_r} X_r) \simeq \bigvee_{q=1}^{r} \Sigma^{\infty}\overline{D}_q\underline{X} \quad \text{and} \quad \Sigma^{\infty}\overline{C}\underline{X} \simeq \bigvee_{q\geq 1} \Sigma^{\infty}\overline{D}_q\underline{X}.$$

For example, with $\mathcal{C} = \mathcal{C}(R^{\infty})$, the case of spaces BG for a topological monoid G gives that $B(\Sigma_{\infty}\int G)$ is stably equivalent to the wedge of the cofibres of the natural maps $B(\Sigma_{q-1}\int G) \to B(\Sigma_q\int G)$.

For the promised analog of Corollary 3.4, we use an approximation of the form

$$\overline{C}(R^n, X) \xleftarrow{\overline{g}} \overline{C}_n(X^+) \xrightarrow{\overline{\alpha}_n} \Omega_0^n\Sigma^n(X^+) .$$

Here X is a connected space and X^+ is the union of X and a disjoint basepoint. The space $\overline{C}_n(X^+)$ is the telescope of a sequence of "right translations"

$$\mathcal{C}_{n,r} \times_{\Sigma_r} X^r \to \mathcal{C}_{n,r+1} \times_{\Sigma_{r+1}} X^{r+1} ,$$

and the map $\overline{\alpha}_n$ is a homology isomorphism constructed at the end of [14, I§5]. While $\overline{\alpha}_n$ is defined there for general X, it is only a homology isomorphism for connected X; in general, the two-variable Browder operations mix components non-trivially in $H_*\Omega_0^n\Sigma^n(X^+)$ but not in $H_*\overline{C}_n(X^+)$. The map $\overline{g}$ is an equivalence analogous to g [17, 3.1].

Corollary 3.8. For all connected based spaces X, there is a natural equivalence in the stable category between $\Sigma^{\infty}\Omega_0^n\Sigma^n(X^+)$ and $\bigvee_{q\geq 1}\Sigma^{\infty}\overline{D}_q(R^n, X)$, $n \geq 2$ or $n = \infty$, and these equivalences are compatible as n varies.

These results by no means exhaust the possibilities of the basic line of argument, and there are various other such splittings known to Cohen, Taylor, and myself but not written down. For example, Joe Neisendorfer reminded us of [36, 6.6], in which I introduced a relative construction $E_n(X, A)$ for a based pair (X, A). When $A \to X$ is a cofibration and A is connected, there is a quasifibering

$$C_nA \to E_n(X, A) \to C_{n-1}(X/A) , \qquad n \geq 1 ,$$

where C_0 is the identity functor [36, 7.3]. There are filtration preserving inclusions $C_nA \subset E_n(X, A) \subset C_nX$ and it is perfectly straightforward to trace through the proof of the stable splitting of C_nX and see that it restricts to give a stable splitting of $E_n(X, A)$.

Theorem 3.9. Let $A \to X$ be a cofibration. For all $r \geq 1$ (including $r = \infty$), and all $n \geq 1$ (including $n = \infty$), there is a natural equivalence

$$\Sigma^{\infty} F_r E_n(X, A) \simeq \bigvee_{q=1}^{r} \Sigma^{\infty}(F_q E_n(X, A)/F_{q-1} E_n(X, A)).$$

These equivalences are compatible as r and n vary and are also compatible with the stable splittings of $C_n A$ and $C_n X$.

The relationship between the splittings of $E_n(X, A)$ and of $C_{n-1}(X/A)$ is unclear and deserves study.

Again, it is a simple matter to give equivariant versions of our splitting theorems, putting actions of a finite group G on all spaces in sight (see [39, §5]), and this in turn is surely a special case of a general categorical version of the argument.

We return to the original splitting theorem and specialize to configuration space coefficient systems $\mathcal{C}(Y)$, the case $Y = R^n$ being of most interest. Actually, we are wholly uninterested in splitting theorems in the rest of the paper, being concerned instead with the analysis of the James maps as a topic of independent interest.

As in section one, think of points of $C(Y,X)$ as pairs (L, λ), where L is a finite subset of Y and $\lambda: L \to X$ is a function. Recall that $B(Y, q) = F(Y, q)/\Sigma_q$. As mentioned above, there are canonical James systems which give rise to James maps

$$j_q: C(Y,X) \to C(B(Y, q), D_q(Y,X)).$$

Explicitly, $j_q(L, \lambda) = (M, \mu)$, where M is the set of all subsets of L with q elements (such a set of q elements of Y being a typical point of $B(Y, q)$) and $\mu: M \to D_q(Y,X)$ sends a point $m \in M$ to the image in $D_q(Y,X)$ of $(m, \lambda|m) \in F_q C(Y,X)$. Of course, it is not immediately apparent that j_q is well-defined. To check this, the more combinatorial description in [16, §5] is perhaps more appropriate. To proceed further, one can assume that $B(Y, q)$ embeds in R^t, say via e_q, and then compose with $C(e_q, 1)$ to obtain a James map

$$j_q: C(Y,X) \to C(R^t, D_q(Y,X)).$$

When $Y = R^n$, we may take $t = 2qn$ (or $(2q-1)n$, by [16, 5.7]).

This functional description of these James maps is due to Koschorke and Sanderson [30], who discovered them independently of Cohen and Taylor. (To see the comparison, their $C^r_m(X)$ is our $C(B(R^m, k), X)$.) Their emphasis is not on the maps and their homotopical implications but rather on their geometrical interpretation. Let V be a smooth manifold without boundary with one-point compactification V_c. Also, let ξ be a vector bundle over some space B, with Thom complex $T\xi$, and let $\xi_{m,k}$ be the evident derived bundle over $B_k = F(R^m, k) \times_{\Sigma_k} B^k$. Consider immersion data consisting of a smooth closed manifold M, an immersion $g_1 : M \to V$ with normal bundle ν, and a bundle map $\overline{g} : \nu \to \xi_{m,k}$ such that $(g_1, g_2) : M \to V \times B(R^m, k)$ is an embedding, where $g_2 : M \to B(R^m, k)$ is the composite of the base space map of $\overline{g}$ and the projection $B_k \to B(R^m, k)$. Let $\mathcal{I}^k_m(V, \xi)$ be the set of bordism classes of such immersions. Koschorke and Sanderson first prove that $C^k_m(T\xi)$ classifies this set,

$$\mathcal{I}^k_m(V, \xi) \cong [V_c, C^k_m(T\xi)],$$

and then explain how to interpret the maps j_q above (for X a Thom space) in terms of certain operations $\mathcal{I}^1_m(V,\xi) \to \mathcal{I}^k_m(V,\xi)$ specified by associating to an immersion $g_1 : M \to V$ with normal bundle ξ an immersion $g^k_1 : M(k) \to V$ with normal bundle mapping appropriately to $\xi_{m,k}$, where $M(k) \subset B(M, k)$ is the manifold of k-tuple self-intersection points of g_1. In this context, they obtain geometric proofs and interpretations of some of the multiplicative properties of James maps we are about to discuss.

In [10], we shall discuss multiplicative properties of James maps in full axiomatic generality. Given suitably related James systems $\mathcal{C} \to \mathcal{C}^{(q)}$ and suitable structure on $\mathcal{C}$ and the $\mathcal{C}^{(q)}$, there is a ring space structure on the infinite product $\underset{q \geq 0}{\times} C^{(q)} D_q X$ and the map

$$(j_q) : CX \to \underset{q \geq 0}{\times} C^{(q)} D_q X$$

is an exponential H-map. Here $D_0 X = S^0$ and j_0 carries CX to $1 \in S^0 \subset C^{(0)} S^0$. For any coefficient system $\mathcal{C}$ with appropriate sums $\mathcal{C}_p \times \mathcal{C}_q \to \mathcal{C}_{p+q}$, the trivial James systems $\mathcal{C} \to \mathcal{N}$ used to prove Theorem 3.1 satisfy the relevant axioms. For any separated $\mathcal{C}$ with sums, the canonical James systems $\mathcal{C} \to \mathcal{C}(\mathcal{B}_q)$ satisfy the axioms. If $\mathcal{C} = \mathcal{C}(Y)$, where Y admits an injection $Y \amalg Y \to Y$ each component of which is homotopic through injections to the identity map, then $\mathcal{C}$ admits sums of the sort required. In particular, this applies to $Y = R^n$. Here the following are all H-maps with respect to the appropriate multiplication on the infinite products:

$$\begin{array}{ccc}
\underset{q\geq 0}{\times}\ C(B(R^n, q), D_q(R^n,X)) & \xrightarrow{\underset{q\geq 0}{\times}\ C(e_q, 1)} & \underset{q\geq 0}{\times}\ C(R^{2qn}, D_q(R^n,X)) \\
\Big\uparrow (j_q) & & \Big\downarrow \underset{q\geq 0}{\times}\ \alpha_{2qn} g^{-1} \\
C(R^n, X) & \overset{(j_q)}{\dashrightarrow} & \underset{q\geq 0}{\times}\ \Omega^{2qn}\Sigma^{2qn} D_q(R^n,X)
\end{array}$$

We continue to write j_q for the composite $\alpha_{2qn} g^{-1} C(e_q, 1) j_q$. We could also have stabilized, replacing $2qn$ by ∞ on the right. The product on the loop space level is induced in an evident way from smash products $\Omega^i Y \times \Omega^j Z \to \Omega^{i+j}(Y \wedge Z)$ and the pairings

$$D_s(R^n, X) \wedge D_t(R^n, X) \to D_{s+t}(R^n, X)$$

induced by the additive H-space structure on $C(R^n, X)$.

We digress to mention an application to Thom spectra in [18]. There we give a simple proof, based solely on use of Steenrod operations, of the following theorem. Let $S^3<3>$ denote the 3-connective cover of S^3.

<u>Theorem 3.10</u>. (i) The Thom spectrum associated to any H-map $\Omega^2 S^3 \to BF$ with non-zero first Stiefel-Whitney class is $K(Z_2, 0)$.
(ii) The Thom spectrum associated to any H-map $\Omega^2 S^3<3> \to BSF$ with non-zero second Stiefel-Whitney class and non-zero first Wu class at each odd prime is $K(Z, 0)$.

Part (i) gives a new proof of Mahowald's result that $M\bar{\eta} = K(Z_2, 0)$. At $p>2$, $D_i(R^2, S^{2q-1}) \simeq 0$ for $1<i<p$ and $D_p(R^2, S^{2q-1}) \simeq M^{2pq-2}$. It follows from the discussion above that

$$j_p : \Omega^2 S^{2q+1} \simeq C_2 S^{2q-1} \to QM^{2pq-1}$$

is a p-local H-map. As explained in [18], with $q = 1$ this easily leads to an H-map as prescribed in part (ii) and so gives Mahowald's result that $K(Z, 0)$ is a Thom spectrum.

Returning to the work in [10], we now head towards the Kahn-Priddy theorem. We follow the ideas of Segal [46], but we work unstably with general spaces X and thus introduce a great deal of new structure into iterated loop space theory. We want first to extend the James maps over $\Omega^n \Sigma^n X$. There is no problem when X is connected, but it is the case $X = S^0$ in which we are most interested. As Segal points out [46], the following obstruction theoretical observation allows use of the exponential H-map property above to extend the j_q simultaneously for all q. Henceforward, all H-spaces are to be homotopy associative and commutative.

Lemma 3.11. Let $g: X \to Y$ be a group completion of H-spaces, where $\pi_0 X$ has a countable cofinal sequence. Then for any grouplike H-space Z and weak H-map $f: X \to Z$, there is a unique weak H-map $\tilde{f}: Y \to Z$ such that $\tilde{f}g$ is weakly homotopic to f.

The "weak" aspect is that we are ignoring $\lim^1$ terms. The interpretation is that, on finite-dimensional CW-complexes A, $g: [A,X] \to [A,Y]$ is universal with respect to natural transformations of monoid-valued functors from $[A,X]$ to group-valued represented functors $[A,Z]$. We take $[\ ,\ ]$ in the sense of based homotopy classes.

By a power series argument, $(1, \times_{q\geq 1} \Omega^{2qn}\Sigma^{2qn}D_q(R^n, X))$ is grouplike; that is, its monoid of components is a group. This gives the following generalization of results of Segal [46]. We assume that $\pi_0 X$ is countable and write $\eta(r,s)$ for the natural inclusion $\Omega^r\Sigma^r X \to \Omega^s\Sigma^s X$, $s \geq r$; $\eta(r,s)$ induces $(s-r)$-fold suspension on homotopy groups.

Theorem 3.12. For $n \geq 2$ and all X, there exist maps

$$j_q : \Omega^n\Sigma^n X \to \Omega^{2nq}\Sigma^{2nq}D_q(R^n, X)$$

such that j_0 is constant at $1 \in S^0$, j_1 is $\eta(n, 2n)$, and

$$j_r(\alpha + \beta) = \sum_{p+q=r} j_p(\alpha) j_q(\beta)$$

for $\alpha, \beta \in [A, \Omega^n\Sigma^n X]$.

Here the sums are loop addition and the products are those specified above. Segal [46] also introduced very special cases of the general maps

$$s_q : C(Z, D_q(Y,X)) \to C(Z \times Y^q, X^{[q]})$$

specified by $s_q(M,\mu) = (N,\nu)$ where if $\mu(m)$ is the image in $D_q(Y,X)$ of an element $(L_m, \lambda_m) \in F_q C(Y,X)$ such that $L_m \subset Y$ has q elements, then

$$N = \bigcup_{m \in M,\, \ell_i \in L_m,\, \sigma \in \Sigma_q} (m, \ell_{\sigma(1)}, \ldots, \ell_{\sigma(q)}) \subset Z \times Y^q$$

and

$$\nu(m, \ell_{\sigma(1)}, \ldots, \ell_{\sigma(q)}) = \lambda_m(\ell_{\sigma(1)}) \wedge \ldots \wedge \lambda_m(\ell_{\sigma(q)}).$$

It is easy to analyze the additive and multiplicative properties of the s_q combinatorially, and we arrive at the following complement to the previous result. Let $X^{[0]} = S^0$.

Theorem 3.13. For $m \geq 2$, $n \geq 1$, and all X, there exist weak H-maps

$$s_q : \Omega^m \Sigma^m D_q(R^n, X) \to \Omega^{m+nq} \Sigma^{m+nq} X^{[q]}$$

such that s_0 is the identity map of $\Omega^m S^m$, s_1 is $\eta(n, m+n)$, and

$$s_r(\beta\gamma) = (p,q) s_p(\beta) s_q(\gamma), \quad r = p+q,$$

for $\beta \in [A, \Omega^{tp}\Sigma^{tp}D_p(R^n, X)]$ and $\gamma \in [A, \Omega^{tq}\Sigma^{tq}D_q(R^n, X)]$, $t \geq 2$. Moreover, for $\alpha \in [A, \Omega^m \Sigma^m X]$,

$$(s_q \circ \Omega^m \Sigma^m \overline{\Delta})(\alpha) = q!\,(\eta(m, m+nq) \circ \Omega^m \Sigma^m \Delta)(\alpha),$$

where $\Delta : X \to Y^{[q]}$ is the diagonal and $\overline{\Delta} : X \to D_q(R^n, X)$ is induced from Δ (via any chosen basepoint in $F(R^n, q)$).

The product $\beta\gamma$ is that above, while that on the right is just smash product of maps. Here s_q is obtained by application of Lemma 3.11 to the additive H-space structures, and the uniqueness clause of that lemma implies the last formula. The passage from the combinatorial level product formula to the loop space level is more subtle and requires use of the following result (the need for which was overlooked in [46]).

Lemma 3.14. Let $g: X \to Y$ and $g': X' \to Y'$ be group completions, where $\pi_0 X$ and $\pi_0 X'$ have countable cofinal sequences. Then for any grouplike H-space Z and weakly homotopy bilinear map $f: X \wedge X' \to Z$ there exists a unique weakly homotopy bilinear map $\tilde{f}: Y \wedge Y' \to Z$ such that $\tilde{f}(g \wedge g')$ is weakly homotopic to f.

Setting $m = 2nq$, we can compose s_q with j_q. To analyze this composite, we need the map

$$k_q : C(Y, X) \to C(Y^q, X^{[q]})$$

specified by $k_q(L, \lambda) = (M, \mu)$ where M is the set of all ordered q-tuples of elements of L and $\mu(\ell_1, \ldots, \ell_q) = \lambda(\ell_1) \wedge \ldots \wedge \lambda(\ell_q)$. Again, easy combinatorics, a power series argument, and use of Lemma 3.11 give the following result.

Theorem 3.15. For $n \geq 2$ and all X, there exist maps

$$k_q : \Omega^n \Sigma^n X \to \Omega^{nq} \Sigma^{nq} X^{[q]}$$

such that k_0 is constant at $1 \in S^0$, k_1 is the identity map, and

$$k_r(\alpha + \beta) = \sum_{p+q=r} (p,q) k_p(\alpha) k_q(\beta)$$

for $\alpha, \beta \in [A, \Omega^n\Sigma^n X]$. Moreover,

$$(\eta(nq, 3nq) \circ k_q)(\alpha) = (s_q \circ j_q)(\alpha).$$

While all this general structure is bound to prove useful, the combinatorics for the last step towards the Kahn-Priddy theorem require $\Delta : X \to X^{[q]}$ to be the identity map, that is to say $X = S^0$. Let c_{iq} be the number of ways of dividing a set of q elements into i unordered subsets. The "therefore" in the following result comes from a purely algebraic argument.

<u>Theorem 3.16.</u> For $n \geq 2$ and $\alpha \in [A, \Omega^n S^n]$,

$$\alpha^q = \sum_{i=1}^{q} c_{iq}(\eta(ni, nq) \circ k_i)(\alpha).$$

Therefore $k_q(\alpha) = \alpha(\alpha-1)\cdots(\alpha-q+1)$ if $A = B^+$, where $r \in [A, \Omega^n S^n]$ is r times the map which sends B to 1. If, further, B is a suspension and α maps B to $\Omega_0^n S^n$, then

$$k_q(\alpha) = (-1)^{q-1}(q-1)!\,\eta(n, nq)(\alpha).$$

The last assertion holds since $\alpha\alpha = 0$ by the standard argument that cup products are trivial for a suspension.

Note that $D_q(R^n, S^0) = B(R^n, q)^+$ and let $\delta : D_q(R^n, S^0) \to S^0$ map 0 to 0 and $B(R^n, q)$ to 1. Let $F_q(n,t)$ be the fibre of $\Omega_0^t\Sigma^t\delta : \Omega_0^t\Sigma^t D_q(R^n, S^0) \to \Omega_0^t S^t$ and note that $F_q(n, \infty) \simeq QB(R^n, q)$. Any choice of basepoint in $B(R^n, q)$ yields $S^0 \to D_q(R^n, S^0)$, and there results a composite equivalence

$$F_q(n,t) \times \Omega^t S^t \to \Omega^t\Sigma^t D_q(R^n, S^0) \times \Omega^t\Sigma^t D_q(R^n, S^0) \to \Omega^t\Sigma^t D_q(R^n, S^0).$$

Let $j_q : \Omega^n S^n \to \Omega^{nq}\Sigma^{nq} D_q(R^n, S^0)$ have components j'_q and j''_q in $F_q(n,t)$ and $\Omega^t S^t$. Theorem 3.15 gives a homotopy commutative diagram

$$\begin{array}{ccc}
\Omega^n S^n & \xrightarrow{(j'_q, j''_q)} & F_q(n, 2nq) \times \Omega^{2nq} S^{2nq} \simeq \Omega^{2nq}\Sigma^{2nq} D_q(R^n, S^0) \\
\downarrow k_q & & \downarrow s_q \\
\Omega^{nq} S^{nq} & \xrightarrow{\eta(nq, 3nq)} & \Omega^{3nq} S^{3nq}
\end{array}$$

On $\pi_r \Omega_0^n S^n$, $r > 0$, Theorems 3.13 and 3.16 yield the formula

$$(s_q j'_q)(\alpha) = (-1)^{q-1}(q-1)!\ \Sigma^{3nq-n}\alpha - q!\ \Sigma^{nq} j''_q(\alpha).$$

This is our unstable version of the Kahn-Priddy theorem. Taking q to be a prime p, we conclude that, up to a constant, $s_q j'_q$ is congruent mod p to the iterated suspension homomorphism. All maps in sight are compatible as n varies. Since $B(R^{\infty}, p) \simeq B\Sigma_p$, we obtain Segal's version [46] of the usual Kahn-Priddy theorem on passage to limits.

Theorem 3.17. The composite $Q_0S^0 \xrightarrow{j'_p} QB\Sigma_p \xrightarrow{s_p} Q_0S^0$ is a p-local homotopy equivalence.

It is not clear to us that s_p is an infinite loop map. According to Adams [1], this is a necessary and sufficient condition that s_p agree with the map used by Kahn and Priddy [28].

By construction, we have the commutative diagram

$$\begin{array}{ccccc} \Omega_0^n S^n & \xrightarrow{j'_2} & QB(R^n, 2) \simeq Q(RP^{n-1}) & & \\ \eta \downarrow & & \downarrow & & \\ Q_0S^0 & \xrightarrow{j'_2} & QB(R^{\infty}, 2) \simeq Q(RP^{\infty}) & \xrightarrow{s_2} & Q_0S^0 . \end{array}$$

Thus stabilization factors through $Q(RP^{n-1})$. This has the following consequence.

Theorem 3.18. If $\alpha \in \pi_r^s$ is a 2-torsion element in the image under stabilization of $\pi_{2n+1+r}S^{2n+1}$, then $2^{n+\varepsilon}\alpha = 0$, where $\varepsilon = 0$ if $n \equiv 0$ or 3 mod 4 and $\varepsilon = 1$ if $n \equiv 1$ or 2 mod 4.

Indeed, Toda [50] proves that the identity of $\Sigma^{2n}RP^{2n}$ has this order.

All of this is quite easy. We close with some remarks on the deeper theory, to appear in [19], which explains what structure the James maps $j_q : C(R^n, X) \to QD_q(R^n, X)$ really carry. As mentioned before, $C(R^n, X)$ is not just an H-space but a $\mathcal{C}_n$-space. Since $H_*C(R^n, X)$ is functorially determined by H_*X via homology operations derived from this structure, one wants to know how this structure behaves with respect to the James maps. Consider the infinite product $\times_{q\geq 0} QD_q(R^n, X)$. We have said that this is a ring space. In fact, it is an E_n ring space (more precisely, it has an equivalent subspace so structured). This means that there is an operad pair $(\mathcal{C}, \mathcal{G})$ in the sense of [37, VI.1.6] such that $\mathcal{C}$ is an E_{∞} operad and $\mathcal{G}$ is an E_n operad (that is, $\mathcal{G}$ is equivalent to $\mathcal{C}_n$) and there is an action in the sense of [37, VI.1.10] of $(\mathcal{C}, \mathcal{G})$ on $\times_{q\geq 0} QD_q(R^n, X)$. (For the aficionados, $\mathcal{C}$ is the little convex bodies operad $\mathcal{K}_{\infty}$ and $\mathcal{G} = \mathcal{C}_n \times \mathcal{L}$, where $\mathcal{L}$ is the linear isometries operad.) The additive action, by $\mathcal{C}$, is the evident product action. The multiplicative action, by $\mathcal{G}$, is a parametrization of the multiplicative

H-space structure described earlier. $\mathcal{J}$ also acts on $C(R^n, X)$ (via the projection $\mathcal{J} \to \mathcal{C}_n$), and the crucial fact is that

$$(j_q): C(R^n, X) \to \times_{q \geq 0} QD_q(R^n, X)$$

is a map of $\mathcal{J}$-spaces. Upon restriction of its target to the unit space (zeroth coordinate 1), the recognition principle of [36] implies that the extension

$$(j_q): \Omega^n \Sigma^n X \to (1, \times_{q \geq 1} QD_q(R^n, X))$$

is actually an n-fold loop map for a suitable n-fold delooping of the target (not, of course, the obvious additive one).

To compute all the j_q on homology, it suffices to determine the multiplicative homology operations on the target. In principle, these are completely determined by the known additive operations and general mixed Cartan and mixed Adem relations for E_n ring spaces like those developed for E_∞ ring spaces in [14, II]. I have no doubt that such calculations will eventually become a powerful tool for the working homotopy theorist, just as the earlier calculations of [14], which once seemed impossibly complicated, are now being assimilated and exploited by many workers in the field.

Bibliography

1. J.F. Adams. The Kahn-Priddy theorem. Proc. Camb. Phil. Soc. 73(1973), 45-55.
2. M.G. Barratt. A free group functor for stable homotopy. Proc. Symp. Pure Math Vol. 22, pp. 31-35. Amer. Math. Soc. 1971.
3. M.G. Barratt and P.J. Eccles. Γ^+ structures III. The stable structure of $\Omega^\infty\Sigma^\infty A$. Topology 13(1974), 199-207.
4. J. Beck. On H-spaces and infinite loop spaces. Springer Lecture Notes in Mathematics Vol. 99, pp. 139-153. 1969.
5. J.M. Boardman and R.M. Vogt. Homotopy invariant algebraic structures on topological spaces. Springer Lecture Notes in Mathematics Vol 347. 1973.
6. E.H. Brown, Jr. and S. Gitler. A spectrum whose cohomology is a certain cyclic module over the Steenrod algebra. Topology 12(1973), 283-295.
7. E.H. Brown, Jr. and F.P. Peterson. On the stable decomposition of $\Omega^2 S^{r+2}$. Trans. Amer. Math. Soc. To appear.
8. E. H. Brown, Jr. and F.P. Peterson. The stable homotopy type of $\Omega^n S^{n+r}$. Quarterly J. Math. To appear.
9. J. Caruso. Thesis. Univ. of Chicago. In preparation.

10. J. Caruso, F.R. Cohen, J.P. May, and L.R. Taylor. James maps, Segal maps, and the Kahn-Priddy theorem. In preparation.

11. J. Caruso and S. Waner. An approximation to $\Omega^n\Sigma^n X$. To appear.

12. F.R. Cohen. Braid orientations and bundles with flat connections. Inventiones Math. 46(1978), 99-110.

13. F. R. Cohen, Little cubes and the classifying space for n-sphere fibrations. Proc. Symp. Pure Math. Vol. 32 Part 2, pp. 245-248. Amer. Math. Soc. 1978.

14. F. R. Cohen, T.J. Lada, and J.P. May. The homology of iterated loop spaces. Springer Lecture Notes in Mathematics Vol 533. 1976.

15. F. R. Cohen, M.E. Mahowald, and R.J. Milgram. The stable decomposition for the double loop space of a sphere. Proc. Symp. Pure Math. Vol 32 Part 2, pp. 225-228. Amer. Math. Soc. 1978.

16. F. R. Cohen, J.P. May, and L.R. Taylor. Splitting of certain spaces $C\underline{X}$. Math. Proc. Camb. Phil. Soc. To appear.

17. F. R. Cohen, J.P. May, and L.R. Taylor. Splitting of some more spaces. Math. Proc. Camb. Phil. Soc. To appear.

18. F. R. Cohen, J.P. May, and L.R. Taylor. $K(Z,0)$ and $K(Z_2,0)$ as Thom spectra. To appear.

19. F. R. Cohen, J.P. May, and L.R. Taylor. James maps and E_n ring spaces. In preparation.

20. F. R. Cohen and L. R. Taylor. Computations of Gelfand-Fuks cohomology, the cohomology of function spaces, and the cohomology of configuration spaces. Springer Lecture Notes in Mathematics Vol. 657, pp. 106-143. 1978.

21. R. L. Cohen. On odd primary stable homotopy theory. Thesis. Brandeis. 1978.

22. R. L. Cohen. The geometry of Ω^2S^3 and braid orientations. In preparation.

23. D. Davis. The antiautomorphism of the Steenrod algebra. Proc. Amer. Math. Soc. 44(1974), 235-236.

24. A. Dold and R. Thom. Quasifaserungen und unendliche symmetrische Produkte. Annals of Math. 67(1958), 239-281.

25. E. Dyer and R.K. Lashof. Homology of iterated loop spaces. Amer. J. Math. 84(1962), 35-88.

26. I.M. James. Reduced product spaces. Annals of Math. 62(1955), 170-197.

27. D.S. Kahn. On the stable decomposition of $\Omega^\infty S^\infty A$. Springer Lecture Notes in Mathematics Vol 658, pp. 206-214. 1978.

28. D.S. Kahn and S.B. Priddy. The transfer and stable homotopy theory. Math. Proc. Camb. Phil. Soc. 83(1978), 103-111.

29. P.O. Kirley. On the indecomposability of iterated loop spaces. Thesis. Northwestern. 1975.

30. U. Koschorke and B. Sanderson. Self intersections and higher Hopf invariants. Topology 17(1978), 283-290.

31. G. Lewis. The stable category and generalized Thom spectra. Thesis. Chicago. 1978.

32. A. Liulevicius. The factorization of cyclic reduced powers by secondary cohomology operations. Memoirs Amer. Math. Soc. 42. 1962.

33. M. Mahowald. A new infinite family in ${}_2\pi_*^s$. Topology 16(1977), 249-256.

34. M. Mahowald and A. Unell. Bott periodicity at the prime 2 and the unstable homotopy of spheres. Preprint.

35. M. Mahowald. Some homotopy classes generated by η_j. These proceedings.

36. J.P. May. The geometry of iterated loop spaces. Springer Lecture Notes in Mathematics Vol 271. 1972.

37. J.P. May (with contributions by F. Quinn, N. Ray, and J. Tornehave). E_∞ ring spaces and E_∞ ring spectra. Springer Lecture Notes in Mathematics Vol 577. 1977.

38. J. P. May. Infinite loop space theory. Bull. Amer. Math. Soc. 83(1977), 456-494.

39. J.P. May. Infinite loop space theory revisited. Proc. conf. alg. top. Waterloo, 1978.

40. D. McDuff. Configuration spaces of positive and negative particles. Topology 14(1975), 91-107.

41. R. J. Milgram. Iterated loop spaces. Annals of Math. 84(1966), 386-403.

42. J. W. Milnor. On the construction FK. In J.F. Adams. Algebraic Topology, a student's guide. London Math. Soc. Lecture Note Series 4, pp. 119-136. 1972.

43. G. Saitati. Loop spaces and K-theory. J. London Math. Soc. 9(1975), 423-428.

44. B. Sanderson. The geometry of Mahowald orientations. These proceedings.

45. G. B. Segal. Configuration-spaces and iterated loop spaces. Invent. Math. 21(1973), 213-222.

46. G. B. Segal. Operations in stable homotopy theory. London Math. Soc. Lecture Note Series 11, pp. 105-110. 1974.

47. V.P. Snaith. A stable decomposition of $\Omega^n S^n X$. J. London Math. Soc. 7(1974), 577-583.

48. V.P. Snaith. On $K_*(\Omega^2 X; Z_2)$. Quarterly J. Math. 26(1975), 421-436.

49. N. E. Steenrod. Cohomology operations and obstructions to extending continuous functions. Advances in Math. 8(1972), 371-416.

50. H. Toda. Order of the identity class of a suspension space. Annals of Math. 78(1963), 300-323.

51. R. J. Wellington. The A-algebra $H^*\Omega_0^{n+1}\Sigma^{n+1}X$, the Dyer-Lashof algebra, and the Λ-algebra. Thesis. Chicago, 1977.

52. S. W. Yang. Thesis. Brandeis. 1978.

Mod p decompositions of H-spaces ; Another approach

by John McCleary

The decomposition of H-spaces into products of simpler spaces has been extensively studied by various authors [5,7,8,11,13,15]. For an arbitrary H-space Y, the problem is to obtain conditions on Y and the prime p such that $H^*(Y;Z_p)$ completely determines the mod p homotopy type of Y, that is, conditions under which Y can be shown to be homotopy-equivalent to a product of spheres and sphere bundles, $B_m(p)$, at the prime p.

The main thrust of this paper is to describe an obstruction theory, based on techniques of Massey and Peterson [9], which is used to prove

Theorem A ([8]) Let Y be an associative mod p H-space where

1) $H^*(Y;Z_p)$ is primitively generated,
2) $H^*(Y;Z_p) = \Lambda(x_{2n_1+1},\ldots,x_{2n_\ell+1})$ where $n_1 \leq n_2 \leq \cdots \leq n_\ell$, and
3) $p \geq n_\ell - n_1 + 2$,

then $Y_{(p)}$ is homotopy-equivalent to $S^{2n_1+1}_{(p)} \times S^{2n_2+1}_{(p)} \times \cdots \times S^{2n_\ell+1}_{(p)}$.

Theorem B Let Y be an associative mod p H-space where

1) $H^*(Y;Z_p)$ is primitively generated,
2) $H^*(Y;Z_p) = \Lambda(x_{2n_1+1},\ldots,x_{2n_\ell+1})$ where $n_1 \leq n_2 \leq \cdots \leq n_\ell$, and
3) $2p > n_\ell - n_1 + 2$ and $p > 5$,

then $Y_{(p)}$ is homotopy-equivalent to the product $\prod_s B_{m_s}(p)_{(p)} \times \prod_t S^{2m_t+1}_{(p)}$

with the numbers m_s and m_t determined by the action of P^1 on $H^*(Y;Z_p)$.

Theorem B includes most cases of theorems proved by Harper [5] and Wilkerson [15]. The condition p > 5 is technical and can be eliminated by other means. The obstruction theory is of independent interest and arises as follows.

Definition. Let M be a module over the mod p Steenrod algebra A(p). We say that M is an <u>unstable module</u> if for p = 2, $Sq^i x = 0$ when dim x < i and for p odd, $P^i x = 0$ when dim x < 2i and $\beta P^i x = 0$ when dim x $\leq$ 2i. An algebra over A(p) is unstable if it is an unstable module and for p = 2, $Sq^i x = x^2$ when dim x = i and for p odd $P^i x = x^p$ when dim x = 2i.

Let UM and UA denote the categories of unstable modules and unstable algebras with degree-preserving maps. It follows that $H^*(\ ;Z_p)$ is a contravariant functor: Top $\longrightarrow$ UA.

The forgetful functor : UA $\longrightarrow$ UM has an adjoint U: UM $\longrightarrow$ UA defined $U(M) = {}^{T(M)}/_{D}$ where T(M) is the infinite tensor product and D is the ideal generated by elements of the form $x \otimes y - (-1)^{\dim x \dim y} y \otimes x$ and $P^i x - x \otimes x \otimes \cdots \otimes x$ (p times) when 2i = dim x. We will call a space <u>very nice</u> (following [2]) if $H^*(Y;Z_p) = U(M_Y)$ for some unstable module M_Y. Examples of such spaces include $K(\pi,n)$'s for π finitely generated, odd-dimensional spheres, most H-spaces and a few projective spaces.

Suppose Y and Y' are very nice spaces and g: $M_{Y'} \longrightarrow M_Y$ is a morphism of unstable modules. Is there a continuous function G: $\tilde{Y} \longrightarrow \tilde{Y}'$ such that $H^*(\tilde{Y};Z_p) = H^*(Y;Z_p)$, $H^*(\tilde{Y}';Z_p) = H^*(Y';Z_p)$ and $G^*|_{M_{Y'}} = g$? If such a function G exists we say that g is <u>realizable</u> by G. The obstruction theory provides a series of obstruction sets, $O_n(g)$, lying in computable groups such that

Theorem There exists a function G: $Y_{(p)} \longrightarrow Y'_{(p)}$ realizing g if and only if $0 \in O_n(g)$ for all n.

This result has been obtained independently by John Harper using the unstable Adams spectral sequence where the obstructions are not as explicitly identified.

In the first section we will provide a thumbnail sketch of the Massey-Peterson theory. The second section is a presentation of the obstruction theory and in the third section we sketch the proofs of theorems A and B. A more detailed account of these results is defered to a later paper.

The results in this paper are part of my Temple University doctoral dissertation written under the direction of Dr. James Stasheff. I am grateful to him for his encouragement and guidance.

§ 1 Massey-Peterson Theory

Let $M \in UM$. We define an endomorphism $\lambda: M \longrightarrow M$ by $\lambda|_{M^n} = Sq^n$ when $p = 2$ and $\lambda|_{M^{2n}} = P^n$ and $\lambda|_{M^{2n+1}} = \beta P^n$ when p is odd. Since λ is an endomorphism it induces an action of $Z_p[\lambda]$ on M. We say that M is a <u>λ-module</u> if $M \in UM$ and M is equipped with this $Z_p[\lambda]$ action.

We call M a <u>free</u> <u>λ-module</u> if M has a homogeneous basis over $Z_p[\lambda]$ or equivalently if for all $x \in M$, $\lambda x = 0$ if and only if $x = 0$. When we consider $Z_p\{\lambda\}$ as a graded algebra on one generator of dimension 1, it follows that submodules of free λ-modules are also free.

The important examples of free λ-modules are $MK(Z_p,n)$, $MK(Z,n)$ and $MK(Z_{p^k},n)$ where $H^*(K(\pi,n);Z_p) \doteq U(MK(\pi,n))$ and $k,n > 1$.

Using the map λ we introduce a functor $\Omega: UM \longrightarrow UM$ defined $(\Omega M)_k = \left(M/_{\lambda M}\right)_{k+1}$. For f: $M \longrightarrow N$ a morphism in UM, f commutes with the action of A(p) and so $f(\lambda M) \subset \lambda N$. Thus $\Omega f: \Omega M \longrightarrow \Omega N$ is well-defined. For π finitely generated, by considering the Cartan basis one can show that $\Omega MK(\pi,n) = MK(\pi,n-1)$. For an Eilenberg-MacLane space, $\Omega K(\pi,n) \simeq K(\pi,n-1)$; this motivates the choice of notation.

Proposition 1.1 If $P \xrightarrow{f} Q \xrightarrow{g} R \longrightarrow 0$ is exact in UM, then $\Omega P \xrightarrow{\Omega f} \Omega Q \xrightarrow{\Omega g} \Omega R \longrightarrow 0$ is also exact. In addition if f is a monomorphism and R is a free λ-module then Ωf is also a monomorphism.

Thus Ω is a right exact functor.

The theorem recorded below is due to Massey and Peterson [9] for the case $p = 2$ and to Barcus [1] for p odd.

Let $\xi_0 = (E_0, p_0, B_0, F)$ be a fibration satisfying

(a) the system of local coefficients of the fibration is trivial.

(b) $H^*(F;Z_p) = U(A)$ where $A \subset H^*(F;Z_p)$ consists of transgressive elements,

(c) E_0 is acyclic and the ideal generated by the extended image of A in $H^*(B_0;Z_p)$ under transgression contains all elements of positive dimension.

By the <u>extended</u> <u>image</u> of A we mean the set $\{y_i\} \cup \{\nu y_i\}$ in $H^*(B_0;Z_p)$ where $\nu: A \longrightarrow A$ is defined $\nu|_{A^{2n}} = 0$ and $\nu|_{A^{2n+1}} = \beta P^n$ and $\{y_i\}$ projects to a basis for Im τ in $H^*(B_0;Z_p)/Q$; Q denotes the indeterminacy of the transgression τ.

Let $f: B \longrightarrow B_0$ be a map and $\xi = (E,p,B,F)$ the induced fibration. Suppose

(d) $H^*(B_0;Z_p) = U(Y)$ and Y is a free λ-module.

(e) $H^*(B;Z_p) = U(Z)$ and $Z = Z_0 \oplus Z_1$ in UM and Z_0 is a free λ-module, and

(f) f is stable, that is, $f^*: H^*(B_0;Z_p) \longrightarrow H^*(B;Z_p)$ is such that $f^*(Y) \subset Z_0$.

Theorem 1.2 (Massey-Peterson-Barcus) Given ξ, ξ_0 and $f: B \longrightarrow B_0$ satisfying (a) through (f) and letting $Z' = \text{coker } f^*|_Y: Y \longrightarrow Z$ and $Y' = \ker f^*|_Y$ then as algebras over Z_p, $H^*(E;Z_p) = U(Z') \otimes U(Y')$ and as algebras over A(p) $H^*(E;Z_p)$ is determined by the short exact sequence in UM,

$$0 \longrightarrow U(Z') \xrightarrow{p^*} N \xrightarrow{i^*} \Omega Y' \longrightarrow 0$$

called the <u>fundamental sequence</u> for ξ, where $i\colon F \longrightarrow E$ is the inclusion and N is an $A(p)$-submodule of $H^*(E;Z_p)$.

For a proof we refer the reader to [9] and [1]. The theorem gives an insight into the algebraic structure of the mod p cohomology of certain fibre spaces and this will be used to obtain a useful interface between algebraic considerations in UM and certain topological constructions.

It is an easy consequence of a theorem of Cartan [3] that the module $MK(Z_p,n)$ is the free unstable module on one generator of dimension n. We also have that $MK(Z_p,n)$ is projective in UM and so we can talk of resolutions of a module in UM. Suppose Y is a very nice space with $H^*(Y;Z_p) = U(M_Y)$ and $X(M_Y)\colon 0 \longleftarrow M_Y \xleftarrow{\varepsilon} X_0 \xleftarrow{d_0} X_1 \xleftarrow{d_1} X_2 \longleftarrow \cdots$ is a (not necessarily projective) resolution of M_Y by modules which are direct sums of $MK(\pi,n)$'s for $\pi = Z$ or Z_{p^k}, k a natural number. Using theorem 1.2 we construct a tower of fibrations that carries the algebraic information contained in $X(M_Y)$.

By a realization, $E(X(M_Y))$, of $X(M_Y)$ we will mean a system of principal fibrations:

$$\begin{array}{ccccccccccccc}
 & & & & F_s & & & & F_3 & & F_2 & & F_1 \\
 & & & & \uparrow f_s & & & & \uparrow f_3 & & \uparrow f_2 & & \uparrow f_1 \\
Y & \longrightarrow & E_s & \xrightarrow[p_s]{s-1} & E_{s-1} & \xrightarrow[p_{s-1}]{s-2} & \cdots & \xrightarrow[p_3]{2} & E_2 & \xrightarrow[p_2]{1} & E_1 & \xrightarrow[p_1]{0} & E_0 \\
 & & \uparrow j_s & & \uparrow j_{s-1} & & & & \uparrow j_2 & & \uparrow j_1 & & \\
 & & \Omega F_{s-1} & & \Omega F_{s-2} & & & & \Omega F_2 & & \Omega F_1 & &
\end{array}$$

that satisfies:

(1) E_0 and F_i are products of $K(\pi,n)$'s, that is, generalized Eilenberg-MacLane spaces (gEMs).

(2) $H^*(E_0;Z_p) = U(X_0)$, $H^*(F_1;Z_p) = U(X_1)$, and $H^*(F_s;Z_p) = U(\Omega^{s-1}X_s)$.

(3) $f_1^* = d_0$, $j_s^* o f_{s+1}^*: \Omega^s X_{s+1} \longrightarrow \Omega^s X_s$ is $\Omega^s d_s$.

(4) The fibration p_s^{s-1} is induced by the path-loop fibration over f_s.

(5) $p_i: Y \longrightarrow E_i$ is the composition $p_{i+1}^i o p_{i+2}^{i+1} o \cdots o p_s^{s-1} o p_s$.

(6) $p_0^*: X_0 \longrightarrow M_Y$ is ε.

Properties (1) through (6) imply

(7) $H^*(E_s;Z_p) = U(M_Y) \otimes U(\Omega^s \ker d_{s-1})$ as algebras over $A(p)$.

Theorem 1.3 Given Y, M_Y and $X(M_Y)$ as above there exists a realization of $X(M_Y)$.

The proof uses theorem 1.2 plus an additional fact in the inductive step; the mapping $p_s: Y \longrightarrow E_s$ splits the fundamental sequence for the fibration p_s^{s-1}:

$$\begin{array}{ccccccccc} 0 & \longrightarrow & U(M_Y) & \xrightarrow{(p_s^{s-1})^*} & N_s & \xrightarrow{j_s^*} & \Omega^s \ker d_{s-1} & \longrightarrow & 0 \\ & & & \diagdown\!\!\diagdown & \downarrow p_s^* & & & & \\ & & & & U(M_Y) & & & & \end{array}$$

Thus $H^*(E_s;Z_p) = U(M_Y) \otimes U(\Omega^s \ker d_{s-1})$ as algebras over $A(p)$. This splitting will play a crucial role in the obstruction theory.

Recall that a graded module is <u>n-connected</u> if $M_k = 0$ for $k \leq n$. Let M be in UM and $X(M): 0 \longleftarrow M \xleftarrow{\varepsilon} X_0 \xleftarrow{d_0} X_1 \xleftarrow{d_1} X_2 \longleftarrow \cdots$ a resolution of M in UM. We will call $X(M)$ <u>convergent</u> if $\Omega^s X_s$ is $f(s)$-connected and $f(s) \rightarrow \infty$ as $s \rightarrow \infty$. Using minimal resolutions and allowing modules $MK(Z,n)$ and $MK(Z_{p^k},n)$ in the construction of resolutions we can guarantee the existence of convergent resolutions for any $M \in$ UM.

Now suppose Y and M_Y are as above and $X(M_Y)$ is a convergent resolution of M_Y. Note $\varinjlim_{r} \Omega^s \ker d_{s-1} \subset \varinjlim_{s} \Omega^s X_s = 0$. Hence $\varinjlim_{s} H^*(E_s;Z_p) =$

$\varinjlim_s U(M_Y) \otimes U(\Omega^s \ker d_{s-1}) = U(M_Y)$. If we let $p_\infty = \varprojlim_s p_s : Y \longrightarrow \varprojlim_s E_s$ then $p_\infty^* : H^*(\varprojlim_s E_s ; Z_p) \xrightarrow{\cong} H^*(Y;Z_p)$. Thus p_∞ induces a homotopy equivalence $(\varprojlim_s E_s)_{(p)} \simeq Y_{(p)}$ where $W_{(p)}$ is the mod p localization of the space W. A realization of a convergent resolution then gives an approximation to the space Y at the prime p.

§ 2 The Obstruction Theory

In this section we will assume that Y and Y' are two very nice spaces with modules M_Y and $M_{Y'}$ in UM such that $H^*(Y;Z_p) = U(M_Y)$ and $H^*(Y';Z_p) = U(M_{Y'})$. Let $X(M_Y)$ and $X(M_{Y'})$ denote resolutions of M_Y and $M_{Y'}$ and $E(X(M_Y))$, $E(X(M_{Y'}))$ realizations of these resolutions as in theorem 1.3. Because we have been liberal in our choices of modules to use in the construction of resolutions we need a definition that provides the analogue of the defining property of projective resolutions. Suppose we have a morphism $g: M_{Y'} \longrightarrow M_Y$ in UM. We will say that g <u>lifts through the resolutions</u> $X(M_{Y'})$ and $X(M_Y)$ if there exist maps $g_i : X_i' \longrightarrow X_i$ such that the following ladder commutes:

$$\begin{array}{ccccccccc}
0 & \longleftarrow & M_{Y'} & \xleftarrow{\varepsilon'} & X_0' & \xleftarrow{d_0'} & X_1' & \xleftarrow{d_1'} & \cdots \\
 & & \Big\downarrow g & & \Big\downarrow g_0 & & \Big\downarrow g_1 & & \\
0 & \longleftarrow & M_Y & \xleftarrow[\varepsilon]{} & X_0 & \xleftarrow[d_0]{} & X_1 & \xleftarrow[d_1]{} & \cdots
\end{array}$$

If $X(M_{Y'})$ is already a projective resolution then any map can be lifted.

The focus of this section will be on the realizability of morphisms in UM. The following theorem indicates the effect of a realizable map on our constructions.

Theorem 2.1 ([9]) Let $k: Y \longrightarrow Y'$ be a map such that $k^*(M_{Y'}) \subset M_Y$ and k^* lifts through the resolutions. Let $\{k_j\}: X(M_{Y'}) \longrightarrow X(M_Y)$ be such a lift. Then there exists a map $\Phi: E(X(M_Y)) \longrightarrow E(X(M_{Y'}))$ realizing the lift of k^* that is, Φ is a collection $\{\phi_i: E_i \longrightarrow E_i',\ \psi_j: F_j \longrightarrow F_j'\}$ satisfying the following

2.1A) $\psi_j^* = U(\Omega^{j-1}k_j): U(\Omega^{j-1}X_j') \longrightarrow U(\Omega^{j-1}X_j)$,

2.1B)
$$\begin{array}{ccc} E_i & \xrightarrow{\phi_i} & E_i' \\ {\scriptstyle p_i^{i-1}}\downarrow & & \downarrow{\scriptstyle 'p_i^{i-1}} \\ E_{i-1} & \xrightarrow[\phi_{i-1}]{} & E_{i-1}' \end{array}$$

2.1C)
$$\begin{array}{ccc} \Omega F_i & \xrightarrow{\Omega\psi_i} & \Omega F_i' \\ {\scriptstyle j_i}\downarrow & & \downarrow{\scriptstyle j_i'} \\ E_i & \xrightarrow[\phi_i]{} & E_i' \end{array}$$

2.1D)
$$\begin{array}{ccc} E_i & \xrightarrow{\phi_i} & E_i' \\ {\scriptstyle f_{i+1}}\downarrow & & \downarrow{\scriptstyle f_{i+1}'} \\ F_{i+1} & \xrightarrow[\psi_{i+1}]{} & F_{i+1}' \end{array}$$

2.1E)
$$\begin{array}{ccc} Y & \xrightarrow{k} & Y' \\ {\scriptstyle p_i}\downarrow & & \downarrow{\scriptstyle p_i'} \\ E_i & \xrightarrow[\phi_i]{} & E_i' \end{array}$$

This theorem demonstrates a kind of naturality for the constructions we have introduced thus far. We note two corollaries to this theorem. First, the maps $\phi_n: E_n \longrightarrow E_n'$ induce mappings $\phi_n^*: N_n' \longrightarrow N_n$ of the extensions in the fundamental sequences for the fibrations $'p_n^{n-1}$ and p_n^{n-1}. Since N_n' and N_n are split extensions, it is natural to ask whether or not ϕ_n respeets this splitting. From the diagrams 2.1 B,D and E and the fundamental sequence we can show

Corrollary 2.2 The mappings $\phi_n: E_n \longrightarrow E_n'$ induce morphisms of split extensions, $\phi_n^*: N_n' \longrightarrow N_n$ in UM for the fundamental sequences of the fibrations $'p_n^{n-1}$ and p_n^{n-1}.

Now suppose Y is a primitively generated mod p H-space. Then the multiplication $m: YxY \longrightarrow Y$ induces $m^*: U(M_Y) \longrightarrow U(M_Y \oplus M_Y)$ such that

$m^*(M_Y) \subset M_Y \oplus M_Y$. In this case theorem 2.1 implies

Corollary 2.3 For Y a primitively generated mod p H-space, the spaces E_n are mod p H-spaces and the maps $f_n : E_{n-1} \longrightarrow F_n$ are H-maps.

Our next theorem obtains a partial converse to theorem 2.1 and provides the basis for the obstruction theory.

Theorem 2.4 Let $g: M_{Y'} \longrightarrow M_Y$ be given such that g lifts through the resolutions $X(M_{Y'})$ and $X(M_Y)$ and let $\{g_i: X_i' \longrightarrow X_i\}$ be such a lifting. Suppose $X(M_{Y'})$ and $X(M_Y)$ are convergent resolutions and $\Phi = \{\phi_i: E_i \longrightarrow E_i'$, $\psi_j: F_j \longrightarrow F_j'\}: E(X(M_Y)) \longrightarrow E(X(M_{Y'}))$ is a map of resolutions satisfying 2.1 A,B,C and D. Then there exists a map $G: Y_{(p)} \longrightarrow Y'_{(p)}$ such that $G^*|_{M_{Y'}} = g$.

The proof uses some facts about localization and a theorem of Cohen [4] concerning the homotopy properties of the inverse limit functor.

We now fix a morphism $g: M_{Y'} \longrightarrow M_Y$ in UM. We will assume that g can be lifted through $X(M_{Y'})$ and $X(M_Y)$ and that the resolutions are convergent. Because we have taken the F_i and F_i' to be gEMs the lifting $\{g_i: X_i' \longrightarrow X_i\}$ gives rise to a collection of maps $\{\psi_i : F_i \longrightarrow F_i'\}$ such that $\psi_i^* = U(\Omega^{i-1} g_i)$. Theorem 2.4 motivates the following

Definition 2.5 Let $\gamma: E_n \longrightarrow E_n'$. We will say that γ is an <u>n-realizer</u> for g if

2.5a_n for $0 \leq i < n$ there exists $\phi_i : E_i \longrightarrow E_i'$ such that ϕ_i is an i-realizer and 2.1B holds.

2.5b_n
$$\begin{array}{ccc} E_n & \xrightarrow{\ \gamma\ } & E_n' \\ {\scriptstyle p_n^{n-1}}\downarrow & (\simeq) & \downarrow{\scriptstyle 'p_n^{n-1}} \\ E_{n-1} & \xrightarrow[\ \phi_i\ ]{} & E_{n-1}' \end{array}$$

2.5c_n
$$\begin{array}{ccc} \Omega F_n & \xrightarrow{\ \Omega\psi_n\ } & \Omega F_n' \\ {\scriptstyle j_n}\downarrow & (\simeq) & \downarrow{\scriptstyle j_n'} \\ E_n & \xrightarrow[\ \gamma\ ]{} & E_n' \end{array}$$

$$2.5d_n \qquad \begin{array}{ccc} E_n & \xrightarrow{\gamma} & E'_n \\ f_{n+1}\big\downarrow & \circlearrowleft & \big\downarrow f'_{n+1} \\ F_{n+1} & \xrightarrow[\psi_{n+1}]{} & F'_{n+1} \end{array}$$

From the definition of a realization of a resolution, everything at the 0-level is a gEMs and so existence of a 0-realizer comes for free. Suppose now that we have an (n-1)-realizer ϕ_{n-1}. By 2.5d_n there exists a homotopy $H: E_{n-1} \times I \longrightarrow F'_n$ such that $H(x,0) = f'_n \circ \phi_{n-1}(x)$ and $H(x,1) = \psi_n \circ f_n(x)$. Recall that $E_n = \{(\lambda,x) \mid \lambda \in PF_n,\ x \in E_{n-1}$ and $\lambda(1) = f_n(x)\}$ and E'_n is the analogous subset of $PF'_n \times E'_{n-1}$. Define $\gamma: E_n \longrightarrow E'_n$ by $\gamma(\lambda,x) = (\lambda_H, \phi_{n-1}(x))$ where λ_H is the path

$$\lambda_H(t) = \begin{cases} \psi_n \circ \lambda(2t), & 0 \le t \le {}^1/_2, \\ H(x,2-2t), & {}^1/_2 \le t \le 1. \end{cases}$$

$\lambda_H(1) = H(x,0) = f'_n(\phi_{n-1}(x))$ implies that $(\lambda_H, \phi_{n-1}(x))$ is in E'_n and hence γ is well-defined. It is easy to show that γ is continuous and satisfies 2.5a_n, b_n and c_n. The splitting of the fundamental sequence gives us the key to condition 2.5d_n for γ.

Theorem 2.6 The obstruction to γ being an n-realizer is the class $[f'_{n+1} \circ \gamma \circ p_n]$ in $[Y, F'_{n+1}]$.

Proof:

$$\begin{array}{ccccc} Y & \xrightarrow{p_n} & E_n & \xrightarrow{\gamma} & E'_n \\ & & f_{n+1}\big\downarrow & & \big\downarrow f'_{n+1} \\ & & F_{n+1} & \xrightarrow[\psi_{n+1}]{} & F'_{n+1} \end{array}$$

From the construction of a realization $f_{n+1} \circ p_n \simeq *$; if $[f'_{n+1} \circ \gamma \circ p_n] \neq 0$, 2.5$d_n$ has no chance of being satisfied. Suppose $[f'_{n+1} \circ \gamma \circ p_n] = 0$. Then $p_n^* \circ \gamma^* \circ (f'_{n+1})^* = 0$ which implies $\gamma^*((f'_{n+1})^*(\Omega^n X'_{n+1}))$ is contained in

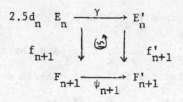

$\ker p_n^*|_{N_n} = \Omega^n \ker d_{n-1}$. Since $(f'_{n+1})^*(\Omega^n X'_{n+1}) = \Omega^n \ker d'_{n-1}$, it follows that $\gamma^*(\Omega^n \ker d'_{n-1}) \subset \Omega^n \ker d_{n-1}$. Now by 2.5$c_n$ and the naturality of the fundamental sequence we get that the following diagram commutes:

$$\begin{array}{ccc} \Omega^n X'_{n+1} & \xrightarrow{(f'_{n+1})^*} & \Omega^n \ker d'_{n-1} \\ {\scriptstyle \Omega^n g_{n+1}} \downarrow & \circlearrowright & \downarrow {\scriptstyle \gamma^*} \\ \Omega^n X_{n+1} & \xrightarrow[(f_{n+1})^*]{} & \Omega^n \ker d_{n-1} \end{array}$$

Since F_{n+1} and F'_{n+1} are gEMs the commutativity of this square implies 2.5d_n and hence γ is an n-realizer.

Observe that $[Y,F'_{n+1}] = H^*(Y;\pi_*(F'_{n+1}))$; this with theorem 2.1 gives

Theorem 2.7 γ is an n-realizer if and only if $[f'_{n+1} \circ \gamma \circ p_n] = 0$ in $H^*(Y;\pi_*(F'_{n+1}))$.

The map γ as constructed above was a single candidate for an n-realizer. Since ${}'p_n^{n-1}: E'_n \longrightarrow E'_{n-1}$ is a principal fibration we can vary γ by the principal action $\mu: \Omega F'_n \times E'_n \longrightarrow E'_n$. That is, if $\zeta \in [E_n, E'_n]$ and $[{}'p_n^{n-1} \circ \zeta] = [{}'p_n^{n-1} \circ \gamma] = [\phi_{n-1} \circ p_n^{n-1}]$ then there exists a w in $[E_n, \Omega F'_n]$ such that $[\mu \circ (w \times \gamma) \circ \Delta] = [\zeta]$ in $[E_n, E'_n]$. If ζ is any map obtained in this manner from γ and the principal action then ζ satisfies 2.5a_n, b_n and c_n and hence theorem 2.6 holds when γ is replaced by ζ.

Define $\Gamma_n: [E_n, \Omega F'_n] \longrightarrow [Y, F'_{n+1}]$ to be the composite

$$[E_n,\Omega F'_n] \xrightarrow[\mu_{\#}(-,\gamma)]{} [E_n,E'_n] \xrightarrow[(f'_{n+1})_{\#}]{} [E_n,F'_{n+1}] \xrightarrow[p_n^{\#}]{} [Y,F'_{n+1}]$$

where $F_{\#}([q]) = [q \circ F]$ and $F^{\#}([q]) = [F \circ q]$. By the previous paragraph the obstructions for all possible candidates for an n-realizer for g must lie in the image of Γ_n in $[Y, F'_{n+1}]$. Let $O_n(g)$ denote the image of Γ_n.

Theorem 2.8 If there is an (n-1)-realizer for g then ther is an n-realizer for g if and only if $0 \in O_n(g) \subset H^*(Y;\pi_*(F'_{n+1}))$.

If an n-realizer exists for all n then, using theorem 2.4 we have that g is realizable. Therefore we conclude

Theorem 2.9 g is realizable if and only if, for all n, $0 \in O_n(g)$.

In [6] Harper proves that the principal action, $\mu: \Omega F'_n \times E'_n \longrightarrow E'_n$ is primitive in the following sense: if $H^*(E;Z_p) = U(N'_n)$ and $y \in N'_n$ then $\mu^*(y) = 1 \otimes y + (j'_n)^*(y) \otimes 1$ in $H^*(\Omega F'_n;Z_p) \otimes H^*(E'_n;Z_p)$. From the definition of a realization of a resolution, the map $f'_{n+1} \circ j'_n: \Omega F'_n \longrightarrow F'_{n+1}$ is determined by $\Omega^n d_n: \Omega^n X_{n+1} \longrightarrow \Omega^n X_n$. Since $\Omega F'_n$ and F'_{n+1} are gEMs, the map $f'_{n+1} \circ j'_n$ determines a primary operation $\Xi_n: H^*(\ ;\pi_*(\Omega F'_n)) \longrightarrow H^*(\ ;\pi_*(F'_{n+1}))$. Utilizing Harper's result we obtain

Theorem 2.10 $O_n(g)$ is the coset $[f'_{n+1} \circ \gamma \circ p_n] + \Xi_n H^*(Y;\pi_*(\Omega F'_n))$ in $H^*(Y;\pi_*(F'_{n+1}))$.

Observe that if Ξ_n is trivial on $H^*(Y;\pi_*(\Omega F'_n))$, then the class $[f'_{n+1} \circ \gamma \circ p_n]$ is the only obstruction to the existence of an n-realizer for g.

§ 3 Applications

It is a consequence of Borel's structure theorem for Hopf algebras that if Y is an H-space without p-torsion, $H^*(Y;Z_p) = \Lambda(x_{2n_1+1},\dots,x_{2n_\ell+1})$ where $\dim x_r = r$. For those primes for which P^1 acts trivially on $H^*(Y;Z_p)$, Y shares the same cohomology as the space $S_p(Y) = S^{2n_1+1} \times \cdots \times S^{2n_\ell^p+1}$. If there is a map $S_p(Y) \longrightarrow Y$ inducing an isomorphism in mod p cohomology then $S_p(Y)_{(p)}$ and $Y_{(p)}$ are homotopy-equivalent and the mod p homotopy information about Y is completely determined. If such a map exists we call the prime regular for Y.

Now consider those primes for which P^1 is the only element of A(p) to act non-trivially on $H^*(Y;Z_p)$. Mimura and Toda [12] have introduced complexes, $B_m(p)$, which are sphere bundles over spheres with cohomology, $H^*(B_m(p);Z_p) = \Lambda(x_{2m+1},P^1x_{2m+1})$. If P^1 acts non-trivially we can ask whether or not Y "looks like" a product of spheres and $B_m(p)$'s at the prime p. That is, if $H^*(Y;Z_p) = \Lambda(x_{2m_1+1},P^1x_{2m_1+1},\ldots,x_{2m_k+1},P^1x_{2m_k+1},x_{2m_{k+1}+1},\ldots,x_{2m_s+1})$ then we wish a map $K_p(Y) \longrightarrow Y$ which induces an isomorphism in mod p cohomology where $K_p(Y) = \prod_{i=1}^{k} B_{m_i}(p) \times \prod_{j=k+1}^{s} S^{2n_j+1}$. If such a map exists, $K_p(Y)_{(p)} \simeq Y_{(p)}$ and we say that p is quasi-regular for Y.

We translate these questions of regularity and quasi-regularity into questions about the realizability of morphisms in UM by observing that $H^*(Y;Z_p) = \Lambda(x_{2n_1+1},\ldots,x_{2n_\ell+1}) = U(M_Y)$ where M_Y is a direct sum of modules $Tr(2n_j+1) = \{x_{2n_j+1}\}$ and $MB_{m_i}(p) = \{x_{2m_i+1},P^1x_{2m_i+1}\}$. As unstable algebras, $H^*(Y;Z_p) = H^*(K_p(Y);Z_p) = U(M_Y)$ so we ask if there is a map $R_p: K_p(Y)_{(p)} \longrightarrow Y_{(p)}$ which realizes the map of modules id: $M_Y \longrightarrow M_Y$. The existence of such a map implies that $K_p(Y)_{(p)} \simeq Y_{(p)}$ as desired.

The strategy of the proofs for theorems A and B will be to employ the obstruction theory to realize each projection $M_Y \longrightarrow Tr(2n_j+1)$ or $MB_{m_i}(p)$ by a map $r_j: S^{2n_j+1}_{(p)} \longrightarrow Y_{(p)}$ or $r_i: B_{m_i}(p)_{(p)} \longrightarrow Y_{(p)}$. We then consider the composite map

$$R_p : B_{m_i}(p)_{(p)} \times \cdots \times B_{m_k}(p)_{(p)} \times S^{2m_{k+1}+1}_{(p)} \times \cdots \times S^{2m_s+1}_{(p)}$$

$$\xrightarrow[r_1 \times \cdots \times r_k \times r_{k+1} \times \cdots \times r_s]{} Y_{(p)} \times Y_{(p)} \times \cdots \times Y_{(p)} \xrightarrow[\xi_s]{} Y_{(p)}$$

where $\xi_s(y_1, y_2, \ldots, y_s) = y_1 y_2 \cdots y_s$, induced by the multiplication on $Y_{(p)}$. It suffices to check $R_p^*: H^*(Y_{(p)}; Z_p) \longrightarrow H^*(K_p(Y)_{(p)}; Z_p)$ on the indecomposables (= the primitives) to determine whether R_p induces the desired isomorphism. Let u be an indecomposable in $H^*(Y_{(p)}; Z_p)$.

$$R_p^*(u) = \text{proj}_1 \otimes \text{proj}_2 \otimes \cdots \otimes \text{proj}_s(\xi_s^*(u))$$

$$= \text{proj}_1 \otimes \text{proj}_2 \otimes \cdots \otimes \text{proj}_s\Big(\sum_{i=1}^{s} 1 \otimes 1 \otimes \cdots \otimes \underset{i^{\text{th}}\text{ place}}{u} \otimes \cdots \otimes 1\Big)$$

$$= u, \text{ the corresponding class in } H^*(K_p(Y)_{(p)}; Z_p).$$

Thus we obtain our desired homotopy equivalence if we can realize each projection $M_Y \longrightarrow Tr(2n_j+1)$ or $MB_{m_i}(p)$.

Now suppose we want a map, $W_{r(p)} \longrightarrow Y_{(p)}$, to realize the projection $M_Y \longrightarrow N_r$ where $W_r = S^{2n_r+1}$ or $B_{m_r}(p)$ and $N_r = Tr(2n_r+1)$ or $MB_{m_r}(p)$. Consider those dimensions in which W_r has non-zero cohomology and those dimensions in which possible obstructions can occur: these dimensions are calculable from knowledge of the direct sum decomposition of M_Y and calculations of modules in convergent resolutions of the summands $Tr(2n_j+1)$ and $MB_{m_i}(p)$. If these two sets of numbers are disjoint then the obstruction theory implies that a map exists realizing the projection. With this in mind we provide the following table which lists the dimensions in which an obstruction might occur when M_Y has the appropriate summand. To obtain the table one computes the first few modules (X_0, X_1,

X_2, and X_3) in a convergent resolution of each summand. The calculations only involve a routine application of the Adem relations and the unstable axioms and so are left to the reader.

Table 1

	Tr(3)-factor	Tr(2n+1)-factor	$MB_1(p)$-factor	$MB_n(p)$-factor
O_1	4p-1, 4p-2	2n+4p-3	4p-1	2n+4p-3
O_2	6p-4	2n+6p-4	6p-3	2n+6p-4

Proof of Theorem A: Recall that the dimension of P^1x_r is $r+2(p-1)$. If $r = 2n_i+1$ then

$$\begin{aligned} r+2(p-1) = 2n_i+1+2(p-1) &\geq 2n_i+1+2(n_\ell-n_1+1) \\ &= 2n_\ell+3+2n_i-2n_1 \\ &> 2n_\ell+1 \end{aligned}$$

since $n_1 \leq n_i$ for all i. The image of a primitive under the action of A(p) is also primitive and since all of the primitives lie in dimensions less than or equal to $2n_\ell+1$, P^1 acts trivially on $H^*(Y;Z_p)$. So $H^*(Y;Z_p) = U(M_Y)$ where $M_Y = Tr(2n_1+1) \oplus Tr(2n_2+1) \oplus \cdots \oplus Tr(2n_\ell+1)$.

Suppose we wish to realize a projection $M_Y \longrightarrow Tr(2n_i+1)$ by a map $S^{2n_i+1}_{(p)} \longrightarrow Y_{(p)}$. From table 1 we see the lowest dimension in which an obstruction may occur is $2n_1+4p-3$. The inequality $p \geq n_\ell-n_1+2$ implies $2n_1+4p-3 > 2n_\ell+1$ and so any obstruction must vanish since the $(2n_i+1)$-sphere has cohomology only in dimension $2n_i+1$. Hence there is a map $r_i\colon S^{2n_i+1}_{(p)} \longrightarrow Y_{(p)}$ realizing each projection $M_Y \longrightarrow Tr(2n_i+1)$. By the discussion in the beginning of the section, this proves the theorem.

Proof of Theorem B: Using corollary 2.3 we first observe the

Lemma 4.1 If Y and Y' are primitively generated mod p H-spaces and very nice spaces and g: $M_{Y'} \longrightarrow M_Y$ a morphism in UM, then the class $[f_2' \circ \gamma \circ p_1] \in O_1(g)$ is primitive.

The spaces $B_{m_i}(p)$ have non-zero cohomology in dimensions $2m_i+1$, $2m_i+1+2(p-1)$ and $2(2m_i+1)+2(p-1)$. When $p > 5$ the spaces $B_{m_i}(p)$ are mod p H-spaces [14] and so we need only consider primitives as O_1 obstructions. The inequality $2p > n_\ell - n_1 + 2$ implies that the first obstructions lie in dimensions larger than $2n_\ell+1$ and hence the O_1 obstructions vanish for dimension reasons.

Now note that the inequality $2p > n_\ell - n_1 + 2$ guarantees that the highest dimension in which a product class $x_{m_i} \smile P^1 x_{m_i}$ can occur is less than 6p-6. Thus the O_2 obstructions all vanish for dimension reasons. Since the higher obstructions lie in still higher dimensions, we have that any projection $M_Y \longrightarrow MB_{m_i}(p)$ can be realized. Similarly any projection $M_Y \longrightarrow Tr(2m_j+1)$ can be realized so Theorem B is proved.

We add that more can be said when the mod p cohomology data for Y is known. In [10], the author obtains a theorem of Mimura and Toda [12] on the quasi-regularity of primes for compact Lie groups without the need of the restriction $p > 5$.

Bibliography

[1] W. D. Barcus, On a theorem of Massey and Peterson, Quart. J. Math.(2) 19 (1968), 33-41.

[2] A. K. Bousfield and D. M. Kan, Pairings and products in the homotopy spectral sequence, Trans.Amer.Math.Soc. 177 (1973), 319-343.

[3] H. Cartan, Séminaire, Algèbres d'Eilenberg-MacLane et Homotopie, Ecole Normale Superiere, Paris(1954/55)

[4] J. M. Cohen, Homotopy groups of inverse limits, Proc. London Math. Soc.(3) 27 (1973), 159-177.

[5] J. Harper, Mod p decompositions of finite H-spaces, LNM No. 428, Springer (1974), 44-51.

[6] _________, H-spaces with torsion, preprint.

[7] P. G. Kumpel, Lie groups and products of spheres, Proc.Amer.Math. Soc. 16(1965), 1350-1356.

[8] _________, Mod p equivalences of mod p H-spaces, Quart.J.Math. 23(1972), 173-178.

[9] W. S. Massey and F. P. Peterson, On the mod 2 cohomology structure of certain fibre spaces, Amer.Math.Soc. Memoirs 74(1967).

[10] J. McCleary, Ph.D. Thesis, Temple University (1978).

[11] M. Mimura, G. Nishida and H. Toda, Mod p decomposition of compact Lie groups, Publ. RIMS, Kyoto Univ. 13(1977), 627-680.

[12] M. Mimura and H. Toda, Cohomology operations and the homotopy of compact Lie groups-I, Topology 9(1970), 317-336.

[13] J. P. Serre, Groupes d'homotopie et classes des groupes abéliens, Ann. of Math. 58 (1953), 258-294.

[14] J. Stasheff, Sphere bundles over spheres as H-spaces mod $p > 2$, LNM No. 249, Springer (1971).

[15] C. Wilkerson, Mod p decompositions of mod p H-spaces, LNM No.428, Springer(1974), 52-57.

Temple University, Philadelphia, PA.

and

Bates College, Lewiston, ME.

Complete Intersections and the Kervaire Invariant

William Browder*
Princeton University

The topology of non-singular complete intersections (i.e. a non-singular variety V of complex dimension n defined by k polynomials in $\mathbb{C}P^{n+k}$) has received considerable attention in recent years. It was observed by Thom in the early 1950's that the degrees of the k polynomials determine the diffeomorphism type. From the Lefschetz Theorem, it follows that for a complete intersection $V^n \subset \mathbb{C}P^{n+k}$, the pair $(\mathbb{C}P^{n+k}, V^n)$ is n-connected, so that the inclusion $i : V \longrightarrow \mathbb{C}P^{n+k}$ induces $i_* : \pi_i(V) \longrightarrow \pi_i(\mathbb{C}P^{n+k})$ which is an isomorphism for $i < n$, and an epimorphism for $i = n$. Thus, in some sense, the topology of V is "concentrated in the middle dimension."

In particular $i_* : H_i(V) \longrightarrow H_i(\mathbb{C}P^{n+k})$ is injective for $i \neq n$, and $(\text{image } i_*)_{2n} = d(H_{2n}(\mathbb{C}P^{n+k}))$ where $d = d_1 \cdot \ldots \cdot d_k$ is the total degree of V, d_i = degree of the i-th polynomial P_i, $P_1, \ldots, P_k$ define V, so that (image i_*) is then completely determined by Poincaré duality.

If s is the largest integer less than $n/2$, we can embed $\mathbb{C}P^s \subset V$, and V will be a bundle neighborhood U of $\mathbb{C}P^s$ with handles $D^r \times D^q$ attached (along $S^{r-1} \times D^q$) with $r \geq n$, $r + q = 2n$. In fact, V can be described as $W \cup U'$ where $W = U \cup$ (n-handles) and U' is another copy of U, and union is along the boundaries. The diffeomorphism type of V is then determined by the attaching maps of the n-handles, and the "gluing" map of $\partial(U')$ to ∂W.

The attaching maps of the n-handles will be closely connected to the middle

*Research partly supported by an NSF Grant.

dimensional intersection form. When n is even, close analysis of this form (as in [Kulkarni-Wood]) leads to interesting results on the topology of V (see also [Wood], [Libgober], [Libgober-Wood]).

When n is odd, the intersection form is skew symmetric and the analysis of the middle dimensional handles relies on more subtle homological information. A basis for the middle dimensional homology $H_n(V)$, when $n > 1$, n odd, can be represented, using Whitney's and Haefliger's embedding theorems, by embedded spheres $S^n_{1i}, S^n_{2i} \subset V$, with $S_{ij} \cap S_{ik} = \emptyset$, any j, k, $S_{1j} \cap S_{2k} = \emptyset$ for $j \neq k$ and $S_{1i} \cap S_{2i}$ = one point, every i, so that $\{S_{ij}\}$ represent a symplectic base for $H_n(V)$. If each S_{ij} could be chosen to have trivial normal bundle, then a neighborhood of $S_{1i} \cup S_{2i}$ would be diffeomorphic to $(S^n \times S^n - (2n\text{-disk}))$ and it would follow that

$$V = (U \cup U') \# \mathop{\#}_{q} S^n \times S^n ,$$

the connected sum of $(U \cup U')$ with q copies of $S^n \times S^n$, where $U \cup U'$ is the "twisted double" of U, i.e. two copies of the disk bundle U over $\mathbb{C}P^s$ $(n = 2s + 1)$, glued by a diffeomorphism of the boundary.

The question of finding a basis for $H_n(V)$ represented by embedded spheres with trivial normal bundle, $(S^n \times D^n \subset V)$, can be studied by the methods of the Kervaire invariant arising in surgery theory. This involves defining a quadratic form $\psi : H_n(V) \longrightarrow \mathbb{Z}/2$, such that $\psi(x) = 0$ if and only if $x \in H_n(V)$ is represented by a $S^n \times D^n \subset V$. We give conditions in terms of the degrees of the defining polynomials of V for such a quadratic form to be defined, and show that <u>when</u> ψ <u>cannot be defined</u>, that any $x \in H_n(V)$ can be represented by $S^n \times D^n$.

When ψ can be defined one can find the sought for basis if and only if the Arf invariant of ψ (called the Kervaire invariant) is zero. We give a formula for computing it in these cases.

Our specific results are as follows:

Let $V \subset \mathbb{C}P^{n+k}$ be a non-singular complete intersection of complex dimension $n = 2s + 1$, defined by k-polynomials of degree $d_1, \ldots, d_k$, and let $d = d_1 \ldots d_k$ (= the degree of V).

<u>Theorem A</u>. Suppose exactly ℓ of the degrees $d_1, \ldots, d_k$ are even (so $k - \ell$ are odd). If the binomial coefficient $\binom{s+\ell}{s+1}$ is odd, and $n \neq 1, 3$ or 7, then there exists a homologically trivial $S^n \subset V$ with non-trivial (stably trivial) normal bundle , and every element $x \in H_n(V)$ can be represented by $S^n \times D^n \subset V$. If $n = 1, 3$ or 7 every embedded $S^n \subset V$ has trivial normal bundle.

This was originally proved by [Morita, H] and [Wood, H] for hypersurfaces and [Wood, CI] for complete intersections.

<u>Theorem B</u>. Notation as in A, if $\binom{s+\ell}{s+1}$ is even, then a quadratic form is defined $\psi : H_n(V) \longrightarrow \mathbb{Z}/2$ such that $x \in H_n(V)$ ($n \neq 1, 3$ or 7) is represented by $S^n \times D^n \subset V$ if and only if $\psi(x) = 0$.

<u>Theorem C</u>. With hypothesis as in B, $H_n(V)$ has a symplectic basis represented by embeddings $S^n \times D^n \subset V$, mutually intersecting exactly as in the intersection matrix if and only if the Arf invariant of ψ (the Kervaire invariant) $K(V) = 0$, ($K(V) \in \mathbb{Z}/2$).

(i) If all $d_1, \ldots, d_k$ are odd,

$$K(V) = \begin{cases} 0 & \text{if } d \equiv \pm 1 \pmod 8 \\ 1 & \text{if } d \equiv \pm 3 \pmod 8 \end{cases}$$

(ii) If some d_i's are even, $K(V) = 1$ if and only if $\ell = 2$, $4 | s$

and $8 \nmid d$.

In C , note that the condition $\binom{s+\ell}{s+1} \equiv 0 \pmod 2$ of Theorem B , imposes a condition on the number of d_i which may be even. For example $\ell \neq 1$, and for $\ell = 2$, $\binom{s+2}{s+1} = s+2$ is even if and only if s is even, ($n = 2s+1$) .

As well as Theorem A , [Morita, H] proved B and C for hypersurfaces ($k = 1$) . This case was also done in [Wood, H] . The author first proved C(i) at that time, and [Wood, CI] gave another proof. [Ochanine] first proved C(ii) in the case where V is a $8q+2$ dimensional Spin manifold, (so $s = 2q$) , which is the only case of even degree when K(V) might equal 1 .

In §1 we discuss the definition of the quadratic form in a general context and prove that when the form is not defined (for framed $M^{2n} \times \mathbb{R}^k \subset W$) there is an embedded sphere $S^n \subset M^{2n}$ ($n \neq 1, 3, 7$) which is homologically trivial (mod 2) in M , with non-trivial (stably trivial) normal bundle. In §2, we consider complete intersections and prove A , B and C . To prove C we invoke a theorem relating the Kervaire invariant of V^n and its hyperplane section V_0^{n-1} which will be proved elsewhere.

§1. Quadratic forms.

In [Browder, K] , a definition of the quadratic form arising in surgery or in framed manifolds was given using functional Steenrod squares. We give here a geometrical version of this definition, and then study when it can be defined for complete intersections, and its meaning.

First note:

(1.1) <u>Proposition</u>. For any $x \in H_n(M^{2n}; \mathbb{Z}/2)$, one can find an embedded $N^n \subset M^{2n}$, with $i_*[N] = x$.

The proof of this is standard as in Thom's proof of representability of homology by maps of manifolds, but using the additional fact that for the canonical n-plane bundle γ^n , the first non-trivial k-invariant occurs in dimension $2n + 1$, so that there is no obstruction to finding a map $f : M^{2n} \longrightarrow T(\gamma^n)$ such that $[M] \cap f^*(U) = x$, ($[M] \cap$ is Poincaré duality, $U \in H^n(T(\gamma^n); \mathbb{Z}/2)$ is the Thom class). Similarly we get:

(1.2) <u>Proposition</u>. If $M^{2n} \times \mathbb{R}^q \subset W^{2n+q}$, W connected, and $y \in H_{n+1}(W, M; \mathbb{Z}/2)$, we can find $N \subset M$ representing ∂y as in (1.1) , with $N = \partial V$, $V \subset W \times [0, 1)$, V connected, with [V] representing y , $[V] \in H_{n+1}(V, N; \mathbb{Z}/2)$ the fundamental class. Further V meets $W \times 0$ transversally in $N \subset M$.

Now the normal bundle of N in $W \times 0$ has a q-frame given by the product $M \times \mathbb{R}^q$ restricted to N . The obstructions to extending this frame to a <u>normal</u> q-frame on $V \subset W \times [0, 1)$ lie in $H^{i+1}(V, N; \pi_i(V_{n+q,q}))$ where $V_{n+q,q}$ is the space of orthogonal q-fames in $\mathbb{R}^{n+q}$, so $V_{n+q,q} = O(n+q) / O(n)$, and is (n-1)-connected. Hence all these obstructions are zero except the last, $\alpha \in H^{n+1}(V, N; \pi_n(V_{n+q,q})) \cong \mathbb{Z}/2$. Evaluating α on [V] we get an element in $\mathbb{Z}/2$, and we would like to define $\psi(x) = \alpha[V]$, (for $x = \partial y$, $y \in H_n(W, M; \mathbb{Z}/2)$) but we have made a number of choices in this process which depend on more than the homology class x, namely the choice of N and the choice of V, with $\partial V = N$.

From the theory of the Stiefel-Whitney class (see [Steenrod]) the first obstruction to finding a q-frame in an $(n+q)$ plane bundle is the Stiefel-Whitney class W_{n+1}, which becomes the ordinary Stiefel-Whitney class w_{n+1} when reduced mod 2. In the relative situation we are discussing, this is in fact the relative Stiefel-Whitney class in the sense of [Kervaire]. Thus it is homologically defined provided that this relative class does not depend on the chocie of V. This will be true provided that any closed manifold $X^{n+1} \subset W^{n+q} \times [0,1)$ admits a normal q-frame, that is, its normal Stiefel-Whitney class $\bar{w}_{n+1}(X) = 0$, so that adding X to V will not change the relative class of V.

If ξ^{n+q} is the normal bundle of X in $W \times [0, 1)$, the normal class $\bar{w}_{n+1}(X)$ is given by the formula

$$\bar{w}_{n+1} \cup U = Sq^{n+1} U \ , \quad U \in H^{n+q}(T(\xi) ; \mathbb{Z}/2)$$

the Thom class. The natural collapsing map $c : \Sigma W = (W \times [0,1]) / \text{boundary} \longrightarrow T(\xi)$ has degree 1 (mod 2), and it follows that:

(1.3) The following are equivalent:

(a) $\bar{w}_{n+1}(X) = 0$ for all $X^{n+1} \subset W \times [0, 1)$

(b) $Sq^{n+1} : H^{n+q-1}(W / \partial W ; \mathbb{Z}/2) \longrightarrow H^{2n+q}(W / \partial W ; \mathbb{Z}/2)$ is zero

(c) $v_{n+1}(W) = 0$ (v_{n+1} = the Wu class).

Thus we get the condition:

(1.4) The obstruction to extending a q-frame over N to V described above defines a quadratic form $\psi : K \longrightarrow \mathbb{Z}/2$ where $K = \text{kernel } H_n(M ; \mathbb{Z}/2) \longrightarrow H_n(W ; \mathbb{Z}/2)$ if and only if $v_{n+1}(W) = 0$.

It is not difficult to translate this relative Stiefel-Whitney class definition into the functional Sq^{n+1} definition of [Browder, K], which shows it defines a quadratic form.

One may prove (1.4) directly as follows:

Since we have shown that the definition is evaluation of a relative Stiefel-Whitney class it follows that the definition depends only on homology class. To show it quadratic, we first prove it in the special case of $S^n \times S^n \times \mathbb{R}^q \subset S^{2n+q} = \partial D^{2n+q+1}$, which may be done directly. It is clear that the function is additive on two non-intersecting manifolds $N_1^n, N_2^n \subset M^{2n}$, (which then bound non-intersecting $V_1^{n+1}, V_2^{n+1} \subset W \times [0, 1)$.

If N_1, N_2 have even intersection number, then if $n > 1$, we may find a bordism of N_1 to N_1' in M, disjoint from N_2', (simply the first few lines of the Whitney process produces the cobordism from each pair of intersection points). Take two intersection points $a, b \in N_1 \cap N_2$ and draw an arc on N_2 joining them. If N_2 were not connected we could first make a bordism of N_2 to a connected submanifold, if M were connected. If M were not connected, we would first take connected sum of its components, without changing the quadratic forms.

A neighborhood of this arc would be of the form $D^1 \times D^{n-1} \times D^n$ where $D^1 \times D^{n-1} \times 0$ is a neighborhood in N_2. Then $D^1 \times 0 \times D^n$ defines a handle which when added to N_1, produces a bordism of N_1 to N_1' which has 2 less intersection points with N_2.

This shows that $\psi(x_1 + x_2) = \psi(x_1) + \psi(x_2)$ whenever $x_1 \cdot x_2$ is even. If $x_1 \cdot x_2$ is odd, let $g_1, g_2 \in H_n(S^n \times S^n)$ be the generators corresponding to the factors, so that $\psi(g_1) = \psi(g_2) = 0$, $\psi(g_1 + g_2) = 1$. Then $(x_1 + g_1) \cdot (x_2 + g_2) = x_1 \cdot x_2 + g_1 \cdot g_2$ is even, so that

(a) $$\psi((x_1 + g_1) + (x_2 + g_2)) = \psi(x_1 + g_1) + \psi(x_1 + g_2) = \psi(x_1) + \psi(x_2).$$

But $(x_1 + x_2) \cdot (g_1 + g_2) = 0$ so

(b) $$\psi((x_1 + x_2) + (g_1 + g_2)) = \psi(x_1 + x_2) + \psi(g_1 + g_2) = \psi(x_1 + x_2) + 1$$

Equating (a) and (b) we get

$$\psi(x_1 + x_2) + 1 = \psi(x_1) + \psi(x_2)$$

and $(x_1 \cdot x_2) \equiv 1 \mod 2$, which completes the proof that ψ is quadratic. □

(1.5) Theorem. Suppose $M^{2n} \times \mathbb{R}^q \subset W^{2n+q}$, $n \neq 0, 1, 3$ or 7, W is 1-connected, (W,M) n-connected, and suppose $v_{n+1}(W) \neq 0$. Then there exists an embedded $S^n \subset M^{2n}$ and $U^{n+1} \subset M^{2n} \times \mathbb{R}^{q+1}$ with $\partial U = S^n$ such that the normal bundle ξ to S^n in M^{2n} is non-trivial, but $\xi + \varepsilon^1$ is trivial. Hence S^n is homologically trivial (mod 2) with non-trivial normal bundle.

Proof: Since $v_{n+1}(W) \neq 0$ we can find an embedding of a closed manifold $j : X^{n+1} \subset W$ whose normal bundle ξ does not admit a q-frame in $W \times [0,1)$. For $v_{n+1}(W) \neq 0$ means there is an $x \in H^{n+q-1}(W/\partial W ; \mathbb{Z}/2)$ such that $(Sq^{n+1} x)[W] \neq 0$, and hence $q \geq 2$. By Thom's theorem, since $n + q - 1 > n$, there is a map $r : (W, \partial W) \longrightarrow (T(\gamma_{n+q-1}), \infty)$ such that $r^*U = x$ (U the Thom class in $H^{n+q-1}(T(\gamma_{n+q-1}) ; \mathbb{Z}/2)$, and the transverse inverse image of the 0-section will be our manifold X^{n+1}, which we may assume connected, by choosing a component with the above property.

Let $X_0 = X - (\text{int } D^{n+1})$ so that $\partial X_0 = S^n$. Since X was connected, X_0 has the homotopy type of a n dimensional complex. Since (W, M) is n-connected, it follows that there is a map $f : X_0 \longrightarrow M$ such that

$$\begin{array}{ccc} X_0 & \xrightarrow{f} & M \\ \cap & & \cap\, i \\ X & \longrightarrow & W \end{array}$$

commutes up to homotopy.

Let g be the composite $S^n = \partial X_0 \subset X_0 \xrightarrow{f} M$, so that $g_*[S^n] = 0$. Since M is 1-connected, the Whitney process will produce a homotopy of g to an embedding (again called g), and we wish to show the normal bundle ζ^n to this embedded sphere $g(S^n)$ is non-trivial.

Using the Whitney general position embedding theorem, we may deform

$X_0 \xrightarrow{f} M \times \mathbb{R}^q \times (-1, 0]$ to an embedding $g_0 : X_0 \longrightarrow M \times \mathbb{R}^q \times (-1, 0]$ extending g which meets $M \times \mathbb{R}^q \times 0$ transversally in $g_0(\partial X_0) = g(S^n) \subset M \subset M \times \mathbb{R}^q \times 0$.

On the other side, the embedding $g : S^n \subset M$ extends to an embedding (using general position) $\bar{g} : D^{n+1} \subset W \times [0, 1)$ meeting $W \times 0$ transversally in $g(S^n) = \bar{g}(\partial D^{n+1}) \subset M \times 0 \subset M \times \mathbb{R}^k \times 0 \subset W \times 0$. The two embeddings $g_0 : X_0 \subset M \times \mathbb{R}^q \times (-1, 0]$, $\bar{g} : D^{n+1} \subset W \times [0, 1)$ together define an embedding $g_1 : X_0 \cup D^{n+1} = X \longrightarrow W \times (-1, 1)$ which is isotopic (by general position) to our original embedding $j : X \subset W \subset W \times (-1, 1)$.

The product structure $M \times \mathbb{R}^q \subset W \times 0$ defines a q-frame in the normal bundle ζ_1^{n+q} of $g(S^n) \subset M \times \mathbb{R}^q \subset W \times 0$ (so that $\zeta^n + \varepsilon^q = \zeta_1^{n+q}$) and let $\theta \in \pi_{n+1}(V_{n+q,q})$ be the obstruction to extending this k-frame over the normal bundle of $\bar{g}(D^{n+1}) \subset W \times [0, 1)$.

(1.6) <u>Lemma</u>. If $n \neq 1, 3$ or 7, the obstruction $\theta = 0$ if and only if ζ^n is trivial.

Proofs of (1.6) can be found in [Wall] or [Browder, S ; (IV 4.2)].

We assume ζ^n is trivial and produce a contradiction. In that case we can find a framed handle $D^{n+1} \times D^n \times \mathbb{R}^q \subset W \times [0, 1)$ (using (1.6)) with $D^{n+1} \times 0 \times 0 = \bar{g}(D^{n+1})$ and $S^n \times D^n \times 0$ a neighborhood of $g(S^n) \subset M \subset M \times \mathbb{R}^q \times 0 \subset W \times [0, 1)$. Let V be the cobordism of M defined by $V = M \times [-1, 0] \cup (D^{n+1} \times D^n)$, so that $V \times \mathbb{R}^q \subset W \times [-1, 1]$, and $g_1(X) \subset (\text{int } V) \times \mathbb{R}^q$.

Hence we have a factorization of the collapsing map $Y = W \times [-1, 1] / \partial(W \times [-1, 1]) \xrightarrow{a} \Sigma^q V/\partial V \xrightarrow{b} T(\xi + \varepsilon^1)$ so that $(ba)^*(U) = \Sigma x \in H^{n+q}(Y ; \mathbb{Z}/2)$ and $(Sq^{n+1}(x))[W] = (Sq^{n+1}(\Sigma x))[Y] \neq 0$, $U \in H^{n+q}(T(\xi) ; \mathbb{Z}/2)$ the Thom class. It follows that $(Sq^{n+1}(b^*U))(\Sigma^q[V]) \neq 0$, so that $Sq^{n+1}(\Sigma^{-q}(b^*U))[V] \neq 0$, where $\Sigma^{-q}(b^*U) \in H^n(V/\partial V ; \mathbb{Z}/2)$, which leads to sought after contradiction since Sq^{n+1} annihilates cohomology of dimension n. This completes the proof of (1.5). □

On the other hand we have:

(1.7) Proposition. If $v_{n+1}(W) = 0$, (W, M) n-connected, $n \neq 1, 3, 7$, n odd, and $\varphi : S^n \subset M^{2n}$ with φ nullhomotopic in W. Then the normal bundle of $\varphi(S^n)$ is trivial if and only $\psi(\varphi_*[S^n]) = 0$ where ψ is the quadratic form of (1.4).

(1.7) follows easily from (1.6) and the definition of ψ.

§2. Complete intersections, their normal bundles and the quadratic form.

In this paragraph, we apply the results of §1 to the case of non-singular complete intersections $V^n \subset \mathbb{C}P^{n+\ell}$, give the conditions for the quadratic form of §1 to be defined, and calculate the Kervaire invariant when it is defined.

Recall that a submanifold $V \subset \mathbb{C}P^{n+k}$ is a non-singular complete intersection if V is the locus of zeros of k homogeneous polynomials $P_1, \ldots, P_k$ where $\dim V = 2n$ (real dimension) and $\operatorname{codim} V = 2k$. The degrees d_i of P_i completely determine V up to diffeomorphism. Thus any question we ask in differential topology about V must have an answer in the form of a formula involving only n, and $d_1, \ldots, d_k$, and we will use the notation $V = V^n(d_1, \ldots, d_k)$.

From the topological point of view it is convenient to view V as a transversal inverse image, to have its normal bundle in $\mathbb{C}P^{n+k}$ evident. We may assume $P_i(1, 0, \ldots, 0) \neq 0$ for all i. Define maps $\overline{P}_i : \mathbb{C}P^{n+k} \longrightarrow \mathbb{C}P^{n+k}$ by $\overline{P}_i(z_0, \ldots, z_{n+k}) = (P_i(z_0, \ldots, z_{n+k}), z_1^{d_i}, \ldots, z_{n+k}^{d_i})$, so $\overline{P}_i^{-1}(z_0 = 0)$ is a hypersurface of degree d_i.

Define $\overline{P} : \mathbb{C}P^{n+k} \longrightarrow \prod_{i=1}^{k} \mathbb{C}P^{n+k}$ by $\overline{P} = \Pi \overline{P}_i$. Then $V = \overline{P}^{-1}\left(\bigcap_{i=1}^{k} (z_0^{(i)} = 0)\right)$, where $z_0^{(i)}$ is the 0-th coordinate in the i-th copy of $\mathbb{C}P^{n+k}$.

Small perturbation of the coefficients (if necessary) will make $\overline{P}$ transversal and V will be a non-singular manifold and we get:

(2.1) <u>Proposition</u>. The non-singular complete intersection V^n defined by the vanishing of $P_1, \ldots, P_k$ on $\mathbb{C}P^{n+k}$ represents the homology class $dx_n \in H_{2n}(\mathbb{C}P^{n+k})$, and the normal bundle ξ of $V \subset \mathbb{C}P^{n+k}$ has a natural bundle map into the bundle $(\alpha^{d_1} + \alpha^{d_2} + \ldots + \alpha^{d_k})$ over $\mathbb{C}P^{n+k}$ where x_n

is the generator dual to $y^k \in H^{2k}(\mathbb{C}P^{n+k})$, $y = c_1(\alpha)$, α the canonical $\mathbb{C}$ bundle over $\mathbb{C}P^{n+k}$, $d = d_1 \dots d_k$ the total degree of V.

We may transform this situation into the situation of §1 by embedding $\mathbb{C}P^{n+k} \subset E$ = the total space of a bundle γ which is stably inverse to $(\alpha^{d_1} + \dots + \alpha^{d_k})$. Then the normal bundle of $V \subset E$ has a bundle map into $(\alpha^{d_1} + \dots + \alpha^{d_k}) + \gamma$ which has a natural trivialization. Hence

(2.2) <u>Proposition</u>. The complete intersection $V^n \subset \mathbb{C}P^{n+k}$ has a natural framing in E, $V \times \mathbb{R}^q \subset E$, where $E = E(\gamma)$, γ a representative of $-(\alpha^{d_1} + \dots + \alpha^{d_k}) \in K(\mathbb{C}P^{n+k})$.

Note that the framing is determined by the structure of V as a complete intersection, and the polynomials $P_1, \dots, P_k$, not simply by the differentiable structure of V.

To apply §1, we need to calculate $v_{n+1}(E(\gamma))$, to see if the quadratic form is well defined. Note that if n is even, $v_{n+1} = 0$ since it lies in a zero group.

(2.3) <u>Theorem</u>. $v_{n+1}(E) \neq 0$ if and only if $\binom{s+\ell}{s+1} \not\equiv 0 \pmod 2$, where ℓ = the number of even integers among the degrees $d_1, \dots, d_k$, $n = 2s+1$.

<u>Proof</u>: The Stiefel-Whitney class

$$W(E) = W(\tau_{\mathbb{C}P^{n+k}} + \gamma) = W((n+k+1)\alpha - (\alpha^{d_1} + \dots + \alpha^{d_k}))$$

$$= \frac{(1+x)^{n+k+1}}{\prod_{i=1}^{k}(1+d_i x)} = \frac{(1+x)^{n+k+1}}{(1+x)^{k-\ell}} = (1+x)^{n+\ell+1} .$$

Hence $W(E) = W(\tau_{\mathbb{C}P^{n+\ell}})$ in dimensions where both cohomologies agree, and hence $v_{n+1}(E) = v_{n+1}(\mathbb{C}P^{n+\ell})$. But $v_{n+1}(\mathbb{C}P^{n+\ell}) \neq 0$ if and only if $Sq^{n+1}: H^{n+2\ell-1}(\mathbb{C}P^{n+\ell}; \mathbb{Z}/2) \longrightarrow H^{2n+2\ell}(\mathbb{C}P^{n+\ell}; \mathbb{Z}/2)$ is non-zero. The group $H^{n+2\ell-1}(\mathbb{C}P^{n+\ell}; \mathbb{Z}/2)$ is generated by $x^{s+\ell}$, where x generates $H^2(\mathbb{C}P^{n+\ell}; \mathbb{Z}/2)$ (since $n = 2s+1$), so $Sq^{n+1}(x^{s+\ell}) = Sq^{2s+2}(x^{s+\ell})$

$= \binom{s+\ell}{s+1} x^{n+\ell}$, which completes the proof.□

In E, x^{s+1} is represented by $X = \mathbb{C}P^{s+1}$, so X_0 in the proof of (1.5) may be taken oriented, and we get:

(2.4) <u>Corollary</u>. (Morita, Wood) If $n = 2s+1$, V non-singular in $\mathbb{C}P^{n+k}$ defined by $P_1, \ldots, P_k$ of degree $d_1, \ldots, d_k$, and if $\binom{s+\ell}{s+1}$ is odd, ℓ = number of even degrees among the d_i's, then there is an embedded $S^n \subset V$ which is homologically trivial, and has a non-trivial normal bundle, (provided $n \neq 1$, 3 or 7).

To calculate the Kervaire invariant (the Arf invariant of ψ) in the other cases (where ψ is well defined) we use the following theorem. The proof will be given in another paper, and it follows from a combination of an additivity theorem for the Kervaire invariant (analogous to Novikov's theorem on index) and the product formula for the Kervaire invariant.

(2.5) <u>Theorem</u>. Let $V^n \subset \mathbb{C}P^{n+k}$ be a non-singular complete intersection, and let $V_0^{n-1} \subset \mathbb{C}P^{n+k-1}$ be a non-singular hyperplane section. If the quadratic forms are defined for both V^n and V_0^{n-1}, then their Kervaire invariants are equal, $K(V) = K(V_0)$.

Note that the definition of $K(V_0)$ may have some extra subtlety as we will see in the calculation.

We can immediately derive the formula for $K(V)$ when $d = d_1 \ldots d_k$ is odd. In that case (2.3) implies that the quadratic forms are defined for all the iterated hyperplane sections $V_0^{n-1} \supset V_1^{n-2} \ldots \supset V_{n-1}^0$ so that (2.5) implies $K(V^n) = K(V_{n-1}^0)$, and we are left with the problem of computing $K(V^0)$ for the zero dimensional complete intersection of degree d, that is, for d similarly oriented points.

This calculation is a special case of that of [Browder, FPK] and is actually equivalent to it using a product formula. We do it explicitly as follows.

Suppose $V^0(d) = d$ disjoint points, embedded in W. A <u>framing</u> of

$V^0(d)$ is simply an orientation on a neighborhood of each point and the condition $v_1(W) = 0$ means that W is orientable. Suppose W is connected, d is odd, and the orientations at all the points are the same.

(2.6) <u>Proposition</u>. $K(V^0(d)) = \begin{cases} 0 & \text{if } d \equiv \pm 1 \mod 8 \\ 1 & \text{if } d \equiv \pm 3 \mod 8 . \end{cases}$

<u>Proof</u>: Since W^m is connected we may assume that $V^0(d) \subset D^m \subset W$ and that the symmetric group Σ_d acts on $V^0(d)$, preserving the framed embedding, so that

$$\psi(\sigma x) = \psi(x) \quad \text{for} \quad \sigma \in \Sigma_d ,$$

$$x \in K_0 = \ker H_0 (V^0(d) ; \mathbb{Z}/2) \longrightarrow H_0 (D^m ; \mathbb{Z}/2) .$$

Now K_0 has a basis $\{x_1 + x_0 , x_2 + x_0 ,\ldots, x_{2s} + x_0\}$, where the d points are $x_0 ,\ldots, x_{2s}$, $d = 2s + 1$. Since Σ_d acts transitively on this basis,

(2.7) $$\psi(x_i + x_0) = \psi(x_j + x_0) \quad \text{for all } i , j .$$

Further the intersection product

(2.8) $$(x_i + x_0) \cdot (x_j + x_0) = \begin{cases} 0 & i = j \\ 1 & i \neq j \end{cases} .$$

Define a module A_s with quadratic form ψ by letting $a_1 , a_2 ,\ldots, a_{2s}$ be a basis for A_s, $\psi(a_i) = 0$ for all i, and

$$(a_i , a_j) = \begin{cases} 0 & \text{if } i = j \\ 1 & \text{if } i \neq j \end{cases}$$

(the opposite of an orthonormal basis). It is easy to check that the bilinear form $(\, , \,)$ is non-singular on A_s and is the associated bilinear form to ψ.

Similarly, define B_s, φ by the basis $b_1 ,\ldots, b_{2s}$, $\varphi(b_i) = 1$, all i, and

$$(b_i , b_j) = \begin{cases} 0 & i = j \\ 1 & i \neq j . \end{cases}$$

(2.9) <u>Lemma.</u> $A_s \cong A_1 + B_{s-1}$

$B_s \cong B_1 + A_{s-1}$

(as modules with quadratic forms).

<u>Proof</u>: Define a new basis for A_s by $a_i' = a_i + a_1 + a_2$, and let $A_1 \subset A_s$ be generated by a_1 , a_2 , $B'_{s-1} \subset A_s$ be generated by $a_3' ,\ldots, a_{2s}'$. Then $A_1 \perp B'_{s-1}$, and it is easy to check that $B'_{s-1} \cong B_{s-1}$.

Similary, define a new basis for B_s by $b_i' = b_i + b_1 + b_2$, let $B_1 \subset B_s$ be generated by b_1 , b_2 , $A'_{s-1} \subset B_s$ generated by $b_3' ,\ldots, b_s'$. Then $B_s \cong B_1 + A'_{s-1}$ as orthogonal direct sum, and $A'_{s-1} \cong A_{s-1}$. Since $\mathrm{Arf}\,(A_1) = 0$, $\mathrm{Arf}\,(B_1) = 1$, we get :

$$\mathrm{Arf}\,(A_s) = \mathrm{Arf}\,(B_{s-1})$$

$$\mathrm{Arf}\,(B_s) = 1 + \mathrm{Arf}\,(A_{s-1})$$

so that

$$\mathrm{Arf}\,(A_s) = 1 + \mathrm{Arf}\,(A_{s-2})$$

$$\mathrm{Arf}\,(B_s) = 1 + \mathrm{Arf}\,(B_{s-2}) .$$

Hence:

(2.10) <u>Proposition.</u> $$\mathrm{Arf}\,(B_s) = \begin{cases} 1 & \text{if } s \equiv 1 \text{ or } 2 \bmod 4 \\ 0 & \text{if } s \equiv 3 \text{ or } 4 \bmod 4 . \end{cases}$$

But in (2.6) , $K_0 \cong B_s$ if $d = 2s + 1$ which completes the proof of (2.6) and the calculation of $K(V^n (d_1 ,\ldots, d_k))$ when $d = d_1 \ldots d_k$ is odd , i.e. Theorem C(i) .

The case of even degree d is more difficult, since the quadratic form may not be defined for the iterated hyperplane sections. However, it is always defined for the first hyperplane section $V_0^{n-1} \subset V^n$, (since the appropriate Wu class v_n lies in a zero group when n is odd). To make the calculation in V_0^{n-1} we need:

(2.11) Proposition. Let $M^{2m} \times \mathbb{R}^k \subset W$, $v_{m+1}(W) = 0$, m even, and M oriented. If $x \in H_m(M;\mathbb{Z})$, $i_*(x_2) = 0$, $i: M \longrightarrow W$ inclusion, $x_2 \in H_m(M;\mathbb{Z}/2)$ the reduction of x mod 2, then $\psi(x_2) \equiv \frac{x \cdot x}{2}$ mod 2.

We sketch a proof, (compare [Morita, P], [Brown]).

First note that if $i_*(x_2) = 0$, then $v_m(M)(x_2) = i^*(v_m(W))(x_2) = v_m(W)(i_* x_2) = 0$, so $x \cdot x \equiv x_2 \cdot x_2$ mod 2, and $x_2 \cdot x_2 = (y \cup y)[M] = (v_m(M) \cup y)[M] = v_m(M)([M] \cap y) = v_m(M)(x_2) = 0$ (where $[M] \cap y = x_2$). Hence $x \cdot x$ is even, so $\varphi(x) = \frac{x \cdot x}{2}$ mod 2 is a well defined quadratic form.

(2.12) Lemma. Let ψ be our usual quadratic form $\psi: K \longrightarrow \mathbb{Z}/2$ ($K = \ker H_m(M^{2m};\mathbb{Z}/2) \longrightarrow H_m(W;\mathbb{Z}/2)$) as in §1, and $\varphi: K \longrightarrow \mathbb{Z}/2$ another quadratic form defined in these circumstances such that $\psi(x) = 0$ implies $\varphi(x) = 0$. Then $\psi = \varphi$.

The proof is similar to that of [Browder, S; (IV. 4.7)]. For the condition implies that on the diagonal $\Delta \in H_m(S^m \times S^m;\mathbb{Z}/2)$ (for $S^m \times S^m \subset S^{2m+1}$), $\varphi(\Delta) = 1 = \psi(\Delta)$. Then by adding $S^m \times S^m$ to M^{2m} and adding Δ to an arbitrary $x \in K$ if necessary we get $\varphi(x) = \psi(x)$ (compare the proof of (1.4)).

Thus to prove (2.11) it suffices to show $\psi(x) = 0$ implies $\varphi(x) = 0$.

(2.13) Lemma. If k is large, $\psi(x) = 0$ implies there exists a framed bordism $U^{2m+1} \times \mathbb{R}^k \subset W \times [0,1]$, $\partial(U \times \mathbb{R}^k) = M \times \mathbb{R}^k \times 0 \cup M' \times \mathbb{R}^k \times 1 \subset W \times \{0,1\}$, and $V^{m+1} \subset U$, $\partial V = N^m \subset M$ with $[N^m] = x_2 \in H_m(M;\mathbb{Z}/2)$.

Proof: As in §1, we can find $N^m \subset M^{2m}$ representing $x \in H_m(M;\mathbb{Z}/2)$ and $V^{m+1} \subset W \times [0,1)$ with $\partial V = N \subset M \times 0$. Then $\psi(x) = 0$ implies that the normal bundle to U admits a k-frame extending that of N (coming from the framing of M in W). The complement of this frame is a D^m bundle over

V which meets M^{2m} in the normal disk bundle of N^m in M, and adding this disk bundle to $M \times [0, \varepsilon]$ clearly defines a framed cobordism of the type required except for the condition $M' \subset W \times 1$, but this may be achieved by an isotopy if k is large.

Now to show $\varphi(x) = 0$ we note that if $[M] \cap \bar{x} = x$, since $x_2 = \partial y$, $y \in H_{m+1}(U, M \cup M'; \mathbb{Z}/2)$, by Poincaré duality, $\bar{x}_2 = i^* \bar{y}$, $\bar{y} \in H^m(U; \mathbb{Z}/2)$, $i'^* \bar{y} = 0$, $i : M \longrightarrow U$, $i' : M' \longrightarrow U$ the inclusions.

Now $x \cdot x = \bar{x}^2 [M] \equiv \mathcal{P}(\bar{x}_2)[M] \pmod 4$ where $\mathcal{P} : H^m(M; \mathbb{Z}/2) \longrightarrow H^{2m}(M; \mathbb{Z}/4)$ is the Pontryagin square (see [Morita, P]). Now $i_* : H_{2m}(M; \mathbb{Z}/2) \cong H_{2m}(U; \mathbb{Z}/2)$ so that $x_2 \cdot x_2 \equiv \mathcal{P}(\bar{x}_2)[M] = \mathcal{P}(i^*\bar{y})[M] = (i^* \mathcal{P}(\bar{y}))[M] = \mathcal{P}(\bar{y})(i_*[M]) \equiv 0 \bmod 2$ and hence $\mathcal{P}(\bar{y}) \in j_* H^{2m}(U; \mathbb{Z}/2)$, where $0 \longrightarrow \mathbb{Z}/2 \xrightarrow{j} \mathbb{Z}/4 \longrightarrow \mathbb{Z}/2 \longrightarrow 0$.

Since $i'^*(\bar{y}) = 0$, it follows that

$$x \cdot x \equiv \mathcal{P}(i^*\bar{y})[M] = (\mathcal{P}(i^*\bar{y}) + \mathcal{P}(i'^*\bar{y}))([M] - [M'])$$

$$= (\bar{i}^*(\mathcal{P}(\bar{y}))\,[\partial U] = \mathcal{P}(\bar{y})\,(\bar{i}_*[\partial U])$$

(all mod 4), $\bar{i} = i \cup i' : M \cup M' = \partial U \longrightarrow U$.

Since $\bar{i}_*[\partial U] \equiv 0 \bmod 2$ it follows that $(\bar{i}_*[\partial U])_4 \in j_* H_{2m}(U; \mathbb{Z}/4)$ $(j : \mathbb{Z}/2 \longrightarrow \mathbb{Z}/4)$. But $\mathcal{P}(\bar{y}) \in j_* H^{2m}(U; \mathbb{Z}/2)$ and $j_* H^{2m}$ and $j_* H_{2m}$ are paired to zero (we get a factor of 2 from each j_* which multiply to become 0 mod 4)

Hence $x \cdot x \equiv 0 \bmod 4$ so $\varphi(x) = 0$, which complete the proof of (2.11) □.

We now proceed to the calculation of $K(V^n(d_1, \ldots, d_k))$ for $d = d_1 \ldots d_k$ even. Recall ℓ = number of even d_i's and $\binom{s+\ell}{s+1} \equiv 0 \bmod 2$ (to have ψ defined), where $n = 2s+1$. By (2.5), $K(V^n(d_1, \ldots, d_k)) = K(V_0^{n-1}(d_1, \ldots, d_k))$ but we must make this statement more precise.

The quadratic form ψ is defined on $L = \ker(H_{n-1}(V_0; \mathbb{Z}/2) \longrightarrow H_{n-1}(\mathbb{C}P^{n+k-1}; \mathbb{Z}/2))$ and since $n-1$ is even and d is even, the associated bilinear form is singular on L. Thus, for the Arf invariant of ψ on L to be defined it is necessary that if $r \in L$ and $(r, x) = 0$ for all $x \in L$, then $\psi(r) = 0$ (see [Browder, FPK]), but this is implicitly included in (2.5).

We will now study the middle dimensional intersection form on V_0^{n-1} and using coefficients in $\mathbb{Z}_{(2)}$ (i.e. introducing all odd denominators) we will put it in a form in which the Arf invariant of φ can be easily computed.

Since $(\mathbb{C}P^{n+k-1}, V_0)$ is $(n-1)$-connected, $i_* : H_{n-1}(V_0) \longrightarrow H_{n-1}(\mathbb{C}P^{n+k-1}) \cong \mathbb{Z}$ is onto, and therefore splits. Hence $i^* : H^{n-1}(\mathbb{C}P^{n+k-1}) \longrightarrow H^{n-1}(V_0)$ also splits and we let $h = i^*(g)$, where g generates $H^{n-1}(\mathbb{C}P^{n+k-1})$.

(2.14) <u>Lemma</u>. The Poincaré dual of L = the annihilator of h under $(\ , \ )$.

<u>Proof</u>. $g(i_*x) = i^*(g)(x) = h(x) = h([V_0] \cap \bar{x}) = (h \cup \bar{x})[V_0]) = (h, \bar{x})$. □

Now let $\beta \in H^{n-1}(V_0)$ be such that $(h, \beta) = 1$, (which is possible since h is indivisible.) Now $(h, h) = (i^*g \cup i^*g)[V_0] = (g \cup g)\, i_*[V_0] = (g \cup g)(d[\mathbb{C}P^{n-1}]) = d$. Hence the quadratic form on A = the submodule generated by h, β, has the matrix

$$\begin{pmatrix} d & 1 \\ 1 & a \end{pmatrix}$$

and hence has <u>odd</u> determinant $ad - 1$, since d is even.

Hence, over $\mathbb{Z}_{(2)}$, we can find a complementary summand B to A so that the matrix for $H^{n-1}(V_0; \mathbb{Z}_{(2)}) = A + B$ becomes

$$\begin{pmatrix} \begin{pmatrix} d & 1 \\ 1 & a \end{pmatrix} & 0 \\ 0 & T \end{pmatrix}$$

Since $B \perp h$, and B is the largest submodule of (annihilator (h)) on which the bilinear form is non-singular (mod 2), it follows that $\text{Arf}\,\varphi = K(V_0)$ is the Arf invariant of the quadratic form $\frac{x \cdot x}{2}$ (mod 2) on B, and T is the matrix for this intersection form.

(2.15) Proposition. The Arf invariant

$$\text{Arf}(B) = \begin{cases} 0 \text{ if } \det T \equiv \pm 1 \bmod 8 \\ 1 \text{ if } \det T \equiv \pm 3 \bmod 8 . \end{cases}$$

See for example [Hirzebruch-Mayer ; (9.3)]. We sketch the proof here.

Over $\mathbb{Z}_{(2)}$, a matrix T with even diagonal entries and odd determinant may be put in form of the sum of 2×2 blocks

$$STS^t = \begin{pmatrix} \begin{pmatrix} a_1 & 1 \\ 1 & b_1 \end{pmatrix} & & 0 \\ & \begin{pmatrix} a_2 & 1 \\ 1 & b_2 \end{pmatrix} & \\ 0 & & \begin{pmatrix} a_t & 1 \\ 1 & b_t \end{pmatrix} \end{pmatrix}$$

For given a generator g of B, since $\det T$ is odd, there is $g' \in B$ such that (g, g') is odd so that over $\mathbb{Z}_{(2)}$ g and g' generate a submodule whose matrix may be made into $\begin{pmatrix} a & 1 \\ 1 & b \end{pmatrix}$ and we may then split off this module (over $\mathbb{Z}_{(2)}$) and proceed by induction.

For $\begin{pmatrix} a & 1 \\ 1 & b \end{pmatrix}$ (a, b even) clearly the Arf invariant is 1 if and only if both $\frac{a}{2}$ and $\frac{b}{2}$ are odd. Then $ab - 1 \equiv \pm 3 \bmod 8$ (i.e. $ab - 1$ is not a quadratic residue mod 8). The result then follows, adding Arf invariants and multiplying determinants of the 2×2 blocks.

Since the bilinear form is unimodular on all of $H^{n-1}(V_0)$, it follows that $\det T = (ad - 1)^{-1}$. The condition $\binom{s+\ell}{s+1} \equiv 0 \bmod 2$ (for defining ψ) implies $\ell \geq 2$ so that $4 \mid d$, (ℓ = number of even d_i's).

Hence $-(ad-1)^{-1} = 1 + ad + a^2d^2 + \ldots \equiv 1 + ad \pmod 8$. Hence, by (2.15), Arf B = 0 if $8 \mid d$ or if a is even. It remains to calculate $a = (\beta . \beta)$.

But $(\beta , \beta) = (v_{n-1}(V_0), \beta) \bmod 2$ so a is odd if and only if $v_{n-1}(V_0)$ is nonzero and equal to $h \pmod 2$. But $v_{n-1}(V_0) = i^*(v_{n-1}(E))$ where $E = E(-(\alpha^{d_1} + \ldots + \alpha^{d_k}))$, the total space of this stable bundle over $\mathbb{C}P^{n+k-1}$, so $a = (\beta , \beta)$ is odd if and only if $v_{n-1}(E) \neq 0$.

As in (2.3) we get that $v_{n-1}(E) \neq 0$ if and only if $v_{n-1}(\mathbb{C}P^{n-1+\ell}) \neq 0$. The latter happens if and only if $\binom{s+\ell}{s} \not\equiv 0 \bmod 2$ which completes the proof of Theorem C. □

Bibliography

W. Browder [K], The Kervaire invariant of framed manifolds and its generalization, Annals of Math 90 (1969), 157-186.

_________ [FPK], Cobordism invariants, the Kervaire invariant and fixed point free involutions, Trans. A.M.S. 178 (1973), 193-225.

_________ [S], Surgery on simply-connected manifolds, Springer Verlag, Berlin 1973.

E. H. Brown, Generalizations of the Kervaire invariant, Annals of Math. 95 (1972) 368-383.

F. Hirzebruch and K.H. Mayer, O(n) - Mannigfaltigkeiten, exotische Sphären und Singularitäten, Springer Lecture Notes No. 57, (1968).

M. Kervaire, Relative characteristic classes, Amer. J. Math. 79 (1957), 517-558.

R. Kulkarni and J. Wood, Topology of non-singular complex hypersurfaces (to appear).

A. Libgober, A geometrical procedure for killing the middle dimensional homology groups of algebraic hypersurfaces, Proc. A.M.S. 63 (1977), 198-202.

__________ and J . Wood, (in preparation.)

S. Morita [H] The Kervaire invariant of hypersurfaces in complex projective space, Commentarii Math. Helv. 50 (1975), 403-419.

__________ [P], On the Pontryagin square and the signature, J. Fac. Sci. Univ. Tokyo, sect IA Math 18 (1971) 405-414.

S. Ochanine, Signature et invariants de Kervaire généralisés. CR Acad. Sci. Paris, 285 (1977) 211-213.

N. Steenrod, Topology of fibre bundles, Princeton Univ. Press, Princeton, NJ 1951.

CTC Wall, Surgery of compact manifolds, Academic Press, New York, 1971.

J. Wood, [H] Removing handles from non-singular algebraic hypersurfaces in $\mathbb{C}P^{n+1}$, Inventiones 31 (1975), 1-6.

_______, [CI] Complete intersections as branched covers and the Kervaire invariant (to appear).

BOUNDS FOR CHARACTERISTIC NUMBERS OF FLAT BUNDLES

Johan L. Dupont

1. A well-known theorem of J. Milnor [8] states that on an oriented surface of genus h any flat $Sl(2,\mathbb{R})$-bundle has Euler number of numerical value at most $h-1$. Here "flat" means that there exists a system of local trivializations for the bundle such that all the transition functions are constant. D. Sullivan [9] has generalized this result by finding bounds for the Euler number of a flat $Sl(2n,\mathbb{R})$-bundle on a 2n-dimensional manifold. In this note we shall generalize Milnor's theorem in a different direction by finding bounds for the 2-dimensional real characteristic numbers of flat G-bundles where G is any connected semi-simple Lie group with finite center. By a real characteristic number we simply mean the evaluation of a real characteristic class (i.e. the pull-back under the classifying map of a class in $H^2(BG,\mathbb{R})$) on a given homology class in the base and we want to estimate this number independently of the flat bundle. Actually it suffices to consider the characteristic numbers of flat bundles over surfaces (see Remark 2 following Proposition 2.2 below) and in this case our results are given by Proposition 2.2 and Theorem 4.1 below.

The results depend on the particular simple description due to Guichardet and Wigner [4] which one has for 2-dimensional continuous cochains on Lie groups and I am indebted to Professor A. Guichardet for informing me about his work.

2. In the following G denotes a connected non-compact Lie group, BG is the classifying space and $c \in H^2(BG,\mathbb{R})$ is any class. Since BG is simply connected $H^2(BG,\mathbb{R}) = H^2(BG,\mathbb{Z}) \otimes \mathbb{R}$ and we shall actually take

$c \in H^2(BG,\mathbb{Z})$. For a principal G-bundle ξ on a space X. The associated characteristic class (with real coefficients) is denoted $c(\xi)$. If ξ is flat its classifying map factors through the natural map $BG_d \to BG$ where G_d is the underlying discrete group of G. Now $H^*(BG_d,\mathbb{R})$ is identified with the Eilenberg-MacLane group cohomology of G_d where a q-cochain is any real valued function on $G \times \ldots \times G$ (q factors) (see e.g. MacLane [7, chapter 4 §5]). Thus the image $\tilde{c} \in H^2(BG_d,\mathbb{R})$ of c is represented by a 2-cochain $f: G \times G \to \mathbb{R}$ and furthermore f can be chosen to satisfy

$$f(x,1) = f(1,x) = f(x,x^{-1}) = 0 \quad \forall x \in G. \tag{2.1}$$

We now introduce the following real number (or $+\infty$):

$$L(c) = \inf\{ \sup_{x_1,x_2 \in G} |f(x_1,x_2)| \,\Big|\, f: G \times G \to \mathbb{R} \text{ represents } \tilde{c} \text{ and satisfies (2.1)}\}.$$

Also for $r \in \mathbb{R}_+$ we use the notation $[r]$ for the largest integer less than r. Then we have:

<u>Proposition 2.2</u>. For ξ a flat G-bundle on a surface X_h of genus h

$$|\langle c(\xi), [X_h]\rangle| \leq [(4h-2)L(c)]$$

<u>Proof</u>. (Implicit in [2, proof of Corollary 4.10] or [1,p.154]). Let Γ be the group generated by $x_1,\ldots,x_{2h}$ satisfying the single relation

$$[x_1,x_2][x_3,x_4]\ldots[x_{2h-1},x_{2h}] = 1 \tag{2.3}$$

Then X_h is homotopy equivalent to $B\Gamma$ and the flat bundle ξ is induced by a homomorphism $\alpha: \Gamma \to G$. The fundamental class $[X_h]$ correspond to 2-cycle

$$\begin{aligned} z = {} & (x_1,x_2) + (x_1x_2,x_1^{-1}) + \ldots + (x_1x_2\ldots x_{2h},x_{2h-1}^{-1}) + \\ & + (1,1) - (x_1,x_1^{-1}) + \ldots + (1,1) - (x_{2h-1},x_{2h-1}^{-1}) \end{aligned} \tag{2.4}$$

in the integral chain complex for the homology of Γ. Now $\langle c(\xi),[X_h]\rangle = \langle\tilde{c},\alpha_* z\rangle$, and since f representing $\tilde{c}$ satisfies (2.1) only the (4h-2) terms in the first line of (2.4) contributes to the characteristic number. The inequality therefore follows from the definition of $L(c)$.

Remark 1. By Proposition 2.2 it remains to estimate $L(c)$ and in fact Milnors result for $Sl(2,\mathbb{R})$ follows once it is proved that $L(c) \leq \frac{1}{4}$ for c the Euler class. Since Milnor in [8] also constructs a flat $Sl(2,\mathbb{R})$-bundle on X_h with Euler number $1-h$ it follows from Proposition 2.2 that in this case $L(c) \geqq \frac{1}{4}$.

Remark 2. The restriction in Proposition 2.2 on the base space to be a surface is unimportant:
In general suppose ξ is a flat G-bundle on any space X and let Π be the fundamental group of X. Then the classifying map for ξ factors as

$$X \xrightarrow{\varphi} B\Pi \xrightarrow{B\alpha} BG$$

where $\alpha: \Pi \to G$ is a homomorphism and φ classifies the universal covering of X. By a classical theorem of H. Hopf [6]

$$(2.5) \qquad H_2(B\Pi,\mathbb{Z}) \cong R \cap [F,F]/[R,F]$$

where $\Pi = F/R$ is a presentation of Π, and under this isomorphism a relation of the form (2.3) corresponds to the cycle given by (2.4). Theorefore for $y \in H_2(X,\mathbb{Z})$ the characteristic number $\langle c(\xi),y\rangle$ is again bounded numerically by $[(4h-2)L(c)]$ if $\varphi_* y$ is represented by a product of h commutators via the isomorphism (2.5).

3. Before estimating $L(c)$ we first calculate $H^2(BG,\mathbb{Z})$:

Lemma 3.1. Let $K \subseteq G$ be a maximal compact subgroup. Then

1) $\quad H^2(BG,\mathbb{Z}) \simeq H^2(BK,\mathbb{Z}) \simeq \mathrm{Hom}_{\mathrm{cont.}}(K,\mathbb{T})$

where $\mathbb{T} \subseteq \mathbb{C}^*$ is the circle group.

2) The free abelian group in 1) has rank equal to the dimension of the center of K.

3) If G is <u>simple</u> with finite center then

$$H^2(BG,\mathbb{Z}) \cong \begin{cases} 0 \\ \mathbb{Z} \end{cases}$$

corresponding to the center of K being either discrete or a circle.

<u>Proof</u>. 1) The first isomorphism is obvious since G/K is contractible. For the second isomorphism let $C \subseteq K$ be the center of K and let $\mathfrak{C} \subseteq \mathfrak{K}$ be the associated Lie algebras. Then $\mathfrak{K} = \mathfrak{C} \oplus [\mathfrak{K}\ \mathfrak{K}]$ (Helgason [5 Chapter II Proposition 6.6]) and we let K' be the analytic subgroup of K with Lie algebra $[\mathfrak{K},\mathfrak{K}]$. Then K' is a semisimple compact Lie group and K/K' is a torus with Lie algebra $\mathfrak{C}$. Hence we get an exact sequence

$$0 \to \pi_1(K') \to \pi_1(K) \to \pi_1(K/K') \to 0$$

and $\pi_1(K')$ is finite. Therefore using the natural isomorphism

$$H^2(BK,\mathbb{Z}) \cong \mathrm{Hom}(\pi_1(K),\mathbb{Z})$$

it follows that the homomorphism $K \to K/K'$ induces an isomorphism

$$H^2(BK,\mathbb{Z}) \xrightarrow{\cong} H^2(B(K/K',\mathbb{Z}).$$

Hence it suffices to prove 1) for K a torus which case is trivial. 2) is implicit in the above since dim K/K' = dim C. 3) follows from 2) and the fact (Helgason [5, Chapter IX, Exercise 2]) that for G simple with finite center dim $C \leq 1$.

<u>Remark</u>. The simple non-compact Lie groups with finite center was classified by E. Cartan (see Helgason [5, p.354]). Up to a finite covering those for which K has non-discrete center are (notation as in

Helgason [5]):

(3.2) $SU(p,q),\ p \leqq q;\quad SO^*(2n),\ n \geqq 2;\quad SO_0(2,q),\ q \neq 2;\quad Sp(n,\mathbb{R});$

plus two exceptional cases.

4. We can now state our main result:

Theorem 4.1. Let G be semi-simple with finite center and $c \in H^2(BG,\mathbb{Z})$.

1) $L(c) < \infty$.

2) If G is one of the simple classical groups in (3.2) and $c \in H^2(BG,\mathbb{Z}) \cong \mathbb{Z}$ is the generator then

$$L(c) \leqq \frac{2^l - 1}{2}, \qquad l = \mathbb{R}\text{-rank}(G).$$

3) For $G = SU(1,q)$ (i.e. $\mathbb{R}$-rank$(G) = 1$) and c the generator we have $L(c) \leqq \frac{1}{4}$.

Remark. The $\mathbb{R}$-rank of G is the dimension of a maximal flat subspace of the symmetric space G/K. For the groups listed in (3.2) the $\mathbb{R}$-ranks are respectively

$$l = p,\ \left[\frac{n}{2}\right],\ 2,\ n\,.$$

For the proof of Theorem 4.1 we shall use the following description due to Guichardet and Wigner [4] of a representative $f\colon G \times G \to \mathbb{R}$ for $\tilde{c} \in H^2(BG_d,\mathbb{R})$. By Lemma 3.1, $c \in H^2(BG,\mathbb{Z})$ is given by a homomorphism $u\colon K \to \mathbb{T}$. Also let $\mathfrak{G} = \mathcal{K} \oplus \mathcal{J}$ be a Cartan decomposition of the Lie algebra of G. Then $G = K\cdot\exp\mathcal{J}$ and the decomposition $g = k\cdot\exp X$, $k \in K$, $X \in \mathcal{J}$, is unique, so one can define $v_0(g) = u(k)$.

Theorem 4.2. There is a representative $f\colon G \times G \to \mathbb{R}$ for $\tilde{c} \in H^2(BG_d,\mathbb{R})$ such that

a) f is continuous.

b) $f(g_1,g_2) = f(k_0 g_1 k_1,\ k_1^{-1} g_2 k_2)$, $g_1,g_2 \in G$, $k_0,k_1,k_2 \in K$.

c) $f(g,g^{-1}) = f(1,g) = f(g,1) = 0 \quad \forall g \in G$.

d) $f(g_1,g_2) = \frac{1}{2\pi} \arg(v_0(g_1)v_0(g_2)v_0(g_1g_2)^{-1})$

Proof of Theorem 4.2. 1) Let $P\colon \mathcal{K} \to \mathbb{R}$ be the K-invariant linear form given by

$$P = \frac{1}{2\pi i}\ u_*$$

where u_* is the differential of u at the identity. Then it is easy to see that $c \in H^2(BK,\mathbb{R})$ is the Chern-Weil image of P. Therefore as in [2. Corollary 1.3] $\tilde{c}$ is represented by the cochain f given by

$$f(g_1,g_2) = \int_{\Delta(g_1,g_2)} P(\Omega), \quad g_1,g_2 \in G \tag{4.3}$$

where $P(\Omega)$ is a certain G-invariant 2-form on G/K and $\Delta(g_1,g_2) \subseteq G/K$ is the "geodesic 2-simplex" with corners o, g_1o, g_1g_2o (where $o = \{K\} \in G/K$). It is clear from (4.3) that f satisfies a), b) and c) and it is shown in [3] that it satisfies d).

Remark. f is of course uniquely determined by a), c) and d). Using this characterization Guichardet and Wigner [4] constructed f without using (4.3).

5. Proof of Theorem 4.1.

1) Let $Z \subseteq G$ be the center. Then $Z \subseteq K$ and clearly $\bar{K} = K/Z$ is the maximal compact subgroup of $\bar{G} = G/Z$, which is semi-simple without center. Using Lemma 3.1, 1) it is easy to see that it suffices to consider the case $Z = \{1\}$ in which case G is a product of simple groups. So it remains to consider G simple and c the generator of $H^2(BG,\mathbb{Z})$ in which case we want to estimate the function $f\colon G \times G \to \mathbb{R}$ given by Theorem 4.2. This can be done for G a general simple group using the description of f given by Guichardet and Wigner [4, §3].

However, this is only interesting for the 2 exceptional cases of G since for the classical groups listed in (3.2) one gets much better estimates by direct computation. Therefore we turn to the proof of

2) Let us restrict to the case $G = SU(p,q)$, $p \leq q$, since the other cases are completely analogous. Every element $g \in G$ we write as a matrix

$$g = \begin{pmatrix} g_{11} & g_{12} \\ g_{21} & g_{22} \end{pmatrix}$$

with respect to the direct sum decomposition $\mathbb{R}^{p+q} = \mathbb{R}^p \oplus \mathbb{R}^q$. The maximal compact subgroup is $K = S(U(p) \times U(q))$ consisting of matrices

$$k = \begin{pmatrix} k_{11} & 0 \\ 0 & k_{22} \end{pmatrix}, \quad k_{11} \in U(p), \ k_{22} \in U(q)$$
$$\det(k_{11})\det(k_{22}) = 1.$$

Also in the Cartan decomposition $\mathfrak{G} = \mathfrak{K} \oplus \mathfrak{P}$, $\mathfrak{P}$ is the set of complex matrices of the form

$$(5.1) \qquad X = \begin{pmatrix} 0 & {}^t\bar{X}_{21} \\ X_{21} & 0 \end{pmatrix}$$

where ${}^t\bar{X}_{21}$ is the transpose conjugate of X_{21}. It is easy to see that the group $\mathrm{Hom}_{cont}(K,\mathbb{T})$ is infinite cyclic with generator u given by

$$u(k) = \det(k_{11})$$

We shall estimate $|f(g_1,g_2)|$, $g_1,g_2 \in G$, where f is given by Theorem 4.2. Notice that by Theorem 4.2, b) we can assume $g_1,g_2 \in \exp(\mathfrak{P})$ in which case Theorem 4.2, d) reads

$$(5.2) \qquad f(g_1,g_2) = -\frac{1}{2\pi} \arg(v_0(g_1g_2)), \ g_1,g_2 \in \exp(\mathfrak{P})$$

Observe that for $g = k \cdot p$ with $k \in K$, $p \in \exp(\mathfrak{P})$ clearly $\det(g_{11}) = \det(k_{11})\det(p_{11})$ where $\det(p_{11})$ is a positive real number. Therefore (5.2) becomes

(5.3) $$f(g_1,g_2) = -\frac{1}{2\pi}\arg(v(g_1g_2)),\quad g_1,g_2 \in \exp(\mathfrak{S}),$$

where $v(g) = \det(g_{11})$ for any $g \in G$ (this observation is also due to Guichardet and Wigner [4]).

Now write $g_2 = \exp(X)$, $X \in \mathfrak{S}$, and observe that X can be changed by the adjoint action of K without changing $f(g_1,g_2)$ (cf. Theorem 4.2, b)), so we can assume X_{21} to be of the form

$$X_{21} = \begin{pmatrix} \alpha_1 & \cdots & 0 \\ \vdots & \ddots & \vdots \\ 0 & \cdots & \alpha_p \\ \vdots & & \vdots \\ 0 & \cdots & 0 \end{pmatrix}$$

where $\alpha_1,\ldots,\alpha_p$ are non-negative reals. Then $g_2 = \exp(X)$ is given by the matrix

$$g_2 = \begin{pmatrix} \cosh A & \sinh A & 0 \\ \sinh A & \cosh A & 0 \\ 0 & 0 & 1 \end{pmatrix}, \quad A = \begin{pmatrix} \alpha_1 & \cdots & 0 \\ \vdots & \ddots & \vdots \\ 0 & \cdots & \alpha_p \end{pmatrix}$$

with respect to the decomposition $\mathbb{R}^{p+q} = \mathbb{R}^p \oplus \mathbb{R}^p \oplus \mathbb{R}^{q-p}$. Hence

(5.4) $$\begin{aligned} v(g_1g_2) &= \det(g_{111}\cosh A + g_{112}\sinh A) \\ &= \sum_{\varepsilon_1,\ldots,\varepsilon_p = \pm 1} \lambda_{\varepsilon_1\ldots\varepsilon_p} e^{\varepsilon_1\alpha_1+\ldots+\varepsilon_p\alpha_p} \end{aligned}$$

where $\lambda_{\varepsilon_1\ldots\varepsilon_p}$ are polynomials in the entries of g_1. Replacing g_2 by $g_2^s = \exp(sX)$, $s \in [0,1]$, we obtain

$$f(g_1,g_2^s) = -\frac{1}{2\pi}\arg(v(g_1g_2^s))$$

where

(5.5) $$v(g_1g_2^s) = \sum_{\varepsilon_1,\ldots,\varepsilon_p} \lambda_{\varepsilon_1\ldots\varepsilon_p} e^{(\varepsilon_1\alpha_1+\ldots+\varepsilon_p\alpha_p)s}$$

Since $f(g_1,g_2^s)$ is continuous in s and is zero for $s = 0$ it is bounded by the number of times which the function given by (5.5) winds around 0 in the complex plane for $s \in [0,1]$. In turn this number is bounded by ½ times the number of zeroes of the function $\operatorname{Im} v(g_1g_2^s)$.

Hence the estimate

$$|f(g_1,g_2)| \leq \frac{2^p-1}{2}$$

follows from the probably well-known

<u>Lemma 5.6</u>. Let ψ be a real function of a real variable given by

$$\psi(s) = \sum_{i=1}^{k} a_i e^{\alpha_i s}$$

where $a_1,\ldots,a_k,\ \alpha_1,\ldots,\alpha_k \in \mathbb{R}$. Then ψ has at most $k-1$ zeroes.

<u>Proof</u>. The zeroes of ψ are clearly the same as those of

$$\varphi(s) = \sum_{i=1}^{k-1} a_i e^{(\alpha_i-\alpha_k)s} + a_k$$

Now by Rolle's theorem φ has at most one more zero than the derivative φ' which is a linear combination of $k-1$ exponentials. Hence the lemma follows by induction.

This proves Theorem 4.1, 2) for the case $G = SU(p,q)$. For the other classical simple groups the proof is exactly analogous using the formula (5.3) where the corresponding functions v are given for each case in Guichardet and Wigner [4, §5] (where they are denoted v').

3) For $G = SU(1,q)$ (5.4) becomes

$$v(g_1g_2) = a \cosh \alpha_1 + b_{11} \sinh \alpha_1$$

where

$$g_1 = \begin{pmatrix} a & b_{11} & \cdots & b_{1q} \\ \bar{b}_{11} & \cdots & & \vdots \\ \vdots & & & \vdots \\ \bar{b}_{1q} & \cdots & & \vdots \end{pmatrix} \qquad g_2 = \begin{pmatrix} \cosh \alpha_1 & \sinh \alpha_1 & 0 & \cdots & 0 \\ \sinh \alpha_1 & \cosh \alpha_1 & 0 & \cdots & \vdots \\ 0 & \cdots & & & \vdots \\ \vdots & & & & \vdots \\ 0 & \cdots & & & 0 \end{pmatrix}$$

Here a is a positive real number and

$$a^2 - |b_{11}|^2 - \ldots - |b_{1q}|^2 = 1$$

so that $a^2 - (\mathrm{Re}\, b_{11})^2 \geqq 1$. Therefore

$$\begin{aligned}\mathrm{Re}(v(g_1 g_2)) &= a \cosh \alpha_1 + (\mathrm{Re}\, b_{11}) \sinh \alpha_1 \\ &= \tfrac{1}{2}(a+\mathrm{Re}\, b_{11}) e^{\alpha_1} + \tfrac{1}{2}(a-\mathrm{Re}\, b_{11}) e^{-\alpha_1} > 0\end{aligned}$$

so that $|f(g_1,g_2)| \leqq \frac{1}{4}$.

Remark 1. The result that on a surface of genus h the Chern number of a flat $SU(1,q)$-bundle is at most $h-1$ has also been proven by D. Toledo (private communication).

Remark 2. I doubt that the estimates in Theorem 4.1 are the best possible except for $G = SU(1,q)$ where it *is* by the Remark 1 following Proposition 2.2. Thus it would be interesting to estimate $L(c)$ from below. E.g. for $G = SU(p,q)$, $p > 1$, products of Milnor's examples just gives $L(c) \geqq \frac{p}{4}$.

Remark 3. In view of the proof of Theorem 4.1 it is natural to ask if all the continuous cochains defined in [2] by integration over geodesic simplices are bounded. However, the above proof in the 2-dimensionale case does not generalize in any obvious way.

REFERENCES

1. J.L. Dupont, Curvature and characteristic classes (Lecture Notes in Mathematics 640), Springer-Verlag, Berlin-Heidelberg-New York.

2. J.L. Dupont, Simplicial de Rham cohomology and characteristic classes of flat bundles, Topology 15(1976), 233-245.

3. J.L. Dupont et A. Guichardet, A propos de L'article "Sur la cohomologie reelle des groupes de Lie simples reels", Ann.Sci. École Norm.Sup., 11(1978), 277-292.

4. A. Guichardet et D. Wigner, Sur la cohomologie reelle des groupes Lie simples reels, Ann.Sci.École Norm.Sup., 11(1978),293-296.

5. S. Helgason, Differential Geometry and Symmetric Spaces, Academic Press, New York - London, 1962.

6. H. Hopf, Fundamentalgruppe und zweite Bettische Gruppe, Comment. Math.Helv.14(1941/42), 257-309.

7. S. MacLane, Homology (Grundlehren Math.Wissensch.114), Springer-Verlag, Berlin-Göttingen-Heidelberg, 1963.

8. J.W. Milnor, On the existence of a connection with curvature zero, Comment.Math.Helv.32(1958), 215-223.

9. D.Sullivan, A generalization of Milnor's inequality concerning affine foliations and affine manifolds, Comment.Math.Helv. 51(1976), 183-189.

/KB

Exotic characteristic classes of Spherical Fibrations

by Friedrich Hegenbarth

Introduction:

F. P. Peterson defined in [1] exotic characteristic classes $e_i \in H^{2^i-1}(BSG, Z_2)$, $i = 2, 3, 4, \ldots$. The definition of e_i is based on the following relation in the Massey-Peterson algebra $A \odot H^*(BG; Z_2)$

$$\sum_{0 \le j \le i} \Theta\, \chi(Sq^{2^i-2^j})\ \Theta\, \chi(Sq^{2^j}) = 0.$$

Here A denotes the Steenrod algebra mod 2 and

$$\Theta : A \to A \odot H^*(BG; Z_2)$$

is the algebra-map which satisfies

$$\Theta(Sq^n) = \Sigma\, Sq^{n-i} \otimes q_i$$

(q_i is the ith Wu-class).

e_2 was already defined by S. Gitler and J. D. Stasheff (see [2]). They also constructed a spherical fibration ξ over S^3 such that $e_2(\xi) \in H^3(S^3; Z_2)$ is the generator.

Later on D. C. Ravenel [3] showed that also the other e_i are non-trivia

Moreover he could show that the e_i are welldefined modulo ordinary characteristic classes.

Following a suggestion by M. Mahowald we are going to prove the followin

Theorem:

There is a stable spherical fibration ξ over $\Omega^2 S^5$ with $e_i(\xi) \neq 0$ for all $i = 2, 3, \ldots$.

The author is grateful to M. Mahowald for suggesting this example and for conversations during the Aarhus Algebraic Topology Symposium in August 1978.

§ 1 The classes e_i.

In this section we will recall some facts from [3] about the e_i. In [3] there is proved that $e_i \neq 0$ in case of Zp-coefficients, where p is an odd prime. In case of Z_2-coefficients (in which we are only interes here) the proof partially differs from that (in particular the proof of Lemma 4.2.2 in [3]). We also need some details. Therefore we will sketch a proof that the $e_i \in H^{2^i-1}(BSG; Z_2)$ are non-trivial.

Before, however, we will recall the homology of SG and BSG. Good references for that are [4] and [5]. There is proved that as Hopf algebras:

$$H_*(SG) = E\{Q^a[1] * [-1] \mid a \geq 1\} \otimes Z_2[Q^a Q^a[1] * [-3] \mid a \geq 3\}$$
$$\otimes Z_2[Q^I[1] * [1-2^{l(I)}] \mid I \text{ allowable, } e(I) > 0,\ l(I) \geq 2]$$

$$H_*(BSG) = H_*(BSO) \otimes E\{\sigma(Q^a Q^a[1] * [-3]) \mid a \geq 1\}$$
$$\otimes Z_2[\sigma(Q^a Q^b[1] * [-3] \mid b < a \leq 2b\}$$
$$\otimes Z_2[\sigma(Q^I[1] * [1-2^{l(I)}) \mid I \text{ allowable } e(I) > 1,\ l(I) > 2]$$

Here $I = \{i_1, i_2, \ldots, i_n\}$, $Q^I = Q^{i_1} \ldots Q^{i_n}$, $l(I) = n$, $e(I) = i_1 - \sum_{j=2}^{n} i_j$.

σ denotes the homology suspension. The algebra structure of $H_*(SG)$ comes from the composition product in $SG \subset \Omega^\infty S^\infty$.

Let now $K_m = K(Z_2, m)$ and consider the following diagram: (cf. diagra on p. 432 in [3])

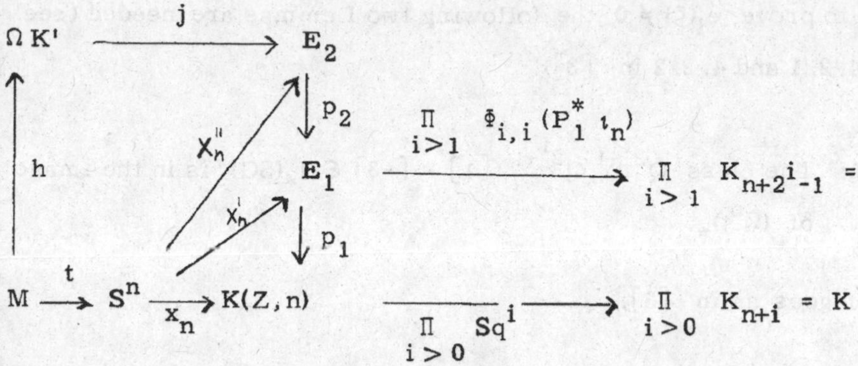

Here E_1 and E_2 are the principal fibrations induced from $\prod_{i>0} Sq^i$ and $\prod_{i>1} \Phi_{i,i}$. $\Phi_{i,i}$ is a secondary operation belonging to the relation $\sum_{0 \leq j \leq i} \chi(Sq^{2^i-2^j})\chi(Sq^{2^j}) = 0$. $x_n \in H^n(S^n;Z)$ is a generator and x_n', x_n'' are liftings of x_n.

M, h and t are defined as the fibre product of j and x_n''. n is a sufficiently large number.

We note that $\Omega^n E_1 \simeq Z \times \Omega^{n+1} K$, $\Omega^n E_2 \simeq Z \times \Omega^{n+1} K' \times \Omega^{n+1} K$ (homotopy-equivalences).

The image of $\Omega^n t : \Omega^n M \to \Omega^n S^n$ is containt in the zero component $(\Omega^n S^n)_0$ of $\Omega^n S^n$ (note that M is n-connected, so $\Omega^n M$ is connected). Taking loop sum with 1 defines a homotopy equivalence

$$*1 : (\Omega^n S^n)_0 \to (\Omega^n S^n)_1 = SG(n)$$

The composition

$$\Omega^n M \xrightarrow{\Omega^n t} (\Omega^n S^n)_0 \simeq SG(n) \to SG$$

defines a spherical fibration ξ over $\Sigma \Omega^n M$. We will denote the composition also by $\Omega^n t$.

In order to prove $e_i(\xi) \neq 0$ the following two Lemmas are needed (see Lemma 4.2.1 and 4.2.2 in [3]).

Lemma 1: The class $Q^{2^i-1} Q^{2^i-1} [1] * [-3] \in H_*(SG)$ is in the image of $(\Omega^n t)_*$.

The proof goes as in [3].

Lemma 2: If $b_{i+1} \in H_{2^{i+1}-2}(\Omega^n E_2)$ is the fundamental class of the factor $K_{2^{i+1}-2}$, then $(\Omega^n \dot{x}''_n)_* (Q^{2^i-1} Q^{2^i-1} [1]*[-3]) = b_i$

If $\tilde{t} : \Sigma^n \Omega^n M \to S^n$ is the adjoint map of $\Omega^n t$ then the mapping cone $C_{\tilde{t}}$ of $\tilde{t}$ is homotopy equivalent to $T(\xi)$, the Thom space of ξ. We consider now the following diagram which is commutative (cf. diagram on p. 433 [3]).

$$
\begin{array}{ccccccc}
\Omega K' & \xrightarrow{j} & E_2 & \xrightarrow{p_2} & E_1 & \xrightarrow{\prod_{i>1} \Phi_{i,i}(p_1^* i_n)} & K' \\
\uparrow h & & \uparrow x''_n & & \uparrow \alpha'' & & \uparrow h' \\
M & \xrightarrow{t} & S^n & \longrightarrow & C_t & \xrightarrow{v} & \Sigma M \\
\uparrow \alpha & & \| & & \uparrow \alpha' & & \uparrow \Sigma\alpha \\
\Sigma^n \Omega^n M & \longrightarrow & S^n & \longrightarrow & T(\xi) & \xrightarrow{\tilde{v}} & \Sigma^{n+1}\Omega^n M
\end{array}
$$

h' is the adjoint map of h and α'' is induced by $x'_n : S^n \to E_1$.

Now by definition (and commutativity of the right squares)

(1) $\quad e_{2^i-1}(\xi) \cup U = [h'_i \circ (\Sigma\alpha) \circ \tilde{v}]$

where $h'_i : \Sigma M \xrightarrow{h'} K' \to K_{n+2^i-1}$

is the composition of h' with the obvious projection. On the other hand it is wellknown (see f. e. [5] p. 127) that

(2) $$e_{2^i-1}(\xi) \cup U = \tilde{v}^*(\Sigma^n)^*(e_{2^i-1}(\xi))$$

($U \in H^n(T\xi)$ is the Thom-class)

Let us now consider the following commutative diagram:

$$\begin{array}{ccccccc}
H^{2^i-2}(\Omega^n S^n) & \xrightarrow{(\Omega^n t)^*} & H^{2^i-2}(\Omega^n M) & \cong & H^{2^i-1}(\Sigma\Omega^n M) & \cong & \\
\uparrow{\scriptstyle (\Omega^n x_n'')^*} & & \uparrow{\scriptstyle (\Omega^n h)^*} & & \uparrow{\scriptstyle g^*} & & \\
H^{2^i-2}(\Omega^n E_2) & \xrightarrow{(\Omega^n j)^*} & H^{2^i-2}(\Omega^{n+1} K') & \xleftarrow{\sigma^*} & H^{2^i-1}(\Omega^n K') & \xleftarrow{(\sigma^*)^n} &
\end{array}$$

$$\begin{array}{ccc}
H^{n+2^i-1}(\Sigma^{n+1}\Omega^n M) & \xrightarrow{(\tilde{v})^*} & H^{n+2^i-1}(T(\xi)) \\
\uparrow{\scriptstyle f^*} & & \\
H^{n+2^i-1}(K') & &
\end{array}$$

where $f = h' \circ \Sigma\alpha$ and $g = h \circ \alpha$.

$(\tilde{v})^*$ is an isomorphism and from the formulae (1) and (2) we see that the fundamental class $\iota_{n+2^i-1} \in H^{n+2^i-1}(K')$ goes to $e_{2^i-1}(\xi)$.

By Lemma 1 we choose $y \in H_{2^i-2}(\Omega^n M)$ with $(\Omega^n t)_*(y) =$
$= Q^{2^{i-1}-1} Q^{2^{i-1}-1}[1] * [-3]$.

Using Lemma 2 we obtain

$$< \Omega^n (j \circ h)^* (\iota_{2^i-2}), y > = < \iota_{2^i-2}, (\Omega^n x''_n)_* (\Omega^n t)_* (y) >$$

$= < \iota_{2^i-2}, b_i > = 1$. This gives us

Corollary 3: $\quad 1 = < \sigma^* (e_{2^i-1}), Q^{2^{i-1}-1} Q^{2^{i-1}-1} [1] * [-3] >$

$$= < e_{2^i-1}, \sigma (Q^{2^{i-1}-1} Q^{2^{i-1}-1} [1] * [-3]) > .$$

As mentioned at the beginning of the section we will sketch a proof of Lemma 2.

Consider $\Omega^n E_1 \simeq Z \times \Omega^{n+1} K$ and the fundamental class

$$h_i \in H_{2^i-1} (\Omega^n E_1)$$

of the factor K_{2^i-1}.

The difference from the Zp-case is the argument which proves the following

Sublemma (Lemma 4.3.1 in [3]):

It is
$$h_i = Q^{2^i-1} [1]$$

Proof: It is sufficient to consider only a part of the fibration $E_1 \to K(n, Z)$, namely

$$\begin{array}{ccccc} K_{n+2^i-1} & \to & E_{1,i} & & \\ & & \downarrow p_{1,i} & & \\ & & K(Z,n) & \xrightarrow{Sq^{2^i}} & K_{n+2^i} \end{array}$$

and its delooping

$$\begin{array}{ccc} K_{2^{i+1}-2} & \xrightarrow{j} & \Omega^{n-2^i+1} E_{1,i} \\ & & \downarrow \pi \\ & K(Z, 2^i-1) & \xrightarrow{Sq^{2^i}} K_{2^{i+1}-1} \end{array}$$

This is trivial and if ψ is the co-multiplication in $H^*(\Omega^{n-2^i+1} E_{1,i})$ then there can be choosen an element

$\gamma \in H^{2^{i+1}-2}(\Omega^{n-2^i+1} E_{1,i})$ such that

(3) $\psi(\gamma) = \gamma \otimes 1 + I \otimes I + 1 \otimes \gamma$

where $j^*(\gamma) = \iota_{2^{i+1}-2}$ and $\pi^*(\iota_{2^i-1}) = I$. (see [6] Lemma 3. 1. 1).

Let $x \in H_{2^i-1}(\Omega^{n-2^i+1} E_{1,i})$ be the dual of I. From (3) follows

$$Q^{2^i-1}(x) = x^2 \neq 0.$$

The Q^m commute with the homology suspension. Delooping the fibration further we obtain

$$Q^{2^i-1}[1] \neq 0 \quad \text{in} \quad H_{2^i-1}(\Omega^n E_{1,i}).$$

But then $Q^{2^i-1}[1] = h_i$ in $H_{2^i-1}(\Omega^n E_1)$ because we have a nth loop map $\Omega^n E_{1,i} \to \Omega^n E_1$.

A similar argument can be used to finish the proof of Lemma 2.

§ 2 Proof of the Theorem.

As mentioned in the introduction S. Gitler and J. Stasheff constructed a spherical fibration over S^3 with $e_2 \neq 0$. It is classified by a map

$$f' : S^3 \to BSG.$$

Under the isomorphism $\pi_3(BSG) \cong \pi_2^S = Z_2$ it corresponds to the generat
Now $BSG \simeq \Omega^2 B^3 SG$ and we can take the adjoint of the above map

$$f : S^5 \to B^3 SG.$$

If we loop it twice we obtain

$$g = \Omega^2 f : \Omega^2 S^5 \to \Omega^2 B^3 SG \simeq BSG.$$

This is the map we will investigate.
The following is wellknown:

Proposition: $H_*(\Omega^2 S^5) = Z_2 [x_i \mid i = 1, 2, 3, \ldots]$

Here $x_i = Q_1 \cdot Q_1 \cdots Q_1 x_1$, $x_1 \in H_3(\Omega^2 S^5)$ is the image of a generato in $H_3(S^3)$ under the natural map $S^3 \to \Omega^2 S^5$.

$$Q_1 : H_m(\Omega^2 S^5) \to H_{2m+1}(\Omega^2 S^5)$$

is the lower Dyer-Lashof operation. Note that $\deg x_i = 2^{i+1} - 1$.
Note also that $g_*(x_1) = e_2$ by the result of Gitler and Stasheff.
Because $g = \Omega^2 f$ we have

$$g_*(x_i) = \hat{Q}_1 \hat{Q}_1 \cdots \hat{Q}_1 (e_2).$$

The $\hat{Q}_r$ are the lower Dyer-Lashof operations in BG which comes from the composition structure in G.

Now $\hat{Q}_1(e_2) = \hat{Q}^4(e_2)$, $\hat{Q}_1 \hat{Q}_1(e_2) = \hat{Q}^8 \hat{Q}^4(e_2)$, ect. One also has $e_2 = \sigma(Q^1 Q^1 [1] * [-3])$ and therefore

$$g_*(x_i) = \hat{Q}^{2^i} \cdots \hat{Q}^4 \sigma(Q^1 Q^1 [1] * [-3])$$
$$= \sigma(\hat{Q}^{2^i} \cdots \hat{Q}^4 (Q^1 Q^1 [1] * [-3]))$$

(The Q^m are the Dyer-Lashof operation extending the loop-sum structure in $\Omega^\infty S^\infty$). We have to use some formulas to simplify the right hand side. Because σ vanishes on products we make our calculation modulo decomposables.

Following I. Madsen [4] we define

D_j to be the set of elements

$$x \in (\tilde{H}_*(\Omega^\infty S^\infty) * \tilde{H}_*(\Omega^\infty S^\infty)) \cap H_*((\Omega^\infty S^\infty)_j)$$

such that $x * [1-j]$ is composition decomposable in $H_*(SG)$.

Let $\mathcal{D} = \bigoplus_j \mathcal{D}j.$

By 4.5 [4] we have mod $\mathcal{D}$

$$\hat{Q}^4 (Q^1 Q^1 [1] * [-3]) \equiv \hat{Q}^4 Q^1 Q^1 [1] * [-3] + Q^4 Q^1 Q^1 [1] * [-7]$$

By Lemma 4.9 [4] for each a it is

$$Q^a [1] \cdot Q^a [1] = Q^a Q^a [1].$$

It follows from 4.6 [4] that

$$\hat{Q}^{2^i}(Q^{2^{i-1}-1}Q^{2^{i-1}-1}[1]) = Q^{2^i-1}Q^{2^i-1}[1]$$

So $\hat{Q}^4 Q^1 Q^1 [1] = Q^3 Q^3 [1]$

By the Adem relation it is

$$Q^{2^i} Q^{2^{i-1}-1} [1] = Q^{2^i-1} Q^{2^i-1} [1].$$

So $Q^4 Q^1 Q^1 [1] = Q^3 Q^2 Q^1 [1] = Q^2 Q^1 [1] * Q^2 Q^1 [1]$.

By Lemma 4.2 [4] $Q^I [1] \cdot Q^I [1]$ is loop decomposable for any sequence $I = \{i_1, \ldots, i_n\}$. From Proposition 4.5 [4] then follows that $Q^I [1] * Q^I [1] \in \mathfrak{D}$.

Hence we have

$$Q^4 Q^1 Q^1 [1] \in \mathfrak{D},$$

that is

$$\hat{Q}^{2^i} \cdots \hat{Q}^4 (\sigma(Q^1 Q^1 [1] * [-3])) =$$
$$\hat{Q}^{2^i} \cdots \hat{Q}^8 (\sigma(Q^3 Q^3 [1] * [-3])).$$

Proceeding in this way we obtain

$$g_*(x_i) = \sigma(Q^{2^i-1} Q^{2^i-1} [1] * [-3])$$

If ξ is the induced fibration by g we obtain by Corollary 3 of § 1

$$e_{2^i-1}(\xi) \neq 0.$$

References

[1] F. P. Peterson: Twisted Cohomology operations and Exotic characteristic Classes, Advances in Math. 4, 81-90 (1970)

[2] S. Gitler, J. D. Stasheff: The first Exotic Class of BF, Topology 4, 257-266 (1965)

[3] D. C. Ravenel: A Definition of Exotic Characteristic Classes of Spherical Fibrations, Comm. Math. Helv. 47, 421-436 (1972)

[4] I. Madsen: On the Action of the Dyer-Lashof Algebra in $H_*(G)$, Pacif. Journ. Math. 60, 235-275 (1975)

[5] J. P. May: The homology of $E\infty$ ring spaces, appeared in The Homology of iterated Loop Spaces, Lecture Notes in Mathematics, Springer-Verlag 1976, pp. 69-207

[6] R. J. Milgram: The Structure over the Steenrod Algebra of some 2-stage Postnikov Systems, Quart. J. Math. Oxford (2) 20 161-169 (1969)

On the (n+1)-tuple points of immersed n-spheres

by Ulrich Koschorke

In this paper we prove the following result.

Theorem. If $n \neq 1,3$ or 7, then any smooth selftransverse immersion of the n-sphere into $\mathbb{R}^{n+1}$ has an even number of $(n+1)$-tuple points.

('selftransverse' means that wherever sheets $V_1,\ldots,V_r$ meet, V_r lies transverse to $V_1 \cap \ldots \cap V_{r-1}$).

Before we start the proof, we express the problem in the language of homotopy theory, and we recall a few known facts.

Given an oriented closed n-manifold M and a smooth immersion $i : M \looparrowright \mathbb{R}^{n+1}$, the tangent map of i determines a stable parallelization $TM \oplus \mathbb{R} \cong \mathbb{R}^{n+1}$, and hence, via the Pontryagin-Thom construction, an element in the stable homotopy group π_n^S of spheres. This procedure allows us to identify π_n^S with the bordism group of oriented n-manifolds smoothly immersed into $\mathbb{R}^{n+1}$. Thus, counting (generic) (n+1)-tuple points we get a homomorphism

$$\Theta = \Theta_n : \pi_n^S \longrightarrow \mathbb{Z}_2 \quad .$$

In [3] M. Freedman conjectured that Θ_n is the stable Hopf invariant (i.e. Θ_n is nontrivial precisely when $n = 0,1,3$, or 7), and he checked this for $n \leq 3$. Then, in [5], I showed among other things that $\Theta_n \equiv 0$ if $n < 2$ $n \neq 0, 1, 3$ or 7; this was done essentially by using Toda's tables and establishing the following two facts.

(i) If $\alpha \in \pi_a^S$, $\beta \in \pi_b^S$ with $a,b > 0$, then

$$\Theta_{a+b}(\alpha \cdot \beta) = 0 \quad .$$

(ii) Given $\alpha \in \pi_a^S$, $\beta \in \pi_b^S$ and $\gamma \in \pi_c^S$ such that $\alpha\cdot\beta = \beta\cdot\gamma = 0$, consider the Toda bracket $<\alpha,\beta,\gamma> = \pm<\gamma,\beta,\alpha>$ as a subset of $\pi_{a+b+c+1}^S$. Assume

$$(*) \qquad 0 < a \leq b+c .$$

If

$$(**) \qquad \Theta_c(\gamma) = 0 \text{ or } c+1 \text{ does not divide } a+b+1 ,$$

then

$$\Theta_{a+b+c+1}(<\alpha, \beta,\gamma>) = \{0\} .$$

Now we are ready to prove the theorem above. It is equivalent, by [4], to the claim that for $n \neq 1,3$ or 7 Θ_n vanishes on the image of the classical J-homomorphism

$$J : \pi_n(SO) \longrightarrow \pi_n^S .$$

We have the following important information concerning this image (see [1] or [2, propositions 4.3, 4.8 and 4.9]). If ρ_j denotes the generator of the 2-primary part of Im J in the $(8j-1)$-stem, $j>1$, then the next four generators of the 2-component of Im J lie in the subsets $\{\eta\rho_j\}$, $\{\eta^2\rho_j\}$, $<2,\eta,\eta^2\rho_j>$ and $<2, 8\sigma,\rho_j>$ of the stems in dimension $8j$, $8j+1$, $8j+3$ and $8j+7$ respectively. Clearly Θ vanishes on the generators $\eta\rho_j$ and $\eta^2\rho_j$, by fact (i). We also want to show that Θ vanishes on the Toda brackets $<2,\eta,\eta^2\rho_j>$ and $<2, 8\sigma,\rho_j>$. Since the dimension assumption (*) of fact (ii) is not satisfied here, we need the following modification in order to complete the proof of our theorem.

Proposition. Fact (ii) still holds if the assumption (*) is replaced by

$$(*') \quad \alpha = 2\alpha' \quad \text{for some} \quad \alpha' \in \pi_a^S .$$

To see this, follow the line of the proof of fact (ii) in [5], § 1. Every element in $<\alpha, \beta, \gamma>$ can be obtained by fitting two bordisms

$$\ell'_+ * k' \; : \; X \times C \longrightarrow \mathbb{R}^{a+b+c+1} \times [0,\infty)$$

and

$$i' * \ell'_- \; : \; A \times Y \longrightarrow \mathbb{R}^{a+b+c+1} \times (-\infty,0]$$

of a certain immersion

$$i' * j' * k' \; : \; A \times B \times C \longrightarrow \mathbb{R}^{a+b+c+1} \times \{0\}$$

together. Condition (**) garantees that $\ell'_+ * k'$ has an even number of (a+b+c+2)-tuple-points. If $\alpha = 2\alpha'$, then we may assume that $i' * \ell'_-$ consists of two disjoint copies of the same immersion; thus its (a+b+c+2)-tuple points occur in pairs.

I would like to thank M. Crabb and K.H. Knapp for useful references. M. Crabb informs me that the theorem above follows also from work of his and from fact (ii) in its original form.

References.

[1] J.F. Adams, On the group J(X) - IV, Topology 5 (1966), 21-71.

[2] S. Feder, S. Gitler and K.Y. Lam, Composition properties of projective homotopy classes, Pacif.Journ.of Math.68(1977),47-61.

[3] M. Freedman, Quadruple points of 3-manifolds in S^4, Comm.Math.Helv. 53 (1978), 385-394.

[4] M. Hirsch, Immersions of manifolds, Trans.Amer.Math. Soc. 93 (1959), 242-276.

[5] U.Koschorke, Multiple points of immersions, and the Kahn-Priddy theorem, Math. Z. 1979.

Isotopy classification of spheres in a manifold

Lawrence L. Larmore

In [4] and [5], a theory of isotopy classification was initiated, based on Haefliger [3]. Similar work has been done by Dax [2], Salomonsen [6]. In this paper, some specific new results are announced, including a geometric interpretation of theorem 1.0.1 of [5].

1. Throughout, let M be a simply connected differentiable $2n$-manifold, $n \geq 4$, and $f : S^n \to M$ an embedding. Fix orientations on S^n and M.

<u>Theorem</u> 1 (Main theorem): Let $[S^n \subset M; f]$ be the set of isotopy classes of embeddings homotopic to f.

1A: $[S^n \subset M; f]$ is an Abelian group where f represents zero.

1B: There is a homomorphism (which does not depend on f) $\psi : \pi_{n+1} M \to H_2(M;G)$, where $G = Z$ if n odd, $G = Z_2$ if n even. $[S^n \subset M; f] = \text{Coker}\,\psi$.

1C: If $n \neq 7$, $2\psi = 0$. (Conjecture: $2\psi = 0$ if $n = 7$, also.)

1D: Let $[S^n \subset M]$ = isotopy classes of embeddings of S^n in M. Then, as a set, $[S^n \subset M] = \pi_n M \times \text{Coker}\,\psi$.

<u>Theorem</u> 2: Let $\alpha \in \pi_{n+1} M$ be represented by an immersion $g : S^{n+1} \to M$ in general position. Let $W \subset M$ be the double point image, a closed surface. If n is odd, W can be canonically oriented. Then $\psi\alpha = [W] \in H_2(M;G)$.

<u>Theorem</u> 3: For any $n \geq 4$, an example M and α can be constructed such that $\psi\alpha \neq 0$.

2. Proofs of theorems 1, 2 and 3. Let $\mathcal{M}_f$ = maps homotopic to f, and $\mathcal{E}_f$ = embeddings homotopic to f. Let $[S^n \subset M]_f = \pi_1(\mathcal{M}_f, \mathcal{E}_f, f)$, an Abelian group, by theorem 2.6.5 of [5]. By theorem 1.0.1 of [5], the Hurewicz theorem, and the universal coefficient theorem, $[S^n \subset M]_f = H_2(M;G)$. Note that $\pi_1(\mathcal{M}_f, f) = \pi_{n+1}M$ and $\pi_0(\mathcal{E}_f, f) = [S^n \subset M;f]$.

We have an action

$$\mu : \pi_{n+1}M \times H_2(M;G) \to H_2(M;G)$$

and $[S^n \subset M;f]$ = the orbits of μ. (See pages 68-69 of [5].) Define

$$\psi\alpha = \mu(\alpha, 0) \text{ for all } \alpha \in \pi_{n+1}M$$

Using §4 of [5], we have, analogous to theorem 3.8.2 of [4], that

$$\mu(\alpha, x) = \psi\alpha + x \text{ for all } x \in H_2(M;G)$$

Results 1.A and 1B follow immediately.

Any map $S^n \to M$ is homotopic to an embedding; since M is simply connected, double points may be eliminated by the Whitney trick. Thus 1D follows from 1B. An interesting question is, can a natural group structure be defined on $[S^n \subset M]$, and is it isomorphic to $\pi_n M \oplus \text{Coker } \psi$? I conjecture that the answers are yes and no, respectively.

To prove 1C, we make use of the following unpublished result.

<u>Theorem</u> 4: If $g : S^m \to M^{2m-k}$ is any map, for $k \leq \frac{1}{2}(m-1)$, then g is homotopic to an immersion unless

(i) $m = 8, \quad k = 2, 3$

(ii) $m = 16, \quad k = 4, 5, 6, 7.$

Furthermore, counter examples exist in all those cases.

The proof of theorem 4 will be the subject of a later paper. It relies on Barratt-Mahowald [1] and Toda [7].

Proof of 1C. Without loss of generality n is odd, since the even case is trivial. Let $n \neq 7$, $\alpha \in \pi_{n+1} M$. By theorem 4, α is represented by an immersion $g : S^{n+1} \to M$. Let $\widetilde{W} \subset S^{n+1}$ and $W \subset M$ be the double point set and its image. Both surfaces can be canonically oriented, and $g_*[\widetilde{W}] = 2[W]$. But $[\widetilde{W}] \in H_2(S^{n+1};Z) = 0$, hence $2[W] = 0$. By theorem 2, $\psi\alpha = [W]$, and we are done.

The motivation for the conjecture is that theorem 4 has a unique counter-example for $n = 7$, $m = 8$, $2m-k = 14$; namely a map $S^8 \to M^{14}$ where M has the homotopy type of $S^8 \vee \Sigma P^2$. Note $H_2(M;Z) = Z_2$. (The uniqueness is up to a natural kind of equivalence.)

In order to prove theorem 2, we need the geometric version of theorem 1.0.1 of [5], namely

Theorem 5. Let $f_t : S^n \to M$, $0 \leq t \leq 1$, be a homotopy, where $f_0 = f$ and f_1 are embeddings. Assume that $F : S^n \times I \to M \times I$ (given by $F(x, t) = f_t x)$ has regular crossings. (This is not a restriction). Let $\widetilde{D} \subset S^n \times I$ be the double point set of F, a closed 1-manifold, and let $\widetilde{W} \subset S^n \times I$ be a surface where $\partial\widetilde{W} = \widetilde{D}$. Then $W = F(\widetilde{W})$ is a closed surface.

If n is odd, $\widetilde{D}$ can be canonically oriented, $\widetilde{W}$ must be chosen to be oriented, and W is oriented. Then the obstruction to deforming $\{f_t\}$ to an isotopy of f_0 with f_1, $\Delta(f_0, f_1; f_t) \in H_2(M;G)$, is equal to $[W]$.

We examine only the odd case. (The even case is similar.) Since M is simply connected, the spectral sequence for $[S^n \subset M]_f$ collapses, and

$$[S^n \subset M]_f = \pi_{2n+1}(M, M^0) \xleftarrow[\iota]{\cong} \pi_2 M$$

where $M^0 = M - \{\text{point}\}$ and ι is Whitehead product with the generator of $\pi_{2n}(M, M^0) = Z$. Careful checking of the constructions of [5] reveals that $[W]$ corresponds to $\Delta(f_0, f_1; f_t)$ under this Whitehead isomorphism.

We need now to show that theorem 5 implies theorem 2. If $\alpha \in \pi_{n+1} M$, let $g : S^{n+1} \to M$ represent α. Let $\overline{f} : S^n \times S^1 \to M$ be given by $\overline{f}(x, t) = f(x)$. Let

$$\overline{f} \# g : S^n \times S^1 \cong S^n \times S^1 \# S^{n+1} \to M$$

where $\#$ denotes connected sum. Finally, pick a homotopy $f_t : S^n \to M$, $0 \le t \le 1$, such that $f_0 = f_1 = f$, and $\overline{F} : S^n \times S^1 \to M$ (given by $\overline{F}(x, t) = f_t x)$ is homotopic to $\overline{f} \# g$.

If g is an immersion, $\{f_t\}$ can be chosen to satisfy the hypotheses of theorem 5, and it can be arranged that the surface W of theorem 5 is the same as the surface W of theorem 2. Thus theorem 5 implies theorem 2.

Proof of theorem 3. Let $n \ge 4$. We construct a simply connected $(2n)$-manifold M and an immersion $g : S^{n+1} \to M$ such that $\psi[g] \neq 0$.

For all n, we choose a double covering of connected surfaces, $\pi : \widetilde{W} \to W$.

If n is even, use $\pi : S^2 \to P^2$, while if n is odd, let $\pi = 2 \times 1 : S^1 \times S^1 \to S^1 \times S^1$. Let $T : \widetilde{W} \to \widetilde{W}$ be the map where $W = \widetilde{W}/T$. Choose 0-codimensional embeddings $i : \widetilde{W} \times D^{n-1} \subset S^{n+1}$ and $i \times 1 : \widetilde{W} \times D^{n-1} \times D^{n-1} \subset S^{n+1} \times D^{n-1}$. Let $M' = (S^{n+1} \times D^{n-1})/\sim$, a (2n)-manifold (ignore corners); where $(\widetilde{w}, x, y) \sim (T\widetilde{w}, y, x)$ for all $\widetilde{w} \in \widetilde{W}$ and $x, y \in D^{n-1}$. Note that, up through dimension n-1, M' is the cofiber of π.

The normal bundle of $W \subset M'$ is $\nu = \oplus_{n-1} (L_\pi \oplus \varepsilon^1)$, where ε^1 is the trivial line bundle and L_π is the line bundle associated with π. In either case, the total space of ν is oriented. Since $\pi_1 W \to \pi_1 M' = Z_2$ is onto, M' is also oriented. Let $M = M' \cup H$, where H is a 1-handle which kills $\pi_1 M'$. Thus M is simply connected. The composition

$$g : S^{n+1} \subset S^{n+1} \times D^{n-1} \to M^1 \subset M$$

has $\widetilde{W}$ and W as its double point set and double point image, respectively, and if n is odd, W is oriented. In either case, $H_2(M';G) = Z_2$ generated by [W], and $H_2(M';G) \hookrightarrow H_2(M;G)$.

This completes the proof of theorem 3.

Bibliography

1. M. G. Barratt and M. E. Mahowald, The Metastable Homotopy of O(n). Bull. AMS 70 (1964), 758-760 MR 31 # 6229.

2. J. P. Dax, Etude Homotopique des Espaces de Plongements, Ann. Sci. École Norm. Sup. (4) 5 (1972), 303-377 MR 47 # 9643.

3. A. Haefliger, Plongements Différentiables dans le Domaine Stable, Comment. Math. Helv. 37 (1961), 57-70 MR 28 # 625.

4. L. L. Larmore, Obstructions to Embedding and Isotopy in the Metastable Range, Rocky Mt. J. Math. 3 (1973), 355-375 MR 50 # 8559.

5. L. L. Larmore, Isotopy Groups, Trans. AMS 239 (1978), 67-97.

6. H. A. Salomonsen, On the Existence and Classification of Differential Embeddings in the Metastable Range, unpublished preprint.

7. Toda, H., Composition Methods in Homotopy Groups of Spheres. Princeton University Press, 1962 MR 26 # 777.

HOMOTOPY RIGIDITY OF STURDY SPACES

Arunas Liulevicius*

1. The main result. Let $U = U(n)$ be the unitary group of the complex n-dimensional vector space C^n with the standard Hermitian inner product. Let $\mu: U \times X \longrightarrow X$ be an action of U on a topological space X. If $\alpha: G \longrightarrow U$ is a representation of a compact topological group into U, then we call the composition $\mu(\alpha \times 1) : G \times X \longrightarrow X$ a linear action of G on X. This G-space is denoted by (X, α). We keep the underlying U-action μ fixed and suppress it in the notation. We wish to show that under certain simple assumptions on the action μ and the topology of the orbit space $B = X/U$ the linear actions of G enjoy a striking homotopy rigidity property.

A U-space X is called sturdy if X is non-empty, completely regular, and all orbits of U in X have the same type U/H, where H is a closed connected subgroup of maximal rank in $U = U(n)$ fixing a unique line in C^n. We also assume that the orbit space $B = X/U$ is simply connected and far from CP^{n-1} in cohomology: if $b \in H^2(B;Z)$ and $b^n = 0$, then $b^{n-1} = 0$ as well.

Theorem 1. Let X be a sturdy space, $\alpha, \beta : G \longrightarrow U$ representations of a compact group G. There exists a G-map $f: (X,\alpha) \longrightarrow (X,\beta)$ such that $f : X \longrightarrow X$ is a homotopy equivalence if and only if there is a linear character $\chi: G \longrightarrow S^1$ such that β or its complex conjugate $\bar{\beta}$ is similar to $\chi\alpha$.

---

* Partially supported by NSF grant # MCS 77-01623.

In the language of Ted Petrie [9] a G-map such that the underlying map is a (nonequivariant) homotopy equivalence is called a G-pseudoequivalence. In general G-pseudoequivalences do not yield an equivalence relation on actions. There is a pleasant surprise in our situation:

Corollary 2. If X is a sturdy space and G is a compact group, then on the set of linear G-actions on X the following three relations coincide: G-equivalence = G-homotopy equivalence = G-pseudoequivalence. Moreover (X, α) is equivalent to (X, β) if and only if β or $\bar{\beta}$ is similar to $\chi\alpha$ as projective representations.

We shall first inspect sturdy spaces a bit more closely. The key cohomology property will be proved using a technique of John Ewing. The proof of Theorem 1 will use K_G-theory. We are grateful to Vic Snaith for a key proof.

2. A closer look at sturdy spaces. Let T be the standard maximal torus of $U = U(n)$ fixing the coordinate axes $L_1, \ldots,$ L_n of C^n. Since we are at liberty to choose for H any representative in its conjugacy class, we may take H so that $T \subset H$ and $L = L_1$ is the line fixed by H in C^n. If $c: U \longrightarrow U$ is complex conjugation, we also may assume that $c(H) = H$.

We let $K = \{u \in U \mid uL = L\}$ and have the inclusions $T \subset H \subset K \subset U$ which induce quotient maps $U/T \xrightarrow{h} U/H$ and $U/H \xrightarrow{k} U/K$. These maps induce monomorphisms in integral cohomology [3] - a fact to be soon exploited. The homogeneous space U/K is of course CP^{n-1}, complex (n-1)-dimensional projective space, and the map k is described by $k(uH) = uL_1$. The maps h and k are U-maps, where the U-action is given by left multiplication in U on coset representatives. The U-maps of U/H into itself are given by $W = N/H$, where N is the normalizer of H in U: $N = \{n \in U \mid nHn^{-1} = H\}$. W acts on the right of U/H as follows: if $w = nH$, then $(uH).w = unH$. Since $c(H) = H$, it follows that $c(N) = N$, and we have induced maps $\underline{c}\colon U/H \longrightarrow U/H$ and $\underline{c}\colon W \longrightarrow W$.

<u>Lemma</u> 3. $\underline{c}\colon W \longrightarrow W$ is the identity map of W.

<u>Proof</u>. Let $N' = \{u \in U \mid uTu^{-1} = T\}$ be the normalizer of T in U. Consider the projection $N \longrightarrow W = N/H$ which takes n to nH. We claim: the restriction of this projection to $N \cap N'$ is still onto W. Let $nHn^{-1} = H$, then nTn^{-1} and T are maximal tori of the compact connected Lie group H, so there exists an $h \in H$ such that $hnTn^{-1}h^{-1} = T$, since any two maximal tori in H are conjugate in H. Then $hn \in N \cap N'$ and $hnH = nH$, as was to be shown. The structure of N' is well known since we took care to choose T as the standard maximal torus of $U = U(n)$. N' consists of elements pt, where p is a permutation matrix and t is an element of T. Since p has real entries, $c(p) = p$. Let now $w \in W$, then $w = pH$ for p a permutation matrix and $\underline{c}(w) = c(p)H = pH = w$. Thus $\underline{c}\colon W \longrightarrow W$ is the identity map.

<u>Corollary</u> 4. $\underline{c}\colon U/H \longrightarrow U/H$ is a W-equivariant map.

<u>Proof</u>. Let $w = nH$, then $(uH).w = unH$ and $\underline{c}((uH).w) = \underline{c}(unH) = c(u)c(n)H = (c(u)H).\underline{c}(w) = \underline{c}(uH).w$, the last step courtesy of Lemma 3.

We now inspect our sturdy space X. Since X is completely regular and has only orbits of type U/H the topological slice theorem (see Bredon [4]) gives us that $X = U/H \times_W E$, where $E = X^H = \{x \in X \mid hx = x \text{ for all } h \in H\}$ is W-free and the orbit projection $p : E \longrightarrow B = X/U = E/W$ is a principal W-bundle. Since $\underline{c} : U/H \longrightarrow U/H$ is a W-map, $\underline{c} \times 1$ induces a map $\hat{c} : X \longrightarrow X$, and $\hat{c}(u.x) = c(u).\hat{c}(x)$ for all $u \in U$ and $x \in X$. If $\beta: G \longrightarrow U$ is a representation then $\bar{\beta} = c\beta$ is the conjugate representation and $\hat{c} : (X, \beta) \longrightarrow (X, \bar{\beta})$ is a G-equivalence.

It is now easy to see why one implication of Theorem 1 is true. We have just seen why $(X, \bar{\beta})$ is G-equivalent to (X, β). Let $C = S^1$ be the center of U, then $C \subset T \subset H$, so C acts trivially on U/H, hence on X. Therefore if $\chi : G \longrightarrow C$ is a linear character then $(X, \chi\alpha) = (X, \alpha)$.

To prove the converse we will have to work a bit more.

<u>Lemma</u> 5. Let $k: U/H \longrightarrow U/K$ be the canonical map. Then k is W-invariant: if $w \in W$ then $k((uH).w) = k(uH)$ for all $u \in U$.

<u>Proof</u>. We have $w = nH$. We notice that $nL = L$, for nL is a line fixed by H in C^n: $hnL = nh'L = L$, where $h' = n^{-1}hn$ is an element of H. Thus $k((uH).w) = k(unH) = unL = uL = k(uH)$, and we are done.

Let U/K have trivial right W-action and B trivial left W-action. The map $k \times p : U/H \times E \longrightarrow U/K \times B$ commutes with the action of W, hence induces a map $q : X = U/H \times_W E \longrightarrow U/K \times B$. We notice that since k was a U-map, q is a U-map. We define an important U-map $r : X \longrightarrow U/K$ by letting $r = \pi_1 q$, where π_1 is the projection on the first factor of $U/K \times B$.

We let h be the Hopf line bundle over U/K and let $y = c_1(h)$ be its first Chern class. We will now find that the class r^*y plays a very central rôle in $H^*(X;Z)$.

Theorem 6. The map $r^*: H^*(U/K;Z) \longrightarrow H^*(X;Z)$ is a monomorphism. If $v \in H^2(X;Z)$ is a class with $v^n = 0$, $v^{n-1} \neq 0$, then $v = ar^*y$ for some a in Z.

The proof of this theorem will involve several steps. We first notice that it is immediate to prove that r^* is a monomorphism, since $q: X \longrightarrow U/K \times B$ is a map of fiber spaces over the simply connected space B inducing $k: U/H \longrightarrow U/K$ on the fiber. Since k^* is a monomorphism by Borel [3] it follows that q^* and hence r^* are monomorphisms.

To prove the second part of the theorem we first have to examine $H^*(U/H;Z)$. We first recall some information about the cohomology of the flag manifold U/T. Probably the most convenient presentation of $H^*(U/T;Z)$ is to give it as the quotient of $H^*(BT;Z)$, namely $H^*(U/T;Z) = Z[x_1,\dots,x_n]/I(n)$, where I(n) is the ideal generated by all symmetric functions in $x_1,\dots,x_n$. The map $h: U/T \longrightarrow U/H$ is a monomorphism in cohomology and $h^*k^*y = x_1 \bmod I(n)$. We will now advertise an important property of $H^*(U/T;Z)$.

Theorem 7. If $u \in H^2(U/T;Z)$ is a class with $u^n = 0$, then $u = ax_i \bmod I(n)$ for some a in Z and some i with $1 \leq i \leq n$.

We are grateful to Ian Macdonald for pointing out that this result appears in D.Monk's thesis [10]. Our favorite proof of the result appears in [5] and uses a trick of John Ewing: the derivation $D = \frac{\partial}{\partial x_1} + \dots + \frac{\partial}{\partial x_n}$ is used to set up an inductive proof.

Corollary 8. If $x \in H^2(U/H;Z)$ with $x^n = 0$, then $x = ak^*y$ for some a in Z.

Proof. Let $u = h^*x$, then $u^n = 0$, so by Theorem 7 we get $u = ax_i$ mod I(n) for some a in Z and some i with $1 \leq i \leq n$. We claim: i = 1 (and here the astute reader will realize that we are in the situation of $n \geq 3$, for n=1 and n=2 are trivial). Since u is in the image of h*, we must have u = z mod I(n), where z is an element of $Z[x_1, \ldots, x_n]$ invariant under the action of the Weyl group of H. This group however fixes x_1 and for all $j \geq 2$ x_j is moved by some element of the Weyl group (remember that we have assumed that H fixes a unique line L in C^n and we chose $L = L_1$). This means that i = 1, and since $h^*k^*y = x_1$ mod I(n) as we have already remarked, it follows that $x = ak^*y$ since h* is a monomorphism.

We can now complete the proof of Theorem 6. Let $i : U/H \longrightarrow X$ be the inclusion of a fiber of $X \longrightarrow B$. Suppose $v \in H^2(X;Z)$ is a class with $v^n = 0$, then of course $(i^*v)^n = 0$ as well, hence by Corollary 8 $i^*v = ak^*y$ for some a in Z. Since $i^*r^*y = k^*y$ and B is simply connected we have $v = ar^*y + b$, where $b \in H^2(B;Z)$. In other words $v = q^*(ay + b)$. Since q* is a monomorphism we have $(ay + b)^n = 0$, so in particular $na^{n-1}y^{n-1}b = 0$. We have $H^2(B;Z)$ torsion free, so either a = 0 or b = 0. The second case is what we want to have, so we want to show that a=0 is untenable. Just think: this means that v = b, so $b^n = 0$, but by hypothesis on B this means that $b^{n-1} = 0$, and this is a contradiction to the assumption that $v^{n-1} \neq 0$.

Corollary 9. If $f : X \longrightarrow X$ is a homotopy equivalence then $f^*r^*y = r^*y$ or $-r^*y$.

Proof. Let $v = f^*r^*y$, then v satisfies the hypotheses of Theorem 6, so $f^*r^*y = ar^*y$, and $a = 1$ or -1 since f^* is an isomorphism.

Here is what happens under a familiar homotopy equivalence:

Lemma 10. Let $\hat{c} : X \longrightarrow X$ be the U-map constructed earlier. We have $\hat{c}^*r^*y = -r^*y$.

Proof. We have the commutative diagram

$$\begin{array}{ccc} X & \xrightarrow{\hat{c}} & X \\ \downarrow r & & \downarrow r \\ U/K & \xrightarrow{\underline{c}} & U/K \end{array}$$

where $\underline{c}(uK) = c(u)K$. Since $\underline{c}^*y = -y$ we have the lemma.

We are now ready to prove Theorem 1.

3. Proof of Theorem 1 completed. We suppose that we are given a G-map $f : (X,\alpha) \longrightarrow (X,\beta)$ such that $f : X \longrightarrow X$ is a homotopy equivalence. If $f^*r^*y = -r^*y$, we replace f by $\hat{c}f$ and β by $\bar{\beta}$. We thus may assume that $f^*r^*y = r^*y$. This means that if we consider the Hopf bundle h on U/K then $f^*r^*h = r^*h$.

At this point we exploit another advantage of linearity. If γ is a representation of G into U, then the Hopf bundle is a G-equivariant bundle $h(\gamma) : (S^{2n-1},\gamma) \longrightarrow (U/K,\gamma)$ and its pullback $r_\gamma^! h(\gamma)$ is a G- equivariant bundle over (X,γ). We now have the situation: $f^! r_\beta^! h(\beta)$ and $r_\alpha^! h(\alpha)$ are both G-equivariant S^1-bundles on (X,α) with the same underlying S^1-bundle r^*h.

Theorem 11. Let a and b be two G-equivariant S^1-bundles on (X,α) with the same underlying S^1-bundle. Then there exists a linear character $\chi : G \longrightarrow S^1$ such that $a = \chi b$.

Proof. The important aspect about X is that it is connected and $H^1(X;Z) = 0$. The technique of proof is Graeme Segal's cohomology of topological groups [13]. Please see [6] for details.

Let $s = h(\alpha)$, $t = h(\beta)$, then we have: $f^! r^!_\beta t = \chi r^!_\alpha s$ for a suitable linear character $\chi: G \longrightarrow S^1$. We now charge ahead to prove that this implies $\beta = \chi\alpha$, completing the proof of Theorem 1.

Theorem 12. The homomorphism

$$q^!: K_G(U/K \times B, \alpha) \longrightarrow K_G(U/H \times_W E, \alpha) = K_G(X, \alpha)$$

makes $K_G(X, \alpha)$ into a free finitely generated $K_G(U/K \times B, \alpha)$-module.

Let us defer the proof of this theorem for a while. We need only the following consequence:

Corollary 13. The homomorphism $r^!_\alpha: K_G(U/K, \alpha) \longrightarrow K_G(X, \alpha)$ is a monomorphism.

This corollary allows us to define a map of R(G)-algebras $\psi: K_G(U/K, \beta) \longrightarrow K_G(U/K, \alpha)$ by setting $\psi = (r^!_\alpha)^{-1} f^! r^!_\beta$. The homomorphism is completely specified by $\psi(t) = \chi s$, since t generates $K_G(U/K, \beta)$ as an R(G)-algebra (see [1] and [12]). Indeed, t satisfies the relation

$$t^n - \beta t^{n-1} + \Lambda^2\beta\, t^{n-2} - \dots + (-1)^n \Lambda^n\beta.1 = 0,$$

so we have the relation

$$\chi^n s^n - \beta\chi^{n-1} s^{n-1} + \quad \dots \quad + (-1)^n \Lambda^n\beta.1 = 0.$$

When we multiply both sides by χ^{-n} we obtain

$$s^n - \beta\chi^{-1} s^{n-1} + \Lambda^2(\beta\chi^{-1}) s^{n-2} - \dots + (-1)^n \Lambda^n(\beta\chi^{-1}).1 = 0.$$

However $K_G(U/K, \alpha)$ is R(G)-free on $1, \dots, s^{n-1}$ with relation

$$s^n - \alpha s^{n-1} + \Lambda^2\alpha\, s^{n-2} - \dots + (-1)^n \Lambda^n\alpha.1 = 0,$$

so comparing coefficients of s^{n-1} we obtain $\beta\chi^{-1} = \alpha$, or $\beta = \chi\alpha$, as was to be shown. The proof of Theorem 1 is complete modulo the proof of Theorem 12 which we present in the next section.

4. Proof of Theorem 12. Both the theorem and its proof are due to Vic Snaith. We begin with a preliminary result which among other things will fix the notation.

Lemma 14. The map $k^!: K_G(U/K, \alpha) \longrightarrow K_G(U/H, \alpha)$ makes $K_G(U/H, \alpha)$ into a free finitely generated $K_G(U/K, \alpha)$-module.

Proof. A geometric argument of J.McLeod [8] shows that $K_G(U/H, \alpha) = R(G) \otimes_{R(U)} R(H)$, where $\alpha^*: R(U) \longrightarrow R(G)$ and $i^*: R(U) \longrightarrow R(H)$ give the R(U)-module structures, $i: H \to U$ being the inclusion. The same is of course true of U/K, and the map $k^!$ is precisely $1 \otimes j^*: R(G) \otimes_{R(U)} R(K) \to R(G) \otimes_{R(U)} R(H)$, where $j: K \longrightarrow H$ is the inclusion. Now R(H) is a free R(K)-module (see Pittie [11] and Steinberg [14]), so $1 \otimes j^*$ makes $R(G) \otimes_{R(U)} R(H)$ into a free $R(G) \otimes_{R(U)} R(K)$ - module.

Steinberg [14] has constructed a natural R(K)-free basis s(1), ... , s(N) of R(H). These elements give rise to $U \times_H s(i)$ in $K_G(U/H, \alpha)$ which form a free basis over $K_G(U/K, \alpha)$.

We use the Steinberg basis in the proof of Theorem 12. Define elements $V(i) = U \times_H s(i) \times_W E$ which restrict to the basis $U \times_H s(i)$ on the fiber U/H. Define a map

$$\bar{q} : \bigoplus_1^N K_G(U/K \times B, \alpha) \longrightarrow K_G(U/H \times_W E, \alpha)$$

by $\bar{q}(a_1, \ldots, a_N) = (q^! a_1)V(1) + \ldots + (q^! a_N)V(N)$. A spectral sequence of G.Segal [12] allows us to conclude that $\bar{q}$ is an isomorphism. This proves Theorem 12.

REFERENCES

1. M.F.Atiyah and G.B.Segal, Lectures on equivariant K-theory, Mimeographed notes, Oxford, 1965.
2. A.Back, Homotopy rigidity for Grassmannians (to appear). Preprint, University of Chicago, 1978.
3. A.Borel, Sur la cohomologie des espaces fibrés principaux et des espaces homogenes de groupes de Lie compacts, Ann. of Math. 57 (1953), 115-207.
4. G.Bredon, Introduction to compact transformation groups, Academic Press, London and New York, 1972.
5. J.Ewing and A.Liulevicius, Homotopy rigidity of linear actions on friendly homogeneous spaces (to appear).
6. A.Liulevicius, Homotopy rigidity of linear actions: characters tell all, Bull. Amer. Math. Soc. 84 (1978), 213 - 221.
7. ______________, Linear actions on friendly spaces, Proceedings of the Waterloo Algebraic Topology Conference, June 1978 (to appear).
8. J.McLeod, The Künneth formula in equivariant K-theory, Proceedings of the Waterloo Algebraic Topology Conference, June 1978 (to appear).
9. A.Meyerhoff and T.Petrie, Quasi-equivalences of G-modules, Topology 15 (1976), 69 - 75.
10. J.D.Monk, The geometry of flag manifolds, Proceedings London Math. Soc. (3) 9 (1959), 253-286.
11. H.V.Pittie, Homogeneous vector bundles on homogeneous spaces, Topology 11 (1972), 199-203.
12. G.B.Segal, Equivariant K-theory, Inst. Hautes Etudes Sci. Publ. Math. No. 34 (1968), 129-151.
13. __________, Cohomology of topological groups, Symposia

Mathematica, vol. IV (INDAM, Rome, 1968/69), 377-387.

14. R. Steinberg, On a theorem of Pittie, Topology 14 (1975), 173-177.

The University of Chicago
Chicago, Illinois

July 1978

The Geometry of Mahowald Orientations

by

Brian J. Sanderson

Mahowald, in [6], has constructed a Thom spectrum equivalent to the mod 2 Eilenberg MacLane spectrum. It follows that there is a description of mod 2 homology in terms of bordism of certain 'Mahowald oriented' manifolds. The Mahowald orientation, for a closed n-manifold M^n, may be roughly described in geometric terms as follows. There is given a generic framed immersion $g:N \to M$, where N is a closed $(n - 1)$-manifold, and g is factored through an embedding in $M \times R^2$. In addition the normal bundle of M is trivial off the image of g and it undergoes a Möbius twist each time g(N) is crossed.

In §1, we describe Mahowald's results, 1.1 and 1.2, and the resulting bordism theory. We also give a homotopical analysis of the Steenrod operation XS^k_q in terms of a k-th Hopf invariant. In §2 there is a discussion of the geometry of generic framed immersions, and the relation with configuration spaces. In §3 we prove that for a Mahowald oriented M, the kth Wu class is represented by the k-tuple points of the immersion, 3.1. We show that the manifold of k-tuple points is itself Mahowald oriented and there follows a geometric interpretation of the operation XS^k_q, 3.5. In §4 we give the precise description of the connection between the immersion $g:N \to M$ and the normal bundle of M, 4.2. The final section, §5, is devoted to proving 4.2, and here we need to look closely at the multiple loop structure of BO.

§1. Mahowald oriented bordism and homology

Let $\eta : \Omega^2 S^3 \to BO$ be $\Omega^2 f$, where $[f] \in \pi_3 B_2 BO$ is a generator - details are given in §5. Let MA denote the Thom spectrum $\eta^* MO$, induced from MO by η. Let $\sigma = \mu \circ \bar{\eta}$, where $\bar{\eta} : MA \to MO$ covers η, $\mu: MO \to H$ is the Thom class, and H is the mod 2 Eilenberg-MacLane spectrum. Thus $H_n(X)$ will mean homology with Z_2 coefficients. Mahowald proved the following theorem. See [6], [5] and [10].

Theorem 1.1 $\sigma : MA \to H$ is an equivalence. ▱

It follows from the theorem that an element $x \in H_n(X)$ may be represented by a triple $(M,f,\hat{h})$, where M is a closed n-manifold, $f : M \to X$, $\hat{h}$ is a stable bundle map from the normal bundle ν_M to the pull back, $\eta^*\gamma$, of the universal bundle, and $f_*[M] = x$. Furthermore the triple representing x is unique up to bordism through such triples.

We refer to $\hat{h}$ as a Mahowald orientation for M. In fact Mahowald proved a refinement of 1.1, which we now describe. Recall that $C_2(S^1)$ the configuration space of 'finite sets of points in R^2 with labels in S^1' is a model for $\Omega^2 S^3$. See [9] and [11]. There is a filtration,

$$\{*\} = F_0 \subset F_1 \subset \cdots \subset F_k \subset \cdots \subset F_\infty = C_2(S^1) \simeq \Omega^2 S^3$$

where F_k contains configurations of no more than k points. Let η_k be the composition $F_k \to \Omega^2 S^3 \to BO$. Define $MA(k) = \eta_k^* MO$. Mahowald's theorem, [6] theorem 4.1, is

Theorem 1.2 The map $MA(k) \to MO \to H$ induces an isomorphism

$$H^*H/\{\chi S_q^i : i > k\} \longrightarrow H^*MA(k).$$ ▱

We are going to identify $\sigma^*\chi S_q^k \in H^k MA$. First we describe the filtration $\{F_k\}_{k=0}^{\infty}$ more carefully. Let

$$\mathcal{C}_{2,k} = \{(z_1,\ldots,z_k) : z_i \in \mathbb{R}^2 \quad z_i \neq z_j \text{ if } i \neq j\}.$$

Then the symmetric group Σ_k acts freely on $\mathcal{C}_{2,k}$ and $\mathcal{C}_{2,k}/\Sigma_k$ may be taken as a classifying space, BB_k, for the braid group B_k. Let Σ_k act on $\mathbb{R}^k$, and on $I^k = [0,1]^k$, by permuting coordinates. There is then a vector bundle ζ^k and a cube bundle $\bar{\zeta}^k$, with total spaces $E(\zeta) = \mathcal{C}_{2,k} \times_{\Sigma_k} \mathbb{R}^k$ and $E(\bar{\zeta}) = \mathcal{C}_{2,k} \times_{\Sigma_k} I^k$ respectively.

By convention $BB_0 = E(\zeta^0) = E(\bar{\zeta}^0) = \{*\}$.

Let $S^1 = I/\partial I$. Then we may identify $C_2(S^1)$ with $\coprod_{k\geq 0} E(\bar{\zeta}^k)/\sim$, where the equivalence relation is generated by:

$$[z_1,\ldots,z_k,t_1,\ldots,t_k] \sim [z_1,\ldots,z_{k-1},t_1,\ldots,t_{k-1}]$$

whenever $t_k \in \partial I$, and $[z_1,t_1] \sim *$ if $t_1 \in \partial I$. Define $F_k = \coprod_{k\geq \ell\geq 0} E(\bar{\zeta}^\ell)/\sim$. Then F_k is obtained from F_{k-1} by attaching $E(\bar{\zeta}^k)$ along the sphere bundle, and F_k/F_{k-1} is the Thom space $T(\zeta^k)$.

Let Q_∞ be the functor which assigns a suspension spectrum to a based space - all our spaces will be compactly generated weakly Hausdorff, and all base points will be non-degenerate.

<u>Proposition 1.3</u> $MA(k)/MA(k-1) \simeq Q_\infty T(\zeta^k)$.

Proof. MA(k)/MA(k-1) is the Thom spectrum of the Whitney sum of ζ^k with the (stable) bundle induced by the composition

$$BB_k \underset{i}{\to} E(\overline{\zeta^k}) \to F_q \underset{q}{\to} \Omega^2 S^3 \to BO$$

where $i[z_1,\dots,z_k] = [z_1,\dots,z_k,1/2,\dots,1/2]$. However $qoi \simeq 0$. The null homotopy is provided by qoi_t, where

$$i_t[z_1,\dots,z_k] = [z_1,\dots,z_k,(1-t)/2,\dots,(1-t)/2]$$

$$\in \mathcal{E}_{2,k} \times_{\Sigma_k} I^k/\sim \; = \; F_k. \qquad \square$$

Proposition 1.4

(i) $H^k(MA/MA(k-1)) \cong Z_2$ generated by x_k, say.

(ii) $j^*(x_k) = \sigma^* \chi S_q^k$ where $j: MA \to MA/MA(k-1)$.

Proof. Consider the exact sequence

$$H^k MA(k-1) \xleftarrow{i^*} H^k MA \xleftarrow{j^*} M^k MA/MA(k-1).$$

From 1.2 we have $i^* \sigma^* \chi S_q^k = 0$ and $\sigma^* \chi S_q^k \neq 0$. Hence for some $x_k \neq 0$ we have $j^* x_k = \sigma^* \chi S_q^k$. From 1.3, the Thom isomorphism, and connectivity of BB_k we have

$$H^k\, MA(k)/MA(k-1) \cong H^k Q_\infty T(\zeta^k) \cong H^o\, BB_k \cong \mathbb{Z}_2.$$

We also have $H^k\, MA/MA(k) \cong H^k\, F_\infty/F_k = 0$. The result now follows from the exact sequence,

$$\begin{array}{ccccc} H^k\, MA(k)/MA(k-1) & \leftarrow & H^k MA/MA(k-1) & \leftarrow & H^k(MA/MA(k)). \\ \wr\wr & & & & \| \\ Z_2 & & & & 0 \end{array}$$

$\square$

There is an equivalence $Q_\infty F_\infty \simeq Q_\infty \bigvee_{k\geq 1} F_k/F_{k-1}$, see [3] and [12]. Projection on the kth factor gives the kth Hopf-invariant

$$h_k : Q_\infty F_\infty \to Q_\infty F_k/F_{k-1}, \quad \text{and}$$

h_k extends the projection $Q_\infty F_k \to Q_\infty F_k/F_{k-1}$.

Let $\phi_k \in \tilde{H}^k(T/\zeta^k))$ be the Thom class and let $\Phi : H^*F_\infty \to H^*MA$ be the Thom isomorphism.

<u>Corollary 1.5</u> $\quad \Phi\, h_k^*\, \phi_k \;=\; \sigma^*\, \chi_{S_q}{}^k$.

<u>Proof</u>. Consider the following diagram.

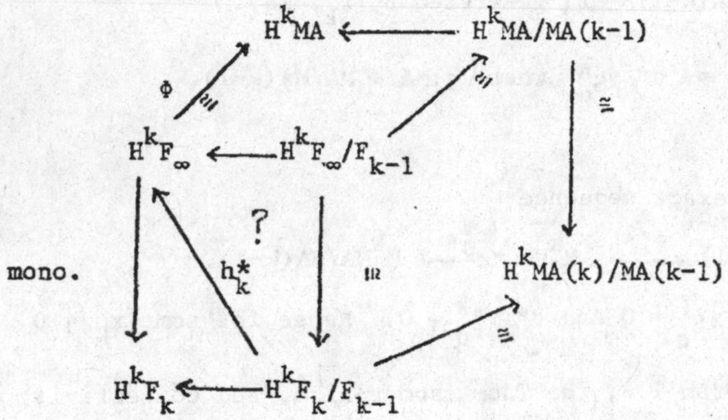

By considering suitable exact sequences we see that the vertical arrows are as indicated. The diagram clearly commutes with the possible exception of the indicated triangle, but this too commutes as there is only one way to factor through a monomorphism. The result follows from 1.4 after a diagram chase. $\square$

§2. $\Omega^2 S^3$ as a classifying space for immersions

In [4] it is shown how configuration spaces may be regarded as classifying spaces for immersions. See also [13]. Here we review this material for the special case, $C_2(S^1)$. In addition we need to go a little deeper into the geometry of self-transverse immersions.

Let N^{n-1} and M^n be closed manifolds. We will describe what we mean for an immersion $\bar{g} : N \times I \to M$ to be in good position. Let $g : N \to M$ be given by $g(x) = \bar{g}(x,1/2)$, then we require g to be self-transverse and we require the k-tuple points of $\bar{g}$ to provide tubular neighbourhoods for the k-tuple points of g - all fitting together nicely. Details follow.

Let $\tilde{N}_k = \{(x_1,\ldots,x_k) : x_i \neq x_j$ if $i \neq j$, and $g(x_1) = g(x_i)$, $1 \leq i \leq k\}$. Then Σ_{k-1} acts freely on $\tilde{N}_k$ by permuting the first k-1 coordinates, and Σ_k acts by permuting all coordinates. Let $N_k' = \tilde{N}_k/\Sigma_{k-1}$ and let $N_k = \tilde{N}_k/\Sigma_k$. Then $p_k : N_k' \to N_k$, given by the identity on representatives, is a p-fold cover. Immersions $g_k' : N_k' \to N$ and $g_k : N_k \to M$ are given by $g_k'[x_1,\ldots,x_k] = x_k$ and $g_k[x_1,\ldots,x_k] = g(x_k)$ respectively. There are cube bundle spaces $\tilde{N}_k \times_{\Sigma_{k-1}} I^{k-1}$ and $\tilde{N}_k \times_{\Sigma_k} I^k$ with 'O'-sections $N_k' \to \tilde{N}_k \times_{\Sigma_{k-1}} I^{k-1}$ and $N_k \to \tilde{N}_k \times_{\Sigma_k} I^k$, given by

$$[x_1,\ldots,x_k] \mapsto [x_1,\ldots,x_k, 1/2,\ldots,1/2].$$

There is a map $\bar{p}_k : \tilde{N}_k \times_{\Sigma_{k-1}} I^{k-1} \to \tilde{N}_k \times_{\Sigma_k} I^k$ given by

$$\bar{p}_k[x_1,\ldots,x_k,t_1,\ldots,t_{k-1}] = [x_1,\ldots,x_k,t_1,\ldots,t_{k-1},1/2].$$

We assume for $1 \leq k \leq n$:

(i) There is a commutative diagram of immersions

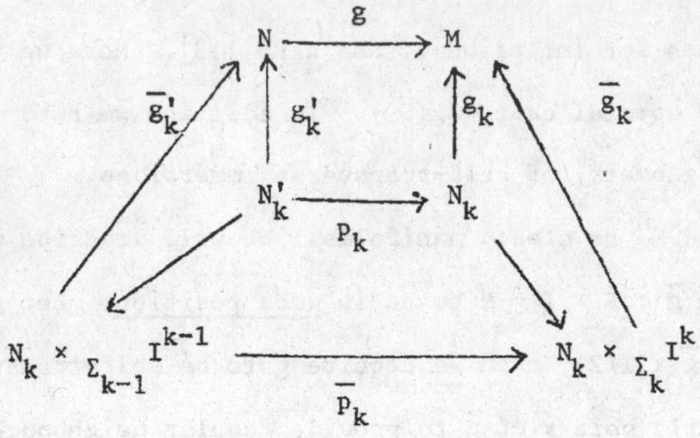

(ii) $\bar g(\bar g'_k([x_1,\ldots,x_k,t_1,\ldots,t_{k-1}]),t) = \bar g_k\,[x_1,\ldots,x_k,t_1,\ldots,t_{k-1},t]$,

and (iii) closure $\{x \in M : |\bar g^{-1}(x)| \geq k\} = \mathrm{im}.(\bar g_k)$.

This completes the description of good position. Now suppose that $\bar g : N \times I \to M$ is an immersion and that $g : N \to M$ is self-transverse. Then there is a regular homotopy of $\bar g$, fixed on $N \times \{1/2\}$, putting $\bar g$ into good position. This is achieved by changing $\bar g$ near the n-tuple points of g, then near the (n-1)-tuple points and so on inductively. We omit the details.

It follows from theorem 1.1 (see also example 1.1) of [4] and the above remarks, that there is a bijection between the set of homotopy classes of maps $M \to F_\infty$, and bordism classes of triples $(M;\bar g,\tilde g)$, where $\bar g : N \times I \to M$ is an immersion in good position and $\tilde g : N \times I \to M \times \mathbb{R}^2$ is an embedding satisfying $\pi_1\tilde g = \bar g$, where $\pi_1 : M \times \mathbb{R}^2 \to M$ is the projection. We also assume that $\pi_2\,\tilde g(x,t) = \pi_2(\tilde g(x,1/2))$, for $(x,t) \in N \times I$, where π_2 is the other projection. In order to describe the map $h : M \to F_\infty$ corresponding to $(M,\bar g,\tilde g)$ we need some more notation. Let $\bar g_o = \mathrm{id}._M$, then M is filtered by the images $\{\text{im}\ \overline{g}_k\}_{k=0}^n$. Let $\underline{M}_k = \text{clos.}\{\text{im.}\ \overline{g}_k \setminus \text{im.}\ \overline{g}_{k+1}\}$ and $\underline{N}_k = \text{clos.}\{\text{im.}\ g_k \setminus \text{im.}\overline{g}_{k+1}\}$. Then $\underline{M}_k$ is a cube bundle space over $\underline{N}_k$, and $h|\underline{M}_k$ factors through a bundle map $\underline{M}_k \to E(\overline{\zeta}^k)$ defined as follows. Suppose $\overline{g}_k|x_1,\ldots,x_k,t_1,\ldots,t_k| = x \in \underline{M}_k$, then

$$h(x) = [\pi_2\tilde{g}(x_1,t_1),\ldots,\pi_2\tilde{g}(x_k,t_k),t_1,\ldots,t_k] \in (\mathcal{Z}_{2,k} \times_{\Sigma_k} I^k)/\sim\ \subset F_\infty.$$

For future reference we also define $\underline{N}_k' = \text{closure}\ \{\text{im}\ g_k' \setminus \text{im}\ \overline{g}_{k+1}'\}$. Then g restricts to give a k-fold cover $\underline{N}_k' \to \underline{N}_k$. Both $\underline{N}_k'$ and $\underline{N}_k$ are manifolds with boundary.

§3. Wu classes and k-tuple points

Let M^n be a closed Mahowald oriented manifold..Then there is a bundle map

$$\begin{array}{ccc} \nu_M & \xrightarrow{\hat{h}} & \eta^*\gamma \\ \downarrow & & \downarrow \\ M & \xrightarrow{h} & F_\infty \end{array},$$

and after the discussion in §2 we may assume that h corresponds to an immersion $\overline{g} : N \times I \to M$ in good position and factored through an embedding $g : N \times I \to M \times \mathbb{R}^2$. Let $[N_k]$ denote the fundamental mod 2 class of N_k.

Theorem 3.1 Let M be a closed Mahowald oriented manifold. Then, in the notation of §2, $g_{k*}[N_k]$ is Poincaré dual to the kth Wu class of $M, v_k(M)$.

We need the following easily proved lemma.

Lemma 3.2 Let M be a closed manifold and let M be a compact codim. 0 submanifold. Then the following diagram commutes,

$$\begin{array}{ccc} H_{n-k}(M) & \underset{\text{P.D.}}{\cong} & H^k(M) \\ \downarrow j_* & & \\ H_{n-k}(M,M_o) & & \downarrow i^* \\ \text{exc.} \cong & & \\ H_{n-k}(M \setminus \text{int.}M, \partial M_o) & \underset{\text{P.D.}}{\cong} & H^k(M \setminus \text{int.}M_o) \end{array}$$

where i and j are inclusions.

□

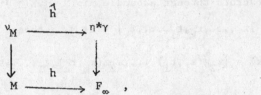

<u>Proof of 3.1.</u> Consider the following commutative diagram

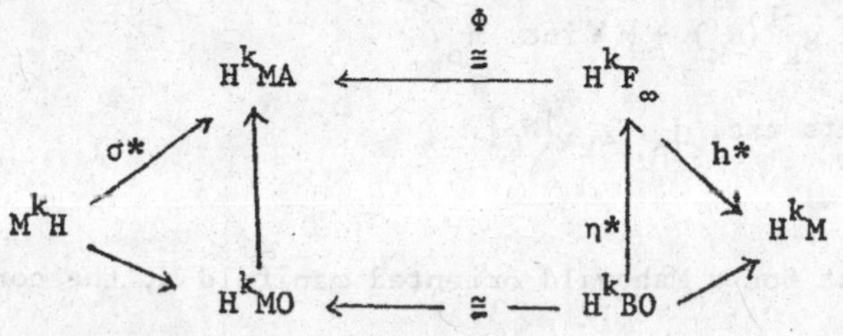

Following $\chi S^k_q \in M^kH$ counter-clockwise we get $\chi S^k_q U$, $V_k(\gamma)$, $V_k(M)$, where U is the universal Thom class. Following χS^k_q clockwise and using 1.5 we get $\sigma^* \chi S^k_q$, $h^*_k \phi_k$, $h^* h^*_k \phi_k$. Hence $h^* h^*_k \phi_k = V_k(M)$. Now let $M_o = \text{im}\, \bar{g}_{k+1}$ and consider the following commutative diagram;

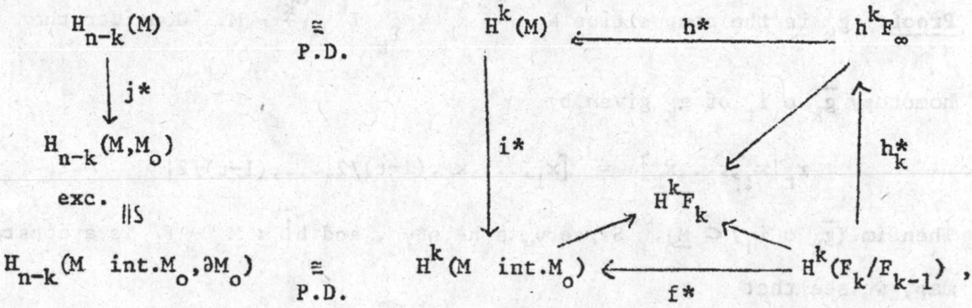

where f is the composition

$$M \setminus \text{int}.M_o \xrightarrow[h|]{} F_k \longrightarrow F_k/F_{k-1}\,.$$

Since M_o has the homotopy type of an n-k-1 complex it follows that j_* is injective. It is now sufficient to prove that

$$\text{exc.}\ j_* \, g_{k_*} [N_k] = \text{P.D.}\ f^* \phi_k .$$

This is easy to see because f is the Thom - Pontrjagin construction for $\underline{M}_k$ over $\underline{N}_k$ and therefore P.D. $f^*\phi_k$ is represented by

$$g_k| : g_k^{-1}(\underline{N}_k) \to M \setminus \text{int. } M_o.$$

But $g_k|$ also represents exc. $j_* \cdot g_{k*}[N_k]$. ▱

We show next that for a Mahowald oriented manifold M, the corresponding N_k are also canonically Mahowald oriented. To see this consider the normal bundle $\nu(N_k)$. We have a (stable) isomorphism

$$\nu(N_k) \cong \nu(g_k) \oplus g_k^*(\nu_M).$$

<u>Lemma 3.3</u>. $g_k^*(\nu_M)$ is trivial.

<u>Proof</u>. g_k is the composition $N_k \xrightarrow{i} \tilde{N}_k \times_{\Sigma_k} I^k \xrightarrow{\bar{g}_k} M$. Consider the homotopy $\bar{g}_k \circ i_t$ of g_k given by

$$i_t[x_1,\dots,x_k] = [x_1,\dots,x_k,(1-t)/2,\dots,(1-t)/2|$$

Then $\text{im.}(\bar{g}_k \circ i_1) \subset \underline{M}_o$. Since $\nu_M \cong h^* \eta^*\gamma$, and $h| : M_o \to F_\infty$ is a constant map, we see that

$$g_k^* \nu_M \cong (\bar{g}_k \circ i_1)^* \nu_M \cong \varepsilon .$$ ▱

It remains to find a stable bundle map $\nu(g_k) \to \eta^*\gamma$. We may identify $E(\nu(g_k))$ with $\tilde{N}_k \times_{\Sigma_k} R^k$. There is a bundle map, $\tilde{N}_k \times_{\Sigma_k} R^k \to \zeta^k$, given by

$$[x_1,\dots,x_k,t_1,\dots,t_k] \to [\pi_2\tilde{g}(x_1,1/2),\dots,\pi_2\tilde{g}(x_k,1/2),t_1,\dots,t_k].$$

Let $B\rho : BB_k \to BO$ classify ζ^k, then from lemma 2.1 of [2] there is a homotopy commutative diagram;

$$\begin{array}{ccc} & & \Omega^2 S^3 \\ & \nearrow & \downarrow \eta \\ BB_k & \xrightarrow{B\rho} & BO \end{array} \qquad (3.4)$$

The lift then provides the required stable bundle map $\zeta^k \to \eta^*\gamma$.

Let $(N_k, \hat{h}_k)$ be the k-tuple manifold, together with its Mahowald orientation, associated with a Mahowald oriented manifold $(M, \hat{h})$. Let $\mathfrak{S} : H^*H \to MA^*MA$ be the isomorphism induced by σ. Then $\mathfrak{S}\chi S_q^k$ may be regarded as a map $MA \to \Sigma^k MA$ which in turn induces, by composition, S-dual operations

$$\chi S_q^k : MA^*(-) \to MA^{*+k}(-) \text{ and}$$

$$\chi S_q^k : MA_*(-) \to MA_{*-k}(-).$$

<u>Theorem 3.5</u> The homology operation $\mathfrak{S}\chi S_q^k$ is given by

$$\mathfrak{S}\,\chi S_q^k[M, f, \hat{h}] = [N_k, f \circ g_k, \hat{h}_k].$$

<u>Proof</u>. It is sufficient, by 1.1, to prove that $\chi S_q^k : H_n(-) \to H_{n-k}(-)$ is given by $\chi S_q^k f_*[M] = (f o g_k)_*[N_k]$. But,

$$(f o g_k)_*[N_k] = f_* g_{k*}[N_k] = f_* \text{ P.D. } V_k(M), \quad \text{by 3.1.}$$

Finally

$$f_* \text{P.D. } V_k(M) = \chi S_q^k f_*[M].$$ ▱

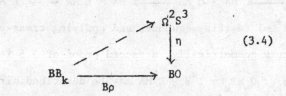

Remark 3.6 It follows from definitions and 1.5 that $\mathfrak{S}\chi S_q^k : MA \to \Sigma^k MA$ is the composition

$$MA \xrightarrow{\Delta} MA \wedge Q_\infty F_\infty \xrightarrow{1\wedge h_k} MA \wedge Q_\infty MB_k \xrightarrow{1\wedge\phi_k} MA \wedge \Sigma^k MA \xrightarrow{m} \Sigma^k MA.$$

By applying $\pi_*(X^+\wedge -)$ to this sequence, and applying transversality to interpret each term geometrically, a second proof of 3.5 is possible. The Thom class $\phi_k : Q_\infty MB_k \to \Sigma^k H \simeq \Sigma^k MA$ may be described directly as the map 'covering' the lift of 3.4.

§4. Construction of the normal bundle of a Mahowald oriented manifold

In §2 we showed how to associate a triple $(M\,\overline{g},\tilde{g})$ with a Mahowald oriented manifold $(M,\hat{h})$. In this section we describe how to reconstruct the stable normal bundle of M from the immersion $\overline{g} : N \times I \to M$.

Lemma 4.1 Suppose given $(M,\overline{g})$ as above, there is then a map $q : N \to \mathbb{R}^{n+1}$ such that if $\overline{g}(x,t) = \overline{g}(x',t')$, for any $(x,t),(x',t') \in N \times I$, then $q(x) \cdot q(x') = 0$. Furthermore, if $i : \mathbb{R}^{n+1} \to \mathbb{R}^{n+2}$ is the inclusion then $i \circ q$ is unique up to homotopy through maps satisfying the condition.

Proof. We construct q by downward induction over the filtration of N by the images $\{\mathrm{im}\,\overline{g}'_k\}_{k=0}^{n}$. For each set of n-points of N having the same image under g choose a total order. Let $(x_1,\ldots,x_n)$ be such an ordered set. Then each x_i has a cube neighbourhood C_i provided by $\overline{g}'_n$. Define $q^{(n)}(x)$ for $x \in C_i$ to be e_i, the ith basis vector of $\mathbb{R}^n$. This defines $q^{(n)} : \mathrm{Im.}\,\overline{g}'_n \to \mathbb{R}^n$ with the required property. Now suppose given $q^{(k+1)} : \mathrm{im.}\,\overline{g}'_{k+1} \to \mathbb{R}^{n+1}$. We show how to construct $q^{(k)}$ extending $q^{(k+1)}$. First we extend over $\underline{N}'_k \subset \mathrm{im.}\,g'_k$, to $p^{(k)}$ say, as follows. Recall from §2 that $\underline{N}'_k$ is a manifold with boundary, and g restricts to a k-fold cover $\underline{N}'_k \to \underline{N}_k$. $p^{(k)}$ is defined on $\partial\underline{N}'_k$ by restriction of $q^{(k+1)}$. We extend over $\underline{N}'_k$ by induction up the skeleta of a triangulation of $\underline{N}_k$. There are obstructions with coefficients in $\pi_i(V_{n+1,k})$ for $i \leq n-k$. Since $\pi_i(V_{n+1,k}) = 0$ in this range, the obstructions vanish. Finally extend $p^{(k)}$ over the whole of $\mathrm{im.}\,\overline{g}'_k$ to $q(k)$ by composing with the projection into $\underline{N}'_k$. This proves existence of q. Uniqueness is proved similarly. $\square$

For $t \in I$ define a line $L_t \subset \mathbb{R}^1 \times \mathbb{R}^1$ by

$$L_t = \{(\lambda \cos \pi t, \lambda \sin \pi t) : \lambda \in \mathbb{R}\}.$$

Given $q : X \to \mathbb{R}^\ell$, define for $x \in X$,

$$q_x : \mathbb{R}^1 \times \mathbb{R}^1 \to R^\ell \times R^\ell \text{ by } q_x(s,t) = (sq(x), tq(x)).$$

Theorem 4.2 Let M^n be Mahowald oriented with immersion $\bar{g} : N \times I \to M$ in good position, and suppose q is as in 4.1. Then the stable normal bundle ν_M of M may be identified with the sub-bundle of the trivial bundle $\varepsilon^{2n+2} = \varepsilon^{n+1} \oplus \varepsilon^{n+1}$ described as follows. Suppose that

$$\bar{g}^{-1}(x) = \{(x_1,t_1),\ldots,(x_k,t_k)\}$$

then the fibre $\nu_x \subset \mathbb{R}^{n+1} \times \mathbb{R}^{n+1}$ is

$$\{(v,o) \in \mathbb{R}^{n+1} \times \{o\} : v, q(x_i) = 0 \quad i = 1,\ldots,k\} \bigoplus_{i=1}^{k} q_{x_i} L_{t_i}.$$

The proof of the theorem occupies the next section. Since $\nu_M : M \to \Omega^2 S^3 \to BO$ we must come to grips with the map $\Omega^2 S^3 \to BO$. For this we need to explain the multiple loop structure of BO.

§5. The multiple loop structure of BO

The Grassmannian of n-planes in (n+m)-space is $G_{n+m,n} = \{P : P$ is an n-dimensional subspace of $\mathbb{R}^n \times \mathbb{R}^m\}$. Then $G_{n+m,n}$ is based by $\mathbb{R}^n \times \{o\}$ and is homeomorphic with the coset space $O(\mathbb{R}^n \times \mathbb{R}^n)/O(\mathbb{R}^n) \times O(\mathbb{R}^m)$. The homeomorphism is given by $[g] \to g(\mathbb{R}^n \times \{o\})$. The inclusions $\mathbb{R}^n \subset \mathbb{R}^{n+1}$, and $\mathbb{R}^m \subset \mathbb{R}^{m+1}$ induce inclusions of Grassmannians and we define $BO_n = \bigcup_m G_{n+m,n}$, and $BO = \bigcup_n BO_n$. Thus a point of BO may be regarded as a certain infinite dimensional subspace of $\mathbb{R}^\infty \times \mathbb{R}^\infty$.

Let $\mathcal{L}$ be the linear isometries operad. See [1], [7] p. 17 and [8]. Then $\mathcal{L}(k)$ is the space of isometric linear maps from $(\mathbb{R}^\infty)^k$ to $\mathbb{R}^\infty$. The action of $\mathcal{L}$ on BO is given by

$$\theta_k : \mathcal{L}(k) \times BO \times \ldots \times BO \to BO$$

$\theta_k(e,Q_1,\ldots,Q_k) = Q$, where

$$Q = \theta_k(e,g_1,\ldots,g_k)\,(\mathbb{R}^\infty \times \{o\}),\quad Q_k = g_k(\mathbb{R}^\infty \times \{o\})$$

and $\theta_k(e,g_1,\ldots,g_k)$ is constructed so as to make the following diagram commute;

$$\begin{array}{ccc} (\mathbb{R}^\infty \times \mathbb{R}^\infty)^k & \xrightarrow{\;e\times e \,\circ\, \text{shuff.}\;} & \mathbb{R}^\infty \times \mathbb{R}^\infty \\ \downarrow{\scriptstyle g_1\times\ldots\times g_k} & & \downarrow{\scriptstyle \theta_k(e,g_1,\ldots,g_i)} \\ (\mathbb{R}^\infty \times \mathbb{R}^\infty)^k & \xrightarrow[\;e\times e \,\circ\, \text{shuff.}\;]{} & \mathbb{R}^\infty \times \mathbb{R}^\infty \end{array},$$

where shuff. is the obvious shuffle and $\theta_k(e,g_1,\ldots,g_k)$ is the identity on the subspace orthogonal to im. (e×e o shuff.).

We need to relate the action of $\mathcal{L}$ with the action of the little cubes operad $\mathcal{C}_\infty^\square$. We assume familiarity with [9] and recall some facts from [3]. The infinite loop structure of a space is intimately related with the action of $\mathcal{C}_\infty^\square$. There is no obvious morphism of operads $\mathcal{L} \to \mathcal{C}_\infty^\square$. In [9] this difficulty is circumvented by considering: $\mathcal{C}_\infty^\square \xleftarrow{\simeq} \mathcal{C}_\infty^\circ \times \mathcal{L} \xrightarrow{\simeq} \mathcal{L}$. In [3] the notion of operad is weakened to the notion of a coefficient system. There is a morphism $\mathcal{L} \to \mathcal{C}_\infty^\circ$ of coefficient systems, and this is appropriate for our purposes. We give some details.

Let Λ be the category of finite based sets $\underline{r} = \{0, 1, 2, \ldots, r\}$, $r \geq 0$, and injective based functions.

A coefficient system is a contravariant functor $\mathcal{C} : \Lambda \to$ spaces, written $\underline{r} \to \mathcal{C}_r$ on objects, and so that $\mathcal{C}_o$ is a single point $*$. For a space X and morphism $\phi : \underline{r} \to \underline{s}$ in Λ there is a map $\phi : X^r \to X^s$ given by $(x_1, \ldots, x_r) = (x'_1, \ldots, x'_s)$, where $x'_{\phi(a)} = x_a$ and $x'_b = *$ if $b \notin \text{im}.\phi$.

Then $C(X) = \coprod_{r \geq 0} \mathcal{C}_r \times X^r/\sim$, where the equivalence relation is given by:

$$(c\ \phi, x) \sim (c, \phi x) \text{ for } c \in \mathcal{C}_s,\ \phi : \underline{r} \to \underline{s}, \text{ and } x \in X^r.$$

<u>Examples 5.1</u> Operads become coefficient systems by forgetting structure. So we have the <u>linear isometries</u> coefficient system $\mathcal{L}$. The <u>little cubes</u> coefficient systems $\mathcal{C}_q^\square$ for $1 \leq q \leq \infty$. The <u>configuration spaces</u> coefficient systems $\mathcal{C}_q$. The space $C_2(S^1)$ was looked at in detail in §1. There are also <u>Stiefel manifolds</u> coefficient systems $\mathcal{V}_q$ given by taking $\mathcal{V}_{q,k}$ to be the space of isometric linear embeddings $\mathbb{R}^k \to \mathbb{R}^q$ i.e. the familiar Stiefel manifold of k-frames in q-space.

Replacing little cubes by their centres determines a morphism of coefficient systems $\kappa : \mathcal{C}_q^{\square} \to \mathcal{C}_q$. There is a morphism $\alpha : \mathcal{L} \to \mathcal{V}_\infty$ given by $\alpha_r(e) = e \circ j_r$, where $j_r : \mathbb{R}^r \to (\mathbb{R}^\infty)^r$ maps the ith basis vector of $\mathbb{R}^r$ to the 1st basis vector of R^∞ in the ith component of the r-fold product. There is a morphism $\beta : \mathcal{V}_\infty \to \mathcal{C}_\infty^{\square}$ given by choosing little cubes of uniform diameter, say 1/100, centred on the ends of the frames.

Now let $\mathcal{C}$ be any E_∞-operad and suppose there is a Λ-morphism $\psi : \mathcal{C} \to \mathcal{C}_\infty^{\square}$, with each $\psi_j : \mathcal{C}_j \to \mathcal{C}_{\infty,j}^{\square}$ a Σ_j-equivariant equivalence. Then for any based X it follows from 2.7 (ii) of [3] that $\Psi : C(X) \to C_\infty^{\square}(X)$ is a homotopy equivalence. Similarly $\psi^1 : C(X) \to D_\infty(X)$ is a homotopy equivalence, where $\mathcal{D}_\infty = \mathcal{C} \times \mathcal{C}_\infty^{\square}$ and ψ^1 is $(id., \psi)$. Thus we have a commutative diagram of homotopy equivalences

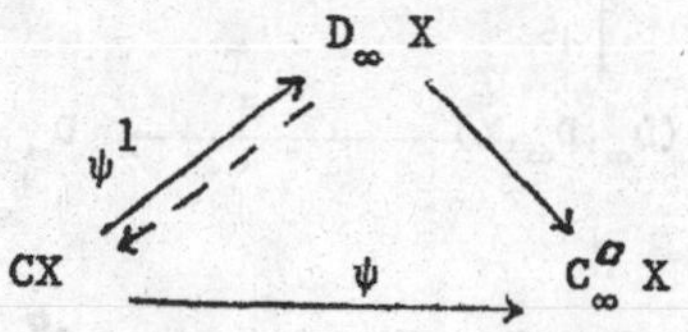

where unlabelled maps are induced by projections. Since $CX \to D_\infty X \to CX$ is the identity we see that ψ^1 is a homotopy inverse for the projection $D_\infty X \to CX$.

For a connected C-space X define $B_i X = \text{ind lim } \Omega^j B(\Sigma^{i+j}, C_{i+j}^{\square} \times C, X)$. there is then a homotopy equivalence $i : X \to B_0 X$, see [8] p. 463.

<u>Proposition 5.2</u> The diagram

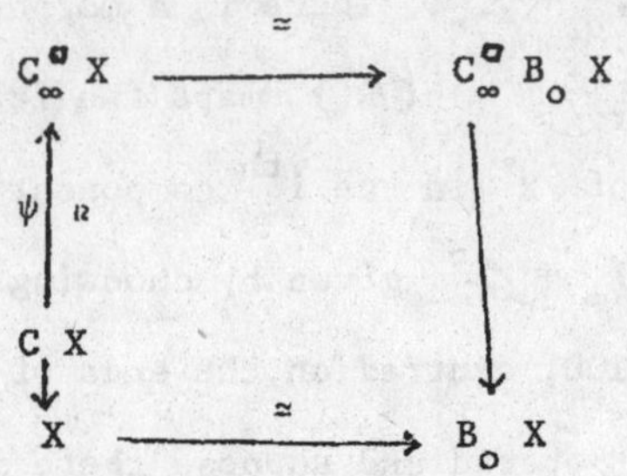

$C_\infty^{\sigma} X$
$C_\infty^{\sigma} B_o X$
ψ
CX
X
$B_o X$

homotopy commutes.

<u>Proof</u>. The result follows from commutativity in the following diagram, the bottom half of which may be found on p.154 of [9].

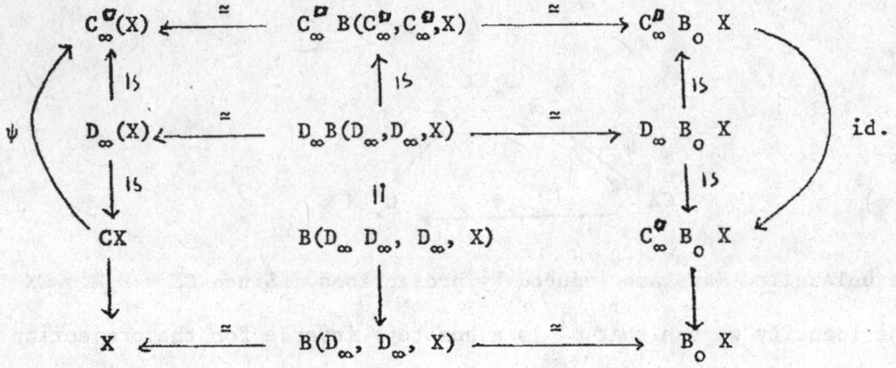

$C_\infty^{\sigma}(X)$
$C_\infty^{\sigma} B(C_\infty^{\sigma}, C_\infty^{\sigma}, X)$
$C_\infty^{\sigma} B_o X$
ψ
$D_\infty(X)$
$D_\infty B(D_\infty, D_\infty, X)$
$D_\infty B_o X$
id.
CX
$B(D_\infty D_\infty, D_\infty, X)$
$C_\infty^{\sigma} B_o X$
X
$B(D_\infty, D_\infty, X)$
$B_o X$

We apply 5.2 below with $\mathcal{C} = \mathcal{L}$, $X = BO$ and $\psi = \beta\circ\alpha$, where β and α are described in 5.1.

$$\Omega^2S^3 \xrightarrow{\Omega^2S^2} \Omega^2S^2BO \xrightarrow{\Omega^2S^2\iota} \Omega^2S^2B_0BO \longrightarrow B_0BO \xleftarrow{\simeq} BO \,.$$

With the aid of 2.7 (ii) of [3], and 5.2 we can factor this map in a form suitable for the proof of 4.2, as follows. There is a homotopy commutative diagram:

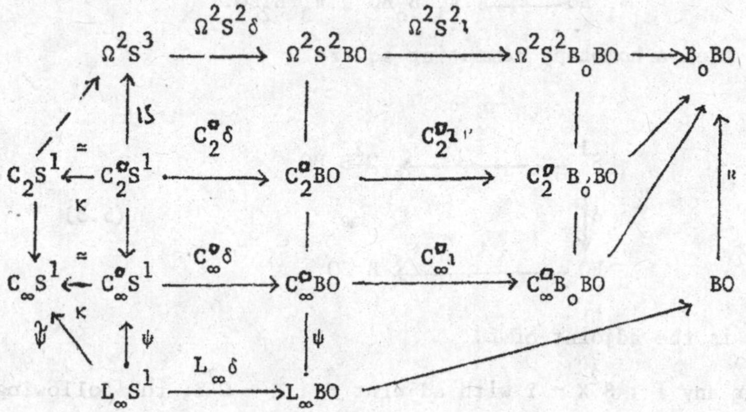

diagram 5.5

<u>Proof of 4.2</u>

From diagram 5.5 we see that it is sufficient to consider the composition

$$M \xrightarrow{h} C_2(S^1) \hookrightarrow C_\infty(S^1) \xleftarrow[\simeq]{\tilde{\psi}} L_\infty(S^1) \xrightarrow{L_\infty\delta} L_\infty(BO) \xrightarrow{\theta} BO$$

where $h : M \to C_2(S^1)$ 'classifies' the triple $(M,\bar{g},\tilde{g})$ with $\bar{g}$ in good position. If we consider h as a map into $C_\infty(S^1)$ then correspondingly we regard $\tilde{g}$ as an embedding in $M \times \mathbb{R}^\infty$, by including $\mathbb{R}^2$ in $\mathbb{R}^\infty$. Since $\tilde{\psi}$ is a homotopy equivalence there is a homotopy h_t of $h : M \to C_\infty(S^1)$ so that h_1 factors through some map $\ell : M \to L_\infty(S^1)$. We construct a

We turn to the description of $\eta : \Omega^2 S^3 \to BO$. Let $\delta : S^1 \to BO$ classify the non-trivial bundle. To be explicit, δ factors through $I/\partial I \to G_{2,1}$, given by $t \to L_t$, where L_t is the line described in §4. Then $[\delta]$ generates $\pi_1 BO$, and the generator $[f]$ of $\pi_3 B_2 BO$ is the image of $[\delta]$ under

$$\pi_1 BO \xrightarrow{\iota_*} \pi_1 B_o BO \cong \pi_3 B_2 BO.$$

We have then a homotopy commuting square

$$\begin{array}{ccc} S^1 & \xrightarrow{\hat{f}} & \Omega^2 B_2 BO \\ \delta \downarrow & & \cong \\ BO & \xrightarrow{\iota} & B_o BO \end{array} \qquad (5.3)$$

where $\hat{f}$ is the adjoint of f.

For any $f : S^2 X \to Y$ with adjoint $\hat{f} : X \to \Omega^2 Y$, the following diagram commutes

$$\begin{array}{ccc} \Omega^2 S^2 X & \xrightarrow{\Omega^2 f} & \Omega^2 Y \\ & {\scriptstyle \Omega^2 S^2 \hat{f}} \searrow \quad \nearrow {\scriptstyle \Omega^2 \phi} & \\ & \Omega^2 S^2 \Omega^2 Y & \end{array} \qquad (5.4)$$

where $\phi : S^2 \Omega^2 Y \to Y$ is the adjunction, given by $\phi(t,h) = h(t)$.

Putting together (5.3) and (5.4) we find that $\eta : \Omega^2 S^3 \to BO$ is the composition:

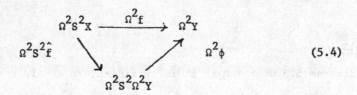

homotopy h_t as follows. To each h_t there will be a corresponding triple $(M,\bar{g},\tilde{g}_t)$ - we do not need to alter $\bar{g}$, and it will be sufficient to describe $\tilde{g}_t$. Lifting our $\tilde{\psi}$ is accomplished in two steps. $\tilde{\psi}$ is the composition

$$L_\infty S^1 \xrightarrow{\alpha} V_\infty S^1 \xrightarrow{\kappa\circ\beta} C_\infty S^1.$$

To lift our $\kappa\circ\beta$ we must change the configuration in $\mathbb{R}^2 \subset \mathbb{R}^\infty$, associated with a point in M, to be orthonormal. This corresponds to the homotopy of $\tilde{g}$. In fact we stay in $\mathbb{R}^{n+1}$, rather than $\mathbb{R}^\infty$, and proceed by induction up the filtration of M by the images $\{\text{im } \bar{g}_k\}_{k=0}^n$. At the kth stage we have $\underline{N}_k \to C_{n+1,k}$ with $\partial\underline{N}_k \to V_{n+1,k} \to C_{n+1,k}$, which is then homotoped rel. $\underline{N}_k$ to a map $\underline{N}_k \to V_{n+1,k}$. There is no obstruction since $\pi_r C_{n+1,k} = 0$ for $r < n$, and $\pi_r V_{n+1,k} = 0$ for $r < n + 1 - k$. Now extend to $\underline{M}_k$ by composing with the projection $\underline{M}_k \to \underline{N}_k$. Finally, lift our α using a similar induction. This time no homotopy is involved.

We now have $h_1 = \tilde{\psi}\circ\ell : M \to C_{n+1}(S^1) \subset C_\infty(S^1)$ and h_1 corresponds to $(M,\bar{g},\tilde{g}_1)$. Now let $q : N \to \mathbb{R}^{n+1}$ be the composition

$$N \xrightarrow{\times 1/2} N \times I \xrightarrow{\tilde{g}_1} M \times \mathbb{R}^{n+1} \xrightarrow{\pi_2} \mathbb{R}^{n+1}.$$

By construction, q has the properties of lemma 2. We now have

$$\nu_M : M \xrightarrow{\ell} L_\infty(S^1) \xrightarrow{L_\infty(\delta)} L_\beta(BO) \xrightarrow{\theta} BO$$

which factors through a map $M \to C_{2n+2,n+1}$, with $x \mapsto \nu_x$ as described in the theorem. ▱

REFERENCES

1. J. M. Boardman and R. M. Vogt, Homotopy-everything H-spaces, Bull. Am. Math. Soc. 74 (1968) 1117-1122.

2. F. R. Cohen, Braid orientations and bundle with flat connections, (to appear).

3. F. R. Cohen, J. P. May, L. Taylor, Splitting of certain spaces C$\underline{X}$, (to appear).

4. U. Koschorke and B. J. Sanderson, Self-intersections and higher Hopf invariants, Topology (to appear).

5. I. Madsen and R. J. Milgram, On spherical fibre bundles and their PL reductions, London Math. Soc. Lecture Notes Series 11 (1974) 43-59.

6. Mark Mahowald, A new infinite family in ${}_2\pi_*^S$, Topology 16 (1977) 249-256.

7. J. P. May, E_∞ Ring Spaces and E_∞ Ring Spectra, Springer Lecture Notes in Mathematics 577, Springer, Berlin (1977).

8. J. P. May, Infinite loop space theory, Bull. Am. Math. Soc. 83 (1977) 456-494.

9. J. P. May, The geometry of iterated loop spaces, Springer Lecture Notes in Mathematics 271. Springer, Berlin (1972).

10. Stewart Priddy, K(Z_2) as a Thom spectrum, (to appear).

11. G. Segal, Configuration spaces and iterated loop spaces, Invent. Math. (1973) 213-221.

12. V. P. Snaith, A stable decomposition of $\Omega^n S^n X$, J. Lond. Math. Soc. 7 (1974) 577-583.

13. P. Vogel, Cobordisme d'immersions, Ann. Sci. de l'Ecole Normale Superieure, 7 (1974) 317-358.

Mathematics Institute,
The University of Warwick,
Coventry CV4 7AL

October 1978

Desuspension in the Symmetric L-groups

by

Gunnar Carlsson

Introduction

In view of the recent strides made in the computation of Wall groups (see e.g. [C-M], [Pardon], [Wall]), the problem of determining which of the surgery obstructions occur as the obstruction of a degree one normal map of closed manifolds has become of increased importance. One approach to this is by product formulae, i.e. formulae which express the surgery obstruction of the degree one normal map

$$M \times N \longrightarrow X \times N$$

in terms of the obstruction of $(M \to X)$ and data derived from N. Morgan has recently analyzed this problem for the case $\pi_1(N) = 0$. (see [Morgan]). The problem reduces to a bordism problem, in fact to the analysis of a bilinear pairing

$$\Omega_*(K(\pi_1(N), 1)) \otimes L_*(\pi_1(X)) \longrightarrow L_*(\pi_1(X \times N)).$$

In [Ranički] it is shown that this pairing actually factors through a pairing

$$L^*(\pi_1(N)) \otimes L_*(\pi_1(X)) \longrightarrow L_*(\pi_1(X \times N)),$$

where the L^*-groups are symmetric versions of the Wall-groups, defined in [Miščenko] and [Ranički]. For purposes of computing product formulae, then, computing the groups L^* becomes of great interest. L^0 turns out to be the Witt group of $\mathbb{Z}\pi$, and L^1 is quite closely tied to the surgery group $L_1(\mathbb{Z}\pi)$. Secondly, there are skew suspension maps relating the high-dimensional L^*-groups to the lower dimensional ones. The approach to calculating the L^*'s, then, is to measure the cokernel of the skew-suspension maps, thereby reducing the problem to a Witt group problem, about which much is kn (see [C]).

The method for analyzing this cokernel is closely related to the method of characteristic elements, which one may use to calculate $W(\hat{\mathbb{Z}}_2)$, and a generalization of which was used in [C] to calculate $W(\hat{\mathbb{Z}}_2)$ for π a 2-group.

§I defines the groups L^*, §II defines the target groups for our invariants, §III proves that the invariants are well-defined, and §IV proves the main theorem, IV.3, which asserts that the defined invariant is the comple obstruction to desuspension.

I. Preliminaries

We recall from [Ranicki] the definition of algebraic Poincaré complexes over a ring Λ with involution and their bordism groups. Given a projective module over Λ, let P^* denote its dual module, $\mathrm{Hom}_\Lambda(P, \Lambda)$, endowed with a Λ-module structure in the usual way.

Definition 1 An n-dimensional ϵ-symmetric complex over Λ is a chain complex of projective Λ-modules, having the chain homotopy type of an n-dimensional chain complex, $\{C_*, \partial_*\}$, together with a collection of Λ-module maps $\Phi = \{\varphi_s \in \mathrm{Hom}_\Lambda(C^{n-r+s}, C_r) | r \in \mathbb{Z}, S \geq 0\}$, so that

$$(*)\ \partial\varphi_s + (-1)^r\varphi_s\partial^* + (-1)^{n+s-1}(\varphi_{s-1} + (-1)^{s+(n-r+s)r}\epsilon\varphi^*_{s-1}) = 0$$

$$: C^{n-r+s-1} \longrightarrow C_r \ (s \geq 0, \varphi_{-1} = 0).$$

(Of course, each φ_s really stands for a collection $\varphi^r_s : C^{n-r+s} \longrightarrow C_r$, $\forall r$. We suppress the superscript for simplicity of notation) Here $C^k = C^*_k$, and ∂^* and φ^*_{s-1} denote the duals to the maps ∂ and φ_{s-1}. Note that φ_0 is thus a chain map from the complex $\{C^{n-*}, \partial^*\}$ to the complex $\{C_*, \partial_*\}$. If φ_0 is a chain equivalence, the symmetric complex is said to be Poincaré.

Definition 2 Let (C_*, Φ) be an n-dimensional Poincaré complex, and let $f : C \to D$ be a chain map. where D is an $(n+1)$ - dimensional chain complex

<u>of projective Λ-modules.</u> <u>Then by surgery data for</u> f <u>we will mean a collection</u> $\Psi = \{\psi_s\}_{s=0}^{s=n+1}$ <u>of Λ-module homomorphisms,</u> $\psi_s : D^{n-r-1} \longrightarrow D_{r+s}$ <u>so that</u>

$$\partial\psi_s + (-1)^r \psi_s \partial^* + (-1)^{n+s} (\psi_{s-1} + (-1)^{s+(n-r-1)(r+s)} \epsilon \psi^*_{s-1})$$

$$+ (-1)^n f\varphi_s f^* = 0.$$

We say that the surgery data Ψ is connected if the map $D^* \longrightarrow MC(f)$ induced by ψ_0, where $MC(f)$ denotes the algebraic mapping cone on f, is surjective in 0-dimensional homology.

<u>Definition 3</u> <u>The Poincaré complex</u> C'_* <u>obtained from</u> C_* <u>by surgery on the map</u> f, <u>using connected surgery data</u> Ψ, <u>is defined by</u>

$$C'_r = D^{n-r+1} \oplus C_r \oplus D_{r+1},$$

$d_{C'}$ is given by the matrix

$$\begin{bmatrix} (-1)^r d_D^* & 0 & 0 \\ (-1)^{n+1} \varphi_0 f^* & d_C & 0 \\ (-1)^r \psi_0 & (-1)^r f & d_D \end{bmatrix}$$

$$\varphi_0' = \begin{bmatrix} 0 & 0 & 0 \\ 0 & \varphi_0 & 0 \\ (-1)^{r(n-r)} & (-1)^{n-r}\epsilon f\varphi_1^* & (-1)^{n-r}\epsilon\psi_1^* \end{bmatrix}$$

$$: (C')^{n-r} = D_{r+1} \oplus C^{n-r} \oplus D^{n-r+1} \longrightarrow C_r = D^{n-r+1} \oplus C_r \oplus D_{r+1}.$$

$$\varphi_s' = \begin{bmatrix} 0 & 0 & \\ 0 & \varphi_s & \\ 0 & (-1)^{n-r}\epsilon f\varphi_{s+1}^* & (-1)^{n-r+s}\epsilon\psi_{s-1}^* \end{bmatrix}$$

$$: (C')^{n-r} = D_{r+1} \oplus C^{n-r} \oplus D^{n-r+1} \longrightarrow C_{r+s} = D^{n-r-s+1} \oplus C_{r+s} \oplus D_{r+s+1}.$$

The equivalence relation generated by all equivalences of the form $(C_*, \Phi) \sim (C_*', \Phi')$, where C_*' is obtained from C_* by surgery, and homotopy equivalence, is called algebraic cobordism. The set of equivalence classes becomes group under direct sum becomes a group under direct sum of Poincaré complexes, and is denoted $L^n(\Lambda, \epsilon)$.

<u>Definition 4</u> <u>The skew-suspension of an n-dimensional ϵ-symmetric Poincaré complex</u> (C_*, Φ) <u>is an</u> (n+2)-<u>dimensional Poincaré complex</u> $(\overline{C}_*, \overline{\Phi})$, <u>where</u>

$$\overline{C_k} = C_{k-1},\ \overline{C_0} = \overline{C_{n+2}} = 0.$$

$$\underline{\overline{\varphi_s} : \overline{C^k} \longrightarrow \overline{C}_{n+2-k+s}}$$

$$= \underline{\varphi_s : C^{k-1} \longrightarrow C_{n+1-k+s}}.$$

It is easily verified that this defines a homomorphism

$$\sigma : L^n(\Lambda, \epsilon) \longrightarrow L^{n+2}(\Lambda, -\epsilon)$$

Remark 1 For the surgery groups $L_n(\mathbb{Z}\pi, \epsilon)$ the analogue to the double skew-suspension $\sigma^2 : L_n(\mathbb{Z}\pi, \epsilon) \longrightarrow L_{n+4}(\mathbb{Z}\pi, \epsilon)$ may be identified with th periodicity isomorphism $L_n(\mathbb{Z}\pi, \epsilon) \xrightarrow{x[\mathbb{C}P^2]} L_{n+4}(\mathbb{Z}\pi, \epsilon)$. In the case of $^{/}L$ however, σ fails to be an isomorphism, and it is this failure we shall analyze.

Remark 2 For complexes C and D of projective Λ-modules, define the complex $\mathrm{Hom}_\Lambda(C, D)$ by $\mathrm{Hom}_\Lambda(C, D)_n = \bigoplus_{q-p=n} \mathrm{Hom}_\Lambda(C_p, D_q)$, $d_{\mathrm{Hom}_\Lambda(C,D)}(f) = d_D f + (-1)^q f d_C$. Note that duality provides an involution on $\mathrm{Hom}_\Lambda(C^*, C_*)$ by $f \to (-1)^{pq} \epsilon f^*$; so the complex $\mathrm{Hom}_\Lambda(C^*, C_*)$ becomes a complex of $\mathbb{Z}[\mathbb{Z}/_2]$-modules. Let W_* denote the standard $\mathbb{Z}[\mathbb{Z}/_2]$-resolution of $\mathbb{Z}$, $W_n = \mathbb{Z}[\mathbb{Z}/_2]$, $\partial e_n = (1 + (-1)^n T)e_{n-1}$, $n \geq 0$, $W_n = 0$ for

$n < 0$. Let $Q^*(C, \epsilon) = \mathrm{Hom}_{\mathbb{Z}[\mathbb{Z}/_2]}(W_*, \mathrm{Hom}_\Lambda(C^*, C_*))$, where $\mathrm{Hom}_\Lambda(C^*, C_*)$ is acted on by $\mathbb{Z}/_2$ by $T\varphi = (-1)^{pq}\,\epsilon\varphi^*$. We may now observe that choosing an ϵ-symmetric structure on a complex C_* amounts to choosing a cycle in $Q^n(C, \epsilon)$. Note also that $Q^n(C, \epsilon)$ is a functor in C_*, since given $f : C_* \longrightarrow D_*$, we may define a map

$$Q^n(C, \epsilon) \xrightarrow{Q^n(f,\epsilon)} Q^n(D, \epsilon)$$

by letting $\hat{f} : \mathrm{Hom}_\Lambda(C^*, C_*) \longrightarrow \mathrm{Hom}_\Lambda(D^*, D_*)$ denote the map $\varphi \longrightarrow f\varphi f^*$, and noting that $\hat{f}$ is $\mathbb{Z}/_2$ - equivariant. Surgery data for the map f consists of a choice of $\psi \in Q^{n+1}(D, \epsilon)$ so that $\partial\psi = Q^n(f, \epsilon)(\varphi)$, where φ is a cycle defining the symmetric structure on C_*.

II The Groups $\mathfrak{W}_n(\Lambda, \epsilon)$

As in the previous section, let Λ be a ring with involution, and let

$$Q^*(C, \epsilon) = \mathrm{Hom}_{\mathbb{Z}[\mathbb{Z}/_2]}(W_*, \mathrm{Hom}_\Lambda(C^*, C_*)),$$

as in remark 2, §I.

Recall that the abelian group

$$H_\epsilon(\mathbb{Z}/_2, \Lambda) = \{\lambda \in \Lambda \mid \lambda = \epsilon\bar{\lambda}\}/\{\lambda + \epsilon\bar{\lambda}, \lambda \in \Lambda\}$$

becomes a Λ-module by

$$\lambda\alpha = \lambda\,\alpha\,\overline{\lambda},$$

for $\lambda \in \Lambda$, $\alpha \in H(\mathbb{Z}/_2, \Lambda)$, and that if $\varphi : M^* \longrightarrow M$ is an ϵ-symmetric Λ-homomorphism (i.e. $\varphi = \epsilon\varphi^*$), we obtain a Λ-map $\hat{\varphi} : M^* \longrightarrow H_\epsilon(\mathbb{Z}/_2, \Lambda)$ $x \longrightarrow \langle x, \varphi x\rangle$, where $\langle\ ,\ \rangle$ denotes the evaluation pairing $\langle\ ,\ \rangle$ $: M^* \otimes M \to \Lambda$.

Let $\mathfrak{R}_* = \mathfrak{R}_*(\Lambda, \epsilon)$ denote a Λ-projective resolution of $H_\epsilon(\mathbb{Z}/_2, \Lambda)$, and let $\mathfrak{R}_*^{(n)}$ denote its n-skeleton. We consider the two complexes $Q(\mathfrak{R}_*^{(n)}, \epsilon)$ and $Q(\mathfrak{R}_*^{(n+1)}, \epsilon)$. Recall that the n-cycles of $Q(\mathfrak{R}_*^{(n)}, \epsilon)$, $Z^n(\mathfrak{R}_*^{(n)}, \epsilon)$ consist of collections $\Phi = \{\varphi_s\}$ of Λ-homomorphisms, satisfyi

$\partial\varphi_s + (-1)^q\varphi_s\partial^* + (-1)^{n+s-1}(\varphi_{s-1} + (-1)^{s+pq}\epsilon\,\varphi^*_{s-1}) = 0 : C^p \longrightarrow C_q$. Therefore, $\varphi_n + (-1)^{n+1}(-1)^{n^2}\epsilon\,\varphi^*_n = 0$, or $\varphi_n = \epsilon\varphi^*_n$. We obtain a homomorphism $\lambda(\Phi) = \varphi_n : \mathcal{R}^*_n \to H_\epsilon(\mathbb{Z}/_2, \Lambda)$. Secondly, φ_0 provides a Λ-homomorphism $\varphi_0 : \mathcal{R}^*_n \longrightarrow \mathcal{R}_0$, which when composed with augmentation map $\eta : \mathcal{R}_0 \to H_\epsilon(\mathbb{Z}/_2, \Lambda$ from the resolution gives a second homomorphism

$$\rho(\Phi) : \mathcal{R}^*_n \longrightarrow H_\epsilon(\mathbb{Z}/_2, \Lambda)$$

These two correspondences define homomorphisms

$$\lambda,\ \rho : Z^n(\mathcal{R}^{(n)}_*, \epsilon) \longrightarrow \mathrm{Hom}_\Lambda(\mathcal{R}^*_n, H_\epsilon(\mathbb{Z}/_2, \Lambda))$$

Define $\widetilde{Z}^n \subseteq Z^n(\mathcal{R}^{(n)}_*, \epsilon)$ by

$$\widetilde{Z}^n = \{x \in Z^n(\mathcal{R}^{(n)}_*, \epsilon) \mid \rho(x) = \lambda(x)\}.$$

We now let B_{n+1} denote the subgroup of the $(n+1)$-chains of $Q(\mathcal{R}^{(n+1)}_*, \epsilon)$ consisting of those chains whose boundary is in the image of $Q(\mathcal{R}^{(n)}_*, \epsilon)$ in $Q(\mathcal{R}^{(n+1)}_*, \epsilon)$ under the natural inclusion. This means that an element of B_{n+1} is a collection $\Psi = \{\psi_s\}$ of Λ-module momomorphisms such that $\psi_{n+1} = \epsilon\,\psi^*_{n+1}$, since

$$\psi_{n+1} + (-1)^{n+2}(-1)^{(n+1)^2} \epsilon \psi^*_{n+1} = 0.$$

This defines a homomorphism

$$\alpha = \psi_{n+1} : B_{n+1} \longrightarrow \mathrm{Hom}_\Lambda(\mathfrak{K}^*_{n+1}, H_\epsilon(\mathbb{Z}/_2, \Lambda))$$

A second homomorphism β is obtained by $\beta(\Psi) = \eta \circ \psi_0$, where $\eta : \mathfrak{K}_0 \longrightarrow H_\epsilon(\mathbb{Z}/_2, \Lambda)$ is the augmentation. Define

$$B_{n+1} = \{\Psi | \alpha(\Psi) = \beta(\Psi)\}$$

<u>Proposition 1</u> $\partial\widetilde{B}_{n+1} \subseteq \widetilde{Z}^n$

<u>Pf.</u> Let $\Phi = \partial\Psi$. Then

(i) $\varphi_0 = \partial\psi_0 + (-1)^q \psi_0 \partial^*$

(ii) $\varphi_n = (-1)^{n+1}(\partial\psi_n + (-1)^{n+1} \psi_n \partial^* - (\psi_{n-1} + \epsilon\psi^*_{n-1}))$

and since $\Psi \in B_{n+1}$,

(iii) $0 = \partial\psi_{n+1} + (-1)^n \psi_{n+1}\partial^* + (\psi_n + (-1)^{n+1} \epsilon \psi^*_n)$

Now,

$$\partial\psi_n + (-1)^{n+1}\psi_n\partial^* = \partial\psi_n + \epsilon\psi^*_n\partial^* + (-1)^{n+1}\partial\psi_{n+1}\partial^*$$

so

$$\varphi_n = (-1)^{n+1}(\partial\psi_n + \epsilon\psi_n^*\partial^* - (\psi_{n-1} + \epsilon\psi_{n-1})) + \partial\psi_{n+1}\partial^*$$

The left hand term in the sum is of the form $\beta + \epsilon\beta^*$, so $\hat{\varphi}_n = \partial\psi_{n+1}\partial^*$.

Equation (i) asserts that $\eta\circ\varphi_0 = \eta\circ\psi_0\circ\partial^*$, since $\eta\circ\partial = 0$, and $H_\epsilon(\mathbb{Z}/_2, \wedge)$ is a $\mathbb{Z}/_2$-vector space. The condition $\alpha(\Psi) = \beta(\Psi)$ guarantees that $\eta\circ\psi_0 = \hat{\psi}_{n+1}$, or $\eta\circ\psi_0\circ\partial^*(x) = \hat{\psi}_{n+1}(\partial^* x) = \partial\psi_{n+1}\partial^*(x) = \hat{\varphi}_n(x)$, so $\eta\circ\varphi_0 = \hat{\varphi}_n$, which implies $\Phi \in \widetilde{Z}^n$. (*)

We now define

$\mathfrak{w}_n(\wedge, \epsilon, \mathfrak{R}) = \widetilde{Z}^n/\partial\widetilde{B}_{n+1}$, and conclude this section by showing that $\mathfrak{w}_n(\wedge, \epsilon, \mathfrak{R})$ is independent of the choice of resolution $\mathfrak{R}$.

<u>Proposition 2</u> <u>If</u> $\mathfrak{R}_*$, $\mathfrak{S}_*$ <u>are two resolutions of</u> $H_\epsilon(\mathbb{Z}/_2, \wedge)$, <u>then</u> $\mathfrak{w}_n(\wedge, \epsilon, \mathfrak{R}) \cong \mathfrak{w}_n(\wedge, \epsilon, \mathfrak{S})$. <u>We then define</u> $\mathfrak{w}_n(\wedge, \epsilon) = \mathfrak{w}_n(\wedge, \epsilon, \mathfrak{R}) = \mathfrak{w}_n(\wedge, \epsilon, \mathfrak{S})$.

<u>Proof</u>. We may assume that there is a chain map $\mathfrak{S} \to \mathfrak{R}$ which is surjective in each degree, since in any event, there is a resolution $\mathfrak{J}$ which maps surjectively in each degree to both $\mathfrak{R}$ and $\mathfrak{S}$. it is then easily seen that $\mathfrak{S}$ is isomorphic to $\mathcal{E} \oplus \mathfrak{R}$, where $\mathcal{E}$ is a contractible complex. Since any sum of elementary complexes

$$\cdots \longrightarrow 0 \longrightarrow 0 \longrightarrow P \xrightarrow{id} P \longrightarrow 0 \longrightarrow 0 \longrightarrow \cdots$$

with P projective, we may assume that $\mathcal{S}$ is obtained from $\mathcal{R}$ by additio[n] with a single elementary complex.

The complex

$\mathrm{Hom}_\Lambda((\mathcal{E} \oplus \mathcal{R})^*, (\mathcal{E} \oplus \mathcal{R}))$ splits as

$\mathrm{Hom}_\Lambda(\mathcal{E}^*, \mathcal{E}) \oplus \mathrm{Hom}_\Lambda(\mathcal{E}^*, \mathcal{R}) \oplus \mathrm{Hom}_\Lambda(\mathcal{R}^*, \mathcal{E}) \oplus \mathrm{Hom}\,(\mathcal{R}^*, \mathcal{R})$,

and the involution preserves the first and fourth summands and permutes the middle two. Thus, $\mathrm{Hom}_{\mathbb{Z}[\mathbb{Z}/_2]}(W_*, \mathrm{Hom}\,((\mathcal{E} \oplus \mathcal{R})^*, \mathcal{E} \oplus \mathcal{R}))$ splits into thr[ee] summands,

$$\mathrm{Hom}_{\mathbb{Z}[\mathbb{Z}/_2]}(W_*, \mathrm{Hom}_\Lambda(\mathcal{E}^*, \mathcal{E})) \oplus$$

$$\mathrm{Hom}_{\mathbb{Z}[\mathbb{Z}/_2]}(W_*, \mathrm{Hom}_\Lambda(\mathcal{E}^*, \mathcal{R}) \oplus \mathrm{Hom}_\Lambda(\mathcal{R}^*, \mathcal{E})) \oplus$$

$$\mathrm{Hom}_{\mathbb{Z}[\mathbb{Z}/_2]}(W_*, \mathrm{Hom}\,(\mathcal{R}^*, \mathcal{R}))$$

Furthermore, the homomorphisms ρ and β vanish identically on the first two of these, and λ and α vanish identically on the middle summand.

It is now easily verified that the middle term contributes nothing to $\mathfrak{w}_n(\Lambda, \epsilon, \mathcal{S})$, since any cycle Z in

$$\mathrm{Hom}_{\mathbb{Z}[\mathbb{Z}/2]}(W_*, \mathrm{Hom}(\mathcal{E}^*, \mathcal{R}_*^{(n)}) \quad \mathrm{Hom}((\mathcal{R}^{(n)})^*, \mathcal{E}))$$

is a boundary in

$$\mathrm{Hom}_{\mathbb{Z}[\mathbb{Z}/2]}(W_*, \mathrm{Hom}(\mathcal{E}^*, \mathcal{R}_*^{(n+1)}) \quad \mathrm{Hom}((\mathcal{R}^{(n+1)})^*, \mathcal{E})),$$

$\mathcal{E}$ being contractible, and the fact that α vanishes identically on this summand guarantees that we may choose the chain x such that $\partial x = z$ with $x \in \tilde{B}_{n+1}$. We must therefore check that the contribution of the first summand is also zero. Let $Z^n(\mathcal{E})$ be the group of n-cycles in $Q(\mathcal{E}, \epsilon)$ and let $\tilde{Z}^n(\mathcal{E}) = \{\Phi \in Z^n(\mathcal{E}) | \varphi_n = \gamma + \epsilon\gamma^*\}$. Also, let $B_{n+1}(\mathcal{E})$ be the group of (n + 1)-chains x in $Q(\mathcal{E}, \epsilon)$ so that $\partial x \in Q(\mathcal{E}^{(n)}, \epsilon)$, and let $\tilde{B}_{n+1}(\mathcal{E}) = \{\Psi \in B_{n+1}(\mathcal{E}) | \psi_{n+1} = \gamma + \epsilon\gamma^*\}$. It is easily seen that $\partial\tilde{B}_{n+1}(\mathcal{E}) \subseteq \tilde{Z}^n(\mathcal{E})$, as in Proposition 1. Moreover, since ρ and β vanish identically on this summand, the contribution of this summand to $\mathfrak{w}_n(\Lambda, \epsilon, \mathcal{S})$ is isomorphic to

$$\tilde{Z}^n(\mathcal{E}) / \partial\tilde{B}_{n+1}(\mathcal{E}) .$$

It is now an easy calculation with the elementary complexes that this group is zero. (*).

III Defining the Invariant

We assume from now on that all Poincaré complexes will in fact be n-dimensional compleses, i.e. that $C_* = 0$ for $* < 0$, $* > n$. This involv no loss of generality since the complexes have the homotopy type of an n-dimensional complex.

Let (C_*, Φ) be an ϵ-symmetric Poincaré complex. From the identity (in the definition of Poincaré complexes, we find

$\partial\varphi_{n+1} + (-1)^r \varphi_{n+1}\partial^* + (-1)(\varphi_n - \epsilon\,\varphi_n^*) = 0 \;:\; C^{2n-r} \to C_r$. Since C_* is n-dimensional, $C^{2n-r} = 0$ for $r < n$, $C_r = 0$ for $r > n$, so the map $\varphi_{n+1} = 0$, and we obtain $\varphi_n = \epsilon\,\varphi_n^*$. Therefore, we have the n-th "Wu class" map $\hat{\varphi}_n : C^n \longrightarrow H_\epsilon(\mathbb{Z}/_2, \wedge)$, as in [Ranički]

<u>Lemma 1</u> <u>Let</u> C_* <u>be a chain complex of projective $\wedge$-modules, bounded below</u> ($C_* = 0$ <u>for</u> $* < 0$) <u>Then any homomorphism</u> $f : C_0 \to M$, <u>where</u> M <u>is a $\wedge$-module, may be extended to a chain map (unique up to chain homotopy)</u> $f : C_* \to \mathbb{R}_*(M)$, <u>where</u> $\mathbb{R}_*(M)$ <u>denotes a resolution of the module</u> M.

<u>Proof</u> The usual argument for maps of resolutions does not use the acyclicit of C_*. (*)

The map $\hat{\varphi}_n$ defines a homotopy class of chain maps

$$W : C^* \longrightarrow \mathbb{R}_*(H_\epsilon(\mathbb{Z}/_2, \wedge)).$$

The invariant we construct will lie in the group

$$\mathbb{W}_n(\wedge, \epsilon)$$

Since (C_*, Φ) is a Poincaré complex, the chain map $\varphi_0 : C^* \to C_*$ is a chain equivalence. We choose $\overline{\varphi}_0$ to be a chain inverse to φ_0(the choice is unique up to chain homotopy).

<u>Proposition 2</u> <u>The element</u> $\{w\overline{\varphi}_0\varphi_s\overline{\varphi}_0^*w^*\} \in Z^n(\mathbb{R}^{(n)}, \epsilon)$ <u>lies in</u> $\widehat{Z^n(\mathbb{R}^{(n)}, \epsilon)}$.

<u>Pf</u>. Let $\Phi = \{w\overline{\varphi}_0\varphi_s\overline{\varphi}_0^*w^*\}$. Then $\lambda(\Phi)(x) = \widehat{w\overline{\varphi}_0\varphi_n\overline{\varphi}_0^*\ w^*}\,(x) = \hat{\varphi}_n(\overline{\varphi}_0^*w^*x)$ Also, $\rho(\Phi)(x) = \eta w\overline{\varphi}_0\varphi_0\overline{\varphi}_0^*w^*(x)$. By the choice of w and $\overline{\varphi}_0$, $\eta w\overline{\varphi}_0\varphi_0 = \hat{\varphi}_n$, so $\rho(\Phi)(x) = \hat{\varphi}_n(\overline{\varphi}_0^*w^*x) = \lambda(\Phi)(x)$. (*)

Let $\xi(C_*, \Phi) \in \mathbb{W}_n(\wedge, \epsilon)$ be defined by $\xi(C_*, \Phi) = \{w\overline{\varphi}_0\varphi_s\overline{\varphi}_0^*w^*\}$.

<u>Proposition 3</u> $\xi(C_*, \Phi)$ <u>is independent of the choice of</u> w <u>and</u> $\overline{\varphi}_0$ <u>within homotopy classes</u>.

<u>Pf</u>. If $w \simeq w'$, $\overline{\varphi}_0 \simeq \overline{\varphi}_0'$, $w\varphi_0 \simeq w'\varphi_0'$, wo we suppose that we have a chain homotopy $h : w\varphi_0 \simeq w'\varphi_0'$

According to [Ranički]

$$\Psi = \{w\overline{\varphi}_0\varphi_s h^* + (-1)^q h\varphi_s\overline{\varphi}_0'^*w'^* + (-1)^{q+1}h\varphi_{s-1}h^*\}$$

satisfies

$$\partial\Psi = \{w\bar{\varphi}_0\varphi_s\bar{\varphi}_0^* w^*\} - \{w'\bar{\varphi}_0'\varphi_s\bar{\varphi}_0'^* w'^*\}$$

We must show that $\Psi \in \widetilde{B}_{n+1} \subseteq B_{n+1}$. To verify this, it will suffice t show $\alpha(\Psi) = \beta(\Psi)$.

$$\alpha(\Psi) = \hat{\psi}_{n+1} = (-1)^{q+1}\widehat{h\varphi_n h^*} = \widehat{h\varphi_n h^*},$$

the last equality since $H_\epsilon(\mathbb{Z}/_2, \Lambda)$ is a $\mathbb{Z}/_2$-vector space.

$\beta(\Psi) = \eta \circ \psi_0 = \eta w\bar{\varphi}_0\varphi_0 h^* + (-1)^q \eta h\varphi_0\bar{\varphi}_0'^* w'^* : \mathfrak{K}_{n+1}^* \to \mathfrak{K}_{n+1}$. The second summand factors through a zero group, hence is zero. By the choice of $\bar{\varphi}_0$ w, we have

$$\eta w\bar{\varphi}_0\varphi_0 h^* = \hat{\varphi}_n \circ h^* = \widehat{h \circ \varphi_n \circ h^*}, \text{ so}$$

$\alpha(\Psi) = \beta(\Psi)$. (*)

Cor. 4 $\xi(C_*, \Phi)$ is independent of the homotopy type of C_*.

Pf. Clear. (*)

Cor. 5. Let (C_*, Φ) and (C_*', Φ') be two Poincaré complexes over Λ. Th

$$\xi(C_* \oplus C_*', \Phi \oplus \Phi') = \xi(C_*, \Phi) \oplus \xi(C_*, \Phi')$$

Pf. Clear, since the homomorphism $\varphi_n \oplus \varphi_n'$ is equal to $\hat{\varphi}_n \oplus \hat{\varphi}_n'$. (*)

IV The Homomorphism $\mathbb{w}_n : L^n(\wedge, \epsilon) \to \mathbb{w}_n(\wedge, \epsilon)$ and Desuspension in the L-groups

In the previous section, it was shown that there is an invariant of the homotopy type of (C_*, Φ), $\xi(C_*, \Phi)$. In this section, we show that $\xi(C_*, \Phi)$ is an invariant of the algebraic cobordism class of (C_*, Φ), and hence induces a homomorphism $\mathbb{w}_n : L^n(\wedge, \epsilon) \longrightarrow \mathbb{w}_n(\wedge, \epsilon)$, in view of corollary III. 5.

Proposition 1 Let (C_*, Φ) be a Poincaré complex, $f : C_* \longrightarrow D_*$ a chain map, and $\Psi = \{\psi_s\}$ surgery data for f. If (C', Φ') denotes the Poincaré complex obtained by surgery on f, then $\xi(C'_*, \Phi') = \xi(C_*, \Phi)$.

Pf. We note that C'_* is obtained by a double mapping cone construction on C_*. That is, we first form the algebraic mapping cone $MC(f)$, and observe that surgery data for f determines a homotopy class om maps $\tilde{f} : D^* \to MC(f)$, together with a Poincaré structure on $MC(\tilde{f})$. In particular, the underlying chain complex of C'_* is $MC(\tilde{f})$. Similarly, C'^* admits D as a subcomplex, as well as $MC(f\varphi_0^*)$. By the definition of the top Wu class of C'_*, $\varphi'_n|D_* = 0$. Therefore, we may choose the chain map w from C'^* to $\mathcal{R}_*(H_\epsilon(\mathbb{Z}/_2, \wedge))$ so that w vanishes on $D_* \subseteq C'^*$. Therefore, there is a splitting of graded $\wedge$-modules (not of chain complexes)

$$C'_* \cong D^* \oplus C_* \oplus D_*$$

$$C'^* \cong D_* \oplus C^* \oplus D^*$$

And the map $w : C'^* \to \mathfrak{R}_*(H_\epsilon(\mathbb{Z}/_2, \wedge))$ has "matrix" $(0, w', w'')$, where w' is a lifting of the n-th Wu class of C_* to $\mathfrak{R}_*(H_\epsilon(\mathbb{Z}/_2, \wedge))$. Consequently, the dual map w^* has matrix

$$\begin{pmatrix} 0 \\ w'^* \\ w''^* \end{pmatrix}$$

Recall from §I that the map φ_0' is given by the matrix

$$\begin{bmatrix} 0 & 0 & (-1)^{q(n-q)}\varepsilon \\ 0 & \varphi_0 & 0 \\ 1 & (-1)^{(-q)+pq} f\varphi_1^* & (-1)^{(n-q)+pq}\epsilon\psi_1 \end{bmatrix}$$

Consequently, if $\overline{\varphi}_0$ is a chain inverse to φ_0, we find that the matrix of chain inverse to φ_0' is given by.

$$\begin{pmatrix} * & * & 1 \\ 0 & \varphi_0 & 0 \\ (-1)^{q(n-q)}\epsilon & 0 & 0 \end{pmatrix}$$

where the *'s represent certain maps, the values of which will not concern us. Now, $w\overline{\varphi_0}$ =

$$(0\ w'\ w'')\begin{pmatrix} * & * & 1 \\ 0 & \overline{\varphi_0} & 0 \\ (-1)^{q(n-q)}\epsilon & 0 & 0 \end{pmatrix}$$

$= ((-1)^{q(n-q)}\epsilon w'', w'\overline{\varphi_0}, 0)$, so $\overline{\varphi}_0^* w^*$ has matrix

$$\begin{pmatrix} w''^* \\ \overline{\varphi}_0^* w'^* \\ 0 \end{pmatrix}$$

Finally, $w\overline{\varphi}_0\varphi_0 = (0, w'\overline{\varphi}_0\varphi_0, w'')$, so $w\overline{\varphi}_0\varphi_0\overline{\varphi}_0^* w^*$ =

$$(0, w'\overline{\varphi}_0\varphi_0, w'')\begin{pmatrix} (-1)^{q(n-q)}\epsilon w'' \\ w'\overline{\varphi}_0 \\ 0 \end{pmatrix}$$

$= w'\overline{\varphi}_0\varphi_0\overline{\varphi}_0^* w^*$. Similarly, $w\overline{\varphi}_0'\varphi_s'\overline{\varphi}_0' w^* = w\varphi_0\varphi_s\varphi_0^* w^*$, so the value of $\xi(C_*, \Phi) = \xi(C_*, \Phi)$. (*)

Applying the definition of the groups $L^n(\Lambda, \epsilon)$ and Corollary II.5, we have defined a homomorphism $\mathfrak{w}_n : L^n(\Lambda, \epsilon) \to \mathfrak{w}_n(\Lambda, \epsilon)$ by $\mathfrak{w}_n(C_*, \Phi) = \xi(C_*, \Phi)$.

<u>Proposition 2</u> $\mathfrak{w}_n$ <u>vanishes on the image of the skew-suspension map</u> σ.

<u>Proof</u>. It is immediate that the n-th Wu class map $\hat{\varphi}_n$ is trivial on a skew-suspension, since it is defined on a trivial group. Thus, the chain ma[p] w may be taken to be zero. (*)

It is shown in [Ranički] that a chain complex is in the image of the skew-suspension if its n-th Wu class vanishes. This allows us to prove th[e] main theorem.

<u>Theorem 3</u>. $x \in L^n(\Lambda, \epsilon)$ <u>is in the image of the skew-suspension if and only if</u> $\mathfrak{w}_n(x) = 0$.

<u>Proof</u>. Consider a representative Poincaré complex (C_*, Φ) for x. We may suppose that the n-th Wu class map $\hat{\varphi}_n$ is onto $H_\epsilon(\mathbb{Z}/_2, \Lambda)$. If not, we may add on some null-cobordant complexes for which $\hat{\varphi}_n$ is onto. Now, if th[e] invariant $\mathfrak{w}_n(x)$ is trivial, there exists $\Psi \in \widetilde{B}_{n+1}$ with $\partial\Psi = \Phi$, where Φ is a cycle representing $\mathfrak{w}_n(x)$. Ψ thus represents surgery data for the chain map $C_* \xrightarrow{w\bar{\varphi}_0} \mathfrak{R}_*^{(n+1)}$. Thus, we form the chain complex (C'_*, Φ') by

surgery on the map w_0. I claim that the n-th Wu class φ'_n is trivial on the n-dimensional cohomology of C'_*. To see this, we analyze $H^n(C)$. In the relevant dimensions, the complex may be represented by

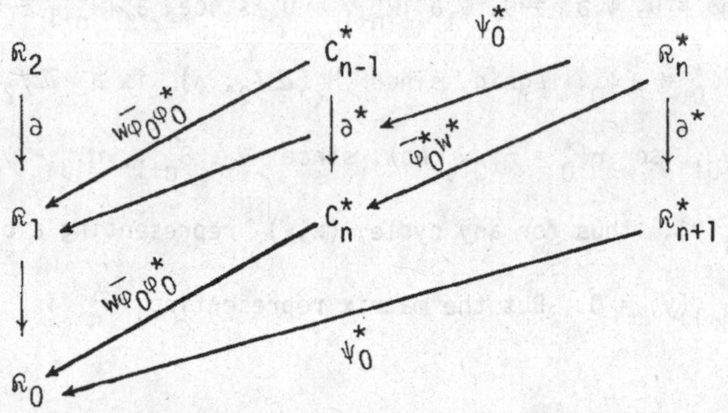

Note that since as a chain map, $\varphi_0 \simeq \varphi^*_0$, and $\bar{\varphi}_0\varphi_0 \simeq \text{id}$, the map $w\bar{\varphi}_0\varphi^*_0$ has the same effect homologically as w. Thus it is surjective in cohomology, and we find

$$H^n(MC(w\bar{\varphi}_0\varphi_0)) \cong \operatorname{Ker}(w_n : H^n(C^*) \longrightarrow H_\epsilon(\mathbb{Z}/_2, \wedge)).$$

The remaining dimensions are unchanged from C^*, since $\mathfrak{K}_*$ is acyclic above dimension 0. We conclude, then, that any cohomology class in $H^n(C'_*)$ is represented by a pair $(x, y) \in C^*_n \oplus \mathfrak{K}^*_{n+1}$, so that $w\bar{\varphi}_0\varphi^*_0(x) = -\psi^*_0(y)$.

Applying the augmentation $\eta : \mathcal{R}_0 \longrightarrow H_\epsilon(\mathbb{Z}/_2, \wedge)$, we find $\eta w\overline{\varphi}_0\varphi_0^*(x) = -\eta\psi^*$

Now by the construction of w and the above remarks about φ_0^*, $\eta w\overline{\varphi}_0\varphi_0^*(x) =$

$\hat{\varphi}_n(x)$. We also note that $\eta\psi_0^* = \eta\psi_0$, for $\psi_0^* = \pm(\psi_0 \pm \partial\psi_1 \pm \psi_1\partial^*) \pm \overline{\varphi}_0\varphi_1\overline{\varphi}_0^*$

$\eta\partial\psi_1 = 0$ since $\eta\partial = 0$, $\psi_1\partial^* = 0$, $\psi_1\partial^*|\mathcal{R}_{n+1}^* = 0$ since $\partial^*|\mathcal{R}_{n+1}^* = 0$, and

$w^*|\mathcal{R}_{n+1}^* = 0$, so $\psi_0^* = \pm\psi_0$. Again, since $H_\epsilon(\mathbb{Z}/_2, \wedge)$ is a $\mathbb{Z}/_2$-vector

space, $\eta\psi_0 = \pm\eta\psi_0$, so $\eta\psi_0^* = \eta\psi_0$. Now, since $\Psi \in \widetilde{B}_{n+1}$, $\eta\psi_0 = \hat{\psi}_{n+1}$,

so $-\eta\psi_0^*(y) = \hat{\psi}_{n+1}(y)$, thus for any cycle (x,y) representing a class in

$H^n(C^*)$, $\hat{\varphi}_n(x) + \hat{\psi}_{n+1}(y) = 0$. But the matrix representing φ_n' is

$$\begin{bmatrix} 0 & 0 & 0 \\ 0 & \varphi_0 & 0 \\ 0 & * & \psi_{n+1} \end{bmatrix}, \quad \text{so } \hat{\varphi}_n'(x,y) = \hat{\varphi}_n(x) + \hat{\psi}_{n+1}(y) = 0.$$

This proves that $\hat{\varphi}_n$ is identically zero on $H^n(C')$. It is shown in [Ranički] that under these circumstances, one may perform a sequence of elementary surgeries to kill $H^n(C)$, leaving a complex C'' with $H^n(C'') =$ Such a complex is the homotopy type of a skew-suspension (again, see [Ranički]). This concludes the proof (*).

Department of Mathematics
University of California at San Diego
La Jolla, California 92037

Bibliography

1.) Carlsson, G. On the Witt Group of a 2-adic Group Ring. (to apper)

2.) Carlsson, G., and Milgram, R.J. The Structure of Odd L-groups. (to appear, Proceedings of Waterloo Conference on Algebraic Topology).

3.) Miščenko, A.S. Homotopy Invariants of Non-Simply Connected Manifolds III. Higher Signatures, Izv. Akad. Nauk. SSSR, ser. mat. 35, pp. 1316-1355 (1971).

4.) Morgan, J. A.M.S. Memoirs, Vol 201.

5.) Pardon, W. The Exact Sequence of a Localization for Witt Groups II: Numerical Invariants of Odd-dimensional Surgery Obstructions (Preprint)

6.) Raniċki, A.A. The Algebraic Theory of Surgery, I.H.E.S. Notes.

7.) Wall, C.T.C. On the Classification of Hermitian Forms VI. Group Rings Ann. of Math. 103, 1-80 (1976).

Product Formulae for Surgery Obstructions.

F.J.-B.J. Clauwens

In this lecture we study the pairing $\Theta : \Omega_n(G,w_G) \times L_m(H,w_H) \to L_{n+m}(G\times H, w_G\, w_H)$ between bordism groups and surgery obstruction groups, arising from the fact that the product of a closed n-manifold P with $\pi_1(P) = G$ and a normal map $f : M \to X$ with $\pi_1(X) = H$ is a well-defined normal map. We reconsider the main theorem proved in [C] for the case $n = 2q$, $m = 2k$.

THEOREM: There exists a free left Z[G] module K and a sesquilinear form σ on K such that $\sigma \otimes \psi$ is a quadratic form representing the obstruction for doing surgery on $id_P \times f$, where ψ is the one for f.

We refer to the introduction of [C] for a more detailed statement. We here give a proof that is both substantially shorter and allows more choice in the construction of σ.

From now on we assume given

i) a subpolyhedron Q of P and a map $\xi : [0,1] \to \mathrm{Diff}(P)$ such that $\xi_0 = 1$ and $\xi_1 Q \cap Q = \emptyset$ and such that $H_i(\tilde{P}-\tilde{Q};\xi_1\tilde{Q})$ vanishes for $2i \neq 2q = n$.

ii) a map $\chi : [0,1] \to \mathrm{Diff}(P)$ homotopic to the one mapping t to $\xi_1^{-1}\xi_t\xi_t$ but such that $\chi_t Q \cap Q = \emptyset$ for all t.

To describe (K,σ) in these terms we note that by Lefschetz duality $H_q(\tilde{P}-\tilde{Q},\xi_1\tilde{Q}) \cong H^q(\tilde{P}-\xi_1\tilde{Q};\tilde{Q})$, and the latter is the dual of $H_q(\tilde{P}-\xi_1\tilde{Q},\tilde{Q})$ as a left Z[G] module. In other words a nonsingular pairing between these two homology groups is defined by the equivariant intersection pairing λ, given by the formula

$\lambda(x,y) = \sum_G (x\cdot g^{-1}y)g \in Z[G]$.

Now we define a homomorphism $H_q(\tilde{P}-\tilde{Q},\xi_1\tilde{Q}) \to H_q(\tilde{P}-\xi_1\tilde{Q},\tilde{Q})$ which maps the class of the chain c to the one of $\xi_1 c - \xi_1\chi([0,1] \times \xi_1^{-1}\partial c)$; it is isomorphic as there exists an inverse, mapping {c} to $\{\xi_1^{-1}c + \chi([0,1] \times \partial c)\}$.

In particular P is mapped to zero by Θ whenever $\xi_1 Q$ is a retract of P-Q. As will be clear from the proof this is also true if n is odd and without the restriction on the homotopy class of χ.

Example 1. One can take Q to be the (q-1) skeleton of some triangulation of P^{2q}; then ξ and χ exist by transversality. In this case we recover the situation of [C]; in particular the relation to the Micenko-Ranicki approach to the product formula is explained in [C].

Example 2. Suppose P is the total space of some smooth fibration $\pi : P \to S^2$. In this case we write S^2 as the union of two discs D^2_+ and D^2_-. We take as our Q the fibre of some point in D^2_+ and define ξ_t and χ_t to be the identity on $\pi^{-1}D^2_-$ for all t. On $\pi^{-1}D^2_+ \cong Q \times D^2_+$ we define them to be the product of id_Q with some suitable diffeomorphisms of D^2_+. Since the inclusion of one fibre into the complement of another is a homotopy equivalence we see that such P are killed by Θ.

Now according to [B] an element of the unoriented bordism group $N_n = \Omega_n(Z^-/2)$ can so be represented

a) for n even, if its mod 2 Euler characteristic w_n vanishes;

b) for n odd, if its de Rham invariant $w_2 w_{n-2}$ vanishes.

We deduce that for $(G,w_G) = Z^-/2$ only a rather small quotient of the bordism ring acts effectively on the L groups.

Example 3. For ℓ odd we consider the action of Z/ℓ on the unit sphere in C^{2a+1} by multiplication with $\exp(^{2\pi i}/\ell)$. Any element of U(2a+1) defines a diffeomorphism of the quotient space L. in particular ξ_1

defined by $\xi_1(z_1 \ldots z_{2a+1}) = (z_{2a+1} \ldots, z_{a+2}, -z_{a+1}, z_a \ldots z_1)$ is homotopic to id; furthermore we take χ_t to be ξ_1 for all t. Now a generalized Morse function f on L is defined by the formula

$$f(z) = -z_1\bar{z}_1 \ldots -z_a\bar{z}_a - \frac{1}{\ell}(z^{\ell}_{a+1} \cdot \bar{z}^{\ell}_{a+1}) + z_{a+2}\bar{z}_{a+2} \ldots + z_{2a+1}\bar{z}_{2a+1}.$$

It has only $+1$, $+2/a$, $-2/a$, -1 as critical levels hence $Q = \{z \in L \mid f(z) \le -\frac{1}{a}\}$ is a retract of the complement of $\xi Q = \{z \in L \mid f(z) \ge \frac{1}{a}\}$. This means that such a lens space of dimension 4a+1 is mapped to zero by Θ. The lens spaces of dimension 3 mod 4 are treated similarly but there we do not need a "middle term" and therefore there is no restriction on the parity of ℓ.

Now $\Omega_*(Z^+/\ell)$ is known to be generated as a $\Omega_*(1)$ module by the point and by lens spaces. We deduce that for $(G, w_G) = Z^+/\ell$ (ℓ odd) the pairing Θ factorizes through $\Omega_*(1)$.

Proof of the Theorem.

We recall the main notations introduced in §1 of [C]:
We may assume that the normal map $f : M \to X$ is k-connected that $K_k(M) = \ker(f_* : H_k(\tilde{M}) \to H_k(\tilde{X}))$ has a basis represented by framed immersions $g_j : S^k \times D^k \to \tilde{M}$ intersecting nicely i.e. not at all except that $g_j(\eta_p(x), y) = \gamma^{-1} g_j(\eta_{p'}(y), x)$ all $x, y \in D^k$ for certain embeddings $\eta_p : D^k \to S^k$. Here and throughout the paper we denote by $S^{k-1}(R)$ and $D^k(R)$ the sphere resp. disc of radius R in $\mathbb{R}^k$; if $R = 1$ we omit R.
We also need a C^∞ function κ with values in $[0,1]$ on the r spheres such that in the above situation κ is 0 on $\eta_p(D^k(\frac{1}{2}))$ and 1 on $\eta_{p'}(D^k(\frac{1}{2}))$ or vice versa and such that κ vanishes outside the η_p-images.
Counting the number of intersections as above satisfying $\kappa(p) = 0$ with multiplicity $\pm\gamma$ (depending on the sign ε_p of $\eta_{p'}$) yields the value $\psi(e_j, e_{j'})$ of a quadratic form ψ on $K_k(M)$ in the sense of [W2], with

invertible symmetrisation $\lambda = \psi + (-1)^k \psi^+$.

By taking a regular neighborhood we may assume that Q is a manifold of dimension n with boundary. The formula $\tilde{\Omega}_j(y,x,v) = (\xi_{\kappa(x)} y, g_j(x,v))$ determines disjoint embeddings $\Omega_j = \pi \circ \tilde{\Omega}_j : Q \times S^k \times D^k(\frac{1}{2}) \to P \times M$. For suppose that $\tilde{\Omega}_j(y,x,v) = (\theta \times \gamma)^{-1} \tilde{\Omega}_{j'}(y',x',v')$; then $g_j(x,v) = \gamma^{-1} g_{j'}(x',v')$ implies that either $j = j'$, $x = x'$, $y = \theta^{-1} y'$ or $x \in \eta_p D^k(\frac{1}{2})$, $x' \in \eta_{p'} D^k(\frac{1}{2})$ and $\{\kappa(x), \kappa(x')\} = \{0,1\}$ for some p, p'; but then $\xi_{\kappa(x)} = \theta^{-1} \xi_{\kappa(x')} y'$ contradicts $Q \cap \xi_1 Q = \emptyset$.

Hence we can use the Ω-s to define a manifold

$$W = P \times M \times [0,1] \cup \bigcup_{j=1}^{r} Q \times D^{k+1} \times D^k(\tfrac{1}{2}),$$

and since Ω_j is homotopic to $(Q \subset P) \times g_j$ we can extend $id_p \times f$ to a map $W \to P \times X$, in fact as a normal map. Now ∂W can be written as a disjoint union $(P \times M \times 0) \cup N$; we are going to show that the induced map $N \to P \times X$ is $(q+k)$-connected.

From now on we write $K_i(W)$ for $\ker(H_i(\tilde{W}) \to H_i(\tilde{P} \times \tilde{X}))$; similarly one has $K_i(P \times M)$, $K_i(N)$ etc.

We can construct a module map $H_i(\tilde{P}) \otimes K_k(M) \to K_{i+k}(P \times M)$ by mapping $\{c\} \otimes e_j$ to the class represented by $c \times g_j(S^k \times 0)$; this is an isomorphism by the Kunneth theorem. Similarly we have a map $H_i(\tilde{Q}) \otimes K_k(M) \to K_{i+k+1}(W, P \times M \times [0,1]) \cong K_{i+k+1}(W, P \times M)$ mapping $\{c\} \otimes e_j$ to the class represented by $c \times D_j^{k+1} \times 0$; this is an isomorphism by excision and the Künneth theorem. Thirdly there is a homomorphism A: $H_i(\tilde{P},\tilde{Q}) \otimes K_k(M) \to K_{i+k}(W)$ mapping $\{c\} \otimes e_j$ to the class represented by $\tilde{\Omega}_j(c \times S^k \times 0) \cup \partial c \times D_j^{k+1} \times 0$; it is an isomorphism by the five lemma since it is consistent with the abovementioned isomorphisms. Finally we note that the union of N and $\bigcup_{j=1}^r Q \times D_j^{k+1} \times D^k(\frac{1}{2})$ is a retract of W; the intersection of these is $\bigcup_{j=1}^r \{Q \times D_j^{k+1} \times S^{k-1}(\frac{1}{2}) \cup \partial Q \times D_j^{k+1} \times D^k(\frac{1}{2})\}$; accordingly we see by retraction, excision and the Kunneth theorem that there is an isomorphism B:

$H_i(\tilde{Q},\partial\tilde{Q}) \otimes K_k(M) \to K_{i+k}(W,N)$ mapping $\{c\} \otimes e_j$ to the class of $c \times 0_j \times D^k_{.}(\frac{1}{2})$; we can replace $(\tilde{Q},\partial\tilde{Q})$ here by $(\tilde{P},\tilde{P}-\tilde{Q})$ using excision.

Lemma 1. The composition

$H_i(\tilde{P},\tilde{Q}) \otimes K_k(M) \cong K_{i+k}(W) \to K_{i+k}(W,N) \cong H_i(\tilde{P},\tilde{P}-\tilde{Q}) \otimes K_k(M)$

is $(-1)^k[(\xi_1^{-1})_* \otimes \tilde{\psi} + (\xi_1)_* \otimes (\tilde{\lambda}-\tilde{\psi})]$, where $\tilde{\psi}$ is defined by the formula $\tilde{\psi}(x) = \sum_{j=1}^{r} \overline{\psi(x,e_j)}e_j$ and $\tilde{\lambda}$ similarly. This uses only the definition of ξ_t.

Corollary. Since χ is a homotopy between ξ_1^{-1} and ξ_1 the above expression becomes $(-1)^k(\xi_1)_* \otimes \tilde{\lambda}$ and so $K_{i+k}(W) \to K_{i+k}(W,N)$ is epi or mono depending on wether $(\xi_1)_*: H_i(\tilde{P},\tilde{Q}) \to H_i(\tilde{P},\tilde{P}-\tilde{Q})$ is. In particular $K_{i+k}(N)$ vanishes whenever $H_i(\tilde{P}-\tilde{Q},\xi_1\tilde{Q})$ does.

Proof. A class in $H_i(\tilde{P},\tilde{Q})$ can be represented by a chain c in $\mathrm{cl}(\tilde{P}-\tilde{Q})$ such that ∂c is in $\partial\tilde{Q}$. Then $A(\{c\} \otimes e_j) = \tilde{\Omega}_j(c \times S^k \times 0) \cup \partial c \times D_j^{k+1} \times 0$; we must consider this relative to $\tilde{N}$ since

$$N = (P \times M - \mathrm{int} \bigcup_{j=1}^{r} \Omega_j(Q \times S^k \times D^k(\tfrac{1}{2})) \cup \bigcup_{j=1}^{r} (\partial Q \times D_j^{k+1} \times D^k(\tfrac{1}{2}) \cup Q \times D_j^{k+1} \times S^{k-1}(\tfrac{1}{2}))$$

and we have to look where $\tilde{\Omega}_j(c \times S^k \times 0) \subset \tilde{P} \times \tilde{M}$ hits an orbit of some $\mathrm{int}\, \tilde{\Omega}_{j'}(Q \times S^k \times D^k(\frac{1}{2}))$.

The equation $\tilde{\Omega}_j(y,x,0) = (\theta \times \gamma)^{-1}\tilde{\Omega}_{j'}(y',x',v')$ leads to

$$g_j(x,0) = \gamma^{-1}g_{j'}(x',v')$$

and $\qquad \xi_{\kappa(x)}y = \xi_{\kappa(x')}y'$.

Hence either $x = x'$ and $y = \theta^{-1}y'$ in contradiction to $y \in \mathrm{cl}(\tilde{P}-\tilde{Q})$ and $y' \in \tilde{Q}$, or else $x \in \eta_p D^k(\frac{1}{2})$ and $x' \in \eta_{p'} D^k(\frac{1}{2})$ for some intersection pair $\{p,p'\}$ and hence $\{\kappa(x),\kappa(x')\} = \{0,1\}$. There are two possibilities.

a) If $\kappa(x) = 0$ and $\kappa(x') = 1$ and therefore $y = \theta^{-1}\xi_1 y'$ we rewrite $\varepsilon_p\tilde{\Omega}_j(c \times \eta_p D^k(\frac{1}{2}) \times 0)$ as $\varepsilon_p\gamma^{-1}\tilde{\Omega}_{j'}(\xi_1^{-1}c \times p' \times D^k(\frac{1}{2}))$ which represents

$\varepsilon_p \gamma^{-1} B(\{\xi_1^{-1} c\} \otimes e_{j'})$. Meanwhile this intersection-pair contributes $\varepsilon_p \gamma^{-1} (-1)^k e_{j'}$ to $\tilde{\psi}(e_j)$.

b) If $\kappa(x) = 1$ and $\kappa(x') = 0$ and therefore $y = \theta^{-1} \xi_1^{-1} y'$ we rewrite the same as $\varepsilon_p \gamma^{-1} \tilde{\Omega}_{j'}(\xi_1 c \times p' \times D^k(\tfrac{1}{2}))$ which represents $\varepsilon_p \gamma^{-1} B(\{\xi_1 c\} \otimes e_{j'})$. The intersection-pair contributes $\varepsilon_p \gamma^{-1} (-1)^k e_{j'}$ to $(\tilde{\lambda} - \tilde{\psi}) e_j$.

Q.E.D.

This lemma justifies the remark preceding the examples; it also suggests the following:

Lemma 2. a) In general there exists an isomorphism

$H_i(\tilde{P}-\tilde{Q}, \xi_1 \tilde{Q}) \otimes K_k(M) \to K_{i+k}(N)$.

b) In the assumed case that $i = q$ and the pair $(\tilde{P}-\tilde{Q}, \xi_1 \tilde{Q})$ is $(q-1)$-connected this yields framed immersions of spheres in $\tilde{N}$.

Proof. We first prove b; we start by some preparations.

By assumption some $H : [0,1] \times [0,1] \to \mathrm{Diff}(\tilde{P})$ exists such that $H(0,t) = \chi_t$, $H(1,t) = \xi_1^{-1} \xi_t \xi_t$, $H(\kappa,0) = \xi_1^{-1}$ and $H(\kappa,1) = \xi_1$. If we define $H' : [0,1] \times [0,1] \to \mathrm{Diff}(\tilde{P})$ by the formula

$H'(\kappa,t) = \xi_{(1-\kappa)t} \xi_{\kappa+(1-\kappa)t}$ then we have $H'(0,t) = \xi_t \xi_t$, $H'(1,t) = \xi_1$, $H'(\kappa,0) = \xi_\kappa$ and $H'(\kappa,1) = \xi_{1-\kappa} \xi_1$. From $\xi_1 H$ and H' we can construct $D : [0,1] \times [0,1] \to \mathrm{Diff}(P)$ such that $D(0,t) = \xi_1 \chi_t$, $D(1,t) = \xi_1$, $D(\kappa,0) = \xi_\kappa$ and $D(\kappa,1) = \xi_{1-\kappa} \xi_1$.

Furthermore we fix an embedding $\upsilon : D^1 \times D^{k-1} \to D^k(1) - \mathrm{int}\, D^k(\tfrac{1}{2})$ such that $1 \times D^{k-1}$ is mapped into $S^{k-1}(\tfrac{1}{2})$.

Finally we remark that any class in $H_q(\tilde{P}-\tilde{Q}, \xi_1 \tilde{Q})$ can be represented by a disc c by the Hurewicz theorem. We may assume it to be collared, framed and embedded i.e. c is an embedding $D^q(2) \times D^q \to P-Q$ such that $(D^q(2) - D^q(1)) \times D^q$ maps into $\xi_1 \tilde{Q}$.

After this preparation we construct a framed immersion of S^{q+k} into $\tilde{N}$. To this end we write S^{q+k} as a union

$D^q \times S^k \cup S^{q-1} \times [1,2] \times S^k \cup S^{q-1} \times D^{k+1}$.

We define a map $\tilde{\Psi}_j$:

$(D^q \times S^k) \times (D^q \times D^1 \times D^{k-1}) \to \tilde{P} \times \tilde{M}$

by the formula

$\tilde{\Psi}_j(y,x,w,s,v) = (\xi_{1-\kappa(x)} c(y,w), g_j(x,\upsilon(s,v)))$.

We also define a map $\tilde{\Gamma}$:

$(S^{q-1} \times [1,2] \times S^k) \times (D^q \times D^1 \times D^{k-1}) \to \tilde{P} \times \tilde{M}$

by the formula

$\tilde{\Gamma}_j(y,t,x,w,s,v) = (D_{\kappa(x),2-t}\xi_1^{-1} c(\alpha y,w), g_j(x,\upsilon(\beta,v)))$, where (α,β) is the embedding of $[1,2] \times [-1,1]$ into itself given by $\alpha(t,s) = 1+(t-1)(2-s)/3$ and $\beta(t,s) = t-ts+2s-1$. In particular $\alpha = 1$ and $\beta = s$ for $t = 1$ and we recover $\tilde{\Psi}_j(y,x,w,s,v)$. For $t = 2$ we get the image under $\tilde{\Omega}_j$ of

$(\xi_1^{-1} c((1+\frac{2-s}{3})y,w), x, \upsilon(1,v)) \in \tilde{Q} \times S^k \times S^{k-1}(\tfrac{1}{2})$.

We now use this last formula to define the image of (y,x,w,s,v) under a map $(S^{q-1} \times D^{k+1}) \times (D^q \times D^1 \times D^{k-1}) \to \tilde{Q} \times D_j^{k+1} \times S^{k-1}(\tfrac{1}{2}) \subset \tilde{N}$.

Here $\tilde{\Psi}_j$ maps into $\tilde{N}$ since

$\tilde{\Psi}_j(y,x,w,s,v) = (\theta \times \gamma)^{-1} \tilde{\Omega}_{j'}(y',x',v')$ means that $g_j(x,\upsilon(s,v)) = \gamma^{-1} g_{j'}(x',v')$ and $\xi_{1-\kappa(x)} c(y,w) = \theta^{-1}\xi_{\kappa(x')} y'$. If $x = x'$ we get $v' = \upsilon(s,v) \in D^k(1) - \operatorname{int} D^k(\tfrac{1}{2})$ in contradiction to $v' \in \operatorname{int} D^k(\tfrac{1}{2})$. If $x \neq x'$ we have $1-\kappa(x) = \kappa(x')$ and we get $\theta^{-1} y' = c(y,w) \in \tilde{P}-\tilde{Q}$ in contradiction to $y' \in \tilde{Q}$.

Similarly $\tilde{\Gamma}_j$ maps into $\tilde{N}$. For $\tilde{\Gamma}_j(y,t,x,w,s,v) = (\theta \times \gamma)^{-1}\tilde{\Omega}_{j'}(y',x',v')$ leads for $x = x'$ to the same contradiction as above and for $x \neq x'$ we have $\{\kappa(x),\kappa(x')\} = \{0,1\}$. Now $\kappa(x) = 0$ leads to $\xi_1\chi_{2-t}\xi_1^{-1} c(\alpha y,w) = \xi_1 y'$ and $\kappa(x) = 1$ leads to $c(\alpha y,w) = y'$; both are contradictions since $c(\alpha y,w) \in \xi_1\tilde{Q}$, $y' \in \tilde{Q}$ and $\chi([0,1] \times \tilde{Q}) \cap \tilde{Q} = \emptyset$.

This concludes the construction of the framed immersions; if we are only interested in their homology classes we can forget about the factor $D^q \times D^1 \times D^{k-1}$ and on the other hand replace the discs c by arbitrary singular chains c i.e. the map in (a) maps $\{c\} \otimes e_j$ to the class of

$$\tilde{\Psi}_j(c \times S^k \times 0) \cup \tilde{\Gamma}_j([1,2] \times \xi_1^{-1}\partial c \times S^k \times 0) \cup \xi_1^{-1}\partial c \times D_j^{k+1} \times \upsilon(0).$$

In particular the induced map $H_i(\tilde{P}-\tilde{Q}) \otimes K_k(M) \to K_{i+k}(N)$ only involves the first $\tilde{\Psi}$ term.

Now we consider the following ladder

$$\begin{array}{ccccccc}
\to H_{i+1}(\tilde{P},\tilde{P}-\tilde{Q}) \otimes K_k(M) & \to & K_i(\tilde{P}-\tilde{Q},\xi_1\tilde{Q}) \otimes K_k(M) & \to & H_i(\tilde{P},\xi_1\tilde{Q}) \otimes K_k(M) & \to & \\
\mathrm{I} \quad \downarrow & & \mathrm{II} \quad \downarrow & & \mathrm{III} \quad \downarrow & & \mathrm{I} \\
\to K_{i+k+1}(W,N) & \longrightarrow & K_{i+k}(N) & \longrightarrow & K_{i+k}(W) & \longrightarrow &
\end{array}$$

where the rows are the obvious ones, the first vertical arrow is $(-1)^k(1 \otimes \tilde{\lambda})B$, the second one is the map constructed above and the last one is $A(\xi_1^{-1} \otimes 1)$.

The square I commutes because of lemma 1; square III commutes because in $\tilde{P} \times \tilde{M} \subset \tilde{W}$ there exists a homotopy from $\tilde{\Psi}_j$ to $\tilde{\Omega}_j$ and from $\tilde{\Gamma}_j$ to a map constant on [1,2]. The commutativity of II follows from that of the following diagram, which is lemma 2c of [C]:

$$\begin{array}{ccc}
H_{i+1}(\tilde{P},\tilde{P}-\tilde{Q}) \otimes K_k(M) & \xrightarrow{\partial} & H_i(\tilde{P}-\tilde{Q}) \otimes K_k(M) \\
\downarrow (-1)^k(1 \otimes \tilde{\lambda})B & & \downarrow \tilde{\Psi} \\
K_{i+k+1}(W,N) & \xrightarrow{\partial} & K_{i+k}(N)
\end{array}$$

Since the rows are exact and the other vertical maps are isomorphisms the proof is finished by an application of the five lemma.

<u>Q.E.D.</u>

<u>Lemma 3</u>. The framings just constructed are consistent with the normal data.

<u>Proof</u>. Even if that were not the case that would affect $\sigma \otimes \psi$ only by changing its ordinary Arf invariant; therefore we only outline the proof.

Since the normal map to $P \times X$ extends to W it is sufficient to check the framed immersions which one gets by composing with the collar $N \times (0,1) \to W$. One can however write down explicitly a regular homotopy from such a framed immersion to the map

$$\Xi_j : (S^k \times D^q \cup [1,2] \times S^k \times S^{q-1} \cup D^{k+1} \times S^{q-1}) \times D^k \times D^q \times (0,1) \to W$$

such that on the three parts

$$\Xi_j(x,y,v,w,z) = (\xi_{\kappa(x)}\xi_1^{-1} c(y,w), g_j(x,\tfrac{1}{2}v), 1-z) \in P \times M \times [0,1] \subset W$$

$$\Xi_j(t,x,y,v,w,z) = (\xi_{\kappa(x)}\xi_1^{-1}\ c((1+(t-1)z)y,w),$$
$$g_j(x,\tfrac{1}{2}v), 1-(2-t)z) \in P \times M \times [0,1] \subset W$$

$$\Xi_j(x,y,v,w,z) = (\xi_1^{-1}\ c((1+z)y,w), x, \tfrac{1}{2}v) \in Q \times D^{k+1} \times D^k(\tfrac{1}{2}) \subset W.$$

Subsequently one notes that $W \times (-2,2)$ is constructed from $P \times M \times [0,1] \times (-2,2)$ by attaching $Q \times D^{k+1} \times D^k(\tfrac{1}{2}) \times (-2,2)$ according to the map $Q \times S^k \times D^k(\tfrac{1}{2}) \times (-2,2) \to P \times M \times 1 \times (-2,2)$ which maps (y,x,v,ρ) to $(\xi_{\kappa(x)}y, g_j(x,v), 1, \rho)$. This embedding however is isotopic to the one mapping (y,x,v,ρ) to $(y, g_j(x,v), 1, \frac{1}{10}\rho+\kappa(x))$. The last one is even defined for all $y \in P$; thus we get an embedding $W \times (-2,2) \subset P \times T$ where $T = M \times [0,1] \times (-2,2) \cup D^{k+1}(2) \times D^k(\tfrac{1}{2}) \times (-2,2)$, identifying $(\tau x, v, \rho)$ with $(g_j(x,v), 2-\tau, \frac{1}{10}\rho+\kappa(x))$

Since the normal map extends to $P \times T$, it is sufficient to check things in $P \times T$. We regularly homotop Ξ_j to the map

$\Xi_j' : S^{k+q} \times D^{k+q+1} \times (-2,2) \to P \times T$ such that on the three parts

$$\Xi_j'(x,y,v,w,z\ \ ) = (\xi_1^{-1}\ c(y,w), g_j(x,\tfrac{1}{2}v), 1-z,$$
$$\tfrac{1}{10}\rho+\kappa(x)) \in P \times M \times [0,1] \times (-2,2) \subset P \times T$$

$$\Xi_j'(t,x,y,v,w,z,\rho) = (\xi_1^{-1}\ c((1+(t-1)z)y,w), g_j(x,\tfrac{1}{2}v),$$
$$1-(2-t)z, \tfrac{1}{10}\rho+\kappa(x)) \in P\times T$$

$$\Xi_j'(x,y,v,w,z,\rho) = (\xi_1^{-1}\ c((1+z)y,w), x, \tfrac{1}{2}v, \rho) \in P\times D^{k+1}\times D^k(\tfrac{1}{2})\times(-2,2) \subset P\times T.$$

Here however we can see that the immersion lands in a disc $\xi_1^{-1}\ c(D^q(2)\times D^q)\times D^{k+1}(2)\times D^k(\tfrac{1}{2})\times(-2,2)$; up to the trivial factors D^q, $D^k(\tfrac{1}{2})$ and $(-2,2)$ we just get the map

$$\Lambda : (S^k\times D^q \cup [1,2]\times S^k\times S^{q-1} \cup D^{k+1}\times S^{q-1})\times(0,1) \to D^q(2)\times D^{k+1}(2)$$

such that

$$\Lambda(x,y,z) = (y,(1+z)x)$$

$$\Lambda(t,x,y,z) = ((1+(t-1)z)y,(1+(2-t)z)x)$$

$$\Lambda(x,y,z) = ((1+z)y,x)$$

respectively. But that is the standard embedded collared sphere, in accordance with the normal data which require a framing which extends over D^{k+q+1}.

Q.E.D.

<u>End of proof of the Theorem</u>. Let c and c' be q-discs in $\tilde{P}-\tilde{Q}$ relative to $\xi_1\tilde{Q}$ as in the proof of lemma 2. To calculate the value of the quadratic form of N on $\{c\}\otimes e_j$ and $\{c'\}\otimes e_{j'}$, we only have to count the intersections of the corresponding framed spheres.

We first fix an ordering of the intersection-pairs by a book-keeping function like κ was for M. For the Ψ term we take the κ of the S^k factor; the Γ term is put 2 lower and the third term we ignore. We can do so because the third terms can only intersect each other, as they are the only ones lying in the handle part of N; but even they do not intersect because c and c' can be assumed to be disjoint.

There are no intersections between the Γ terms because $\tilde{\Gamma}_j(y,t,x,0,0,0) = (\theta\times\gamma)^{-1}\tilde{\Gamma}_{j'}(y',t',x',0,0,0)$ leads to $g_j(x,\upsilon(\beta,0)) = \gamma^{-1}g_{j'}(x',\upsilon(\beta',0))$.

If $x = x'$ we must have $\beta(t,0) = \beta(t',0)$ hence $t = t'$ so $y = y'$. If $x \neq x'$ we must have $\{\kappa(x), \kappa(x')\} = \{0,1\}$ and so we get $(\xi_1 \chi_t)(\xi_1^{-1} y) = \xi_1(\xi_1^{-1} y')$ contradicting $\chi_t Q \cap Q = \emptyset$ or the same with y, y' interchanged. The intersections between Ψ_j and $\Gamma_{j'}$ do not contribute to the quadratic form due to the above ordering convention.

To count the intersections between the Ψ terms we have to solve $\widetilde{\Psi}_j(y,x,0,0,0) = (\theta \times \gamma)^{-1} \widetilde{\Psi}_{j'}(y',x',0,0,0)$, hence $g_j(x,0) = \gamma^{-1} g_{j'}(x',0)$ and $\xi_{1-\kappa(x)} y = \theta^{-1} \xi_{1-\kappa(x')} y'$. For $x = x'$ we get $y = \theta^{-1} y'$ which does not count; otherwise we are only interested in the case $\kappa(x) = 0$, $\kappa(x') = 1$ which gives $\xi y = y'$. So counting multiplicities we get contributions from these intersections corresponding to the terms of $(-1)^{kq} \lambda(\xi c, c') \otimes \psi(e_j, e_{j'})$.

Finally we intersect Γ_j with $\Psi_{j'}$; the order condition is here always satisfied. We have to solve the equations $g_j(x, \mu(\beta,0)) = \gamma^{-1} g_{j'}(x', \mu(0))$ and $D_{\kappa(x),2-t} \xi_1^{-1} y = \theta^{-1} \xi_{1-\kappa(x')} y'$. Again $x = x'$ does not count. For $\kappa(x) = 1$, $\kappa(x') = 0$ we get $\xi_1 \xi_1^{-1} y = \theta^{-1} \xi_1 y'$ which contradicts the fact that $\xi_1 y \in \widetilde{Q}$ and $y' \in \mathrm{int}(\widetilde{P} - \widetilde{Q})$. For $\kappa(x) = 0$, $\kappa(x') = 1$ we get $\xi_1 \chi_{2-t} \xi_1^{-1} y = y'$ and we get contributions corresponding to the terms of $-(-1)^{kq} \lambda(\xi_1 \chi([0,1] \times \xi_1^{-1} \partial c), c') \otimes \psi(e_j, e_{j'})$.

This proves the theorem. Notice that the tensor-product was meant in a graded sense: an extra factor $(-1)^{kq}$.

Computational results

The foregoing theorem and a similar one for the odd-dimensional case have been applied to calculate Θ on smooth bordism of cyclic groups. The details about this will be published elsewhere.

1. For $(G,w_G) = Z^-$ one gets the bordism theory W_*, which is according to [W3] a polynomial algebra over Z/2 on generators x_j for each j with j+1 not a power of 2. For $t,s \geq 1$ one can represent $x_{2^t(2s+1)}$ by the quotient of $S^1 \times S^{2^t-1} \times CP(2^t s)$ under the identifications $(-I) \times \begin{pmatrix}-1 & 0\\ 0 & I\end{pmatrix} \times (I)$ and $(I) \times (-I) \times$ (conjugation); unless t = 1 and s is odd we can choose Q,ξ,χ in such a way that σ is of the type $\sigma(e,e) = \sigma(f,f) = 1$, $\sigma(e,f) = \sigma(f,e) = 0$ on some basis {e,f} i.e. these manifolds behave as 0 under Θ. In this way we see that x_n acts as

1 if n is a power of 2

x_6 if n = 6 mod 8

x_5 if n = 5 mod 8

0 otherwise.

So the action of $\Omega_n(Z^-)$ on L theory is trivial for n = 3 mod 4 and factorises through some Z/2 for n = 0,1 or 2 mod 4 which is detected by the mod 2 Euler characteristic w_n, de Rham invariant w_2w_{n-2}, and codimension one de Rham invariant $w_1w_2w_{n-3}$ respectively.

2. For $(G,w_G) = Z^-/2$ one gets unoriented bordism theory, which is a free module over $\Omega_*(Z^-) = W_*$ generated by the even dimensional real projective spaces $\mathbb{RP}(2q)$. As described in [C] these have $\sigma = (1)$ if q is even and (generator of Z/2) if q is odd. Thus we recover the results of example 2. Note that the image of $x_6 \in W_6$ acts now as 0.

3. Since the behavior of $\Omega_*(1)$ is known, that of $\tilde{\Omega}_*(Z^+/2)$ follows from that of the reduced bordism group $\tilde{\Omega}_*(Z^+/2)$. However there exists a homomorphism $E : \Omega_{n-1}(Z^-/2) \to \tilde{\Omega}_n(Z^+/2)$ which maps the manifold P/T where T is a fix point-free orientation reversing involution to the quotient of $S^1 \times P$ under (reflection) $\times$ T. This is an isomorphism, the inverse map being the Smith homomorphism. By a commutative diagram for Θ the results now follow from 2. and knowledge of the analoguous map (not

isomorphic) $E : L_{m-1}(Z^-/2 \times H) \to L_m(H \to Z^+/2 \times H) \subset L_m(Z^+/2 \times H)$. Since $E(\mathbb{R}P(2)) = \mathbb{R}P(3)$ acts trivially according to example 3 we only get a contribution (factorising through Z/2) for n = 1 or 2 mod 4.

4. The foregoing determines the action on L groups of the part of $\tilde{\Omega}(Z^+/2^\ell)$ which is in the image of $\tilde{\Omega}(Z^+/2)$ under the inclusion map $Z^+/2 \to Z^+/2^\ell$. According to [K] the quotient is generated as an $\Omega_*(1)$ module by lens spaces $(S^{2a+1})/2^\ell$. As we have seen in example 3 these act as 0 for a odd; similarly one can prove that they act as S^1 for a even. The result is that the action of $P \in \tilde{\Omega}_{4a+1}(Z^+/2^\ell)$ is determined by the characteristic number βL_a in $Z/2^\ell$ where $\beta \in H^1(P,Z/2^\ell) = \mathrm{Hom}(\pi_1(P),Z/2^\ell)$ is the canonical element and $L_a \in H^{4a}(P,Z/2^\ell)$ is the Hirzebruch polynomial in the Pontrjagin classes of P.

5. As an $\Omega(Z^-)$ algebra $\Omega(Z^-/2^\ell)$ is freely generated by 1, the Klein bottle and manifolds P of dimension 2^q (where $q \geq 2$) for which one can take the equator f = 0 (see example 3) of a lens space. These have $\sigma = (1)$ for $q \geq 3$ and $(T^{2^{\ell-1}})$ for q = 2; here T is a generator of the cyclic group.

6. For m odd the action of $\Omega_*(Z/m2^\ell)$ on the L groups factorises through the canonical projection $\Omega_*(Z/m2^\ell) \to \Omega_*(Z/2^\ell)$. This is the case since $\Omega_*(Z/m2^\ell)$ is the tensor product over $\Omega_*(1)$ of $\Omega_*(Z/2^\ell)$ and $\Omega_*(Z/m)$ according to [H] (and a similar argument for the orientation reversing case) and since $\tilde{\Omega}_*(Z/m)$ is generated over $\Omega_*(1)$ by lens spaces [CF] and hence acts as 0.

References

[B] R.L.W. Brown, "Cobordism and bundles over spheres", Michigan Math. J. 16 (1969), 315-320.

[C] F.J.B.J. Clauwens, "Surgery on products", to appear in Indag.Math.

[CF] P.E. Conner and E.E. Floyd, "Differentiable Periodic Maps", Erg.d.Math. Bd.33, Springer Verlag 1964.

[H] M. Nouredine Hassani, "Sur le bordisme des groupes cycliques", C.R.Acad.Sci. Paris, t.272 (1971) Serie A, 776-778.

[K] Y. Katsube, "Principal oriented bordism algebra $\Omega_*(Z_{2^k})$", Hiroshima Math.J. 4 (1974), 265-277.

[R] A. Ranicki, "An algebraic theory of surgery", preprint, Cambridge 1975.

[S] K. Shibata, "Oriented and weakly complex bordism algebra of free periodic maps", Trans.Amer.Math.Soc. 177 (1973), 199-220.

[W1] C.T.C. Wall, "Surgery on compact manifolds" Academic Press, 1970, London.

[W2] C.T.C. Wall, "On the axiomatic foundation of the theory of Hermitian forms", Proc.Cambridge Philos.Soc. 67 (1970), 243-250.

[W3] C.T.C. Wall, "Determination of the cobordism ring", Ann. of Math. 72 (1960), 292-311.

F.J.-B.J. Clauwens
Department of Mathematics
Catholic University
Toernooiveld
6525 ED Nijmegen
The Netherlands

ALGEBRAIC K-THEORY AND FLAT MANIFOLDS

Jean-Claude HAUSMANN

Statement of the main results

Let Λ be a subring of the real number field $\mathbb{R}$. We denote by $SL_n^\delta(\Lambda)$ the special linear group over Λ endowed with the discrete topology, while $SL_n(\Lambda)$ (used only for $\Lambda = \mathbb{R}$) is reserved for the topology induced from those natural on $\mathbb{R}$. For a topological group G, the classifying space for principal G-bundles is denoted by BG. The continuous homomorphism $SL_n^\delta(\Lambda) \longrightarrow SL_n(\mathbb{R})$ determines a map $BSL_n^\delta(\Lambda) \longrightarrow BSL_n(\mathbb{R})$.

When $n \geq 3$ or $\frac{1}{6} \in \Lambda$, $\pi_1(BSL_n^\delta(\Lambda)) = SL_n^\delta(\Lambda)$ is perfect and thus the plus construction of Quillen [H-H] produces a functorial map $BSL_n^\delta(\Lambda) \longrightarrow BSL_n^\delta(\Lambda)^+$ which induces an isomorphism on homology and $\pi_1(BSL_n^\delta(\Lambda)^+) = 1$. As $BSL_n(\mathbb{R})$ is simply connected, the natural map $BSL_n^\delta(\Lambda) \longrightarrow BSL_n(\mathbb{R})$ factors through $BSL_n^\delta(\Lambda)^+$ [H-H, Proposition 3.1], i.e. can be expressed as the composition of two functorial maps $BSL_n^\delta(\Lambda) \longrightarrow BSL_n^\delta(\Lambda)^+ \longrightarrow BSL_n(\mathbb{R})$

As we are interested in vector bundles only up to isomorphism, we shall identify the concept of a n-dimensional (always orientable) vector bundle ξ over a CW-complex X and of its characteristic map $\xi : X \longrightarrow BSL_n(\mathbb{R})$ Let us denote by $< \tau S^n >$ the subgroup of $\pi_n(BSL_n(\mathbb{R}))$ generated by the tangent bundle of the standard sphere S^n. On the other hand, define $I_n(\Lambda)$ as the image $\mathrm{im}(\pi_n(BSL_n^\delta(\Lambda)^+ \longrightarrow \pi_n(BSL_n(\mathbb{R})))$. Our first results give information on some $I_n(\Lambda)$. ($\mathbb{Z}_m$ denotes the cyclic group of order m) :

Theorem 1

a) $I_2(\Lambda) = \pi_2(BSL_2(\mathbb{R}))$, when $\frac{1}{6} \in \Lambda$.

b) For $n = 2k \geq 4$ and $\frac{1}{2} \in \Lambda$, one has

$$2^{k-2} < \tau S^n > \quad \subset \quad I_n(\Lambda)$$

and $$I_n(\Lambda) \subset < \tau S^n > \quad \text{if } n \neq 8r+2 .$$

In particular $I_4(\Lambda) = < \tau S^4 >$ for $\frac{1}{2} \in \Lambda$. Recall that one has a natural decomposition $\pi_{8r+2}(BSL_{8r+2}(\mathbb{R})) = < \tau S^{8r+2} > \oplus \mathbb{Z}_2$ (used in the statement of Theorem 2 below ; for a proof, see Lemma 4.5). Then one has $2I_{8r+2}(\Lambda) \subset < \tau S^{8r+2} >$, but we were not able to get rid of the factor 2 . Theorem 1 shows that $I_{2k}(\Lambda)$ is never trivial when $\frac{1}{2} \in \Lambda$ ($\frac{1}{6} \in \Lambda$ if $2k = 2$). This contrasts with the following result, which will be deduced from a theorem of Sullivan [Su] :

<u>Theorem 2</u>

a) $I_{2k}(\mathbb{Z}) = 0$ when $2k \neq 8r+2$

b) $I_{8r+2}(\mathbb{Z}) \subset 0 \times \mathbb{Z}_2$ for $r \geq 1$.

Let M^n be a closed C^∞-manifold of dimension n. A cobordism (W^{n+1}, M, M_-) is called a semi-s-cobordism if the inclusion $M \subset W$ is a simple homotopy equivalence [H-V 1] . One says that M is semi-s-cobordant to M_- . A manifold M^n is called Λ-flat if its tangent bundle is Λ-flat, i.e. $\tau M : M \longrightarrow BSL_n(\mathbb{R})$ admits a lifting through $BSL_n^\delta(\Lambda)$. Theorem 1 together with results of [H-V 2] will permit us to prove the following :

<u>Theorem 3</u> :

Let M^{2k} be a stably parallelizable closed manifold with $2k \geq 6$. If the Euler characteristic $\chi(M)$ of M is a multiple of 2^{k-1}, then M is semi-s-cobordant to a $\mathbb{Z}[\frac{1}{2}]$-flat manifold.

Let (W, M, M_-) be a semi-s-cobordism. By Poincaré Duality, the homomorphisms induced by the inclusions $H_*(M_-;B) \to H_*(W;B) \leftarrow H_*(M;B)$ for any $\mathbf{Z}\pi_1(W)$-module B are all isomorphisms. As a consequence, the Euler characteristics $\chi(M)$ and $\chi(M_-)$ are equal. As any even integer occurs as the Euler characteristic of a stably parallelizable manifold M^{2k} (take connected sums of copies of $S^2 \times S^{2k-2}$ and of $S^1 \times S^{2k-1}$), Theorem 3 gives the following corollary :

Corollary 4 :

Let $n = 2k \geq 6$ and $c = 2^{k-1}d$ with $d \in \mathbf{Z}$. Then there exists a $\mathbf{Z}[\frac{1}{2}]$-flat manifold V^n with Euler characteristic $\chi(V) = c$.

Let M^{2k} $(2k \geq 6)$ be a stably closed manifold which is semi-s-cobordant to a $\mathbf{Z}[\frac{1}{2}]$-flat manifold. We shall prove in Section 5 that, under these assumptions, there exists a semi-s-cobordism (V, M, M_-) such that $\pi_1(M_-) = \pi_1(M) \times St_{2k}(\mathbf{Z}[\frac{1}{2}])$, where $St_n(\Lambda)$ denotes the Steinberg group of rank n over Λ [Mi 2, p. 39]. Also the $\mathbf{Z}[\frac{1}{2}]$-flat structure on the tangent bundle of M_- comes from the natural epimorphism $\pi_1(M_-) \xrightarrow{proj} St_{2k}(\mathbf{Z}[\frac{1}{2}]) \longrightarrow SL_{2k}(\mathbf{Z}[\frac{1}{2}])$. Finally, M_- is always stably parallelizable.

Simillarly to Theorem 2, the result for $\Lambda = \mathbf{Z}$ as follows :

Theorem 5 :

A stably parallelizable closed manifold M^{2k} is semi-s-cobordant to a $\mathbf{Z}$-flat manifold if and only if it is parallelizable.

Finally, let $e^\delta \in H^{2k}(BSL^\delta_{2k}(\mathbf{R}))$ be the flat Euler class, i.e. the image of the Euler class $e \in H^{2k}(BSL_{2k}(\mathbf{R}))$ under the natural homomorphism $H^{2k}(BSL_{2k}(\mathbf{R})) \longrightarrow H^{2k}(BSL^\delta_{2k}(\mathbf{R}))$. We shall prove that the flat Euler class behaves somewhat like the standard Euler class, namely :

Proposition 6 :

a) e^{δ} is indecomposable, i.e. is not a sum of cup products of classes of strictly lower dimensions.

b) e^{δ} is not in the image of $H^{2k}(BSL^{\delta}_{\infty}(\mathbb{R})) \longrightarrow H^{2k}(BSL^{\delta}_{2k}(\mathbb{R}))$.

D. Quillen has proved (but not yet published) that $H_n(BGL^{\delta}_i(F)) \to H_n(BGL^{\delta}_{i+1}(F))$ is an isomorphism for $n < i$, when $F \neq \mathbb{Z}_2$ is a field. Thus Proposition 6 shows that this stability result of Quillen is the best possible for fields in general.

The paper is organized as follows :

Section 2 : one gives an improvement (known by specialists) of a Theorem of J. Milnor concerning the existence of some flat bundles over surfaces.

Section 3 : Low dimension computations are made for the algebraic K-theory of the ring $\mathbb{Z}[\frac{1}{2}]$.

Section 4 : One uses results of Section 2 and 3 to prove Theorem 1 and 2.

Section 5 : Theorem 3 and 5 are proven together with some improvements.

Section 6 : Results are given on flat bundles over homology spheres.

Section 7 : proof of Proposition 6 .

2. Flat bundles over surfaces with structure group $SL_2(\mathbb{Z}[\frac{1}{2}])$.

In [Mi 1], J. Milnor constructed $\mathbb{R}$-flat 2-dimensional bundles over surfaces with non-zero Euler class. It has been observed that the methode of Milnor can actually produce $\mathbb{Z}[\frac{1}{2}]$-flat bundles. In fact, we shall use several time the following result, to whose proof this section in devoted.

Theorem 2.1 :

There exists a $\mathbb{Z}[\frac{1}{2}]$-flat vector bundle η of dimension 2 over a closed surface T_3 of genus 3 with Euler class $e(\eta)$ satisfying $e(\eta)[T_3]=$

Proof : We invite the reader to follow with us the argument of Milnor [Mi 1, pp. 218-220] . The notations of that paper are used without explanation throughout this proof.

Let $G[\frac{1}{2}]$ denote the group $SL_2(\mathbb{Z}[\frac{1}{2}])$ included in $G = GL_2^+(\mathbb{R})$ (matrices with positive determinant). Define $\widetilde{G}(\frac{1}{2})$ by the pull-back diagram :

$$\begin{array}{ccc} \widetilde{G}(\frac{1}{2}) & \longrightarrow & \widetilde{G} \\ \downarrow & & \downarrow p \\ G(\frac{1}{2}) & \longrightarrow & G \end{array}$$

Define :

$$\gamma_0 = \begin{pmatrix} 2 & 0 \\ 0 & \frac{1}{2} \end{pmatrix} \in G(\tfrac{1}{2})$$. Chose the lifting and

chose the lifting $\widetilde{\gamma}_0$ of [Mi 1, p.219] belonging to $\widetilde{G}(\frac{1}{2})$.

Let $K(\frac{1}{2})$ and $\widetilde{K}(\frac{1}{2})$ denote $K \cap G(\frac{1}{2})$ and $\widetilde{K} \cap \widetilde{G}(\frac{1}{2})$ respectively. One begins by improving [Mi 1, Lemma 4] by establishing that any element of $(\text{Exp} \pm \pi)\widetilde{K}(\frac{1}{2})$ can be expressed as a product of two elements in

$\widetilde{K}(\frac{1}{2})$. Indeed, the liftings Γ_1 and Γ_2 of [Mi 1, proof of Lemma 4] can be chosen in $\widetilde{K}(\frac{1}{2})$. Using the equalities :

$$\begin{pmatrix} -\frac{5}{2} & -\frac{9}{2} \\ -3 & 5 \end{pmatrix} = \begin{pmatrix} -2 & 3 \\ 2 & 2 \end{pmatrix}^{-1} \gamma_0 \begin{pmatrix} -2 & 3 \\ 2 & 2 \end{pmatrix}$$

and

$$\begin{pmatrix} -5 & -9 \\ -\frac{3}{2} & -\frac{5}{2} \end{pmatrix} = \begin{pmatrix} 1 & -2 \\ -1 & 3 \end{pmatrix}^{-1} \gamma_0 \begin{pmatrix} 1 & -2 \\ -1 & 3 \end{pmatrix}$$

one deduces that $\Gamma_1 \Gamma_2 \in (\mathrm{Exp} \pm n\pi)\widetilde{K}(\frac{1}{2})$ for some odd integer n. Using that the map $\theta : \widetilde{G}(\frac{1}{2}) \longrightarrow \mathbb{R}$ factors through the connected group G, one deduces as in [Mi 1] that $n = \pm 1$ and then every element in $(\mathrm{Exp} \pm (\frac{1}{2}))$can be expressed as a product of a conjugate of Γ_1 and of a conjugate of Γ_2 (possibly Γ_1 and Γ_2 have to replaced by Γ_2^{-1} and Γ_1^{-1} respectively).

Then one deduces, as in [Mi 1, (7)] that every element in $(\mathrm{Exp} \pm \pi)\widetilde{K}(\frac{1}{2})$ is a commutator of elements of $\widetilde{K}(\frac{1}{2})$. For, one does the same argument as in [Mi 1] together with the observation that

$$\gamma_0^{-1} = \begin{pmatrix} 0 & 1 \\ 1 & 0 \end{pmatrix} \gamma_0 \begin{pmatrix} 0 & 1 \\ 1 & 0 \end{pmatrix}$$

Chose now elements $\overline{\Gamma}_1$ and $\overline{\Gamma}_2$ in $\widetilde{K}(\frac{1}{2})$ such that $\overline{\Gamma}_1\overline{\Gamma}_2 = (\mathrm{Exp}(-\pi))\Gamma_0$. Setting $\overline{\Gamma}_3 = \overline{\Gamma}_0^{-1}$ one gets $\overline{\Gamma}_1 \overline{\Gamma}_2 \overline{\Gamma}_3 = \mathrm{Exp}(-\pi)$. Using this relation as in [Mi 1, end of proof of Theorem 2] one constructs a homomorphism

$\pi_1(T_3) \longrightarrow G(\frac{1}{2})$ (T_3 denotes the surface of genus 3) such that the characteristic number of the corresponding bundle is -1. This proves Theorem 2.1 .

3. K-theory computation

The algebraic K-theory groups $K_i(\Lambda)$ of a ring Λ have been defined by D. Quillen as

$$K_i(\Lambda) = \pi_i(BGL_\infty{}^\delta(\Lambda)^+)$$

where $GL_\infty(\Lambda)$ is the inductive limit of the groups $GL_n(\Lambda)$. For $i = 1$ or 2, these K_i's coincide with those used in [Mi 2] . If Λ is a subring of $\mathbb{R}$, $SL_n(\Lambda)$ is the maximal perfect subgroup of $GL_n(\Lambda)$ $(n \geq 3)$ and thus $BSL_n{}^\delta(\Lambda)^+$ is the universal covering of $BGL_n^\delta(\Lambda)^+$ (see [H-H]). In this section, we use K-theory computations to obtain results on $H_2(SL_n(\mathbb{Z}[\frac{1}{2}]))$ and $H^2(SL_n(\mathbb{Z}[\frac{1}{2}]) ; \mathbb{Z}_2)$. These results will be useful for proving Theorem 1.

Lemma 3.1 :

For $n \geq 4$, one has $H_2(SL_n(\mathbb{Z}[\frac{1}{2}])) = \mathbb{Z}_2$ and the natural homomorphism $H_2(SL_n(\mathbb{Z}[\frac{1}{2}])) \longrightarrow H_2(SL_{n+1}(\mathbb{Z}[\frac{1}{2}]))$ is an isomorphism.

Proof :

One first proves that $K_2(\mathbb{Z}[\frac{1}{2}]) = \mathbb{Z}_2$. This can be done by comparing the localization exact sequence [Q, Theorem 6] :

$$\begin{array}{ccccccc}
\bigoplus_p K_2(\mathbb{Z}/p\mathbb{Z}) & \longrightarrow & K_2(\mathbb{Z}) & \longrightarrow & K_2(\mathbb{Q}) & \longrightarrow & \bigoplus_p K_1(\mathbb{Z}/p\mathbb{Z}) \\
\downarrow & & \downarrow & & \downarrow & & \downarrow \\
\bigoplus_{p\neq 2} K_2(\mathbb{Z}[\frac{1}{2}]/p\mathbb{Z}[\frac{1}{2}]) & \rightarrow & K_2(\mathbb{Z}[\frac{1}{2}]) & \rightarrow & K_2(\mathbb{Q}) & \longrightarrow & \bigoplus_{p\neq 2} K_1(\mathbb{Z}[\frac{1}{2}]/p\mathbb{Z}[\frac{1}{2}])
\end{array}$$

where p ranges over all primes. The two left groups are zero, since

K_2 (finite field) = 0 [M 2, Corollary 9.13] . For a field F, $K_1(F)$ is the group of units of F [Mi 2, p. 28] . Thus, the right hand vertical arrow is an isomorphism. Therefore $K_2(\mathbf{Z}) \xrightarrow{\sim} K_2(\mathbf{Z}[\frac{1}{2}])$. But $K_2(\mathbf{Z}) = \mathbf{Z}_2$ [Mi 2, Corollary 10.2] .

Let $St_n(\Lambda)$ denote the Steinberg group over Λ [Mi 2, p.39] . By [vdK-S, (2.6)], and [Ke p. 224], one has $H_2(St_n(\mathbf{Z}[\frac{1}{2}])) = 0$ for $n \geq 4$ and thus the kernel $K_2(n,\mathbf{Z}[\frac{1}{2}])$ of the homomorphism $St_n(\mathbf{Z}[\frac{1}{2}]) \to SL_n(\mathbf{Z}[\frac{1}{2}])$ is isomorphic to $H_2(SL_n(\mathbf{Z}[\frac{1}{2}]))$ [Ke] . The last assertion of Lemma 3.1 comes then from the stability of $K_2(n,\mathbf{Z}[\frac{1}{2}])$ for $n \geq 3$ [vdK, Theorem 1].

Since $H_1(SL_n(\mathbf{Z}[\frac{1}{2}])) = 0$ for $n \geq 3$, Lemma 3.1 implies that $H^2(SL_n(\mathbf{Z}[\frac{1}{2}]);\mathbf{Z}_2) = \mathbf{Z}_2$ when $n \geq 4$.

Lemma 3.2 :

For $n \geq 4$, the generator of $H^2(SL_n(\mathbf{Z}[\frac{1}{2}]) = \mathbf{Z}_2$ is the second Stiffel-Whitney class $w_2(\rho)$ of the natural bundle $\rho : BSL_n{}^{\delta}(\mathbf{Z}[\frac{1}{2}]) \longrightarrow BSL$

Proof :

As $H^2(BSL_n{}^{\delta}(\mathbf{Z}[\frac{1}{2}])^+ ; \mathbf{Z}_2) = \mathbf{Z}_2$, it suffices to prove that $w_2(\rho) \neq$ But his is true since $H^*(BSL_n(\mathbf{R}); \mathbf{Z}_2) \longrightarrow H^*(BSL_n{}^{\delta}(\mathbf{Z}[\frac{1}{2}]); \mathbf{Z}_2)$ is injective by [Bo 1, Theorem 22.7]. One may also see this fact using the bundle of Theorem 2.1.

By Lemma 3.1 and [Ha, Proposition 7.1.3], one has a fibration

$$K(\mathbf{Z}_2,1) \longrightarrow BSt_n{}^{\delta}(\mathbf{Z}[\tfrac{1}{2}])^+ \longrightarrow BSL_n{}^{\delta}(\mathbf{Z}[\tfrac{1}{2}])^+ \qquad (n \geq 4) .$$

Since $\pi_1(BSL_n(\mathbf{Z}[\frac{1}{2}]) = H_1(SL_n(\mathbf{Z}[\frac{1}{2}])) = 1$, this fibration is induced by a fibration

$$BSt_n{}^{\delta}(\mathbf{Z}[\tfrac{1}{2}])^+ \longrightarrow BSL_n{}^{\delta}(\mathbf{Z}[\tfrac{1}{2}])^+ \xrightarrow{w} K(\mathbf{Z}_2, 2)$$

and the map w is not trivial, for $\pi_1(BSt_n(Z[\frac{1}{2}])) = 1$ and thus the fibration is not trivial. By Lemma 3.2, one deduces that $w = w_2(\rho^+)$ where $\rho^+ : BSL_n^{\delta}(Z[\frac{1}{2}])^+ \longrightarrow BSL_n(\mathbb{R})$ is the map induced by ρ.

Thus, one deduces the following corollary

Corollary 3.3 :

Let $\xi : X \longrightarrow BSL_n^{\delta}(Z[\frac{1}{2}])^+$ be a map, with $n \geq 4$. Then ξ lifts through $BSt_n^{\delta}(Z[\frac{1}{2}])^+$ if and only if $w_2(\rho^+ \circ \xi) = 0$.

Finally, using that the diagram

$$\begin{array}{ccc} BSt_n^{\delta}(\Lambda) & \longrightarrow & BSL_n^{\delta}(\Lambda) \\ \downarrow & & \downarrow \\ BSt_n^{\delta}(\Lambda)^+ & \longrightarrow & BSL_n^{\delta}(\Lambda)^+ \end{array} \qquad (n \geq 4 \text{ and } \tfrac{1}{2} \in \Lambda)$$

is a pull-back diagram (use [Ha, Proposition 7.1.3] as above), one can remove the "+" in the statement of Corollary 3.3, namely :

Corollary 3.4 :

Let $\xi : X \longrightarrow BSL_n^{\delta}(Z[\frac{1}{2}])$ be a map, with $n \geq 4$. Then ξ lifts through $BSt_n^{\delta}(Z[\frac{1}{2}])$ if and only if $w_2(\rho \circ \xi) = 0$.

4. Proofs of Theorems 1 and 2

Proof of Theorem 1 a) :

The space $BSL_2^{\delta}(\Lambda)^+$ is simply connected when $\frac{1}{6} \in \Lambda$ (since $SL_2(\Lambda)$ is then perfect), as well as $BSL_2(\mathbb{R})$ of course. Therefore, by the Hurewicz isomorphism theorem, it suffices to prove that the homomorphis $H_2(BSL_2^{\delta}(\Lambda)^+) \longrightarrow H_2(BSL_2(\mathbb{R}))$ is surjective. The Euler class $e \in H^2(BSL_2(\mathbb{R});\mathbb{Z})$ gives an isomorphism $H_2(BSL_2(\mathbb{R})) \xrightarrow{\sim} \mathbb{Z}$. Hence it suffices to prove the existence of a class $\alpha \in \mathrm{Im}(H_2(BSL_2^{\delta}(\Lambda)^+ \longrightarrow H_2(BSL_2(\mathbb{R}))$ such that $e(\alpha) = 1$, which is clear when $\frac{1}{2} \in \Lambda$ using the $\mathbb{Z}[\frac{1}{2}]$-flat bundle of Theorem 2.1.

Remark : For $n \geqslant 3$, $BSL_n^{\delta}(\Lambda)^+$ is simply connected and $H_2(BSL_n(\mathbb{R})) \simeq \mathbb{Z}$ (isomorphism given by w_2). Thus, the same argument together with Lemma 3.1 and 3.2 gives the following proposition :

Proposition 4.1 :

For $n \geqslant 4$ and $\frac{1}{2} \in \Lambda$, the homomorphism $\pi_2(BSL_n^{\delta}(\Lambda)^+) \longrightarrow \pi_2(BSL_n(\mathbb{R}$ $= \mathbb{Z}_2$ is surjective and admits a section. In particular, one gets a split surjection $K_2(\Lambda) \twoheadrightarrow \mathbb{Z}_2$.

Proposition 4.2 :

For $4 \leqslant n = 2k \neq 8r+2$, one has $I_n(\mathbb{R}) \subset \langle \tau S^n \rangle$.

Proof :

It suffices to prove that a bundle ξ: $S^n \to BSL_n^{\delta}(\mathbb{R})^+ \longrightarrow BSL_n(\mathbb{R}$ is stably trivial. This clear for $n = 8r+6$, since $\pi_{8r+6}(BSL_{\infty}(\mathbb{R})) = 0$ (Bott periodicity ; see [Ka, p. 157]) . For $n = 4s$, observe that the rational Pontrjagin classes $p_i(\xi)$ are all zero. Indeed, p_i vanishes in $H^{4i}(BSL_n{}^{\delta}(R); Q)$ [Mi 1, Theorem 3] and the homomorphism $H^*(BSL_n{}^{\delta}(\Lambda)^+;Q) \to$ $\longrightarrow H^*(BSL_n{}^{\delta}(\Lambda); Q)$ is an isomorphism [H-H, Remark 1.6]. But the stable bundles over S^{4s} are classified by their rational Pontrjagin classes classes [Hr, Corollary 9.8].

In the next proof and later we shall make use several times of the following classical lemma (for a proof, see [Sm, Proposition 4]):

Lemma 4.3 :

Let ξ_1 and ξ_2 be two stably trivial oriented 2k-vector bundle over a closed manifold M^{2k}. If $e(\xi_1) = e(\xi_2)$ then ξ_1 and ξ_2 are isomorphic.

Lemma 4.4 :

For $n = 2k \geq 4$, one has $2^{k-2} < \tau S^n > \subset I_n(Z[\frac{1}{2}])$.

Proof :

We first prove Lemma 4.4 for $n = 4$ and then for $n = 2k \geq 6$ by an induction argument. Throughout this proof, we denote by A the ring $Z[\frac{1}{2}]$.

The case n = 4 :

Let $\eta : T_3 \longrightarrow BSL_2{}^{\delta}(A)$ be the A-flat bundle of Theorem 2.1. One considers the product flat bundle $\eta.\eta : T_3 \times T_3 \longrightarrow BSL_4{}^{\delta}(A)$.

Its second Stiffel-Whitney class $w_2(\eta.\eta)$ satisfies $w_2(\eta.\eta) =$ $= w_2(\eta) \otimes 1 + 1 \otimes w_2(\eta)$. Thus $w_2(\eta.\eta)$ is represented by a map $w : T_3 \times T_3 \to K(Z_2,2)$ which admits a factorization $T_3 \times T_3 \xrightarrow{\alpha} K(Z_2;2)\times K(Z_2;2) \xrightarrow{\beta} K(Z_2;2)$, where $\alpha = w_2(\eta) \times w_2(\eta)$ and β is the H-space multiplication on $K(Z_2;2)$. As $[T_3 \times T_3] = [T_3] \otimes [T_3]$, the induced homomorphism $\omega_* : H_4(T_3 \times T_3) \longrightarrow H_4(K(\mathbb{Z}_2;\ 2))$ sends $[T_3 \times T_3]$ onto an element of order 2 of $H_4(K(\mathbb{Z}_2;2))$. Since $H_1(St_4(A)) = H_2(St_4(A)) = 0$ [vdK-S, 2.6] , the space $BSt_4{}^{\delta}(A)^+$ is 2-connected and the spectral sequence of the fibration

$$BSt_4{}^{\delta}(A)^+ \longrightarrow BSL_4{}^{\delta}(A)^+ \xrightarrow{w_2} K(\mathbb{Z}_2,2) \quad \text{(see Section 3)}$$

gives rise to an exact sequence

$$H_4(BSt_4{}^{\delta}(A)^+) \xrightarrow{i} H_4(BSL_4{}^{\delta}(A)^+) \longrightarrow H_4(K(\mathbb{Z}_2,2)) \ .$$

Therefore, there is a class $\alpha \in H_4(BSt_4^{\delta}(A)^+)$ such that $i(\overline{\alpha}) = 2(\eta.\eta)_*([T_3 \times T_3])$ in $H_4(BSL_4{}^{\delta}(A)^+)$. By the Hurewicz theorem, $\overline{\alpha}$ can be represented by a map $\alpha: S^4 \longrightarrow BSt_4{}^{\delta}(A)^+$ ($BSt_4{}^{\delta}(A)^+$ is 2-connected). The Euler class $e(\alpha)$ of the corresponding bundle satisfies $e(\alpha)([S^4]) = 2e(\eta.\eta)([T_3 \times T_3]) = 2$. By Proposition 4.2, α must be a multiple of τS^4 and then $\alpha = \tau S^4$.

The induction step :

One assumes by induction that $2^{k-3} < \tau S^{n-2} > \subset I_{n-2}(A)$ for $n = 2k \geq 6$. This means that the bundle

$$S^{n-2} \xrightarrow{\alpha} S^{n-2} \xrightarrow{\tau S^{n-2}} BSL_{n-2}(\mathbb{R}) \qquad (\alpha \text{ of degree } 2^{k-3})$$

admits a lifting $\xi_0 : S^{n-2} \longrightarrow BSL_{n-2}{}^{\delta}(A)^+$. Let $\tilde{\eta} : T \longrightarrow BSL_2(A)$ be the composition of a two-fold covering $T \to T_3$ with the bundle $\eta : T_3 \longrightarrow BSL_2{}^{\delta}(A)$ of Theorem 2.1 . The bundle

$$\xi_1 : S^{n-2} \times T \xrightarrow{\xi_0 \times \tilde{\eta}} BSL_{n-2}{}^{\delta}(A)^+ \times BSL_2{}^{\delta}(A) \longrightarrow BSL_n{}^{\delta}(A)^+$$

satisfies by construction $e(\xi_1)[S^{n-2} \times T] = 2^{k-1}$. As $e(\tilde{\eta})[T] = 2$, the bundle $\tilde{\eta}$ is stably trivial [Sm, Proposition 2]; therefore ξ_1 admits a lifting (call it ξ_1 again) $\xi_1 : S^{n-2} \times T \longrightarrow F$, where F is the fiber of the map $BSL_n{}^{\delta}(A)^+ \longrightarrow BSL_\infty(\mathbb{R})$. The space $BSL_\infty(\mathbb{R})$ is homotopy equivalent to BSO and Corollary 3.3 and its context show that the following homotopy commutative diagram :

$$\begin{array}{ccc} BSt_n{}^{\delta}(A)^+ & \longrightarrow & BSpin \\ \downarrow & & \downarrow \\ BSL_n{}^{\delta}(A)^+ & \longrightarrow & BSO \end{array} \qquad (n \geq 4)$$

is a pull-back diagram. Hence F is homotopy equivalent to the fiber of $BSt_n(A)^+ \longrightarrow BSpin$ which is 2-connected.

As $S^{n-2} \times T$ is a stably parallelizable manifold of dimension $n \geq 6$ having non-zero homology groups only in dimension 0,1,2,n-2,n-1 and n, one can perform surgeries of index 2 and 3 so that the resulting manifold M is a homotopy sphere. In other words, there is a cobordism $(W^{n+1}, S^{n-2} \times T, M)$ such that W is obtained from $S^{n-2} \times T \times I$ by attaching handles of index 2 and 3 and there exists a homotopy equivalence $\beta : S^n \to M$. As F is 2-connected, the map ξ_1 extends to $\xi_1 : W \longrightarrow F$. Then we get a bundle $\xi : S^n \longrightarrow F$ given by $\xi = (\xi_1|M)\,\beta$. One has $e(\xi)[S^n] = e(\xi_1)[S^{n-2} \times T] = 2^{k-1}$. By Lemma 4.3, the map ξ is a lifting of $2^{k-2}\tau S^n$ which proves our induction step.

<u>Proof of Theorem 1b</u>) Theorem 1 b) follows directly from Proposition 4.2 and Lemma 4.4, at least when $n \neq 8r+2$. For this latest case, one uses also the following lemma :

<u>Lemma 4.5</u> : For $m = 8r+2 \geq 10$, one has a natural decomposition :

$$\pi_m(BSL_m(\mathbb{R})) = < \tau S^m > \oplus \pi_m(BSL_\infty(\mathbb{R})) = \mathbb{Z} \oplus \mathbb{Z}_2 .$$

<u>Proof</u> : From the fibration $S^m \xrightarrow{\tau S^m} BSL_m(\mathbb{R}) \longrightarrow BSL_{m+1}(\mathbb{R})$ one draws the following exact sequence :

$$\pi_m(S^m) \xrightarrow{\tau^*} \pi_m(BSL_m(\mathbb{R})) \longrightarrow \pi_m(BSL_{m+1}(\mathbb{R})) \longrightarrow 0$$
$$\pi_m(BSL_{m+1}(\mathbb{R})) = m(BSL_\infty(\mathbb{R})) = \mathbb{Z}_2$$

The composition $\pi_m(S^m) \xrightarrow{\tau_*} {}_m(BSL_m(\mathbb{R})) \xrightarrow{\partial} \pi_{m-1}(S^{m-1})$ is the multiplication by 2 [Br, Lemma IV.1.9], where ∂ is the boundary of the homotopy exact sequence of the fibration $S^{m-1} \longrightarrow BSL_{m-1}(\mathbb{R}) \longrightarrow BSL_m(\mathbb{R})$. Thus τ_* is injective and $\pi_m(BSL_m(\mathbb{R})) = \mathbb{Z}$ or $\mathbb{Z} \oplus \mathbb{Z}_2$. But ∂ is not surjective [Br, Theorem IV.1.10], which implies that $\pi_m(BSL_m(\mathbb{R})) = \mathbb{Z} \oplus \mathbb{Z}_2$. In such a group, there is only one element of order 2, so the decomposition is natural.

Proof of Theorem 2 : Let $\alpha : S^n \longrightarrow BSL_n(\mathbb{R})$ represent $[\alpha] \in I_n(\mathbb{Z})$. Suppose that $n \neq 8r+2$. By Proposition 4.2, $[\alpha] = k[\tau S^n]$ for some $k \in \mathbb{Z}$ and thus $[\alpha]$ is determined by its Euler class $\alpha^*(e_{\mathbb{Q}})$, with $e_{\mathbb{Q}} \in H^n(BSL_n(\mathbb{R});\mathbb{Q})$. But $e_{\mathbb{Q}}$ goes to zero in $H^n(BSL_n{}^{\delta}(\mathbb{Z})^+;\mathbb{Q}) = H^n(BSL_n{}^{\delta}(\mathbb{Z});\mathbb{Q})$ by [Su]. Then $[\alpha] = 0$. The same argument using Lemma 4.5 gives the corresponding result for $n = 8r+2 \geq 10$.

5. Proof of Theorems 3 and 5 :

Throughout this section, A denotes the ring $\mathbb{Z}[\frac{1}{2}]$.

Let $M^n (n = 2k \geq 6)$ be a stably parallelizable closed manifold with $\chi(M) = 2^{k-1}d$. By Lemma 4.3, the tangent map τM is represented by the composition :

$$M \xrightarrow{\alpha} S^n \xrightarrow{\beta} BSL_n(\mathbb{R})$$

where α is a map of degree d and β represents $2^{k-2}\tau S^n$. By Theorem 1b), β lifts through $BSL_n\delta(A)^+$ and so τM lifts through $BSL_n{}^{\delta}(A)^+$. By [H-V 2, Theorem 1.1] one deduces that M is semi-s-cobordant to a A-flat manifold M, using the same argument as for Theorem 1.3 of [H-V 2]. Theorem 3 is thus proved.

Let us now emphasize some properties of the manifold M.

Lemma 5.1 :

The manifold M_- is stably parallelizable.

Proof :

Let (W,M,M_-) be a semi-s-cobordism. The stable tangent bundle $\tau^S(M \amalg M_-) : (M \amalg M_-) \longrightarrow BSL_\infty(\mathbb{R})$ extends to $\tau^S W : W \longrightarrow BSL_\infty(\mathbb{R})$. As $\tau^S M$ is homotopic to a constant map and $H_*(W,M) = 0$ one has by obstruction theory that $\tau^S W$ is homotopic to a constant map. This proves Lemma 6.1.

Proposition 5.2 :

Let $M^n (N = 2k \geq 6)$ be a stably parallelizable closed manifold which is semi-s-cobordant to a Λ-flat manifold M_-, where Λ is a subring of $\mathbb{R}$ which is finitely generated as $\mathbb{Z}$-algebra. Then there exists a semi-s-cobordism (W,M,M_-) such that $\pi_1(M_-) = \pi_1(M) \times St_n(\Lambda)$ and such that τM_- admits a Λ-flat structure given by the composed map

$$M_- \longrightarrow B\pi_1(M_-) = B\pi_1(M) \times BSt_n{}^{\delta}(\Lambda) \xrightarrow{\text{proj.}} BSt_n{}^{\delta}(\Lambda) \longrightarrow BSL_n{}^{\delta}(\Lambda)$$

Proof :

By the hypotheses on M and [H-V, Corollary 1.2], the tangent bundle τM of M admits a lifting $\tau M : M \longrightarrow BSL_n{}^{\sigma}(\Lambda)^+$. As M is stably parallelizable, one has $w_2(\tau M) = 0$ and, by Corollary 3.3, the map τM factors through a map (call it τM again) $\tau M : M \longrightarrow BSt_n{}^{\delta}(\Lambda)^+$. Then the map $M \xrightarrow{j\times\tau M} B\pi_1(M) \times BSt_n{}^{\delta}(\Lambda)^+$ is 2-connected (where $M \xrightarrow{j} B\pi_1(M)$ is of course the classifying map for the universal covering of M). Take the pull-back diagram :

$$\begin{array}{ccc} X & \longrightarrow & B\pi_1(M) \times BSt_n{}^{\delta}(\Lambda) \\ \downarrow & & \downarrow \\ M & \xrightarrow{j\times\tau M} & B\pi_1(M) \times BSt_n{}^{\delta}(\Lambda)^+ \end{array}$$

By Theorem 6.3 of [H-H] and its proof, the map $X \longrightarrow M$ is an acyclic map and $\ker(\pi_1(X) \longrightarrow \pi_1(M))$ acts trivially on $\pi_i(X)$ for $i \geq 2$. As $j\times\tau M$ is 2-connected, the homomorphism

$\pi_1(X) \longrightarrow \pi_1(M) \times St_n(\Lambda)$ is an isomorphism (five Lemma applied to the homotopy ecact sequence of the two vertical fibrations ; see detail in [Ha, proof of Lemma 4.1]). Now $St_n(\Lambda)$ is finitely presented [R-S]. By [Ha, Theorem 3.1], there exists a semi-s-cobordism (W,M,M_-) such that the restriction to M_- of the map $W \longrightarrow M \longrightarrow B\pi_1(M) \times BSt_n{}^{\delta}(\Lambda)^+$ lifts through $\gamma : M_- \longrightarrow B\pi_1(M)\times BSt_n{}^{\delta}(\Lambda)$ and $\pi_1\gamma$ is an isomorphism (X is a Poincaré complex over $\mathbb{Z}\pi_1(M)$ and $\gamma \in \mathcal{S}^s{}_{DIFF}(X; \mathbb{Z}\pi_1(M))$ corresponds to $id_M \in \mathcal{S}^s{}_{DIFF}(M)$ via the bijection of [Ha, Theorem 3.1]). The map

$$\mu : M_- \longrightarrow X \longrightarrow M \xrightarrow{\tau M} BSL_n(\mathbb{R}) \quad \text{is homotopic to}$$

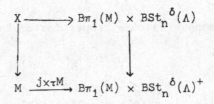

$M_- \longrightarrow B\pi_1(M) \longrightarrow BSt_n{}^\delta(\Lambda) \longrightarrow ESt_n{}^\delta(\Lambda) \longrightarrow BSL_n(\mathbb{R})$ and produces a bundle such that $e(\mu)[M_-] = e(\tau M)[M] = \chi(M) = \chi(M_-)$. As M_- is stably parallelizable (Lemma 5.1), one has $\mu = \tau M_-$ by Lemma 4.3.

Proof of Theorem 5 :

Let M^{2K} be a stably parallelizable closed manifold which is semi-s-cobordant to a $\mathbf{Z}$-flat manifold M_-. Then $\chi(M_-) = 0$ by [Su]. Thus $\chi(M) = 0$ and M is parallelizable by Lemma 4.3.

6. Flat bundles over homology spheres

A homology sphere Σ^n is a smooth manifold such that $H_*(\Sigma^n) = H_*(S^n$
A vector bundle $\xi : X \longrightarrow BSL_n(\mathbf{R})$ is called Λ-flat (Λ a subring of $\mathbf{R}$) if there is a lifting $\xi : X \longrightarrow BSL_n{}^{\delta}(\Lambda)$ of ξ .

Proposition 6.1 :

For $8r+2 \neq n = 2k \geq 6$, the following two conditions are equivalent :

a) There is a Λ-flat vector bundle ξ of rank n over a homology sphere Σ^n such that $e(\xi)[\Sigma^n] = d \in \mathbf{Z}$.

b) $\frac{d}{2}\ \tau S^n \in I_n(\Lambda)$

If $n = 8r+2 \geq 6$, Condition b) implies Condition a) . On the other hand, Condition a) implies Condition b) provided ξ is stably trivial.

Proof :

Suppose a) holds and apply the plus construction to ξ :

$$\begin{array}{ccc} \Sigma^n & \xrightarrow{\ \xi\ } & BSL_n^{\delta}(\Lambda) \\ \downarrow & & \downarrow \\ (\Sigma^n)^+ & \xrightarrow{\ \xi^+\ } & BSL_n^{\delta}(\Lambda)^+ \end{array}$$

One has $(\Sigma^n)^+ = S^n$. This implies b) when $n \neq 8r+2$ by Lemma 4.2. If ξ is stably trivial, so is ξ^+ by obstruction theory which proves the second assertion for the case $n = 8r+2$. The fact that b) implies a) follows directly from [H-V 2, Theorem 1.1] or [H-V 1, Theorem 4.1].

Corollary 6.2 :

Let $n = 2k \geq 6$, $c \in \mathbf{Z}$, and $A = \mathbf{Z}[\frac{1}{2}]$. There exists a homology sphere Σ^n with $\pi_1(\Sigma^n) = St_n(A)$ and such that the natural map $\Sigma^n \longrightarrow B\pi_1(\Sigma^n) \longrightarrow BSt_n^{-\delta}(A) \longrightarrow BSL_n^{\delta}(A)$ produces a A-flat bundle ξ over Σ^n with $e(\xi)[\Sigma^n] = 2^{k-1} c$.

Proof :

By Proposition 6.1 and Theorem 1, there is a A-flat bundle ξ_0 over a homology sphere Σ_0^n with $e(\xi_0)[\Sigma_0^n] = 2^{k-1} c$. The existence of the required manifold can then be deduced from those of Σ_0 in the same way as in the proof of Proposition 5.2.

Question : Do there exist flat homology spheres ?

We were not able to give an answer to that question. So far, we can only give the following comment on it :

Proposition 6.3 :

For $n = 2k \geq 6$, and Λ a subring of $\mathbb{R}$, the following three conditions are equivalent :

1) Any stably parallelizable manifold M of dimension n is semi-s-cobordant to a Λ-flat manifold .

2) There exists a Λ-flat homology sphere of dimension n

3) $\tau S^n \in I_n(\Lambda)$.

Proof :

1) implies 2) by taking $M = S^n$. Condition 2) implies Condition 3) by Proposition 6.1 . Let us prove that 3) implies 1) .

If M^n is stably parallelizable, the Euler characteristic $\chi(M)$ is even ($e(\tau M)$ reduces mod 2 to $w_n(\tau m)$ which is zero). So, by Lemma 4.3, the tangent bundle M is represented by $M \xrightarrow{\alpha} S^n \xrightarrow{\tau S^n} BSL_n(\mathbb{R})$, where α is of degree $\chi(M)/2$. Condition 1) follows then from Condition 3) by the argument used to prove Theorem 3 .

7. Proof of Proposition 6 :

As $H^*(BSL_n{}^\delta(\mathbb{R})^+ ;\mathbb{Z}) \longrightarrow H^*(BSL_n{}^\delta(\mathbb{R}); \mathbb{Z})$ is an isomorphism [H-H, Remark 1.6], it suffices to prove Proposition 6 for the class e_+^δ which is the image of e in $H^n(BSL_n{}^\delta(\mathbb{R})^+;\mathbb{Z})$.

Proof of a) :

By Theorem 1, e_+^δ takes a non-zero value on a spherical homology class (representing $2^{k-2}\ S^n$) . Thus e_+^δ cannot be a non-trivial sum of cup products.

Proof of b) :

Let $A = \mathbb{Z}[\ \frac{1}{2}\ ]$. For $n \in \mathbb{N}$, and $i \in \mathbb{N} \cup \{\infty\}$, define p_i^n as the composed homomorphism :

$$H^n(BSL_i^\delta(\mathbb{R})^+;\mathbb{Z}) \longrightarrow H^n(BSL_i^\delta(A)^+;\mathbb{Q}) \longrightarrow \mathrm{Hom}(H_n(BSL_i^\delta(A)^+);\mathbb{Q}) \longrightarrow$$

$$\longrightarrow \mathrm{Hom}(\pi_n(BSL_i^\delta(A)^+);\mathbb{Q})$$

where the last homomorphism is the dual of the Hurewicz homomorphism. One has the commutative diagram :

$$\begin{array}{ccc} H^n(BSL_\infty^\delta(\mathbb{R})^+;\mathbb{Z}) & \xrightarrow{p_\infty^n} & \mathrm{Hom}(\pi_n(BSL_\infty^\delta(A)^+);\mathbb{Q}) \\ \downarrow \alpha & & \downarrow \\ H^n(BSL_n^\delta(\mathbb{R})^+;\mathbb{Z}) & \xrightarrow{p_n^n} & \mathrm{Hom}(\pi_n(BSL^\delta(A)^+);\mathbb{Q}) \end{array}$$

Let us now suppose that n = 2k. Theorem 1 implies that $p^n(e_+^\delta) \neq 0$. Therefore e_+^δ is not in the image of α because $\mathrm{Hom}(\pi_n(BSL_\infty^\delta(A)^+;\mathbb{Q}) = 0$. Indeed, $\pi_n(BSL_\infty^\delta(A)^+) = K_n(A)$ for $n \geq 2$. By the localization exact sequence [Q, Theorem 6] and the fact that the algebraic K-theory groups of a finite field are finite [Q, Theorem 4] one deduces that

$K_n(A) \otimes \mathbb{Q} = K_n(\mathbb{Z}) \otimes \mathbb{Q}$. If $n = 2k$, one has $K_n(\mathbb{Z}) \otimes \mathbb{Q} = 0$ [Bo 2, Proposition 12.2] .

REFERENCES

[Bo 1] A. BOREL Topics in the homotopy thepry of fibre bundles, Springer Lect. Notes 36 (1967) .

[Bo 2] ---- Stable real cohomology of arithmetic groups, Ann. Ec. Norm. Sup. 4e série, t.7 (1974), p. 235-272.

[Br] W. BROWDER Surgery on simply connected manifolds, Springer-Verlag 1972.

[Ha] J.-Cl. HAUSMANN - Manifolds with a given homology and fundamental group Comm. Math. Helv. 53 (1978) p. 113-134.

[H-H] J.-Cl. HAUSMANN- D. HUSEMOLLER - Acyclic maps. L'Enseign.Math., to appear .

[H-V 1] J.-Cl.HAUSMANN- P. VOGEL - The plus construction and lifting maps from manifolds. Proceedings of Symposia in Pure Math. 32 (1977) p. 417-426.

[H-V 2] ---- Reduction of structure on manifolds by semi-s-cobordism . Monographie de l'Enseign. Math. dedicated to B. Eckmann , p. 117-124.

[Hr] D. HUSEMOLLER - Fibre bundles. 2e édition Springer-Verlag 1975.

[Ka] M. KAROUBI K-theory, an Introduction. Springer-Verlag 1978.

(vdK] W. van der KALLEN - Injective stability for K_2. Algebraic K-theory, Evanston 1976, Springer Lect. Notes Nb 551 .

[vdK-S] W. van der KALLEN-M. STEIN - On the Schur multipliers of Steinberg and Chevalley Groups. Math. Zeitsch. 155 (1977) 83-94.

[Ke] M. KERVAIRE : Multiplicateurs de Schur et K-théorie. Essays on Topology and related Topics, Springer-Verlag 1970, p. 212-225.

[Mi 1] J. MILNOR : On the existence of a connection with curvature zero. Comm. Math. Helv. 32 (1958) p. 215-223.

[Mi 2] ---- Introduction to algebraic K-theory, Princeton Univ. Press 1971.

[Q] D. QUILLEN : Higher Algebraic K-theory. Proc. Int. Congress of Math. Vancouver 1974, 171-176.

[R-S] U. REHMANN- Ch. SOULE : Finitely presented groups of matrices. Algebraic K-theory, Evanston 1976, Springer Lect. Notes , p. 551 .

[Sm] J. SMILLIE : Flat manifolds with non-zero Euler characteristics Comm. Math. Helv. 52 (1977) 453-455.

[Su] D. SULLIVAN : La classe d'Euler réelle d'un fibré à groupe structural $SL_n(Z)$ est nulle. Comptes rendus Ac. Sc. Paris, Série A, t. 281 (1975) p. 17-18.

University of Geneva, Switzerland
The Institute for Advanced Study, Princeton .

TOPOLOGICAL CLASSIFICATIONS OF $S\ell_2(\mathbb{F}_p)$ SPACE FORMS

Erkki Laitinen and Ib Madsen

§0. Introduction.

Let p be a prime and $\mathbb{F}_p$ the field with p elements. The group $S\ell_2(\mathbb{F}_p)$ of 2×2 matrices with determinant 1 admits free actions on spheres, and free linear actions only if $p \le 5$. More precisely, let $d = p-1$ if $p \equiv 1(4)$ and $d = 2(p-1)$ if $p \equiv 3(4)$. Then $S\ell_2(\mathbb{F}_p)$ acts freely on S^{nd-1} for each positive integer n. The orbit space of such an action is called a (spherical) $S\ell_2(\mathbb{F}_p)$ space form, and is often denoted $M(S\ell_2(\mathbb{F}_p))$.

Much is known about which homotopy types can contain a space form and about the set of simple homotopy types of space forms (cf. [W6], [M], [Mg]). In this paper we consider the relative classification problem of counting the topological manifolds contained in a fixed simple homotopy type $M = M^{4n-1}(\pi)$, $\pi = S\ell_2(\mathbb{F}_p)$.

We follow the standard approach from surgery theory. Let $\mathcal{S}(M)$ be the structure set of homeomorphism classes of pairs (N,f) consisting of a topological manifold and a simple homotopy equivalence $f\colon N \to M$. The surgery obstruction group $L_0^s(\pi)$ acts on $\mathcal{S}(M)$ but not freely. Indeed, $L_0^s(1) \subset L_0^s(\pi)$ acts trivially so we consider only the action of the reduced group $\tilde{L}_0^s(\pi)$.

__Theorem A__. The group $\tilde{L}_0^s(\pi)$ acts freely on $\mathcal{S}(M^{4n-1})$, and the cardinality of the orbit space is $p^s(p^2-1)^{n-1}$, $s = 4n/p-1 - 1$.

There are two well understood invariants of $\mathcal{S}(M)$: The ρ-invariant $\rho\colon \mathcal{S}(M) \to \tilde{RO}(\pi) \otimes \mathbb{Q}$ (cf.[W ,ch.13]) and the (index part of) 2-local normal invariant $N\colon \mathcal{S}(M) \to \oplus H^{4i}(M;\mathbb{Z}_{(2)})$, $\tilde{RO}(\pi) = \mathrm{cok}\{i_*\colon RO(1) \to RO(\pi)\}$.

__Theorem B__. The number of topological manifolds contained in the s-type of M^{4m-1} and with fixed N- and ρ-invariants is bounded by the cardinality of $\mathrm{Tor}\, L_0^s(\pi)$.

These results are proved in §2. They are, however, of limited scope since Tor $L_o^s(\pi)$ is extremely hard to get a hold on. Explicit results depend on Tor $Wh(\pi)$ which at present (for $\pi = S\ell_2(\mathbb{F}_p)$) seems only to be known for $p = 3,5$.

In order to get more explicit results we retreat and ask for a classification modulo weak simple h-cobordism, that is, h-cobordisms whose torsions are torsion. We start by fixing a weak simple homotopy type $M = M^{4n-1}(\pi)$. This is specified by the Reidemeister torsion invariant $\Delta(M) \in K_1(Q\pi)/\langle -1\rangle \oplus \pi/[\pi,\pi]$.

Theorems A and B remain valid when we substitute for $\mathcal{S}(M)$ the set $\mathcal{S}'(M)$ of weak h-cobordism classes of pairs (N,f) with f a weak simple homotopy equivalence and for $L_o^s(\pi)$ the 'intermediate' groups $L_o'(\pi)$. In §3 and 4 we follow a proceedure developed by Wall (cf.[W5]) to determine $L_o'(S\ell_2(\mathbb{F}_p))$ modulo classical number theoretic questions. In §5 we give explicit results for $p < 47$. In this range Tor $L_o'(S\ell_2(\mathbb{F}_p)$ $\neq 0$ only for $p = 17, 41$, so we have

Corollary C. If $p < 47$ and $p \neq 17,41$ then the invariants Δ, N and ρ uniquely specify the weak simple h-cobordism type of $S\ell_2(\mathbb{F}_p)$ space forms. For $p = 17$ and $p = 41$ one needs further invariants.

We have limited ourselves to $S\ell_2(\mathbb{F}_p)$ mostly to save space, however, the 2-adic calculations of §4 become more complicated for certain groups with periodic cohomology. These questions are taken up in a forthcoming paper [LM].

§1. The subgroups of $S\ell_2(\mathbb{F}_p)$.

We briefly recall the group theory of $S\ell_2(\mathbb{F}_p)$. Details may be found in [H].

There are 3 basic cyclic subgroups C_1, C_2 and C_3 of $S\ell_2(\mathbb{F}_p)$, whose conjugates cover $S\ell_2(\mathbb{F}_p)$. The subgroup $C_1 = \mathbb{Z}/2p$ consists of all matrices $\pm\begin{pmatrix}1 & a\\ 0 & 1\end{pmatrix}$, $a \in \mathbb{F}_p$; $C_2 = \mathbb{Z}/p-1$ is the group of diagonal matrices $\begin{pmatrix}a & 0\\ 0 & a^{-1}\end{pmatrix}$, $a \in \mathbb{F}_p^\times$. Finally, $C_3 = \mathbb{Z}/p+1$ is the kernel of the norm homomorphism $N: \mathbb{F}_{p^2}^{\times} \to \mathbb{F}_p^{\times}$, where $\mathbb{F}_{p^2}$ is the field with p^2 elements. Consider $\mathbb{F}_{p^2}$ as a 2-dimensional vector space over $\mathbb{F}_p$. Then any element $\lambda \in \mathbb{F}_{p^2}^{\times}$ gives an element of $G\ell_2(\mathbb{F}_p)$ with determinant $N(\lambda) = \lambda \cdot \lambda^p$. Thus $C_3 = \ker N$ embeds into $S\ell_2(\mathbb{F}_p)$.

We also need the normalizers of C_i:, $N(C_1)$ consists of matrices $\begin{pmatrix} b & a \\ 0 & b^{-1} \end{pmatrix}$ and it defines a split extension $1 \to C_1 \to N(C_1) \to \mathbb{F}_p^{\times}/\langle -1 \rangle \to 1$. Here $b \in \mathbb{F}_p^{\times}/\langle -1 \rangle$ acts on C_1 by sending $\pm\begin{pmatrix} 1 & a \\ 0 & 1 \end{pmatrix}$ to $\pm\begin{pmatrix} 1 & b^2a \\ 0 & 1 \end{pmatrix}$. Both C_2 and C_3 have index 2 in their normalizers.

The (generalized) quaternion group with 2^k elements is denoted by $H2^k$, and is presented as $H2^k = \langle X, Y \mid X^{2^{k-2}} = Y^2,\ YXY^{-1} = X^{-1} \rangle$. There are 3 maximal conjugacy classes of 2-hyperelementary subgroups in $S\ell_2(\mathbb{F}_p)$, τ_1, τ_2 and τ_3, where

$$1 \to \mathbb{Z}/p \to \tau_1 \to \mathbb{Z}/2^{\ell} \to 1$$

1.1 $$1 \to \mathbb{Z}/m \to \tau_2 \to \mathbb{Z}/4 \to 1$$

$$1 \to \mathbb{Z}/n \to \tau_3 \to H2^k \to 1$$

Here $\mathbb{Z}/2^{\ell}$ is the Sylow 2-subgroup of $\mathbb{F}_p^{\times}$, m and n are odd and $\{2m, 2^{k-1}n\} = \{p-1, p+1\}$, so $\tau_2 \subset N(C_3)$ and $\tau_3 \subset N(C_2)$ if $p \equiv 1\,(4)$, and otherwise $\tau_i \subset N(C_i)$.

<u>Lemma 1.2</u>. The Sylow subgroups π_ℓ of $\pi = S\ell_2(\mathbb{F}_p)$ are cyclic for ℓ odd and quaternion for $\ell = 2$. The only element of order 2 is the central $\begin{pmatrix} -1 & 0 \\ 0 & -1 \end{pmatrix}$. Each 2-hyperelementary subgroup is conjugate to a subgroup of τ_1, τ_2 or τ_3.

Each $H8$ in $S\ell_2(\mathbb{F}_p)$ is contained in a binary tetrahedral group T^*(cf. [MTW]), and T^* is a split extension $1 \to H8 \to T^* \to \mathbb{Z}/3 \to 1$, where $\mathbb{Z}/3$ acts on $H8$ by cyclic permutation of X, Y, XY.

<u>Lemma 1.3</u>. The group $\pi = S\ell_2(\mathbb{F}_p)$ has periodic cohomology with period $p-1$ if $p \equiv 1(4)$ and period $2(p-1)$ if $p \equiv 3(4)$. More precisely

$$H^*(\pi) = \bigoplus_{\ell \neq p} H^{4*}(\pi_\ell) \oplus H^{(p-1)*}(\pi_p),$$

where H^* denotes integral group cohomology.

Proof. We shall use the exact sequence

$$1 \to H^*(\pi) \to \bigoplus_{\ell} H^*(\pi_\ell) \to \bigoplus_{\ell,g} H^*(g\pi_\ell g^{-1})$$

of Cartan-Eilenberg. Let us call the image of $H^*(\pi)$ in $H^*(\pi_\ell)$ the stable elements and denote it by $H^*(\pi_\ell)^{st}$. If ℓ is odd then π_ℓ is abelian and the stable elements are $H^*(\pi_\ell)^{\Phi(\pi_\ell)}$, where $\Phi(\pi_\ell) = N(\pi_\ell)/Z(\pi_\ell)$ [M, 1.8]. An automorphism of $\mathbb{Z}/\ell^n$ is multiplication by some $a \in (\mathbb{Z}/\ell^n)^\times$. It is well known that $H^*(\pi_\ell)$ is a polynomial algebra over $\mathbb{Z}/\ell^n$ with a 2-dimensional generator y, and a induces $y_\ell \to ay_\ell$.

If $\ell \neq p,2$ then $\Phi(\pi_\ell) = \langle -1\rangle$ and $H^*(\pi_\ell)^{st} = \mathbb{Z}/\ell^n[y^2] = H^{4*}(\pi_\ell)$. If $\ell = p$, then $\Phi(\pi_\ell) = \mathbb{F}_p^\times/\langle -1\rangle$, and the invariant part is $\mathbb{Z}/p\,[y_p^m] = H^{(p-1)*}(\pi_p)$, $m = \frac{p-1}{2}$.

Finally, by [M, 1.10] an element of $H^*(\pi_2)$ comes from $H^*(\pi)$ iff its image in $H^*(H8)$ comes from $H^*(T^*)$ for every $H8 < \pi_2$. From [CE p.254] we have

$$H^n(H2^k) = \begin{cases} \mathbb{Z}/2 \oplus \mathbb{Z}/2\,, & n \equiv 2(4) \\ \mathbb{Z}/2^k\,, & n \equiv 0(4) \\ 0\,, & \text{otherwise} \end{cases}$$

and the $4k+2$-dimensional classes are detected by the $H8 \subset \pi_2$ [MTW p.379]. But $H8 \triangleleft T^*$, so the image of $H^*(T^*)$ is $H^*(H8)^{\mathbb{Z}/3} = H^{4*}(H8)$. It follows that $H^*(\pi)$ maps onto $H^{4*}(\pi_2)$.

There are two conjugacy classes of $H8$ in $H2^k$, namely $\langle X^{2^{k-3}},Y\rangle$ and $\langle X^{2^{k-3}},XY\rangle$. In §2 we shall need

Lemma 1.4. In dimensions 1 and 2 (mod 4) the cohomology groups $H^*(H2^k;\mathbb{Z}/2)$ are detected on the two conjugacy classes of $H8$ in $H2^k$.

§2. The structure set.

Set $\pi = S\ell_2(\mathbb{F}_p)$ and fix a topological space form $M = M^{4n-1}(\pi)$. In this section we determine the set $\mathcal{S}(M(\pi))$ of simple homotopy manifold structures on $M^{4n-1}(\pi)$ modulo knowledge of $L_o^s(\pi)$.

Elements $\{f_o\}$ of $\mathcal{S}(M)$ are represented by simple homotopy equivalences $f_o\colon M_o \to M$, where M_o is a topological manifold. Two such f_o, f_1 are counted equal if f_0 is homotopic to $f_1 \circ d$ for some homeomorphism $d\colon M_o \to M_1$. Each $\{f_o\} \in \mathcal{S}(M)$ defines a topological surgery problem $(f_o, \hat{f}_o)$, classified by a homotopy class of maps from M to G/TOP. This gives the 'normal invariant', $N\colon \mathcal{S}(M) \to [M, G/TOP]$.

Recall from [W , ch.5 and 10], that for $\{f_o\} \in \mathcal{S}(M)$ and $\alpha \in L_o^s(\pi)$ there exists a normal cobordism $(F, \hat{F})$ with $F\colon W \to M \times I$, $\partial W = M_o \cup -M_1$, $F \mid M_o = f_o$, $F \mid M_1$ a simple homotopy equivalence and such that $(F, \hat{F})$ has surgery obstruction α. The restriction $f_1 = F \mid M_1$ gives a new element of $\mathcal{S}(M)$, and $\{f_1\} = \alpha \cdot \{f_o\}$ defines an action of $L_o^s(\pi)$ on $\mathcal{S}(M^{4n-1}(\pi))$. The main theorem of surgery asserts exactness of

2.1 $$* \to \mathcal{S}(M(\pi))/L_o^s(\pi) \to [M(\pi), G/TOP] \overset{\lambda}{\to} L_3^s(\pi)$$

We begin by evaluating $[M, G/TOP]$. Since G/TOP is an infinite loop space we have the decomposition

$$[M(\pi), G/TOP] = \bigoplus_{\ell} [M(\pi_\ell), G/TOP]^{st}$$

from [M, 1.7] (cf. Lemma 1.3). Here $M(\pi_\ell)$ is the cover of $M(\pi)$ with fundamental group π_ℓ, and $[M(\pi_\ell), G/TOP]$ is ℓ-local. We can then use Sullivan's theorems (cf. [MM]):

$$[M(\pi_\ell), G/TOP] = \widetilde{KO}(M(\pi_\ell)), \quad \ell \text{ odd}$$

$$[M(\pi_2), G/TOP] = \bigoplus_{i=1}^{n-1} H^{4i-2}(M(\pi_2), \mathbb{Z}/2) \oplus H^{4i}(M(\pi_2), \mathbb{Z}_{(2)})$$

For odd ℓ, the spectral sequence $H^*(M(\pi_\ell), \pi_*(BO)) \Rightarrow \widetilde{KO}(M(\pi_\ell))$ collapses for dimensional reasons and gives the order of the $[M(\pi_\ell), G/TOP]^{st}$,

$$
\begin{aligned}
|[M(\pi_\ell),BO]^{\Phi(\pi_\ell)}| &= |\pi_\ell|^{n-1} \quad \text{for} \quad \ell \neq p \\
&= p^s,\ s = 4n/p-1 - 1, \quad \text{for} \quad \ell = p
\end{aligned}
$$

From 1.3 we see that $H^{4i-2}(M(\pi_2);\mathbb{Z}/2)^{st} = 0$. Since $H^{4i}(M(\pi_2);\mathbb{Z})^{st} = H^{4i}(M(\pi_2);\mathbb{Z})$ and $|\pi| = p(p^2-1)$ we have proved

<u>Proposition 2.2</u>. The group of normal invariants $[M^{4n-1}(\pi),G/TOP]$ has order $(p^2-1)^{n-1}\cdot p^s$, $s = 4n/p-1 - 1$.

<u>Remark 2.3</u>. We have not listed the odd-primary group structure of $[M^{4n-1}(\pi),G/TOP]$, and there are indeed non-trivial extensions in the spectral sequence. For each odd ℓ, $M^{4n-1}(\pi_\ell)$ is homotopy equivalent to some linear space form $S(V)/\pi_\ell$ so $\widetilde{KO}(M(\pi_\ell)) = RO(\pi_\ell)/\langle\lambda_{-1}(V)\rangle$. This, at least abstractly, solves the extension problems (cf. [W, p.208]).

For any closed n-manifold M^n with $\pi_1(M) = \pi$ we have a commutative diagram

$$
\begin{array}{ccc}
[M,G/TOP] & \xrightarrow{\ \lambda\ } & L_n^s(\pi) \\
\big\downarrow h & \nearrow \lambda & \\
\Omega_*(B\pi \times G/TOP) & &
\end{array}
$$

Here $h([M,g]) = [M,f\times g]$ where $f\colon M \to B\pi$ classifies the universal cover of π. As $L_n^s(\pi) \to L_n^s(\pi)\otimes\mathbb{Z}_{(2)}$ is injective, λ is determined completely by $\lambda_{(2)} = \lambda\otimes\mathbb{Z}_{(2)}$.

More precisely, in [W1] Wall gives a formula for $\lambda_{(2)}[M,f\times g]$:

2.4
$$
\begin{aligned}
\lambda_{(2)}[M,f\times g] &= \eta_o(f_*(Lg^*(\ell)\cap[M])) + \beta_* C(M;f,g) \\
C(M;f,g) &= (V^2 f^*(\alpha_3) + VSq^1V\cdot f^*(\alpha_o))g^*(k)[M].
\end{aligned}
$$

Here L is the Hirzebruch class, V is the Wu class, ℓ and k are graded characteristic classes of G/TOP in dimensions congruent to 0 and 2 (mod 4); α_o and α_3 are (unspecified) graded cohomology classes in $H^*(B\pi; L_{n+1}^s(\pi;\mathbb{Z}/2))$ and $\beta\colon L_{n+1}^s(\pi;\mathbb{Z}/2) \to L_n^s(\pi)$ is the L-theoretic Bockstein. Finally, $\eta_o\colon H_*(B\pi;\mathbb{Z}_{(2)}) \to L_n^s(\pi)\otimes\mathbb{Z}_{(2)}$ is a homomorphism, related to α_o by the formula $\beta_*(\alpha_o) = \eta_o\circ Sq^1_*$.

Proposition 2.5. With the notation of 2.1, λ vanishes identically on $[M^{4n-1}(\pi), G/TOP]$.

Proof. Let $\pi_2 \subset \pi$ be the Sylow 2-subgroup and let $i: \Omega_*(B\pi_2 \times G/TOP) \to \Omega_*(B\pi \times G/TOP)$. Then λ vanishes iff $\lambda \circ i$ vanishes (cf. [M, 4.2] so we can use 2.4 for $M = M^{4n-1}(\pi_2)$. The term $\beta_* C(M;f,g)=0$ because $g^*(k) = 0$, and we are left with $\eta_o(f_*(L\cdot g^*(\ell) \cap [M]))$. Let $c: K^8 \to S^8$ be the Milnor surgery problem, and consider $\lambda_o: \Omega_*(B\pi_2) \otimes \mathbb{Z}_{(2)} \to L_3^s(\pi_2) \otimes \mathbb{Z}_{(2)}$ which to a singular manifold $f: N \to B\pi_2$ of dimension $4n-1$ associates the surgery obstruction of $1 \times c: N \times K^8 \to N \times S^8$. Then (by definition) $\lambda_o(M,f) = \eta_o\sigma(M,f)$, where $\sigma(M,f) = f_*(L \cap [M])$. But σ is onto and Stein shows in [S] that $\lambda_o=0$.

Remark 2.6. Stein states the result used in 2.5 only for the obstruction in $L_3^h(\pi_2)$. However, the argument based on the bundle $S^1 \to S^3/H2^k \to \mathbb{RP}^2$ [S,p.493] obviously carries over to $L_3^s(\pi_2)$, cf.[A ,p.175].

We have left to examine the action of $L_o^s(\pi)$ on $\mathcal{S}(M^{4n-1}(\pi))$. Let $G \subset L_o^s(\pi)$ be the isotropy subgroup of an $\{f_o\} \in \mathcal{S}(M(\pi))$. I.Hambleton pointed out to us

Lemma 2.7. The transfer $i^*: L_o^s(\pi) \to L_o^s(\pi_2)$ associated with a Sylow 2-subgroup maps G monomorphically.

Proof. Let $(F,\hat{F}), F: W \to M(\pi) \times I$ be the normal cobordism realizing the action of $\gamma \in L_o^s(\pi)$ on $\{f_o\}$. Then $\partial W = M_o + (-M_1)$ with M_o and M_1 homeomorphic. Let V be the closed manifold obtained from W by gluing together the two ends of W. The normal map $(F,\hat{F})$ induces a normal map from V to $M \times S^1$ with surgery obstruction $\beta \in L_o^s(\pi \times \mathbb{Z})$. The image of β in $L_o^s(\pi)$ under the projection $\pi \times \mathbb{Z} \to \pi$ is the original γ. Thus $\gamma \in \mathrm{Image}(\Omega_{4n}(B\pi \times G/TOP) \xrightarrow{\lambda} L_{4n}^s(\pi))$. But this whole image is mapped monomorphically to $L_o^s(\pi_2)$, cf. [M ,4.2].

In our case $\pi_2 = H2^k$, and we can get further information about $G \subset L_o^s(\pi_2)$ from the cohomological formula 2.4. Since $\tilde{H}_{4*}(H2^k;\mathbb{Z}) = 0$

the formula reduces to

2.8 $$\lambda_{(2)}\ [V^{4n},f,g] = Lg^*(\ell)[V]+(V^2f^*(\beta_*\alpha_3)+VSq^1Vf^*(\beta_*\alpha_o))g^*(k)[$$

Here $\beta_*(\alpha_3)$, $\beta_*(\alpha_o)$ are homomorphisms from $H_*(B\pi_2;\mathbb{Z}/2)$ into $L_o^s(\pi_2)$, and they are both concentrated in degrees $4i+2$. This is obviou for $\beta_*(\alpha_3)$ and follows for $\beta_*(\alpha_o)$ by the formula $\beta_*(\alpha_o) = \eta_o \circ Sq^1_*$.

Since we work in the topological category, $L_o^s(1) \subset L_o^s(\pi)$ acts trivially on $\mathcal{S}(M(\pi))$. Let $\tilde{L}_o^s(\pi)$ be the kernel of $L_o^s(\pi) \to L_o^s(1)$. Then $L_o^s(\pi) = L_o^s(1) \oplus \tilde{L}_o^s(\pi)$ and we have

<u>Theorem 2.9</u>. For each space form $M^{4n-1}(\pi)$, $\pi = S\ell_2(\mathbb{F}_p)$, $\tilde{L}_o^s(\pi)$ acts freely on $\mathcal{S}(M^{4n-1}(\pi))$ and the orbit space is $[M(\pi), G/TOP]$.

<u>Proof</u>. It suffices to prove that $\tilde{L}_o^s(\pi_2)$ acts freely on $\mathcal{S}(M(\pi_2))$; the rest of 2.9 then follows from 2.5 and 2.7. Recall from 1.4 that the homology of π_2 in dimensions congruent to 2(mod 4) is detected on the subgroups $H8 \subset \pi_2$. Since $\beta_*(\alpha_3)$, $\beta_*(\alpha_o)$ are natural with respect to induced maps we are reduced to consider the action of $\tilde{L}_o^s(H8)$ on $\mathcal{S}(M^{4n-1}(H8))$.

It is known that Wh(H8) has no torsion. Thus $L_o^s(H8) = L_o'(H8)$ an $L_o'(H8)$ is torsion free by [W5 ,p.69]. It follows from the proof of 2. and 2.8 that an isotropic element of $L_o^s(H8)$ must have the form $Lg^*(\ell)[V] \in L_o^s(1)$. Thus $\tilde{L}_o^s(H8)$ acts freely as claimed.

Let $\tilde{R}(\pi) = R(\pi)/\langle\mathbb{C}\pi\rangle$ be the representation ring modulo the ideal generated by the regular representation. There is an invariant $\rho\colon \mathcal{S}(M^{4n-1}(\pi)) \to \tilde{R}(\pi) \otimes Q$, related to the action of $\tilde{L}_o^s(\pi)$ by the formula $\rho(\alpha\cdot\{f\}) = \rho(\{f\}) - \text{sign}(\alpha)$, where sign: $\tilde{L}_o^s(\pi) \to \tilde{R}(\pi) \otimes Q$ is the multisignature (cf.[W , ch.14]). Now sign is injective on the torsi free part of $\tilde{L}_o^s(\pi)$, so ρ detects the action of $\tilde{L}_o^s(\pi)/\text{Tor}$ inside each orbit.

The odd-primary part of the normal invariant N is determined by ρ: this follows from [W,p.212] and induction. The 2-primary part $N_{(2}$ of N is not determined by ρ, but it is not independent of ρ either

Corollary 2.10. The 'kernel' of

$$\rho \times N_{(2)} : \mathcal{S}(M^{4n-1}(\pi)) \to (\tilde{R}(\pi) \otimes \mathbb{Q}) \times (\bigoplus_{i=1}^{n-1} H^{4i}(\pi_2; \mathbb{Z}_{(2)}))$$

is contained in the torsion subgroup of $\tilde{L}^s_o(\pi)$.

Remark 2.11. Let $\mathrm{Out}^s(M(\pi))$ denote the group of homotopy classes of simple homotopy equivalences $M(\pi) \to M(\pi)$. It acts on $\mathcal{S}(M(\pi))$ and the orbit space is the set of homeomorphism types of manifolds simply homotopy equivalent to $M(\pi)$. We have $\mathrm{Out}^s(M(\pi)) \subseteq \mathrm{Out}(\pi) = \mathbb{Z}/2$, cf. [M, §2]. If $M(\pi)$ is a space form whose 2-hyperelementary and cyclic covers are orthogonal space forms (cf. [M1]) then $\mathrm{Out}^s(\pi)$ acts trivially. In general, however, we have not been able to determine the action.

§3. The intermediate L-groups, $L'_o(S\ell_2(\mathbb{F}_p))$.

In a series of papers Wall has shown how to effectively calculate the socalled intermediate L-groups $L'_*(\pi)$. They are related to the groups $L^s_*(\pi)$ via a long exact sequence whose third terms $H^*(SK_1(\mathbb{Z}\pi))$ seem exceedingly hard to compute. In this section we outline the calculation of $L'_o(S\ell_2(\mathbb{F}_p))$. We assume some familarity with the basic paper [W5].

The induction theorem of A.Dress [D] implies that $L'_*(S\ell_2(\mathbb{F}_p))$ is calculable in terms of $L'_*(\tau)$ where τ runs through the 2-hyperelementary subgroups. A 2-hyperelementary group τ admits a projection $\tau \to \tau_{(2)}$ on its Sylow 2-subgroup and $L'_*(\tau) = L'_*(\tau_{(2)}) \oplus L'_*(\tau)_{od}$ where $L'_*(\tau)_{od} = \mathrm{Ker}(L'_*(\tau) \to L'_*(\tau_{(2)}))$.

The conjugacy classes of hyperelementary subgroups in $S\ell_2(\mathbb{F}_p)$ were listed in §1. There were three 2-hyperelementary subgroups, namely τ_1, τ_2 and τ_3, where τ_1 and τ_2 are metacyclic and τ_3 is of quaternion type. Induction now gives

Lemma 3.1. $L'_*(S\ell_2(\mathbb{F}_p)) = L'_*(H2^k)^{st} \oplus L'_*(\tau_1)_{od}^{\mathbb{F}_p^\times} \oplus L'_*(\tau_2)_{od} \oplus L'_*(\tau_3)_{od}$ where $\mathbb{F}_p^\times$ acts on $L'_*(\tau_1)$ through its quotient $\mathbb{F}_p^\times/(\mathbb{F}_p^\times)_{(2)}$, and $H2^k \subset S\ell_2(\mathbb{F}_p)$ is the Sylow 2-subgroup, $k = \nu_2(p^2-1)$.

The stable part of $L_*^{!}(H2^k)$ is also easy to describe, explicitly

3.2 $$L_*^{!}(H2^k)^{st} = \{x \in L_*(H2^k) \mid i^*(x) \in L_*(H8)^{\mathbb{Z}/3} \text{ for each } i: H8 \subset H2^k\}.$$

(The argument is similar to [M, 1.10]).

Two of the groups $L_*(\tau_i)$ are metacyclic, and are calculated in [W5] (cf. §5 below). We can thus concentrate on the third, which is

3.3 $$1 \to \mathbb{Z}/n \to \tau \to H2^k \to 1, \quad H2^k = \tau_{(2)}$$

where $H2^k = \langle X, Y \mid X^{2^{k-2}} = Y^2, \ Y X Y^{-1} = X^{-1}\rangle$ acts on the generator $T \in \mathbb{Z}/n$ by $XTX^{-1} = T$, $YTY^{-1} = T^{-1}$, and $|\tau| = 2(p-1), 2(p+1)$, when $p \equiv 1,3 \pmod{4}$

The splitting $L_*^{!}(\tau) = L_*(\tau_{(2)}) \oplus L_*^{!}(\tau)_{od}$ used above can be refined to a splitting

3.4 $$L_*^{!}(\tau) = L_*^{!}(\tau_{(2)}) \oplus \coprod_{d>1} L_*^{!}(\tau)(d)$$

where d runs through all divisors of n, $d > 1$. This is obvious when n is a product of distinct primes - the splitting is then induced from endomorphisms of τ - but otherwise highly non-trivial, cf. [W5 , §4.1]. The single components $L_*^{!}(\tau)(d)$ are calculated via an exact sequence, analogous to the Mayer-Vietoris sequence of algebraic K-theory. Before we give some details of this we briefly recall the algebraic setup for L-theory.

An antistructure (R,α,u) consists of a ring R, an anti-involution $\alpha: R \to R$ and a (central) unit $u \in R^\times$ such that $\alpha(u) = u^{-1}$. Usually u will be ±1 but in the next section we shall need the more general case.

To each α-invariant subgroup $X(R)$ of $K_1(R)$ Wall defines L-groups $L_i^{X(R)}(R,\alpha,u)$, $i = 0,1$, and 4-periodic groups by $L_{i+2}^{X(R)}(R,\alpha,u) = L_i^{X(R)}(R,\alpha,-u)$. (cf. [W2]). The groups for varying superscripts are related by Rothenberg type exact sequences

3.5 $$\cdots \to H^{i+1}(Y(R)/X(R)) \to L_i^{X(R)}(R,\alpha,u) \to L_i^{Y(R)}(R,\alpha,u) \to H^i(Y(R)/X(R)) \to \cdots$$

Here $H^i(\ )$ are the Tate cohomology groups of $\mathbb{Z}/2$ with coefficients in $Y(R)/X(R)$. The $\mathbb{Z}/2$-module structure on $Y(R)/X(R)$ is inherited from the one on $K_1(R)$ and corresponds to the transpose conjugate (w.r.t. α) on representing matrices.

The groups $L_*^{X(R)}(R,\alpha,u)$ are related to the surgery obstruction groups as follows

$$L^s_{2k}(\pi) = L^U_{2k}(\mathbb{Z}\pi,\alpha,1),\ L'_{2k}(\pi) = L^Y_{2k}(\mathbb{Z}\pi,\alpha,1), L^h_{2k}(\pi) = L^K_{2k}(\mathbb{Z}\pi,\alpha,1)$$

where $\alpha(\Sigma a_g g) = \Sigma a_g g^{-1}$, $U = \{\pm 1\} \oplus \pi/[\pi,\pi] \in K_1(\mathbb{Z}\pi)$, $Y = U \oplus SK_1(\mathbb{Z}\pi)$ and $K = K_1(\mathbb{Z}\pi)$, and similarly for L_{2k+1}, cf. [W5 , §5.4].

In the rest of the paper we write $L_*(R,\alpha,1)$ or just $L_*(R)$ for the groups $L_*^{X(R)}(R,\alpha,1)$ where $X(R) = SK_1(R) = \mathrm{Ker}\{K_1(R) \to K_1(R \otimes \mathbb{Q})\}$.

We begin with rational calculations, and first we shall decompose the antistructure $(\mathbb{Q}\tau,\alpha,1)$ into its simple components, where τ is the group from 3.3.

The standard decomposition $\mathbb{Q}[\mathbb{Z}/n] = \Pi\, \mathbb{Q}(\zeta_d)$, $d|n$, of $\mathbb{Q}[\mathbb{Z}/n]$ in cyclotomic fields induces a decomposition $\mathbb{Q}\tau = \Pi\, S(d)$. Here $S(d) = \mathbb{Q}(\zeta_d)^t[H2^k]$ is the twisted group ring with $X\zeta_d X^{-1} = \zeta_d$ and $Y\zeta_d Y^{-1} = \zeta_d^{-1}$. But $S(d)$ splits further. Let E_j be the real field

$$E_j = \mathbb{Q}(\zeta_{2^j d}) \cap \mathbb{R} = \mathbb{Q}(\zeta_{2^j d} + \zeta_{2^j d}^{-1})$$

where $\zeta_{2^j d}$ is a primitive $2^j d$'th root of 1.

<u>Lemma 3.6</u>. $S(d) = \mathbb{H}(E_{k-1}) \times \prod_{j=0}^{k-2} M_2(E_j)$ where $M_2(E)$ is the ring of 2×2 matrices over E and $\mathbb{H}(E)$ is the usual quaternion algebra over E, $\mathbb{H}(E) = E\cdot 1 \oplus E\cdot i \oplus E\cdot j \oplus E\cdot k$.

(The proof of 3.6 is a tedious computation which we leave for the reader to carry out).

The standard anti-involution α on $\mathbb{Q}\tau$ induces an anti-involution on each $S(d)$ which respects the further decomposition in 3.6 in the sense that each simple component is preserved by α. We want a decomposition of antistructures and must specify α on each component in

3.6. This can be quite difficult to carry out explicitly, but fortunatel it is not necessary either, as we shall now explain.

Call two antistructures (R,α,u) and (R,β,v) equivalent if $\beta(r) = c\alpha(r)c^{-1}$ and $v = c\alpha(c)^{-1}u$ for some unit $c \in R^{\times}$. Equivalent ant structures have isomorphic L-theory,

$$L_*(R,\alpha,u) \cong L_*(R,\beta,v).$$

The isomorphism is induced from scaling the quadratic form by c. To make the most out of this we suppose further that R is a central simpl algebra over the (number) field E. The Skolem-Noether theorem and Hilbert's Theorem 90 implies that there is at most one equivalence class of antistructures (R,α,u) with $\alpha|E \neq \mathrm{id}_E$, and we call such an equivaler ce class for <u>type U</u>. There are two (if any) equivalence classes of ant structures (R,α,u) with $\alpha|E = \mathrm{id}_E$. If $u = \pm 1$ then (R,α,u) has <u>type O</u> if $\dim_E R^{\alpha} = \frac{1}{2}(n^2+un)$ and (R,α,u) has <u>type Sp</u> if $\dim_E R^{\alpha} = \frac{1}{2}(n^2-un)$. Here $R^{\alpha} \subset R$ is the fix ring of α and $\dim_E R = n^2$.

<u>Addendum 3.7</u>. In the decomposition of $(S(d),\alpha,1)$ from 3.6 the first factor $(\mathbb{H}(E_{k-1}),\alpha,1)$ has type Sp, so α can be taken to be t standard conjugation, and the other factors have type O.

The L-groups are additive with respect to product decompositions of antistructures, and they satisfy Morita equivalence,

3.8 $$L_*(M_n(E),\alpha,\pm 1) \cong L_*(E,\alpha|E,u)$$

where $u = \pm 1$ is determined such that $(M_n(E),\alpha,\pm 1)$ and $(E,\alpha|E,u)$ have the same type. Thus the computation of $L_*(\mathbb{Q}\tau,\alpha,1)$ is reduced to calculating $L_*(E_j,1,1)$ and $L_*(\mathbb{H}(E_{k-1}),\alpha,1)$.

We next recall the exact sequence from [W5, §4.1]. Let $\mathbb{Q}_A$ be the adele ring of $\mathbb{Q}$, $\mathbb{Q}_A = \hat{\mathbb{Q}} \oplus \mathbb{R}$ and let (S,α,u) be an antistructure on the $\mathbb{Q}$-algebra S. We define $CL_i(S) = CL_i(S,\alpha,u)$ to be the cokernel in

$$0 \to L_i(S,\alpha,u) \to L_i(S_A,\alpha,u) \to CL_i(S) \to 0, \quad S_A = S \otimes_{\mathbb{Q}} \mathbb{Q}_A$$

For each divisor $d > 1$ of n there is an exact sequence

3.9 $$\to \prod_{p\nmid d} L_1(\hat{R}_p(d)) \oplus L_1(T(d)) \xrightarrow{\gamma_1} CL_1(S(d)) \to L'_o(\tau)(d) \to$$
$$\prod_{p\nmid d} L_o(\hat{R}_p(d)) \oplus L_o(T(d)) \xrightarrow{\gamma_0} CL_o(S(d)) \to$$

where $R(d) = \mathbb{Z}(\zeta_d)^t[H2^k] \subset S(d)$, $\hat{R}_p(d) = \hat{\mathbb{Z}}_p \otimes_{\mathbb{Z}} R(d)$ and $T(d) = \mathbb{R} \otimes_Q S(d)$.

Let E be a number field. We write $P(E)$ for the set of all primes of E and $P_\infty(E) \subset P(E)$ for the set of infinite primes. Let $A \subset E$ denote the ring of integers, and define for each finite set of primes $\Omega \supseteq P_\infty(E)$

$$E_A(\Omega) = \Pi\{\hat{E}_y \mid y \in \Omega\} \times \Pi\{\hat{A}_y \mid y \notin \Omega\}$$

Here $\hat{E}_y$, $\hat{A}_y$ denote the completion at $y \in P(E)$. We have $E_A = Q_A \otimes_Q E = \lim\limits_{\rightarrow} E_A(\Omega)$. The idele class group is $C(E) = E_A^\times / E^\times$ where $E^\times \subset E_A^\times$ is the diagonal embedding. The global square theorem implies that the subgroup ${}_2C(E)$ of elements of order 2 is ${}_2C(E) = \prod\{<\pm 1> \mid y \in P(E)\}/<\pm 1>$.

From [W4, 5,5] and 3.7 we get

3.10 $$CL_1(S(d)) \cong \prod_{j=0}^{k-2} C(E_j)/2C(E_j)$$
$$CL_o(S(d)) \cong {}_2C(E_{k-1}) \times \prod_{j=0}^{k-2} \mathbb{Z}/2$$

Recall that the signature invariants define isomorphisms $\sigma\colon L_o(\mathbb{H},\alpha,1) \cong 2\mathbb{Z}$ and $\sigma\colon L_o(\mathbb{R},1,1) \cong 4\mathbb{Z}$. The $\mathbb{R}$-algebra $T(d)$ decomposes in copies of $\mathbb{H} = \mathbb{H}(\mathbb{R})$ and $M_2(\mathbb{R})$ with one copy of $\mathbb{H}$ for each $y \in P_\infty(E_{k-1})$ and one copy of $M_2(\mathbb{R})$ for each $y \in P_\infty(E_j)$, $j = 0,\dots,k-2$. Hence we have

3.11 $$L_o(T(d),\alpha,1) = \prod\{2\mathbb{Z} \mid y \in P_\infty(E_{k-1})\} \times \prod_{j=0}^{k-2} \prod\{4\mathbb{Z} \mid y \in P_\infty(E_j)\}$$

At various point later we shall use the maps in the Rothenberg sequence relating L_*^K and L_*, and it is convenient to name them:

3.12
$$\delta_i : L_i^K(E,1,1) \to H^i(E^\times)$$
$$\pi_i : H^{i+1}(E^\times) \to L_i(E,1,-1)$$

Here δ_o is the discriminant; it is surjective except if char(E) = 2. The homomorphism π_o is always surjective since every skew symmetric form has a symplectic basis, and consequently $L_o^K(E,1,-1) = 0$. In fact π_o is an isomorphism except if char(E) = 2. This follows from 3.5 and the following general result from [W 2],

Theorem 3.13. Let (D,α,u) be an antistructive on the division algebra D. Then $L_1^K(D,\alpha,u) = 0$ except if $(D,\alpha,u) = (E,1,1)$ where $L_1^K(E,1,1) \cong \mathbb{Z}/2$. The isomorphism is given by δ_1 except if char(E) = 2 where $\delta_1 = 0$.

Note as a special case of 3.13 that $L_1(T(d),\alpha,1) = \prod_{j=0}^{k-2} \Pi\{\mathbb{R}^\times/\mathbb{R}^{\times 2} | y \in P_\infty(E_j)\}$.

If p is an odd prime then all idempotent elements used in 3.6 are also present in $\hat{R}_p(d)$, so we have an analogous splitting

$$\hat{R}_p(d) = \mathbb{H}(A_{k-1}\otimes\hat{\mathbb{Z}}_p) \times \prod_{j=0}^{k-2} M_2(A_j\otimes\hat{\mathbb{Z}}_p)$$
$$= \prod_{y|p} \mathbb{H}(\hat{A}_{k-1,y}) \times \prod_{j=0}^{k-2} \prod_{y|p} M_2(\hat{A}_{j,y})$$

Moreover, $\mathbb{H}(\hat{A}_{k-1,y}) = M_2(\hat{A}_{k-1,y})$ since y is odd. Again, α preserves the splitting and we claim the equivalences of antistructures

$$(\mathbb{H}(\hat{A}_{k-1,y}),\alpha,1) \sim (M_2(\hat{A}_{k-1,y}),{}^t(\ ),-1)$$
$$(M_2(\hat{A}_{j,y}),\alpha,1) \sim (M_2(\hat{A}_{j,y}),{}^t(\ ),+1)$$

Indeed, this follows by reducing to the residue field and there using th concept of types: the rational type and the residue type agree. From 3.8 w

3.14 $$L_*(\hat{R}_p(d),\alpha,1) = \prod_{y|p} L_*(\hat{A}_{k-1,y},1,-1) \times \prod_{j=0}^{k-2} \prod_{y|p} L_*(\hat{A}_{j,y},1,1).$$

To obtain further information we invoke the following fundamental reduction theorem from [W3]:

Theorem 3.15. Let $J \subset R$ be a 2-sided ideal and suppose $R = \varprojlim R/J^n$. Then the natural projection induces isomorphism $L_*^K(R,\alpha,u) \cong L_*^K(\bar{R},\bar{\alpha},\bar{u})$, $\bar{R} = R/J$.

We can apply 3.15 to $R = \hat{A}_{j,y}$ and take J to be the maximal ideal. Then $\bar{R}$ is a finite field $F_{j,p}$ of odd characteristic and $A_{j,y}^\times \to F_{j,p}^\times$ induces isomorphism on the Tate cohomology groups. Thus we may combine 3.5 and 3.15 to get $L_*(\hat{A}_{j,y}) = L_*(F_{j,p})$, and the remarks following 3.12 imply

$$\pi_0 : {}_2A_{k-1,y}^\times \xrightarrow{\cong} L_0(\hat{A}_{k-1,y},1,-1)$$

3.16 $$\pi_3 : \hat{A}_{j,y}^\times / \hat{A}_{j,y}^{\times 2} \xrightarrow{\cong} L_1(\hat{A}_{j,y},1,1)$$

$$L_0(\hat{A}_{j,y},1,1) = 0, \quad L_1(\hat{A}_{k-1,y},1,-1) = 0$$

Together with 3.14 this evaluates the groups $L_i(\hat{R}_p(d))$, $i=0,1$ for p an odd prime. The groups $L_i(\hat{R}_2(d))$ will be deferred to §4 below.

We now determine the restrictions $\bar{\gamma}_i$ of γ_i in 3.9,

$$\bar{\gamma}_0 : \prod_{p \nmid 2d} L_0(\hat{R}_p(d)) \oplus L_0(T(d)) \to CL_0(S(d))$$

$$\bar{\gamma}_1 : \prod_{p \nmid 2d} L_1(\hat{R}_p(d)) \oplus L_1(T(d)) \to CL_1(S(d))$$

Clearly, each map breaks up in components according to the splittings in 3.6, 3.11 and 3.14, $\bar{\gamma}_0 = \prod_{j=0}^{k-1} \gamma_0(E_j)$, $\bar{\gamma}_1 = \prod_{j=0}^{k-1} \gamma_1(E_j)$

We have from 3.10, 3.11 and 3.16,

$$\gamma_0(E_{k-1}) : \prod\{2\mathbb{Z} \mid y \in P_\infty(E_{k-1})\} \times \prod_{y \nmid 2d} {}_2\hat{A}_{k-1,y}^\times \to {}_2C(E_{k-1})$$

This is the natural inclusion on the second factor and it sends an element $\{2x_y\}$ of the first factor to $\{(-1)^x y\}, y \in P_\infty(E_{k-1})$, in ${}_2C(E_{k-1})$. Similarly, for $j < k-1$

$$\gamma_0(E_j) : \prod\{4\mathbb{Z} \mid y \in P_\infty(E_j)\} \to \mathbb{Z}/2$$

sends $\{4x_y\}$ to $\prod(-1)^x y$. It follows that $\operatorname{Ker} \bar{\gamma}_0 = \Sigma(d)$, where $\Sigma(d)$ is a direct sum of $4\mathbb{Z}$ and $8\mathbb{Z}$, with one copy of $8\mathbb{Z}$ for each type

O component of $S(d)$.

For coker $\bar{\gamma}_1$ it is only the type O summands which contribute (cf. 3.10) and

$$\gamma_1(E_j): \prod_{y \nmid 2d} \hat{A}^{\times}_{j,y}/\hat{A}^{\times 2}_{j,y} \times \prod_{y=\infty} \mathbb{R}^{\times}/\mathbb{R}^{\times 2} \to C(E_j)/2C(E_j),\ j=0,\dots,k-$$

is again the natural map (3.13, 3.16). We can express the kernel and cokernel of $\gamma_1(E_j)$ in terms of the class group $\Gamma(E_j) = \tilde{K}_o(A_j)$. This is clear from the idelic definition of $\Gamma(E)$ as the cokernel of $E^{\times}_A(P_\infty) \to C(E)$. Indeed, a simple diagram chasing ('snake lemma') gives the exact sequence (cf. [W5, p.47])

$$0 \to \operatorname{Ker}\gamma_1(E_j) \to E_j^{(2)}/E_j^{\times 2} \to \prod_{y|2d} \hat{A}^{\times}_{j,y}/\hat{A}^{\times 2}_{j,y} \to \operatorname{cok}\gamma_1(E_j) \to \Gamma(E_j)/2\Gamma(E_j) \to 0$$

Here $E^{(2)} \subset E^{\times}$ consists of elements with even valuation at all (finite) primes. There is an exact sequence

3.17 $$1 \to A^{\times}/A^{\times 2} \to E^{(2)}/E^{\times 2} \to {}_2\Gamma(E) \to 1,$$

so if $\Gamma(E)$ has odd order then $A^{\times}/A^{\times 2} = E^{(2)}/E^{\times 2}$.

§4. The 2-adic calculation.

In this paragraph we complete the calculation of the exact sequence 3.9 by evaluating $L_*(\hat{R}_2(d))$ and the remaining maps.

Conjugation with Y gives a Galois automorphism σ on $B = \mathbb{Z}(\zeta_d)[\mathbb{Z}/2^{k-1}]$, $\sigma(\zeta_d) = \zeta_d^{-1}$, $\sigma(X) = X^{-1}$, in the sense of Galois extensions of commutative rings [AG, Appendix]. Thus $R(d) = \mathbb{Z}(\zeta_d)[\mathbb{Z}/2^{k-1}]^t[Y]$ can be viewed as a cyclic algebra $(B/B^{\sigma}, \sigma, X^{2^{k-2}})$ over the fixed ring B^{σ} of B. Note that B^{σ} is a free module over $\mathbb{Z}[\zeta_d+\zeta_d^{-1}]$ with basis

4.1 $$1,\ X^{2^{k-2}},\ X^i+X^{-i},\ \zeta_d X^i+\zeta_d^{-1}X^{-i},\ 1 \le i \le 2^{k-2}-1$$

We are interested primarily in the 2-adic completion $C_{k-1}(d) = \hat{\mathbb{Z}}_2 \otimes B^{\sigma}$. However, many computations will be made in the integral basis.

As a cyclic algebra, $\hat{R}_2(d)$ is an Azumaya algebra over $C_{k-1}(d)$,

[AG,A.12]. The residue ring $C_{k-1}(d)/\mathrm{Rad}$ is a direct sum of finite fields and the reduction of $\hat{R}_2(d)$ over it is a matrix ring by Wedderburn's theorem. The matrix idempotents can be lifted back to $\hat{R}_2(d)$ to show that $\hat{R}_2(d) \cong M_2(C_{k-1}(d))$. It follows that the antistructure $(\hat{R}_2(d),\alpha,1)$ is Morita equivalent to $(C_{k-1}(d),1,u)$ for some unit u, which can be determined from the previous rational calculations. In fact

Lemma 4.2. The antistructure $(\hat{R}_2(d),\alpha,1)$ is Morita equivalent to $(C_{k-1}(d),1,X^{2^{k-2}})$ where $C_{k-1}(d) = \hat{\mathbb{Z}}_2 \otimes \mathbb{Z}(\zeta_d)[\mathbb{Z}/2^{k-1}]^\sigma$ and σ is conjugation by Y.

The elements $\varepsilon_1 = 1-X^{2^{k-2}}$ and $\varepsilon_2 = 1+X^{2^{k-2}}$ satisfy $\varepsilon_i^2 = 2\varepsilon_i$, $\varepsilon_1\varepsilon_2 = 0$ and $\varepsilon_1 + \varepsilon_2 = 2$, and $C_{k-2}(d) \cong C_{k-1}(d)/(\varepsilon_1)$. Moreover, $\varepsilon_1 C_{k-1}(d) \cong 2C_{k-1}(d)/(\varepsilon_2)$ and $C_{k-1}(d)/(\varepsilon_2)$ is by 4.1 isomorphic to $A_{k-1} = \hat{\mathbb{Z}}_2 \otimes$ (integers in $\mathbb{Q}(\zeta_{2^{k-1}d})_{re}$), $\mathbb{Q}(\zeta_{2^{k-1}d})_{re} = \mathbb{Q}(\zeta_{2^{k-1}d}) \cap \mathbb{R}$. Summarising, we have an exact sequence

$$0 \to 2A_{k-1} \to C_{k-1}(d) \to C_{k-2}(d) \to 0 \tag{4.3}$$

Since $\varepsilon_1^2 \in 2C_{k-1}(d)$ the ideal (ε_1) is contained in the radical of $C_{k-1}(d)$ and by 3.15 $L_*^K(C_{k-1}(d)) \cong L_*^K(C_{k-2}(d))$. We can iterate to obtain a sequence of epimorphisms (inducing isomorphism on L_*^K)

$$C_{k-1}(d) \to C_{k-2}(d) \to \ldots \to C_o(d) \tag{4.4}$$

Here, $C_o(d) = \hat{\mathbb{Z}}_2 \otimes \mathbb{Z}(\zeta_d)_{re} = \prod_{i=1}^{g} \hat{\mathbb{Z}}_2(\zeta_d)_{re}$, where g is the number of dyadic primes in $\mathbb{Z}(\zeta_d)_{re} = \mathbb{Z}(\zeta_d+\zeta_d^{-1})$, that is, $g = |(\mathbb{Z}/d)^\times/\langle -1\rangle:\langle 2\rangle|$. The radical of $\mathbb{Z}_2(\zeta_d)_{re}$ is $2\mathbb{Z}_2(\zeta_d)_{re}$ and the residue field is $\mathbb{F}_2(\zeta_d)_{re} = \mathbb{F}_{2^f}$, $f = |\langle 2\rangle|$. Again from 3.15,

$$L_i^K(C_{k-1}(d),1,X^{2^{k-2}}) \cong g\cdot L_i^K(\mathbb{F}_{2^f},1,1) \cong g\cdot(\mathbb{Z}/2). \tag{4.5}$$

We shall use 3.5 and 4.5 to calculate $L_*(C_{k-1}(d),1,X^{2^{k-2}})$. Now, $K_1(C_{k-1}(d)) = C_{k-1}(d)^\times$ and we need information about $H^*(C_{k-1}(d)^\times)$. This will follow from 4.3 and

Lemma 4.6. Let A be the ring of integers in a 2-adic field with prime element π, residue field F and ramification index e (over $\hat{Q}_2$). Then

$$H^1((1+2A)^\times) \cong \langle\pm 1\rangle, \quad H^0((1+2A)^\times) \cong \mathbb{Z}/2 \oplus e\cdot F$$

Here $\mathbb{Z}/2$ is generated by the non-zero element β in the cokernel of $p: F \to F$, $p(x) = x+x^2$. Moreover, if $\alpha_1,\ldots,\alpha_f$ is a vector space basis for F over $\mathbb{F}_2$, then $1+2\pi^i\alpha_j$ $(0 \le i < e,\ 1 \le j \le f)$ and $1+4\beta$ form a basis for H^0.

Proof. We calculate $H^i((1+2A)^\times)$ from the exact sequence

$$1 \to (1+2\pi A)^\times \to (1+2A)^\times \to A/\pi A \to 0$$

The first group is isomorphic under the 2-adic logarithm to $2\pi A^+ \cong A^+$ so $H^1(1+2\pi A^\times) = 0$ and $H^0(1+2\pi A^\times) = A/2A$. We get the exact sequence

$$0 \to H^1(1+2A^\times) \to H^1(F) \xrightarrow{\delta} A/2A \to H^0(1+2A^\times) \to H^0(F) \to 0$$

Now $\delta(f) = (1+2f)^2 = 1+4(f+f^2)$ corresponds to $\frac{2}{\pi}(f+f^2) \in A$ under the 2-adic logarithm, and

$$A/2A = A/(\pi^e) \cong F \oplus F\pi \oplus \ldots \oplus F\pi^{e-1}.$$

Under this identification δ has the same kernel and cokernel as $p: F \to F$ mapping onto the last summand. This completes the proof.

Proposition 4.7. Let $g = |(\mathbb{Z}/d)^\times/\langle -1\rangle : \langle 2\rangle|$ and $f = |\langle 2\rangle|$ denote the decomposition number and residue number of 2 in $Q(\zeta_d)_{re}$. Let $\hat{A}_i$ be the integers in $\hat{Q}_2(\zeta_{2^i d})_{re}$ and set $\hat{C}_\ell(d) = (\hat{\mathbb{Z}}_2(\zeta_d)[\mathbb{Z}/2^\ell])^\sigma$. Then $C_\ell(d)$ splits in g equal copies of $\hat{C}_\ell(d)$. The cohomology of each one is

(i) $H^1(\hat{C}_o(d)^\times) = \langle -1\rangle$, $H^1(\hat{C}_\ell(d)^\times) = \langle -1\rangle \oplus \langle X^{2^{\ell-1}}\rangle$, $\ell \ge 1$

(ii) $H^0(\hat{C}_\ell(d)^\times) = \bigoplus_{i=2}^{\ell} H^0((1+2\hat{A}_i)^\times) / \langle 1+2\gamma\rangle \oplus H^0((1+2\hat{A}_1)^\times) \oplus H^0((1+2\hat{A}_o)^\times)$

where $\gamma = 2(\zeta_d+\zeta_d^{-1})^{-2}-1$ is non-trivial in $H^0((1+2\hat{A}_i)^\times)$, $i \ge 2$.

Proof. The splitting follows from that of $\hat{\mathbb{Z}}_2 \otimes \mathbb{Z}(\zeta_d)_{re}$; it makes 4.3 to a sum of g copies of the corresponding local sequence $\hat{4.3}$.

For the unramified 2-ring $\hat{C}_o(d) = \hat{\mathbb{Z}}_2(\zeta_d)_{re}$ the result follows from 4.6 and the long exact cohomology sequence associated with $1 \to (1+2A_o)^\times \to C_o(d)^\times \to F^\times \to 1$ $(H^*(F^\times) = 0!)$.

We suppose inductively that $H^1(\hat{C}_{\ell-1}(d)^\times) = \langle -1 \rangle \oplus \langle X^{2^{\ell-2}} \rangle$, $\ell \geq 2$. The sequence $\hat{4.3}$ gives the exact sequence

$$1 \to (1+2\hat{A}_\ell)^\times \to \hat{C}_\ell(d)^\times \to \hat{C}_{\ell-1}(d)^\times \to 1$$

and $\langle -1 \rangle = H^1((1+2A_\ell)^\times)$ corresponds to $\langle X^{2^{\ell-1}} \rangle$. Thus we get

$$0 \to \langle X^{2^{\ell-1}} \rangle \to H^1(\hat{C}_\ell(d)^\times) \to H^1(\hat{C}_{\ell-1}(d)) \overset{\delta}{\to} H^o((1+2\hat{A}_\ell)^\times)$$

$$\to H^o(\hat{C}_\ell(d)^\times) \to H^o(\hat{C}_{\ell-1}(d)^\times) \to 0,$$

and we must evaluate δ. Clearly, $\delta(-1) = 1$ and a little calculation gives

$$\delta(X^{2^{\ell-2}}) = \left(\frac{\zeta_d X^{2^{\ell-2}} + \zeta_d^{-1} X^{2^{\ell-2}}}{\zeta_d + \zeta_d^{-1}}\right)^2 = 1 + \gamma(1 - X^{2^{\ell-1}})$$

where $\gamma = 2(\zeta_d + \zeta_d^{-1})^{-2} - 1$. Thus $\delta(X^{2^{\ell-2}}) = 1 + 2\gamma \in H^o((1+2\hat{A}_\ell)^\times)$ and the class is non-trivial since it projects non-trivially under $1+2\hat{A}_\ell \to \hat{A}_\ell/\pi\hat{A}_\ell = F$. The calculation of $H^o(\hat{C}_\ell(d)^\times)$ is quite similar.

Corollary 4.8. With the notation of 4.7,

(i) $L_o(\hat{R}_2(d), \alpha, 1) = g \cdot (\mathbb{Z}/2)$

(ii) $L_1(\hat{R}_2(d), \alpha, 1) = H^o(C_{k-1}(d)^\times)/g \cdot (\mathbb{Z}/2)$

Proof. We have $L_i(\hat{R}_2(d), \alpha, 1) = L_i(C_{k-1}(d), 1, X^{2^{k-2}})$ by 4.2 and we can use 4.5, 4.7 and the Rothenberg sequence 3.5 to obtain the result. This requires that we determine the maps

$$\delta_o : L_o^K(C_{k-1}(d), 1, X^{2^{k-2}}) = g \cdot (\mathbb{Z}/2) \to H^o(C_{k-1}(d)^\times)$$

$$\delta_1 : L_1^K(C_{k-1}(d), 1, X^{2^{k-2}}) = g \cdot (\mathbb{Z}/2) \to H^1(C_{k-1}(d)^\times)$$

$$\delta_2\colon L^K_2(C_{k-1}(d),1,X^{2^{k-2}}) = g\cdot(\mathbb{Z}/2)\to H^o(C_{k-1}(d)^\times).$$

All three maps are injective. This is easy for, δ_o cf.[W3 ,Theorem 11] we give the argument for δ_1 and δ_2. Consider a single summand in the splitting $C_{k-1}(d) = g\,\hat{C}_{k-1}(d)$. The non-trivial element $\tau\in L^K_1(\hat{C}_{k-1}(d),1$ $X^{2^{k-2}}) = \mathbb{Z}/2$ is represented by the automorphism $\tau(e) = f$, $\tau(f) = X^{2^{k-2}}e$ of the hyperbolic plane $\{e,f\}$. It has determinant $-X^{2^{k-2}}\in H^1(\hat{C}_{k-1}(d)^\times)$, which is non-xero by 4.7(i). Then δ_1 is injective. The non-trivial class in $L^K_2(\hat{C}_{k-1}(d),1,X^{2^{k-1}})$ is represented by the quadratic plane $\{e,f\}$ with non-zero Arf invariant: The matrix of its quadratic form is $Q = \begin{pmatrix}1 & 1\\ 0 & b\end{pmatrix}$, where $b\in\hat{\mathbb{Z}}_2(\zeta_d)_{re}$ maps to $\beta\in F$ (cf. 4.6). The associated bilinear form has matrix $B = Q - X^{2^{k-1}}Q^t$ and discriminant $1 - 2b(1-X^{2^{k-1}})\in H^o(\hat{C}_{k-1}(d)^\times)$. This element maps to $1 - 4b$ $\in H^o((1+2\hat{A}_{k-1})^\times)$ in the splitting 4.7(ii) and $1 - 4b \neq 0$ in this group. This completes the proof.

We can now complete the computation of the exact sequence 3.9. Indeed, arguments similar to the paragraphs following 3.16 together with 4.7, 4.8 and the calculation of $L'_o(H2^k)$ from [W5] yield

<u>Theorem 4.9</u>. For the group in 3.3 we have

$$L'_o(\tau) = \Sigma \oplus \Pi\{\mathrm{cok}\,\gamma^d_1 \mid d|n,\ d>1\}$$

where Σ is a free group of rank $2^{k-2}n+3$. The torsion part $\Pi\,\mathrm{cok}\,\gamma^d_1$ is an elementary 2-group. Each factor $\mathrm{cok}\,\gamma^d_1$ decomposes further as

$$\mathrm{cok}\,\gamma^d_1 = \Pi\,\mathrm{cok}\,\gamma^d_1(E_j(d)),\ 0\le j\le k-2$$

where the components fit into exact sequences

(i) $E^{(2)}_j/E^{\times 2}_j \to \prod_{y|d} H^o(\hat{A}^\times_{j,y}) \to \mathrm{cok}\,\gamma^d_1(E_j) \to \Gamma(E_j)\otimes\mathbb{Z}/2\to 0$, $j=0,1$

(ii) $E^{(2)}_j/E_j{}^2 \to \prod_{y|2}\hat{C}_{j,y}\times\prod_{y|d} H^o(\hat{A}^\times_{j,y}) \to \mathrm{cok}\,\gamma^d_1(E_j)\to\Gamma(E_j)\otimes\mathbb{Z}/2\to 0$, $j\geq$

Here $E_j = E_j(d)$ is the field $Q(\zeta_{2^jd})_{re}$ with integers A_j, $H^o(\hat{A}^\times_{j,y}) =$

$\hat{A}^{\times}_{j,y}/\hat{A}^{\times 2}_{j,y}$ and $\hat{C}_{j,y} = \operatorname{cok}\{H^{o}((1+2\hat{A}_{j,y})^{\times}) \to H^{o}(\hat{A}^{\times}_{j,y})\}$.

Since $L'_o(\pi)$ is torsion free for cyclic groups and quaternion 2-groups, 3.1 implies a splitting

4.10 $\quad \operatorname{Tor} L'_o(S\ell_2(\mathbb{F}_p)) = \operatorname{Tor} L'_o(\tau_1)^{\mathbb{F}_p^{\times}} \oplus \operatorname{Tor} L'_o(\tau_2) \oplus \operatorname{Tor} L'_o(\tau_3)$.

The groups $L'_o(\tau_1)$, $L'_o(\tau_2)$ were calculated in [W5, ch.4] Torsion in L'_o has exponent 2. More precisely, $L'_o(\tau_1)$ is torsion free for $p \equiv 3(4)$, and for $p \equiv 1(4)$ $\operatorname{Tor} L'_o(\tau_1)$ is determined from

4.11 $\quad E^{(2)}/E^{\times 2} \to A_p^{\times}/A_p^{\times 2} \to \operatorname{Tor} L'_o(\tau_1) \to \Gamma(E) \otimes \mathbb{Z}/2 \to 0$

where $E = Q(\zeta_p)^{\mathbb{Z}/2^{\ell-1}}$ and $\mathbb{Z}/2^{\ell}$ is the Sylow 2-subgroup of $\mathbb{F}_p^{\times}$. For τ_2 we have $L'_o(\tau_2) = \Pi\{L'_o(\tau_2)(d) \mid d>1,\ d|m\}$, and there are exact sequences

4.12 $\quad E^{(2)}/E^{\times 2} \to \prod_{y|d} \hat{A}^{\times}_y/\hat{A}^{\times 2}_y \to \operatorname{Tor} L'_o(\tau_2)(d) \to \Gamma(E) \otimes \mathbb{Z}/2 \to 0$

where $E = E(d)$ is the field $Q(\zeta_d)_{re}$. The details of these calculations are quite similar, only easier than for τ_3.

Quite generally, the multisignature $\operatorname{sign}\colon L'_o(S\ell_2(\mathbb{F}_p)) \to R_{\mathbb{R}}(S\ell_2(\mathbb{F}_p))$ maps the torsion free part monomorphically. Its cokernel is a 2-group of exponent 8. Let $\Sigma = \Sigma(S\ell_2(\mathbb{F}_p))$ be the image of sign. The calculations in §3 and §4 show that $\Sigma(\tau_3)$ is a sum of copies of $4\mathbb{Z}$ and $8\mathbb{Z}$ with one copy of $8\mathbb{Z}$ for each type O factor of $Q\tau_3$, see in particular the final part of §3. Similar results hold for $L'_o(\tau_1)$, $L'_o(\tau_2)$, and induction calculates $L'_o(S\ell_2(\mathbb{F}_p))$. The representation theory of $S\ell_2(\mathbb{F}_p)$ was done in [Sch]. Let $r_{\mathbb{R}}$ (resp. r_Q) be the rank of $R_{\mathbb{R}}(S\ell_2(\mathbb{F}_p))$ (resp. $R_Q(S\ell_2(\mathbb{F}_p))$). Then

$$r_{\mathbb{R}} = p+4 \text{ if } p \equiv 1(4),\quad r_{\mathbb{R}} = p+2 \text{ if } p \equiv 3(4)$$

$$r_Q = \#\{\text{divisors of } p-1\} + \#\{\text{divisors of } p+1\}$$

and the number of type O summands r^{O}_Q is given by

$r_Q^o =$ even divisors if $p \equiv 1(4)$, $r_Q^o + 1 =$ even divisors if $p \equiv 3(4)$

Thus we have

Theorem 4.13. The signature group Σ is a sum of r_Q^o copies of $8\mathbb{Z}$ and $r_{\mathbb{R}} - r_Q^o$ copies of $4\mathbb{Z}$.

§5. Number Theory.

This section is devoted to an explicit determination of the torsion in $L_o'(S\ell_2(\mathbb{F}_p))$. It splits in 3 pieces according to 4.10, and for each piece $\mathrm{Tor}\, L_o'(\tau_i)$ there are two separate number theoretic sources for torsion, namely the class group part and the cokernel of the map from $E^{(2)}/E^{\times 2}$ to local units. They are, however, intimately related (cf.3.17). The class numbers $h(E) = |\Gamma(E)|$ are hard to calculate, and we shall restrict ourselves to cases where we can bypass this problem.

Proposition 5.1. If the fields involved have odd class number, then torsion in $L_o(\tau)$ is the sum of cokernels of

$$(\tau_1) \quad A^\times/A^{\times 2} \to \hat{A}_p^\times/\hat{A}_p^{\times 2}, \qquad A = \text{integers in } Q(\zeta_p)^{\mathbb{Z}/2^{\ell-1}}, \; p \equiv 1(4)$$

$$(\tau_2) \quad A^\times/A^{\times 2} \to \prod_{y|d} \hat{A}_y^\times/\hat{A}_y^{\times 2}, \qquad A = \text{integers in } Q(\zeta_d)_{re}, \; 1<d|m$$

$$(\tau_3) \quad A^\times/A^{\times 2} \to \prod_{y|2} \hat{C}_y \times \prod_{y|d} \hat{A}_y^\times/\hat{A}_y^{\times 2}, \; A = \text{integers in } Q(\zeta_{2^i d})_{re}, \; 1<d|n, \; 0 \le i \le k-2$$

This holds for τ_1 when $p < 197$, for τ_2 when $p < 131$ and for τ_3 when $p < 113$.

Proof. The first claim follows since $\Gamma(E) \otimes \mathbb{Z}/2$ vanishes in 4.9 4.11 and 4.12 and because $E^{(2)}/E^{\times 2} = A^\times/A^{\times 2}$, by 3.17.

As to τ_1, consider the extension $Q(\zeta_p)^{\mathbb{Z}/2^{\ell-1}} \subset Q(\zeta_p)_{re}$. It is cyclic of degree $2^{\ell-2}$ and there is only one ramified prime, the p-adic one, which ramifies totally. It is a simple consequence of the existence of Hilbert's class field that under these conditions $h(Q(\zeta_p)^{\mathbb{Z}/2^{\ell-1}})$ is odd if and only if $h(Q(\zeta_p)_{re})$ is odd, where h is the class number ([I

For imaginary cyclic number fields E, the relative class number of E/E_{re} can be odd only when $h(E_{re})$ is odd [Ha, Satz 45]. The explicit computations of Kummer now show that $h(Q(\zeta_p)_{re})$ is odd for $p = 4n+1$, $p < 197$ (cf. [K], [Mn, p.413]).

For odd prime powers $d = p^n$ we can use the same result to conclude that $h(Q(\zeta_d)_{re})$ is odd by appealing to the relative class number tables of [Ha]. For composite d we have to refer to recent computer runs for real cyclic fields announced in [B]. The bound for τ_2 comes from the first non cyclic case. For τ_3 the first three non-cyclic fields $Q(\zeta_{24})_{re}$, $Q(\zeta_{40})_{re}$ and $Q(\zeta_{48})_{re}$ ($p = 47, 79$ and 97) have class number one [MMo]; $Q(\zeta_{56})_{re}$ ($p = 113$) is the first unsettled case. This proves Proposition 5.1.

There is a relation between the units and the class number for (real) cyclic number fields E which we now recall. Choose f smallest possible with $E \subset Q(\zeta_f)$ and write $E = Q(\zeta_f)^H$, $H \leq (\mathbb{Z}/f)^\times$. Let s be any generator of $(\mathbb{Z}/f)^\times/H$ and define

5.2
$$\eta_s = \prod_{a \in H/\langle -1 \rangle} \frac{\zeta_{2f}^{sa} - \zeta_{2f}^{-sa}}{\zeta_{2f}^{a} - \zeta_{2f}^{-a}}$$

This is a unit of E, called the fundamental cyclotomic unit. If we take one for each subfield $E' \subset E$, their conjugates generate a subgroup of index $h(E)$ in the group of units $A^\times$, [Ha, Satz 9]. Especially if $h(E)$ is odd, it is enough in 5.1 to check the images of the cyclotomic units. For f odd, we can take ζ_f instead of ζ_{2f} as this only affects the sign.

<u>Theorem 5.3</u>. For $p < 197$, the torsion in $L_0'(\tau_1)$ is 0 when $p \equiv 1,3,7\ (8)$ and $\mathbb{Z}/2$ when $p \equiv 5\ (8)$.

<u>Proof</u>. If $p \equiv 3\ (4)$, then $\tau_1 = \mathbb{Z}/2p$ has torsion free L_0'. Otherwise consider $E = Q(\zeta_p)^{\langle X \rangle}$, $\langle X \rangle = \mathbb{Z}/2^{\ell-1} \leq (\mathbb{Z}/p)^\times$. Here $A_p \to \mathbb{F}_p$ induces an isomorphism $\hat{A}_p^\times/\hat{A}_p^{\times 2} \to \mathbb{F}_p^\times/\mathbb{F}_p^{\times 2}$, so we must evaluate the image of η_s in $\mathbb{F}_p^\times$, where s is a generator of $(\mathbb{Z}/p)^\times$. As the conjugates have the same image, it suffices to compute

$$\frac{\zeta_p^{sa}-\zeta_p^{-sa}}{\zeta_p^{a}-\zeta_p^{-a}} = \zeta_p^{(s-1)a} + \zeta_p^{(s-3)a} + \ldots + \zeta_p^{(1-s)a} \equiv s .$$

modulo the prime ideal $(1-\zeta_p)$ in $Q(\zeta_p)$. Thus the reduction of η_s in $\mathbb{F}_p$ is

$$\bar{\eta}_s = \bar{s}^n, \quad n = |\langle X\rangle : \langle -1\rangle| = 2^{\ell-2}$$

which is a square if $\ell \geq 3$, i.e. $8|p-1$, and a non-square otherwise.

The first example of torsion is $p = 17$, where $E = Q(\zeta_{17})^{\mathbb{Z}/8} = Q(\sqrt{17})$ has units $\pm\varepsilon^n$, $\varepsilon = 4+\sqrt{17}$. From $\varepsilon^2-8\varepsilon-1 = 0$ we see that $\bar{\varepsilon} = 4 \in \mathbb{F}_{17}^2$. Note that the same method shows that $A^\times/A^{\times 2} \to \mathbb{F}_p^\times/\mathbb{F}_p^{\times 2}$ is onto for $Q(\zeta_{p^n})_{re}$.

Example 5.4. We give the calculation of $A^\times/A^{\times 2} \to \prod_{y|d} \hat{A}_y^\times/\hat{A}_y^{\times 2}$ for $E = Q(\zeta_{15})_{re}$, of degree 4 over Q. The ramification indices and residue class degrees are $e_3 = 2$, $f_3 = 2$ and $e_5 = 4$ (note that $E \subset Q(\zeta_{15})$ is unramified). Also, the norm induces an isomorphism $\mathbb{F}_9^\times/\mathbb{F}_9^{\times 2} \to \mathbb{F}_3^\times/\mathbb{F}_3^{\times 2}$

In Q we have the unit $\eta_1 = -1$ with reduction $(1,1)$ in $\mathbb{F}_9^\times/\mathbb{F}_9^{\times 2} \oplus \mathbb{F}_5^\times/\mathbb{F}_5^{\times 2}$. In $Q(\sqrt{5})$ the fundamental unit $\varepsilon = \frac{1+\sqrt{5}}{2}$ satisfies $\varepsilon^2 - \varepsilon - 1 = 0$, so the reduction in $\mathbb{F}_5$ is $3 \notin \mathbb{F}_5^2$. The reduction in $\mathbb{F}_9$ is not in $\mathbb{F}_3$ but has norm $-1 \notin \mathbb{F}_3^2$. Hence $\eta_2 = \varepsilon$ goes to $(-1,-1)$ in $\mathbb{F}_9^\times/\mathbb{F}_9^{\times 2} \oplus \mathbb{F}_5^\times/\mathbb{F}_5^{\times 2}$.
Finally, the cyclotomic generating unit of $Q(\zeta_{15})_{re}$ is

$$\eta_3 = \frac{\zeta_{15}^2-\zeta_{15}^{-2}}{\zeta_{15}-\zeta_{15}^{-1}} = \zeta_{15} + \zeta_{15}^{-1}$$

If we choose $\zeta_{15} = \zeta_3\zeta_5$, then $\eta_3 \equiv \zeta_3 + \zeta_3^{-1} = -1 \mod 5$, $\eta_3 \equiv \zeta_5+\zeta_5^{-1} = \varepsilon-1 \mod 3$. As $\varepsilon-1$ satisfies $(\varepsilon-1)^2 + (\varepsilon-1) - 1 = 0$, it has norm $-1 \notin \mathbb{F}_3^2$. Thus

$$\eta_3 \mapsto (-1,1) \in \mathbb{F}_9^\times/\mathbb{F}_9^{\times 2} \oplus \mathbb{F}_5^\times/\mathbb{F}_5^{\times 2}.$$

and we conclude that $A^\times/A^{\times 2} \to \hat{A}_3^\times/\hat{A}_3^{\times 2} \oplus \hat{A}_5^\times/\hat{A}_5^{\times 2}$ is onto.

Similar computations can be carried further to show

Proposition 5.5. For $p < 100$, $L_0'(\tau_2)$ is torsion free.

The analysis of $L_0'(\tau_3)$ is a bit harder. The problems begin when the quaternionic Sylow subgroup has order at least 16, so that one has to take the (ramified) 2-adic units into account. We do not give further details but record

Summary 5.6. For the primes $p \leq 43$, $L_0'(S\ell_2(\mathbb{F}_p))$ is torsion free except in the cases

$$\text{Tor } L_0'(S\ell_2(\mathbb{F}_{17})) = \mathbb{Z}/2\,, \quad \text{Tor } L_0'(S\ell_2(\mathbb{F}_{41})) = (\mathbb{Z}/2)^3\,,$$

where τ_1 gives $\mathbb{Z}/2$ in both and τ_3 gives $\mathbb{Z}/2 \oplus \mathbb{Z}/2$ in the last one.

Remark 5.7. The torsion in the above examples is unit torsion, as we restricted ourselves to the cases where $\Gamma(E)$ has odd order. The class group torsion will show up, however. Indeed, Kummer knew that $h(\mathbb{Q}(\zeta_{937})_{re})$ is even [K, p.944], so for $p = 937$, $L_0'(\tau_1)$ has class group torsion. Actually T. Metsänkylä informs us that already $p = 277$ is an example of the same phenomenon. We don't know whether the bounds in 5.1 are the best possible, e.g. is $h(\mathbb{Q}(\zeta_{197})_{re})$ even?

REFERENCES

[A] D.A. Anderson, The Whitehead torsion of a fiber homotopy equivalenc Michigan Math.J.21 (1974), 171-180.

[AG] M.Auslander and O.Goldman, The Brauer group of a commutative ring, Trans.Amer.Math.Soc. 97 (1960), 367-409.

[B] H.Bauer, Numerische Bestimmung von Klassenzahlen reeller zyklischer Zahlkörper, J.Number Theory 1 (1969), 161-162.

[CE] H.Cartan and S.Eilenberg, Homological Algebra, Princeton University Press 1956.

[D] A.Dress, Induction and structure theorems for orthogonal representations of finite groups, Ann.of Math. 102 (1975), 291-325.

[Ha] H.Hasse, Über die Klassenzahl abelscher Zahlkörper, Akademie-Verlag Berlin 1952.

[H] B.Huppert, Endliche Gruppen I, Die Grundlehren der mathematischen Wissenschaften 134, Springer-Verlag, Berlin Heidelberg New York 1967.

[I] K.Iwasawa, A note on class numbers of algebraic number fields, Abh. Math.Sem.Univ.Hamburg 20 (1956), 257-258.

[K] E.E.Kummer, Über eine Eigenschaft der Einheiten der aus den Wurzeln der Gleichung $\alpha^{\lambda} = 1$ gebildeten complexen Zahlen,und über den Zweiten Factor der Klassenzahl, pp.919-944 in Collected Papers Vol.I, Springer-Verlag, Berlin Heidelberg New York 1975.

[M] I.Madsen, Smooth spherical space forms, pp.303-352 in Geometric applications of homotopy theory I, Proceedings Evanston 1977, Lecture Notes in Mathematics 657, Springer-Verlag, Berlin Heidelberg New York 1978.

[M1] I.Madsen, Spherical space forms, Proceedings of the International Congress of Mathematicians 1978, Helsinki.

[MM] I.Madsen and R.J.Milgram, The cobordism of topological and PL-manifolds, to appear as an Annals of Math.Study, Princeton University Press.

[MTW] I.Madsen, C.B. Thomas and C.T.C.Wall, The topological spherical space form problem II. Existence of free actions, Topology 15 (1976), 375-382.

[MMo] J.M.Masley and H.L.Montgomery, Cyclotomic fields with unique factorization, J.Reine Angew.Math. 286/287 (1976), 248-256.

[Mg] R.J.Milgram, Evaluating the Swan finiteness obstruction for periodic groups, these proceedings.

[Mn] J.Milnor, Whitehead torsion, Bull.Amer.Math.Soc. 72 (1966), 358-426.

[Sch] I.Schur, Untersuchungen über die Darstellung der endlichen Gruppen durch gebrochene lineare Substitutionen, J.Reine Angew. Math. 132 (1907), 85-137.

[S] E.Stein, Surgery on products with finite fundamental group, Topology 16 (1977), 473-493.

[W] C.T.C.Wall, Surgery on Compact Manifolds, Academic Press, London & New York 1970.

[W1] C.T.C.Wall, Formulae for surgery obstructions, Topology 15 (1976), 189-210 Corrigendum, Topology 16 (1977), 495-496.

[W2] C.T.C.Wall, Foundations of algebraic L-theory, pp.266-300 in Algebraic K-theory III, Lecture Notes in Mathematics 343, Springer-Verlag, Berlin Heidelberg New York 1973.

[W3] C.T.C.Wall, On the classification of Hermitian forms. III Complete semilocal rings, Invent.Math. 19 (1973), 59-71.

[W4] C.T.C.Wall, On the classification of Hermitian forms. IV Adele rings, Invent.Math.23 (1974), 241-260.

[W5] C.T.C.Wall, Classification of Hermitian forms. VI Group rings, Ann. of Math. 103 (1976), 1-80.

[W6] C.T.C.Wall, Free actions of finite groups on spheres, pp.115-125 in Proceedings of Symposia in pure Mathematics 32, AMS 1978.

E. Laitinen
Department of Mathematics
University of Helsinki
Hallituskatu 15
SF-00100 Helsinki 10
Finland

I. Madsen
Matematisk Institut
Aarhus Universitet
DK-8000 Aarhus C
Denmark

E. Laitinen was partially supported by the Emil Aaltonen Foundation.

C*-ALGEBRAS AND K-THEORY

A.S. Mishchenko

This paper is devoted to the review of results mainly obtained by Moscow mathematicians,in the domain of functional methods used in algebraic topology, to be more precise, in K-theory. These are generally results concerning the problem of effectively calculating the invariants of smooth manifolds which describe and distinguish the smooth structure of the manifolds. While for the simply connected manifolds the smooth structure classification problem was solved in a general way by W.Browder and S.Novikow in 1962-1964, for nonsimply connected manifolds the solution has proceeded in steps, and is far from completion even now.

Now the problem of the smooth structure classification for nonsimply connected manifolds is divided into some independent problems, each having its own methods of investigation and each connected with other problems.

We shall especially dwell on the application of the representation theory to investigate the homotopy invariants of the nonsimply connected manifolds, in particular, the signature formulae of Hirzebruch type.

§1. C*-algebras and vector bundles

Let Λ be any C*-algebra with unit. Making use of the algebra Λ as a scalar ring, consider, following M. Karoubi [1], the category $Vect_{\Lambda}(X)$ of the vector bundles over X with the finite generated projective Λ-module P as the fiber. Denote the Grothendick group of that category by $K_\Lambda(X)$, and the bigraded cohomology theory by $K^{p,q}(X,Y;\Lambda)$. It is convenient also to consider the relative K-theory for the ring homomorphism $\varphi: \Lambda \longrightarrow \Lambda'$, which we denote by $K^{p,q}(X,Y;\varphi)$. All these K-theories satisfy the Bott periodicity and have the Chern character

$$ch_\Lambda: K^{**}(X,Y;\Lambda) \to H^*(X,Y;K^{**}(\Lambda) \otimes Q)$$

which is multiplicative up to the tensor product of the algebras. In general, the ring homomorphism in the same way can be included into the bigraded category of the representations, whose Grothendick group is denoted in terms of $R^{p,q}(\Lambda,\Lambda')$. Then we have the following pairing:

$$R^{p,q}(\Lambda,\Lambda') \times K^{p',q'}(X,Y;\Lambda) \to K^{p+p',q+q'}(X,Y;\Lambda)$$

At last, if $\varphi: \Lambda' \to \Lambda''$ is a ring homomorphism, then one can define the relative Grothendick groups $R^{p,q}_{rel}(\Lambda,\varphi)$.

In particular, the Fredholm representations of the algebra Λ from [2] are the elements of $R^{o,o}_{rel}(\Lambda,\varphi)$, where $\varphi: B(H) \to B(H)/K(H)$ is a projection of the algebra $B(H)$ of the bounded operators of the Hilbert space H to factor algebra modulo the ideal $K(H)$ of the compact operators.

It is useful to consider the hermitian Λ-bundles with non-degenerate hermitian Λ-form in each fiber. If the Λ-module P is free, then the hermitian form λ is identified with an element of the algebra $End(n,\Lambda)$. The form $\lambda \in End(n,\Lambda)$ is called positively determined $(\lambda > 0)$ iff $Spec\ \lambda > \varepsilon > 0$, with the equivalent definition: $Spec\ \lambda(x,x) > 0$ for any $x \in P$. The last variant extends to the projective Λ-modules.

Theorem 1. (a) Any two nondegenerate positively determined forms on the free Λ-module are equivalent.

(b) Any finitely generated projective Λ-module is self-conjugated and determined by the self-conjugated projector.

(c) Any two nondegenerate positively determined forms on the finitely generated projective Λ-module are equivalent.

(d) Any nondegenerate form λ on the finitely generated projective Λ-module P decomposes into a direct sum

$$(P,\lambda) = (P_1,\lambda_1) \oplus (P_2,\lambda_2),$$

and $\lambda_1 > 0$, $\lambda_2 > 0$.

Let us denote the Grothendick group for the category of nondegenerate forms on the finitely generated projective Λ-module by $K_h^{p,q}(X,Y;\Lambda)$. Then by theorem 1

$$K_h^{p,q}(X,Y;\Lambda) = K^{p,q}(X,Y;\Lambda).$$

§2. The index of elliptic operators over C*-algebras

The numerous variants of elliptic theory of which we are aware are included in a general construction of elliptic operators in which some C*-algebra Λ is used as the scalar ring.

Let P be a finitely generated projective Λ-module with fixed nondegenerate positively determined hermitian form λ. Denote the value of λ by (x,y). Let $l_2(P)$ be the space of sequences $x = \{x_k\}_{k=1}^{\infty}$, $x_k \in P$, such that $\|\sum_{k=1}^{\infty} (x_k,x_k)\| < \infty$.

The space $l_2(P)$ is a Λ-module and a complete Banach space with respect to the norm

$$\|x\|^2 = \| \sum_{k=1}^{\infty} (x_k, x_k) \| .$$

Denote the subspace of the elements $x \in l_2(P)$ by $[l_2(P)]^n$ such that $x_1 = x_2 = \cdots = x_n = 0$.

<u>Definition</u>. The bounded Λ-operator $A: l_2(P) \to l_2(P)$ is called a compact operator iff $\lim_{n\to\infty} \|A|[l_2(P)]^n\| = 0$. The bounded Λ-operator A is called Fredholm operator iff there exist bounded Λ-operators B, B' such that $(B'A-1)$ and $(AB-1)$ are compact.

<u>Theorem 2</u>. (a) If $A: l_2(P) \to l_2(P)$ is Fredholm Λ-operator, then there exist decompositions of the image and preimage $l_2(P) = M_1 \oplus N_1$, $l_2(P) = M_2 \oplus N_2$ such that the matrix of A equals

$$A = \begin{pmatrix} A_1 & 0 \\ 0 & A_2 \end{pmatrix},$$

where A_1 is invertible, N_1, N_2 are finitely generated projective Λ-modules.

(b) The element $\text{index}\, A = [N_1]-[N_2] \in K^0(\Lambda)$ does not depend on the decompositions.

(c) The function index is locally constant on the space of the Fredholm Λ-operators.

(d) If K is a compact Λ-operator, then AK, KA are compact, $A+K$ is a Fredholm operator and $\text{index}\, A = \text{index}\, (A+K)$.

Let us consider the compact smooth manifold X, the cotangent bundle T^*X, the projection $\pi: T^*X \to X$, two locally trivial hermitian vector Λ-bundles E_1, E_2 over X. The symbol of pseudodifferential Λ-operator of degree n is a Λ-homomorphism

$$\sigma: \pi^*(E_1) \to \pi^*(E_2)$$

such that

$$\left\| \frac{\partial^{|\alpha+\beta|}\sigma}{\partial_x^{\alpha}\partial\xi^{\beta}} \right\| \le C_{\alpha,\beta}(1+|\xi|)^{n-|\beta|}, \quad \xi \in T_x^*X, \quad x \in X.$$

Using the standard construction for the symbol σ one can define a pseudodifferential operator

$$\sigma(\mathcal{D}): \Gamma(E_1) \to \Gamma(E_2)$$

which acts in the space of the sections of Γ-bundles E_1, E_2. Let us fix some Riemannian metrix on X and the Laplace operator Δ which acts in the space of the sections $\Gamma(E_i)$, $i = 1,2$. Let $\|u\|_s^2 = \|\int_X (u,(1+\Delta)^s u \, d\mu\|$ be the Sobolev norm in the spaces $\Gamma(E_i)$, where μ is a measure associated with metric. Denote the completion of $\Gamma(E_i)$ due to the Sobolev norm by $H^s(E_i)$. The pseudodifferential Λ-operator $\sigma(D)$ is called elliptic iff $\sigma(x,\xi)$ is an isomorphism when $\xi \to \infty$. Then $[\sigma]$ will mean the element of $K(T^*X;\Lambda)$.

Theorem 3. (a) The Λ-operator $\sigma(D)$ is bounded in the following norms:

$$\sigma(D): H^s(E_1) \to H^{(s-n)}(E_2).$$

(b) The spaces $H^s(E_i)$ are Λ-isomorphic to $l_2(P_i)$, where P_i is the fiber of E_i.

(c) The inclusion $H^{(s+1)}(E_i) \to H^s(E_i)$ is a compact Λ-operator and

$$\text{index } \sigma(D) = (-1)^n < T(X)ch[\sigma] ,[T^*X]> \in K^o(\Lambda) \otimes Q \quad (1)$$

Examples. 1. The family of pseudodifferential operators in the sense of Shih Weishu [3] can be interpreted as a single elliptic operator over the ring $C(Y)$ of the continuous functions on the compact Y. Since $K^o(C(Y)) = K(Y)$, the Shih Weishu formula is a particular case of (1).

2. Let G be a compact Lie group, E_1, E_2 vector G-bundles over manifold X, G acts in a trivial way on X. Then any equivariant operator in the sense of [4] can be interpreted as Λ-operator where $\Lambda = C^*[G]$ is the group C^*-algebra of the group G. Since the ring $R(G)$ of the virtual representations of the group G has a natural mapping into $K^o(C^*[G])$, the index formula for the elliptic equivariant operators is a particular case of (1).

3. Let E_1, E_2 be infinite dimensional bundles over the manifolds X with the Hilbert fiber H, $\sigma: \pi^*(E_1) \to \pi^*(E_2)$ the elliptic symbol in the sense of G.Luke [5]. In particular it means that $\sigma(x,\xi)$ is the Fredholm operator for any $(x,\xi) \in T^*(X)$. Then one can interprete the index formula for the operator $\sigma(D)$ as a

particular case of (1) for the relative K-functor of the ring homomorphisms $\Lambda \to \Lambda'$, where $\Lambda = B(H)$ is the algebra of the bounded operators, $\Lambda' = \Lambda/K$ is the factoralgebra modulo the ideal of compact operators.

4. In [6] the result of M.F. Atiyah and I.M. Singer on the index for the case of the Λ-bundles, when Λ is a factor of type II_1 is described. In that case the group $K^0(\Lambda)$ is isomorphic to the group of real numbers. Therefore the index formula of [6] is a particular case of (1).

§3. The analogue of the Hirzebruch formula

Let X be a compact smooth nonsimply connected manifold, $\pi_1(X) = \pi$, $f: X \to B\pi$ a classifying mapping. In differential topology the investigation of the problem of describing the homotopy invariant rational characteristic number of the singular bordism f is of great importance.

Theorem 4. Any homotopy invariant rational characteristic number of nonsimply connected manifolds X has the following form

$$\operatorname{sign}_x(X) = \langle L(X) f^*(x), [X] \rangle , \qquad (2)$$

where $x \in H^*(B\pi;Q)$, L is the Pontrjagin-Hirzebruch characteristic class of X.

The numbers of form (2) are called the "higher signatures" of the manifold X.

For instance, G. Lusztig [7] gave a direct proof of homotopy invariance for all higher signatures for free abelian fundamental group. For that he used the family of all characters of fundamental group π.

In general, to prove the homotopy invariance of the higher signatures (2) it is sufficient to represent the right hand side of (2) by a priori homotopy invariants. Such invariants are the generalized nonsimply connected signatures of the nonsimply connected manifolds, which were defined by the author in [8]. This generalisation is founded on the investigation of the special algebraic object - algebraic Poincaré complexes. The algebraic Poincaré complex (APC) of dimension n is such a chain graded complex of free Λ-modules (C,d) with the homomorphism $\xi: C^* \to C$, $\deg \xi = n$, $\xi^* = \xi$, where ξ induces an isomorphism in homology. There is a natural definition of bordism of APC: the bordism group of APC is denoted by $\Omega_n(\Lambda)$. The case of algebras over integral numbers was considered by the author [9] and investigated completely by A.A. Ranicki [10].

Let $\Lambda = C^*[\pi]$ be the group C^*-algebra.

Theorem 5. ([11]). (a) There exists a natural homomorphism $\sigma: \Omega_*^{SO}(B\pi) \to \Omega_*(\Lambda)$ which has the same value for homotopy equivalent manifolds.

(b) $\Omega_*(\Lambda)$ is isomorphic to $K_h^*(\Lambda) \cong K^*(\Lambda)$,

(c) Let ξ be a canonical Λ-bundle over $B\pi$, induced by inclusion $\pi \subset \Lambda$. Then the following Hirzebruch formula holds:

$$\sigma(X) = 2^n\langle L(X)f^*ch_\Lambda\xi, [X]\rangle \in K^o(\Lambda) \otimes Q \qquad (3)$$

(d) If the kernel of inclusion $L_*(\mathbb{Z}[\pi]) \to K^*(\Lambda)$ has no elements of infinite order then (3) gives all homotopy invariant higher signatures.

The Lusztig formula from [7] is a particular case of (3), if one considers the canonical representation of the group $\pi = \mathbb{Z}^n$ into the ring $\Lambda = C^*(T^n)$ of the continuous functions on the torus.

The proof of (3) can be obtained in two ways. One of them is founded on the interpretation of the nonsimply connected signature $\sigma(X)$ of the manifold X as the index of Hirzebruch Λ-operator on X, as well as on application of the index formula (1). The second is founded on the construction and investigation of the homotopy properties of the universal space for the hermitian K-theory.

§4. The universal space of the hermitian K-theory

Let us consider the cell complex K, such that each cell with its boundary is included by some homeomorphism into K. The set of the closed subcomplexes with their inclusions forms the category $\underline{K}$. Let us consider splitting functors Π from $\underline{K}$ into the category of the graded complexes of projective Λ-modules, that is, functors such that $K_1 \subset K_2$ implies $\Pi(K_1) \subset \Pi(K_2)$ and $\Pi(K_1)$ is a direct summand, $\Pi(K_1 \cup K_2) = \Pi(K_1) + \Pi(K_2)$, $\Pi(K_1 \cap K_2) = \Pi(K_1) \cap \Pi(K_2)$. For any cell let us consider homomorphisms $\xi(\sigma)\colon \Pi(\sigma)^* \to \Pi(\sigma)$,

$$\deg \xi(\sigma) = \dim(\sigma), \qquad \xi(-\sigma) = -\xi(\sigma) \quad ,$$

such that the triad $(\Pi(\sigma),\Pi(\partial\sigma),\ \xi(\sigma))$ is APC with boundary, and a boundary homomorphism

$$\xi(\sigma)d^* + d\xi(\sigma) = \sum_{\sigma'} \xi(\sigma'),$$

σ' runs through all cells $\sigma' \subset \partial\sigma$, $\dim \sigma' = \dim \sigma - 1$, The pair (Π,ξ) is called a bundle of APC over K. For the bundles of APC the operations of restriction, preimage and tensor product are defined. Then let us denote the space of the final bundle of APC Π^W by $W(\Lambda)$. If Λ is a Banach algebra, and all bundles of APC have bases, the space $W(\Lambda)$ possesses weak topology, which is denoted by $\widetilde{W}(\Lambda)$.

<u>Theorem 6</u>. (a) Let X be an orientable PL-manifold, Π - the bundle of APC over X. Then the pair $(\Pi(X),\xi)$ is APC, where $\xi = \sum_{\sigma} \xi(\sigma)$, σ runs through all the cells of maximal dimension.

(b) $\pi_n(W(\Lambda)) \simeq \Omega_n(\Lambda)$ and is the direct summand of $\Omega_n(W(\Lambda))$.

(c) For a Banach algebra Λ the natural inclusion $W(\Lambda) \to \widetilde{W}(\Lambda)$ is a weak homotopy equivalence.

For C^*-algebras Λ the structure of bundle of APC can be defined on the Λ-bundles over X. Let us denote the final object in the category of Λ-bundles with the structure of the bundle of APC by $V(\Lambda)$.

<u>Theorem 7</u>. (a) The natural inclusion $W(\Lambda) \to V(\Lambda)$ is a weak homotopy equivalence.

(b) The natural inclusion $BGL(\Lambda) \to V(\Lambda)$ is a weak homotopy equivalence modulo 2-torsion.

(c) Let X be nonsimply connected compact closed manifold, $f: X \to \widetilde{W}(\Lambda)$ the canonical map constructed by means of a PL-structure on the X, $g: X \to BGL(\Lambda)$ - the map classifying the canonical bundle ξ, generated by the inclusion $\pi \subset \Lambda$. Then f and g are homotopic in $V(\Lambda)$.

The theorem 7(c) presents another way of proving formula (3).

§5. The higher signatures

Choosing suitable representations of the fundamental group in the Banach algebras one can point out sufficiently many sets of the homotopy invariant higher signatures by means of the formula (3).

Consider the Fredholm representation ρ of the fundamental group π, that is $\rho \in R^{o}(\varphi)$, where $\varphi: B(H) \to B(H)/K(H)$ is the quotient map. Under the pairing $K^{o}(B\pi;\Lambda) \times R^{o}(\varphi) \to K(B\pi,\varphi)$ the canonical bundle $\xi \in K^{o}(B\pi;\Lambda)$ maps to $\xi_{\rho} \in K^{o}(B\pi,\varphi) \simeq$ $\simeq K^{1}(B\pi;B(H)/K(H)) \simeq K^{o}(B\pi)$. Thus the formula (3) gives a homotopy invariant number

$$\mathrm{sign}_{\rho}(\sigma(X)) = 2^{n}\langle L(X)\mathrm{ch}\xi_{\rho}, [X]\rangle \in Q.$$

If some set of representations $\{\rho\}$ is such that the cohomology elements $\mathrm{ch}\xi_{\rho} \in H^{*}(B\pi;Q)$ generate the whole cohomology group, we'll say that the set $\{\rho\}$ is large enough.

Theorem 8. (a) ([2],[12]). Let $B\pi$ be a complete riemannian manifold of nonnegative curvature. Then π has a large enough set of the Fredholm representations.

(b) ([13]). Let π be a discrete subgroup of a product of semisimple Lie groups or linear algebraic groups over local locally compact fields. Then there exists a large enough set of the Fredholm representations.

(c) For the groups as in (a) or (b) each higher signature is homotopy invariant.

Ju.P. Solovjov [13] proved, that the propositions of the theorem 8 are true for the rational homology manifolds by means of the final object method.

§6. Problems

1. Can one distinguish the elements from $K^o(\Lambda)$ by means of the Fredholm representations ?

2. In (1) to substitute the Fredholm representation for the representation to some algebra Λ', such that $K^o(\Lambda') \subset \mathbb{R}$. (For instance, Λ' is a factor of the kind I_n or II_1).

3. Let $B\pi$ be a manifold. Does there exist a complete riemannian metric of nonnegative curvature on the $B\pi$?

4. To describe K^o and K^1 for the von Neuman algebras.

REFERENCES

[1] M. Karoubi, Algebres de Clifford et K-thêorie, Ann.Sci. Ecole Norm.Super., t.1, N2 (1968), 161-270.

[2] A. C. Мищенко, Бесконечномерные предсдавления дискретных групп и высше сигнатуры. Изв. АН СССР, сер. Матем., 38, 1974, 81 - 106

[3] Shih Weishu, Fiber cobordism and the index of a family of elliptic differential operators, Bull. Amer.Math. Soc., v.72, N 6 (1966).

[4] M.F. Atiyah, I.M. Singer, The index of elliptic operators, I, Ann.Math.,Ser.2, 87:3 (1968), 484-530.

[5] G.Luke, Pseudo-differential operators on Hilbert bundles, J.Diff.Equations, 12:3 (1972), 566-589.

[6] I.M. Singer, Some remarks on operator theory and index theory, Lecture Notes in Math., N 575 (1977), 128-138.

[7] G. Lusztig, Novikov's higher signature and families of elliptic operators, J.Diff.Geometry, 7 (1971), 229-256.

[8] A. C. Мищенко, Гомотопические инварианты неодносвязных многообразний, I Изв. АН СССР, сер. Матем., 34, 1970, 501 - 514

[9] A. C. Мищенко, Гомотопические инварианты неодносвязных многообрвзний, 3 Изв. АН СССК, сер. Матем. 35, 1971, 1316 - 1355

[10] A. A. Ranicki, preprint

[11] Ф. С. Мищенко, Ю. П. Соловьев, О бесконечномерных предстаилениях Фундаментальных групп и формулы типв Хирцебруха, ДВН СССР, 234, No. 4, 1977

[12] A. C. Мищенко, О фредгольмовых представлениях дискретных групп, Функц. анализ, 9:2V 1975, 36 - 41

[13] Ю. П. Соловьев, Гомотопические инварианты рациональных гомологиеских многообразий, ДАН СССР, 230:I, 1976, 41 - 43

The total surgery obstruction

by Andrew Ranicki, Princeton University

Let $n \geq 5$.

According to the Browder-Novikov-Sullivan-Wall theory of surgery ([B1],[B2],[N],[Su1],[W1]) a finite n-dimensional Poincaré complex X is homotopy equivalent to a compact topological manifold if and only if

i) the Spivak normal fibration $\nu_X : X \longrightarrow BG(k)$ $(k \gg n)$ admits a topological reduction $\widetilde{\nu}_X : X \longrightarrow BTOP(k)$, in which case topological transversality applied to a degree 1 map $\rho_X : S^{n+k} \longrightarrow T(\nu_X)$ gives a topological manifold $M^n = \rho_X^{-1}(X) \subset S^{n+k}$ and a map of topological bundles $b : \widetilde{\nu}_M \longrightarrow \widetilde{\nu}_X$ covering the degree 1 map $f = \rho_X| : M \longrightarrow X$, and hence a surgery obstruction $\theta(f,b) \in L_n(\pi_1(X))$

ii) there exists a topological reduction $\widetilde{\nu}_X$ such that $\theta(f,b) = 0$, in which case the normal map $(f,b) : M \longrightarrow X$ is normal bordant to a homotopy equivalence.

The theory was initially developed in the smooth and PL categories; the extension to the topological category is due to Kirby and Siebenmann ([KS]).

We present here the preliminary account of a theory which replaces the two-stage obstruction with a single invariant, 'the total surgery obstruction'.

We shall only consider the oriented case, but in principle there exists an unoriented version involving twisted coefficients. For the sake of the s-cobordism theorem we shall be working with simple homotopy types and the Wall L^s-groups, but there is also an ordinary homotopy version which we discuss briefly at the end. Thus Poincaré complexes will be finite, simple and oriented; manifolds will be compact, topological and oriented.

The invariant lies in one of the groups $\mathcal{S}_*(X)$ (defined for any space X) appearing in an exact sequence of abelian groups

$$\dots \longrightarrow H_n(X;\underline{\mathbb{L}}_0) \xrightarrow{\sigma_*} L_n(\pi_1(X)) \longrightarrow \mathcal{S}_n(X) \longrightarrow H_{n-1}(X;\underline{\mathbb{L}}_0) \longrightarrow \dots ,$$

where $\underline{\mathbb{L}}_0$ is a 1-connective Ω-spectrum with 0th space homotopy equivalent to G/TOP and σ_* is a universal assembly map. Both $\underline{\mathbb{L}}_0$ and σ_* were originally constructed by

Quinn ([Q1],[Q2]) using geometric methods. Here, $\underline{\mathbb{L}}_0$ and σ_* are constructed using algebraic methods, and the groups $\mathcal{S}_*(X)$ are the relative homotopy groups of a map of simplicial Ω-spectra $\sigma_*: X_+ \wedge \underline{\mathbb{L}}_0 \longrightarrow \underline{\mathbb{L}}_0(\pi_1(X))$ inducing the assembly maps $\sigma_*: H_*(X;\underline{\mathbb{L}}_0) = \pi_*(X_+ \wedge \underline{\mathbb{L}}_0) \longrightarrow \pi_*(\underline{\mathbb{L}}_0(\pi_1(X))) = L_*(\pi_1(X))$ $(X_+ = X \cup \{pt.\})$.
There are also defined relative groups $\mathcal{S}_*(X,Y)$ for pairs (X,Y), to fit into an exact sequence of abelian groups

$$\ldots \longrightarrow H_n(X,Y;\underline{\mathbb{L}}_0) \xrightarrow{\sigma_*} L_n(\pi_1(Y) \to \pi_1(X)) \longrightarrow \mathcal{S}_n(X,Y) \longrightarrow H_{n-1}(X,Y;\underline{\mathbb{L}}_0) \longrightarrow$$

The functor $\mathcal{S}_*$ satisfies the first five of the seven Eilenberg-Steenrod axioms for a homology theory, failing excision and dimension:

$$\mathcal{S}_*(\text{pushout square}) = \text{Cappell's } \mathrm{Unil}_* \ , \ \mathcal{S}_*(\text{pt.}) = 0 \ .$$

<u>Theorem 1</u> An n-dimensional Poincaré complex X determines an element $s(X) \in \mathcal{S}_n(X)$, the <u>total surgery obstruction</u> of X, such that $s(X) = 0$ if and only if X is simple homotopy equivalent to a closed topological manifold. The image of $s(X)$ in $H_{n-1}(X;\underline{\mathbb{L}}_0)$ is the obstruction to a topological reduction of the Spivak normal fibration $\nu_X: X \longrightarrow BSG$.

[]

There are also relative versions (and even n-ad versions) of Theorem 1:

<u>Theorem 1 (rel)</u> An n-dimensional Poincaré pair (X,Y) determines an element $s(X,Y) \in \mathcal{S}_n(X,Y)$ such that $s(X,Y) = 0$ if and only if (X,Y) is simple homotopy equivalent to a manifold with boundary.

[]

<u>Theorem 1 (rel ∂)</u> An n-dimensional Poincaré pair (X,Y) with manifold boundary Y determines an element $s_\partial(X,Y) \in \mathcal{S}_n(X)$ such that $s_\partial(X,Y) = 0$ if and only if (X,Y) is simple homotopy equivalent to a manifold with boundary by an equivalence which restricts to a homeomorphism of the boundaries.

[]

The obstruction theory of Sullivan [Su1] for the problem of deforming a homotopy equivalence of manifolds to a homeomorphism has a natural expression as a total surgery obstruction:

Corollary 1 A simple homotopy equivalence of closed n-dimensional manifolds $f: M \longrightarrow X$ determines an element $s(f) \in \mathcal{S}_{n+1}(X)$ such that $s(f) = 0$ if and only if f is homotopic to a homeomorphism.

Proof: Let W be the mapping cylinder of f, so that $(W, M \cup -X)$ is an (n+1)-dimensional Poincaré pair with manifold boundary. Define

$$s(f) = s_\partial(W, M \cup -X) \in \mathcal{S}_{n+1}(W) \; (= \mathcal{S}_{n+1}(X) \text{ by the homotopy invariance of } \mathcal{S}_*) .$$

By Theorem 1 (rel ∂) $s(f) = 0$ if and only if there exists a topological s-cobordism $(W'; M', X')$ simple homotopy equivalent to $(W; M, X)$ by an equivalence which restricts to homeomorphisms of the boundary components. Now apply the topological s-cobordism theorem (in dimension $n+1 \geqslant 6$).

[]

There are also relative versions, Corollary 1 (rel) and Corollary 1 (rel ∂).

Given an n-dimensional Poincaré complex X let $\mathcal{S}^{TOP}(X)$ be the topological manifold structure set of X, defined as usual to be the set of equivalence classes of pairs

(closed n-dimensional topological manifold M,

orientation preserving simple homotopy equivalence $f: M \longrightarrow X$)

under the relation

$(M, f) \sim (M', f')$ if there exist a homeomorphism $h: M \longrightarrow M'$ and a homotopy $f'h \simeq f: M \longrightarrow X$.

Define similarly structure sets $\mathcal{S}^{TOP}(X, Y)$ for Poincaré pairs (X, Y), and also $\mathcal{S}_\partial^{TOP}(X, Y)$ for Poincaré pairs (X, Y) with manifold boundary Y.

Corollary 2 If X is a closed n-dimensional manifold the function

$$s : \mathcal{S}^{TOP}(X) \longrightarrow \mathcal{S}_{n+1}(X) \; ; \; (f:M \longrightarrow X) \longmapsto s(f)$$

is a bijection, and there is a natural identification of the Sullivan-Wall surgery exact sequence

$$\cdots \longrightarrow \mathcal{S}^{TOP}_{\partial}(X\times\Delta^1, \partial(X\times\Delta^1)) \longrightarrow [X\times\Delta^1, \partial(X\times\Delta^1); G/TOP, *] \xrightarrow{\theta} L_{n+1}(\pi_1(X))$$
$$\longrightarrow \mathcal{S}^{TOP}(X) \longrightarrow [X, G/TOP] \xrightarrow{\theta} L_n(\pi_1(X))$$

with the exact sequence

$$\cdots \longrightarrow \mathcal{S}_{n+2}(X) \longrightarrow H_{n+1}(X;\mathbb{L}_0) \xrightarrow{\sigma_*} L_{n+1}(\pi_1(X))$$
$$\longrightarrow \mathcal{S}_{n+1}(X) \longrightarrow H_n(X;\mathbb{L}_0) \xrightarrow{\sigma_*} L_n(\pi_1(X)) \quad .$$

In particular,

$$\mathcal{S}^{TOP}_{\partial}(X\times\Delta^k, \partial(X\times\Delta^k)) = \mathcal{S}_{n+k+1}(X) \quad , \; [X\times\Delta^k, \partial(X\times\Delta^k); G/TOP, *] = H_{n+k}(X;\mathbb{L}_0) \quad (k \geqslant 0) \; .$$

[]

Again, there are relative versions, Corollary 2 (rel) and Corollary 2 (rel∂). If $(X,\partial X)$ is an n-dimensional manifold with boundary there are natural identifications

$$\begin{aligned} \mathcal{S}^{TOP}(X\times\Delta^k, \partial(X\times\Delta^k)) &= \mathcal{S}_{n+k+1}(X,\partial X) \\ \mathcal{S}^{TOP}_{\partial}(X\times\Delta^k, \partial(X\times\Delta^k)) &= \mathcal{S}_{n+k+1}(X) \end{aligned} \quad (k \geqslant 0) \; .$$

We shall only sketch a proof of Theorem 1 here. There are 4 main ingredients:

i) the Browder-Novikov-Sullivan-Wall theory in the topological category

ii) the isomorphisms $\theta : \pi_*(G/TOP) \longrightarrow L_*(1)$ defined by the surgery obstruction

iii) transversality in Quinn's category of normal spaces and spherical fibrations

iv) the algebraic theory of surgery.

We start with a brief account of iv) - the first two instalments of a full account are due to appear shortly ([R2]).

Given a group π and a (left) $\mathbb{Z}[\pi]$-module chain complex C let $T\in\mathbb{Z}_2$ act on $C\otimes_{\mathbb{Z}[\pi]}C = C\otimes_{\mathbb{Z}}C/\{x\otimes gy - g^{-1}x\otimes y \mid x,y\in C, g\in\pi\}$ by $T(x\otimes y) = (-)^{|x||y|}y\otimes x$, and define the

$\mathbb{Z}_2$-hypercohomology / $\mathbb{Z}_2$-hyperhomology groups $\begin{cases} Q^n(C) = H_n(\mathrm{Hom}_{\mathbb{Z}[\mathbb{Z}_2]}(W, C\otimes_{\mathbb{Z}[\pi]}C)) \\ Q_n(C) = H_n(W\otimes_{\mathbb{Z}[\mathbb{Z}_2]}(C\otimes_{\mathbb{Z}[\pi]}C)) \end{cases}$ with W the free $\mathbb{Z}[\mathbb{Z}_2]$-module resolution of $\mathbb{Z}$ $W : \ldots \longrightarrow \mathbb{Z}[\mathbb{Z}_2] \xrightarrow{1+T} \mathbb{Z}[\mathbb{Z}_2] \xrightarrow{1-T} \mathbb{Z}[\mathbb{Z}_2] \longrightarrow 0.$

An element $\begin{cases} \varphi\in Q^n(C) \\ \psi\in Q_n(C) \end{cases}$ is an equivalence class of collections $\begin{cases} \{\varphi_s \in (C\otimes_{\mathbb{Z}[\pi]}C)_{n+s} \mid s\geq 0\} \\ \{\psi_s \in (C\otimes_{\mathbb{Z}[\pi]}C)_{n-s} \mid s\geq 0\} \end{cases}$ such that

$$\begin{cases} d(\varphi_s) + (-)^{n+s-1}(\varphi_{s-1} + (-)^s T\varphi_{s-1}) = 0 \in (C\otimes_{\mathbb{Z}[\pi]}C)_{n+s-1} & (s\geq 0, \varphi_{-1} = 0) \\ d(\psi_s) + (-)^{n-s-1}(\psi_{s+1} + (-)^{s+1}T\psi_{s+1}) = 0 \in (C\otimes_{\mathbb{Z}[\pi]}C)_{n-s-1} & (s\geq 0). \end{cases}$$

The $\begin{cases}\text{symmetric}\\ \text{quadratic}\end{cases}$ L-groups $\begin{cases} L^n(\pi) \\ L_n(\pi) \end{cases}$ $(n\geq 0)$ are defined to be the algebraic Poincaré cobordism groups of n-dimensional $\begin{cases}\text{symmetric}\\ \text{quadratic}\end{cases}$ Poincaré complexes over $\mathbb{Z}[\pi]$ $\begin{cases} (C, \varphi\in Q^n(C)) \\ (C, \psi\in Q_n(C)) \end{cases}$, with $C: C_n \xrightarrow{d} C_{n-1} \longrightarrow \ldots \longrightarrow C_1 \xrightarrow{d} C_0$ a based f.g. free $\mathbb{Z}[\pi]$-module chain complex such that slant product with the cycle $\begin{cases} \varphi_0 \in (C\otimes_{\mathbb{Z}[\pi]}C)_n \\ (1+T)\psi_0 \in (C\otimes_{\mathbb{Z}[\pi]}C)_n \end{cases}$ defines a simple chain equivalence $C^{n-*} = \mathrm{Hom}_{\mathbb{Z}[\pi]}(C, \mathbb{Z}[\pi])_{n-*} \longrightarrow C$. The quadratic L-groups are 4-periodic, $L_n(\pi) = L_{n+4}(\pi)$, being just the Wall surgery obstruction groups. The symmetric L-groups were introduced by Mishchenko [Mi]; they are not in general 4-periodic, $L^n(\pi) \neq L^{n+4}(\pi)$. There are defined symmetrization maps

$$1+T : L_n(\pi) \longrightarrow L^n(\pi) \ ; \ (C,\psi) \longmapsto (C, (1+T)\psi_0)$$

which are isomorphisms modulo 8-torsion. The cobordism classes of (n-1)-dimensional quadratic Poincaré complexes with an n-dimensional symmetric Poincaré null-cobordism define hyperquadratic L-groups $\hat{L}^n(\pi)$ $(n\geq 1)$ of exponent 8 which fit into a long exact sequence of abelian groups

$$\ldots \longrightarrow L_n(\pi) \xrightarrow{1+T} L^n(\pi) \xrightarrow{J} \hat{L}^n(\pi) \xrightarrow{H} L_{n-1}(\pi) \xrightarrow{1+T} L^{n-1}(\pi) \longrightarrow \ldots .$$

For example, $\begin{cases} L^0(\pi) \\ L_0(\pi) \end{cases}$ is the Witt group of non-singular $\begin{cases}\text{symmetric}\\ \text{quadratic}\end{cases}$ forms over $\mathbb{Z}[\pi]$.

The L-groups of the trivial group $\pi = \{1\}$ are given by

$$L^n(1) = \begin{cases} \mathbb{Z} & \text{(signature)} \\ \mathbb{Z}_2 & \text{(deRham)} \\ 0 & \\ 0 & \end{cases}, \quad L_n(1) = \begin{cases} \mathbb{Z} & (\tfrac{1}{8}(\text{signature})) \\ 0 & \\ \mathbb{Z}_2 & \text{(Arf)} \\ 0 & \end{cases}$$

$$\hat{L}^n(1) = \begin{cases} \mathbb{Z}_8 & \text{(signature (mod 8))} \\ \mathbb{Z}_2 & \text{(deRham)} \\ 0 & \\ \mathbb{Z}_2 & \text{(Arf)} \end{cases} \quad \text{if } n \equiv \begin{cases} 0 \\ 1 \\ 2 \\ 3 \end{cases} \pmod 4$$

An n-dimensional geometric Poincaré complex X is an n-dimensional finite CW complex together with a fundamental homology class $[X] \in H_n(X)$ such that cap product defines a simple chain equivalence of based f.g. free $\mathbb{Z}[\pi_1(X)]$-module chain complexes

$$[X] \cap - : C(\widetilde{X})^{n-*} \longrightarrow C(\widetilde{X}) ,$$

with $C(\widetilde{X})$ the cellular chain complex of the universal cover $\widetilde{X}$. Applying $H_*(\mathbb{Z} \otimes_{\mathbb{Z}[\pi_1(X)]} -)$ to a diagonal approximation $\Delta : C(\widetilde{X}) \longrightarrow \mathrm{Hom}_{\mathbb{Z}[\mathbb{Z}_2]}(W, C(\widetilde{X}) \otimes_{\mathbb{Z}} C(\widetilde{X}))$ and evaluating $\Delta : H_n(X) \longrightarrow Q^n(C(\widetilde{X}))$ defines an n-dimensional symmetric Poincaré complex over $\mathbb{Z}[\pi_1(X)]$ $(C, \varphi) = (C(\widetilde{X}), \Delta[X])$, and hence a symmetric signature geometric Poincaré bordism invariant

$$\sigma^*(X) = (C(\widetilde{X}), \Delta[X]) \in L^n(\pi_1(X))$$

(which was introduced by Mishchenko [Mi]). Given a group morphism $\pi_1(X) \longrightarrow \pi$ we shall denote the image of $\sigma^*(X) \in L^n(\pi_1(X))$ in $L^n(\pi)$ also by $\sigma^*(X)$. For example, if $n = 4k$ $\sigma^*(X) \in L^{4k}(1) = \mathbb{Z}$ is just the ordinary signature of X.

An n-dimensional geometric Poincaré complex X carries a stable equivalence class of Spivak normal structures

$$(\nu_X : X \longrightarrow BSG(k), \rho_X : S^{n+k} \longrightarrow T(\nu_X)) ,$$

such as arise from an embedding $X \subset S^{n+k}$ $(k \gg n)$ by taking a closed regular neighbourhood W of X in S^{n+k} and setting

$$S^{k-1} \longrightarrow \partial W \xrightarrow{\nu_X} W$$

$$\rho_X : S^{n+k} \xrightarrow{\text{collapse}} S^{n+k}/\overline{S^{n+k} - W} = W/\partial W = T(\nu_X) .$$

s usual, $[X] = h(\rho_X) \cap U_{\nu_X} \in H_n(X)$, with h = Hurewicz map : $\pi_{n+k}(T(\nu_X)) \longrightarrow \dot{H}_{n+k}(T(\nu_X))$, (ν_X) = Thom space of ν_X, U_{ν_X} = Thom class $\in \dot{H}^k(T(\nu_X))$, $\dot{H}$ = reduced (co)homology.

A normal map of n-dimensional geometric Poincaré complexes

$$(f,b) : (M,\nu_M,\rho_M) \longrightarrow (X,\nu_X,\rho_X)$$

s a map $f:M \longrightarrow X$ of degree 1, $f_*[M] = [X] \in H_n(X)$, together with specified Spivak ormal structures (ν_M,ρ_M), (ν_X,ρ_X) and a stable map of spherical fibrations $:\nu_M \longrightarrow \nu_X$ covering f such that $T(b)_*(\rho_M) = \rho_X \in \pi^S_{n+k}(T(\nu_X))$. Such a normal map etermines an n-dimensional quadratic Poincaré complex over $\mathbb{Z}[\pi_1(X)]$ (C,ψ), and here is defined a quadratic signature normal map bordism invariant

$$\sigma_*(f,b) = (C,\psi) \in L_n(\pi_1(X))$$

such that

$$(1+T)\sigma_*(f,b) = \sigma^*(M) - \sigma^*(X) \in L^n(\pi_1(X)) .$$

Here, $C = C(f^!)$ is the algebraic mapping cone of the Umkehr $\mathbb{Z}[\pi_1(X)]$-module chain map

$$f^! : C(\tilde{X}) \xrightarrow[\sim]{([X]\cap -)^{-1}} C(\tilde{X})^{n-*} \xrightarrow{\tilde{f}^*} C(\tilde{M})^{n-*} \xrightarrow[\sim]{([M]\cap -)} C(\tilde{M})$$

with $\tilde{M}$ the cover of M induced from the universal cover $\tilde{X}$ of X by f, and Ψ is defined as follows. Let $\nu_{\tilde{M}}:\tilde{M} \longrightarrow M \xrightarrow{\nu_M} BSG(k)$, $\nu_{\tilde{X}}:\tilde{X} \longrightarrow X \xrightarrow{\nu_X} BSG(k)$, so that b lifts to a stable map $\tilde{b}:\nu_{\tilde{M}} \longrightarrow \nu_{\tilde{X}}$ covering $\tilde{f}:\tilde{M} \longrightarrow \tilde{X}$. The induced map of Thom spaces $T(\tilde{b}):T(\nu_{\tilde{M}}) \longrightarrow T(\nu_{\tilde{X}})$ has an equivariant S-dual stable $\pi_1(X)$-equivariant map $F:\Sigma^\infty\tilde{X}_+ \longrightarrow \Sigma^\infty\tilde{M}_+$ $(\tilde{X}_+ = \tilde{X}\cup\{pt.\})$ inducing $f^!:C(\tilde{X}) \longrightarrow C(\tilde{M})$ on the chain level, and such that $(\Sigma^\infty\tilde{f})F \simeq 1:\Sigma^\infty\tilde{X}_+ \longrightarrow \Sigma^\infty\tilde{X}_+$. The evaluation of the composite

$$\Psi_F : H_n(X) \xrightarrow{(\text{adjoint } F)_*} \dot{H}_n(\Omega^\infty\Sigma^\infty\tilde{M}_+/\pi) \xrightarrow{\text{projection}} Q_n(C(\tilde{M})) \xrightarrow{e_\%} Q_n(C(f^!))$$

on the fundamental class $[X] \in H_n(X)$ defines $\Psi = \Psi_F[X] \in Q_n(C(f^!))$, where $e_\%$ is induced by the natural projection $e:C(\tilde{M}) \longrightarrow C(f^!)$ and $\pi = \pi_1(X)$. The standard map $\coprod_{k\geq 0} E\Sigma_k \times_{\Sigma_k} (\prod_k \tilde{M})/\pi \longrightarrow \Omega^\infty\Sigma^\infty\tilde{M}_+/\pi$ is a group completion in homology, so that $\dot{H}_n(\Omega^\infty\Sigma^\infty\tilde{M}_+/\pi) = \mathbb{Z}[\mathbb{Z}] \otimes_{\mathbb{Z}[\mathbb{N}]} (\bigoplus_{k\geq 1} H_n(E\Sigma_k \times_{\Sigma_k} (\prod_k \tilde{M})/\pi))$ contains $H_n(E\Sigma_2 \times_{\Sigma_2} (\tilde{M}\times\tilde{M})/\pi) = Q_n(C(\tilde{M}))$ as a direct summand.

An n-dimensional normal map $(f,b):M \longrightarrow X$ in the sense of Browder and Wall is a degree 1 map $f:M \longrightarrow X$ from an n-dimensional manifold M to an n-dimensional geometric Poincaré complex X, together with a stable map $b:\nu_M \longrightarrow \nu$ of topological bundles covering f, where $\nu_M:M \longrightarrow BSTOP(k)$ is the normal bundle of an embedding $M \subset S^{n+k}$. The algebraic theory of surgery identifies the surgery obstruction of (f,b) with the quadratic signature of the underlying normal map of geometric Poincaré complexes $(f,Jb):(M,J\nu_M,\rho_M) \longrightarrow (X,J\nu_X,\rho_X)$

$$\theta(f,b) = \sigma_*(f,Jb) \in L_n(\pi_1(X)) .$$

An n-dimensional normal space is a triple

$$(X,\nu_X:X \longrightarrow BSG(k),\rho_X:S^{n+k} \longrightarrow T(\nu_X))$$

consisting of an n-dimensional finite CW complex X, an oriented spherical fibration ν_X and a map ρ_X. There are evident notions of normal pair, normal bordi normal space n-ad. Given a normal space (X,ν_X,ρ_X) it is possible to construct a stable $\pi_1(X)$-equivariant map $G:\Sigma^\infty Z \longrightarrow \Sigma^\infty \widetilde{X}_+$ inducing $[X] \cap -:C(\widetilde{X})^{n-*} \longrightarrow C(\widetilde{X})$ on the chain level, with Z an equivariant S-dual of $T(\widetilde{\nu_X})$ and $[X] = h(\rho_X) \cap U_{\nu_X} \in H_n$ The quadratic construction now applies to define a hyperquadratic signature normal bordism invariant

$$\hat{\sigma}^*(X) \in \hat{L}^n(\pi_1(X))$$

(where $\hat{\sigma}^*(X)$ is short for $\hat{\sigma}^*(X,\nu_X,\rho_X)$) such that $H\hat{\sigma}^*(X) = (C,\psi) \in L_{n-1}(\pi_1(X))$, with C the algebraic mapping cone of $[X] \cap -:C(\widetilde{X})^{n-*} \longrightarrow C(\widetilde{X})$. An n-dimensional geometric Poincare complex X is essentially the same as an n-dimensional normal space (X,ν_X,ρ_X) such that $[X] \cap -:C(\widetilde{X})^{n-*} \longrightarrow C(\widetilde{X})$ is a chain equivalence, in whic case (ν_X,ρ_X) is a Spivak normal structure, $Z = \widetilde{X}_+$, $G = 1$ and

$$\hat{\sigma}^*(X) = J\sigma^*(X) \in \hat{L}^n(\pi_1(X)) \quad , \quad H\hat{\sigma}^*(X) = 0 \in L_{n-1}(\pi_1(X)) \ ,$$

If (X,Y) is an (n+1)-dimensional normal pair with Poincaré boundary Y there is defined a quadratic signature (normal, Poincaré)-bordism invariant

$$\sigma_*(X,Y) = (C,\psi) \in L_n(\pi_1(X))$$

such that C is the algebraic mapping cone of $[X] \cap -:C(\widetilde{X})^{n+1-*} \longrightarrow C(\widetilde{X},\widetilde{Y})$ and

$$(1+T)\sigma_*(X,Y) = \sigma^*(Y) \in L^n(\pi_1(X)) .$$

he mapping cylinder W of a normal map of n-dimensional geometric Poincaré complexes $(f,b):(M,\nu_M,\rho_M)\longrightarrow(X,\nu_X,\rho_X)$ defines an (n+1)-dimensional normal pair $(W,M\cup -X)$ ith Poincaré boundary $M\cup -X$, such that

$$\sigma_*(W,M\cup -X) = \sigma_*(f,b)\in L_n(\pi_1(X)) .$$

he various signature maps fit together to define a natural transformation of xact sequences of abelian groups (for any space K)

$$\begin{array}{ccccccccc}
\dots\longrightarrow & \Omega^N_{n+1}(K) & \longrightarrow & \Omega^{N,P}_{n+1}(K) & \longrightarrow & \Omega^P_n(K) & \longrightarrow & \Omega^N_n(K) & \longrightarrow\dots \\
 & \downarrow\hat{\sigma}^* & & \downarrow\sigma_* & & \downarrow\sigma^* & & \downarrow\hat{\sigma}^* & \\
\dots\longrightarrow & \hat{L}^{n+1}(\pi_1(K)) & \xrightarrow{H} & L_n(\pi_1(K)) & \xrightarrow{1+T} & L^n(\pi_1(K)) & \xrightarrow{J} & \hat{L}^n(\pi_1(K)) & \longrightarrow\dots ,
\end{array}$$

ith $\left\{\begin{array}{l}\Omega^P_*(K)\\ \Omega^N_*(K)\\ \Omega^{N,P}_*(K)\end{array}\right.$ the bordism groups of $\left\{\begin{array}{l}\text{geometric Poincaré complexes}\\ \text{normal spaces}\\ \text{(normal,Poincaré) pairs}\end{array}\right.$ mapping to K.

I should like to thank Frank Quinn for inventing normal spaces ([Q3]), and or suggesting that they should have a hyperquadratic invariant. Unfortunately, he results and constructions of [Q1],[Q2],[Q3] have not yet been fully documented. he theory announced here is independent of Quinn's (although evidently influenced y its philosophy), with the following two exceptions:

i) Normal space transversality: given a spherical fibration $\nu:K\longrightarrow BSG(k)$ over finite CW complex K and a map $\rho:S^{n+k}\longrightarrow T(\nu)$ to the Thom space $T(\nu)$ it is ossible to deform ρ by a homotopy to a map (also called ρ) for which $X=\rho^{-1}(K)\subset S^{n+k}$ as the structure of an n-dimensional normal space (X,ν_X,ρ_X) with

$$\nu_X : X\xrightarrow{\rho|}K\xrightarrow{\nu}BSG(k) \quad , \quad \rho_X : S^{n+k}\xrightarrow{\text{collapse}}S^{n+k}/\overline{S^{n+k}-W} = W/\partial W\longrightarrow T(\nu_X)$$

for some closed regular neighbourhood W of X in S^{n+k}, and with

$$\rho : S^{n+k}\xrightarrow{\rho_X}T(\nu_X)\longrightarrow T(\nu) .$$

Along with the relative normal transversality for maps of n-ads. It follows that he maps

$$\Omega^N_n(K)\longrightarrow H_n(K;\underline{MSG}) \ ; \ (X,\nu_X,\rho_X)\longmapsto (S^{n+k}\xrightarrow{\rho_X}T(\nu_X)\xrightarrow{\Delta}X_+\wedge T(\nu_X)\longrightarrow K_+\wedge MSG(k))$$

are isomorphisms, by analogy with the Pontrjagin-Thom isomorphisms for smooth bordism $\Omega^{SO}_n(K)\xrightarrow{\simeq}H_n(K;\underline{MSO})$ obtained by smooth transversality. (I am indebted to Norman Levitt for an elementary handle exchange argument establishing normal space transversality).

ii) Poincaré surgery: in the starred discussion surrounding Theorem 4 below (and Theorem 4 itself) we shall make use of the geometric Poincaré surgery theory initiated by Browder [B3], and developed further by Levitt [Le], Jones [J1] and Quinn [Q3]. Some details of the theory still remain obscure, especially in the non-simply-connected case. The main result of this theory is an exact sequence

$$\ldots \longrightarrow L_n(\pi_1(K)) \longrightarrow \Omega_n^P(K) \longrightarrow \Omega_n^N(K) \xrightarrow{H\hat{\sigma}^*} L_{n-1}(\pi_1(K)) \longrightarrow \ldots ,$$

or equivalently that the quadratic signature maps $\sigma_*:\Omega_{n+1}^{N,P}(K) \longrightarrow L_n(\pi_1(K))$ are isomorphisms, for any space K. It is immediate from the Wall realization theorem for surgery obstructions that the quadratic signature maps are split surjective, so that $\Omega_{n+1}^{N,P}(K) = L_n(\pi_1(K))\oplus ?$, but it it is not so easy to see that ? = 0 (althoug almost certainly true). In particular, <u>the proof of Theorem 1 makes no use of geometric Poincaré surgery</u>, relying instead on the algebraic Poincaré surgery of [Assuming ? = 0 it is in fact possible to give an alternative proof of Theorem 1 which makes no use of algebraic Poincaré surgery, relying instead on geometric Poincaré surgery. (Follow the same steps as in the proof below, but with the $\left\{\begin{matrix}\text{geometric Poincaré}\\ \text{normal space}\end{matrix}\right.$ bordism spectrum $\left\{\begin{matrix}\underline{\Omega}^P(K(\pi,1))\\ \underline{\Omega}^N(K(\pi,1))\end{matrix}\right.$ in place of the $\left.\begin{matrix}\text{symmetric}\\ \text{hyperquadratic}\end{matrix}\right\}$ $\mathbb{L}$-spectrum $\left\{\begin{matrix}\underline{\mathbb{L}}^0(\pi)\\ \underline{\hat{\mathbb{L}}}^0(\pi)\end{matrix}\right.$. If ? = 0 the quadratic signature map $\sigma_*:\underline{\Omega}^{N,P}(K(\pi,1)) \longrightarrow \Sigma\underline{\mathbb{L}}_0($ to the suspension of the 1-connective quadratic $\mathbb{L}$-spectrum is a homotopy equivalen

The original simply-connected surgery theory of Browder and Novikov was reformulated in terms of classifying spaces for normal maps (such as G/O, G/PL, G/TO by Sullivan [Su1] and Casson, and the non-simply-connected surgery theory of Wall was reformulated in terms of geometric classifying spaces by Quinn [Q1], see Rourke [Ro] and §17A of Wall [W1]. We shall now outline an algebraic constructi of surgery classifying spaces, leading to an algebraic formulation of surgery.

Given an abelian group G let $\underline{K}(G)$ be the Ω-spectrum with kth term the Eilenberg-MacLane space K(G,k). Given a connective spectrum $\underline{R}$ let $\underline{R}_{\S}$ denote the 1-connective covering of $\underline{R}$, i.e. the fibre of the evident map $\underline{R} \longrightarrow \underline{K}(\pi_0(\underline{R}))$.

Let π be a group. A $\begin{cases}\text{symmetric}\\ \text{quadratic}\end{cases}$ Poincaré n-ad over $\mathbb{Z}[\pi]$ is an n-ad of chain complexes of based f.g. free $\mathbb{Z}[\pi]$-modules, together with a simple $\begin{cases}\text{symmetric}\\ \text{quadratic}\end{cases}$ Poincaré duality. (See §0 of Wall [W1] for the general properties of n-ads). For example, an algebraic Poincaré 1-ad (resp. 2-ad) is the same as an algebraic Poincaré complex (resp. pair). The $\begin{cases}\text{symmetric}\\ \text{quadratic}\end{cases}$ Poincaré n-ads over $\mathbb{Z}[\pi]$ are the simplexes of (k+1)-connected Kan complexes $\begin{cases}\mathbb{L}^k(\pi)\\ \mathbb{L}_k(\pi)\end{cases}$ $(k\in\mathbb{Z})$ such that

$$\begin{cases}\Omega\mathbb{L}^k(\pi) = \mathbb{L}^{k+1}(\pi)\ , & \pi_n(\mathbb{L}^k(\pi)) = L^{n+k}(\pi)\\ \Omega\mathbb{L}_k(\pi) = \mathbb{L}_{k+1}(\pi)\ , & \pi_n(\mathbb{L}_k(\pi)) = L_{n+k}(\pi)\end{cases}\quad (k\in\mathbb{Z}, n+k\geqslant 0)\ .$$

Thus $\begin{cases}\underline{\mathbb{L}}^0(\pi) = \{\mathbb{L}^{-k}(\pi)\,|\,k\geqslant 0\}\\ \underline{\mathbb{L}}_0(\pi) = \{\mathbb{L}_{-k}(\pi)\,|\,k\geqslant 0\}\end{cases}$ is a connective Ω-spectrum such that

$$\begin{cases}\pi_n(\underline{\mathbb{L}}^0(\pi)) = L^n(\pi)\\ \pi_n(\underline{\mathbb{L}}_0(\pi)) = L_n(\pi)\end{cases}\quad (n\geqslant 0)\ .$$

The cofibre of the symmetrization map $1+T:\underline{\mathbb{L}}_0(\pi)_{\S}\longrightarrow\underline{\mathbb{L}}^0(\pi)$ is a connective Ω-spectrum $\underline{\hat{\mathbb{L}}}^0(\pi) = \{\hat{\mathbb{L}}^{-k}(\pi)\,|\,k\geqslant 0\}$ such that

$$\pi_n(\underline{\hat{\mathbb{L}}}^0(\pi)) = \begin{cases}\hat{L}^n(\pi) & (n\geqslant 1)\\ L^0(\pi) & (n=0)\ ,\end{cases}$$

which fits into a fibration sequence of spectra

$$\underline{\mathbb{L}}_0(\pi)_{\S}\xrightarrow{1+T}\underline{\mathbb{L}}^0(\pi)\xrightarrow{J}\underline{\hat{\mathbb{L}}}^0(\pi)\xrightarrow{H}\Sigma\underline{\mathbb{L}}_0(\pi)_{\S}\xrightarrow{1+T}\Sigma\underline{\mathbb{L}}^0(\pi)\ .$$

The tensor product of chain complex n-ads defines pairings of $\mathbb{L}$-spectra

$$\otimes:\underline{\mathbb{L}}^0(\pi)\wedge\underline{\mathbb{L}}^0(\rho)\longrightarrow\underline{\mathbb{L}}^0(\pi\times\rho)\ ,\quad \otimes:\underline{\mathbb{L}}^0(\pi)\wedge\underline{\mathbb{L}}_0(\rho)\longrightarrow\underline{\mathbb{L}}_0(\pi\times\rho)$$
$$\otimes:\underline{\hat{\mathbb{L}}}^0(\pi)\wedge\underline{\hat{\mathbb{L}}}^0(\rho)\longrightarrow\underline{\hat{\mathbb{L}}}^0(\pi\times\rho)$$

for any groups π,ρ. On the L-group level the tensor product of chain complexes defines pairings

$$\otimes: L^m(\pi)\otimes_{\mathbb{Z}}L^n(\rho)\longrightarrow L^{m+n}(\pi\times\rho)\ ,\quad \otimes: L^m(\pi)\otimes_{\mathbb{Z}}L_n(\rho)\longrightarrow L_{m+n}(\pi\times\rho)$$
$$\otimes:\hat{L}^m(\pi)\otimes_{\mathbb{Z}}\hat{L}^n(\rho)\longrightarrow\hat{L}^{m+n}(\pi\times\rho)\ .$$

We shall write $\begin{cases} \underline{\mathbb{L}}^0(1) = \underline{\mathbb{L}}^0 \\ \underline{\mathbb{L}}_0(1)_{\S} = \underline{\mathbb{L}}_0 \\ \underline{\hat{\mathbb{L}}}^0(1) = \underline{\hat{\mathbb{L}}}^0 \end{cases}$. Both $\underline{\mathbb{L}}^0$ and $\underline{\hat{\mathbb{L}}}^0$ are ring spectra; every algebraic $\underline{\mathbb{L}}$-spectrum above is an $\underline{\mathbb{L}}^0$-module spectrum. There is defined a commutative braid of fibration sequences of spectra

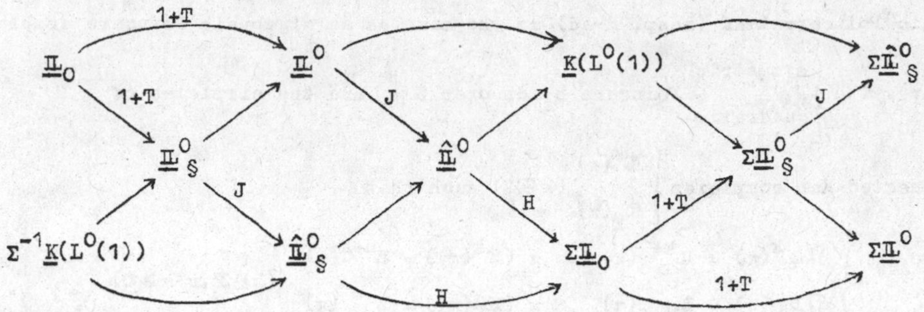

Given a space K let $\begin{cases} \underline{\Omega}^P(K) \\ \underline{\Omega}^N(K) \end{cases}$ be the connective Ω-spectrum of Kan complexes of maps $f:X \longrightarrow K$ from $\begin{cases} \text{geometric Poincaré} \\ \text{normal space} \end{cases}$ n-ads X to K such that

$\begin{cases} \pi_n(\underline{\Omega}^P(K)) = \Omega_n^P(K) \\ \pi_n(\underline{\Omega}^N(K)) = \Omega_n^N(K) \end{cases}$ $(n \geqslant 0)$ is the nth $\begin{cases} \text{geometric Poincaré} \\ \text{normal space} \end{cases}$ bordism group of K.

The cofibre of the forgetful map $\underline{\Omega}^P(K) \longrightarrow \underline{\Omega}^N(K)$ is denoted by $\underline{\Omega}^{N,P}(K)$, so that $\pi_n(\underline{\Omega}^{N,P}(K)) = \Omega_n^{N,P}(K)$ $(n \geqslant 0)$ is the nth (normal, Poincaré) pair bordism group of K. The cartesian product of topological n-ads defines pairings of spectra

$$\begin{cases} \otimes : \underline{\Omega}^P(K) \wedge \underline{\Omega}^P(L) \longrightarrow \underline{\Omega}^P(K \times L) \\ \otimes : \underline{\Omega}^N(K) \wedge \underline{\Omega}^N(L) \longrightarrow \underline{\Omega}^N(K \times L) \\ \otimes : \underline{\Omega}^P(K) \wedge \underline{\Omega}^{N,P}(L) \longrightarrow \underline{\Omega}^{N,P}(K \times L) \end{cases}$$

for any spaces K,L. We shall write $\underline{\Omega}^Q(\text{pt.}) = \underline{\Omega}^Q$ $(Q = P, N, (N,P))$. Let $\underline{K}_+$ be the suspension spectrum of $K_+ = K \cup \{\text{pt.}\}$, with kth term $\Sigma^k K_+ = S^k \wedge K_+$. A singular simpl in K is a particular example of a $\begin{cases} \text{geometric Poincaré} \\ \text{normal space} \end{cases}$ n-ad mapping to K, so there is defined a forgetful map $\begin{cases}\sigma^*:\underline{K}_+ \longrightarrow \underline{\Omega}^P(K)\\ \hat{\sigma}^*:\underline{K}_+ \longrightarrow \underline{\Omega}^N(K)\end{cases}$. The composites

$$\begin{cases}\sigma^* : \underline{K}_+\wedge\underline{\Omega}^P \xrightarrow{\sigma^*\wedge 1} \underline{\Omega}^P(K)\wedge\underline{\Omega}^P \xrightarrow{\otimes} \underline{\Omega}^P(K)\\ \hat{\sigma}^* : \underline{K}_+\wedge\underline{\Omega}^N \xrightarrow{\hat{\sigma}^*\wedge 1} \underline{\Omega}^N(K)\wedge\underline{\Omega}^N \xrightarrow{\otimes} \underline{\Omega}^N(K)\\ \sigma_* : \underline{K}_+\wedge\underline{\Omega}^{N,P} \xrightarrow{\sigma^*\wedge 1} \underline{\Omega}^P(K)\wedge\underline{\Omega}^{N,P} \xrightarrow{\otimes} \underline{\Omega}^{N,P}(K)\end{cases}$$

induce the assembly maps appearing in the natural transformation of exact sequences

$$\begin{array}{ccccccccc}\cdots\longrightarrow & H_{n+1}(K;\underline{\Omega}^{N,P}) & \longrightarrow & H_n(K;\underline{\Omega}^P) & \longrightarrow & H_n(K;\underline{\Omega}^N) & \longrightarrow & H_n(K;\underline{\Omega}^{N,P}) & \longrightarrow\cdots\\ & \downarrow{\sigma_*} & & \downarrow{\sigma^*} & & \downarrow{\hat{\sigma}^*} & & \downarrow{\sigma_*} & \\ \cdots\longrightarrow & \Omega^{N,P}_{n+1}(K) & \longrightarrow & \Omega^P_n(K) & \longrightarrow & \Omega^N_n(K) & \longrightarrow & \Omega^{N,P}_n(K) & \longrightarrow\cdots\end{array}$$

The assembly maps $\hat{\sigma}^*:H_n(K;\underline{\Omega}^N)\longrightarrow\Omega^N_n(K)$ are isomorphisms inverse to the natural maps $\Omega^N_n(K)\longrightarrow H_n(K;\underline{MSG}) = H_n(K;\underline{\Omega}^N)$, identifying $\underline{MSG} = \underline{\Omega}^N$ by normal transversality. (The Pontrjagin-Thom isomorphisms $H_n(K;\underline{MSO})\xrightarrow{\sim}\Omega^{SO}_n(K)$ have a similar expression as assembly maps).

The chain complex of the universal cover $\widetilde{X}$ of a geometric Poincaré n-ad X defines a symmetric Poincaré n-ad over $\mathbb{Z}[\pi_1(|X|)]$ $(C(\widetilde{X}),\Delta[X])$, so there is defined a map of Ω-spectra

$$\sigma^* : \underline{\Omega}^P(K)\longrightarrow\underline{\mathbb{L}}^0(\pi_1(K))$$

inducing the symmetric signature $\sigma^*:\Omega^P_n(K)\longrightarrow L^n(\pi_1(K))$ in the homotopy groups.

Similarly for $\begin{cases}\text{normal space}\\ \text{(normal,Poincaré) pair}\end{cases}$ n-ads, with a map of Ω-spectra

$$\begin{cases}\hat{\sigma}^* : \underline{\Omega}^N(K)\longrightarrow\underline{\hat{\mathbb{L}}}^0(\pi_1(K))\\ \sigma_* : \underline{\Omega}^{N,P}(K)\longrightarrow\Sigma\underline{\mathbb{L}}_0(\pi_1(K))\end{cases}$$

inducing the $\begin{cases}\text{hyperquadratic}\\ \text{quadratic}\end{cases}$ signature $\begin{cases}\hat{\sigma}^*:\Omega^N_n(K)\longrightarrow L^n(\pi_1(K))\\ \sigma_*:\Omega^{N,P}_{n+1}(K)\longrightarrow L_n(\pi_1(K))\end{cases}$. The pairings $\otimes$ defined for the Ω-spectra correspond to $\otimes$ for the $\mathbb{L}$-spectra. In particular, $\begin{cases}\sigma^*:\underline{\Omega}^P\longrightarrow\underline{\mathbb{L}}^0\\ \hat{\sigma}^*:\underline{\Omega}^N\longrightarrow\underline{\hat{\mathbb{L}}}^0\end{cases}$ is a morphism of ring spectra.

For an Eilenberg-MacLane space $K = K(\pi,1)$ the composite

$$\sigma^* : \underline{K}_+ \xrightarrow{\sigma^*} \underline{\Omega}^P(K) \xrightarrow{\sigma^*} \underline{\mathbb{L}}^0(\pi)$$

can be defined algebraically, using the standard simplicial model for $K(\pi,1)$. On the 1-skeleton $K(\pi,1)^{(1)} = \pi$ σ^* sends $g\in\pi$ to the 1-dimensional symmetric Poincaré complex over $\mathbb{Z}[\pi]$ $\sigma^*(g) = (C,\varphi\in Q^1(C))$ defined by

$$\begin{array}{ccccc} C^{1-*} & : C^0 = \mathbb{Z}[\pi] & \xrightarrow{d^* = 1-g^{-1}} & C^1 = \mathbb{Z}[\pi] \\ \varphi_0 \downarrow & \varphi_0 = 1 \downarrow & \varphi_1 = 1 \swarrow & \downarrow \varphi_0 = -g \\ C & : C_1 = \mathbb{Z}[\pi] & \xrightarrow{d = 1-g} & C_0 = \mathbb{Z}[\pi] \end{array} \quad .$$

This is the symmetric Poincaré complex corresponding to the simple automorphism $g:(\mathbb{Z}[\pi],1) \longrightarrow (\mathbb{Z}[\pi],1)$ of the non-singular symmetric form over $\mathbb{Z}[\pi]$ $(\mathbb{Z}[\pi],1)$. For the generator $g\in\pi = \mathbb{Z}$ $\sigma^*(g)$ is just the symmetric Poincaré complex $\sigma^*(S^1)$ of $K(\mathbb{Z},1) = S^1$.

Given a space X use the composite

$$\sigma^* : \underline{X}_+ \longrightarrow \underline{K}(\pi_1(X,1)_+ \xrightarrow{\sigma^*} \underline{\mathbb{L}}^0(\pi_1(X))$$

(which is also the composite $\sigma^*:\underline{X}_+ \xrightarrow{\sigma^*} \underline{\Omega}^P(X) \xrightarrow{\sigma^*} \underline{\mathbb{L}}^0(\pi_1(X))$) to define assembl maps of spectra

$$\begin{array}{l} \sigma^* : \underline{X}_+ \wedge \underline{\mathbb{L}}^0 \xrightarrow{\sigma^*\wedge 1} \underline{\mathbb{L}}^0(\pi_1(X))\wedge\underline{\mathbb{L}}^0 \xrightarrow{\otimes} \underline{\mathbb{L}}^0(\pi_1(X)) \\ \sigma_* : \underline{X}_+ \wedge \underline{\mathbb{L}}_0 \xrightarrow{\sigma^*\wedge 1} \underline{\mathbb{L}}^0(\pi_1(X))\wedge\underline{\mathbb{L}}_0 \xrightarrow{\otimes} \underline{\mathbb{L}}_0(\pi_1(X))_S \\ \hat{\sigma}^* : \underline{X}_+ \wedge \hat{\underline{\mathbb{L}}}^0 \xrightarrow{\sigma^*\wedge 1} \underline{\mathbb{L}}^0(\pi_1(X))\wedge\hat{\underline{\mathbb{L}}}^0 \xrightarrow{\otimes} \hat{\underline{\mathbb{L}}}^0(\pi_1(X)) \end{array} ,$$

and hence a natural transformation of exact sequences of abelian groups

$$\begin{array}{ccccccccc} \cdots \longrightarrow & H_n(X;\underline{\mathbb{L}}_0) & \xrightarrow{1+T} & H_n(X;\underline{\mathbb{L}}^0) & \xrightarrow{J} & H_n(X;\hat{\underline{\mathbb{L}}}^0) & \xrightarrow{H} & H_{n-1}(X;\underline{\mathbb{L}}_0) & \longrightarrow \cdots \\ & \sigma_* \downarrow & & \sigma^* \downarrow & & \hat{\sigma}^* \downarrow & & \sigma_* \downarrow & \\ \cdots \longrightarrow & L_n(\pi_1(X)) & \xrightarrow{1+T} & L^n(\pi_1(X)) & \xrightarrow{J} & \hat{L}^n(\pi_1(X)) & \xrightarrow{H} & L_{n-1}(\pi_1(X)) & \longrightarrow \cdots . \end{array}$$

Define the quadratic $\mathcal{S}$-groups $\mathcal{S}_*(X)$ of a space X by

$$\mathcal{S}_n(X) = \pi_n(\sigma_*:X_+\wedge\underline{\mathbb{L}}_0 \longrightarrow \underline{\mathbb{L}}_0(\pi_1(X))_S) \quad ,$$

to fit into an exact sequence of abelian groups

$$\cdots \longrightarrow H_n(X;\underline{\mathbb{L}}_0) \xrightarrow{\sigma_*} L_n(\pi_1(X)) \longrightarrow \mathcal{S}_n(X) \longrightarrow H_{n-1}(X;\underline{\mathbb{L}}_0) \longrightarrow \cdots .$$

The construction of the algebraic assembly maps σ_* and of the groups $\mathcal{S}_*(X)$ was motivated by Quinn's analysis of the surgery exact sequence in terms of geometric assembly maps ([Q1],[Q2]), and by the higher Whitehead groups $Wh_*(X)$ of Waldhausen [Wa]. Loday [Lo] has obtained similar maps in the context of Karoubi's hermitian K-theory, and also in algebraic K-theory. The maps σ_* are L-theoretic analogues of the maps $H_*(X;\underline{\underline{K}}(\mathbb{Z})) \longrightarrow K_*(\mathbb{Z}[\pi_1(X)])$ used to define $Wh_*(X)$ to fit into an exact sequence

$$\ldots \longrightarrow H_n(X;\underline{\underline{K}}(\mathbb{Z})) \longrightarrow K_n(\mathbb{Z}[\pi_1(X)]) \longrightarrow Wh_n(X) \longrightarrow H_{n-1}(X;\underline{\underline{K}}(\mathbb{Z})) \longrightarrow \ldots ,$$

with $\underline{\underline{K}}(\mathbb{Z})$ the spectrum of the algebraic K-theory of $\mathbb{Z}$, $\pi_*(\underline{\underline{K}}(\mathbb{Z})) = K_*(\mathbb{Z})$. For example, $Wh_1(K(\pi,1)) = Wh(\pi)$, $Wh_0(K(\pi,1)) = \tilde{K}_0(\mathbb{Z}[\pi])$. The groups $\mathcal{S}_*(X)$ are thus L-theoretic analogues of $Wh_*(X)$.

Transversality in the $\begin{cases}\text{topological}\\ \text{normal}\end{cases}$ category allows us to replace the Thom spectrum $\begin{cases}\underline{MSTOP}\\ \underline{MSG}\end{cases}$ by the homotopy equivalent Ω-spectrum $\begin{cases}\underline{\Omega}^{STOP}\\ \underline{\Omega}^{N}\end{cases}$ of Kan complexes of $\begin{cases}\text{topological manifold}\\ \text{normal space}\end{cases}$ n-ads. (It may be objected that we have ignored the absence of topological transversality in dimension 4, but there is at least enough of it to define a forgetful map $\underline{MSTOP} \longrightarrow \underline{\Omega}^P$, which is all we need. See Scharlemann [Sch]). Let $\underline{MS(G/TOP)}$ be the fibre of the forgetful map $\underline{MSTOP} \longrightarrow \underline{MSG}$, the spectrum with kth space $MS(G(k)/TOP(k))$, the homotopy-theoretic fibre of $MSTOP(k) \longrightarrow MSG(k)$. Then $\Sigma\underline{MS(G/TOP)}$ is homotopy equivalent to $\underline{\Omega}^{N,STOP}$, the cofibre of $\underline{\Omega}^{STOP} \longrightarrow \underline{\Omega}^N$.

The $\begin{cases}\text{symmetric}\\ \text{hyperquadratic}\\ \text{quadratic}\end{cases}$ signature map $\begin{cases}\sigma^*:\Omega_n^{STOP}(K) \longrightarrow L^n(\pi_1(K))\\ \hat{\sigma}^*:\Omega_n^{N}(K) \longrightarrow \hat{L}^n(\pi_1(K))\\ \sigma_*:\Omega_{n+1}^{N,STOP}(K) \longrightarrow L_n(\pi_1(K))\end{cases}$ factors through the algebraic assembly map

$$\begin{cases}\sigma^* : \Omega_n^{STOP}(K) = H_n(K;\underline{MSTOP}) \xrightarrow{\sigma^*} H_n(K;\underline{\mathbb{L}}^0) \xrightarrow{\sigma^*} L^n(\pi_1(K))\\ \hat{\sigma}^* : \Omega_n^{N}(K) = H_n(K;\underline{MSG}) \xrightarrow{\hat{\sigma}^*} H_n(K;\underline{\hat{\mathbb{L}}}^0) \xrightarrow{\hat{\sigma}^*} \hat{L}^n(\pi_1(K))\\ \sigma_* : \Omega_{n+1}^{N,STOP}(K) = H_n(K;\underline{MS(G/TOP)}) \xrightarrow{\sigma_*} H_n(K;\underline{\mathbb{L}}_0) \xrightarrow{\sigma_*} L_n(\pi_1(K))\end{cases}$$

.

(These factorizations can be interpreted in terms of characteristic numbers, in particular for the surgery obstructions of normal maps of manifolds, which ca then be used to determine the homotopy types of the $\mathbb{L}$-spaces, following the work of Sullivan [Su1] and Morgan and Sullivan [MS] in the simply-connected case. See Wall [W3], Jones [J2], Taylor and Williams [TaW] for generalizations to the non-simply-connected case. In [TaW] it is shown that the algebraic $\mathbb{L}$-spectra become generalized Eilenberg-MacLane spectra localized at 2, and wedges of bo-coefficient spectra localized away from 2).

Given a ring Ω-spectrum $\underline{R} = \{R_k = \Omega R_{k+1}, \otimes : R_j \wedge R_k \longrightarrow R_{j+k}, 1_k : S^k \longrightarrow R_k\}$ let $B\underline{R}G$ be the classifying space for stable $\underline{R}$-oriented spherical fibrations over finite CW complexes, and let $R_\otimes$ be the component of $1 \in \pi_0(\underline{R})$ in R_0. If $\pi_0(\underline{R}) = \mathbb{Z}$ the morphism $\underline{R} \longrightarrow \underline{K}(\mathbb{Z})$ induces a forgetful map $B\underline{R}G \longrightarrow B\underline{K}(\mathbb{Z})G = BSG$, and ther is defined a fibration sequence of spaces

$$R_\otimes \longrightarrow B\underline{R}G \longrightarrow BSG\ .$$

In particular, we have defined a commutative braid of fibration sequences

$$\begin{array}{ccccccccc}
\mathbb{L}_0 & & \longrightarrow & & B\underline{\mathbb{L}}^0G & & \longrightarrow & & BSG \\
& \searrow^{1+T} & & \nearrow & & \searrow^{J} & & \nearrow & \\
& & \mathbb{L}^0_\otimes & & & & B\hat{\underline{\mathbb{L}}}^0G & & \\
& & & \searrow^{J} & & \nearrow & & & \\
& & & & \hat{\mathbb{L}}^0_\otimes & & & &
\end{array}$$

with $\mathbb{L}_0$ the 0th term of $\underline{\mathbb{L}}_0 = \underline{\mathbb{L}}_0(1)_{\S}$, i.e. the connected Kan complex of quadratic Poincaré n-ads over $\mathbb{Z}$ such that $\pi_n(\mathbb{L}_0) = L_n(1)\ (n \geqslant 1)$.

We have defined a commutative square of ring spectra

$$\begin{array}{ccc}
\underline{MSTOP} & \xrightarrow{\sigma^*} & \underline{\mathbb{L}}^0 \\
{\scriptstyle J}\downarrow & & \downarrow{\scriptstyle J} \\
\underline{MSG} & \xrightarrow{\hat{\sigma}^*} & \hat{\underline{\mathbb{L}}}^0
\end{array} \quad .$$

An oriented $\left\{\begin{array}{l}\text{topological bundle } \alpha : K \longrightarrow BSTOP(k) \\ \text{spherical fibration } \beta : K \longrightarrow BSG(k)\end{array}\right.$ over a finite CW complex K has a

canonical $\left\{\begin{array}{l}\underline{MSTOP}\text{-} \\ \underline{MSG}\text{-}\end{array}\right.$ orientation $\left\{\begin{array}{l}U(\alpha) \in \dot{H}^k(T(\alpha);\underline{MSTOP}) \\ \hat{U}(\beta) \in \dot{H}^k(T(\beta);\underline{MSG})\end{array}\right.$, and hence also a canonical

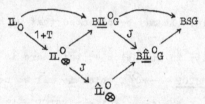

$\left\{\begin{array}{l}\underline{\mathbb{L}}^0\text{-}\\ \hat{\underline{\mathbb{L}}}^0\text{-}\end{array}\right.$ orientation $\left\{\begin{array}{l}\sigma^*U(\alpha)\in \dot{H}^k(T(\alpha);\underline{\mathbb{L}}^0)\\ \hat{\sigma}^*\hat{U}(\beta)\in \dot{H}^k(T(\beta);\hat{\underline{\mathbb{L}}}^0)\end{array}\right.$. There is induced a morphism of fibrations

$$\begin{array}{ccccc} G/TOP & \longrightarrow & BSTOP & \xrightarrow{J} & BSG \\ \downarrow{\scriptstyle \sigma_*} & & \downarrow{\scriptstyle \sigma^*U} & & \downarrow{\scriptstyle \hat{\sigma}^*\hat{U}} \\ \mathbb{L}_0 & \longrightarrow & B\underline{\mathbb{L}}^0G & \xrightarrow{J} & B\hat{\underline{\mathbb{L}}}^0G \end{array}$$

with $\sigma_*:G/TOP \longrightarrow \mathbb{L}_0$ the map associating to each singular simplex $\Delta \longrightarrow G/TOP$ the quadratic Poincaré n-ad $\sigma_*(f,b)$ over $\mathbb{Z}$ of the normal map of manifold n-ads $(f,b):M \longrightarrow \Delta$ that it classifies. Now $\sigma_*:G/TOP \longrightarrow \mathbb{L}_0$ induces the surgery obstruction isomorphisms

$$\sigma_* = \theta : \pi_*(G/TOP) \longrightarrow \pi_*(\mathbb{L}_0) = L_*(1) \ ,$$

so that it is a homotopy equivalence by J.H.C.Whitehead's theorem. The right hand square is thus a homotopy-theoretic pullback, and for any spherical fibration $\beta:K \longrightarrow BSG(k)$ there is an identification of sets of equivalence classes

$\{$stable topological reductions $\tilde{\beta}:K \longrightarrow BSTOP$ of $\beta:K \longrightarrow BSG(k)\}$

$= \{$pairs (V,h) consisting of a map $V:T(\beta) \longrightarrow \underline{\mathbb{L}}^{-k}$ and a homotopy

$h : JV \simeq \hat{V} : T(\beta) \longrightarrow \hat{\underline{\mathbb{L}}}^{-k}\}$

for some fixed map $\hat{V}:T(\beta) \longrightarrow \hat{\underline{\mathbb{L}}}^{-k}$ representing the canonical $\underline{\mathbb{L}}^0$- orientation $\hat{\sigma}^*\hat{U}(\beta) \in \dot{H}^k(T(\beta);\hat{\underline{\mathbb{L}}}^0) = [T(\beta),\hat{\underline{\mathbb{L}}}^{-k}]$. We thus have an equivalence of categories

$\{$stable oriented topological bundles (over finite CW complexes)$\}$

$\approx \{$stable spherical fibrations with an $\underline{\mathbb{L}}^0$- orientation lifting the canonical $\hat{\underline{\mathbb{L}}}^0$- orientation$\}$.

Localizing away from 2 we have the Sullivan [Su2] characterization of stable topological bundles as $KO[\frac{1}{2}]$-oriented spherical fibrations, with

$$\mathbb{L}_0[\tfrac{1}{2}] = BO[\tfrac{1}{2}] \quad , \quad \underline{\mathbb{L}}^0[\tfrac{1}{2}] = \underline{bo}[\tfrac{1}{2}] \quad , \quad \hat{\underline{\mathbb{L}}}^0[\tfrac{1}{2}] = \underline{K}(\mathbb{Z})[\tfrac{1}{2}] \ .$$

I should like to thank Graeme Segal and Frank Quinn for discussions pertaining to the L-theoretic characterization of topological bundles. (It is in fact equivalent to the Levitt-Morgan-Brumfiel characterization of stable topological bundles as spherical fibrations with geoemtric Poincaré transversality [LeM],[BM]. Unstably, the result $G(k)/TOP(k) \simeq G/TOP$ $(k \geqslant 3)$ of Rourke and Sanderson [RS] applies to show that there is an equivalence of categories

$$\{\text{oriented topological k-block bundles (over finite CW complexes)}\}$$
$$\approx \{\text{(k-1)-spherical fibrations with an } \underline{\mathbb{L}}^0\text{-orientation lifting the canonical } \hat{\underline{\mathbb{L}}}^0\text{- orientation}\} .$$

The homotopy equivalence $\sigma_*:G/TOP \longrightarrow \mathbb{L}_0$ is not an H-map from the H-space structure on G/TOP defined by the Whitney sum of bundles to the H-space structure on $\mathbb{L}_0$ defined by the direct sum of quadratic Poincaré n-ads. The latter is equivalent to the Quinn disjoint union of surgery problems addition, and also to the Sullivan characteristic variety addition in G/TOP. The former is expressed in terms of the latter by $(a,b) \longmapsto a{\oplus}b{\oplus}(a\otimes b)$. Madsen and Milgram [MM] show that there exists no (2-local) homotopy equivalence $B(G/TOP) \longrightarrow \mathbb{L}_{-1}$ extending the above diagram to the right by a commutative square

$$\begin{array}{ccc} BSG & \longrightarrow & B(G/TOP) \\ {\scriptstyle \hat{\sigma}*\hat{U}}\downarrow & & \downarrow \\ B\hat{\underline{\mathbb{L}}}^0G & \longrightarrow & \mathbb{L}_{-1} \end{array} .$$

Here, $\mathbb{L}_{-1}$ is the 1st term of $\underline{\mathbb{L}}_0$, the delooping of $\mathbb{L}_0$ defined by the universal cover of the connected Kan complex $\mathbb{L}_{-1}(1)$ of quadratic Poincaré n-ads over $\mathbb{Z}$ such that $\pi_n(\mathbb{L}_{-1}(1)) = L_{n-1}(1)$ $(n \geqslant 1)$. Localizing at 2 we have

$$\mathbb{L}_0(1)_{(2)} = \prod_{i=0}^{\infty}(K(\mathbb{Z}_{(2)},4i)\times K(\mathbb{Z}_2,4i+2)) \ , \ \mathbb{L}_{-1}(1)_{(2)} = \prod_{i=0}^{\infty}(K(\mathbb{Z}_{(2)},4i+1)\times K(\mathbb{Z}_2,4i+$$

$$\underline{\mathbb{L}}^0{}_{(2)} = \prod_{i=0}^{\infty}(\Sigma^{4i}\underline{K}(\mathbb{Z}_{(2)})\times\Sigma^{4i+1}\underline{K}(\mathbb{Z}_2)) \ ,$$
$$\hat{\underline{\mathbb{L}}}^0{}_{(2)} = \underline{K}(\mathbb{Z}_{(2)})\times\prod_{i=0}^{\infty}(\Sigma^{4i+1}\underline{K}(\mathbb{Z}_2)\times\Sigma^{4i+3}\underline{K}(\mathbb{Z}_2)\times\Sigma^{4i+4}\underline{K}(\mathbb{Z}_8)) \ ).$$

Given an oriented spherical fibration $\beta:K \longrightarrow BSG(k)$ over a finite CW complex K define

$$t(\beta) = H\hat{\sigma}*\hat{U}(\beta) \in \dot{H}^{k+1}(T(\beta);\underline{\mathbb{L}}_0) \ ,$$

the image of the canonical $\hat{\underline{\mathbb{L}}}^0$- orientation $\hat{\sigma}*\hat{U}(\beta) \in \dot{H}^k(T(\beta);\hat{\underline{\mathbb{L}}}^0)$ under the map H appearing in the exact sequence

$$\cdots \longrightarrow \dot{H}^k(T(\beta);\underline{\mathbb{L}}_0) \xrightarrow{1+T} \dot{H}^k(T(\beta);\underline{\mathbb{L}}^0) \xrightarrow{J} \dot{H}^k(T(\beta);\hat{\underline{\mathbb{L}}}^0) \xrightarrow{H} \dot{H}^{k+1}(T(\beta);\underline{\mathbb{L}}_0) \longrightarrow \cdots$$

By the above, β admits a stable topological reduction $\tilde{\beta}:K \longrightarrow BSTOP$ if and only if $t(\beta) = 0$. (We have that $t(\beta)$ is a torsion element, and that

$$\underline{\mathbb{L}}_0[\tfrac{1}{2}] = \underline{bso}[\tfrac{1}{2}] \ , \quad \underline{\mathbb{L}}_{0(2)} = \prod_{i=0}^{\infty}(\Sigma^{4i+2}\underline{K}(\mathbb{Z}_2)\times\Sigma^{4i+4}\underline{K}(\mathbb{Z}_{(2)})) \ .$$

Localized at 2 $t(\beta)$ can be expressed as a stable characteristic class

$$t(\beta)_{(2)} \in \prod_{i=1}^{\infty} H^{4i-1}(K;\mathbb{Z}_2)\oplus \mathrm{im}(H^{4i}(K;\mathbb{Z}_8)\longrightarrow H^{4i+1}(K;\mathbb{Z}_{(2)})) \ .$$

Away from 2 $t(\beta)$ is the obstruction to a $KO[\frac{1}{2}]$-orientation of β

$$t(\beta)[\tfrac{1}{2}] = \widetilde{KSO}^{k+1}(T(\beta))[\tfrac{1}{2}] \quad).$$

Given an n-dimensional geometric Poincaré complex X let $\mathcal{T}^{TOP}(X)$ be the topological normal map bordism set of X, defined as usual to be the set of equivalence classes of normal maps $(f,b):M\longrightarrow X$ in the sense of Browder and Wall, under the relation

$(f,b)\sim(f',b')$ if there exists a normal map

$$((g;f,f'),(c;b,b')) \ : \ (N;M,M') \longrightarrow (X\times I;X\times 0,X\times 1) \ .$$

The surgery obstruction function

$$\theta \ : \mathcal{T}^{TOP}(X)\longrightarrow L_n(\pi_1(X)) \ ; \ (f,b)\longmapsto \sigma_*(f,Jb)$$

fits into the Sullivan-Wall surgery exact sequence of sets

$$L_{n+1}(\pi_1(X))\longrightarrow \mathcal{S}^{TOP}(X)\longrightarrow \mathcal{T}^{TOP}(X)\xrightarrow{\ \theta\ } L_n(\pi_1(X)) \ .$$

In the case $\mathcal{T}^{TOP}(X)\neq\emptyset$ (i.e. if the Spivak normal fibration $\nu_X:X\longrightarrow BSG$ admits a topological reduction) we shall express θ in terms of the assembly map $\sigma_*:H_n(X;\underline{\mathbb{L}}_0)\longrightarrow L_n(\pi_1(X))$.

Let $G(k)/TOP(k)$ denote the homotopy-theoretic fibre of the forgetful map $J:BSTOP(k)\longrightarrow BSG(k)$, as usual, and let $MS(G(k)/TOP(k))$ be the homotopy-theoretic fibre of the forgetful map of Thom spaces $J:MSTOP(k)\longrightarrow MSG(k)$ $(k\geqslant 0)$. The canonical topological bundle $\eta_k:G(k)/TOP(k)\longrightarrow BSTOP(k)$ has a canonical fibre homotopy trivialization $h_k:J\eta_k\simeq J\varepsilon^k:G(k)/TOP(k)\longrightarrow BSG(k)$. The canonical $\underline{MSTOP}$-orientation $U(\eta_k)\in \dot{H}^k(T(\eta_k);\underline{MSTOP})$ is represented by the induced map of Thom spaces

$$U(\eta_k) \ : \ T(\eta_k) = \Sigma^k(G(k)/TOP(k))_+\longrightarrow MSTOP(k) \ ,$$

using h_k to identify $T(\eta_k) = T(\varepsilon^k) = \Sigma^k(G(k)/TOP(k))_+$. The canonical

MSTOP-orientation $U(\varepsilon^k) \in \dot{H}^k(T(\varepsilon^k);\underline{MSTOP})$ of the trivial topological bundle $\varepsilon^k: G(k)/TOP(k) \longrightarrow BSTOP(k)$ is represented by the composite

$$U(\varepsilon^k) : T(\varepsilon^k) = \Sigma^k(G(k)/TOP(k))_+ \xrightarrow{\text{collapse}} \Sigma^k(S^0) = S^k \xrightarrow{1_k} MSTOP(k) .$$

The fibre homotopy $h_k: J\eta_k \simeq J\varepsilon^k: G(k)/TOP(k) \longrightarrow BSG(k)$ determines a homotopy

$$T(h_k) : JU(\eta_k) \simeq JU(\varepsilon^k) : \Sigma^k(G(k)/TOP(k))_+ \longrightarrow MSG(k) ,$$

and hence a map

$$\Gamma_k : G(k)/TOP(k) \longrightarrow \Omega^k MS(G(k)/TOP(k))$$

such that

$$\text{adjoint}\, U(\eta_k) - \text{adjoint}\, U(\varepsilon^k) : G(k)/TOP(k) \xrightarrow{\Gamma_k} \Omega^k MS(G(k)/TOP(k)) \longrightarrow \Omega^k MSTOP(k)$$

(up to homotopy). The maps Γ_k $(k \geqslant 0)$ fit together to define a map

$$\Gamma = \varinjlim_k \Gamma_k : G/TOP = \varinjlim_k G(k)/TOP(k) \longrightarrow \Omega^\infty MS(G/TOP) = \varinjlim_k \Omega^k MS(G(k)/TOP(k))$$

Now $\Omega^\infty MS(G/TOP)$ is the infinite loop space corresponding to the (normal,manifold) bordism spectrum with a dimension shift, $\underline{MS(G/TOP)} = \Sigma^{-1}\underline{\Omega}^{N,STOP}$, and so can be regarded as a Kan complex of (normal,manifold)-pair n-ads. The quadratic signature of such n-ads defines a map

$$\sigma_* : \Omega^\infty MS(G/TOP) \longrightarrow \mathbb{L}_0 .$$

The map $\Gamma: G/TOP \longrightarrow \Omega^\infty MS(G/TOP)$ sends a singular simplex in G/TOP to the mapping cylinder of the normal map of manifold n-ads that it classifies. The composite

$$\sigma_* : G/TOP \xrightarrow{\Gamma} \Omega^\infty MS(G/TOP) \xrightarrow{\sigma_*} \mathbb{L}_0$$

is the homotopy equivalence defined previously.

Let X be an n-dimensional geometric Poincaré complex, and let

$$(\nu_X: X \longrightarrow BSG(k), \rho_X: S^{n+k} \longrightarrow T(\nu_X))$$

be a Spivak normal structure. The composite

$$\alpha_X : S^{n+k} \xrightarrow{\rho_X} T(\nu_X) \xrightarrow{\Delta} X_+ \wedge T(\nu_X)$$

is an S-duality map between X_+ and $T(\nu_X)$, so that for any spectrum $\underline{R} = \{R_k, \Sigma R_k \longrightarrow R_{k+1}\}$ there are defined isomorphisms

$$\alpha_X : \dot{H}^*(T(\nu_X);\underline{R}) = \varinjlim_j [\Sigma^j T(\nu_X), R_{j+*}] \xrightarrow{\sim} H_{n+k-*}(X;\underline{R}) = \varinjlim_j \pi_{n+j+k-*}(X_+ \wedge R_j) ;$$

$$\{g_j : \Sigma^j T(\nu_X) \longrightarrow R_{j+*}\} \longmapsto \{S^{n+j+k} \xrightarrow{\Sigma^j \alpha_X} X_+ \wedge \Sigma^j T(\nu_X) \xrightarrow{1 \wedge g_j} X_+ \wedge R_{j+*}\} .$$

Any two Spivak normal structures on X (ν_X, ρ_X), (ν'_X, ρ'_X) are related by a stable fibre homotopy equivalence $c : \nu_X \longrightarrow \nu'_X$ over $1 : X \longrightarrow X$ such that $T(c)_*(\rho_X) = \rho'_X \in \pi^S_{n+k'}(T(\nu'_X))$, and any two such fibre homotopy equivalences are related by a stable fibre homotopy. The Browder-Novikov transversality construction of normal maps identifies

$\mathcal{T}^{TOP}(X)$ = the set of equivalence classes of topological normal structures

$$(\nu_X : X \longrightarrow BSTOP(k), \rho_X : S^{n+k} \longrightarrow T(\nu_X)) .$$

Thus if $\mathcal{T}^{TOP}(X) \neq \emptyset$ and $x_0 = ((f_0, b_0) : M_0 \longrightarrow X) \in \mathcal{T}^{TOP}(X)$ is the normal map bordism class associated to some topological normal structure $(\nu_0 : X \longrightarrow BSTOP(k_0)$, $\rho_0 : S^{n+k_0} \longrightarrow T(\nu_0))$ we have the usual bijections (depending on x_0)

$\mathcal{T}^{TOP}(X) \approx$ the set of equivalence classes of stable topological reductions

$$\tilde{\nu}_0 : X \longrightarrow BSTOP \text{ of } J\nu_0 : X \longrightarrow BSG(k_0) ,$$

and

$$x_0 : \mathcal{T}^{TOP}(X) \xrightarrow{\sim} [X, G/TOP] ; ((f_1, b_1) : M_1 \longrightarrow X) \longmapsto (\nu_1 - \nu_0, c) ,$$

with $(\nu_1 : X \longrightarrow BSTOP(k_1), \rho_1 : S^{n+k_1} \longrightarrow T(\nu_1))$ a topological normal structure associated to $(f_1, b_1) \in \mathcal{T}^{TOP}(X)$. Let $\alpha_0 : S^{n+k_0} \xrightarrow{\rho_0} T(\nu_0) \xrightarrow{\Delta} X_+ \wedge T(\nu_0)$ be the S-duality map determined by (ν_0, ρ_0). The image of the canonical MSTOP-orientation $U(\nu_0) \in \dot{H}^k(T(\nu_0);\underline{MSTOP})$ under the S-duality isomorphism

$$\alpha_0 : \dot{H}^{k_0}(T(\nu_0);\underline{MSTOP}) \xrightarrow{\sim} H_n(X;\underline{MSTOP}) = \Omega_n^{STOP}(X)$$

is the MSTOP-orientation $[X]_0 = (M_0, f_0) \in \Omega_n^{STOP}(X)$ of X determined by $(f_0, b_0) \in \mathcal{T}^{TOP}(X)$. For any MSTOP-module spectrum $\underline{R} = \{R_j, \Sigma R_j \longrightarrow R_{j+1}, \otimes : MSTOP(j) \wedge R_k \longrightarrow R_{j+k}\}$ there is defined an $\underline{R}$-coefficient Thom isomorphism

$$- \cup U(\nu_0) : H^0(X;\underline{R}) \xrightarrow{\sim} \dot{H}^{k_0}(T(\nu_0);\underline{R}) ;$$

$$\{g_j : \Sigma^j X_+ \longrightarrow R_j\} \longmapsto \{\Sigma^j T(\nu_0) \xrightarrow{\Delta} T(\nu_0) \wedge \Sigma^j X_+ \xrightarrow{U(\nu_0) \wedge g_j} MSTOP(k_0) \wedge R_j \xrightarrow{\otimes} R_{j+k_0}\} ,$$

so that the composite

$$[X]_0 \cap - : H^0(X;\underline{R}) \xrightarrow[\sim]{U(\nu_0) \cup -} \dot{H}^{k_0}(T(\nu_0);\underline{R}) \xrightarrow[\sim]{\alpha_0} H_n(X;\underline{R})$$

is an $\underline{R}$-coefficient Poincaré duality isomorphism. (This point of view derives fr G.W.Whitehead's treatment of orientability with respect to extraordinary (co)homology theories, and from Atiyah's reformulation of Thom's smooth cobordism theory in terms of $\underline{MSO}$-orientations). In particular, $\underline{MSTOP}$ and $\underline{MS(G/TOP)}$ are $\underline{MSTOP}$-module spectra. Let $\Phi : G/TOP \longrightarrow \Omega^\infty MSTOP = \varinjlim_k \Omega^k MSTOP(k)$ be the map which restricts to the adjoints $(G(k)/TOP(k))_+ \longrightarrow \Omega^k MSTOP(k)$ of the canonical $\underline{MSTOP}$-orientations $U(\gamma_k) : \Sigma^k (G(k)/TOP(k))_+ \longrightarrow MSTOP(k)$, so that

$$\Phi - 1 : G/TOP \xrightarrow{\Gamma} \Omega^\infty MS(G/TOP) \longrightarrow \Omega^\infty MSTOP \ .$$

Given a topological bundle $\eta : X \longrightarrow BSTOP(j)$ and a fibre homotopy trivialization $h : J\eta \simeq \varepsilon^j : X \longrightarrow BSG(j)$ there is defined a topological normal structure

$$(\nu_1 = \eta \oplus \nu_0 : X \longrightarrow BSTOP(k_1), \rho_1 : S^{n+k_1} \xrightarrow{\Sigma^j \rho_0} \Sigma^j T(\nu_0) = T(\varepsilon^j \oplus \nu_0) \xrightarrow[\simeq]{T(h\oplus 1)^{-1}} T(\nu_1)) \ ,$$

where $k_1 = j + k_0$. The image of the classifying map $(\eta, h) : X \longrightarrow G/TOP$ under the bijection $x_0^{-1} : [X, G/TOP] \xrightarrow{\simeq} \mathcal{T}^{TOP}(X)$ is the bordism class of the normal map $(f_1, b_1) : M_1 \longrightarrow X$ associated to (ν_1, ρ_1). The composite

$$\begin{array}{l} [X,G/TOP] \xrightarrow{\Phi} [X_+, \Omega^\infty MSTOP] = H^0(X;\underline{MSTOP}) \xrightarrow[\simeq]{-\cup U(\nu_0)} \dot{H}^{k_0}(T(\nu_0);\underline{MSTOP}) \\ (=[X_+,G/TOP]) \\ \qquad \xrightarrow[\simeq]{\Sigma^j} \dot{H}^{k_1}(T(\varepsilon^j \oplus \nu_0);\underline{MSTOP}) \xrightarrow[\simeq]{T(h\oplus 1)^*} \dot{H}^{k_1}(T(\nu_1);\underline{MSTOP}) \end{array}$$

sends $(\eta, h) \in [X, G/TOP]$ to the canonical MSTOP-orientation $U(\nu_1) \in \dot{H}^{k_1}(T(\nu_1);\underline{MSTOP})$ The composite

$$\begin{array}{l} \alpha_1 : \dot{H}^{k_1}(T(\nu_1);\underline{MSTOP}) \xrightarrow[\simeq]{T(h\oplus 1)^{*-1}} \dot{H}^{k_1}(T(\varepsilon^j \oplus \nu_0);\underline{MSTOP}) \\ \qquad \xrightarrow[\simeq]{\Sigma^{-j}} \dot{H}^{k_0}(T(\nu_0);\underline{MSTOP}) \xrightarrow[\simeq]{\alpha_0} H_n(X;\underline{MSTOP}) \end{array}$$

is the S-duality isomorphism determined by (ν_1, ρ_1). The composite

$$[X,G/TOP] \xrightarrow{\Gamma} [X, \Omega^\infty MS(G/TOP)] = H^0(X;\underline{MS(G/TOP)}) \xrightarrow[\simeq]{[X]_0 \cap -} H_n(X;\underline{MS(G/TOP)}) = \Omega_{n+1}^{N,STOP}$$

sends $(\eta, h) \in [X, G/TOP]$ to $(W_1 \cup_X -W_0, M_1 \cup -M_0) \in \Omega_{n+1}^{N,STOP}(X)$, where W_i is the mapping cylinder of $f_i : M_i \longrightarrow X$ $(i = 0,1)$. Let $\sigma^*[X]_0 \in H_n(X;\underline{L}^0)$ be the $\underline{L}^0$-orientation of determined by $[X]_0 \in H_n(X;\underline{MSTOP})$, so that there is defined a commutative diagram

$$\begin{array}{ccccc}
[X,G/TOP] & \xrightarrow{\Gamma} & [X,\Omega^{\infty}MS(G/TOP)] = H^0(X;\underline{MS(G/TOP)}) & \xrightarrow{[X]_0\cap -} & H_n(X;\underline{MS(G/TOP)}) = \Omega^{N,STOP}_{n+1}(X) \\
\sigma_* \downarrow \wr & & \downarrow \sigma_* & & \downarrow \sigma_* \\
[X,\mathbb{L}_0] & \xrightarrow[\sim]{1} & H^0(X;\underline{\mathbb{L}}_0) & \xrightarrow[\sim]{\sigma^*[X]_0\cap -} & H_n(X;\underline{\mathbb{L}}_0)
\end{array} .$$

Furthermore, there is defined a commutative diagram

$$\begin{array}{ccc}
\Omega^{N,STOP}_{n+1}(X) & \longrightarrow & \Omega^{N,P}_{n+1}(X) \\
\sigma_* \downarrow & & \downarrow \sigma_* \\
H_n(X;\underline{\mathbb{L}}_0) & \xrightarrow{\sigma_*} & L_n(\pi_1(X))
\end{array} ,$$

and

$$(W_1\cup_X -W_0, M_1\cup -M_0) = (W_1, M_1\cup -X) - (W_0, M_0\cup -X) \in \Omega^{N,P}_{n+1}(X) .$$

Thus the surgery obstruction $\theta(f_1,b_1) = \sigma_*(W_1,M_1\cup -X) \in L_n(\pi_1(X))$ of $(f_1,b_1)\in \mathcal{J}^{TOP}(X)$ is given by

$$\begin{aligned}
\theta(f_1,b_1) &= \sigma_*(W_1\cup_X -W_0, M_1\cup -M_0) + \sigma_*(W_0, M_0\cup -X) \\
&= \sigma_*(x_1) + \theta(f_0,b_0) \in L_n(\pi_1(X)) ,
\end{aligned}$$

where $\sigma_*(x_1)\in L_n(\pi_1(X))$ is the image of (f_1,b_1) under the composite

$$\mathcal{J}^{TOP}(X) \xrightarrow[\sim]{x_0} [X,G/TOP] \xrightarrow[\sim]{\sigma^*[X]_0\cap -} H_n(X;\underline{\mathbb{L}}_0) \xrightarrow{\sigma_*} L_n(\pi_1(X)) .$$

We now define the total surgery obstruction $s(X)\in \mathcal{S}_n(X)$ of an n-dimensional geometric Poincaré complex X, as follows. Let $(\nu_X : X \longrightarrow BSG(k), \rho_X : S^{n+k} \longrightarrow T(\nu_X))$ be a Spivak normal structure of X, and let $\alpha_X : S^{n+k} \xrightarrow{\rho_X} T(\nu_X) \xrightarrow{\Delta} X_+ \wedge T(\nu_X)$ be the corresponding S-duality map. Consider the commutative diagram

$$\begin{array}{ccccc}
\dot{H}^k(T(\nu_X);\underline{\hat{\mathbb{L}}}^0) & \xrightarrow[\sim]{\alpha_X} & H_n(X;\underline{\hat{\mathbb{L}}}^0) & \xrightarrow{\hat{\sigma}^*} & \hat{L}^n(\pi_1(X)) \\
H \downarrow & & H \downarrow & & H \downarrow \\
\dot{H}^{k+1}(T(\nu_X);\underline{\mathbb{L}}_0) & \xrightarrow[\sim]{\alpha_X} & H_{n-1}(X;\underline{\mathbb{L}}_0) & \xrightarrow{\sigma_*} & L_{n-1}(\pi_1(X))
\end{array} .$$

The canonical $\underline{\hat{\mathbb{L}}}^0$-orientation $\hat{V} = \hat{\sigma}^*\hat{U}(\nu_X) \in \dot{H}^k(T(\nu_X);\underline{\hat{\mathbb{L}}}^0)$ is such that

i) $H(\hat{V}) = t(\nu_X) \in \dot{H}^{k+1}(T(\nu_X);\underline{\mathbb{L}}_0)$ is the obstruction to a stable topological reduction of ν_X

ii) $\hat{\sigma}^*\alpha_X(\hat{V}) = \hat{\sigma}^*(X) = J\sigma^*(X) \in \hat{L}^n(\pi_1(X))$ is the hyperquadratic signature of X, with $\sigma^*(X)\in L^n(\pi_1(X))$ the symmetric signature of X.

Thus $\sigma_*(\alpha_X H(\hat{V})) = HJ\sigma^*(X) = 0 \in L_{n-1}(\pi_1(X))$, and working on the $\mathbb{L}_0(\pi_1(X))$-space level we can use the $\mathbb{Z}[\pi_1(X)]$-coefficient Poincaré duality on the chain level to obtain an explicit null-homotopy of a simplex representing $\sigma_*(\alpha_X H(\hat{V})) \in L_{n-1}(\pi_1(X$ and hence an element $s(X) \in \pi_n(\sigma_*: X_+ \wedge \underline{\mathbb{L}}_0 \longrightarrow \underline{\mathbb{L}}_0(\pi_1(X))_{\S}) = \mathcal{S}_n(X)$. The image of $s(X)$ in $H_{n-1}(X;\underline{\mathbb{L}}_0)$ is the S-dual of $t(\nu_X) \in \dot{H}^{k+1}(T(\nu_X);\underline{\mathbb{L}}_0)$. If $t(\nu_X) = 0$ choose a stable topological reduction $\nu_0 : X \longrightarrow BSTOP$ of ν_X, let $x_0 = (f_0, b_0) \in \mathcal{T}^{TOP}(X)$ be t corresponding normal map, and let $[X]_0 = \alpha_X(\sigma^* U(\nu_0)) \in H_n(X;\underline{\mathbb{L}}^0)$ denote the $\underline{\mathbb{L}}^0$-orientation of X determined by the canonical $\underline{\mathbb{L}}^0$-orientation of ν_0 $\sigma^* U(\nu_0) \in \dot{H}^k(T(\nu_X);\underline{\mathbb{L}}^0)$. By the above, the surgery obstruction function is given b

$$\theta : \mathcal{T}^{TOP}(X) \longrightarrow L_n(\pi_1(X)) \ ; \ x_1 \longmapsto \sigma_*(x_1) + \theta(x_0) \ ,$$

where $\sigma_*(x_1)$ is the evaluation of the composite

$$\mathcal{T}^{TOP}(X) \xrightarrow[\sim]{x_0} [X, G/TOP] \xrightarrow[\sim]{\sigma_*} [X, \mathbb{L}_0] = H^0(X;\underline{\mathbb{L}}_0) \xrightarrow[\sim]{[X]_0 \cap -} H_n(X;\underline{\mathbb{L}}_0) \xrightarrow{\sigma_*} L_n(\pi_1(X)) \ .$$

The composite $\mathcal{T}^{TOP}(X) \xrightarrow{\theta} L_n(\pi_1(X)) \longrightarrow \mathcal{S}_n(X)$ sends every element $x_1 \in \mathcal{T}^{TOP}(X)$ to $s(X) \in \mathcal{S}_n(X)$, and the inverse image of $s(X)$ in $L_n(\pi_1(X))$ is precisely the coset of the subgroup $\mathrm{im}(\sigma_* : H_n(X;\underline{\mathbb{L}}_0) \longrightarrow L_n(\pi_1(X)))$ consisting of the surgery obstruction $\theta(x_1) \in L_n(\pi_1(X))$ of all the elements $x_1 \in \mathcal{T}^{TOP}(X)$. The surgery exact sequence has been extended to the right

$$L_{n+1}(\pi_1(X)) \longrightarrow \mathcal{S}^{TOP}(X) \longrightarrow \mathcal{T}^{TOP}(X) \xrightarrow{\theta} L_n(\pi_1(X)) \longrightarrow \mathcal{S}_n(X) \longrightarrow H_{n-1}(X;\underline{\mathbb{L}}_0) \longrightarrow \ldots$$

with $s(X) = 0 \in \mathcal{S}_n(X)$ if and only if there exists a normal map $x_1 = (f_1, b_1) \in \mathcal{T}^{TOP}$ with surgery obstruction $\theta(f_1, b_1) = 0 \in L_n(\pi_1(X))$, i.e. if and only if X is simpl homotopy equivalent to a closed topological manifold.

This completes the sketch of the proof of Theorem 1.

In order to identify $\mathcal{S}^{TOP}(X) = \mathcal{S}_{n+1}(X)$ for an n-dimensional manifold X note that an element $x \in \mathcal{S}_{n+1}(X)$ is defined by a pair (y,z) consisting of a normal map bordism class $y \in H_n(X;\underline{\mathbb{L}}_0) = \mathcal{T}^{TOP}(X)$ such that $\sigma_*(y) = \theta(y) = 0 \in L_n(\pi_1(X))$, together with a particular solution z of the associated surgery problem. Such a pair (y,z) is essentially the same as a homotopy triangulation $(f:M \longrightarrow X) \in \mathcal{S}^{TOP}(X)$. The function $\mathcal{S}_{n+1}(X) \longrightarrow \mathcal{S}^{TOP}(X)$; $x = (y,z) \longmapsto (f:M \longrightarrow X)$ is an inverse for the total surgery obstruction function $s:\mathcal{S}^{TOP}(X) \longrightarrow \mathcal{S}_{n+1}(X)$.

The identification of the structure sets $\mathcal{S}_\partial^{TOP}(X\times\Delta^k, \partial(X\times\Delta^k))$ $(k \geqslant 0)$ for an n-dimensional manifold with boundary $(X,\partial X)$ with a sequence of universally defined abelian groups $\mathcal{S}_{n+k+1}(X)$ is implicit in Quinn's identification ([Q2]) of the surgery obstruction function

$$\theta : \mathcal{T}_\partial^{TOP}(X\times\Delta^k, \partial(X\times\Delta^k)) = [X\times\Delta^k, \partial(X\times\Delta^k); G/TOP, *] \longrightarrow L_{n+k}(\pi_1(X))$$

with the restrictions of universally defined abelian group morphisms

$$A : H_{n+k}(X;\underline{\mathcal{L}}) \longrightarrow L_{n+k}(\pi_1(X))$$

to $\mathrm{im}(H_{n+k}(X;\underline{\mathcal{L}}_S) \hookrightarrow H_{n+k}(X;\underline{\mathcal{L}}))$. See the forthcoming Princeton Ph.D. thesis of Andrew Nicas for induction theorems for the structure sets which exploit this group structure. (I am indebted to Larry Siebenmann for the following description of the assembly map A. Given a finite CW complex X let W be the closed regular neighbourhood of X for some embedding $X \subset S^q$ $(q \gg \dim X)$. Then $(W,\partial W)$ is a framed q-dimensional manifold with boundary, enjoying universal Poincaré duality. Let $\underline{\mathcal{L}} = \{\mathcal{L}_{-k} = \Omega\mathcal{L}_{-k-1} \mid k \in \mathbb{Z}\}$ be the connective Ω-spectrum with kth space $\mathcal{L}_{-k}$ the Kan complex of normal maps of manifold n-ads such that $\pi_{n+k}(\mathcal{L}_{-k}) = L_n(1)$ $(n, n+k \geqslant 0)$ i.e. Quinn's surgery spectrum, with $\mathcal{L}_0 \simeq L_0(1) \times G/TOP$ [Q1]. Define

$$A : H_n(X;\underline{\mathcal{L}}) = H_n(W;\underline{\mathcal{L}}) = H^{q-n}(W,\partial W;\underline{\mathcal{L}}) = [W,\partial W;\mathcal{L}_{n-q},*] \longrightarrow L_n(\pi_1(X))$$

by sending a simplicial map $(W,\partial W) \longrightarrow (\mathcal{L}_{n-q},*)$ to the surgery obstruction $\sigma_*(f,b) \in L_n(\pi_1(X))$ of the n-dimensional normal map $(f,b):M \longrightarrow N$ obtained by glueing together ("assembling") the normal maps classified by the composites $\Delta^q \hookrightarrow W \longrightarrow \mathcal{L}_{n-q}$, which comes equipped with a reference map $N \longrightarrow W \simeq X$. The quadratic signature map $\sigma_*:\underline{\mathcal{L}} \longrightarrow \underline{\mathbb{L}}_0(1)$ is a homotopy equivalence, and

$$\sigma_* : H_n(X;\underline{\mathbb{L}}_0) = H_n(X;\underline{\mathcal{L}}_S) \longrightarrow H_n(X;\underline{\mathbb{L}}_0(1)) = H_n(X;\underline{\mathcal{L}}) \xrightarrow{A} L_n(\pi_1(X)) \text{).}$$

Any simple homotopy invariant of an n-dimensional geometric Poincaré complex X which vanishes if X has the simple homotopy type of a manifold can now be expressed in terms of the total surgery obstruction $s(X)\in\mathcal{S}_n(X)$. We have alread dealt with the obstruction to a topological reduction of the Spivak normal fibrat the image of s(X) in $H_{n-1}(X;\mathbb{L}_0)$. Examples of geometric Poincaré complexes without topological reduction were first obtained by Gitler and Stasheff [GS], and Wall - of course, at the time it was only clear there was no PL reduction, but the subsequent computation $TOP/PL \simeq K(\mathbb{Z}_2,3)$ implied that there was also no topological reduction. (The Hambleton-Milgram [HM] geometric Poincaré splitting obstruction for a double cover of a 2m-dimensional geometric Poincaré complex X (which need not be oriented) is a part of the topological reducibility obstruction, being the image of $s(X)\in\mathcal{S}_{2m}(X^w)$ under the composite

$$\mathcal{S}_{2m}(X^w)\longrightarrow H^w_{2m-1}(X;\mathbb{L}_0)\xrightarrow{p_*} H^w_{2m-1}(B\Sigma_2;\mathbb{L}_0)\xrightarrow{c} L_{2m-2}(\mathbb{Z}_2^w)=\mathbb{Z}_2\ ,$$

where w refers to homology and L-theory with orientation-twisted coefficients, $p:X\longrightarrow B\Sigma_2$ is the classifying map of the covering, and c is the codimension 1 Arf invariant). The symmetric signature $\sigma^*(X)\in L^n(\pi_1(X))$ is a simple homotopy invariant of X such that $\sigma^*(X)\in \mathrm{coker}(\sigma^*:H_n(X;\mathbb{L}^0)\longrightarrow L^n(\pi_1(X)))$ vanishes if X has the simple homotopy type of a manifold. We shall express this invariant in terms of s(X) in Theorem 2 below. For example, if n = 2m and $\pi_1(X)\longrightarrow\pi$ is a morphism to a finite group π, the image of this invariant in $\mathrm{coker}(\sigma^*:H_n(K(\pi,1);\mathbb{L}^0)\longrightarrow L^n(\pi))\otimes\mathbb{Z}[\frac{1}{2}]$ is the corresponding multisignature of X reduced modulo the multisignatures of closed manifolds, i.e. those with equal components (cf. p.175 of Wall [W1]). The 4-dimensional geometric Poincaré complexe X of Wall [W2] such that $\pi_1(X)=\mathbb{Z}_p$, $\sigma^*(\widetilde{X})\neq p\sigma^*(X)\in L^4(1)=\mathbb{Z}$ are thus detecte by this invariant. (There is no problem in defining the total surgery obstruction $s(X)\in\mathcal{S}_n(X)$ for $n\leqslant 4$, or in showing that s(X) = 0 if X has the simple homotopy type of a manifold. However, the usual difficulties with low-dimensional geometric surgery prevent us from deducing the converse).

The construction of the assembly map $\sigma_*:X_+\wedge\underline{\mathbb{L}}_0 \longrightarrow \underline{\mathbb{L}}_0(\pi_1(X))_{\S}$ generalizes to a natural transformation of commutative braids of fibration sequences of spectra

$$\sigma : X_+\wedge\underline{\mathbb{L}} \longrightarrow \underline{\mathbb{L}}(\pi_1(X))$$

(for any space X), from

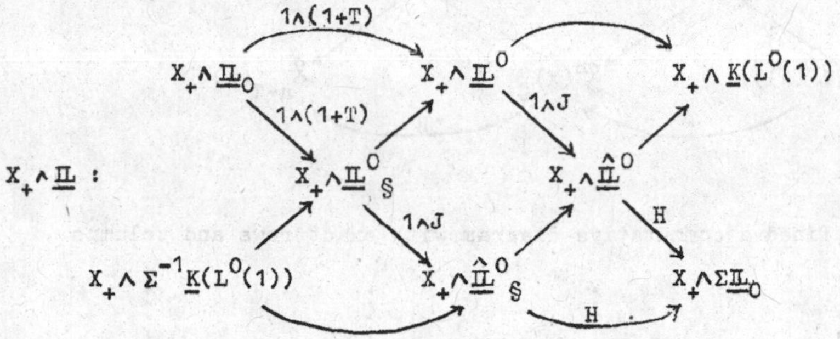

to

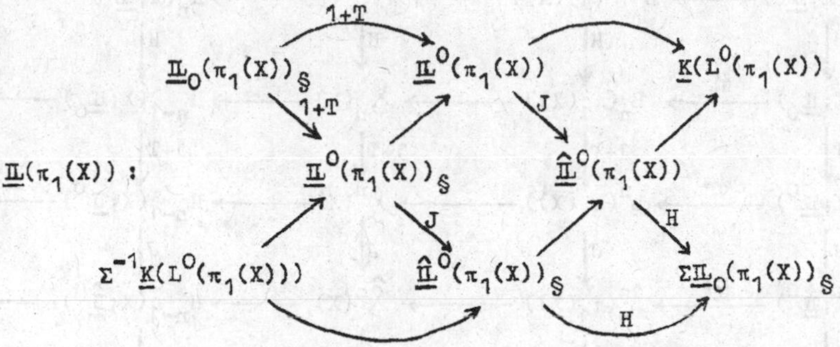

The relative homotopy groups of all the maps appearing in $\sigma:X_+\wedge\underline{\mathbb{L}} \longrightarrow \underline{\mathbb{L}}(\pi_1(X))$ define a commutative braid of exact sequences of abelian groups

$$\mathcal{S}(X):\quad \begin{array}{ccccccc}
\mathcal{S}_n(X) & & \xrightarrow{1+T} & & \mathcal{S}^n(X) & & \longrightarrow \; H_{n-1}(X;L^0(1)) \\
 & \searrow^{1+T} & & \nearrow & & \searrow^{J} & \nearrow \\
 & & \mathcal{S}^n(X)_{\S} & & & \hat{\mathcal{S}}^n(X) & \\
 & \nearrow & & \searrow^{J} & \nearrow & & \searrow^{H} \\
H_n(X;L^0(1)) & & \longrightarrow & & \hat{\mathcal{S}}^n(X)_{\S} & \xrightarrow{H} & \mathcal{S}_{n-1}(X)
\end{array}$$

,

and there are defined a commutative diagram with exact rows and columns

$$\begin{array}{ccccccccc}
 & \vdots & & \vdots & & \vdots & & \vdots & \\
 & \downarrow & & \downarrow & & \downarrow & & \downarrow & \\
\cdots \longrightarrow & H_{n+1}(X;\hat{\mathbb{L}}^0) & \xrightarrow{\hat{\sigma}^*} & \hat{L}^{n+1}(\pi_1(X)) & \longrightarrow & \hat{\mathcal{S}}^{n+1}(X) & \longrightarrow & H_n(X;\hat{\mathbb{L}}^0) & \longrightarrow \cdots \\
 & \downarrow{\scriptstyle H} & & \downarrow{\scriptstyle H} & & \downarrow{\scriptstyle H} & & \downarrow{\scriptstyle H} & \\
\cdots \longrightarrow & H_n(X;\mathbb{L}_0) & \xrightarrow{\sigma_*} & L_n(\pi_1(X)) & \longrightarrow & \mathcal{S}_n(X) & \longrightarrow & H_{n-1}(X;\mathbb{L}_0) & \longrightarrow \cdots \\
 & \downarrow{\scriptstyle 1+T} & & \downarrow{\scriptstyle 1+T} & & \downarrow{\scriptstyle 1+T} & & \downarrow{\scriptstyle 1+T} & \\
\cdots \longrightarrow & H_n(X;\mathbb{L}^0) & \xrightarrow{\sigma^*} & L^n(\pi_1(X)) & \longrightarrow & \mathcal{S}^n(X) & \longrightarrow & H_{n-1}(X;\mathbb{L}^0) & \longrightarrow \cdots \\
 & \downarrow{\scriptstyle J} & & \downarrow{\scriptstyle J} & & \downarrow{\scriptstyle J} & & \downarrow{\scriptstyle J} & \\
\cdots \longrightarrow & H_n(X;\hat{\mathbb{L}}^0) & \xrightarrow{\hat{\sigma}^*} & \hat{L}^n(\pi_1(X)) & \longrightarrow & \hat{\mathcal{S}}^n(X) & \longrightarrow & H_{n-1}(X;\hat{\mathbb{L}}^0) & \longrightarrow \cdots \\
 & \downarrow & & \downarrow & & \downarrow & & \downarrow & \\
 & \vdots & & \vdots & & \vdots & & \vdots &
\end{array}$$

and the corresponding diagram with $\begin{cases} \mathbb{L}^0{}_{\S}, \mathcal{S}^*(X)_{\S} \\ \hat{\mathbb{L}}^0{}_{\S}, \hat{\mathcal{S}}^*(X)_{\S} \end{cases}$ in place of $\begin{cases} \mathbb{L}^0, \mathcal{S}^*(X) \\ \hat{\mathbb{L}}^0, \hat{\mathcal{S}}^*(X) \end{cases}$.

If X is an n-dimensional geometric Poincaré complex the image of the total surgery obstruction $s(X) \in \mathcal{S}_n(X)$ in $H_{n-1}(X;\mathbb{L}_0)$ is the image under H of the canonical $\hat{\mathbb{L}}^0$- orientation $[\hat{X}] \in H_n(X;\hat{\mathbb{L}}^0)$.

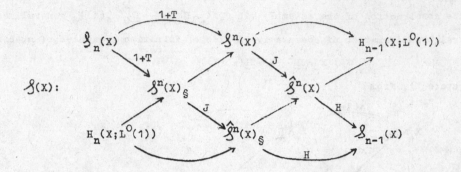

For any space X there is defined a commutative exact braid

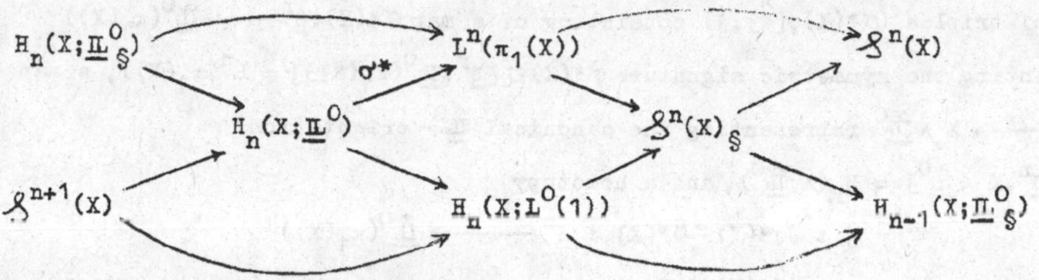

giving rise to the exact sequence

$$\ldots \longrightarrow H_n(X;\underline{\mathbb{L}}^0) \longrightarrow L^n(\pi_1(X)) \oplus H_n(X;L^0(1)) \longrightarrow \mathcal{S}^n(X)_{\S} \longrightarrow H_{n-1}(X;\underline{\mathbb{L}}^0) \longrightarrow \ldots .$$

Theorem 2 Let X be an n-dimensional geometric Poincaré complex, with total surgery obstruction $s(X) \in \mathcal{S}_n(X)$.

i) The symmetrization $(1+T)s(X)_{\S} \in \mathcal{S}^n(X)_{\S}$ is the image of

(symmetric signature $\sigma^*(X)$, fundamental class $[X]) \in L^n(\pi_1(X)) \oplus H_n(X;L^0(1))$,

so that $(1+T)s(X)_{\S} = 0$ if and only if X has an $\underline{\mathbb{L}}^0$-orientation $[X] \in H_n(X;\underline{\mathbb{L}}^0)$ which assembles to $\sigma^*([X]) = \sigma^*(X) \in L^n(\pi_1(X))$.

ii) The image of $(1+T)s(X)_{\S} \in \mathcal{S}^n(X)_{\S}$ in $H_{n-1}(X;\underline{\mathbb{L}}^0_{\S})$ is the obstruction to an $\underline{\mathbb{L}}^0$- orientation of X, or equivalently of the Spivak normal fibration $\nu_X : X \longrightarrow BSG$.

iii) The symmetrization $(1+T)s(X) \in \mathcal{S}^n(X)$ is the image of $\sigma^*(X) \in L^n(\pi_1(X))$, so that $(1+T)s(X) = 0$ if and only if $\sigma^*(X) \in \mathrm{im}(\sigma^* : H_n(X;\underline{\mathbb{L}}^0) \longrightarrow L^n(\pi_1(X)))$.

[]

It should be noted that the symmetrization maps

$$1+T : \mathcal{S}_n(X) \longrightarrow \mathcal{S}^n(X)_{\S}$$

are isomorphisms modulo 8-torsion (for any space X), since the hyperquadratic L-groups $\hat{L}^*(\pi_1(X))$ are of exponent 8, and hence so are $\pi_*(\hat{\underline{\mathbb{L}}}^0_{\S}) = \hat{L}^*(1)$, $\hat{\mathcal{S}}^*(X)_{\S}$. Thus if X is an n-dimensional geometric Poincaré complex $s(X)[\frac{1}{2}] = 0 \in \mathcal{S}_n(X)[\frac{1}{2}]$ if and only if X has a $KO[\frac{1}{2}]$-orientation $[X] \in KO_n(X)[\frac{1}{2}]$ which assembles to the symmetric signature away from 2 $\sigma^*[X] = \sigma^*(X)[\frac{1}{2}] \in L^n(\pi_1(X))[\frac{1}{2}]$. Here, we can identify the assembly map $\sigma^* : H_n(X;\underline{\mathbb{L}}^0) \longrightarrow L^n(\pi_1(X))$ localized away from 2 with the composite $KO_n(X)[\frac{1}{2}] \longrightarrow KO_n(K(\pi_1(X),1))[\frac{1}{2}] \xrightarrow{1_\pi^!} L_n(\pi_1(X))[\frac{1}{2}] = L^n(\pi_1(X))[\frac{1}{2}]$, where $1_\pi^!$ is as defined on p.265 of Wall [W1], and $\underline{\mathbb{L}}^0[\frac{1}{2}] = \underline{bo}[\frac{1}{2}]$ as before.

An n-dimensional geometric Poincaré complex X carries an equivalence class of triples $(\sigma^*(X),[\hat{X}],j)$ consisting of a map $\sigma^*(X):\underline{S}^n \longrightarrow \underline{\mathbb{L}}^0(\pi_1(X))$ representing the symmetric signature $\sigma^*(X) \in [\underline{S}^n,\underline{\mathbb{L}}^0(\pi_1(X))] = L^n(\pi_1(X))$, a map $[\hat{X}]:\underline{S}^n \longrightarrow X_+ \wedge \hat{\underline{\mathbb{L}}}^0$ representing the canonical $\hat{\underline{\mathbb{L}}}^0$- orientation $[\hat{X}] \in [\underline{S}^n, X_+ \wedge \hat{\underline{\mathbb{L}}}^0] = H_n(X;\hat{\underline{\mathbb{L}}}^0)$, and a homotopy

$$j : J\sigma^*(X) \simeq \hat{\sigma}^*[\hat{X}] : \underline{S}^n \longrightarrow \hat{\underline{\mathbb{L}}}^0(\pi_1(X)) .$$

Fixing one such triple $(\sigma^*(X),[\hat{X}],j)$ we can express the original two-stage obstruction theory for X to be simple homotopy equivalent to a manifold entirely in terms of the algebraic $\mathbb{L}$-spectra: $\mathcal{S}^{TOP}(X) \neq \emptyset$ if and only if

i) $[\hat{X}] \in \operatorname{im}(J:H_n(X;\underline{\mathbb{L}}^0) \longrightarrow H_n(X;\hat{\underline{\mathbb{L}}}^0))$, in which case a choice of map $[X]:\underline{S}^n \longrightarrow X_+ \wedge \underline{\mathbb{L}}^0$ and homotopy $g:J[X] \simeq [\hat{X}]:\underline{S}^n \longrightarrow X_+ \wedge \hat{\underline{\mathbb{L}}}^0$ together with j determine an element $\theta([X],g) \in L_n(\pi_1(X))$ with images $s(X) \in \mathcal{S}_n(X)$, $\sigma^*([X]) - \sigma^*(X) \in L^n(\pi_1(X))$

ii) there exists a pair $([X],g)$ such that $\theta([X],g) = 0$.

(In geometric terms $([X],g)$ corresponds to a topological reduction $\tilde{\nu}_X:X \longrightarrow BSTOP$ of the Spivak normal fibration $\nu_X:X \longrightarrow BSG$, and if $(f,b):M \longrightarrow X$ is the associated normal map then $\theta([X],g) = \theta(f,b) \in L_n(\pi_1(X))$ is the surgery obstruction and $[X] = f_*[M] \in H_n(X;\underline{\mathbb{L}}^0)$ is the image of the canonical $\underline{\mathbb{L}}^0$-orientation $[M] \in H_n(M;\underline{\mathbb{L}}^0)$ of the manifold M, so that $\sigma^*([X]) = \sigma^*(M) \in L^n(\pi_1(X))$).

The invariant $(1+T)s(X)_{\S} \in \mathcal{S}^n(X)_{\S}$ is the primary obstruction of a distinct two-stage theory: $\mathcal{S}^{TOP}(X) \neq \emptyset$ if and only if

i)' there exists an $\underline{\mathbb{L}}^0$- orientation $[X] \in H_n(X;\underline{\mathbb{L}}^0)$ such that $\sigma^*([X]) = \sigma^*(X) \in L^n(\pi_1(X))$, in which case a choice of representative map $[X]:\underline{S}^n \longrightarrow X_+ \wedge \underline{\mathbb{L}}^0$ and of a homotopy $h:\sigma^*(X) \simeq \sigma^*[X]:\underline{S}^n \longrightarrow \underline{\mathbb{L}}^0(\pi_1(X))$ together with j determine an element $\hat{s}([X],h)_{\S} \in \hat{\mathcal{S}}^{n+1}(X)_{\S}$ with images $s(X) \in \mathcal{S}_n(X)$, $J[X] - [\hat{X}] \in H_n(X;\hat{\underline{\mathbb{L}}}^0{}_{\S})$

ii)' there exists a pair $([X],h)$ such that $\hat{s}([X],h)_{\S} = 0$.

(In the previous theory the primary obstruction $t(\nu_X) \in \hat{H}^{k+1}(T(\nu_X);\underline{\mathbb{L}}_0) = H_{n-1}(X;\underline{\mathbb{L}}_0$ is a torsion element, with the 2-primary torsion of exponent 8. In this theory the secondary obstruction $\hat{s}([X],h)_{\S} \in \hat{\mathcal{S}}^{n+1}(X)_{\S}$ is 2-primary torsion of exponent 8). Combining the two approaches we have that $\mathcal{S}^{TOP}(X) \neq \emptyset$ if and only if there exists a quadruple $([X],g,h,i)$ consisting of a map $[X]:\underline{S}^n \longrightarrow X_+ \wedge \underline{\mathbb{L}}^0$, homotopies $g:J[X] \simeq [\hat{X}]:\underline{S}^n \longrightarrow X_+ \wedge \hat{\underline{\mathbb{L}}}^0$, $h:\sigma^*(X) \simeq \sigma^*[X]:\underline{S}^n \longrightarrow \underline{\mathbb{L}}^0(\pi_1(X))$, and a homotopy of homotopies $i:(\hat{\sigma}^*g)(Jh) \simeq j : J\sigma^*(X) \simeq \hat{\sigma}^*[\hat{X}] : \underline{S}^n \longrightarrow \hat{\underline{\mathbb{L}}}^0(\pi_1(X))$.

An n-dimensional manifold X carries an equivalence class of such quadruples $([X],g,h,i)$, with $[X] \in H_n(X;\underline{\mathbb{L}}^0)$ the canonical $\underline{\mathbb{L}}^0$-orientation, $J[X] = [\hat{X}] \in H_n(X;\hat{\underline{\mathbb{L}}}^0)$ the canonical $\hat{\underline{\mathbb{L}}}^0$- orientation, and $\sigma^*([X]) = \sigma^*(X) \in L^n(\pi_1(X))$ the symmetric signature. Conversely, an n-dimensional geometric Poincaré complex X is simple homotopy equivalent to a manifold if and only if it admits such a quadruple $([X],g,h,i)$. (In geometric terms $([X],g)$ corresponds to a particular topological reduction of the Spivak normal fibration ν_X, and (h,i) to a particular solution of the associated surgery problem). We can thus identify:

$$\mathcal{S}^{TOP}(X) = \text{the set of equivalence classes of quadruples } ([X],g,h,i) ,$$

and if $\mathcal{S}^{TOP}(X) \neq \emptyset$ (i.e. if $s(X) = 0 \in \mathcal{S}_n(X)$) then choosing one manifold structure on X as a base point of $\mathcal{S}^{TOP}(X)$ we have the bijection of Corollary 2 to Theorem 1

$$s : \mathcal{S}^{TOP}(X) \longrightarrow \mathcal{S}_{n+1}(X) \ ; \ (f:M \longrightarrow X) \longmapsto s(f) .$$

This defines an equivalence of categories

{compact n-dimensional topological manifolds,
homotopy classes of homeomorphisms}

$\approx$ {n-dimensional geometric Poincaré complexes with extra structure $([X],g,h,i)$,
homotopy classes of simple homotopy equivalences preserving
the extra structure} .

By the above, an n-dimensional geometric Poincaré complex X is simple homotopy equivalent to a closed topological manifold if and only if there exists an element $[X] \in H_n(X;\underline{\mathbb{L}}^0)$ such that

i) $J[X] = [\hat{X}] \in H_n(X;\hat{\underline{\mathbb{L}}}^0)$ is the canonical $\hat{\underline{\mathbb{L}}}^0$-orientation of X, in which case $[X] \in H_n(X;\underline{\mathbb{L}}^0)$ is an $\underline{\mathbb{L}}^0$-orientation (since $\pi_0(\underline{\mathbb{L}}^0) = \pi_0(\hat{\underline{\mathbb{L}}}^0) = L^0(1)$)

ii) $\sigma^*([X]) = \sigma^*(X) \in L^n(\pi_1(X))$ is the symmetric signature of X

iii) the relations i) and ii) are compatible on the $\mathbb{L}$-space level, i.e. ca be realized by a quadruple $([X],g,h,i)$.

In certain cases we can ensure that condition iii) is redundant:

Theorem 3 Let X be an n-dimensional geometric Poincaré complex such that the hyperquadratic signature map $\hat{\sigma}^*: H_{n+1}(X;\hat{\underline{\mathbb{L}}}^0) \longrightarrow \hat{L}^{n+1}(\pi_1(X))$ is onto. Then X is simple homotopy equivalent to a closed topological manifold if and only if there exists an $\underline{\mathbb{L}}^0$-orientation $[X] \in H_n(X;\underline{\mathbb{L}}^0)$ such that $J[X] = [\hat{X}] \in H_n(X;\hat{\underline{\mathbb{L}}}^0)$ and $\sigma^*([X]) = \sigma^*(X) \in L^n(\pi_1(X))$.

Proof: Given such an $\underline{\mathbb{L}}^0$-orientation $[X]$ there are defined homotopies $g: J[X] \simeq [\hat{X}] : \underline{S}^n \longrightarrow X_+ \wedge \hat{\underline{\mathbb{L}}}^0$, $h: \sigma^*(X) \simeq \sigma^*([X]) : \underline{S}^n \longrightarrow \underline{\mathbb{L}}^0(\pi_1(X))$. These determi an element $\hat{\sigma}([X],g,h) \in \hat{L}^{n+1}(\pi_1(X))$, the obstruction to the existence of a homoto of homotopies $i: (\hat{\sigma}^*g)(Jh) \simeq j: J\sigma^*(X) \simeq \hat{\sigma}^*[\hat{X}] : \underline{S}^n \longrightarrow \hat{\underline{\mathbb{L}}}^0(\pi_1(X))$. Now $H\hat{\sigma}^*([X],g,h) = \theta([X],g) = \theta(f,b) \in L_n(\pi_1(X))$ is the surgery obstruction of the normal map $(f,b): M \longrightarrow X$ associated to the topological reduction of ν_X determin by $([X],g)$. By assumption $\hat{\sigma}([X],g,h) \in \mathrm{im}(\hat{\sigma}^*: H_{n+1}(X;\hat{\underline{\mathbb{L}}}^0) \longrightarrow \hat{L}^{n+1}(\pi_1(X)))$, so tha $\theta(f,b) \in \mathrm{im}(\sigma_*: H_n(X;\underline{\mathbb{L}}_0) \longrightarrow L_n(\pi_1(X)))$ and there exists a topological reduction with 0 surgery obstruction.

[]

In particular, suppose that π is a group such that $K(\pi,1)$ is an n-dimensional geometric Poincaré complex for which $\sigma^*: H_n(K(\pi,1);\underline{\mathbb{L}}^0) \longrightarrow L^n(\pi)$ is an isomorphism and $\hat{\sigma}^*: H_{n+1}(K(\pi,1);\hat{\underline{\mathbb{L}}}^0) \longrightarrow \hat{L}^{n+1}(\pi)$ is onto. Then $K(\pi,1)$ is simple homotopy equivalent to a closed topological manifold if and only if the composit $L^n(\pi) \xrightarrow[\simeq]{\sigma^{*-1}} H_n(K(\pi,1);\underline{\mathbb{L}}^0) \xrightarrow{J} H_n(K(\pi,1);\hat{\underline{\mathbb{L}}}^0)$ sends the symmetric signature $\sigma^*(K(\pi,1)) \in L^n(\pi)$ to the canonical $\hat{\underline{\mathbb{L}}}^0$-orientation $[K(\hat{\pi},1)] \in H_n(K(\pi,1);\hat{\underline{\mathbb{L}}}^0)$.

(The hypothesis of Theorem 3 is not satisfied in general: the infinitely generat subgroup $\mathbb{Z}_2^\infty \subseteq \mathrm{Unil}_{4k+2}(1;\mathbb{Z},\mathbb{Z}_2) = \mathrm{coker}(L_{4k+2}(\mathbb{Z}) \oplus L_{4k+2}(\mathbb{Z}_2) \longrightarrow L_{4k+2}(\mathbb{Z}*\mathbb{Z}_2))$ constructed by Cappell [C] can be used to detect an infinitely generated subgrou $\mathbb{Z}_2^\infty \subseteq \mathrm{coker}(\hat{\sigma}^*: H_{4k+3}(K(\mathbb{Z}*\mathbb{Z}_2,1);\hat{\underline{\mathbb{L}}}^0) \longrightarrow \hat{L}^{4k+3}(\mathbb{Z}*\mathbb{Z}_2))$. This also shows that the hyperquadratic signature map $\hat{\sigma}^*: \Omega_n^N(K) \longrightarrow \hat{L}^n(\pi_1(K))$ is not onto in general).

For any space K there is defined a natural transformation of exact sequences

$$\begin{array}{ccccccccc}
\cdots \longrightarrow \Omega^N_{n+1}(K) & \longrightarrow & \Omega^{N,P}_{n+1}(K) & \longrightarrow & \Omega^P_n(K) & \longrightarrow & \Omega^N_n(K) & \longrightarrow & \cdots \\
\downarrow H\hat{\sigma}^* & & \downarrow \sigma_* & & \downarrow s & & \downarrow H\hat{\sigma}^* & & \\
\cdots \longrightarrow H_n(K;\underline{\mathbb{L}}_0) & \xrightarrow{\sigma_*} & L_n(\pi_1(K)) & \longrightarrow & \mathcal{S}_n(K) & \longrightarrow & H_{n-1}(K;\underline{\mathbb{L}}_0) & \longrightarrow & \cdots
\end{array}$$

with $\sigma_*:\Omega^{N,P}_{n+1}(K) \longrightarrow L_n(\pi_1(K))$ the quadratic signature map and

$$H\hat{\sigma}^* : \Omega^N_{n+1}(K) = H_{n+1}(K;\underline{\Omega}^N) \xrightarrow{\hat{\sigma}^*} H_{n+1}(K;\underline{\hat{\mathbb{L}}}^0) \xrightarrow{H} H_n(K;\underline{\mathbb{L}}_0)$$

$$s : \Omega^P_n(K) \longrightarrow \mathcal{S}_n(K) \ ; \ (f:X \longrightarrow K) \longmapsto f_*s(X) \ .$$

In particular, the quadratic signature $\sigma_*(f,b) = \sigma_*(W,M\cup -X) \in L_n(\pi_1(X))$ of a normal map of n-dimensional geometric Poincaré complexes

$$(f,b) : (M,\nu_M,\rho_M) \longrightarrow (X,\nu_X,\rho_X)$$

has image

$$[\sigma_*(f,b)] = f_*s(M) - s(X) \in \mathcal{S}_n(X) \ ,$$

where W is the mapping cylinder of f, $(W,M\cup -X) \in \Omega^{N,P}_{n+1}(X)$.

For any space K define a morphism of abelian groups

$$L_n(\pi_1(K)) \longrightarrow \Omega^P_n(K) \ ; \ x \longmapsto (f:X \longrightarrow K)$$

as follows. Let Y be an (n-1)-dimensional manifold (possibly with boundary) equipped with a map $Y \longrightarrow K$ inducing an isomorphism $\pi_1(Y) \xrightarrow{\sim} \pi_1(K)$. By Wall's realization theorem every element $x \in L_n(\pi_1(K))$ is the surgery obstruction $x = \sigma_*(F,B)$ of a normal map of manifold triads

$$(F,B) : (Z;Y,Y') \longrightarrow (Y \times I; Y \times 0, Y \times 1)$$

such that $F| = 1 : Y \longrightarrow Y \times 0$ and $F| = h : Y' \longrightarrow Y \times 1$ is a simple homotopy equivalence. Define $X = Z/Y \underset{h}{\simeq} Y'$ to be the n-dimensional geometric Poincaré complex obtained from Z by glueing Y to Y' by h, let $g:X \longrightarrow Y \times S^1$ be the degree 1 map obtained from F, and define $f:X \longrightarrow K$ to be the composite

$$f : X \xrightarrow{g} Y \times S^1 \xrightarrow{\text{projection}} Y \longrightarrow K \ .$$

Now g is covered by a bundle map of topological reductions of the Spivak normal fibrations such that the quadratic signature $\sigma_*(g,c) \in L_n(\pi_1(Y \times S^1))$ of the corresponding normal map of geometric Poincaré complexes

$$(g,c) : (X,\nu_X,\rho_X) \longrightarrow (Y\times S^1, \nu_{Y\times S^1}, \rho_{Y\times S^1})$$

has image

$$\sigma_*(g,c) = x \in L_n(\pi_1(K)) \ .$$

By the above

$$[\sigma_*(g,c)] = g_*s(X) - s(Y\times S^1) \in \mathcal{S}_n(Y\times S^1) \ ,$$

and $s(Y\times S^1) = 0$, so that

$$[x] = f_*s(X) \in \mathcal{S}_n(K) \ .$$

(Incidentally, the image $[x]\in\mathcal{S}_n(Y)$ is the obstruction to deforming the simple homotopy equivalence $h:Y' \longrightarrow Y$ to a homeomorphism, $[x] = s(h) \in \mathcal{S}_n(Y)$, cf. Corollary 1 to Theorem 1 above). The composite

$$L_n(\pi_1(K)) \longrightarrow \Omega_n^P(K) \xrightarrow{\ s\ } \mathcal{S}_n(K)$$

is thus the canonical map $L_n(\pi_1(K)) \longrightarrow \mathcal{S}_n(K)$.

****

We have the following extension of the Levitt-Jones-Quinn geometric Poinca surgery exact sequence [Le],[J1],[Q3]

$$\ldots \longrightarrow \Omega^N_{n+1}(K) \longrightarrow L_n(\pi_1(K)) \longrightarrow \Omega^P_n(K) \longrightarrow \Omega^N_n(K) \longrightarrow \ \ldots \ .$$

<u>Theorem 4</u> For any space K there is defined a commutative braid of exact sequence of abelian groups

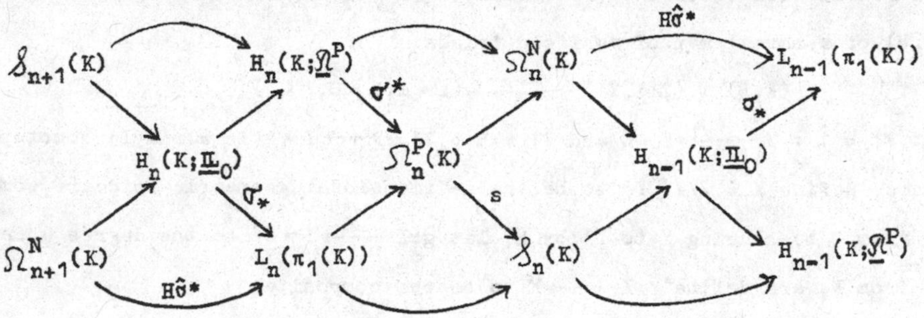

[]

For example, $\Omega^P_n(T^n) = H_n(T^n;\underline{\Omega}^P)\oplus\mathcal{S}_n(T^n)$, $\mathcal{S}_n(T^n) = L_0(1)$

(since $\sigma_*:H_n(T^n;\underline{\mathbb{L}}_0) = \bigoplus_{i=1}^{n}\binom{n}{i}L_i(1) \hookrightarrow L_n(\pi_1(T^n)) = \bigoplus_{i=0}^{n}\binom{n}{i}L_i(1)$).

From the point of view of geometric Poincaré surgery theory there are defined equivalences of categories

$\{$stable oriented topological bundles (over finite CW complexes)$\}$

$\approx \{$stable spherical fibrations with an $\underline{\Omega}^P$-orientation lifting the canonical $\underline{\Omega}^N$-orientation$\}$,

$\{$compact oriented n-dimensional topological manifolds$\}$

$\approx \{$n-dimensional geometric Poincaré complexes X with an $\underline{\Omega}^P$-orientation $[X] \in H_n(X;\underline{\Omega}^P)$ which assembles to $\sigma^*([X]) = (1:X \longrightarrow X) \in \Omega_n^P(X)\}$.

****

Product with the symmetric signature $\sigma^*(\mathbb{CP}^2) \in L^4(1)$ $(=1 \in \mathbb{Z})$ of the complex projective plane $\mathbb{CP}^2$ defines the periodicity isomorphisms in the quadratic L-groups

$$\sigma^*(\mathbb{CP}^2)\otimes - : L_n(\pi) \longrightarrow L_{n+4}(\pi) \quad (n \geqslant 0)$$

for any group π. For any space K there is defined a commutative braid of exact sequences of abelian groups

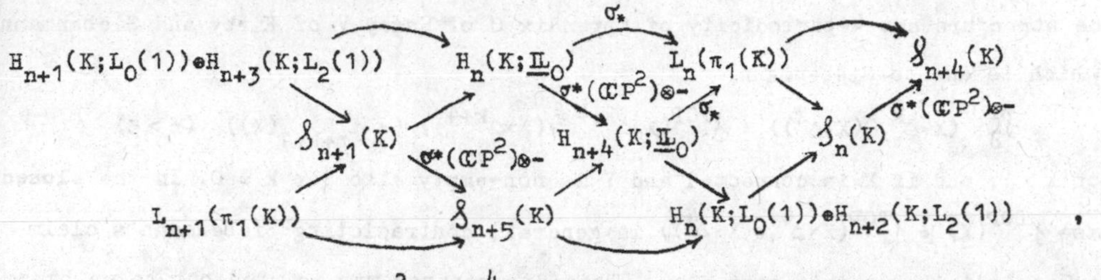

involving the products $\sigma^*(\mathbb{CP}^2)\otimes - : \Sigma^4\underline{\mathbb{L}}_0 \longrightarrow \underline{\mathbb{L}}_0$ and the homotopy-theoretic analysis

$$\underline{\mathbb{L}}_0[\tfrac{1}{2}] = \underline{\mathrm{bso}}[\tfrac{1}{2}] \quad , \quad \underline{\mathbb{L}}_{0(2)} = \prod_{i=1}^{\infty} \Sigma^{4i}\underline{K}(L_0(1)_{(2)}) \times \Sigma^{4i-2}\underline{K}(L_2(1)) .$$

The maps $H_{n+4}(K;\underline{\mathbb{L}}_0) \longrightarrow H_n(K;L_0(1)) \oplus H_{n+2}(K;L_2(1))$ have odd torsion cokernel.

(More generally, we have that the $\left\{\begin{matrix}\text{symmetric}\\ \text{quadratic}\end{matrix}\right.$ signature of a product is given by

$$\left\{\begin{aligned} &\sigma^*(M^m \times N^n) = \sigma^*(M) \otimes \sigma^*(N) \in L^{m+n}(\pi_1(M\times N)) \\ &\sigma_*(1\times(f,b):M^m \times N^n \longrightarrow M\times X) = \sigma^*(M) \otimes \sigma_*(f,b) \in L_{m+n}(\pi_1(M\times X)) \end{aligned}\right.$$

. Product with the canonical $\underline{\mathbb{L}}^0$-orientation $[M] \in H_m(M;\underline{\mathbb{L}}^0)$ of an m-dimensional manifold M defines a map $[M]\otimes - : \mathcal{S}_n(X) \longrightarrow \mathcal{S}_{m+n}(M\times X)$ (for any space X) compatible with the product map $\sigma^*(M)\otimes - : L_n(\pi_1(X)) \longrightarrow L_{m+n}(\pi_1(M\times X))$ $(\sigma^*(M) = \sigma^*([M]) \in L^m(\pi_1(M)))$. If X is an n-dimensional geometric Poincaré complex $s(M\times X) = [M]\otimes s(X) \in \mathcal{S}_{m+n}(M\times X)$. The maps appearing above are $\sigma^*(\mathbb{CP}^2)\otimes - : \mathcal{S}_n(K) \xrightarrow{[\mathbb{CP}^2]\otimes -} \mathcal{S}_{n+4}(\mathbb{CP}^2\times K) \xrightarrow{\text{proj.}_*} \mathcal{S}_{n+4}(K)$).

<u>Theorem 5</u> i) If X is a connected n-dimensional geometric Poincaré complex there are defined periodicity isomorphisms

$$\sigma^*(\mathbb{C}P^2)\otimes - : \mathcal{S}_{n+k}(X) \longrightarrow \mathcal{S}_{n+k+4}(X) \quad (k \geq 2)$$

and an exact sequence

$$0 \longrightarrow \mathcal{S}_{n+1}(X) \xrightarrow{\sigma^*(\mathbb{C}P^2)\otimes -} \mathcal{S}_{n+5}(X) \longrightarrow L_0(1) \longrightarrow \mathcal{S}_n(X) \xrightarrow{\sigma^*(\mathbb{C}P^2)\otimes -} \mathcal{S}_{n+4}(X) \longrightarrow \ldots .$$

ii) If (X,Y) is an n-dimensional geometric Poincaré pair with X connected and Y non-empty there are defined periodicity isomorphisms

$$\sigma^*(\mathbb{C}P^2)\otimes - : \mathcal{S}_{n+k}(X) \longrightarrow \mathcal{S}_{n+k+4}(X) \quad (k \geq 1)$$

and an exact sequence

$$0 \longrightarrow \mathcal{S}_n(X) \xrightarrow{\sigma^*(\mathbb{C}P^2)\otimes -} \mathcal{S}_{n+4}(X) \longrightarrow H_{n-1}(X;L_0(1))$$
$$\longrightarrow \mathcal{S}_{n-1}(X) \xrightarrow{\sigma^*(\mathbb{C}P^2)\otimes -} \mathcal{S}_{n+3}(X) \longrightarrow \ldots .$$

[]

In particular, if (X,Y) is an n-dimensional manifold with boundary we have the structure set 4-periodicity of Appendix C of Essay V of Kirby and Siebenmann [(which is due to Siebenmann)

$$\mathcal{S}^{TOP}_\partial(X\times\Delta^k,\partial(X\times\Delta^k)) = \mathcal{S}^{TOP}_\partial(X\times\Delta^{k+4},\partial(X\times\Delta^{k+4})) \ (= \mathcal{S}_{n+k+1}(X)) \quad (n \geq 5)$$

for $k \geq 1$, and if X is connected and Y is non-empty also for k = 0. In the closed case $\mathcal{S}^{TOP}(X) \neq \mathcal{S}^{TOP}_\partial(X\times\Delta^4,\partial(X\times\Delta^4))$ in general, contradicting Siebenmann's claim for periodicity in this case also. (This discrepancy was pointed out to me by Andrew Nicas). For example,

$$\mathcal{S}_{n+k}(S^n) = \begin{cases} L_{k-1}(1) & \text{if } k \geq 2 \\ 0 & \text{if } k = 0,1 \end{cases} \quad (n \geq 2)$$

so that

$$\mathcal{S}^{TOP}_\partial(S^n\times\Delta^4,\partial(S^n\times\Delta^4)) = \mathcal{S}_{n+5}(S^n) = L_4(1) \neq \mathcal{S}^{TOP}(S^n) = \mathcal{S}_{n+1}(S^n) = 0 \quad (n \geq 5).$$

On the other hand,

$$\mathcal{S}_{n+k}(T^n) = \begin{cases} 0 & \text{if } k \geq 1 \\ L_0(1) & \text{if } k = 0 \end{cases} \quad (n \geq 1)$$

so that

$$\mathcal{S}^{TOP}_\partial(T^n\times\Delta^k,\partial(T^n\times\Delta^k)) = \mathcal{S}^{TOP}_\partial(T^n\times\Delta^{k+4},\partial(T^n\times\Delta^{k+4})) = \mathcal{S}_{n+k+1}(T^n) = 0 \quad (k \geq 0, n \geq$$

In conclusion, we note that it is also possible to define quadratic $\mathcal{S}$-groups $\begin{cases} \mathcal{S}^h_*(X) \\ \mathcal{S}^p_*(X) \end{cases}$ appropriate to $\begin{cases} \text{finite} \\ \text{infinite} \end{cases}$ homotopy types and the $\begin{cases} \text{free} \\ \text{projective} \end{cases}$ L-groups $\begin{cases} L^h_*(\pi) \\ L^p_*(\pi) \end{cases}$, which fit into a commutative braid of exact sequences of abelian groups

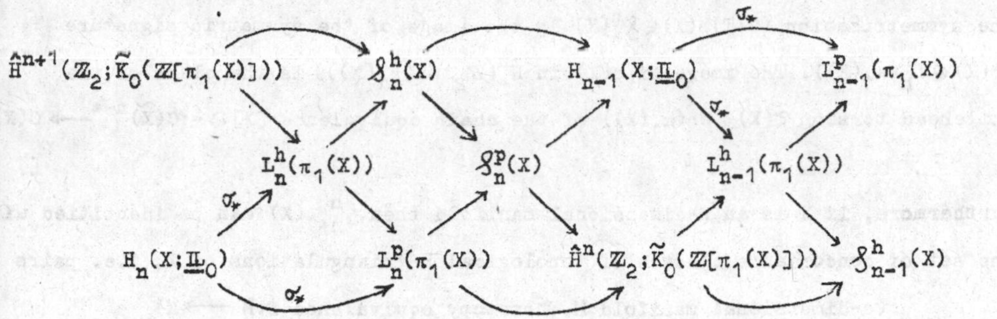

$\hat{H}^{n+1}(\mathbb{Z}_2;\tilde{K}_0(\mathbb{Z}[\pi_1(X)]))$
$\mathcal{S}^h_n(X)$
$H_{n-1}(X;\underline{\mathbb{L}}_0)$
$L^p_{n-1}(\pi_1(X))$
$L^h_n(\pi_1(X))$
$\mathcal{S}^p_n(X)$
$L^h_{n-1}(\pi_1(X))$
$H_n(X;\underline{\mathbb{L}}_0)$
$L^p_n(\pi_1(X))$
$\hat{H}^n(\mathbb{Z}_2;\tilde{K}_0(\mathbb{Z}[\pi_1(X)]))$
$\mathcal{S}^h_{n-1}(X)$
σ_*
σ_*
σ_*
σ_*

involving the Tate $\mathbb{Z}_2$-cohomology groups of the duality involution $[P] \longmapsto [P^*]$ ($P^* = \mathrm{Hom}_A(P,A)$, $A = \mathbb{Z}[\pi_1(X)]$) on the reduced projective class group $\tilde{K}_0(\mathbb{Z}[\pi_1(X)])$. There is a similar braid relating $\mathcal{S}^h_*(X)$ and $\mathcal{S}^s_*(X) = \mathcal{S}_*(X)$, involving the duality involution $\tau(f:P \longrightarrow Q) \longmapsto \tau(f^*:Q^* \longrightarrow P^*)$ on the Whitehead group $Wh(\pi_1(X))$. The free symmetric L-groups $L^*_h(\pi)$ are related to the projective symmetric L-groups $L^*_p(\pi)$ by an exact sequence

$$\ldots \longrightarrow \hat{H}^{n+1}(\mathbb{Z}_2;\tilde{K}_0(\mathbb{Z}[\pi])) \longrightarrow L^n_h(\pi) \longrightarrow L^n_p(\pi) \longrightarrow \hat{H}^n(\mathbb{Z}_2;\tilde{K}_0(\mathbb{Z}[\pi]))$$
$$\longrightarrow L^{n-1}_h(\pi) \longrightarrow \ldots$$

(which actually connects with the quadratic L-group sequence for $L^h_*(\pi), L^p_*(\pi)$ on setting $L^n(\pi) = L_{n+4k}(\pi)$ ($n \leqslant -3$, $n+4k \geqslant 0$), see [R2]) and similarly for $L^*_s(\pi) \equiv L^*(\pi)$, $L^*_h(\pi)$, $Wh(\pi)$. Thus it is also possible to define symmetric $\mathcal{S}$-groups $\begin{cases} \mathcal{S}^*_h(X) \\ \mathcal{S}^*_p(X) \end{cases}$ with properties analogous to those of $\mathcal{S}^*_s(X) \equiv \mathcal{S}^*(X)$, $\begin{cases} \mathcal{S}^h_*(X) \\ \mathcal{S}^p_*(X) \end{cases}$.

The hyperquadratic L-groups are such that

$$\hat{L}^*_p(\pi) = \hat{L}^*_h(\pi) = \hat{L}^*_s(\pi) \equiv \hat{L}^*(\pi) ,$$

and accordingly we define

$$\hat{\mathcal{S}}^*_p(X) = \hat{\mathcal{S}}^*_h(X) = \hat{\mathcal{S}}^*_s(X) \equiv \hat{\mathcal{S}}^*(X) .$$

Similarly for $\mathcal{S}^*(X)_{\S}$, $\hat{\mathcal{S}}^*(X)_{\S}$.

Theorem 1(h) A finite n-dimensional geometric Poincaré complex X determines an element $s(X) \in \mathcal{S}_n^h(X)$ such that $s(X) = 0$ if and only if X is homotopy equivalent to a closed topological manifold. The image of $s(X)$ in $H_{n-1}(X;\underline{\mathbb{L}}_0)$ is the obstruction to a topological reduction of the Spivak normal fibration $\nu_X : X \longrightarrow BSG$. The symmetrization $(1+T)s(X) \in \mathcal{S}_h^n(X)$ is the image of the symmetric signature $\sigma^*(X) \in L_h^n(\pi_1(X))$. The image of $s(X)$ in $\hat{H}^n(\mathbb{Z}_2;Wh(\pi_1(X)))$ is the class of the Whitehead torsion $\tau(X) \in Wh(\pi_1(X))$ of the chain equivalence $[X] \cap - : C(\widetilde{X})^{n-*} \longrightarrow C(\widetilde{X})$

[]

Furthermore, if X is an n-dimensional manifold then $\mathcal{S}_{n+1}^h(X)$ can be identified wit the set of concordance classes of topological h-triangulations of X, i.e. pairs

(n-dimensional manifold M, homotopy equivalence $f:M \longrightarrow X$)

with $(M,f) \sim (M',f')$ if there exist an h-cobordism $(W;M,M')$ and a homotopy equival

$$(g;f,f') : (W;M,M') \longrightarrow (X \times I; X \times 0, X \times 1) .$$

Theorem 1(p) A finitely dominated n-dimensional geometric Poincaré complex X determines an element $s(X) \in \mathcal{S}_n^p(X)$ such that $s(X) = 0$ if and only if $X \times S^1$ is homotopy equivalent to a closed topological manifold. The image of $s(X)$ in $H_{n-1}(X;\underline{\mathbb{L}}_0)$ is the obstruction to a topological reduction of the Spivak normal fibration $\nu_X : X \longrightarrow BSG$. The symmetrization $(1+T)s(X) \in \mathcal{S}_p^n(X)$ is the image of the symmetric signature $\sigma^*(X) \in L_p^n(\pi_1(X))$. The image of $s(X)$ in $\hat{H}^n(\mathbb{Z}_2;\widetilde{K}_0(\mathbb{Z}[\pi_1(X)]))$ is the class of the Wall finiteness obstruction $[C(\widetilde{X})] \in \widetilde{K}_0(\mathbb{Z}[\pi_1(X)])$.

[]

Theorem 1(p) is the special case of Theorem 1(h) obtained by first noting that $X \times S^1$ has the homotopy type of a finite complex and then applying the algebraic aplitting theorem $L_{n+1}^h(\pi \times \mathbb{Z}) = L_{n+1}^h(\pi) \oplus L_n^p(\pi)$ ([R1]) to identify

$$s(X \times S^1) = (0, s(X)) \in \mathcal{S}_{n+1}^h(X \times S^1) = \mathcal{S}_{n+1}^h(X) \oplus \mathcal{S}_n^p(X) .$$

(The definitive version of the non-compact manifold surgery theories of Taylor [Ta and Maumary [Ma] should interpret $s(X) \in \mathcal{S}_n^p(X)$ as the total obstruction to X being homotopy equivalent to a topological manifold allowed a certain degree of non-compactness, such as an end).

The invariant $s(X)\in\mathcal{S}_n^p(X)$ may be of interest in the classification of free ctions of finite groups on spheres, the "topological spherical space form problem" cf. Swan [Sw], Thomas and Wall [ThW], Madsen, Thomas and Wall [MTW]) since its efinition does not presuppose a vanishing of the finiteness obstruction. If π is a inite group with cohomology of period dividing n+1, to every generator $g\in H^{n+1}(K(\pi,1))$ here is associated a finitely dominated n-dimensional geometric Poincaré complex X_g equipped with an isomorphism $\pi_1(X_g)\xrightarrow{\sim}\pi$, a homotopy equivalence $\widetilde{X}_g\xrightarrow{\sim}S^n$, and first k-invariant $g\in H^{n+1}(K(\pi,1))$. Ultimately, it might be possible to give a direct description of $s(X_g)\in\mathcal{S}_n^p(X_g)$. In this connection, it should also be mentioned that the $\mathcal{S}$-groups (in each of the categories s,h,p) behave well with respect to finite covers $p:\overline{X}\longrightarrow X$, with transfer maps defining a natural transformation of exact sequences of abelian groups

$$\begin{array}{ccccccccc}
\cdots\longrightarrow & H_n(X;\underline{\mathbb{L}}_0) & \xrightarrow{\sigma_*} & L_n(\pi_1(X)) & \longrightarrow & \mathcal{S}_n(X) & \longrightarrow & H_{n-1}(X;\underline{\mathbb{L}}_0) & \longrightarrow\cdots \\
 & \downarrow p^! & & \downarrow p^! & & \downarrow p^! & & \downarrow p^! & \\
\cdots\longrightarrow & H_n(\overline{X};\underline{\mathbb{L}}_0) & \xrightarrow{\sigma_*} & L_n(\pi_1(\overline{X})) & \longrightarrow & \mathcal{S}_n(\overline{X}) & \longrightarrow & H_{n-1}(\overline{X};\underline{\mathbb{L}}_0) & \longrightarrow\cdots ,
\end{array}$$

using the canonical S-map $p^!:\Sigma^\infty X_+\longrightarrow\Sigma^\infty\overline{X}_+$ to define

$$p^! : H_n(X;\underline{\mathbb{L}}_0)\longrightarrow H_n(\overline{X};\underline{\mathbb{L}}_0) ,$$

and the restriction of $\pi_1(X)$-action to $\pi_1(\overline{X})$-action to define

$$p^! : L_n(\pi_1(X))\longrightarrow L_n(\pi_1(\overline{X})) ; (C,\psi)\longmapsto(p^!C,p^!\psi) .$$

If $\begin{cases} X \\ (f,b):M\longrightarrow X\end{cases}$ is an n-dimensional $\begin{cases}\text{geometric Poincaré complex}\\ \text{normal map}\end{cases}$ then so is $\begin{cases}\overline{X}\\ (\overline{f},\overline{b}):\overline{M}\longrightarrow\overline{X}\end{cases}$, and

$$\begin{cases} s(\overline{X}) = p^!s(X)\in\mathcal{S}_n(\overline{X}) \ , \ \sigma^*(\overline{X}) = p^!\sigma^*(X)\in L^n(\pi_1(\overline{X})) \\ \sigma_*(\overline{f},\overline{b}) = p^!\sigma_*(f,b)\in L_n(\pi_1(\overline{X})) \end{cases} .$$

Similarly for the $\begin{cases}\text{symmetric}\\ \text{hyperquadratic}\end{cases}$ $\mathcal{S}$-groups $\begin{cases}\mathcal{S}^*(X),\ \mathcal{S}^*(X)_{\S} \\ \hat{\mathcal{S}}^*(X),\ \hat{\mathcal{S}}^*(X)_{\S}\end{cases}$.

References

[B1] W.Browder Homotopy type of differentiable manifolds
Århus Colloquium on Algebraic Topology, 42-46 (1962)

[B2] Surgery on simply-connected manifolds
Ergebnisse der Mathematik 65, Springer (1972)

[B3] Poincaré Spaces, Their Normal Fibrations and Surgery
Inventiones math. 17, 191-202 (1972)

[BM] G.Brumfiel and J.Morgan
Homotopy-theoretic consequences of N.Levitt's obstruction theory to transversality for spherical fibrations
Pacific J. Math. 67, 1-100 (1976)

[C] S.Cappell Splitting obstructions for hermitian forms and manifolds with $\mathbb{Z}_2 \subset \pi_1$ Bull. Amer. Math. Soc. 79, 909-913 (1973)

[GS] S.Gitler and J.Stasheff
The first exotic class of BF Topology 4, 257-266 (1965)

[HM] I.Hambleton and J.Milgram
Poincaré transversality for double covers
Canadian J. Math. XXX, 1319-1330 (1978)

[J1] L.Jones Patch spaces: a geometric representation for Poincaré spaces
Ann. Math. 97, 306-343 (1973)

[J2] The nonsimply connected characteristic variety theorem
Proceedings of Symposia in Pure Mathematics 32, 131-140 (1978)

[KS] R.Kirby and L.Siebenmann
Foundational essays on topological manifolds, smoothings, and triangulations Ann. Math. Study 88 (1977)

[Le] N.Levitt Poincaré duality cobordism Ann. Math. 96, 211-244 (1972)

[LeM] and J.Morgan
Transversality structures and PL structures on spherical fibrations Bull. Amer. Math. Soc. 78, 1064-1068 (1972)

[Lo] J.-L.Loday K-theorie algébrique et représentations de groupes
Ann. sci. Éc. Norm. Sup. (4) 9, 309-377 (1976)

[MM] I.Madsen and J.Milgram
On spherical fiber bundles and their PL reductions
New Developments in Topology, Lond. Math. Soc. LN 11, 43-59 (1

[MTW] , C.B.Thomas and C.T.C.Wall
The topological spherical space form problem - II Existence of free actions Topology 15, 375-382 (1976)

[Ma] S.Maumary Proper surgery groups and Wall-Novikov groups
Proceedings of 1972 Battelle Conference on Algebraic K-theory
Vol. III, Springer LN 343, 526-539 (1973)

[Mi] A.S.Mishchenko Homotopy invariants of non-simply-connected manifolds
III. Higher signature Izv. Akad. Nauk SSSR, ser. mat. 35, 1316-1355 (1971)

[MS] J.Morgan and D.Sullivan
The transversality characteristic class and linking cycles in surgery theory Ann. Math. 99, 463-544 (1974)

[N] S.P.Novikov Homotopy equivalent smooth manifolds I.
Izv. Akad. Nauk SSSR, ser. mat. 28, 365-474 (1964)

[Q1] F.Quinn A geometric formulation of surgery Proceedings of 1969 Georgia Conference on Topology of Manifolds, 500-512 (1972)
Princeton Ph.D. thesis (1969)

[Q2] $\underline{B_{(TOP)}}^{\sim}_{n}$ and the surgery obstruction
Bull. Amer. Math. Soc. 77, 596-600 (1971)

[Q3] Surgery on Poincaré and normal spaces
Bull. Amer. Math. Soc. 78, 262-267 (1972)

[R1] A.A.Ranicki Algebraic L-theory II. Laurent extensions
Proc. Lond. Math. Soc. (3) 27, 126-158 (1973)

[R2] The algebraic theory of surgery I.Foundations,
II. Applications to topology
to appear in Proc. Lond. Math. Soc.

[Ro] C.P.Rourke The Hauptvermutung according to Casson and Sullivan
Warwick notes (1972)

[RS] and B.J.Sanderson
On topological neighbourhoods Comp. Math. 22, 387-424 (1970)

[Sch] M.G.Scharlemann Transversality theories at dimension 4
Inventiones math. 33, 1-14 (1976)

[Su1] D.Sullivan Triangulating and smoothing homotopy equivalences and homeomorphisms Princeton Geometric Topology Seminar notes (1967) Princeton Ph.D. thesis (1965)

[Su2] Geometric Topology I. Localization, periodicity and Galois symmetry MIT notes (1970)

[Sw] R.G.Swan <u>Periodic resolutions for finite groups</u>
Ann. Math. 72, 267-291 (1960)

[Ta] L.R.Taylor <u>Surgery on paracompact manifolds</u> Berkeley Ph.D. thesis (1972

[TaW] and B.Williams
<u>Surgery spaces: formulae and structure</u> to appear in the Proceedings of 1978 Waterloo Conference, Springer LN

[ThW] C.B.Thomas and C.T.C.Wall
<u>The topological spherical space-form problem I.</u>
Comp. math. 23, 101-114 (1971)

[Wa] F.Waldhausen <u>Algebraic K-theory of generalized free products I.,II.</u>
Ann. Math. 108, 135-256 (1978)

[W1] C.T.C.Wall <u>Surgery on compact manifolds</u> Academic Press (1970)

[W2] <u>Poincaré complexes I.</u> Ann. Math. 86, 213-245 (1970)

[W3] <u>Formulae for surgery obstructions</u> Topology 15, 189-210 (1976

April, 1979

ON THE EQUIVALENCE OF THE TWO DEFINITIONS OF THE ALGEBRAIC K-THEORY OF A TOPOLOGICAL SPACE

Mark Steinberger*

Massachusetts Institute of Technology
Cambridge, Massachusetts 02139

Waldhausen, in [5], constructs the algebraic K-theory of a topological space X, which we shall denote W(X), together with a fibre sequence

$$F(X) \longrightarrow W(X) \longrightarrow Wh(X),$$

which is functorial in X. Here, the homotopy groups of F(X) are isomorphic to the homology of X with coefficients in a generalized homology theory whose homotopy groups are those of W of a point, and, if X is a finite P.L. space, Wh(X) is a delooping of Hatcher's higher simple homotopy functor [1]. Waldhausen conjectured, [5], that W(X) could be constructed, for based, connected spaces X, by application of a K-theory functor on rings up to homotopy to $Q(|GX|_+)$, the free infinite loop space on the union of |GX| with a disjoint basepoint, where |GX| is the geometric realization of the Kan loop group of the total singular complex of X. Waldhausen defined a K-theoretic functor, which we shall denote K^h, on topological rings, and noted that the truth of the above conjecture

Author partially supported by NSF Grant MCS 78-02315.

ought to lead to a line of proof for the following theorem.

Theorem 1. Let X be a connected, based space. Then there is a rational equivalence between $W(X)$ and $K^h|Z[GX]|$, where $Z[GX]$ is the simplicial integral group ring of the simplicial group GX.

Burghelea, Hsiang and Staffeldt have developed techniques to compute the rational homotopy of $K^h|Z[GX]|$, and have used these, modulo Theorem 1 and the presumed equivalence of the second loop space of $Wh(X)$ with the stable concordance space of X, to compute the rational homotopy of the stable diffeomorphism groups of discs and other spaces.

May, in [2], developed a conceptual framework for the definition of the K-theory of an A_∞ ring space, such as $Q(|GX|_+)$. His definition (with suitable corrections) will be recalled below. Following May, we shall write $A(X)$ for the K-theory of $Q(|GX|_+)$ and shall sketch a proof of the following theorem.

Theorem 2. The functors $W(X)$ and $A(X)$ are naturally equivalent.

May, [2], also constructed a rational equivalence between $A(X)$ and $K^h|Z[GX]|$. Thus, our Theorem 2 does imply Theorem 1.

In section 1, we explain the idea behind May's construction, define Waldhausen's construction, and show that Theorem 2 will follow from the construction of deloopings for certain natural equivalences. In section 2, we describe May's construction in greater detail. In section 3, we show how to deloop the maps defined in section 1.

The details of our assertions will be presented in a later work.

I wish to thank Peter May and Bob Thomason for helpful

conversations regarding the material presented here, and to thank Dan Burghelea, Wu-Chung Hsiang, Kiyoshi Igusa, Dick Lashof and Friedhelm Waldhausen for conversations on related topics.

§1. Basic definitions and a comparison map.

Let R be any ring space. That is, R has two homotopy associative H-space structures (each with its own basepoint), a homotopy commutative "additive" structure and a "multiplicative" structure, such that the usual distributivity diagram homotopy commutes. We shall also assume that the additive structure is group-like, so that the zero-th homotopy "group" $\pi_0 R$ is a ring. The space of $n \times n$ matrices in R, $M_n R$, inherits a natural ring space structure, with $\pi_0 M_n R$ isomorphic as a ring to the ring of $n \times n$ matrices with coefficients in $\pi_0 R$. We define $\widehat{Gl}_n R$ to be the subspace of multiplicative "units" of $M_n R$. Thus, $\widehat{Gl}_n R$ is the pullback of the following diagram

$$\begin{array}{c} M_n R \\ \Big\downarrow \text{componentization} \\ Gl_n \pi_0 R \subset M_n \pi_0 R \, . \end{array}$$

The notion of A_∞ ring space parametrizes the additive and multiplicative structures of R by E_∞ and A_∞ operads respectively, in such a way that we may define deloopings $B\widehat{Gl}_n R$, together with maps $B\widehat{Gl}_n R \to B\widehat{Gl}_{n+1} R$ (after inversion of certain equivalences) corresponding to the natural inclusion of matrices. We then take $BGl_\infty R$ to be the telescope of these maps, and define the K-theory of

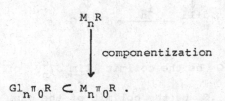

R, KR, to be $(B\widehat{Gl}_\infty R)^+ \times Z$, the product of the plus construction on $B\widehat{Gl}_\infty R$ with the discrete space of integers.

Before making the above theory precise, we shall define W(X) and explain the line of proof of Theorem 2.

Waldhausen's construction has several equivalent definitions. We shall present one of these, and shall prove the relevant equivalences in a later work.

Fix a grouplike topological monoid G . One should think of G as a suitable model for the loopspace on a space X.

Definition 1.1. Let $Gl_n^d G$ be the topological monoid of self G-homotopy equivalences of $G_+ \wedge \bigvee^n S^d$, which is a G-space via left translation in G. Let $\sigma : Gl_n^d G \to Gl_n^{d+1} G$ be the suspension map defined by the smash product with the identity map of the 1-sphere, and let $i_n : Gl_n^d G \to Gl_{n+1}^d G$ be defined by the wedge product with the identity map of $G_+ \wedge S^d$. We have a commutative diagram of inclusions

$$\begin{array}{ccc} Gl_n^d G & \xrightarrow{i_n} & Gl_{n+1}^d G \\ \downarrow \sigma & & \downarrow \sigma \\ Gl_n^{d+1} G & \xrightarrow{i_n} & Gl_{n+1}^{d+1} G . \end{array}$$

Let $Gl_n G = Gl_n^\infty G$ be the colimit $\varinjlim_d Gl_n^d G$ of the maps σ, and let $i_n : Gl_n G \to Gl_{n+1} G$ be the colimit of the maps i_n. Let $BGl_\infty G$ be the telescope of the maps Bi_n. We shall see below that $\pi_1 BGl_\infty G$ is isomorphic to $Gl_\infty Z[\pi_0 G]$, so that the plus construction $(BGl_\infty G)^+$ is defined. We define the geometric K-theory, $K^g G$, of G to be $(BGl_\infty G)^+ \times Z$. Waldhausen's algebraic K-theory of the space X, W(X) is defined to be equal to $K^g|GX|$.

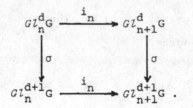

The comparison of this definition with higher simple homotopy theory is rather convoluted, but its comparison with A(X) is intuitively clear. We shall prove the following more general form of Theorem 2.

Theorem 1.2. Let G be a grouplike topological monoid. Then there is a natural equivalence between K^gG and the K-theory of the A_∞ ring space $Q(G_+)$.

Very briefly, the idea is to first construct a natural homotopy equivalence $f_n : Gl_n G \to \widehat{Gl}_n Q(G_+)$ and then put enough structure on this map to deloop it. We carry out the first step here.

The maps f_n are easy to construct. Let $M_n^d G$ be the monoid of all G-equivariant self-maps of $G_+ \wedge \bigvee^n S^d$. Then restriction to the subspace $e_+ \wedge \bigvee^n S^d$ defines a homeomorphism

$$g_n : M_n^d G \to F(\bigvee^n S^d, G_+ \wedge \bigvee^n S^d) .$$

Here e is the identity element of G, and $F(X,Y)$ is the space of based maps between the based spaces X and Y. Now using the universal property of wedges, together with the commutation formulae for wedges and suspensions with respect to the smash product, we obtain a homeomorphism

$$h_n : F(\bigvee^n S^d, G_+ \wedge \bigvee^n S^d) \to \prod^n \Omega^d \Sigma^d (\bigvee^n (G_+))$$

between the latter space and the product of n copies of the free d-fold loop space on the wedge of n copies of G_+. The maps $h_n \circ g_n$ commute with suspension, and provide a homeomorphism

$$\bar{g}_n : M_n^\infty G \to \prod^n Q(\bigvee^n (G_+)).$$

Regarding the coordinates of the target space as columns, let

$$k_n : \prod^n Q(\bigvee^n (G_+)) \to M_n Q(G_+)$$

be obtained from the maps by which Q converts wedges to products, namely the maps obtained by application of the functor Q to the projections $\bigvee^n G_+ \subset \prod^n G_+ \to G_+$. Of course, k_n is a homotopy equivalence. We now check that the product on $\pi_0 M_n^\infty G$ induced by the monoid structure corresponds under $k_n \circ \bar{g}_n$ to matrix multiplication in $\pi_0 M_n Q(G_+)$, which we already know to be isomorphic to $M_n Z[\pi_0 G]$. Since $Gl_n G$ is the "unit space" of the monoid $M_n^\infty G$, $k_n \circ \bar{g}_n$ restricts to a homotopy equivalence

$$f_n : Gl_n G \to \widehat{Gl}_n Q(G_+) \ .$$

Note also that the maps f_n are inclusions.

We shall deloop the maps f_n in section 3. Theorem 1.2 will then follow, since our constructions commute with the maps i_n.

§2. <u>The K-theory of an A_∞ ring space.</u>

To begin with, we must examine the notion of A_∞ ring space. Recall, [3], that an operad pair (C,G) consists of an "additive" operad C, a "multiplicative" operad G, and action maps

$$\lambda : G(k) \times C(j_1)\times\ldots\times C(j_k) \to C(j_1\ldots j_k),$$

which satisfy certain diagrams. A (C,G)-space (R,θ,ξ) consists of a space R, together with a C-action θ and a G-action ξ (with different basepoints) such that a certain distributivity diagram commutes. For simplicity, we shall also assume that R is a grouplike C-space, and therefore a ring space under our previous definition. If C is an E_∞ operad and G is an A_∞ operad, we say that (C,G) is an A_∞ operad pair and that R is an A_∞ ring space.

We shall adopt the convention that all A_∞ operads are to be considered in their non-Σ form (without permutations [4]). Thus, G consists of contractible spaces $G(j)$ for $j \geq 0$, with $G(0) = *$, an element $1 \in G(1)$, and associative and unital maps

$$\gamma : G(k) \times G(j_1)\times\ldots\times G(j_k) \to G(j_1+\ldots+j_k) .$$

In particular, "forget permutations" forms a functor from E_∞ operads to A_∞ operads, and hence from E_∞ operad pairs to A_∞ operad pairs. We shall call particular attention to the A_∞ operad pair (K_∞, L) obtained in this manner, where K_∞ is the (partial) operad of little convex bodies and L is the linear isometries operad [3]. This pair is particularly useful, because it plays a universal role in May's theory. For our purposes it is important, being the operad pair which acts on $Q(G_+)$ [2].

Now fix an A_∞ operad pair (C,G). We would like to find an A_∞ operad which parametrizes the multiplicative H-space structure of M_nR, for R a (C,G)-space. At this point we note that the combinatorics in May's paper are not quite right. The first change we must make is to allow the zero-th space of an A_∞ operad G, $G(0)$ to be a contractible based space rather than a point. In particular, the maps

$$\gamma : G(k) \times G(0)^k \to G(0)$$

then give $G(0)$ the structure of a G-space. This has no effect on the resulting classifying space theory.

We wish now to associate to an ordinary A_∞ operad pair $(\mathcal{C},\mathcal{G})$ a generalized A_∞ operad, $\mathcal{H}_n(\mathcal{C},\mathcal{G})$, of this type, which will parametrize multiplication of $n\times n$ matrices. For $j > 0$, we set

$$\mathcal{H}_n(\mathcal{C},\mathcal{G})(j) = (M_n\mathcal{C}(n^{j-1})) \times \mathcal{G}(j),$$

the product of $n\times n$ matrices with entries in $\mathcal{C}(n^{j-1})$ with the j-th space of $\mathcal{G}$. We define $\mathcal{H}_n(\mathcal{C},\mathcal{G})(0)$ to be matrices whose diagonal entries lie in $\mathcal{C}(1)$, while all other entries are $* = \mathcal{C}(0)$. The action of $\mathcal{H}_n(\mathcal{C},\mathcal{G})$ on $n\times n$ matrices with coefficients in the $(\mathcal{C},\mathcal{G})$-space R is given by parametrized matrix multiplication

$$\psi_j : \mathcal{H}_n(\mathcal{C},\mathcal{G})(j) \times (M_nR)^j \to M_nR .$$

In more detail, for $A_1,\ldots,A_j \in M_nR$, $C \in M_n\mathcal{C}(n^{j-1})$ and $g \in \mathcal{G}(j)$, one obtains the (r,s)-th entry of $\psi_j((C,g);A_1,\ldots,A_j)$ by taking the n^{j-1}-fold sum specified by the (r,s)-th entry of C of the n^{j-1} summands of the formal matrix product of $A_1,\ldots,A_j$ specified by the j-fold product g. The structure maps, γ, of $\mathcal{H}_n(\mathcal{C},\mathcal{G})$ are forced by the requirement that ψ_j specify an action.

The projection map $\pi_2 : \mathcal{H}_n(\mathcal{C},\mathcal{G}) \to \mathcal{G}$ onto the second component is a map of operads. We may regard $\mathcal{H}_n(\mathcal{C},\mathcal{G})$ as something of a semidirect product of $\mathcal{G}$ with matrices in $\mathcal{C}$. Since $\mathcal{C}(j)$ and $\mathcal{G}(j)$ are contractible for all $j \geq 0$, $\mathcal{H}_n(\mathcal{C},\mathcal{G})$ is an A_∞ operad. We obtain the following definition.

<u>Definition 2.1</u>. Let R be a $(\mathcal{C},\mathcal{G})$-space. Then $B\widehat{Gl}_nR$ is the delooping of $\widehat{Gl}_nR$ with respect to its $\mathcal{H}_n(\mathcal{C},\mathcal{G})$ action.

We wish now to compare $B\widehat{Gl}_nR$ and $B\widehat{Gl}_{n+1}R$. In fact, the following proposition was stated in [2].

Proposition 2.2. For $m > 0$ and $n > 0$, there is a natural morphism of operads

$$\tau_{n,m} : \mathcal{H}_{n+m}(C,G) \to \mathcal{H}_n(C,G) \times \mathcal{H}_m(C,G)$$

which provides the Whitney sum

$$M_n R \times M_m R \xrightarrow{\oplus} M_{n+m} R$$

with the structure of an $\mathcal{H}_{n+m}(C,G)$ map.

Intuitively, $\tau_{n,m}$ is obtained by restricting the additive parameters to act on those addends which would not be automatically zero in the j-fold product of Whitney sums. We obtain the following definitions.

Definition 2.3. Let R be a (C,G)-space. Then $i_n : B\widehat{Gl}_n R \to B\widehat{Gl}_{n+1} R$ is the homotopy class of the comparison induced from the $\mathcal{H}_{n+1}$-maps

$$\widehat{Gl}_n R \xleftarrow{\pi_1} \widehat{Gl}_n R \times \mathcal{H}_1(0) \xrightarrow{1\times\psi_0} \widehat{Gl}_n R \times \widehat{Gl}_1 R \xrightarrow{\oplus} \widehat{Gl}_{n+1} R .$$

Here, π_1 is an equivalence, and modulo a little care [2], $B\widehat{Gl}_\infty R$ is the telescope of the resulting maps i_n. Since $\pi_1 B\widehat{Gl}_\infty R$ is isomorphic to $Gl_\infty \pi_0 R$, we may define $KR = (B\widehat{Gl}_\infty R)^+ \times Z$.

§3. Delooping the maps f_n.

Let $\mathcal{H}_n = \mathcal{H}_n(\mathcal{K}_\infty, \mathcal{L})$. Recall that $Q(G_+)$ is a $(\mathcal{K}_\infty, \mathcal{L})$-space, so that $\widehat{Gl}_n Q(G_+)$ is an $\mathcal{H}_n$-space. We shall first put an $\mathcal{H}_n$ structure on $Gl_n G$, and show that its $\mathcal{H}_n$ delooping is equivalent to the classical bar construction. We shall then show how to construct "higher homotopies" between the actions of $\mathcal{H}_n$ on $Gl_n G$ and on $\widehat{Gl}_n Q(G_+)$. In particular, we obtain $\mathcal{H}_n$-maps

$$Gl_n G \xleftarrow{\varepsilon} B(H_n, H_n, Gl_n G) \xrightarrow{\bar{f}_n} \widehat{Gl}_n Q(G_+) .$$

Here $B(H_n, H_n, Gl_n G)$ is the two-sided bar construction [4], obtained from the action of the monad H_n associated to $\mathcal{H}_n$ on $Gl_n G$, ε is the natural map induced by this action, and the composite of $\bar{f}_n$ with the natural homotopy inverse to ε is the map f_n.

Recall that the projection $\pi_2 : \mathcal{H}_n \to \mathcal{L}$ is a map of operads. The action of $\mathcal{H}_n$ on $Gl_n G$ will be induced by pullback from an $\mathcal{L}$ action we shall define below.

Recall that $\mathcal{I}$ is the category of finite or countably infinite real inner product spaces and their linear isometries. As shown in [3], most known E_∞ $\mathcal{L}$-spaces arise from continuous functors T from $\mathcal{I}$ to based spaces, together with commutative and associative natural transformations

$$\oplus : TV \times TW \to T(V \oplus W),$$

such that the restriction of $\oplus$ to $TV \times *$ is an inclusion which coincides with $T(i : V \subset V \oplus W)$, and TV is the colimit of TW as W ranges over the finite dimensional subspaces of V. T is uniquely determined by its behavior on finite dimensional inner product spaces and their linear isometric isomorphisms [3]. The $\mathcal{L}$-space associated to T is $T(R^\infty)$, with L-action specified by

$$L(j) \times T(R^\infty)^j \xrightarrow{1\times\oplus} I(R^{\infty j}, R^\infty) \times T(R^{\infty j}) \longrightarrow T(R^\infty) .$$

There is an analogous notion of a non-Σ I-functor with $\oplus$ only required to be associative. The same recipe now provides a non-Σ action of L on $T(R^\infty)$. We specify such a functor, Gl_nG, by setting $Gl_nG(V)$ equal to the monoid of self G-homotopy equivalences of $G_+ \wedge \overset{n}{\bigvee} tV$, where tV is the one-point compactification of the finite inner product space V. If $g : V \rightarrow V'$ is an isometric isomorphism, then $g_* : Gl_nG(V) \rightarrow Gl_nG(V')$ is specified by conjugation with $1 \wedge \overset{n}{\bigvee} tg$, and the suspension $\sigma : Gl_nG(V) \rightarrow Gl_nG(V \oplus W)$ is obtained by smash product with the identity map of tW. The direct sum is the composite

$$Gl_nG(V) \times Gl_nG(W) \xrightarrow{\sigma\times\sigma} Gl_nG(V \oplus W) \times Gl_nG(V \oplus W) \xrightarrow{\mu} Gl_nG(V \oplus W),$$

where μ is the monoid product. Of course, $Gl_nG(R^\infty)$ is the previously defined monoid Gl_nG.

The bar construction classifying space of a grouplike monoid is naturally equivalent to the May delooping of its M-space structure. We shall construct a diagram of A_∞ operads

$$\begin{array}{ccc} & & L \\ & & \downarrow \chi \\ M & \xrightarrow{\phi} & \hat{L} \end{array}$$

such that the actions of L and M on Gl_nG are obtained by pull-back from an action by $\hat{L}$. From this it will follow that the standard classifying space of Gl_nG used in the construction of K^gG is equivalent to its May delooping as an L-space (or H_n-space by pullback). Here, $\hat{L}$ is the "product operad" associated to L, meaning that $\hat{L}(j) = L(1)^j$ and that the structure map of $\hat{L}$ is obtained from suitable choices of multiplications in the monoid $L(1)$, precisely as in the case of the little cubes operads. We may also consider $\hat{L}$ to be the operad whose j-th space consists of linear maps from $R^{\infty j}$ to R^{∞} whose restriction to the p-th summand is an isometry for $1 \leq p \leq j$. Thus, $\chi_j(g) = (g\circ\iota_1,\ldots,g\circ\iota_j)$, where ι_p is the inclusion of the p-th summand for $1 \leq p \leq j$. Of course, ϕ is defined by $\phi_j(*) = 1^j$, and the action of $\hat{L}$ on Gl_nG is defined by

$$\theta_j((g_1,\ldots,g_j);x_1,\ldots,x_j) = \mu(g_{1*}x_1,\ldots,g_{j*}x_j) ,$$

where μ is the j-fold iterated monoid product and g_*x is the effect of the linear isometry g on $x \in Gl_nG(R^\infty)$.

The definition of the higher homotopies requires greater care. We wish to find maps

$$\bar{f}_{n,p} : H_n^{p+1}Gl_nG \times \Delta^p \to \widehat{Gl}_nQ(G_+) ,$$

compatible with the faces and degeneracies used to define the two-sided bar construction, such that $\bar{f}_{n,0}$ is the composite

$$H_nGl_nG \xrightarrow{H_nf_n} H_n\widehat{Gl}_nQ(G_+) \xrightarrow{\psi} \widehat{Gl}_nQ(G_+) ,$$

and the restriction of $\bar{f}_{n,p}$ to $H_n^{p+1}Gl_nG \times u$ is an H_n-map for each $u \in \Delta^p$.

We must carefully examine the two H_n-actions and the map f_n.

Suppose that $x \in \mathcal{Gl}_n G$ is concentrated on $G_+ \wedge \overset{n}{\vee} tV$ for the finite dimensional $V \subset R^\infty$. Let $x(r,s)$ be the composite

$$G_+ \wedge tV \xrightarrow{1\wedge\iota_s} G_+ \wedge \overset{n}{\vee} tV \xrightarrow{x} G_+ \wedge \overset{n}{\vee} tV \xrightarrow{1\wedge\pi_r} G_+ \wedge tV,$$

where ι_s is the inclusion of the s-th summand of $\overset{n}{\vee} tV$ and π_r is the projection onto the r-th summand of $\overset{n}{\vee} tV$. We see that $f_n x$ is the matrix whose (r,s)-th coordinate is the suspension over R^∞ of the restriction of $x(r,s)$ to $tV = e_+ \wedge tV$. Now let $x_p \in \mathcal{Gl}_n G(V_p)$ for $1 \leq p \leq j$ and let $\sigma : \mathcal{Gl}_n G(V_p) \to \mathcal{Gl}_n G(V_1 \oplus \ldots \oplus V_j)$ be the suspension. Fixing r and s, let $U = (u_0, \ldots, u_j)$ be a sequence of integers $1 \leq u_p \leq n$ with $u_0 = r$ and $u_j = s$. Consider the composite

$$x_U = \sigma x_1(u_0, u_1) \circ \ldots \circ \sigma x_j(u_{j-1}, u_j) .$$

We claim the following

(1) Let $v \in tV$. Then there is at most one sequence U such that the restriction of x_U to $G \times v$ is nontrivial.

(2) The (r,s)-th coordinate of the sum $x_1 \oplus \ldots \oplus x_j$ is the "sum" over all such sequences U of the maps x_U. That is, if the restriction of x_U to $G \times v$ is nontrivial, then its effect is equal to that of $(x_1 \oplus \ldots \oplus x_j)(r,s)$ on that subspace.

(3) Let $(C,g) \in \mathcal{H}_n(j)$ and let $C(r,s)$ be the collection $\langle c_U \rangle$ of little convex bodies, indexed on such sequences U. Suppose that each c_U is concentrated on some finite dimensional subspace $W \subset R^\infty$ containing $g(V_1 \oplus \ldots \oplus V_j)$ and let $c_U^{-1} : tW \to tW$ be the map

induced by inversion of c_U. Then the (r,s)-th coordinate of $\psi_j((C,g);f_nx_1,\ldots,f_nx_j)$ is the suspension over R^∞ of the sum in the sense of (2) of the maps

$$\sigma(g_* x_u) \circ \iota_+ \wedge c_U^{-1} ,$$

where $\iota : e \subset G$ is the inclusion of the identity element and $\sigma : Gl_n G(g(V_1 \oplus \ldots \oplus V_j)) \subset Gl_n G(W)$ is the suspension.

Thus, if we could homotop the c_U's to the identity map of R^∞ in such a way that the sum of the maps $\sigma(g_* x_U) \circ \iota_+ \wedge c_U(t)^{-1}$ is defined for each t in the unit interval, we would obtain a homotopy from the upper composite to the lower composite in the following diagram

$$\begin{array}{ccc} H_n(j) \times Gl_n G^j & \xrightarrow{1\times f_n^j} & H_n(j) \times \widehat{Gl}_n Q(G_+)^j \\ \downarrow \theta_j & & \downarrow \psi_j \\ Gl_n G & \xrightarrow{f_n} & \widehat{Gl}_n Q(G_+) . \end{array}$$

The higher homotopies could be defined similarly.

Little convex bodies were set up in such a way that for each little convex body d, the linear homotopy, $d_t(v) = (1-t)d(v) + tv$, from d to the identity map, is a homotopy through convex bodies. But the convex bodies $c_{U,t}$ will not be disjoint for all t, so we require more explicit information concerning the summability of $\sigma(g_* x_U) \circ \iota_+ \wedge c_U^{-1}$, for $\langle c_U \rangle$ a not necessarily disjoint collection of convex bodies. We already know that the identity map poses no problem, and we notice the following.

<u>Lemma 3.1</u>. If $W = g(V_1 \oplus \ldots \oplus V_j) \oplus W'$ and each c_U is concentrated

on W', then

$$\sigma(g_* x_U) \circ \iota_+ \wedge c_U^{-1}(v,w) = g_* x_U(e,v) \wedge c_U^{-1}(w) ,$$

for $v \in g(V_1 \oplus \ldots \oplus V_j)$ and $w \in W$.

Thus, under the hypothesis of Lemma 3.1, the summability of the maps x_U guarantees the summability of the maps $\sigma(g_* x_U) \circ \iota_+ \wedge c_U^{-1}$ regardless of the nature of the collection $\langle c_U \rangle$. We notice at this point that we may as well define $B(H_n, H_n, Gl_n G)$ in terms of a partial action of the partial operad H_n on $Gl_n G$ [3]. We can define composability in H_n, and a partial action on $Gl_n G$, in a way that codifies the disjunction requirement of Lemma 3.1. We can now find simple combinatorial formulae for the maps $\bar{f}_{n,p}$. The rest is mere detail.

Bibliography

1. A. Hatcher. Higher simple homotopy theory. Annals of Math. 102 (1975), 101-137.

2. J.P. May, A_∞ ring spaces and algebraic K-theory, Springer Lecture Notes in Math. Vol. 658, 240-315, 1978.

3. J.P. May (with contributions by N. Ray, F. Quinn, and J. Tornehave). E_∞ Ring Spaces and E_∞ Ring Spectra. Springer Lecture Notes in Math. Vol. 577, 1977.

4. J.P. May, The Geometry of Iterated Loop Spaces. Springer Lecture Notes in Math. Vol. 271, 1972.

5. F. Waldhausen, Algebraic K-theory of topological spaces, I. Proc. Symposia in Pure Math. Vol. 32, part 1, 35-60. Amer. Math. Soc. 1978.

FIRST QUADRANT SPECTRAL SEQUENCES IN ALGEBRAIC K-THEORY

by

R.W. Thomason*

Department of Mathematics
Massachusetts Institute of Technology
Cambridge, Massachusetts 02139
U.S.A.

By algebraic K-theory I understand the study of the following process: one takes a small category $\underline{\underline{S}}$ provided with a "direct sum" operation, $\oplus$; "group completes" the monoid structure induced by $\oplus$ on the classifying space $B\underline{\underline{S}}$; and then takes the homotopy groups of the resulting space. For R a ring, this process applied to the category of finitely generated projective R-modules yields Quillen's $K_*(R)$. Karoubi's L-theory is also a special case of this generalized algebraic K-theory.

My aim in this paper is to show how K-theory may be axiomatized as a generalized homology theory on the category of such categories $\underline{\underline{S}}$, and to give a construction that yields the K-theoretic analogues of mapping cones, mapping telescopes and the like. Even if one is interested only in the K-theory of rings, these results for generalized K-theory should be useful technical tools. In particular the mapping cone construction of §6 may be useful in situations where appeal to

* Author partially supported by NSF Grant MCS77-04148.

Quillen's Theorem B fails. Two examples are given in §6 to illustrate this point.

§1. Definitions of some algebraic structures on categories.

K-theory assigns a graded abelian group to each small symmetric monoidal category. Recall this is a category $\underline{\underline{S}}$ together with a selected object $0 \in \underline{\underline{S}}$, a bifunctor $\oplus : \underline{\underline{S}} \times \underline{\underline{S}} \longrightarrow \underline{\underline{S}}$, and natural isomorphisms

$$\alpha : (A \oplus B) \oplus C \xrightarrow{\sim} B \oplus A$$

$$\gamma : A \oplus B \xrightarrow{\sim} B \oplus A$$

$$\lambda : A \xrightarrow{\sim} 0 \oplus A$$

These natural isomorphisms are subject to "coherence conditions;" we require a certain five diagrams to commute, and these imply all (generic) diagrams made up of α's, γ's, and λ's commute. Check [2], II, §1, III, §1; or [6], 3.3; or [9], VII, §1, VII, §7; for details. The word "small" means only that $\underline{\underline{S}}$ has a set of objects, rather than a proper class of them; this is a technicality required so the classifying space [14] $B\underline{\underline{S}}$ exists.

The standard example of a symmetric monoidal category is any additive category with $\oplus$ given by direct sum. The subcategory of isomorphisms in an additive category has an induced symmetric monoidal structure.

A permutative category is a symmetric monoidal one where the α's and λ's are required to be identity natural transformations: thus $\oplus$ is strictly associative and unital. Every symmetric monoidal category is equivalent to a permutative one ([7], 1.2; [11], 4.2), so we may assume things are permutative whenever convenient.

Let $\underline{\underline{S}}$, $\underline{\underline{T}}$ be symmetric monoidal categories. A <u>lax symmetric monoidal functor</u> $F : \underline{\underline{S}} \longrightarrow \underline{\underline{T}}$ is a functor together with natural transformations

$$\tilde{f} : O_T \longrightarrow FO_S \qquad\qquad \bar{f} : FA \oplus FB \longrightarrow F(A \oplus B)$$

such that the following diagrams commute

$$\begin{array}{ccc} FA \oplus FB & \xrightarrow{\ \bar{f}\ } & F(A \oplus B) \\ \gamma \downarrow & & \downarrow F\gamma \\ FB \oplus FA & \xrightarrow[\ \bar{f}\ ]{} & F(B \oplus A) \end{array}$$

$$\begin{array}{ccccc} (FA \oplus FB) \oplus FC & \xrightarrow{\ \bar{f} \oplus 1\ } & F(A \oplus B) \oplus FC & \xrightarrow{\ \bar{f}\ } & F((A \oplus B) \oplus C) \\ \alpha \downarrow & & & & \downarrow F\alpha \\ FA \oplus (FB \oplus FC) & \xrightarrow{\ 1 \oplus \bar{f}\ } & FA \oplus F(B \oplus C) & \xrightarrow{\ \bar{f}\ } & F(A \oplus (B \oplus C)) \end{array}$$

$$\begin{array}{ccc} FA & \xrightarrow{\ F\lambda\ } & F(O_S \oplus A) \\ \lambda \downarrow & & \uparrow \bar{f} \\ O_T \oplus FA & \xrightarrow{\ \tilde{f} \oplus 1\ } & FO_S \oplus FA \end{array}$$

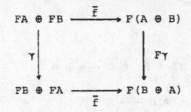

F is a <u>strong</u> symmetric monoidal functor (the usual notion in infinite loop space theory) if in addition $\tilde{f}$ and $\bar{f}$ are isomorphisms. F is a <u>strict</u> symmetric monoidal functor if $\tilde{f}$ and $\bar{f}$ are the identity natural transformation. F is a lax (strong, strict)

permutative functor if its source and target are permutative.

If F, G are lax symmetric monoidal functors, a *symmetric monoidal natural transformation* $\eta : F \Rightarrow G$ is a natural transformation making the following diagrams commute:

$$\begin{array}{ccc} & & FO_S \\ & \nearrow^{\tilde{f}} & \\ O_T & & \downarrow \eta \\ & \searrow_{\tilde{g}} & \\ & & GO_S \end{array} \qquad \begin{array}{ccc} FA \oplus FB & \xrightarrow{\bar{f}} & F(A \oplus B) \\ \downarrow \eta \oplus \eta & & \downarrow \eta \\ GA \oplus GB & \xrightarrow{\bar{g}} & G(A \oplus B) \end{array}$$

As an example, any additive functor between two additive categories is a strong symmetric monoidal functor. Any natural transformation between two additive functors is a symmetric monoidal natural transformation.

For any of the above concepts, we have a corresponding nonunital version obtained by dropping any condition relating to the unit O.

§2. *Passage to the associated spectrum and definition of algebraic K-theory*.

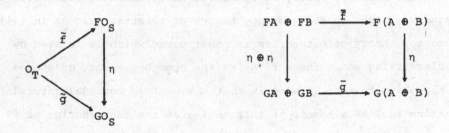

Let *Sym Mon* be the category of small symmetric monoidal categories and lax symmetric monoidal functors. By [17], 4.2.1, there is a functor, $Spt : \underline{Sym\ Mon} \longrightarrow \underline{Spectra}$, into the category of (connective) spectra, and a natural transformation $B\underline{S} \longrightarrow Spt_0(\underline{S})$, which exhibits the zeroth space of the spectrum $Spt(\underline{S})$ as the group completion of the classifying space of the category $\underline{S}$. This extends the usual functors defined on the subcategory of strong symmetric monoidal functors by [11], [15]. The proof of [13] is easily adapted to show Spt is uniquely determined up to natural equivalence by the above properties.

A symmetric monoidal natural transformation $\eta : F \Rightarrow G$ induces a homotopy of maps of spectra from Spt F to Spt G. Thus one may construct a homotopy theory of symmetric monoidal categories which is related via Spt to the homotopy theory of spectra, just as in [14] a homotopy theory of categories is constructed which is related by the classifying space functor B to the homotopy theory of spaces. The key thing to keep in mind is that a symmetric monoidal natural transform is like a homotopy; this motivates the construction of §3 and its relation to the homotopy colimit constructions of §4.

I define K-theory as a functor from <u>Sym</u> <u>Mon</u> to graded abelian groups by $K_*(\underline{\underline{S}}) = \pi_*^S(\text{Spt } \underline{\underline{S}})$, the reduced stable homotopy groups of the spectrum Spt $\underline{\underline{S}}$. As by spectrum I mean what used to be called an Ω-spectrum, this is equivalent to $\pi_*(\text{Spt}_0\ \underline{\underline{S}})$, the homotopy groups of the group completion of $B\underline{\underline{S}}$. Thus if $\underline{\underline{S}}$ is the subcategory of isomorphisms in an additive category $\underline{\underline{\mathcal{A}}}$, my $K_*(\underline{\underline{S}})$ is $K_*(\underline{\underline{\mathcal{A}}})$ as defined by Quillen's plus construction or group completion method.

I prefer to think of K_* as π_*^S Spt rather than in terms of homotopy groups, as the former is more "homological."

§3. <u>The fundamental construction and its first quadrant K-theory spectral sequence</u>.

I will now present a construction on diagrams in <u>Sym</u> <u>Mon</u>, show it has a reasonable universal mapping property (so it is not an ad hoc construction), and then give a spectral sequence for its K-groups.

Let $\underline{\underline{L}}$ be a small category. For simplicity, consider a diagram of permutative categories; i.e., a functor $F : \underline{\underline{L}} \longrightarrow \underline{\text{Perm}}$ into the category of permutative categories.

Let Perm-hocolim F be the permutative category with objects $n[(L_1,X_1),\ldots,(L_n,X_n)]$ where $n = 0,1,2,\ldots$; L_i is an object of $\underline{\underline{L}}$, and X_i is an object of $F(L_i)$. A morphism $n[(L_1,X_1),\ldots,(L_n,X_n)] \longrightarrow k[(L_1',X_1'),\ldots,(L_k',X_k')]$ consists of data $(\psi;\ell_i;x_i)$

1) a surjection of sets $\psi : \{1,\ldots,n\} \twoheadrightarrow \{1,\ldots,k\}$
2) maps $\ell_i : L_i \longrightarrow L'_{\psi(i)}$ in $\underline{\underline{L}}$
3) maps $x_j : \bigoplus_{i\in\psi^{-1}(j)} F(\ell_i)(X_i) \longrightarrow X_j'$ in L_j'

There is a straightforward rule giving the composition of morphisms. I will not explicitly give it, but it is implicitly determined by the universal mapping property below.

The unit of Perm-hocolim F is $0[\ ]$, and $\oplus$ is given by

$n[(L_1,X_1),\ldots,(L_n,X_n)] \oplus m[(L_{n+1},X_{n+1}),\ldots,(L_{n+m},X_{n+m})] =$
$n+m[(L_1,X_1),\ldots,(L_{n+m},X_{n+m})]$.

The universal mapping property is given by

<u>Lemma</u>: Strict permutative functors $G : \text{Perm hocolim } F \longrightarrow \underline{\underline{T}}$ correspond bijectively with systems consisting of non-unital lax permutative functors $G_L : F(L) \longrightarrow \underline{\underline{T}}$ for each $L \in \underline{\underline{L}}$, and non-unital permutative natural transformations $G_\ell : G_L \Rightarrow G_{L'} \cdot F(\ell)$ for each $\ell : L \longrightarrow L'$ in $\underline{\underline{L}}$; which must satisfy the conditions $G_1 = \text{id}$ and $G_\ell \cdot G_{\ell'} = G_{\ell\ell'}$.

<u>Proof</u>: Let $J_L : F(L) \longrightarrow \text{Perm hocolim } F$ be given by $J_L(X) = 1[(L,X)]$. This J_L is a non-unital lax permutative functor in an obvious way, and for $\ell : L \longrightarrow L'$ there is an obvious choice of $J_\ell : J_L \Rightarrow J_{L'} \cdot F(\ell)$.

Given this system, the bijective correspondence sends G to the system $G_L = G \cdot J_L$, $G_\ell = G \cdot J_\ell$. It is tedious but easy to see this works.

Given a diagram $F : \underline{\underline{L}} \longrightarrow$ Sym Mon, there is an analogous Sym-Mon-hocolim F with the corresponding universal mapping property. All I say below about Perm-hocolim F applies to it as well. The explicit description of the objects and morphisms of Sym-Mon-hocolim F differs slightly from that given above.

This construction turns out to have good properties with respect to K-theory.

Theorem: Let $F : \underline{\underline{L}} \longrightarrow$ Perm be a diagram. Then there is a natural first quadrant spectral sequence

$$E^2_{p,q} = H_p(\underline{\underline{L}};K_qF) \Rightarrow K_{p+q}(\text{Perm-hocolim } F)$$

Here $H_*(\underline{\underline{L}};K_qF)$ is the homology of the category $\underline{\underline{L}}$ with coefficients in the functor $L \longmapsto K_qF(L)$. A convenient source for information on this is [5], IX §6 or [14], §1. I'll identify the E^2 term with more familiar objects for the examples of §4.

This theorem is an immediate corollary of the theorem of §5 and the proposition of §4.

§4. Facts about and examples of homotopy colimits.

To prepare the way for the statement of the fundamental theorem of §5, and to explain the strange-looking name Perm-hocolim, I will review the homotopy colimit (homotopy direct limit) construction of Bousfield and Kan [1]. Some version of this exists for every category admitting a reasonable homotopy theory, e.g., Sym Mon and Spectra; but [1] concentrates on the category of simplicial sets. I'll give some of their results translated for the category of topological spaces, Top. One can also read Vogt [20] for this material.

Let $F : \underline{L} \longrightarrow \text{Top}$ be a diagram. Associated naturally to F is a space Top-hocolim F, the homotopy colimit of F. It is characterized by a universal mapping property ([1], XII, 2.3) establishing a bijective correspondence between maps g : Top-hocolim $F \longrightarrow X$ and a system of maps $g_L : F(L) \longrightarrow X$ and homotopies relating them. With the philosophy of §2 that symmetric monoidal natural transforms are like homotopies, this universal mapping property of Top-hocolim F is much like that of Perm-hocolim F given in the lemma of §3 (cf. [18], 1.3.2).

For any generalized homology theory E_* on Top, there is a first quadrant spectral sequence [1], XII, 5.7

$$E^2_{p,q} = H_p(\underline{\underline{L}};E_qF) \Rightarrow E_{p+q}(\text{Top-hocolim } F)$$

This construction subsumes many well-known constructions.

Example 1: Let $F : \underline{L} \longrightarrow \text{Top}$ be the diagram

$$\begin{array}{ccc} A & \longrightarrow & B \\ \downarrow & & \\ C & & \end{array}$$

Then Top-hocolim F is the double mapping cyclinder on $A \longrightarrow B$ and $A \longrightarrow C$. In this case the spectral sequence collapses to the long exact Mayer-Vietoris sequence

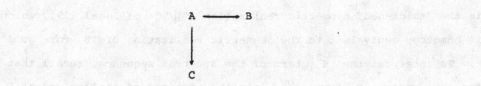

$$\cdots \xrightarrow{\partial} E_q(A) \longrightarrow E_q(B) \oplus E_q(C) \longrightarrow E_q(\text{double mapping cylinder}) \xrightarrow{\partial} \cdots$$

For C a point, the Top-hocolim is the mapping cone of $A \longrightarrow B$; and for B and C points, it is the suspension of A.

Example 2: Let $\underline{\underline{L}}$ be the category of the positive integers as a partially ordered set. Then $F : \underline{\underline{L}} \longrightarrow \underline{\text{Top}}$ is a diagram: $F(1) \longrightarrow F(2) \longrightarrow F(3) \longrightarrow \cdots$, and Top-hocolim F is the mapping telescope. In the spectral sequence $H_p(\underline{\underline{L}};E_qF) = 0$ if $p > 0$, and $H_0(\underline{\underline{L}};E_qF) = \varinjlim_n E_qF(n)$.

Example 3: Let $\underline{\underline{L}}$ be a group G considered as a category with one object $*$, and morphisms being the elements of G. A functor $F : \underline{\underline{L}} \longrightarrow \underline{\text{Top}}$ is a homomorphism $G \longrightarrow \text{Aut}(F(*))$; that is, an action of G on $F(*)$. If EG is a free acyclic G-complex, Top-hocolim F is $EG \times_G F(*)$. The spectral sequence is identified to the usual one

$$H_p(G,E_qF(*)) \Rightarrow E_{p+q}(EG \times_G F(*)) .$$

Example 4: Let $\underline{\underline{L}}$ be Δ^{op}. Then $F : \Delta^{op} \longrightarrow \underline{\text{Top}}$ is just a simplicial space. It follows from [1], XII, 3.4 that Top-hocolim F is the "thickened" geometric realization "$\|\ \|$" of Segal [15], which is homotopy equivalent to the geometric realization of F for "good" F. To interpret the E^2 term of the spectral sequence, recall that for any functor E from Δ^{op} into the category of abelian groups, i.e., for E a simplicial abelian group, $H_*(\Delta^{op};E)$ is the homology of the chain complex which in degree p is E_p, and has differential $\partial = \sum(-1)^i d_i$. This follows from [1], XII, 5.6 and [12], 22.1.

One has analogous results in many categories admitting a homotopy theory. In particular, consider a functor $F : \underline{\underline{L}} \longrightarrow$ Spectra. One may define Spectra-hocolim F as follows. Let $F_n : \underline{\underline{L}} \longrightarrow$ Top be the diagram of nth spaces of the spectra. Form Top-hocolim F_n. As homotopy colimits in Top commute with suspensions, we get maps

$$\Sigma \text{ Top-hocolim } F_n \xrightarrow{\cong} \text{Top-hocolim } \Sigma F_n \longrightarrow \text{Top-hocolim } F_{n+1}$$

induced by the maps $\Sigma F_n \longrightarrow F_{n+1}$ adjoint to the structure maps $F_n \longrightarrow \Omega F_{n+1}$. Passing to the adjoints again, we get maps Top-hocolim $F_n \longrightarrow \Omega$Top-hocolim F_{n+1}. These maps are not in general equivalences; so the sequence of spaces Top-hocolim F_n is not a spectrum, but only a prespectrum. To this prespectrum one canonically associates an equivalent spectrum [10]; this spectrum is our Spectra-hocolim F. As above, we have

<u>Proposition</u>: For any connective generalized homology theory E_* on <u>Spectra</u>, there is a first quadrant spectral sequence natural in $F : \underline{\underline{L}} \longrightarrow$ <u>Spectra</u>

$$E^2_{p,q} = H_p(\underline{\underline{L}}; E_q F) \Rightarrow E_{p+q}(\text{Spectra-hocolim } F).$$

<u>Proof</u>: Use the fact $E_*(\text{Spectra-hocolim } F) = \varinjlim_n E_{k+n}(\text{Top-hocolim } F_n)$ and the spectral sequences for Top-hocolim F_n. Here one regards E_* as a generalized homology theory on spaces in the usual way, via the suspension spectrum functor.

For special diagrams, we may identify the E^2 term as in the examples above. In particular, for a diagram of spectra

$$\begin{array}{ccc} A & \longrightarrow & B \\ \downarrow & & \\ * & & \end{array}$$

the Spectra-hocolim is the mapping cone or cofibre spectrum of $A \longrightarrow B$, and the spectral sequence degenerates into the long exact cofibre sequence.

One may also consider homotopy colimits in Cat, the category of small categories. This is treated in [17], [18]. It is shown there that the classifying space functor $B : \underline{\text{Cat}} \longrightarrow \underline{\text{Top}}$ commutes with homotopy colimits up to homotopy equivalence. This is an essential ingredient at several points in the proof of the theorem of §5.

§5. Homotopy colimits are preserved by Spt.

Theorem: Let $F : \underline{\underline{L}} \longrightarrow \underline{\text{Perm}}$ be a functor. There is a natural equivalence of spectra

$$\text{Spectra-hocolim (Spt } F) \simeq \text{Spt (Perm-hocolim } F)$$

Sketch of proof: The universal mapping property of a homotopy colimit gives a natural map Spectra-hocolim (Spt F) $\longrightarrow$ Spt (Perm-hocolim F). This map will be the equivalence. The proof uses the resolution technique of [19].

Use the monad T on Cat which sends a category to the free permutative category over it to construct a Kliesli standard simplicial resolution of F. This is a simplicial object in the category of diagrams of permutative categories, $n \longmapsto T^{n+1}F$, with face and degeneracy operators induced by the action of T on F, the multiplication of T, and the unit of T. This simplicial object is augmented to F via the action $TF \longrightarrow F$.

Applying Spectra-hocolim (Spt ?) $\longrightarrow$ Spt (Perm-hocolim ?), one gets a map of simplicial objects in <u>Spectra</u>. One may "geometrically realize" such simplicial spectra. The augmentation induces a map from the realization of $n \longmapsto$ Spectra-hocolim (Spt $T^{n+1}F$) to Spectra-hocolim (Spt F), and one from the realization of $n \longmapsto$ Spt (Perm-hocolim $T^{n+1}F$) to Spt (Perm-hocolim F). The first map is a homotopy equivalence by general nonsense; the second, by a calculation given below. Granted this, one is reduced to showing that Spectra-hocolim (Spt $T^{n+1}F$) $\longrightarrow$ Spt (Perm-hocolim $T^{n+1}F$) is an equivalence, using the usual fact that a simplicial map which is an equivalence in each degree has a geometric realization which is an equivalence.

One next reduces to the theorem [18], 1.2 relating homotopy colimits in <u>Top</u> and <u>Cat</u>. Recall if $T\underline{\underline{C}}$ is the free permutative category on $\underline{\underline{C}}$, Spt $T\underline{\underline{C}}$ is equivalent to $\Sigma^{\infty}B\underline{\underline{C}}$, the suspension spectrum on the classifying space of the category $\underline{\underline{C}}$. Thus for $G : \underline{\underline{L}} \longrightarrow \underline{\text{Cat}}$, one has equivalences: Spectra-hocolim (Spt TG) $\simeq$ Spectra-hocolim $\Sigma^{\infty}BG \simeq \Sigma^{\infty}$ Top-hocolim BG. On the other hand, recall from [18] that <u>Cat</u> has homotopy colimits given by the Grothendieck construction $\underline{\underline{L}} \int G$ on $G : \underline{\underline{L}} \longrightarrow \underline{\text{Cat}}$, and that there is a natural equivalence Top-hocolim BG $\simeq B(\underline{\underline{L}} \int G)$. One finds functors and natural transforms giving inverse homotopy equivalences between Perm-hocolim TG and $T(\underline{\underline{L}} \int G)$, so there are equivalences Spt (Perm-hocolim TG) $\simeq$ Spt $(T(\underline{\underline{L}} \int G) \simeq \Sigma^{\infty}(\underline{\underline{L}} \int G) \simeq \Sigma^{\infty}$(Top-hocolim BG). Combining the two series of equivalences, we get Spectra-hocolim (Spt $T^{n+1}F$) $\simeq$ Spt (Perm-hocolim $T^{n+1}F$) as required.

It remains only to indicate why the map of the realization of $n \longmapsto$ Spt (Perm-hocolim $T^{n+1}F$) to Spt (Perm-hocolim F) is an equivalence. The functor Spt factors as the composite of an infinite loop space machine and a functor which regards the classifying space of a permutative category as an E_∞-space, as in [11]. As the machine commutes with geometric realization, it suffices to show the realization of the simplicial E^∞-space $n \longmapsto B(\text{Perm-hocolim } T^{n+1}F)$ is equivalent as a space to $B(\text{Perm-hocolim } F)$. But by the preceding paragraph, this simplicial space is equivalent to $n \longmapsto B(T(\underline{\underline{L}}\int T^n F))$, and so by [18] its realization is equivalent to the classifying space of $\Delta^{op}\int n \longrightarrow T(\underline{\underline{L}}\int T^n F)$, or even $[\Delta^{op}\int n \longrightarrow [T(\underline{\underline{L}}\int T^n F)]^{op}]^{op}$. There is a functor from this last category to Perm-hocolim F which induces a homotopy equivalence of classifying spaces by an argument similar to the proofs of [17], VI, 2.3, VI, 3.4. This completes the sketch.

To produce an honest proof, a few technical tricks must be applied. For example, at various points there are problems with basepoints of spaces and with units of permutative categories. One deals with this by noting that if one adds a new disjoint unit 0 to a permutative category, the associated spectrum doesn't change up to homotopy. Also, the above proof works only in the case where $F : \underline{\underline{L}} \longrightarrow \underline{\text{Perm}}$ is such that for each morphism ℓ in $\underline{\underline{L}}$, $F(\ell)$ is a strict permutative functor. The general case is deduced from this special case. Finally, to avoid trouble with the universal mapping property of Spectra-hocolim, part of the argument must be done in the category of prespectra. A fully detailed honest proof will appear elsewhere, someday.

§6. A simplified mapping cone and how to use it.

The category Perm-hocolim F is generally somewhat complicated, although not impossibly so. In certain situations of interest it may be replaced by a simpler homotopy equivalent construction. I will indicate how to do this in the case of mapping cones. This construction should be useful for producing exact sequences in the K-theory of rings. It can be employed in place of Quillen's Theorem B ([14], §1) for this purpose. It has the advantage that its hypotheses are easier to satisfy in practice than those of Theorem B. As with Quillen's theorem, it leaves one with the problem of identifying what it gives as the third terms in a long exact sequence of K-groups with what one wants there. This problem is generally that of showing some functor induces a homotopy equivalence of classifying spaces; and may be attacked by the methods of [14], essentially by cleverness and the use of Quillen's Theorem A. These points should become clear in the two examples below.

To make the simplified double mapping cylinder consider a diagram of symmetric monoidal categories and strong symmetric monoidal functors

$$\begin{array}{ccc} \underline{\underline{A}} & \xrightarrow{V} & \underline{\underline{B}} \\ {\scriptstyle U}\big\downarrow & & \\ \underline{\underline{C}} & & \end{array}$$

Suppose every morphism of $\underline{\underline{A}}$ is an isomorphism. Let $\underline{\underline{P}}$ be the category with objects (C,A,B) where C is an object of $\underline{\underline{C}}$; A, of $\underline{\underline{A}}$; and B, of $\underline{\underline{B}}$. A morphism in $\underline{\underline{P}}$, $(C,A,B) \to (C',A',B')$ is given by an equivalence class of data:

1) $\psi : A \longrightarrow A_1 \oplus A' \oplus A_2$ an isomorphism in $\underline{\underline{A}}$;

2) $\psi_1 : C \oplus UA_L \longrightarrow C'$ a morphism in $\underline{\underline{C}}$

3) $\psi_2 : VA_2 \oplus B \longrightarrow B'$ a morphism in $\underline{\underline{B}}$.

Equivalent data are obtained by changing A_1 and A_2 up to isomorphism; thus if $a : A_1 \xrightarrow{\cong} A_1'$, $b : A_2 \xrightarrow{\cong} A_2'$ are isomorphisms, the above data (ψ, ψ_1, ψ_2) is equivalent to $(a \oplus A' \oplus b \cdot \psi,\ \psi_1 \cdot C \oplus Ua^{-1},\ \psi_2 \cdot Vb^{-1} \oplus B)$.

This $\underline{\underline{P}}$ is the simplified double mapping cylinder of the diagram. It has the obvious symmetric monoidal structure with $(C,A,B) \oplus (C',A',B') = (C \oplus C', A \oplus A', B \oplus B')$. There are strong symmetric monoidal functors $\underline{\underline{B}} \longrightarrow \underline{\underline{P}}$, $\underline{\underline{C}} \longrightarrow \underline{\underline{P}}$, and $\underline{\underline{P}}$ has a simple universal mapping property. In the special case where $\underline{\underline{C}} = \underline{\underline{O}}$ is a point, the construction of $\underline{\underline{P}}$ yields the simplified mapping cone on $\underline{\underline{A}} \longrightarrow \underline{\underline{B}}$.

One uses Theorem A of [14] to show the canonical map from the double mapping cylinder in the sense of §3 to the simplified version $\underline{\underline{P}}$ is a homotopy equivalence. This justifies the name "double mapping cylinder" for $\underline{\underline{P}}$ and yields:

<u>Proposition</u>: In the situation above, there is a long exact Mayer-Vietoris sequence

$$\longrightarrow K_{i+1}(\underline{\underline{P}}) \xrightarrow{\partial} K_i(\underline{\underline{A}}) \longrightarrow K_i(\underline{\underline{B}}) \oplus K_i(\underline{\underline{C}}) \longrightarrow K_i(\underline{\underline{P}}) \xrightarrow{\partial} \cdots$$

As an example of how to use this construction, I will give a quick proof of Quillen's theorem that his two definitions of K-theory agree [3]. Let $\mathcal{Q}$ be an additive category, and $\underline{\underline{A}} = \text{Iso}\ \mathcal{Q}$ the category of isomorphisms in $\mathcal{Q}$. As remarked above, $\underline{\underline{A}}$ is symmetric monoidal with $\oplus$ given by direct sum. For $\mathcal{Q}$ the category of finitely generated projective R-modules, $\coprod_n GL_n(R)$ is cofinal in $\underline{\underline{A}}$, so $K_i(\underline{\underline{A}}) = \pi_i^S \operatorname{Spt}(\underline{\underline{A}}) = \pi_i \operatorname{Spt}_0(\underline{\underline{A}}) \cong \pi_i(BGL(R)^+)$ for $i > 0$. Thus $K_*(\underline{\underline{A}})$ coincides with Quillen's "plus construction" or group completion definition of K-theory.

Consider now the suspension of $\underline{\underline{A}}$, $\Sigma\underline{\underline{A}} = \underline{\underline{P}}$ obtained as the simplified double mapping cylinder from the diagram where both $\underline{\underline{B}}$ and $\underline{\underline{C}}$ are points. By the Proposition above we have an isomorphism $\partial : K_{i+1}(\Sigma\underline{\underline{A}}) \cong K_i(\underline{\underline{A}})$, and $K_{i+1}(\Sigma\underline{\underline{A}}) = \pi_{i+1}^S \operatorname{Spt}(\Sigma\underline{\underline{A}}) \cong \pi_{i+1} \operatorname{Spt}_0(\Sigma\underline{\underline{A}}) \cong \pi_{i+1} B\Sigma\underline{\underline{A}}$, the last isomorphism being due to the fact $B\Sigma\underline{\underline{A}}$ is connected, hence group complete.

Now $\Sigma\underline{\underline{A}}$ has objects $(0,A,0)$, which I abbreviate to A. A morphism $A \to A'$ in $\Sigma\underline{\underline{A}}$ consists of an equivalence class of data, which reduces to giving an isomorphism $\psi : A \xrightarrow{\cong} A_1 \oplus A' \oplus A_2$, up to isomorphism in A_1 and A_2. From ψ, one constructs a diagram of epimorphisms and monomorphisms in $\underline{\underline{\mathcal{A}}}$, with choices of splittings $A \cong A_1 \oplus A' \oplus A_2 \rightleftarrows A' \oplus A_2 \rightleftarrows A'$. In fact, $\Sigma\underline{\underline{A}}$ is easily seen to be isomorphic to the category $Q^s\mathcal{A}$ ([16], §3) whose objects are those of $\mathcal{A}$; and whose morphisms $A \to A'$ are equivalence classes of data $A \rightleftarrows E \rightleftarrows A'$, where the indicated arrows are splittable mono- and epimorphisms, together with a choice of splitting (shown as dotted arrows). Changing E and all arrows by the same isomorphism $E \cong E'$ gives the equivalence relation. Composition is induced by taking pullbacks as in Quillen's $Q\mathcal{A}$ [14], and there is a functor $\Sigma\underline{\underline{A}} \cong Q^s\mathcal{A} \to Q\mathcal{A}$ that forgets the choice of splittings. (Actually, this $Q\mathcal{A}$ is the opposite category of the one in [14].) As $K_i(\underline{\underline{A}}) \cong \pi_{i+1} BQ^s\mathcal{A}$ by the above, to show $K_i(A) = \pi_{i+1} BQ\mathcal{A}$; i.e., that the group completion definition of K-theory agrees with the Q-construction definition, it remains only to show $Q^s\mathcal{A} \to Q\mathcal{A}$ is a homotopy equivalence.

To see this, consider the category $Q^{se}\mathcal{A}$ defined like $Q^s\mathcal{A}$, but where the morphisms are classes of $A \leftarrowtail E \rightleftarrows A'$, with a choice of splitting made only for the epimorphism. The functor

$Q^s\mathcal{a} \to Q\mathcal{a}$ factors as $Q^s\mathcal{a} \to Q^{se}\mathcal{a} \to Q\mathcal{a}$. I'll show $\rho : Q^{se}\mathcal{a} \to Q\mathcal{a}$ is a homotopy equivalence; the proof that $Q^s\mathcal{a} \to Q^{se}\mathcal{a}$ is, is similar. To show ρ is a homotopy equivalence, by Theorem A I need only show the categories ρ/A are contractible for each A in $Q\mathcal{a}$ [14]. But ρ/A contains as a reflexive subcategory, and so as a deformation retract, the full subcategory whose objects are splittable epimorphisms $j : B \twoheadrightarrow A$, and whose morphisms from $j : B \twoheadrightarrow A$ to $j' : B' \twoheadrightarrow A$ are epimorphisms $g : B \overset{\twoheadrightarrow}{\leftarrow\!\!-\!\!-} B'$ with a choice of splitting, and such that $j'g = j$. This subcategory has a symmetric monoidal structure induced by pullback over A, so $(B \twoheadrightarrow A) \otimes (B' \twoheadrightarrow A) = B \times_A B \twoheadrightarrow A$. Using this structure, the proof of [3], p. 227 that "For any C in P, $\langle S, E_C \rangle$ is contractible" applies to show this subcategory, and so ρ/A, is contractible, as required. This argument is the clever part of Quillen's proof given in [3]; the virtue of my machinery here is merely that it gets one down to this crux very quickly.

As a second example, I will indicate how one can approach the Lichtenbaum conjecture. I will reduce the problem to showing that a certain functor is a homotopy equivalence, but I have been unable to complete the proof by showing this is so.

Recall that the conjecture states that if $\bar{k}$ is an algebraically closed field in characteristic p, and $\bar{\mathbb{F}}_p$ is the algebraic closure of the prime field, there is a short exact sequence

$$0 \longrightarrow K_*(\bar{\mathbb{F}}_p) \longrightarrow K_*(\bar{k}) \longrightarrow K_*(\bar{k}) \otimes \mathbb{Q} \longrightarrow 0$$

Proof of this is crucial to understanding K-theory in characteristic p. By the work of Howard Hiller [4], this conjecture is equivalent to the conjecture of Quillen: Let $\phi^q : \bar{k} \longrightarrow \bar{k}$ be the Frobenius map $x \longmapsto x^q$, then $BGL(\mathbb{F}_q)^+$ is the homotopy fibre of $1-\phi^q : BGL(\bar{k})^+ \to BGL(\bar{k})^+$. Equivalently, there should be a fibre sequence of infinite loop spaces, or cofibre sequence of spectra. I'll produce the cofibre of $BGL(\mathbb{F}_q)^+ \to BGL(\bar{k})^+$; to prove the conjecture one then wants to show it's $BGL(\bar{k})^+$.

Let $\underline{\underline{A}} = \coprod_n GL_n(\mathbb{F}_q)$, $\underline{\underline{B}} = \coprod_n GL_n(\bar{k})$, with symmetric monoidal structure induced by direct sum. Thus $Spt_0(\underline{\underline{A}}) \simeq \mathbb{Z} \times BGL(\mathbb{F}_q)^+$, $Spt_0(\underline{\underline{B}}) \simeq \mathbb{Z} \times BGL(\bar{k})^+$. Consider the simplified mapping cone $\underline{\underline{P}}$ on the functor $\bar{k} \otimes_{\mathbb{F}_q} : \underline{\underline{A}} \to \underline{\underline{B}}$. This $\underline{\underline{P}}$ has objects (A,B), with A a vector space over $\mathbb{F}_q$ and B a vector space over $\bar{k}$. A morphism $(A,B) \to (A',B')$ is a class of data: $\psi : A \cong A_1 \oplus A' \oplus A_2$, $\psi_2 : B \oplus (\bar{k} \otimes A_2) \xrightarrow{\cong} B'$.

Let $\underline{\underline{B}}^{-1}\underline{\underline{B}}$ be Quillen's group completed category, as described in [3]. Then $Spt_0(\underline{\underline{B}}^{-1}\underline{\underline{B}}) \simeq \mathbb{Z} \times BGL(\bar{k})^+$, and as $\pi_0(\underline{\underline{B}}^{-1}\underline{\underline{B}})$ is a group, $Spt_0(\underline{\underline{B}}^{-1}\underline{\underline{B}}) \simeq B(\underline{\underline{B}}^{-1}\underline{\underline{B}})$. Let $(\underline{\underline{B}}^{-1}\underline{\underline{B}})_0$ be the connected component of $(0,0)$. Then $B(\underline{\underline{B}}^{-1}\underline{\underline{B}})_0 \simeq BGL(\bar{k})^+$.

There is a symmetric monoidal functor $\rho : \underline{\underline{P}} \to (\underline{\underline{B}}^{-1}\underline{\underline{B}})_0$ given on objects by $\rho(A,B) = (B, \phi^q B)$, where $\phi^q B$ is the $\bar{k}$ vector space B with the $\bar{k}$-module structure changed by $\phi^q : \bar{k} \to \bar{k}$. On a morphism $(A,B) \to (A',B')$ of $\underline{\underline{P}}$ given by data as above, ρ is given by the morphism in $(\underline{\underline{B}}^{-1}\underline{\underline{B}})_0$ determined ([3]) by the object $\bar{k} \otimes A_2$, and the pair of isomorphisms

$$B \oplus (\bar{k} \otimes_{\mathbb{F}_q} A_2) \xrightarrow{\psi_2} B'$$

$$\phi^q B \oplus (\bar{k} \otimes_{\mathbb{F}_q} A_2) \cong \phi^q B \oplus \phi^q(\bar{k} \otimes_{\mathbb{F}_q} A_2) \xrightarrow{\phi^q(\psi_2)} \phi^q B'$$

where one uses the canonical isomorphism $\bar{k} \otimes_{\mathbb{F}_q} A_2 \cong \phi^q \bar{k} \otimes_{\mathbb{F}_q} A_2$ which is the identity on the subgroup $\mathbb{F}_q \otimes_{\mathbb{F}_q} A_2$.

The diagram

$$
\begin{array}{ccccc}
\mathbb{Z} \times BGL(\mathbb{F}_q)^+ & \longrightarrow & \mathbb{Z} \times BGL(\bar{k})^+ & \longrightarrow & BGL(\bar{k})^+ \\
\wr\| & & \wr\| & & \wr\| \\
Spt_0(\underline{\underline{A}}) & \longrightarrow & Spt_0(\underline{\underline{B}}) & \xrightarrow{1-\phi^q} & Spt_0((\underline{\underline{B}}^{-1}\underline{\underline{B}})_0) \\
\| & & \| & & \uparrow Spt_0(\rho) \\
Spt_0(\underline{\underline{A}}) & \longrightarrow & Spt_0(\underline{\underline{B}}) & \longrightarrow & Spt_0(\underline{\underline{P}})
\end{array}
$$

commutes, and the bottom row is a fibre sequence, as the sequence of zeroth spaces of a cofibre sequence of spectra. Thus Quillen's conjecture is equivalent to the statement that $Spt_0(\rho)$ is an equivalence. As both $\underline{\underline{P}}$ and $(\underline{\underline{B}}^{-1}\underline{\underline{B}})_0$ are connected and so group complete, this is in fact equivalent to the functor $\rho : \underline{\underline{P}} \longrightarrow (\underline{\underline{B}}^{-1}\underline{\underline{B}})_0$ being a homotopy equivalence.

So far, I have been unable to show this, but there are signs that it is true. One could try to appeal to Quillen's Theorem A. I know the fibre $(0,0)/\rho$ is contractible by Lang's Theorem [8] that all torsors for the Frobenius action on $GL_n(\bar{k})$ are trivial with trivialization unique up to $GL_n(\mathbb{F}_q)$. Unfortunately, $(B,B)/\rho$ is in general disconnected, although I suspect each component is contractible. One could hope to show $H^*(\rho)$ is an isomorphism by considering the Grothendieck spectral sequence $H^p((\underline{\underline{B}}^{-1}\underline{\underline{B}})_0,\ H^q(\ /\rho)) \Rightarrow H^{p+q}(\underline{\underline{P}})$ and proving it collapses by analysis of $H^*(\ /\rho)$ and the action of the $GL_n(\bar{k})$ on it. Something like Tits buildings seems to play a role here. The interested reader is invited to try to make sense of this.

§7. Axiomitization of K-theory as a generalized homology theory on Sym Mon.

I will give an axiomatiziation characterizing the functor K_*. The axioms are reminiscent of the usual axioms for a generalized homology theory on Top if one accepts the idea that the appropriate Sym-Mon-hocolim is the analogue of the mapping cone. This gives a reassuring picture of K-theory as a generalized homology theory on Sym Mon, and suggests that analogues of the usual theorems about homology theories on Spectra ought to be true for K-theory.

I do not see how to characterize K-theory restricted to rings or exact categories in any similar fashion.

Consider the following four axioms on a functor K_* from Sym Mon to the category of non-negatively graded abelian groups.

I. (Homotopy axiom). If $F : \underline{\underline{A}} \longrightarrow \underline{\underline{B}}$ is a morphism that induces an isomorphism on homology with $\mathbb{Z}$-coefficients after group completion, $\pi_0^{-1}H_*(F) : \pi_0^{-1}H_*(\underline{\underline{A}}) \xrightarrow{\cong} \pi_0^{-1}H_*(\underline{\underline{B}})$; then $K_*(F)$ is an isomorphism.

Here $H_*(\underline{\underline{A}})$ is homology of the category $\underline{\underline{A}}$, as in [5], IX, §6, or [14]. By [14], $H_*(\underline{\underline{A}})$ is isomorphic to $H_*(B\underline{\underline{A}})$, the homology of the classifying space. By $\pi_0^{-1}H_*(\underline{\underline{A}})$, I mean the homology localized with respect to the multiplicative subset $\pi_0\underline{\underline{A}} \subseteq H_0(\underline{\underline{A}})$. As is well known, $\pi_0^{-1}H_*(\underline{\underline{A}})$ is isomorphic to $H_*(\mathrm{Spt}_0\ \underline{\underline{A}})$, so this axiom is equivalent to the statement that $K_*(F)$ is an isomorphism if $\mathrm{Spt}_0(F)$ is a homotopy equivalence.

II. (Cofibre sequence axiom). For $F : \underline{\underline{A}} \longrightarrow \underline{\underline{B}}$ a morphism, let

$\underline{\underline{B}} \longrightarrow \underline{\underline{P}}$ be the mapping cone on F. Then there is a long exact sequence

$$\longrightarrow K_{i+1}(\underline{\underline{P}}) \xrightarrow{\partial} K_i(\underline{\underline{A}}) \longrightarrow K_i(\underline{\underline{B}}) \longrightarrow K_i'(\underline{\underline{P}}) \xrightarrow{\partial} \cdots \longrightarrow K_0(\underline{\underline{P}}) \longrightarrow 0$$

Here the mapping cone $\underline{\underline{P}}$ is the Sym Mon-hocolim of a diagram as in §4, Example 1.

III (Continuity axiom). If $\underline{\underline{A}}_i$, $i \in I$ is a directed system of symmetric monoidal categories, $\varinjlim K_*(\underline{\underline{A}}_i) \cong K_*(\varinjlim \underline{\underline{A}}_i)$.

IV (Normalization axiom). Let $\underline{\underline{\mathbb{Z}}}$ be the category whose objects are integers $\underline{n}$, and whose morphisms are all identity morphisms. Let $\underline{\underline{\mathbb{Z}}}$ have the symmetric monoidal structure $\underline{n} \oplus \underline{m} = \underline{n+m}$. Then $K_0(\underline{\underline{\mathbb{Z}}}) = \mathbb{Z}$, $K_i(\underline{\underline{\mathbb{Z}}}) = 0$ for $i > 0$.

<u>Theorem</u>: If K_* is any functor from <u>Sym</u> <u>Mon</u> to non-negatively graded abelian groups satisfying the above four axioms, then K_* is naturally isomorphic to algebraic K-theory, π_*^S Spt.

<u>Idea of Proof</u>: Because of the homotopy axiom, K_* induces a functor out of the homotopy category of <u>Sym</u> <u>Mon</u> (obtained by formally inverting all $F : \underline{\underline{A}} \longrightarrow \underline{\underline{B}}$ such that Spt(F) is an equivalence) into the homotopy category of <u>Spectra</u>. Suppose first one knew the induced map of homotopy categories was an equivalence. Then one shows K_* is stable homotopy. Using axiom II and the tower of higher connected coverings of a system, which is a sort of upside-down Postnikov tower, one can reduce to checking K_* is π_*^S on Eilenberg-MacLane spectra. Using axiom II again, one can shift dimensions until one is dealing with

$K(\pi,0)$-spectra. By axiom III, reduce to the case π is finitely generated; by axiom II, to the case π is cyclic; and by axiom II again to the case $\pi = \mathbb{Z}$; which holds by axiom IV.

While I do not know the two homotopy categories are the same, I can show the homotopy category of Sym Mon is a retract of that of Spectra, and that the retraction does not change the homotopy type of Spt_0. Proof of this involves 2-category theory, and a generalization of the theorem of §5, so I'll say no more about it. This relation between the homotopy categories is strong enough to make possible an argument along the lines of the first paragraph.

Bibliography

[1] Bousfield, A.K., and Kan, D.M.: Homotopy Limits, Completions, and Localizations, Springer Lecture Notes in Math., Vol. 304, (1972).

[2] Eilenberg, S., and Kelly, G.M.: "Closed categories"; in Proceedings of the Conference on Categorical Algebra: La Jolla 1965, pp. 421-562 (1966).

[3] Grayson, D.: "Higher algebraic K-theory: II (after Quillen)," in Algebraic K-Theory: Evanston 1976, Springer Lecture Notes in Math., Vol. 551, pp. 217-240 (1976).

[4] Hiller, H.: "Fixed points of Adams operations," thesis, MIT (1978).

[5] Hilton, P., and Stammbach, U.: A Course in Homological Algebra, Springer Graduate Texts in Math. Vol. 4 (1971).

[6] Kelly, G.M.: "An abstract approach to coherence," in Coherence in Categories, Springer Lecture Notes in Math., Vol. 281, pp. 106-147 (1972).

[7] Kelly, G.M.: "Coherence theorems for lax algebras and for distributive laws," in Category Seminar, Springer-Lecture Notes in Math., Vol. 420, pp. 281-375 (1974).

[8] Lang, S.: "Algebraic groups over finite fields," Amer. J. Math. Vol. 78, no. 3, pp. 555-563, (1956).

[9] MacLane, S.: Categories for the Working Mathematician, Springer Graduate Texts in Math., vol. 5, (1971).

[10] May, J.P.: "Categories of spectra and infinite loop spaces," in Category Theory, Homology Theory, and Their Applications III, Springer Lecture Notes in Math., vol. 99, pp. 448-479 (1969).

[11] May, J.P.: "E_∞ spaces, group completions, and permutative categories," in New Developments in Topology, London Math. Soc. Lecture Notes, no. 11, pp. 61-94 (1974).

[12] May, J.P.: "Simplicial Objects in Algebraic Topology," D. VanNostrand Co., (1967).

[13] May, J.P.: "The spectra associated to permutative categories," preprint.

[14] Quillen, D.: "Higher algebraic K-theory: I," in Higher K-Theories, Springer Lecture Notes in Math., vol. 341, pp. 85-147 (1973).

[15] Segal, G.: "Categories and cohomology theories," Topology, vol. 13, pp. 293-312 (1974).

[16] Segal, G.: "K-homology theory and algebraic K-theory," in K-Theory and Operator Algebras: Athens 1975, Springer Lecture Notes in Math., vol. 575, pp. 113-127 (1977).

[17] Thomason, R.W., "Homotopy colimits in Cat, with applications to algebraic K-theory and loop space theory," thesis, Princeton, (1977).

[18] Thomason, R.W.: "Homotopy colimits in the category of small categories," to appear, Math. Proc. Cambridge Phil. Soc.

[19] Thomason, R.W.: "Uniqueness of delooping machines," preprint.

[20] Vogt, R.M.: "Homotopy limits and colimits," Math. Z., vol. 134, no. 1, pp. 11-52 (1973).

Algebraic K-theory of topological spaces. II

Friedhelm Waldhausen

The purpose of this paper is to explore the relation between stable homotopy theory and the functor $A(X)$ of the title. The relation turns out to be very simple: The former splits off the latter.

This splitting of $A(X)$ is an unexpected phenomenon. Consider the case where

$$X = * ,$$

a point. In this case we may (and will) take as the definition

$$A(*) = Z \times \left(\varinjlim_{n,k} B\, Aut(\vee^k S^n) \right)^+$$

where

$\vee^k S^n$ = wedge of k spheres of dimension n

$Aut(..)$ = simplicial monoid of pointed homotopy equivalences

$B\, Aut$ = its classifying space

$(...)^+$ = the + construction of Quillen

$\varinjlim_{n,k}$: by suspension, and by wedge with identity maps, respectively.

The artificial factor Z is required to avoid disagreement with other definitions of $A(*)$. Thanks to a theorem of Barratt-Priddy, Quillen, and Segal on the other hand stable homotopy is definable in terms of the symmetric groups,

$$\Omega^\infty S^\infty \simeq Z \times \left(\varinjlim_{k} B\, \Sigma_k \right)^+ .$$

Since $\Sigma_k \approx Aut(\vee^k S^0)$, the map

$$Aut(\vee^k S^0) \longrightarrow \varinjlim_{n} Aut(\vee^k S^n)$$

therefore induces a map $\Omega^\infty S^\infty \to A(*)$. It is this map for which the splitting theorem provides a left inverse, up to homotopy.

Let us compare with known facts from algebraic K-theory. There is a map from

$A(*)$ to the algebraic K-theory of the ring of integers,

$$K(Z) = Z \times (\lim_{\vec{k}} B\, GL_k(Z))^+ ,$$

it is induced from

$$Aut(V^k S^n) \longrightarrow Aut(H_n(V^k S^n)) \approx GL_k(Z)$$

This map is a rational homotopy equivalence [14] (an easy consequence of the finiteness of the stable homotopy groups of spheres π_i^S, $i > 0$).

The composite map

$$\Omega^\infty S^\infty \longrightarrow A(*) \longrightarrow K(Z)$$

is the usual map resulting from identification of a symmetric group with a group of permutation matrices. This map has been studied by Quillen [10]. The main result is that

$$\pi_{4k+3}^S \longrightarrow K_{4k+3}(Z)$$

is injective on the image of the J-homomorphism, the subgroup $\mathrm{Im}J_{4k+3}$; in fact, the map is split injective on the odd torsion, and also on the 2-torsion in half the cases (k odd). In the other half it is not. For, Lee and Szczarba [5] have computed $K_3(Z)$ and as a result the map $\mathrm{Im}J_{4k+3} \to K_{4k+3}(Z)$ is, for $k = 0$, the inclusion

$$Z/24 \approx \mathrm{Im}J_3 \approx \pi_3^S \longrightarrow K_3(Z) \approx Z/48 .$$

Browder [3] has deduced from this that the map is not split for all even k. It also follows from the Lee-Szczarba computation that $\pi_i^S \to K_i(Z)$ is not in general injective, and specifically [3] that

$$Z/2 \approx \pi_6^S \longrightarrow K_6(Z)$$

is the zero map. To sum up, the relation between π_i^S and $K_i(Z)$ is very interesting, but apparently also very complicated. Certainly the map $\Omega^\infty S^\infty \to K(Z)$ does not split.

One may wonder here how possibly a result can be provable in the 'non-linear' case (the splitting theorem for $A(X)$) but fail to hold in the 'linear' case (algebraic K-theory). The answer is of course that the proof does not really break down in the linear case, it just proves a different result. This result will be discussed at the end of the paper.

Returning to the splitting theorem, to prove it we must in fact prove a stronger result involving the *stabilization of* $A(X)$,

$$A^S(X) = \lim_{\vec{m}} \Omega^m \text{ fibre}(A(S^m \wedge X_+) \to A(*))$$

where

X_+ = X with a disjoint basepoint added

fibre(..) = the homotopy theoretic fibre

Ω^m = the m-th loop space,

and where the direct system involves certain naturally defined maps.

There is a natural transformation

$$A(X) \longrightarrow A^S(X)$$

of which one should think of being induced from the identification of $A(X)$ with the 0-th term in the direct system defining $A^S(X)$. (There is a technical point here The definition of $A(X)$ we use requires that X be connected. So the 0-th term in the direct system is not defined. So the map

$$A(X) \longrightarrow \Omega \text{ fibre}(\, A(S^1 \wedge X_+) \to A(*)\,)$$

must be artificially produced. We have to introduce the external pairing for that purpose).

<u>Theorem</u>. There is a natural map, well defined up to weak homotopy,

$$A^S(X) \longrightarrow \Omega^\infty S^\infty(X_+)$$

so that the diagram

$$\begin{array}{ccccc} \Omega^\infty S^\infty(X_+) & & & & \\ \downarrow & & \overset{=}{\searrow} & & \\ A(X) & \longrightarrow & A^S(X) & \longrightarrow & \Omega^\infty S^\infty(X_+) \end{array}$$

commutes up to (weak) homotopy.

Recall that two maps are called weakly homotopic if their restrictions to every compactum are homotopic. 'Weak homotopy' is the price we have to pay for working with stable range arguments.

To produce the required map on $A^S(X)$ is equivalent more or less, in view of the definition of A^S, to producing for highly connected Y a map, defined in a stable range,

$$A(Y) \longrightarrow \Omega^\infty S^\infty(Y) \; .$$

It is not obvious that such a map should exist, and considerable work goes into its construction.

Our method to produce the map is to first manipulate $A(Y)$ in a stable range (section 3). A curious construction of simplicial objects is needed here which will be referred to as the *cyclic bar construction*. The idea for this construction comes from unpublished work of K. Dennis (talk at Evanston conference, January 1976), in fact, the *Hochschild homology* that Dennis uses may be regarded as a linear version of the cyclic bar construction. General facts relating to the cyclic bar construction are assembled in section 2.

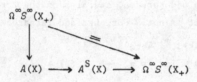

Given the manipulation of $A(Y)$ in the stable range, a map $A(Y) \to \Omega^\infty S^\infty(Y)$, defined in a stable range, may simply be written down (section 4, there are however some technicalities involved here) and it is entirely obvious that this map admits *some* section.

We are then left to show (section 5) that the section is what we want it to be. This requires some preparatory material which is scattered through earlier sections, particularly section 1 which gives a review of some general properties of $A(X)$ and of material involved in the Barratt-Priddy-Quillen-Segal theorem.

§1. Review of $A(X)$ and stable homotopy.

Let X be a simplicial set. We assume X is connected and pointed, so the *loop group* G(X) in the sense of Kan [4] is defined. The geometric realization $|G(X)|$ is a topological group which will be called G for short.

Letting G_+ denote G with a disjoint basepoint added, and V^kS^n the wedge of k spheres of dimension n, we form the G-space

$$V^kS^n \wedge G_+ \qquad (\approx V^kS^n \times G \, / \, {*}\times G)$$

which should be thought of as a *free pointed G-cell complex with k G-cells of dimension n.*

We consider the simplicial set (= singular complex of the topological space) of G-equivariant pointed maps

$$M_k^n(G) \;=\; Map_G(V^kS^n \wedge G_+,\; V^kS^n \wedge G_+)$$

which may be given the structure of a simplicial monoid, by composition of maps. Further we consider the simplicial monoid of G-equivariant pointed weak homotopy equivalences

$$H_k^n(G) \;=\; Aut_G(V^kS^n \wedge G_+) .$$

There is a stabilization map from n to $n+1$, by suspension, hence we can form the direct limit with respect to n. We can also consider a stabilization map from k to $k+1$; in the case of $H_k^n(G)$ it is given by adding the identity map on a new summand in the wedge.

Using the identity element of G we have a canonical map $S^0 \to G_+$. By restriction along this map we obtain an isomorphism

$$Map_G(V^kS^n \wedge G_+,\; V^kS^n \wedge G_+) \xrightarrow{\;\approx\;} Map(V^kS^n,\; V^kS^n \wedge G_+) .$$

This isomorphism in turn restricts to an isomorphism from the underlying simplicial set of $H_k^n(G)$ to a union of connected components of $Map(V^kS^n,\; V^kS^n \wedge G_+)$.

It is suggestive to think of $M_k^n(G)$ as a space of $k\times k$ matrices of some kind. The suggestion is particularly attractive in the limiting case $n = \infty$, for in this case $M_k^\infty(G)$ is actually homotopy equivalent, in the obvious way, to the product of $k\times k$ copies of

$$M_1^\infty(G) = \varinjlim_n Map(S^n, S^n \wedge G_+) = \Omega^\infty S^\infty(G_+) ,$$

and the composition law on $M_k^\infty(G)$ corresponds, under the homotopy equivalence, to matrix multiplication.

Let $NH_k^n(G)$ denote the nerve (or bar construction) of the simplicial monoid $H_k^n(G)$; it is the simplicial object

$$[m] \longmapsto H_k^n(G) \times \ldots \times H_k^n(G) \quad \text{(m factors)}$$

with the usual face structure. Let $B\, H_k^n(G) = |NH_k^n(G)|$ be its geometric realization. Then, by definition,

$$A(X) = Z \times (\varinjlim_{n,k} B\, H_k^n(G))^+$$

where $(..)^+$ denotes the + construction of Quillen [9] (recall that G denotes the geometric realization of the loop group of X).

This definition is essentially the same as the first definition of $A(X)$ in [14] To make the translation one verifies that the space $BH_k^n(G)$ used here is homotopy equivalent to the classifying space of the category used there (this is the content of [14, lemma 2.1], essentially). The requisite arguments are probably well known, a detailed account will be in [15].

The above construction can also be made for any finite n, giving a kind of unstable approximation to $A(X)$. In particular, the case $n = 0$ gives stable homotopy. Indeed, $H_1^0(G) \approx S(G)$ (the singular complex of G) and in general

$$H_k^0(G) \approx \Sigma_k \int S(G)$$

(wreath product with the symmetric group on k letters). Hence the theorem of Barratt-Priddy, Quillen, and Segal [11] gives a homotopy equivalence

$$(\Omega^\infty S^\infty |X_+| \simeq)\ \Omega^\infty S^\infty (BS(G)_+) \simeq Z \times (\varinjlim_k B\, H_k^0(G))^+ .$$

The map

$$H_k^0(G) \longrightarrow \varinjlim_n H_k^n(G)$$

therefore induces

$$\Omega^\infty S^\infty |X_+| \longrightarrow A(X) .$$

We will need a different description of this map, in a stable range.

<u>Lemma</u> 1.1. The following diagram commutes up to weak homotopy (homotopy on compacta) in which the homotopy equivalence on the right is that of the Barratt-Priddy-Quillen-Segal theorem and the map on the bottom is the natural stabilization map:

$$\begin{array}{ccc} BH_1^0(G) \longrightarrow BH_\infty^0(G) & \longrightarrow & Z \times BH_\infty^0(G)^+ \\ \downarrow \wr & & \uparrow \wr \\ |X| & \longrightarrow & \Omega^\infty S^\infty |X_+| \end{array}$$

The lemma is, essentially, a quotation from Segal [11]. Before making this explicit we review some material on Γ-spaces. We do this in some detail as the material will also be needed for other purposes, particularly the treatment of pairings below.

(1.2). Γ-spaces. Our reference is Segal [11]; cf. also Anderson [1] for some reformulation. Let $\underline{s}$ denote the basepointed set with s non-basepoint elements 1, ..., s. We recall that a (special) Γ-*space* is a covariant functor F from the category of finite pointed sets to the category of spaces (respectively, the category of (multi-)simplicial sets in our case) which satisfies that $F(\underline{0}) = *$, and which takes sums to products, up to homotopy; this means, if $p_1: X_1 \vee X_2 \to X_1$ is the retraction which takes X_2 to the basepoint, and p_2 similarly, then

$$(p_{1*}, p_{2*}): \; F(X_1 \vee X_2) \longrightarrow F(X_1) \times F(X_2)$$

is a weak homotopy equivalence. The space $F(\underline{1})$ is called the *underlying space* of the Γ-space F.

In our present situation we have for every $n = 0, 1, \ldots,$ or $= \infty$, a Γ-space F_G^n whose underlying space is

$$F_G^n(\underline{1}) \;=\; \coprod_k \; NH_k^n(G) \;.$$

The higher terms can be obtained by a general procedure of Segal [11, section 2]; the next term is

$$F_G^n(\underline{2}) \;=\; \coprod_{k,l} \;(\, EH_k^n(G) \times EH_l^n(G) \times EH_{k+l}^n(G)\,) \,/\, H_k^n(G) \times H_l^n(G)$$

where E denotes a universal bundle (one-sided bar construction) and '/' means quotienting out of the action, and the general term is

$$F_G^n(\underline{s}) \;=\; \coprod_{k_1,..,k_s} \;(\, \textstyle\prod_{\sigma \subset \underline{s}} EH_{k_\sigma}^n(G)\,) \,/\, H_{k_1}^n(G) \times .. \times H_{k_s}^n(G)$$

where $k_\sigma = \Sigma_{r \in \sigma} k_r$.

Returning to the general notion of Γ-space, we can extend the functor F, by direct limit and degreewise extension, to a functor defined on the category of pointed simplicial sets. For example if the original functor took values in the category of simplicial sets, the extended functor will take values in the category of bisimplicial sets.

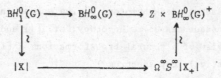

In the special case of a Γ-space which is 'group-valued' (for example this holds if the underlying space is connected) the extended functor is a (reduced) homology theory; that is, it preserves weak homotopy equivalences, and it takes cofibration sequences to fibration sequences up to homotopy, cf. [1] and e.g. [13] for a more detailed account. In view of a natural transformation $X\wedge F(Y) \to F(X\wedge Y)$ it therefore gives rise to a (connective) loop spectrum

$$|F(\underline{1})| \xrightarrow{\simeq} \Omega|F(S^1)| \;, \quad |F(S^1)| \xrightarrow{\simeq} \Omega|F(S^2)| \;, \;\ldots$$

Our Γ-spaces F^n_G are not group valued in the above sense. In this general case the list of properties must be weakened a bit, namely the extended functor F will not in general produce a fibration sequence from a cofibration sequence unless the latter involves connected spaces only. Thus the spectrum $m \mapsto F(S^m)$ is a loop spectrum only after the first map. The space $F(S^1)$ is equivalent to the underlying space of the Γ-space which in Segal's notation would be called BF, and one of the main general results about Γ-spaces says that it is computable by means of the + construction. Specifically in our situation we have

$$\Omega|F^n_G(S^1)| \simeq Z \times (\lim_{\vec{k}} B\, H^n_k(G))^+ .$$

Thus in the cases $n = 0$ and $n = \infty$ we recover $\Omega^\infty S^\infty|X_+|$ and $A(X)$, respectively.

Remark. The latter homotopy equivalence is well defined up to weak homotopy only (for it is obtained by means of an isomorphism of homotopy functors on the category of finite CW complexes [11]). This kind of ambiguity (weak homotopy instead of homotopy) arises frequently in connection with the + construction. It would be tempting to avoid the ambiguity by avoiding the + construction, and specifically by not using the universal property. We could indeed avoid the + construction altogether. But the effort would be in vain. For the stable range arguments that we have to use later on, would re-introduce the ambiguity.

<u>Proof</u> <u>of</u> <u>lemma</u> 1.1. This is a corollary of Segal's proof of the homotopy equivalence of infinite loop spaces

$$\Omega^\infty S^\infty|X_+| \simeq \Omega|F^0_{G(X)}(S^1)| .$$

In [11, proofs of propositions 3.5 and 3.6] Segal does in fact exhibit a specific map of spectra from the suspension spectrum of $|X_+|$ to the spectrum $m \mapsto \Omega|F^0_{G(X)}(S^{m+1})|$ which he then shows is a weak homotopy equivalence of spectra. Since the receiving spectrum is a loop spectrum this map is characterized by the map of first terms which is the composite map

$$BS|G(X)|_+ \longrightarrow \coprod_k BH^0_k(G) = |F^0_G(S^0)| \longrightarrow \Omega|F^0_G(S^1)|$$

$$BS|G(X)| \xrightarrow{\approx} BH^0_1(G)$$

$$* \xrightarrow{\approx} BH^0_0(G) .$$

It is immediate from this that there is a version of lemma 1.1 in which $Z \times BH^0_\infty(G)^+$

has been replaced by $\Omega|F^0_G(S^1)|$. To translate into the form stated, one has to take into account the way the homotopy equivalence between these two spaces arises [11, section 4] and particularly the way that $Z \times BH^0_\infty(G)$ arises as the telescope of $\coprod_k BH^0_k(G)$ and a shift map. □

(1.3). Pairings. Smash product induces a pairing $H^n_k(G) \times H^{n'}_{k'}(G') \longrightarrow H^{n+n'}_{k\cdot k'}(G\times G')$ and therefore also a pairing of Γ-spaces (resp. of their extensions described above)

$$F^n_G(Y) \wedge F^{n'}_{G'}(Y') \longrightarrow F^{n+n'}_{G\times G'}(Y\wedge Y') .$$

The pairing is compatible with the natural transformation $Y'' \wedge F^n_G(Y) \longrightarrow F^n_G(Y''\wedge Y)$.

Taking Y and Y' to be spheres, we have in particular

$$F^n_G(S^m) \wedge F^{n'}_{G'}(S^{m'}) \longrightarrow F^{n+n'}_{G\times G'}(S^{m+m'})$$

which defines a pairing of spectra because of the compatibility with the structure map $S^1 \wedge F^n_G(S^m) \longrightarrow F^n_G(S^{m+1})$.

Using that, for $m > 0$, we have $A(X) \simeq \Omega^m|F^\infty_{G(X)}(S^m)|$, and using that the weak homotopy equivalence $G(X\times X') \to G(X)\times G(X')$ induces one

$$F^n_{G(X\times X')}(Y) \xrightarrow{\ \sim\ } F^n_{G(X)\times G(X')}(Y) ,$$

we thus obtain a pairing, well defined up to (weak) homotopy,

$$A(X) \wedge A(X') \longrightarrow A(X\times X') .$$

Note that the pairing could also have been defined more directly in terms of the definition of $A(X)$ by the + construction (similarly to the pairing in K-theory in [6]); with the present definition any desired naturality properties of the pairing are essentially obvious.

The pairing formally implies others. Let $\widetilde{A}(X)$ be the *reduced part* of $A(X)$,

$$\widetilde{A}(X) = \text{fibre}(\, A(X) \to A(*)\,) .$$

Taking the difference (with respect to the H-space structure) of the identity map on $A(X)$ and the composite map $A(X) \to A(*) \to A(X)$, one obtains the required map in a splitting

$$A(X) \simeq A(*) \times \widetilde{A}(X) .$$

There is a pairing

$$\widetilde{A}(X) \wedge A(Y) \longrightarrow \widetilde{A}(X\wedge Y_+)$$

which is definable as the composite map

$$\widetilde{A}(X) \wedge A(Y) \longrightarrow A(X) \wedge A(Y) \longrightarrow A(X\times Y) \longrightarrow \widetilde{A}(X\times Y/*\times Y) ;$$

it satisfies that the following diagram is (weakly) homotopy commutative

$$\begin{array}{ccc} A(X) \wedge A(Y) & \longrightarrow & A(X\times Y) \\ \downarrow & & \downarrow \\ \widetilde{A}(X) \wedge A(Y) & \longrightarrow & \widetilde{A}(X\wedge Y_+) \end{array} .$$

Similarly there is a pairing

$$\widetilde{A}(X) \wedge \widetilde{A}(Y) \longrightarrow \widetilde{A}(X\wedge Y) .$$

There are analogous pairings involving (reduced and/or unreduced) stable homotopy, resp. stable homotopy and $A(X)$. For uniformity of notation we let

$$(\Omega^\infty S^\infty |X_+| \simeq) \quad Q(X) \;=\; Z \times (\varinjlim_k B\, H_k^0(G))^+$$

<u>Lemma</u> 1.4. There is a map $A(X) \to \Omega\widetilde{A}(S^1\wedge X_+)$ so that the diagram

$$\begin{array}{ccc} Q(X) & \xrightarrow{\simeq} & \Omega\widetilde{Q}(S^1\wedge X_+) \\ \downarrow & & \downarrow \\ A(X) & \longrightarrow & \Omega\widetilde{A}(S^1\wedge X_+) \end{array}$$

commutes up to homotopy.

<u>Proof</u>. Let $S^1 \to Q(S^1) \to \widetilde{Q}(S^1)$ be the Hurewicz map from homotopy to stable homotopy (the first map is that of lemma 1.1). Using the above pairings we have a diagram

$$\begin{array}{ccccc} S^1 \wedge Q(X) & \longrightarrow & \widetilde{Q}(S^1) \wedge Q(X) & \longrightarrow & \widetilde{Q}(S^1\wedge X_+) \\ \downarrow & & \downarrow & & \downarrow \\ S^1 \wedge A(X) & \longrightarrow & \widetilde{Q}(S^1) \wedge A(X) & \longrightarrow & \widetilde{A}(S^1\wedge X_+) \end{array}$$

and the adjoint of the composite map on the bottom will have the required property if we can show that the adjoint map

$$Q(X) \longrightarrow \Omega\widetilde{Q}(S^1\wedge X_+)$$

is a homotopy equivalence.

We note here that in treating this $Q(X)$ the necessity of having X connected and pointed is of course an illusion. For

$$N(\Sigma_k \int G(X)) \;\approx\; E\Sigma_k \times^{\Sigma_k} NG(X)^k \;\simeq\; E\Sigma_k \times^{\Sigma_k} X^k$$

so that we are in the situation of [11] and the term on the right is quite generally defined. Furthermore the pairing extends to this more general situation. Therefore

$$Q(X) \longrightarrow \Omega\widetilde{Q}(S^1\wedge X_+)$$

is in fact a natural transformation from stable homotopy theory to itself, and it suffices to show it is a homotopy equivalence in the case $X = *$.

Since $Q(*) \to \Omega\widetilde{Q}(S^1)$ extends to a map of spectra it suffices in fact to show that it induces an isomorphism on π_0; equivalently, that its adjoint is surjective on π_1. But from the explicit description of the Hurewicz map (lemma 1.1) we see that the composite map

$$\begin{array}{ccccccc} S^1 \wedge S^0 & \longrightarrow & S^1 \wedge Q(*) & \longrightarrow & Q(S^1) \wedge Q(*) & \longrightarrow & Q(S^1) \\ & & & & \downarrow & & \downarrow \\ & & & & \widetilde{Q}(S^1) \wedge Q(*) & \longrightarrow & \widetilde{Q}(S^1) \end{array}$$

is itself the Hurewicz map, and we are done. □

§2. Simplicial tools.

(2.1). The realization lemma. This asserts that a map of simplicial objects which is a weak homotopy equivalence locally (i.e., the partial map in every degree is a weak homotopy equivalence) is also one globally. We need a version of this for finite connectivity.

We say a map is k-*connected* (or is a k-*equivalence*, by abuse of language) if it induces an isomorphism on π_j for $j < k$, and an epimorphism on π_k.

Lemma 2.1.1. Let $X.. \to Y..$ be a map of bisimplicial sets. Suppose that for every n the map of simplicial sets $X._n \to Y._n$ is k-connected. Then the map $X.. \to Y..$ is also k-connected.

Indeed, recall the argument in the case $k = \infty$, cf. e.g. [16]. One considers the 'skeleton filtration' $X_{(n)}$ of $|X..|$ induced from the second simplicial direction, that is, $X_{(n)}$ is the geometric realization of the bisimplicial subset of $X..$ generated by $X._n$. Then one proves inductively that $X_{(n)} \to Y_{(n)}$ is a k-equivalence using the *gluing lemma*. The same argument works in the case of finite k in view of the following version of the gluing lemma.

Lemma 2.1.2. In the commutative diagram

$$\begin{array}{ccccc} X_1 & \leftarrow & X_0 & \rightarrow & X_2 \\ \downarrow & & \downarrow & & \downarrow \\ Y_1 & \leftarrow & Y_0 & \rightarrow & Y_2 \end{array}$$

let the two left horizontal maps be cofibrations, and suppose that all the vertical maps are k-connected. Then the map of pushouts $X_1 \cup_{X_0} X_2 \to Y_1 \cup_{Y_0} Y_2$ is also k-connected. □

(2.2). Partial monoids. This notion, due to Segal [12], allows a concise description of certain simplicial objects. By definition, a *partial monoid* is a set E together with a partially defined composition law

$$E \times E \supset E_2 \longrightarrow E$$

which is associative in the sense that if one of $(e_1e_2)e_3$ and $e_1(e_2e_3)$ is defined then so is the other and the two are equal. Further there must be a two-sided identity element $*$ and multiplication by $*$ must be everywhere defined, that is,

$$E \vee E \subset E_2 .$$

The simplicial set associated to the partial monoid (we refer to it as the *nerve* of E, notation NE) is given by

$$[n] \longmapsto E_n = \text{set of composable n-tuples}$$

with face and degeneracy maps given in the usual way by composition, resp. by insertion of the identity.

Similarly one has the notion of a *simplicial partial monoid*; its nerve is a bisimplicial set.

For example [12] a pointed simplicial set X can be considered as a simplicial partial monoid in a trivial way, with $X_2 = X \vee X$. The nerve in this case is the simplicial object

$$[n] \longmapsto \underbrace{X \vee \ldots \vee X}_{n}$$

whose diagonal simplicial set is a suspension of X .

Other examples arise in the following way. Let M be a monoid and A a submonoid of M. Then we can manufacture a partial monoid by declaring that two elements of M shall be composable if and only if at least one of them belongs to the submonoid. Thus $M_2 = M{\times}A \cup A{\times}M$, and M_n is what we will refer to as a *generalized wedge*,

$$V^n(M,A) = \text{set of n-tuples of elements in } M, \text{ with at least } (n-1) \text{ elements in } A .$$

Similarly this construction can be made with a simplicial monoid M and a simplicial submonoid A of M .

Lemma 2.2.1. In this situation, if $A \to M$ is (k-1)-connected then the inclusion of simplicial objects

$$[n] \longmapsto (V^n(M,A) \longrightarrow M^n)$$

is (2k-1)-connected.

Proof. In view of the realization lemma (2.1.1.) it suffices to show that for every n the inclusion $V^n(M,A) \to M^n$ is (2k-1)-connected. This is certainly true if n

is either 0 or 1 as the inclusion is an isomorphism in those cases. The case $n = 2$ follows from the following remark.

A map of simplicial sets is $(k-1)$-connected if and only if its geometric realization is homotopy equivalent to an inclusion of CW complexes $K \to X$ so that $X \setminus K$ has no cells of dimension $< k$. Let similarly $Y \setminus L$ have no cells of dimension $< l$. Then $X \times Y \setminus X \times L \cup K \times Y$ has no cells of dimension $< k+l$, and therefore the map $X \times L \cup K \times Y \to X \times Y$ is $(k+l-1)$-connected.

The general case follows inductively by factoring the inclusion suitably and using the same remark and the gluing lemma. □

Finally we will need to consider, in this framework of partial monoids, the notion of *semi-direct product*.

Suppose first that F is a monoid (which we think of as multiplicative) and that E is another (which we think of as additive). Let F act from both sides, and compatibly, on E (in other words, if F^{op} denotes the opposite monoid of F then $F \times F^{op}$ acts on E from the left, say). In this situation, the semi-direct product

$$F \ltimes E$$

is the monoid of pairs (f,e) with multiplication given by the formula

$$(f,e)(f',e') = (ff', ef' + fe')$$

Remark. In case this looks unfamiliar, consider the case where F is a group. Here one can rewrite in the usual form, as follows. Write

$$(f,e) = (f, f\tilde{e})$$

where $\tilde{e} = f^{-1}e$. Then

$$\begin{aligned}(f,f\tilde{e})(f',f'\tilde{e}') &= (ff', f\tilde{e}f' + ff'\tilde{e}') \\ &= (ff', (ff')f'^{-1}\tilde{e}\, f' + (ff')\tilde{e}')\end{aligned}$$

and hence with $[f,\tilde{e}] = (f,f\tilde{e})$ the multiplication is given by the formula

$$[f,\tilde{e}][f',\tilde{e}'] = [ff', f'^{-1}\tilde{e}f' + \tilde{e}']$$

This ends the remark. □

Suppose now that E is a partial monoid on which the monoid F acts compatibly from both sides. We need to assume that E is *saturated* with respect to the action in the sense that the following condition is satisfied: for every pair (e,e') whose sum is defined, and for every f, the sums of the four pairs

$$(fe,e') , \quad (ef,e') , \quad (e,fe') , \quad (e,e'f)$$

must also be defined (they need not however be related in any particular way). Under this assumption the formula $(f,e)(f',e') = (ff',ef'+fe')$ carries over to define a partial monoid $F \ltimes E$ with underlying set $F \times E$ and with $(F \ltimes E)_2 \approx F \times F \times E_2$.

We will be especially concerned with the particular case where E is a pointed set X considered as a partial monoid in a trivial way. In this case $(F \ltimes E)_n$ is the generalized wedge

$$(F \ltimes X)_n = V^n(F{\times}X, F{\times}*) \approx F^n \times (X\vee\ldots\vee X) .$$

In particular $(F \ltimes X)_2 \approx F{\times}F{\times}(X\vee X)$, and the partial composition law is given by the case distinction

$$(f,x)(f',*) = (ff', xf')$$

$$(f,*)(f',x) = (ff', fx) .$$

All of the above extends to (and will be used in) a simplicial framework.

(2.3). <u>The cyclic bar construction</u>. Let F be a monoid which acts on a set X both from the left and the right, and compatibly. The *cyclic bar construction* is defined to be the simplicial set

$$N^{cy}(F,X) , \qquad [k] \longmapsto \underbrace{F \times \ldots \times F}_{k} \times X$$

with face maps

$$d_0(f_1,\ldots,f_k, x) = (f_2,\ldots,f_k, xf_1)$$

$$d_i(f_1,\ldots,f_k, x) = (f_1,\ldots,f_if_{i+1},\ldots,f_k, x) \qquad \text{if } 0 < i < k$$

$$d_k(f_1,\ldots,f_k, x) = (f_1,\ldots,f_{k-1}, f_kx)$$

Similarly if F is a simplicial monoid and X a simplicial set, the cyclic bar construction is defined in the same way, giving a bisimplicial set.

The cyclic bar construction may be regarded as a generalization of the two-sided bar construction. Indeed, the latter may be identified to the special case of the former where X is the product of two factors of which the first has a left F-structure and the second a right F-structure, respectively.

As another example consider the case of a (simplicial) group acting on its underlying (simplicial) set from either side by multiplication. Then the map which in degree k is

$$(g_1,\ldots,g_k, g) \longmapsto (g_1,\ldots,g_k, g(g_1\cdots g_k))$$

defines an isomorphism from $N^{cy}(G,G)$ to the one-sided bar construction of G acting on itself by conjugation. The latter represents the free loop space of NG .

The case of main concern to us arises in the situation where a (simplicial) monoid F acts on a (simplicial) partial monoid E in such a way that the semi-direct product $F \ltimes E$ is defined. In this situation F will also act on the nerve NE in such a way that the cyclic bar construction $N^{cy}(F,NE)$ is defined. We denote by $\mathrm{diag}N^{cy}(F,NE)$ the simplicial (resp. bisimplicial) set resulting from diagonalizing the two N-directions of the latter.

Lemma 2.3.1. There is a natural map

$$u\colon \mathrm{diag}\, N^{cy}(F,NE) \longrightarrow N(F \ltimes E)\ .$$

The map u is an isomorphism if F acts invertibly. If $\pi_0 F$ is a group then u is a weak homotopy equivalence.

Proof. In the formulas to follow we will suppose for simplicity of notation that F and E are a monoid and partial monoid, respectively, rather than a simplicial monoid and simplicial partial monoid. In the general case the formulas are exactly the same except that a dummy index has to be added everywhere.

By definition, $\mathrm{diag}\, N^{cy}(F,NE)$ is the simplicial set (resp. simplicial object in the general case)

$$[n] \longmapsto \underbrace{F\times\ldots\times F}_{n} \times E_n \subset \underbrace{F\times\ldots\times F}_{n} \times \underbrace{E\times\ldots\times E}_{n}$$

with face maps taking $(f_1,\ldots,f_n;\ e_1,\ldots,e_n)$ to

$$d_0(..) = (f_2,..,f_n;\ e_2f_1,..,e_nf_1)$$

$$d_i(..) = (f_1,..,f_if_{i+1},..,f_n;\ e_1,..,e_i+e_{i+1},..,e_n)\ , \qquad 0 < i < n$$

$$d_n(..) = (f_1,..,f_{n-1};\ f_ne_1,..,f_ne_{n-1})\ ,$$

while $N(F \ltimes E)$ is given by

$$[n] \longmapsto (F \ltimes E)_n \subset \underbrace{F\times E \times \ldots \times F\times E}_{n}$$

with face maps taking $(f_1,e_1;\ldots;\ f_n,e_n)$ to

$$d_0(..) = (f_2,e_2;..;\ f_n,e_n)$$

$$d_i(..) = (f_1,e_1;..;\ f_if_{i+1},\ e_if_{i+1}+f_ie_{i+1};\ldots;\ f_n,e_n)\ , \quad 0 < i < n$$

$$d_n(..) = (f_1,e_1;..;\ f_{n-1},e_{n-1})\ .$$

We define $u_n(f_1,..,f_n;\ e_1,..,e_n)$ to be

$$(\ f_1,\ (f_1..f_n)e_1(f_1)\ ;\ f_2,\ (f_2..f_n)e_2(f_1f_2)\ ;\ \ldots\ ;\ f_n,\ (f_n)e_n(f_1..f_n)\ )\ ,$$

then the collection of maps u_n forms a simplicial map u as one checks. Here is the situation for face maps: evaluating on $(f_1,..,f_n;\ e_1,..,e_n)$ we obtain

$(d_0u_n:)\quad (\ f_2,\ (f_2..f_n)e_2(f_1f_2)\ ;\ \ldots\ ;\ f_n,\ (f_n)e_n(f_1..f_n)\ )$

$(u_{n-1}d_0:)\quad (\ f_2,\ (f_2..f_n)(e_2f_1)(f_2)\ ;\ldots;\ f_n,\ (f_n)(e_nf_1)(f_2..f_n)\ )$

and similarly with $d_n u_n$ and $u_{n-1} d_n$, further if $0 < i < n$ then

$$(d_i u_n:) \quad (..;\ f_i f_{i+1},\ (f_i..f_n)e_i(f_1..f_i)f_{i+1} + f_i(f_{i+1}..f_n)e_{i+1}(f_1..f_{i+1})\ ;\ ..)$$
$$(u_{n-1} d_i:) \quad (..;\ f_i f_{i+1},\ ((f_i f_{i+1})f_{i+2}..f_n)(e_i+e_{i+1})(f_1..f_{i-1}(f_i f_{i+1}))\ ;\ ..)\ ,$$

thus the identities for iterated face maps are satisfied.

If the two actions of F on E are invertible then each of the maps u_n is an isomorphism, therefore u is an isomorphism in this case.

Suppose now that $\pi_0 F$ is a group. Then any action of F is homotopy invertible that is, if F acts, from the left say, on X then the shearing map

$$F \times X \longrightarrow F \times X$$
$$(f,x) \longmapsto (f,fx)$$

is a weak homotopy equivalence. Therefore in order to show the map u_n is a weak homotopy equivalence it suffices to write it as a composite of maps each of which is isomorphic to a shearing map. But u_n is isomorphic to the composite map

$$F\times\ldots\times F \times E_n \xrightarrow{u_n} (F \times E)_n \xrightarrow{\approx} F\times\ldots\times F \times E_n$$

and the latter may be factored

$$l_1 \ldots l_n r_n \ldots r_2 r_1 \qquad \text{(composition from right to left)}$$

where r_i is the restriction of the map

$$F \times\ldots\times F \times E \times\ldots\times E \longrightarrow F \times\ldots\times F \times E \times\ldots\times E$$
$$(f_1,\ldots,f_n;\ e_1,\ldots,e_n) \longmapsto (f_1,\ldots,f_n;\ e_1,\ldots,e_{i-1},\ e_i f_i,\ldots,e_n f_i)$$

and where l_i is similarly defined using the left action.

Thus each of the maps u_n is a weak homotopy equivalence. In view of the realization lemma therefore the entire map u is a weak homotopy equivalence, too. The proof is complete. □

§3. Manipulation in a stable range.

In the theorem below we will suppose that X is highly connected and, for technical reasons, that it actually be given as a suspension. While there is no canonical way to suspend a simplicial set, a choice can of course be made universally. Our present choice is to be made so that $G(SX)$ is the free simplicial group generated by the non-basepoint simplices of X [4]. The geometric realization of the canonical map $X \to G(SX)$ then represents $|X| \to \Omega S|X|$ and is $(2m-1)$-connected if X is $(m-1)$-connected.

If V, W are pointed topological spaces we denote $Map(V,W)$ the pointed simplicial set (= the singular complex of the topological space) of pointed maps from V to W, and $H(V)$ the simplicial monoid of pointed (weak) self-homotopy equivalences of V. In a context of G-equivariant maps the analogous notions are indicated by a subscript G.

The simplicial monoid $H(V^kS^n)$ acts from the left on the pointed simplicial set $Map(V^kS^n, V^kS^n \wedge |X|)$, by composition of maps. But it also acts from the right in view of the canonical map

$$\begin{aligned} H(V^kS^n) &\longrightarrow H(V^kS^n \wedge |X|) \,, \\ h &\longmapsto h \wedge id_{|X|} \end{aligned}$$

and the two actions are compatible. Hence the cyclic bar construction, cf. (2.3),

$$N^{cy}(\, H(V^kS^n),\, Map(V^kS^n, V^kS^n \wedge |X|)\,)$$

is defined.

<u>Theorem</u> 3.1. Let X be a pointed simplicial set which is m-connected, $m \geqslant 0$. Let SX be its suspension. Then the two spaces

$$N\, H_{|G(SX)|}(V^kS^n \wedge |G(SX)|_+)$$

and

$$N^{cy}(\, H(V^kS^n),\, Map(V^kS^n, V^kS^n \wedge |SX|)\,)$$

are naturally q-equivalent, where

$$q \;=\; \min(n-2, 2m+1) \;;$$

that is, there is a chain of natural maps connecting these two spaces, and all the maps in the chain are q-connected.

Naturality here refers to n and k, and the X variable. We will also need a further piece of naturality which we record in the following addendum.

<u>Addendum</u> 3.2. There is a chain of $(2m+1)$-equivalences between $NG(SX)$ and SX, and a transformation from this chain to the one of the theorem with the property that the first map in the transformation is the composite of $NG(SX) \xrightarrow{\sim} NH^0_1(|G(SX)|)$ with the inclusion $NH^0_1(|G(SX)|) \to NH^n_k(|G(SX)|)$ (cf. lemma 1.1); and the last map in the transformation is given by the composite map

$$SX \xrightarrow{\simeq} S|SX| \xrightarrow{\approx} Map(V^1S^0, V^1S^0 \wedge |SX|) \longrightarrow Map(V^kS^n, V^kS^n \wedge |SX|)$$

together with the identification of the latter space with the term in degree 0 of $N^{cy}(..)$.

The proof of the theorem will occupy this section. The addendum will be noted as we go along. The chain of maps will consist of five maps; it could be reduced to four as the first two maps are composable. Each of the maps will be described in its own subsection.

(3.3). The first map. The simplicial monoid of the theorem,

$$H_{|G(SX)|}(V^k S^n \wedge |G(SX)|_+) ,$$

can be considered as a simplicial partial monoid by declaring that multiplication of elements in a fixed degree is possible if and only if at most one of them is outside the simplicial submonoid

$$H(V^k S^n) .$$

Thus the nerve of the simplicial monoid contains as a simplicial subobject the nerve of that simplicial partial monoid (the situation of lemma 2.2.1). The inclusion map will be our first map.

To verify the asserted connectivity, and also for its own sake, we do some rewriting now. As pointed out in section 1, the canonical map $S^0 \to |G(SX)|_+$ induces an isomorphism from the underlying simplicial set of the simplicial monoid to a union of connected components of the simplicial set of maps

$$Map(V^k S^n, V^k S^n \wedge |G(SX)|_+) ;$$

we denote this union of components by

$$\overline{Map}(..) .$$

Clearly the isomorphism is compatible with the left and right actions of $H(V^k S^n)$. Further the inclusion of the underlying simplicial set of the simplicial submonoid $H(V^k S^n)$ corresponds, under the isomorphism, to the natural inclusion

$$H(V^k S^n) \longrightarrow \overline{Map}(V^k S^n, V^k S^n \wedge |G(SX)|_+)$$

induced from $S^0 \to |G(SX)|_+$.

But it is only those two bits of structure, the latter inclusion and the left and right actions of $H(V^k S^n)$, which matter in the structure of the simplicial partial monoid. Therefore its nerve may be described as the simplicial object given by *generalized wedges* (cf. (2.2) for this notation),

$$[p] \longmapsto V^p(\overline{Map}(V^k S^n, V^k S^n \wedge |G(SX)|_+), H(V^k S^n)) .$$

The inclusion into the nerve of the original simplicial monoid is (2m+1)-connected by lemma 2.2.1, for the inclusion

$$H(V^k S^n) \longrightarrow \overline{Map}(V^k S^n, V^k S^n \wedge |G(SX)|_+)$$

is m-connected since $S^0 \to |G(SX)|_+$ is.

This finishes the account of the first map. Concerning the addendum, the first map in that chain is given by the analogous inclusion

$$[p] \longmapsto (V^p(G(SX), G(*)) \longrightarrow G(SX)^p) .$$

(3.4). *The second map*. The inclusion $X \to G(SX)$ induces one

$$\overline{Map}(V^kS^n, V^kS^n \wedge |X|_+) \longrightarrow \overline{Map}(V^kS^n, V^kS^n \wedge |G(SX)|_+)$$

where we are continuing to denote by $\overline{Map}$ a suitable union of connected components of Map, and the latter inclusion is (2m+1)-connected since the former is.

The inclusion is compatible with the left and right actions of $H(V^kS^n)$. It is also compatible with the inclusion of the underlying simplicial set of $H(V^kS^n)$, for the natural map $S^0 \to |X|_+$ given by the basepoint of X satisfies that

$$\begin{array}{ccc} S^0 & \longrightarrow & |X|_+ \\ & \searrow \quad \swarrow & \\ & |G(SX)|_+ & \end{array}$$

commutes.

Therefore the nerve of the simplicial partial monoid considered before, contains another,

$$[p] \longmapsto V^p(\,\overline{Map}(V^kS^n, V^kS^n \wedge |X|_+),\ H(V^kS^n)\,)\ .$$

The inclusion is our second map.

To show the map is (2m+1)-connected it suffices, by the realization lemma, to show this in each degree p. The case $p = 1$ was noted before. It implies the general case in view of the gluing lemma (2.1.2) and induction.

This finishes the account of the second map. Concerning the addendum, the second map in that chain is given by a similar inclusion, namely

$$[p] \longmapsto (\ V^p(X,*) \longrightarrow V^p(G(SX),G(*))\ )\ .$$

(3.5). *The third map*. The pointed simplicial set $Map(V^kS^n, V^kS^n \wedge |X|)$ can be considered as a simplicial partial monoid in a trivial way, and the simplicial monoid $H(V^kS^n)$ acts on it from both sides, and compatibly. Hence the product

$$H(V^kS^n) \times Map(V^kS^n, V^kS^n \wedge |X|)$$

can be given the structure of a simplicial partial monoid, namely the semi-direct product in the sense of (2.2).

The pair of maps $|X|_+ \to S^0$, $|X|_+ \to |X|$ induces a map of simplicial partial monoids whose underlying map of simplicial sets is

$$\overline{Map}(V^kS^n, V^kS^n \wedge |X|_+) \longrightarrow H(V^kS^n) \times Map(V^kS^n, V^kS^n \wedge |X|)\ .$$

We show this map is (n-2)-connected. Indeed, since X is connected (we assumed this in the theorem) this map is the restriction to a union of connected components of the map

$$Map(V^kS^n, V^kS^n \wedge |X|_+) \longrightarrow Map(V^kS^n) \times Map(V^kS^n, V^kS^n \wedge |X|)\ ,$$

so it suffices to show the latter map is (n-2)-connected. We treat the case $k = 1$ first.

Lemma. The map $\Omega^n S^n(|X|\cup *) \to \Omega^n S^n \times \Omega^n S^n|X|$ is (n-2)-connected.

Proof. The long exact sequence of stable homotopy groups of the cofibration sequence

$$S^n(S^0) \longrightarrow S^n(|X|\cup *) \longrightarrow S^n|X|$$

decomposes into split short exact sequences. As $\pi_i S^n Y \to \pi_i^S S^n Y$ is an isomorphism for $i \leq 2n-2$ it follows that $S^n(|X|\cup *) \to S^n \times S^n|X|$ induces an isomorphism on homotopy groups for $i \leq 2n-2$. The assertion results by taking loop spaces. □

The case $k = 1$ being established, the case of general k now follows from the isomorphism

$$Map(V^k S^n, Y) \xrightarrow{\approx} (\ Map(S^n, Y)\ )^k$$

and the (n-1)-equivalence

$$Map(S^n, V^k S^n \wedge Y') \longrightarrow (\ Map(S^n, S^n \wedge Y')\ )^k$$

induced from the (2n-1)-equivalence

$$(S^n \wedge Y') \vee \ldots \vee (S^n \wedge Y') \longrightarrow (S^n \wedge Y') \times \ldots \times (S^n \wedge Y') .$$

The map of simplicial partial monoids induces a map of their nerves. In the notation of generalized wedges, this is a map from

$$[p] \longmapsto V^p(\ \overline{Map}(V^k S^n, V^k S^n \wedge |X|_+),\ H(V^k S^n)\ )$$

to

$$[p] \longmapsto V^p(\ H(V^k S^n) \times Map(V^k S^n, V^k S^n \wedge |X|),\ H(V^k S^n) \times *\ ) .$$

This map is (n-2)-connected for every p (the gluing lemma reduces the assertion to the case $p = 1$ which was verified above) and therefore the entire map is also (n-2)-connected by the realization lemma. This is our third map.

Concerning the addendum, the third map in that chain is the identity map on

$$[p] \longmapsto V^p(|X|, *) .$$

(3.6). The fourth map. Considering the pointed simplicial set $Map(V^k S^n, V^k S^n \wedge |X|)$ as a simplicial partial monoid in a trivial way, and forming the nerve of the latter, we obtain the simplicial object

$$[p] \longmapsto \underbrace{Map(V^k S^n, V^k S^n \wedge |X|) \vee \ldots \vee Map(V^k S^n, V^k S^n \wedge |X|)}_{p}$$

which we denote by

$$\Sigma\ Map(V^k S^n, V^k S^n \wedge |X|) .$$

It inherits compatible left and right actions of the simplicial monoid $H(V^kS^n)$, so we can form the cyclic bar construction

$$N^{cy}(\, H(V^kS^n),\ \Sigma\, Map(V^kS^n, V^kS^n \wedge |X|)\ ),$$

a trisimplicial set. Our fourth map is provided by lemma 2.3.1. It is the weak homotopy equivalence whose source is

$$\text{diag}\ N^{cy}(\, H(V^kS^n),\ \Sigma\, Map(V^kS^n, V^kS^n \wedge |X|)\ )$$

(diagonal along the N- and Σ-directions) and whose target is identical to the target of the third map, namely the nerve of the simplicial partial monoid given by the semi-direct product of $H(V^kS^n)$ acting on $Map(V^kS^n, V^kS^n \wedge |X|)$.

Concerning the addendum, the fourth map in that chain is again the identity map on

$$(\ [p] \longmapsto V^p(|X|,*)\ ) \qquad (\ = \ \Sigma\ |X|\ )\ .$$

(3.7). <u>The fifth map</u>. Partial geometric realization takes the bisimplicial set

$$\Sigma\, Map(V^kS^n, V^kS^n \wedge |X|)$$

to the simplicial topological space

$$S^1 \wedge Map(V^kS^n, V^kS^n \wedge |X|)$$

and the canonical map from the latter to

$$Map(V^kS^n, V^kS^n \wedge S^1 \wedge |X|) \quad \approx \quad Map(V^kS^n, V^kS^n \wedge |SX|)$$

is (2m+1)-connected. The induced map from (the partial geometric realization of)

$$N^{cy}(\, H(V^kS^n),\ \Sigma\, Map(V^kS^n, V^kS^n \wedge |X|)\ )$$

to

$$N^{cy}(\, H(V^kS^n),\ Map(V^kS^n, V^kS^n \wedge |SX|)\ )$$

is therefore also (2m+1)-connected, by the realization lemma. This is our fifth map.

Concerning the addendum, the fifth map in that chain is the isomorphism from (the geometric realization of)

$$\Sigma\ |X|$$

to

$$|SX|\ .$$

The proof of the theorem and its addendum are now complete. □

§4. The stabilization of $A(X)$.

We will need the following elementary properties of the functor $A(X)$. Namely, it

(i) takes n-equivalences to n-equivalences if n is at least 2,

(ii) satisfies a version of homotopy excision, namely for $m, n \geq 2$, $k \leq m+n-2$, it preserves (m,n)-*connected* k-*homotopy cartesian squares*, that is, commutative squares

$$\begin{array}{ccc} V & \longrightarrow & W \\ \downarrow & & \downarrow \\ X & \longrightarrow & Y \end{array}$$

in which the horizontal (resp. vertical) arrows are m-connected (resp. n-connected) and the map $\text{fibre}(V \to X) \to \text{fibre}(W \to Y)$, or equivalently the map $\text{fibre}(V \to W) \to \text{fibre}(X \to Y)$, is (k+1)-connected.

These properties are propositions 2.3 and 2.4 of [14]. Their proofs are actually easiest with the definition of $A(X)$ used here.

We note here that $A(X)$ can be a functor on the nose, not just up to homotopy. In our present context we may simply point to the possibility of performing the + construction uniformly (for example by attaching a single 2-cell and 3-cell to $BH(V^5 S^0)$). In particular the above maps of homotopy fibres are well defined.

Let S^m denote a suitable simplicial set representing the m-sphere, and let

$$D_1^m \cup_{S^{m-1}} D_2^m \xrightarrow{\approx} S^m$$

be a decomposition into hemispheres. Then for any X the diagram

$$\begin{array}{ccc} S^{m-1} \wedge X_+ & \longrightarrow & D_1^m \wedge X_+ \\ \downarrow & & \downarrow \\ D_2^m \wedge X_+ & \longrightarrow & S^m \wedge X_+ \end{array}$$

being (m-1,m-1)-connected, is (2m-4)-homotopy cartesian by the homotopy excision theorem. In view of the above therefore the map

$$\text{fibre}(\,A(S^{m-1} \wedge X_+) \to A(D_2^m \wedge X_+)\,) \longrightarrow \text{fibre}(\,A(D_1^m \wedge X_+) \to A(S^m \wedge X_+)\,)$$
$$\simeq \Omega\, \text{fibre}(\,A(S^m \wedge X_+) \to A(*)\,)$$

is (2m-3)-connected. Thus we have a spectrum

$$m \longmapsto \text{fibre}(\,A(S^m \wedge X_+) \to A(*)\,)\,,$$

and we define $A^S(X)$ to be its telescope,

$$A^S(X) = \varinjlim \Omega^m\, \text{fibre}(\,A(S^m \wedge X_+) \to A(*)\,)\,.$$

The map $|X| \to A(X)$ (lemma 1.1) is a natural transformation if we write it in the form $|NG(X)| \to A(X)$, therefore it is compatible with the stabilization process and induces a map

$$\Omega^\infty S^\infty |X_+| \longrightarrow A^S(X) .$$

<u>Theorem</u> 4.1. There is a map

$$A^S(X) \longrightarrow \Omega^\infty S^\infty |X_+| ,$$

well defined up to weak homotopy, so that the composite map

$$\Omega^\infty S^\infty |X_+| \longrightarrow A^S(X) \longrightarrow \Omega^\infty S^\infty |X_+|$$

is weakly homotopic to the identity map.

The proof of the theorem will occupy this section. The first step is to rewrite $A^S(X)$ in terms of the cyclic bar construction. We abbreviate

$$C_k^n(X) = N^{cy}(H(V^k S^n), Map(V^k S^n, V^k S^n \wedge |X|))$$

$$C(X) = \varinjlim_{n,k} C_k^n(X) .$$

<u>Lemma</u> 4.2. The chain of maps of theorem 3.1 induces a homotopy equivalence between

$$A^S(X)$$

and

$$\varinjlim_m \Omega^m \text{ fibre}(C(S^m \wedge X_+) \to C(*))$$

where the maps in the latter direct system are, up to homotopy, given by Ω^{m-1} applied to the vertical homotopy fibres of the stabilization diagram

$$\begin{array}{ccc} |C(S^{m-1} \wedge X_+)|^+ & \longrightarrow & |C(D_1^m \wedge X_+)|^+ \\ \downarrow & & \downarrow \\ |C(D_2^m \wedge X_+)|^+ & \longrightarrow & |C(S^m \wedge X_+)|^+ . \end{array}$$

The homotopy equivalence itself is well defined up to weak homotopy.

<u>Proof</u>. In order to get theorem 3.1 to apply to all the terms in the stabilization diagram, we replace the variables $S^m \wedge X_+$, $D_1^m \wedge X_+$, etc., by their suspensions $S(S^m \wedge X_+)$, $S(D_1^m \wedge X_+)$, etc. This can be accounted for in the end by passing to loop spaces.

In view of the naturality with respect to n, k, and the X variable, theorem 3.1 induces, for every m, a chain of natural transformations of stabilization diagrams before the + construction. By performing the + construction uniformly (for example, by attaching a 2-cell and 3-cell to $|N\, H(V^5 S^0)|$ which is contained in everything in sight) we obtain from this another chain of natural transformations of stabilization diagrams, and all the diagrams involved are still strictly commutative. So the requisite maps of homotopy fibres are well defined, and we obtain a chain of transformations connecting the m-th map of the original direct system to the m-th map of the new direct system.

By splicing these, for varying m, we obtain a chain of transformations between the original direct system and the new one. As the connectivity of the transformations increases with m, we obtain in the limit a chain of weak homotopy equivalences. To show the latter is well defined up to weak homotopy, it suffices to show that the chain of maps is well defined up to homotopy if everything in sight is replaced by a term in its Postnikov tower. But if we replace by the m-th terms in the Postnikov towers then our original direct system becomes essentially constant (the maps are weak homotopy equivalences from number $m+3$ on). Consequently, in view of the connectivity of the transformations, the other direct systems also become essentially constant. So the chain of maps between those terms in the Postnikov towers comes from a chain of maps at some finite stage, and this is well defined up to homotopy. □

We note that the addendum 3.2 provides a description of the map $\Omega^\infty S^\infty|X_+| \to A^S(X)$ in terms of our new definition of $A^S(X)$.

Before proceeding we state a lemma which will be needed presently.

<u>Lemma</u> 4.3. Suppose that Y is $(m-1)$-connected. Then the map

$$Map(V^k S^n, S^{n+m}) \wedge Map(S^{n+m}, S^{n+m}\wedge Y) \longrightarrow Map(V^k S^n, S^{n+m}\wedge Y)$$

given by composition, is $(3m-1)$-connected. Similarly, in the case $k = 1$, we obtain a $(3m-1)$-connected map if we compose the other way, that is, consider the map

$$Map(S^{n+m}, S^{n+m}\wedge Y) \wedge Map(S^n, S^{n+m}) \longrightarrow Map(S^{n+m}, S^{n+2m}\wedge Y)$$

obtained by stabilizing the second factor to $Map(S^{n+m}\wedge Y, S^{n+2m}\wedge Y)$, and composing.

<u>Proof</u>. The first map is isomorphic to the upper horizontal map in the commutative diagram

$$\begin{array}{ccc}
Map(S^n, S^{n+m})^k \wedge Map(S^{n+m}, S^{n+m}\wedge Y) & \longrightarrow & Map(S^n, S^{n+m}\wedge Y)^k \\
\uparrow & & \uparrow \\
V^k S^0 \wedge S^m \wedge Map(S^{n+m}, S^{n+m}\wedge Y) & & \\
\uparrow & & \\
V^k S^0 \wedge S^m \wedge Map(S^m, S^m\wedge Y) & \longrightarrow & V^k S^0 \wedge (S^m \wedge Y) \\
& \nwarrow & \uparrow \| \\
& & (V^k S^0 \wedge S^m) \wedge Y
\end{array}$$

The arrow on the right is $(4m-1)$-connected. Each of the two arrows on the left and the diagonal arrow on the bottom is the smash product of a $(2m-1)$-connected map with the identity on an $(m-1)$-connected space, hence $(3m-1)$-connected. So we must have the asserted connectivity of the first map.

The second map is part of the commutative diagram

$$\begin{array}{ccc}
Map(S^{n+m},S^{n+m}\wedge Y)\wedge Map(S^{n},S^{n+m}) & \longrightarrow & Map(S^{n+m},S^{n+2m}\wedge Y)\\
\uparrow & & \\
Map(S^{n+m},S^{n+m}\wedge Y)\wedge Map(S^{0},S^{m}) & & \uparrow\\
\uparrow & & \\
Map(S^{0},Y)\wedge Map(S^{0},S^{m}) & \xrightarrow{\ \approx\ } & Map(S^{0},S^{m}\wedge Y)
\end{array}$$

and the same kind of connectivity considerations apply as before. □

Returning to the proof of the theorem, we will proceed in two steps. In the first step (4.4 below) we represent, in a stable range, the asserted map by a chain of two maps of which one is highly connected and has to be inverted. By taking into account some more data it will be immediate that the map is a retraction up to homotopy in that range. In the second step (4.5 below) we discuss the stabilization procedure.

(4.4). The representative in the stable range. The relevant data are displayed on the following diagram. The diagram shows the part in degree p of a commutative diagram of simplicial objects. Two of these simplicial objects are given by the cyclic bar construction (the upper and middle terms in the left column), the four others are trivial simplicial objects.

$$\begin{array}{ccc}
 & & Map(S^{n},S^{n+2m}\wedge|X_+|)\\
 & \swarrow & \uparrow\\
H(V^kS^n)^p\times Map(V^kS^n,V^kS^{n+2m}\wedge|X_+|) & & \\
\uparrow & & Map(S^{n},S^{n+m})\wedge Map(S^{n+m},S^{n+2m}\wedge|X_+|)\\
 & \swarrow & \\
H(V^kS^n)^p\times\big(\,Map(V^kS^n,S^{n+m})\wedge Map(S^{n+m},V^kS^{n+2m}\wedge|X_+|)\,\big) & & \downarrow\\
\downarrow & & Map(S^{n+m},S^{n+3m}\wedge|X_+|)\\
 & \overset{=}{\swarrow} & \\
Map(S^{n+m},S^{n+3m}\wedge|X_+|) & &
\end{array}$$

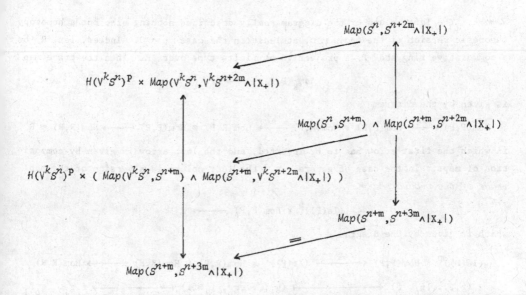

Two of the maps require comment, these are the lower vertical maps in the diagram. The one on the right is given by composition of maps *after switch of factors*. The one on the left similarly involves a switch of factors. It is the unique map of quotient spaces induced by the following sequence of maps,

$$H(V^kS^n) \times \ldots \times H(V^kS^n) \times Map(V^kS^n, S^{n+m}) \times Map(S^{n+m}, V^kS^{n+2m} \wedge |X_+|)$$

$$\Big\downarrow \text{(switch of factors)}$$

$$Map(S^{n+m}, V^kS^{n+2m} \wedge |X_+|) \times H(V^kS^n) \times \ldots \times H(V^kS^n) \times Map(V^kS^n, S^{n+m})$$

$$\Big\downarrow \text{(smash product with identity maps)}$$

$$Map(S^{n+m}, V^kS^{n+2m} \wedge |X_+|) \times H(V^kS^{n+2m} \wedge |X_+|) \times \ldots \times Map(V^kS^{n+2m} \wedge |X_+|, S^{n+3m} \wedge |X_+|)$$

$$\Big\downarrow \text{(composition of maps)}$$

$$Map(S^{n+m}, S^{n+3m} \wedge |X_+|)$$

The map is compatible with the structure maps of the cyclic bar construction. This fact, indeed, is the reason why we are using the cyclic bar construction.

Remark. The left column of the diagram really describes nothing else but a homotopy theoretic version of the *trace map*, at least in the case $p = 0$. Indeed, let R be a commutative ring and P a projective of finite type over R. Then the trace map

$$\mathrm{Hom}_R(P,P) \longrightarrow R$$

is given by the diagram

$$\mathrm{Hom}(P,P) \xleftarrow{\approx} \mathrm{Hom}(P,R) \otimes \mathrm{Hom}(R,P) \xrightarrow{\approx} \mathrm{Hom}(R,P) \otimes \mathrm{Hom}(P,R) \longrightarrow \mathrm{Hom}(R,R) \approx R$$

in which the first arrow has to be inverted, and the last arrow is given by composition of maps. In the case of general p, the left column is a version of the map *trace of the product matrix*

$$(\mathrm{Is}(P))^p \times \mathrm{Hom}(P,P) \longrightarrow R$$

which is given by the diagram

$$\begin{array}{ccccc} (\mathrm{Is}(P))^p \times \mathrm{Hom}(P,P) & \xleftarrow{\approx} & (\mathrm{Is}(P))^p \times \mathrm{Hom}(P,R) \otimes \mathrm{Hom}(R,P) & \longrightarrow & \mathrm{Hom}(R,R) \\ (g_1,\ldots,g_p,\ f) & \longleftarrow\!\!\shortmid & (g_1,\ldots,g_p,\ f_1 \otimes f_2) & \longmapsto & f_2 g_1 \ldots g_p f_1 \end{array}$$

This ends the remark.

Concerning the relevance of the diagram of simplicial objects described, we will eventually have to pass to loop spaces, namely the (2m)-th loop spaces. Thus any required connectivities must increase faster than 2m. This is indeed the case. *The map on the upper left is* (3m-1)-*connected*: The realization lemma reduces us to showing this in every degree p in which case it is the content of the first part of lemma 4.3. Thus the left column does represent a map, defined in a stable range, from top to bottom. *This map is a retraction up to homotopy, in that range.* This information is provided by the rest of the diagram since the two vertical maps on the right are (3m-1)-connected by lemma 4.3 again. *The coretraction involved* (the upper horizontal map) *is a representative* (before the + construction, in a stable range) *of the map* $\Omega^\infty S^\infty|S^{2m}\wedge X_+| \to A(S^{2m}\wedge X_+)$. As noted before, this is the content of the addendum 3.2.

Passing to geometric realization and performing the + construction to the terms on the left, we obtain the diagram

$$\begin{array}{ccc}
 & & |Map(S^n, S^{n+2m}\wedge|X_+|)| \\
 & \swarrow & \uparrow \\
|N^{cy}(H(V^kS^n), Map(V^kS^n, V^kS^{n+2m}\wedge|X_+|))|^+ & & \\
\uparrow & & |Map(S^n,S^{n+m})\wedge Map(S^{n+m},S^{n+2m}\wedge|X_+|)| \\
 & \swarrow & \\
|N^{cy}(H(V^kS^n), Map(V^kS^n,S^{n+m})\wedge Map(S^{n+m},V^kS^{n+2m}\wedge|X_+|))|^+ & & \downarrow \\
\downarrow & & |Map(S^{n+m},S^{n+3m}\wedge|X_+|)| \\
 & \swarrow & \\
|Map(S^{n+m},S^{n+3m}\wedge|X_+|)|^+ & &
\end{array}$$

The + construction is possible if k is at least 5 and it can be done uniformly with regard to the upper and middle space on the the left by attaching a pair of cells to the common subspace $|N(H(V^5S^0))|$. It preserves the connectivity of the upper map on the left (by the gluing lemma). The + construction on the bottom term on the left refers to the induced attaching of the pair of cells (a pushout). As the original term had abelian fundamental group, the + construction does not change the homotopy type. Its sole purpose is to keep the whole diagram strictly commutative.

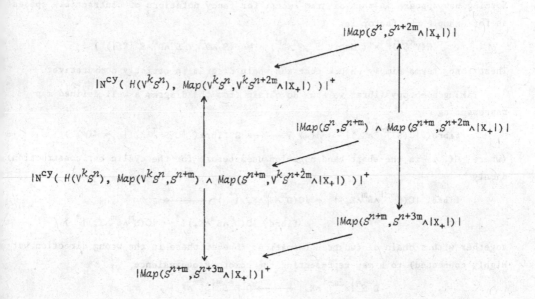

Everything we have done so far is natural with respect to n and k, so we may pass to the direct limit in those variables (recall that stabilizing with respect to k involves wedge with an identity map on the $H(..)$ part, but wedge with a trivial map on the *Map*(..) part).

(4.5). The stabilization procedure. This must be adjusted to the needs of the preceding subsection. Namely the two factors $\mathcal{S}^m$ in $\mathcal{S}^m \wedge \mathcal{S}^m \wedge |X_+|$ play rather different roles, so we must stabilize in both of these factors. To do this we just alternate in stabilizing either the first or the second.

In order to stabilize in the first $\mathcal{S}^m$, say, we must write down (or better, contemplate) a large diagram involving four versions of the diagram of (4.4), one for each of the terms in

$$\begin{array}{ccc} S^{m-1} & \longrightarrow & D_1^m \\ \downarrow & & \downarrow \\ D_2^m & \longrightarrow & S^m \end{array} .$$

Nothing new appears in this diagram except for fancy notations of contractible spaces as for example the factor in

$$H(V^k \mathcal{S}^n)^p \times \left(\mathit{Map}(V^k \mathcal{S}^n, \mathcal{S}^n \wedge D_1^m) \wedge \mathit{Map}(\mathcal{S}^n \wedge D_1^m, V^k \mathcal{S}^n \wedge D_1^m \wedge \mathcal{S}^m |X_+|) \right) .$$

These fancy terms simply ensure that the whole diagram is strictly commutative.

Taking homotopy fibres we thus do obtain from the diagram a well defined map representing

$$\mathrm{fibre}(\ |C(S^{2m-1} \wedge X_+)|^+ \to A(*)\ ) \longrightarrow \Omega\ \mathrm{fibre}(\ |C(S^{2m} \wedge X_+)|^+ \to A(*)\ )$$

(where $C(..)$ is the short hand notation used before for the cyclic bar construction), namely

$$\mathrm{fibre}(\ |C(S^{m-1} \wedge S^m \wedge X_+)|^+ \to |C(D_1^m \wedge S^m \wedge X_+)|^+\ ) \longrightarrow$$
$$\mathrm{fibre}(\ |C(D_2^m \wedge S^m \wedge X_+)|^+ \to |C(S^m \wedge S^m \wedge X_+)|^+\ ) ,$$

together with a chain of two transformations (one of these in the wrong direction but highly connected) to a map representing the homotopy equivalence

$$\Omega^\infty \mathcal{S}^\infty |S^{2m-1} \wedge X_+| \longrightarrow \Omega\ \Omega^\infty \mathcal{S}^\infty |S^{2m} \wedge X_+| .$$

We apply Ω^{2m-1} to all this. Then we may splice, for varying m, to obtain a chain of transformations of direct systems. Passing to the limit we obtain what we are after; the appropriate concluding remarks here are similar to the proof of lemma 4.2. This completes the argument.

§5. The splitting of $A(X)$.

Let $\Omega^\infty S^\infty|X_+| \to A(X)$ be the map given by the Barratt-Priddy-Quillen-Segal theorem (section 1), and let $A(X) \to A^S(X)$ be the stabilization map (it will be defined in lemma 5.2 below).

Theorem 5.1. There is a map $A^S(X) \to \Omega^\infty S^\infty|X_+|$ so that the diagram

$$\begin{array}{ccccc} \Omega^\infty S^\infty|X_+| & & & \overset{=}{\searrow} & \\ \downarrow & & & & \\ A(X) & \longrightarrow & A^S(X) & \longrightarrow & \Omega^\infty S^\infty|X_+| \end{array}$$

is weakly homotopy commutative.

Proof. This results from theorem 4.1 in view of the following lemma. □

Lemma 5.2. There is a natural stabilization map $A(X) \to A^S(X)$. Its composition with $\Omega^\infty S^\infty|X_+| \to A(X)$ is weakly homotopic to the map used in theorem 4.1.

Proof. Letting $\widetilde{A}(X)$ denote the factor in the natural splitting (section 1)

$$A(X) \simeq \widetilde{A}(X) \times A(*)$$

we define a direct system

$$A(X) \longrightarrow \Omega\,\widetilde{A}(S^1 \wedge X_+) \longrightarrow \Omega^2\,\widetilde{A}(S^2 \wedge X_+) \longrightarrow$$

in which the first map is provided by lemma 1.4, and the other maps are given by the maps of vertical homotopy fibres in the appropriate stabilization diagrams (as described in the beginning of section 4). The map from the initial term of the system to its telescope gives the required map $A(X) \to A^S(X)$.

To make the asserted comparison we consider the map of direct systems

$$\begin{array}{ccccccc} \Omega^\infty S^\infty|X_+| & \longrightarrow & \Omega\,\Omega^\infty S^\infty|S^1 \wedge X_+| & \longrightarrow & \Omega^2\,\Omega^\infty S^\infty|S^2 \wedge X_+| & \longrightarrow \\ \downarrow & & \downarrow & & \downarrow & \\ A(X) & \longrightarrow & \Omega\,\widetilde{A}(S^1 \wedge X_+) & \longrightarrow & \Omega^2\,\widetilde{A}(S^2 \wedge X_+) & \longrightarrow \end{array}$$

where the vertical maps are the natural ones (the weak homotopy commutativity of the first square is due to lemma 1.4). The maps in the upper direct system are homotopy equivalences: the first map by lemma 1.4, and the other maps by the excision property of stable homotopy. The maps in the direct system defining $A^S(X)$ are eventually highly connected (cf. the beginning of section 4). So it will suffice to compare the vertical maps in the diagram with the map used in theorem 4.1, and to show these coincide in a stable range.

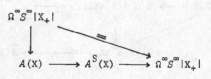

The diagram of inclusions (section 1)

$$\begin{array}{ccc} H_1^0(G) & \longrightarrow & H_\infty^0(G) \\ & \searrow & \downarrow \\ & & H_\infty^\infty(G) \end{array}$$

with $G = |G(S^m \wedge X_+)|$, induces the left part of the following diagram.

$$\begin{array}{ccccc} |S^m \wedge X_+| & \longrightarrow & \Omega^\infty S^\infty |(S^m \wedge X_+)_+| & \longrightarrow & \Omega^\infty S^\infty |S^m \wedge X_+| \\ & \searrow & \downarrow & & \downarrow \\ & & A(S^m \wedge X_+) & \longrightarrow & \widetilde{A}(S^m \wedge X_+) \end{array}$$

The vertical map on the right is, up to de-looping, the same as the m-th vertical map in the diagram above, and the composite map on the bottom is an approximation to the map used in theorem 4.1. The composite map on top is the Hurewicz map from homotopy to stable homotopy (lemma 1.1), hence it is $(2m-1)$-connected. So the two maps in question do agree in a stable range, and the proof is complete. □

Remark 5.3. The maps in theorem 5.1 are maps of infinite loop spaces, and the diagram is weakly homotopy commutative as a diagram of infinite loop spaces.

Here is an indication of proof for the first assertion, the second involves similar considerations. Two of the maps are clearly infinite loop maps, namely the map $\Omega^\infty S^\infty |X_+| \to A(X)$ as it is the map of underlying spaces of a map of Γ-spaces, and the map $A^S(X) \to \Omega^\infty S^\infty |X_+|$ of theorem 4.1 as it was defined as the telescope of a map of spectra.

The remaining map $A(X) \to A^S(X)$ is also a map of infinite loop spaces provided that we use a possibly different infinite loop structure on $A^S(X)$. For the stabilization diagram

$$\begin{array}{ccc} A(S^{m-1} \wedge X_+) & \longrightarrow & A(D_1^m \wedge X_+) \\ \downarrow & & \downarrow \\ A(D_2^m \wedge X_+) & \longrightarrow & A(S^m \wedge X_+) \end{array}$$

is in fact the diagram of underlying spaces of a diagram of Γ-spaces. Therefore there is a Γ-space of which $A^S(X)$ is the underlying space, and the map

$$\Omega \text{ fibre}(\, A(S^1 \wedge X_+) \to A(*)\,) \longrightarrow A^S(X)$$

is a map of underlying spaces of Γ-spaces. The map

$$A(X) \longrightarrow \Omega \text{ fibre}(\, A(S^1 \wedge X_+) \to A(*)\,)$$

of lemma 1.4, too, is a map of underlying spaces of Γ-spaces. Hence so is the composite map $A(X) \to A^S(X)$.

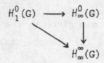

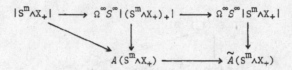

It remains to be seen that the two infinite loop structures on $A^S(X)$ are equivalent. In view of their definitions these infinite loop structures are compatible in the following sense. They are definable in terms of spectra (obtainable from the Γ-structure, resp. from stabilization in the X-variable) and the two spectra can be combined into a *double spectrum*. Further both spectra are connective. But this implies they are equivalent (the argument is probably well known; cf. [13, section 16] for a detailed account in a particular case).

Remark 5.4. The maps in theorem 5.1 are compatible with pairings.

Here is an indication of why this is so. In the case of $\Omega^\infty S^\infty |X_+| \to A(X)$ it is immediate from the definition of the pairings.

To treat the case of the map $A(X) \to A^S(X)$ one shows that the stabilization map

$$\mathrm{fibre}(\, A(S^m \wedge X_+) \to A(*)\,) \longrightarrow \Omega\ \mathrm{fibre}(\, A(S^{m+1} \wedge X_+) \to A(*)\,)$$

is the same, up to homotopy, as the adjoint of the composite map (cf. section 1)

$$S^1 \wedge \tilde{A}(S^m \wedge X_+) \longrightarrow \tilde{Q}(S^1) \wedge \tilde{A}(S^m \wedge X_+) \longrightarrow \tilde{A}(S^{m+1} \wedge X_+)$$

where $S^1 \to \tilde{Q}(S^1)$ is the Hurewicz map (to prove this one has to use that $A(X)$ is definable in a more general context than we are using here, i.e., for X which are not necessarily pointed nor connected - cf. a similar point in the proof of lemma 1.4). Thus stabilization itself is definable in terms of the pairing, and so the pairing on $\tilde{A}$ induces one on A^S and the required compatibility holds.

To treat the case of the map $A^S(X) \to \Omega^\infty S^\infty |X_+|$ one redefines $A^S(X)$ in terms of the cyclic bar construction (section 4). One notes that the smash product also induces a pairing in terms of the cyclic bar construction, and that this pairing is (obviously) compatible with the one on stable homotopy via the two maps of theorem 4.1 of which the map in question is one. To finish one has to chase the pairing through the chain of maps of theorem 3.1 in order to compare with the pairing formerly used. This ends the indication.

§6. Appendix: The stabilization of *K*-theory.

The stabilization of $A(X)$ to $A^S(X)$ may be mimicked with *K*-theory provided that one works with a suitably extended notion of *K*-theory in the framework of simplicial rings [14, section 1]. The extended notion of *K*-theory is needed even in the treatment of the stabilized *K*-theory of an ordinary ring.

We need some notation. If A is an abelian group and X a set we denote A[X] the direct sum of A with itself indexed by the elements of X. Similarly A[X] is defined if A is a simplicial abelian group and X a simplicial set, and is a bisimplicial abelian group (which we may diagonalize if we wish to a simplicial abelian group). If R is a (simplicial) ring and G a (simplicial) group then R[G] may be equipped with a multiplication in the usual way so that it is a 'group ring'. For pointed X we let $\tilde{A}[X] = A[X]/A[*]$. If A has an R-module structure then so have A[X] and $\tilde{A}[X]$, respectively.

The set of connected components $\pi_0 R$ is a ring in a natural way (the exotic case $1 = 0$ in $\pi_0 R$ may be ignored, for in this case R is contractible (multiply by a path from 1 to 0) and such an R is without interest to us); we let $K_0(\pi_0 R)$ denote its projective class group, as usual.

If A is a simplicial abelian group we denote $M_k(A)$ the simplicial abelian group of $k \times k$ matrices in A. If R is a simplicial ring then so is $M_k(R)$ and we denote $\widehat{GL}_k(R)$ the multiplicative simplicial monoid of homotopy units in $M_k(R)$ (the matrices in the connected components indexed by the elements of $GL_k(\pi_0 R) \subset M_k(\pi_0 R)$). The *K*-theory of the simplicial ring R is, by definition,

$$K(R) = K_0(\pi_0 R) \times \varinjlim B\widehat{GL}_k(R)^+$$

(in [14] the factor $K_0(\pi_0 R)$ was replaced by Z in order to simplify the comparison with $A(X)$).

The functor $R \mapsto K(R)$ is a homotopy functor in a suitable sense (cf. the properties of $A(X)$ stated in the beginning of section 4). In particular if $R \to R'$ is a weak homotopy equivalence then so is $K(R) \to K(R')$. It extends the *K*-theory of Quillen in the sense that it reduces to the latter in the case of a ring considered as a simplicial ring in a trivial way.

The *stabilized K-theory* of R is defined to be

$$K^S(R) = \varinjlim_m \Omega^m \operatorname{fibre}(K(R[G(S^m)]) \to K(R))$$

where S^m denotes a simplicial set representing the m-sphere and G(..) is Kan's loop group functor; the maps in the direct system are defined as in section 4. It is natural, in fact, to consider a slight generalization, the functor of two variables R and X (a simplicial set)

$$K^S(X,R) \;=\; \varinjlim_m \Omega^m \text{ fibre}(\, K(R[G(S^m \wedge X_+)]) \to K(R)\,)$$

In detail, the terms in the direct system are defined for $m > 0$, and the maps are (loops of) the maps of vertical homotopy fibres of $K(R[G(\;?\;)])$ applied to the stabilization diagram

$$\begin{array}{ccc} S^{m-1}\wedge X_+ & \longrightarrow & D^m_1 \wedge X_+ \\ \downarrow & & \downarrow \\ D^m_2 \wedge X_+ & \longrightarrow & S^m \wedge X_+ \end{array}$$

This $K^S(X,R)$ is a *homology theory* in the X variable [14], the *coefficients* of the homology theory are given by $K^S(*,R) \approx K^S(R)$.

Here are some remarks about the numerical significance of stabilized K-theory. Let R be a ring (not simplicial ring), let $K^S_i(R) = \pi_i K^S(R)$. There is a spectral sequence (with trivial action in the E^2 term)

$$H_p(GL(R), K^S_q(R)) \;\Rightarrow\; H_{p+q}(GL(R), M(R)) \;,$$

with abutment the homology of $GL(R)$ acting by conjugation on $M(R)$, the essentially finite matrices in R. This is proved by the method of [14, lemma 1.5]: to deduce the existence of the spectral sequence in a stable range, one compares the spectral sequence for stable homotopy of the map $B\widehat{GL}(R[G(S^m)]) \to BGL(R)$ with that of the corresponding map after the + construction. After a suitable dimension shift the latter spectral sequence has the desired E^2 term, while the former one collapses and gives the desired abutment (everything in a stable range).

Stabilized K-theory may be 'computed' in the following way. Let again R be a ring (not simplicial ring). Let $F(R)$ be the homotopy fibre

$$F(R) \;=\; \text{fibre}(\, BGL(R) \to BGL(R)^+\,)\,.$$

Then $F(R)$ is an acyclic space with $\pi_1 F(R) \approx St(R)$ (the Steinberg group), and $\pi_i F(R) \approx K_{i+1}(R)$ if $i > 1$.

Denoting the homotopy fibre of the map $B\widehat{GL}(R[G(S^m)]) \to BGL(R)^+$ by U, one shows that after the + construction one obtains a homotopy equivalence

$$U^+ \;\simeq\; \text{fibre}(\, B\widehat{GL}(R[G(S^m)])^+ \to BGL(R)^+\,)\,.$$

On the other hand U may be identified to the homotopy pullback of the diagram

$$F(R) \longrightarrow BGL(R) \longleftarrow B\widehat{GL}(R[G(S^m)])\,.$$

As $U \to U^+$ is an acyclic map, the spectral sequence of a generalized homology theory h_* for the map $U \to F(R)$ therefore gives a spectral sequence

$$H_p(F(R),\, h_q\text{fibre}(\, B\widehat{GL}(R[G(S^m)]) \to BGL(R)\,)) \;\Rightarrow\; h_{p+q}\text{fibre}(\, B\widehat{GL}(R[G(S^m)])^+ \to BGL(R)^+\,)\,.$$

The fibre involved in the E^2 term may be identified, in a stable range, with the Eilenberg-Mac Lane space $BM(\widetilde{R}[S^{m-1}])$. Taking h_* to be the stable homotopy groups one obtains hence that, in a stable range, the stable groups can be identified to the actual ones and the spectral sequence collapses. Whence the isomorphism

$$K_i^S(R) \approx H_i(F(R), M(R))$$

where, as one checks, the homology involves the action of $\pi_1 F(R)$ on $M(R)$ pulled back from the conjugation action of $GL(R)$. In particular,

$$K_0^S(R) \approx H_0(St(R), M(R)) \approx R/[R,R] , \qquad K_1^S(R) \approx H_1(St(R), M(R)) .$$

It will be indicated now how the results on $A(X)$ described in the earlier sections can be adapted to K-theory.

The heart of the matter is to recast the definition of stabilized K-theory in terms of the cyclic bar construction. Let Y be an m-connected simplicial set, $m \geq 0$, and let SY be its suspension. As in section 3 one constructs a natural chain of maps (five of them, just as in theorem 3.1) between $N\widehat{GL}_k(R[G(SY)])$ and $N^{cy}(\widehat{GL}_k(R), M_k(\widetilde{R}[SY]))$ satisfying that each of the maps in the chain is (2m+1)-connected. One deduces from this a homotopy equivalence

$$K^S(X,R) \simeq \varinjlim_{k,m} \Omega^m \text{ fibre}(|N^{cy}(\widehat{GL}_k(R), M_k(\widetilde{R}[S^m \wedge X_+]))|^+ \to |N\widehat{GL}_k(R)|^+) .$$

Let us insert here as a parenthesis how to go from this homotopy equivalence to an interesting new definition of stabilized K-theory which we do not have occasion to use, though. If R is a ring and A an R-bimodule (resp. simplicial ring and simplicial bimodule) then $R \oplus A$ can be considered as a ring (resp. simplicial ring) by giving A trivial multiplication. Now suppose that A is connected. Then there is a natural isomorphism

$$\widehat{GL}_k(R \oplus A) \approx \widehat{GL}_k(R) \ltimes M_k(A)$$

where the term on the right is the semi-direct product in the sense of (2.2). Hence lemma 3.1 gives a homotopy equivalence

$$\text{diag } N^{cy}(\widehat{GL}_k(R), NM_k(A)) \xrightarrow{\simeq} N\widehat{GL}_k(R \oplus A) .$$

On the other hand, $NM_k(A) \approx M_k(NA) \approx M_k(\widetilde{A}[S^1])$, and so we can conclude

$$K^S(X,R) \simeq \varinjlim_m \Omega^m \text{ fibre}(K(R \oplus \widetilde{R}[S^{m-1} \wedge X_+]) \to K(R)) .$$

Notice in particular that $\widetilde{R}[S^{m-1}]$ is just an Eilenberg-Mac Lane group, and

$$K^S(R) \simeq \varinjlim_m \Omega^m \text{ fibre}(K(R \oplus \widetilde{R}[S^{m-1}]) \to K(R)) .$$

This ends the parenthesis. □

Let $h(X,R)$ denote the (unreduced) homology of X with coefficients in R, it is represented by $|R[X]|$. There is a natural map $h(X,R) \to K^S(X,R)$. It arises from the homotopy equivalence $h(X,R) \simeq \varinjlim \Omega^m |\tilde{R}[S^m \wedge X_+]|$ together with the identification of $\tilde{R}[S^m \wedge X_+]$ with the part in degree 0 of $N^{cy}(\hat{GL}_1(R), M_1(\tilde{R}[S^m \wedge X_+]))$.

Proposition 6.1. If R is commutative then $h(X,R) \to K^S(X,R)$ is a coretraction, up to weak homotopy.

This is the analogue of theorem 4.1. Concerning the proof, if A is an R-module (resp. simplicial R-module) considered as a bimodule in a trivial way (both the left and the right structure are given by the original module structure) then the trace map

$$\begin{array}{ccc} \hat{GL}_k(R)^p \times M_k(A) & \longrightarrow & A \\ (g_1,..g_p,\ a) & \longmapsto & tr(g_1..g_p a) \end{array}$$

is insensitive to cyclic rearrangement of the factors. Therefore it is compatible with the face maps of the cyclic bar construction and defines a map

$$N^{cy}(\hat{GL}_k(R), M_k(A)) \longrightarrow A$$

which is a retraction with section as described. To complete the proof one has to check naturality with regard to stabilization, as in section 4.

One constructs a natural transformation $K(R[G(X)]) \to K^S(X,R)$ by producing artificially a map $K(R[G(X)]) \to \Omega\ \text{fibre}(K(R[G(S^1 \wedge X_+)]) \to K(R))$ as in lemma 1.4, using pairings.

The inclusion of the 'monomial matrices', $\Sigma_k \int G(X) \to \hat{GL}_k(R[G(X)])$, induces a map, as usual, $\Omega^\infty S^\infty |X_+| \to K(R[G(X)])$.

Let $\Omega^\infty S^\infty |X_+| \to h(X,R)$ be the Hurewicz map from stable homotopy to R-homology.

Proposition 6.2. The diagram of the above maps commutes up to weak homotopy,

$$\begin{array}{ccc} \Omega^\infty S^\infty |X_+| & \longrightarrow & h(X,R) \\ \downarrow & & \downarrow \\ K(R[G(X)]) & \longrightarrow & K^S(X,R) \end{array} .$$

Putting this together with the preceding result we obtain for commutative R an analogue of the splitting theorem 5.1, a diagram

$$\begin{array}{ccccc} \Omega^\infty S^\infty |X_+| & \longrightarrow & h(X,R) & & \\ \downarrow & & \downarrow & \overset{=}{\searrow} & \\ K(R[G(X)]) & \longrightarrow & K^S(X,R) & \longrightarrow & h(X,R) \end{array}$$

that commutes up to weak homotopy and whose maps have the naturality properties indicated in section 5: they are infinite loop maps and compatible with the respective pairings.

Proposition 6.2 is the analogue of lemma 5.2, and the proof of the latter may be adapted. One can also deduce it from lemma 5.2 because of the following naturality property: there is a natural transformation

$$A(X) \longrightarrow K(R[G(X)]) \ ,$$

it induces a corresponding transformation of the stabilized theories, and

$$\begin{array}{ccccccc} & \nearrow & A(X) & \longrightarrow & A^S(X) & \longleftarrow & \Omega^\infty S^\infty |X_+| \\ \Omega^\infty S^\infty |X_+| & & \downarrow & & \downarrow & & \downarrow \\ & \searrow & K(R[G(X)]) & \longrightarrow & K^S(X,R) & \longleftarrow & h(X,R) \end{array}$$

commutes up to (weak) homotopy, and finally in the case of commutative R so does

$$\begin{array}{ccc} A^S(X) & \longrightarrow & \Omega^\infty S^\infty |X_+| \\ \downarrow & & \downarrow \\ K^S(X,R) & \longrightarrow & h(X,R) \ . \end{array}$$

Using the notion of 'Hochschild homology' one can give a variant of the map $K^S(X,R) \to h(X,R)$ which is more generally defined. We no longer assume that R is commutative, but we do assume that R is given as an algebra (resp. simplicial algebra) over some commutative ring (resp. simplicial ring) k, and that it is flat over k (resp. degreewise flat).

Let A be a (simplicial) R-bimodule, over k. Following K. Dennis, one defines the *Hochschild homology*

$$H(R/k,A)$$

as the additive version of the cyclic bar construction, the simplicial object

$$[p] \longmapsto \underbrace{R \otimes_k \ldots \otimes_k R}_{p} \otimes_k A$$

(degreewise tensor product) with face and degeneracy maps as in the cyclic bar construction. We will need the fact, due to Dennis [talk at Evanston conference, January 1976, unpublished], that the Hochschild homology is *Morita invariant* in the sense of the following lemma.

Recall that two rings are called Morita equivalent if their module categories are equivalent categories. This relation is equivalent [2, chapter II] to the following property which in our present more general situation we will take as the definition.

We say that R is Morita equivalent over k with a (simplicial) k-algebra R' if there exist (simplicial) bimodules ${}_{R}E_{R'}$, ${}_{R'}F_{R}$ over k which are (degreewise) projective both from the left and the right, so that

$$E \otimes_{R'} F \approx R , \qquad F \otimes_{R} E \approx R'$$

as (simplicial) R-bimodules, resp. R'-bimodules.

<u>Lemma</u> (K. Dennis). In this situation there is a natural homotopy equivalence

$$\mathcal{H}(R/k,\ A) \simeq \mathcal{H}(R'/k,\ F \otimes_R A \otimes_R E) .$$

<u>Proof</u>. Letting $B = F \otimes_R A$ we may reformulate the assertion as a homotopy equivalence

$$\mathcal{H}(R/k,\ E \otimes_{R'} B) \simeq \mathcal{H}(R'/k,\ B \otimes_R E) .$$

To prove this it suffices to consider the case of rings rather than simplicial rings and establish the homotopy equivalence by a chain of two natural maps. The general case then follows in view of the realization lemma. So we assume R, R' are rings, not simplicial rings.

The common source of the two maps to be constructed will be the following bisimplicial object. The object in bidegree (p,q) is given by

$$\begin{array}{ccc} & \longleftarrow p \longrightarrow & \\ & R \otimes \ldots\ldots \otimes R & \\ \otimes & & \otimes \\ E & & B \\ \otimes & & \otimes \\ & R' \otimes \ldots \otimes R' & \\ & \longleftarrow q \longrightarrow & \end{array} \qquad (\mathcal{H})_{p,q}$$

(tensor products over k), and the way this has been written as a circle is to suggest in which way the various face maps are given by multiplication at the appropriate tensor product signs.

Let $\mathcal{H}(E,R'/k,B)$ be the simplicial object

$$[q] \longmapsto E \otimes \underbrace{R' \otimes \ldots \otimes R'}_{q} \otimes B$$

(a 'two-sided bar construction'). It maps to the trivial simplicial object $E \otimes_{R'} B$ by the map which in degree q multiplies together all the factors. This map is a homotopy equivalence. Indeed, using the right projectivity of E over R' we can reduce the assertion to the case where $E = R'$. But this case is clear (the simplicial object is a 'cone').

The bisimplicial object $(\mathcal{H})$ may be identified to one

$$\mathcal{H}(R/k,\ \mathcal{H}(E,R'/k,B)\ )$$

(a combination of the cyclic bar construction and the two-sided bar construction) and the map described just before, induces a map from this bisimplicial object to the simplicial object $H(R/k, E\otimes_R B)$. The latter map is a homotopy equivalence degreewise in the p-direction. Indeed this follows from the homotopy equivalence established just before in view of the flatness of R over k. In view of the realization lemma it therefore follows that (H) maps by homotopy equivalence to $H(R/k, E\otimes_R B)$.

By identifying (H) to a bisimplicial object $H(R'/k, H(B,R/k,E))$ one similarly sees that (H) maps by homotopy equivalence to $H(R'/k, B\otimes_R E)$. This completes the proof of the lemma. □

The lemma applies to the case where $R' = M_k(R)$, the $k\times k$ matrices in R. The required (simplicial) bimodules are given in this case by the 'row vectors' and 'column vectors', respectively. Hence we have a homotopy equivalence

$$H(R/k, A) \simeq H(M_k(R)/k, F\otimes_R A\otimes_R E) \simeq H(M_k(R)/k, M_k(A)) .$$

This homotopy equivalence is compatible with stabilization (stabilization is given on $M_k(R)$, resp. $M_k(A)$, by adding 1, resp. 0, in the lower right corner), one sees this by comparing stabilization with the maps involved in the lemma.

The map from $N^{cy}(\widehat{GL}_k(R), M_k(A))$ to $H(M_k(R)/k, M_k(A))$ given by

$$\underbrace{\widehat{GL}_k(R) \times \ldots \times \widehat{GL}_k(R)}_{p} \times M_k(A) \longrightarrow \underbrace{M_k(R) \otimes_k \ldots \otimes_k M_k(R)}_{p} \otimes_k M_k(A)$$

therefore induces a map

$$K^S(X,R) \longrightarrow H(R/k, R[X]) .$$

This map is the promised generalization of the map $K^S(X,R) \to h(X,R)$ constructed earlier. For, as one may check, it reduces to the latter in the case where R is commutative and $k = R$.

Remark. Maps like the ones here, from (unstabilized) K-theory to group homology, resp. Hochschild homology, have been constructed earlier by K. Dennis [talk at Evanston conference, January 1976, unpublished]. Dennis' constructions are somewhat different from the ones here. It remains to be seen if the maps are equivalent.

Concluding remark. It has been stressed that the material on *K*-theory described in this appendix is an analogue of the splitting theorem for $A(X)$. However the connection is more than just an analogy, both of these results may be considered as special cases of one and the same general result. To formulate this result one needs a common framework for $A(X)$ and *K*-theory.

One such common framework is a *K*-theory of 'rings up to homotopy'. This was indicated in [14] as a means of how to think about $A(X)$ in terms of what one is accustomed to from *K*-theory. In fact, it is a useful way to think about $A(X)$, occasionally: the splitting theorem for $A(X)$ was found that way (and for a while it even required the *K*-theory of rings up to homotopy in its proof - the only result about $A(X)$ so far which ever did that). In the long run the *K*-theory of rings up to homotopy may hopefully turn out to be useful as a computational tool.

The *K*-theory of rings up to homotopy does involve serious technical problems. The prime one is to give sense to the classifying space of the homotopy monoid of homotopy invertible matrices. May [8] has made a start in dealing with these problems, in particular he has given a definition and verified a few of the elementary properties. However as May states, there is difficulty in showing his definition is the correct one in the sense that it produces $A(X)$ from the appropriate ring up to homotopy. (There is an alternative framework in which to handle those technical problems, a notion of ring up to homotopy elaborating on one proposed by Segal [11, section 5]. Here that particular difficulty does not arise).

In this framework of rings up to homotopy and their *K*-theory, the general result referred to is simply propositions 6.1 and 6.2, with the abuse of allowing R to be a 'ring up to homotopy', resp. 'commutative ring up to homotopy'. Note how this explains the difference of why we get a splitting theorem in the case of $A(X)$ but not in the case of *K*-theory. We get a splitting theorem only if the map $\Omega^\infty S^\infty |X_+| \to h(X,R)$ is a homotopy equivalence. For this to hold, 'R-homology' must be stable homotopy, so R must be $\Omega^\infty S^\infty$ and we must be dealing with $A(X)$.

References.

1. D.W. Anderson, *Chain functors and homology theories*, Sympos. Algebraic Topology, Lecture Notes in Math., vol. 249, Springer Verlag, 1971, pp. 1 - 12.

2. H. Bass, *Algebraic K-theory*, Benjamin, New York (1968).

3. W. Browder, *Algebraic K-theory with coefficients Z/p*, Geometric Applications of Homotopy Theory I, Lecture Notes in Math., vol. 657, Springer Verlag, 1978, pp. 40 - 8

4. D.M. Kan, *A combinatorial definition of homotopy groups*, Ann. of Math. 67 (1958), 282 - 312.

5. R. Lee and R.H. Szczarba, *The group $K_3(Z)$ is cyclic of order forty-eight*, Ann. of Math. 104 (1976), 31 - 60.

6. J.L. Loday, *K-théorie algébrique et représentations des groupes*, Ann. Sc. Ec. Norm Sup., t. 9 (1976), 309 - 377.

7. ———, *Homotopie des espaces de concordances*, Séminaire Bourbaki, 30e année, 1977/78, n° 516.

8. J.P. May, *A_∞ ring spaces and algebraic K-theory*, Geometric Applications of Homotopy Theory II, Lecture Notes in Math., vol. 658, Springer Verlag, 1978, pp. 240 - 315

9. D.G. Quillen, *Cohomology of groups*, Actes, Congrès Intern. Math., 1970, t. 2, pp. 47 - 51.

10. ———, (Letter to Milnor), Algebraic K-theory, Lecture Notes in Math., vol. 551 Springer Verlag, 1976, pp. 182 - 188.

11. G. Segal, *Categories and cohomology theories*, Topology 13 (1974), 293 - 312.

12. ———, *Configuration spaces and iterated loop spaces*, Inventiones math. 21 (1973), 213 - 221.

13. F. Waldhausen, *Algebraic K-theory of generalized free products*, Ann. of Math. 108 (1978), 135 - 256.

14. ———, *Algebraic K-theory of topological spaces. I*, Proc. Symp. Pure Math., Vol. 32, 1978, pp. 35 - 60.

15. ———, *Algebraic K-theory of spaces*, in preparation.

16. M. Zisman, *Suite spectrale d'homotopie et ensembles bisimpliciaux*, preprint.

Pseudo-free actions, I.

by Sylvain E. Cappell[1] and Julius L. Shaneson[1]

Contents

§0. Introduction

This paper begins a study of pseudo-free group actions on spheres. By definition, an action of a finite group on a manifold is said to be pseudo-free if its restriction to the complement of a finite invariant set of points has the property that no non-trivial element of the group fixes any point; i.e. the restriction of the action to the complement of a finite set the ("singular set") is free. Future papers will be devoted to the existence and classification questions for general pseudo-free actions in the smooth, piecewise linear (P.L.), and topological category. For example, the groups admitting piecewise linear (P.L.) pseudo-free actions will be characterized. In addition to groups that can act freely (and hence pseudo-freely by suspension), it turns out that only metacyclic groups of

[1]Both authors supported by NSF Grants

order 2n, n odd can act pseudo-freely on higher dimensional spheres. In low dimensions, some examples also arise from polyhedral symmetry groups.

This paper studies piecewise linear pseudo-free actions of cyclic groups on spheres. It is not too hard to see that in this case there can be at most two singular points, and that if there is a fixed point, then the action is semi-free. Free actions have been studied extensively by (e.g. [BL], [LM], [BPW], [W1], [MTW] and references of these) and semi-free actions are (at least in principle) amenable to the methods of [R], [BP].

The present paper gives a complete classification of P.L. conjugacy class (also called "equivalence class") of P.L. actions of cyclic groups Z_m, on spheres, that are pseudo-free but neither free nor have a fixed point. Necessarily, $m = 2N$ and the sphere on which Z_{2N} acts will have even dimension. One way of constructing such actions is the followi[ng]: let γ be a <u>free</u> action of Z_{2N} on S^{2k-1}. The unique non-trivial homomorphism $\psi: Z_{2N} \to \{\pm 1\} = S^0$ determines an action of Z_{2N} on S^0. The <u>join</u> of these two actions, denoted $\Sigma\gamma$, and called the <u>twisted suspension</u> of γ will be a pseudo-free P.L. action of Z_{2N} on S^{2k}. Recall that the join is defined as the union of all (abstract) lines from points in S^{2k-1} to points in S^0, and that the action is obtained by linear extension from γ and the action on S^0. One consequence of our results is the <u>twisted desuspension theorem</u>:

<u>Theorem[1] 5.1.</u> <u>For</u> $k \geq 3$, <u>every pseudo-free action without a fixed point of</u> Z_{2N} <u>on</u> S^{2k} <u>is P.L. equivalent to a twisted suspension.</u>

To explain the complete classification, let α be a pseudo-free action of Z_{2N} on S^{2k}; then $Z_N \subset Z_{2N}$ has the singular points as fixed points. The action of Z_N on the boundary of an invariant disk neighborhood of a singular point is a free action on S^{2k-1}; call its

[1]This theorem does not hold for the quaternion group of order 8.

P.L. equivalence class the local type. In §4 (see also §7), a torsion invariant $\delta(\alpha)$ is associated to the quotient space S^{2k}/α; it lies in $V_N/\{\pm 1\}$ where V_N is the subgroup of the units of the form $u(-T)u(T)^{-1}$ of $Z[Z_{2N}]/(1+T+\ldots+T^{2N-1})$. The space S^{2k}/α is actually a manifold with an isolated singularity. In §4 it is shown that the non-singular part of this space has the (simple) homotopy type of a (fake) lens space. From the (normal invariant of) the homotopy equivalence one obtains (see §4) some Z_2-invariants $\tau_{4r}(\alpha)$, $1 \le r \le (k-1)/2$, for N even. For N odd, $\tau_1(\alpha) \in Z_4$ and $\tau_{2r}(\alpha) \in Z_2$, $2 \le r \le \begin{cases} k-1 \\ k-2 \end{cases}$ for k $\begin{cases} \text{odd} \\ \text{even} \end{cases}$ are needed. A further invariant $\nu(\alpha) \in 8Z/16Z$ is defined for k even in §6, as is an invariant $\lambda(\alpha) \in 4Z/8Z$ for N even and k odd. The classification theorem 7.1 can then be paraphrased as follows:

Theorem. A pseudo-free P.L. action of Z_{2N} on S^{2k} without a fixed point is determined up to equivalence by the local type and the invariants $\tau_{4r}(\alpha)$ for N even and $\tau_{2r}(\alpha)$ for N odd, $\delta(\alpha)$, $\nu(\alpha)$ for k even, and $\lambda(\alpha)$ for k odd and N even.

A companion theorem (7.2) determines which invariants actually arise. Here is a paraphrase of that result:

Theorem. The invariants τ_{2r} or τ_{4r}, λ and ν, where appropriate, may be varied at will. If M is a fake lens space with $\pi_1 M = Z_{2N}$ and $\delta \in V_N/\{\pm 1\}$, then M and δ are local types and torsion of an action α iff and only if there is $u(T) \in (Z[Z_{2N}]/(1+T+\ldots+T^{2N-1}))^x$ with $\Delta(M) = (T^2-1)^k u(T)u(-T)$ and $\delta = u(-T)u(T)^{-1}$.

These two results are our classifiction. Since much is known about units in $Z[Z_{2N}]/(1+T+\ldots+T^{2N-1})$, it seems possible to reduce our classification to a completely numerical form, but we do not make this explicit here. We do obtain the following numerical results:

Theorem (see 7.3 and 7.4). With a given local type and simple homotopy type, there are exactly[1] $\begin{cases} 2[\frac{1}{2}(k+1)] \\ 2^{k-1} \end{cases}$ distinct equivalence classes of pseudo free actions of Z_{2N} on S^{2k} without fixed points, for $\begin{cases} N \text{ even} \\ N \text{ odd} \end{cases}$ The simple homotopy types with a given homotopy type and local type are in 1-1 correspondence with a free abelian group of rank

$$\sum_r \phi(r),$$

where ϕ is the Euler ϕ-function and r ranges over natural numbers dividing $2N$ but not N.

In describing these results as classifications, one is of course assuming that fake lens spaces with $\pi_1 = Z_N$ have been classified. This is true for N odd, [BPW] [W1] and many partial results are given in [W1] for N even. In fact, it seems clear that the classification for N even can be completed using among other things, some of the methods to follow.

From the above results one sees that the "desuspension" of a pseudo-free action is far from unique. It is possible to give a complete analysis of the non-uniqueness, but we do not do so here.

An important ingredient of the current work is a new structure sequence for non-trivial line bundles or, more generally, for manifolds (or Poincare complexes) with boundary so that the fundamental group of the boundary has index two in that of the entire space. For the case of the total space $E = E(\zeta)$ of the $[-1,1]$-bundle associated to a non-trivial line bundle ζ over a closed orientable manifold, X^n, the first few ter

[1] $[x]$ = greatest integer in x.

of the sequence look like this:

$$[\Sigma X;G/K]\oplus L^{\psi}_{n+1}(H) \to L_{n+1}(\pi) \to S_K(E) \to [X;G/K]\oplus L^{\psi}_{n}(H) \to L_n(\pi).$$

Here K is TOP (topological category), P.L. (P.L. category) or O (smooth category) and G/K is the classifying space for stable trivializations as fiber spaces of stable K-bundle. Also, $K = \pi_1 X$ and $\psi:\pi \to \{\pm 1\}$ is the homomorphism determined by the first Stiefel-Whitney class of ζ, and $H = \psi^{-1}(1)$. The groups $L_n(\pi)$ are the usual L-groups [W1], and $L^{\psi}_{n+1}(H)$ are Browder-Livesay type groups. $S_K(E)$ denotes the structure set; i.e. equivalence classes of simple homotopy equivalences

$$h:(W,\partial W) \to (E,\partial E),$$

where two such h and h' are equivalent if there is a K-isomorphism $\phi:(W,\partial W) \to (W',\partial W')$ with $h'\phi$ homotopic to h. It is easy to envision many other potential applications of the new structure sequence.

As for L-groups, $L^{\psi}_{i} = L^{\psi}_{i+4}$. To apply the sequence to the study of group actions of cyclic groups, it is necessary to calculate $L^{\psi}_{n}(Z_N)$,

$$\psi:Z_{2N} \to \{\pm 1\}$$

the surjective map. This is done in §3 (see 3.3, 3.4, 3.5, 3.6). The results may be partially summzrized as follows:

__Theorem__. __For__ n __odd, the groups__ $L^{\psi}_{n}(Z_N)$ __vanish__. __For__ N __odd__, $L^{\psi}_{2}(Z_N)$ __has 2-torsion__ Z_2; __otherwise__ $L^{\psi}_{even}(Z_N)$ __is a free abelian group__. __The rank of__ $L^{\psi}_{2k}(Z_N)$ __is__ $N-[N/2]$ __for__ k __even and__ $N-[(N-1)/2]-1$ __for__ k __odd__.

The result on $L^{\psi}_{odd}(Z_N)$ has already been proved by F. Hegenbarth [H] by completely different means. In general, odd Browder-Livesay groups do not vanish; e.g. for ψ the quaternionic extension of Z_4, $L_3^{\psi}(Z_4,)$ is non-trivial, as are some odd groups of cyclic extensions, with the non-trivial orientation.

Browder-Livesay groups are of use in the important problem of determining which elements in surgery groups arise from normal maps of closed manifolds. There is a homomorphism $L_n(\pi,w) \overset{\bar{\sigma}}{\to} L_{n-2}(H,\psi w)$, $\psi:\pi \to \{\pm 1\}$ any non-trivial map, $H = \psi^{-1}(1)$ with the following property

Theorem (special case of 1.5). *If* $x \in L_n(\pi,w)$ *is the surgery obstruction of a normal map of closed manifolds, then* $\bar{\sigma}(x) = 0$.

Actually $\bar{\sigma}$ in this result is the composite of σ_1 of §1 and a natural map $L_n(\pi,w) \to L_n(\pi,H,w)$. Also, w has been replaced by ψw $(\psi^2 w = w)$.

It is a general fact that if an element x of $L_n(\pi,w)$ ever acts trivially on any element of $S_K(M)$ for any manifold M^n with $(\pi_1 M, w_1 M) = (\pi,w)$, then it is represented by a normal map of closed manifolds in the category K. To see this, just glue up the ends of a normal cobordism from an element of $S_K(M)$ to itself with obstruction x to obtain a normal map of closed manifolds with $\pi_1 = \pi\times Z$. Then perform surgery on an appropriate circle in the domain and range to obtai a normal map with the fundamental group π and obstruction x. Thus one has:

Corollary. *If* $\bar{\sigma}(x) \neq 0$ *for* $x \in L_n(\pi,w)$, *then* x *never acts triviall on any element of* $S_K(M)$, $(\pi_1 M, w_1 M) = (\pi,w)$.

A more subtle application of Browder-Livesay groups sometimes allows one to show that certain non-trivial obstructions are realized by closed manifolds. See [CS2].

Finally we mention another structure sequence involving Browder-Livesay groups. Let M^n be a closed manifold, and let ζ be a non-trivial line bundle classified by $\psi_\zeta : \pi = \pi_1 M \to \{\pm 1\}$. Then there is a sequence $(w = w_1 M)$:

$$\ldots \to S_K(E\times I/\partial) \to L^{\psi}_{n+1}(\pi, w) \to S_K(M) \to S_K(E(\zeta)) \to L^{\psi}_n(\pi, w).$$

Analogous exists for manifolds with boundary as well, using relative Browder-Livesay groups.

We also note that R. Kulkarni [K] has recently determined which groups act pseudo-freely and orientably on integral homology manifolds which have the homology of sphere.

§1. Browder-Livesay groups

Let H be a group and throughout this section let

$$1 \to H \to \pi \overset{\psi}{\to} \{\pm 1\} \to 1$$

be an extension by the group $\{\pm 1\}$ of order two. Let $w:\pi \to \{\pm 1\}$ be a homomorphism (possibly trivial). Let $Z[H]$ denote the integral group ring of H. Let $t \in \pi$ with $\psi(t) = -1$. For $h \in H$, let

$$\beta_t(h) = w(h)t^{-1}h^{-1}t \in \pm H \subset Z[H],$$

and extend linearly to obtain $\beta_t : Z[H] \to Z[H]$. Then the triple $(Z[H], \beta_t, w(t)t^2)$ is an <u>anti-structure</u>; i.e. $\beta_t(\beta_t(x)) = t^{-2}ht^2$ and $\beta_t(w(t)t^2) = (w(t)t^2)^{-1}$. Therefore, its algebraic L-groups ([W2], [R1]) are defined. Thus let $L_i(Z[H], \beta_t, w(t)t^2)$ be the group (properly) denoted $L_i^{\{\pm H\}}(Z[H], \beta_t, w(t)t^2)$ in [W2] for i even, and this group modulo the element represented by $\begin{pmatrix} 0 & 1 \\ (-1)^k w(t)t^2 & 0 \end{pmatrix}$ for $i = 2k+1$.

For example, $L_0(Z[H], \beta_t, w(t)t^2)$ is as usual a reduced Grothendiec group of unimodular quadratic forms, on stably based $Z[H]$ modules, whose bilinearizations satisfy

$$\phi(x,y) = w(t)t^2\beta_t(\phi(y,x)).$$

The reduction is accomplished by requiring <u>hyperbolic</u> forms to represent the trivial element.

<u>Proposition 1.1</u>. <u>Let</u> $s, t \in \pi$ <u>with</u> $\psi(s) = \psi(t) = -1$. <u>Then</u> $(Z[H], \beta_t, w(t)t^2)$ <u>and</u> $(Z[H], \beta_s, w(s)s^2)$ <u>are Morita equivalent</u>.

<u>Proof</u>. Let $s = th$, $h \in H$. Then for $g \in H$,

$$\beta_s(g) = h^{-1}\beta_t(g)h,$$

and

$$w(s)s^2 = w(t)t^2\beta_t(h)^{-1}h.$$

Therefore the two anti-structures are Morita equivalent.

It is easy to see that the Morita equivalence of Proposition 1.1 gives an isomorphism of L-groups. Therefore we define $L_n^{\psi}(H,w)$ to be the subset of the product $\prod_{\psi(t)=-1} L_n(Z[H],\beta_t,w(t)t^2)$ consisting of elements whose components agree under the isomorphisms induced by the Morita equivalences of Proposition 1.1. Thus projection on a co-ordinate gives a canonical isomorphism

$$(1.2) \qquad L_n^{\psi}(H,w) \cong L_n(Z[H],\beta_t,w(t)t^2)$$

for each choice of t. In practice we often fix t and view (1.2) as an identity. Also, if w is trivial it will usually be omitted entirely.

For $H = \{e\}$ and $w = (-1)^{n-1}$, $L_n^{\psi}(H,w)$ are precisely the obstruction groups introduced by Browder and Livesay for desuspending free involutions on spheres. For ψ a split central extension, $L_n^{\psi}(H,w)$ is just the usual L-group $L_{n+1-w(t)}(H,w|H)$; this is the case explicitly considered by Wall in [W1]. Groups $L_n^{\psi,h}(H,w)$, $L_n^{\psi,p}(H,w)$, $L_n^{\psi,A}(H,w)$, exist, $A \subset Wh(H)$ any subgroup that is closed under the outer automorphism of H determined by ψ. For example, for the first type one omits consideration of Whitehead torsion considerations, for the second one allows the underlying modules to be projective. The various Browder-Livesay groups are related by the analogues of [Sh 1,4.1].

These Browder-Livesay groups are related to L-groups by the following exact sequence:

$$(1.3) \quad \to L_n(\pi,w) \xrightarrow{\psi^!} L_{n+1}(\pi,H;w\psi) \xrightarrow{\sigma_1} L_{n-1}^{\psi}(H,w) \xrightarrow{j} L_{n-1}(\pi,w) \to$$

Here the second term is a relative L-group with the indicated product as orientation character.

A geometric proof of (1.3) can be derived from material in Chapter 9, 11 and 12 of [W2]. Here we shall merely discuss the definitions and some properties of the various maps. From these and basic facts about surgery theory, it is actually not too difficult for the reader t supply a direct proof of (1.3). Alternatively, one might formulate and prove everything in (1.3) algebraically, as in [R2] for ordinary surgery theory. This would have the virtue that it would apply in more general algebraic settings.

The map $\psi^!$ has the following property: Let $(X^n, \partial X^n)$ be a manifold or simple Poincare pair for which $(\pi_1 X, w_1 X)$ is identified wi (π, w). Let (f,b)

$$f:(M,\partial M) \to (X,\partial X)$$

be a degree one normal map, with $f|\partial M:\partial M \to \partial X$ a simple homotopy equivalence. (Actually a simple homology equivalence over $Z\pi_1 X$ will do; see [CS1].) Let $E(\xi)$ be the total space of the $[-1,1]$-bundle ξ determined by ψ; i.e. ξ is induced from the universal line bundle over $\mathbb{RP}^\infty = K(\{\pm 1\},1)$ by the homotopy class of maps inducing ψ on π_1 Let $S(\xi)$ be the associated S^0-bundle. Then (f,b) is covered by an induced normal map $(\hat{f},\hat{b})$,

$$\hat{f}:(E(f^*\xi),S(f^*\xi),E(f^*\xi|\partial M)) \to (E(\xi),S(\xi),E(\xi|\partial X)).$$

Further, on the third entries $\hat{f}$ induces a simple homotopy equivalence. Clearly $\pi_1 S(\xi) = H$, and $\pi_1 E(\xi) = \pi_1 X = \pi$. Therefore the relative surgery obstruction $\sigma(\hat{f},\hat{b})$ is defined, and one has

$$\psi^!(\sigma(f,b)) = \sigma(\hat{f},\hat{b}) \in L_{n+1}(\pi,H;w\psi).$$

To describe the map σ_1, one must first describe the problem that the obstruction groups $L_*^{\psi}(H,w)$ solve. Let V^n be a connected manifold (or even a Poincare complex of suitable type), with $(\pi_1 V, w_1 V) = (\pi, w\psi)$. Let $(Y,\partial Y) \subset (V,\partial V)$ be a (locally flat) codimension one submanifold representing the element of $H_{n-1}(V,\partial V;Z_2)$ that corresponds under Poincare duality to $\psi \in \mathrm{Hom}(\pi_1 V;\{\pm 1\}) = H^1(V;Z_2)$. Assume also that inclusion induces an isomorphism $\pi_1 Y \cong \pi_1 V$. It follows by Van-Kampen's theorem that $\pi_1(V-Y) = H$. Also, $w_1 Y = w$.

Let $h:(W,\partial W) \to (V,\partial V)$ induce a simple homotopy equivalence of W and V and a simple homology equivalence over $Z\pi$ of ∂W and ∂V. Assume that h and $h|\partial W$ are transverse to Y and ∂Y resp., and let $N = h^{-1}(Y)$. Assume also that $h|\partial N:\partial N \to \partial Y$ is a simple homology equivalence over $Z\pi$, and that $h|\partial W-\partial N:\partial W-\partial N \to \partial V-\partial Y$ is a simple homology equivalence over $Z[H]$. In this case there is a well-defined (Browder-Livesay) obstruction $\beta(h) \in L_{n-1}^{\psi}(H,w)$, that depends only upon V, ψ, and the homotopy class of h.

<u>Theorem 1.4</u>. <u>The following are equivalent for</u> $n \geq 6$ ($a \Longrightarrow b$ <u>all</u> n):

a) h <u>is homotopic relative to the boundary to</u> f, <u>transverse to</u> Y, <u>where</u> $f|f^{-1}Y:f^{-1}Y \to Y$ <u>is a simple homotopy equivalence and</u> $f|W-f^{-1}Y:W-f^{-1}Y \to V-Y$ <u>is a simple homotopy equivalence over</u> $Z[H]$; <u>and</u>

b) $\beta(h) = 0$.

<u>Note</u>: One may have $\partial Y = \phi$. Also, in a) (and similarly in the hypotheses on $h|\partial N$), the statement about $f|f^{-1}Y$ actually follows from the one about $f|W-f^{-1}Y$ and the fact that f is a simple homotopy equivalence.

The map σ_1 can be explained as follows: Let $\gamma \in L_{n+1}(\pi,H;w\psi)$.

Let $(V^n, \partial V)$ be a manifold pair, but assume further that ∂V is connected, that $\pi_1(\partial V) = H$, and that $H \subset \pi$ is the map induced by inclusion. Let (F,B),

$$F:(U,\partial_-U,\partial_+U;\partial_0U) \to (V\times[0,1],V\times 0,V\times 1;\partial V\times[0,1]),$$

with $F|\partial_-U:\partial_-U \to V\times 0$ and $F|\partial_+U:\partial_+U \to V\times 1$ simple homotopy equivalenc and $F|(\partial_0U,\partial(\partial_0U)):(\partial_0U,\partial(\partial_0U)) \to (\partial V\times[0,1],\partial V\times\{0,1\})$ a simple homology equivalence over $Z[H]$. Then $\sigma(F,B) \in L_{n+1}(\pi,H,w\psi)$ is defined, and by [W1 ,10.4], we can find (F,B) with $\sigma(F,B) = \gamma$. Then if $h_1 = F|\partial_+U:\partial_+U \to V = V\times 1$, and $h_0 = F|\partial_-U:\partial_-U \to V = V\times 0$,

$$\sigma_1(\gamma) = \beta(h_1)-\beta(h_0).$$

It is not difficult to give a geometric proof that the indicated definition is well-defined and gives a homomorphism of groups.

As to the proof of 1.4, it is essentially contained in [W1]. Alternatively, given an algebraic proof of 1.3, 1.4 would follow formally

To discuss j, let N, Y, and

$$h:(W,\partial W) \to (V,\partial V)$$

be as in 1.4 and the preceeding discussion. Let $b:\nu_W \to \eta$ be a stable bundle map covering h. Then the restriction

$$(h,b)|N: N \to Y$$

is also a normal map, and

$$j(\beta(h)) = \sigma((h,b)|N) \in L_{n-1}(H,w|H);$$

the right side is just the usual (abstract) surgery obstruction. In fact, one can take this formula as the definition.

However, $j : L_m^{\psi}(H,w) \to L_m(\pi,w)$ has a simple algebraic definition as well. For $m \equiv 0 \pmod 2$, an element of $L_m^{\psi}(H,w)$ is represented by a quadratic module (M,q) over the anti-structure $(Z[H], \beta_t, (-1)^{m/2} w(t)t^2)$, with suitable usual properties (e.g. M is stably based, the associated bilinear form is unimodular, etc. ...;

Let $M_1 = M \underset{Z[H]}{\otimes} Z[\pi]$ and set $q_1(x) = q(x)t^{-1} \in Z[\pi]$ for $x \in M = M \otimes 1 \subset M_1$. This extends uniquely to yield a stably based unimodular quadratic module (M_1, q_1) over $(Z\pi, \alpha, 1)$, $\alpha(g) = w(g)g^{-1}$ for $g \in \pi$. In terms of matrices, the associated bilinearization of q_1 is obtained from that of q by multiplying each entry on the right by t^{-1}. The map j then satisfies

$$j(x) = x_1,$$

where x and x_1 are represented by $[M,q]$ and $[M_1,q_1]$ respectively.

For m odd, an element x of $L_m^{\psi}(H,w)$ is represented by an automorphism $\alpha : (E_r, q_r) \to (E_r, q_r)$, some r, where E_r is a free module over $Z[H]$ with basis $\{x_1, \ldots, x_r, y_1, \ldots, y_r\}$ and q_r is the unique quadratic structure over $(Z[H], \beta_t, \pm w(t)t^2)$ satisfying $q_r(x_i) = q_r(y_j) = 0$ and $b_{q_r}(x_i, y_j) = \delta_{ij}$, b_{q_r} the bilinearization of q_r. Clearly $\alpha \otimes 1$ will be an automorphism of $(E_{r_1}, (q_r)_1)$. But when the basis $\{x_1 \otimes 1, \ldots, y_r \otimes 1\}$ is replaced by $\{x_1 \otimes 1, \ldots, x_r \otimes 1, t^{-1}(y_1 \otimes 1), \ldots, t^{-1}(y_r \otimes 1)\}$, $((E_r)_1, (q_r)_1)$ becomes a standard kernel over $(Z\pi, \alpha, 1)$; thus $\alpha \otimes 1$ represents an element x_1 of $L_m(\pi,w)$, and $j(x) = x_1$.

This completes the description of the maps in (1.3). As mentioned above, it is not hard to use the geometric definitions to supply a proof of (1.3). In addition one has the following vanishing property of σ_1 on surgery obstructions realized by closed manifolds:

Theorem 1.5. Let $(X^{n+1},\partial X)$ be a connected manifold pair, with $\pi_1\partial X = H$, $\pi_1 X = \pi$, $w_1 X = w\psi$, and with $H \subset \pi$ induced by inclusion $\partial X \subset X$. Let (f,b), $f:(W,\partial W) \to (X,\partial X)$ be a degree one normal map, and let $\gamma = \sigma(f,b) \in L_{n+1}(\pi,H;\psi w)$. Then

$$\sigma_1(\gamma) = 0.$$

Proof. Let V be as in the discussion of σ_1, and consider $(g,c) = (f,b) \cup id_{V\times[0,1]}$. By [W1 , Chapter 9] this normal map is cobordant in the unrestricted sense to (F,B), relative $V\times\{0,1\}$, (F,B) a normal map in $V\times[0,1]$, and

$$\sigma(F,B) = \sigma(g,c) + \sigma(id_{V\times[0,1]}) = \gamma.$$

But if $F:(W,\partial_-W,\partial_+W;\partial_0W) \to (V\times[0,1],V\times 0,V\times 1,\partial V\times[0,1])$, $F|\partial_\pm W$ is the identity of $V\times 0$ or $V\times 1$; thus from the above

$$\sigma(\gamma) = 0.$$

§2. A structure sequence for line bundles

Let X^n, $n \geq 5$, be a simple Poincare complex without boundary. Let $\pi = \pi_1 X$, $w = w_1 X$. Let ζ be a non-trivial line bundle over X; ζ is determined by its 1st Stiefel-Whitney class, which may be viewed as a homomorphism

$$\psi = \psi_\sigma : \pi \to \{\pm 1\},$$

and any homomorphism is ψ_ζ for some ζ. Let $E = E(\zeta)$ be the total space of the associated $[-1,1]$ bundle. Thus if $H = \psi^{-1}(1)$, $\pi_1(\partial E) = H$. For K = Diff, PL, TOP, let $S_K(E)$ be the structure set of E; its elements are equivalence classes of simple homotopy equivalences $h:(W,\partial W) \to (E,\partial E)$, where h_0 and h_1 are equivalent if there is a K-isomorphism $\phi:(W_0,\partial W_0) \to (W_1,\partial W_1)$ with $h_1\phi$ homotopic (as a map of pairs) to h_0. In this case, note that $\beta(h_0) = \beta(h_1)$; i.e. β is well-defined on $S_K(E)$.

The usual theory of surgery says that $S_K(X) \neq \phi$ if there is a degree one normal map into X, in the category K, with trivial surgery obstruction.

<u>Theorem 2.1</u>. $S_K(E) \neq \Phi$ <u>if and only if there is a degree 1 normal map</u>

$$\begin{array}{ccc} \nu_M & \xrightarrow{\ b\ } & \xi \\ \downarrow & & \downarrow \\ M & \xrightarrow{\ f\ } & X \end{array}$$

<u>such that</u> $\sigma(f,b) \in L_n(\pi,w)$ <u>is in the image of</u> $j:L_n^\psi(\pi,w) \to L_n(\pi,w)$.

Next suppose X has a manifold structure. Let $\eta:S_K(E) \to [E,G/K] \cong_{i*} [X;G/K]$, $i:X \subset E$ the 0-section, be the usual normal invariant [B1], [W2] G/K the classifying space for fibre homotopy trivializations of stable orthogonal, P.L., or TOP bundles as appropriate Let

$$\tau : L_{n+1}(\pi,w) \times S_K(E) \to S_K(E)$$

be the action one obtains by first applying

$$\psi^! : L_{n+1}(\pi,w) \to L_{n+2}(\pi,H,\psi w)$$

and then letting the result act as in the usual surgery sequence (e.g. see [W2,§10]). Let s always denote a surgery obstruction map in the usual surgery sequence.

<u>Theorem 2.2</u>. (Structure sequence for line bundles) <u>The sequence</u>

$$[\Sigma X;G/K]\oplus L^{\psi}_{n+1}(H,w) \xrightarrow{s-j} L_{n+1}(\pi,w) \xrightarrow{\tau} S_K(E) \xrightarrow{(i^*\eta,\beta)} [X;G/K]\oplus L^{\psi}_{n}(\pi,w) \xrightarrow{s-j} L_n($$

<u>is exact</u>.

As usual exactness at $S_K(E)$ means that the orbits of the action τ are the inverse images of points under $(i^*\eta,\beta)$, to the left of $S_K(E)$ one has groups and homomorphisms, and sets to the right.

The reader will no doubt see how to continue the sequence of 2.2 to the left; e.g. the next term would be $S_K(E\times I, E\times\partial I)$. The reader may also ask about the case where X has not the homotopy type of a manifold, but $S_K(E) \neq \phi$. In fact, there actually is a structure sequence

$$\to S_K(W^n\times I,\partial W\times I) \to [\Sigma W;G/K]\oplus L^{\psi}_{n+1}(H,w) \to L_{n+1}(\pi,w) \to$$
$$\to S_K(W) \to [W;G/K]\oplus L^{\psi}_{n}(h,w) \to L_n(\pi,w)$$

whenever $(W,\partial W)$ is a connected manifold pair and $(\pi_1 W,\pi_1(\partial W),w_1(W)) = (\pi,H,w)$. Theorem 2.2 is essentially just the special case with $W = E(\zeta)$. Since 2.2 is all that will be applied in this paper, it is all that will be proved. The proof of the general result does involve some extra difficulties.

To prove 2.2, we start at the point $[X;G/K]\oplus L_n^{\psi}(\pi,w)$. We first assert that

$$(2.3) \qquad j\circ\beta = s\circ i^*\circ\eta$$

In fact, given $h:(W,\partial W) \to (E,\partial E)$, both sides applied to $[h] \in S(E)$ are easily seen to be just the surgery obstruction of the normal map obtained by making h transverse to the 0-section $X\subset E$ and restricting to the inverse image of X.

Thus $(j-s)\circ(\beta,i^*\eta) = 0$. To prove the converse, let $x \in [M;G/K]$, $y \in L_n^{\psi}(\pi,w)$, and suppose $j(y) = s(x)$. Then $\psi^! s(x) = \psi^! j(y) = 0$ by 1.3. However, the diagram

$$\begin{array}{ccc} [E;G/K] & \xrightarrow{\ s\ } & L_{n+1}(\pi,H;\psi w) \\ \uparrow{\scriptstyle p^*} & & \uparrow{\scriptstyle \psi^!} \\ [X;G/K] & \xrightarrow{\ s\ } & L_n(\pi) \end{array}$$

commutes, $p:E \to X$ the bundle projection. Hence by the usual surgery sequence there is $z \in S_K(E)$ with $\eta(z) = p^*x$; i.e. $x = i^*\eta(z)$.

Let $\gamma \in L_{n+2}(\pi,H,\psi w)$ and write $\gamma\cdot z$ for the action of γ on z as in the usual surgery sequence. Then $i^*\eta(\gamma\cdot z) = i^*\eta(z) = x$. By properties of σ_1 in §1,

$$(2.4) \qquad \beta(\gamma\cdot z) = \sigma_1(\gamma) + \beta(z).$$

But $j\beta(z) = si^*\eta(z) = s(x) = j(y)$; i.e. $j(y-\beta(z)) = 0$. Hence by 1.3 there is γ with $\sigma_1(\gamma) = y-\beta(z)$. Thus $\beta(\gamma\cdot z) = y$. Thus $(\beta,i^*\eta)(z) = (y,x)$.

Next we show exactness at $S_K(E)$. By definition $\tau(\delta,z) = (\psi^!\delta)\cdot z$. By (2.4),

$$\beta((\psi^! \delta)\cdot z) = \sigma_1(\psi^! \delta) + \beta(z),$$

and by (1.3), $\sigma_1(\psi^! \delta) = 0$. So $\beta(\tau(\delta,z)) = \beta(z)$. From the usual surger sequence, $\eta((\psi^! \delta)\ z) = \eta(z)$.

Suppose on the other hand that $\beta(z_0) = \beta(z_1)$ and $i^*\eta(z_0) = i^*\eta(z_1$ $z_0, z_1 \in S_K(E)$. Then there is $\gamma \in L_{n+2}(\pi,H,\psi w)$ with $\gamma\cdot z_0 = z_1$. By (2.4) again $\beta(z_1) = \sigma_1(\gamma) + \beta(z_0)$. Hence $\sigma_1(\gamma) = 0$. By (1.3), $\gamma = \psi^! \delta$; i.e. $z_1 = \tau(\delta, z_0)$.

Finally, let $\delta \in L_{n+1}(\pi,w)$, and suppose first $\delta = s(x)-j(y)$. Then $\psi^!\delta = \psi^! s(x)-\psi^! j(y) = \psi^! s(x)$ by 1.3. The diagram

$$\begin{array}{ccc} [\Sigma E;G/K] & \xrightarrow{\ s\ } & L_{n+2}(\pi,H;\psi w) \\ \uparrow p^* & & \uparrow \psi^! \\ [\Sigma X;G/K] & \xrightarrow{\ s\ } & L_{n+1}(\pi,w) \end{array}$$

commutes. Therefore $\psi^!\delta = s(u)$, $u \in [\Sigma E;G/K]$. Thus from the usual surgery sequence (see next remark)

$$\tau(\delta,z) = (\psi^!\delta)\cdot z = z, \quad \text{all } z \text{ in } S_K(E).$$

Now, suppose that $\tau(\delta, [id]) = [id]$. Then $(\psi^!\delta)\cdot[id] = [id]$, so by t surgery exact sequence (see next remark) $\psi^!\delta = s(u)$, $u \in [\Sigma E;G/K]$. From the last diagram above, $s(u) = \psi^! s(i^*u)$, as $i^* = (p^*)^{-1}$; i.e. $\psi^!\delta = \psi^!(s(i^*u))$. So by 1.3, $\delta - s(i^*u) = jx$; i.e. $\delta = s(i^*u)-j(-x)$, completing the proof.

Remark. The last part showed the following strong version of exactness at $L_{n+1}(\pi,w)$:

Addendum to 2.2. *If* $\delta \in$ *Image* $(s-j)$, *then* $\tau(\delta,z) = z$ *for all* z *in* $S_K(E)$. *If* $\tau(\delta,z_0) = z_0$, z_0 *represented by the identity of* E, *then, conversely*, $\delta \in$ Image $(s-j)$.

Of course, the proof referred to the analogous statement for the usual surgery sequence

$$[\Sigma E;G/K] \xrightarrow{s} L_{n+2}(\pi,H,\psi w) \xrightarrow{\partial} S_K(E)$$

for E. The anologous strengthening of the usual statement ∂^{-1} [id] = Image s seems usually to be taken for granted without explict mention. (Note that ∂^{-1} [id] = Im s is what is proved in [Wl, 10.8]). However, it is easy to prove it using basic naturality properties of surgery obstructions and the obvious fact that a homotopy equivalence $E' \to E$ induces an isomorphism $[\Sigma E;G/K] \to [\Sigma E';G/K]$.

What does not seem to hold <u>in general</u> is the strongest possible version of exactness, namely that, in addition to the above, $\delta \in \mathrm{Im}\,(s\text{-}j)$ <u>whenever there is some</u> z, <u>not neccessarily equal to</u> z_0, <u>with</u> $\tau(\delta,z) = z$. In other words, the action of $L_{n+1}(\pi,w)$ may not in general be uniform.

§3. Calculation of some Browder-Livesay groups

Let N be an integer, and let ψ be the cyclic extension by $\{\pm 1\}$ of the cyclic group $Z_N = Z/NZ$. The aim of this section is to calculate $L_i^{\psi}(Z_N)$. The arguments will refer to the following diagram:

(3.1)
$$\begin{array}{ccccccccc}
 & & L_{i+2}(Z_{2N},\psi) & & & & L_{i+1}(Z_{2N},\psi \\
 & & \downarrow \alpha & & & & \downarrow \alpha \\
\xrightarrow{j} L_{i+1}(Z_{2N}) & \xrightarrow{\psi^!} & L_{i+2}(Z_{2N},Z_N;\ ) & \xrightarrow{\sigma_1} & L_i(Z_N) \xrightarrow{j} L_i(Z_{2N}) & \xrightarrow{\psi^!} & L_{i+1}(Z_{2N},Z \\
 & \searrow t & \downarrow \partial & & & \searrow t & \downarrow \partial \\
 & & L_{i+1}(Z_N) & & & & L_i(Z_N) \\
 & & \downarrow i_* & & & & \downarrow i_* \\
 & & L_{i+1}(Z_{2N},\psi) & & & & L_i(Z_{2N};\psi)
\end{array}$$

In (3.1) t denotes the transfer homomorphism (geometrically, this corresponds to passing the double cover .) The horizontal sequence is (1.3), the vertical ones the exact sequence for relative L-groups, and the triangles are easily seen to commute.

<u>Lemma 3.2. The transfer map</u>

$$t: L_i(Z_{2N}) \to L_i(Z_N)$$

<u>is surjective for</u> $i \not\equiv 2 \pmod 4$ <u>and is also injective for</u> i <u>odd and</u> N <u>even. For</u> $i \equiv 2 \pmod 4$, <u>the image of</u> t <u>is the kernel of the Arf-invariant</u> $c: L_2(Z_N) \to Z_2$.

To prove 3.2 let i first be odd. If $i \equiv 1 \pmod 4$ or if N is odd, $L_i(Z_N) = 0$ [L] [B1] [W2] and there is nothing to prove. In the remaining case we have a commutative diagram

$$\begin{array}{ccc}
Z_2 \cong L_3(Z) & \longrightarrow & L_3(Z_{2N}) \\
\downarrow t & & \downarrow t \\
L_3(2Z) & \longrightarrow & L_3(Z_N)
\end{array}$$

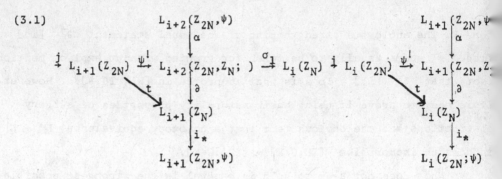

and the horizontal maps are isomorphisms (see [W2 , Chapter 14].) But the diagram

$$\begin{array}{ccc} L_{4i+3}(Z) & \longleftarrow & L_{4i+2}(e) = Z_i \\ \downarrow & \swarrow & \\ L_{4i\times 3}(2Z) & & \end{array}$$

obviously commutes, where the two unlabelled arrows are defined by taking products with S^1. By [Sh 1], the unlabelled maps are isomorphisms.

To discuss the case i even, we need appropriate formulations of some invariants and calculations of L-groups; these will also be useful for other purposes. For a finite group G, let $R(G)$ denote the complex representation ring, and let

$$\chi : L_{2k}(G) \to R(G)$$

be the map defined in [P,W2]. For μ an irreducible representation and $x \in L_{2k}(G)$, let $\chi_\mu(x)$ denote the coefficient of μ in $\chi(x)$. In case μ is one-dimensional (which will always be the case if G is abelian) it can be seen that $\chi_\mu(x)$ can be calculated as follows: let x be represented by the quadratic module (M,q) over $(Z[G],\alpha,1)$ ($\alpha(g) = g^{-1}$), and let b_q be the bilinearization of q. Then, if $M_C = M \otimes_{Z[G]} C$, C = complex numbers, let

$$\bar{b}_q(x\otimes 1, y\otimes 1) = \mu(b_q(x,y)).$$

Since $\mu(g)$ is a root of unity, $\bar{b}_q$ extends uniquely to a $(-1)^k$-skew-Hermitian complex form, also denoted $\bar{b}_q$. Then $\chi_\mu(x)$ is just the usual signature of $\bar{b}_q$; i.e. $\dim M_C^+ - \dim M_C^-$, where M_C^+ and M_C^- are maximal subspaces of M_C on which $\bar{b}_q(x,x) > 0$ or $\bar{b}_q(x,x) < 0$, respectively, k even, or $-i\bar{b}_q(x,x) > 0$ or $-i\bar{b}_q(x,x) < 0$, respectively, k odd.

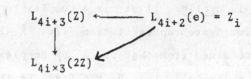

For example, for k even the coefficient of the trivial representation is the usual signature, and for k odd it is always zero.

Now assume G is <u>cyclic</u>. If λ is a complex representation of G, let $\bar{\lambda}$ be the conjugate representation (i.e. $\bar{\lambda}$ is the representati[on] whose character is obtained from that of λ by complex conjugation. In this case, $\bar{\lambda}(g) = \lambda(g^{-1})$.) Let $R_k(\pi) \subset R(\pi)$ be the subset of $\lambda \in R(\pi)$ satisfying the following conditions:

(i) $\lambda = (-1)^k \bar{\lambda}$,

(ii) $\lambda \in 4R(\pi) \subset R(\pi)$

(iii) if μ is a real one-dimensional representation, (i.e. $\mu = \bar{\mu}$) the coefficient of μ in λ is $\begin{cases} \equiv 0 \pmod 8 \\ \text{zero} \end{cases}$ if k is $\begin{cases} \text{even} \\ \text{odd} \end{cases}$.

Then the results of [W2] imply that for k even χ maps $L_{2k}(G)$ isomorphically to $R_k(\pi)$ and that for k odd, χ maps $L_{2k}(G)$ onto $R_k(\pi)$ and has kernel $L_{2k}(e) = Z_2 \subset L_{2k}(G)$.

To complete the proof of (3.2), note that the following diagram obviously commutes:

$$\begin{array}{ccc} L_{2k}(Z_{2N}) & \xrightarrow{\chi} & R(Z_{2N}) \\ \downarrow t & & \downarrow r \\ L_{2k}(Z_N) & \xrightarrow{\chi} & R(Z_N) \end{array}$$

where r is the restriction map.

If μ is an irreducible representation of Z_N, then obviously $\mu = \mu_1 | Z_N$, μ_1 an irreducible representation of Z_{2N} (e.g. if $\mu(T^2) = e^{2\pi i m/N}$, T a generator Z_{2N}, set $\mu_1(T) = e^{\pi i m/N}$.) Thus $4\mu + (-1)^k 4\bar{\mu} = r(4\mu_1 + (-1)^k 4\bar{\mu}_1)$. If μ is not real, neither is μ_1. If μ is the unique non-trivial real irreducible representation, then $\mu_1(T) = i$ is complex and, for k even,

$$8\mu = r(4\mu_1 + 4\bar{\mu}_1).$$

If μ is trivial, then μ_1 is real $(\mu_1(T) = -1)$ and, for k even,

$$8\mu = r(8\mu_1).$$

Thus $r(R_k(Z_{2N})) = r(R_k(Z_N))$, k even or odd. This proves 3.2 for $i = 2k \equiv 0 \pmod 4$. For $i \equiv 2 \pmod 4$, it remains only to see that $c\circ t = 0$, $c:L_2(Z_N) \to Z_2$ the Arf-invariant. But the Arf-invariant $c:L_2(Z_{2N},\psi) \to Z_2$ is easily seen to be surjective, and i_* in (3.1) respects the Arf-invariant. So $c\circ t = c\circ i_*\circ t = c\circ i_*\circ\partial\circ\psi^! = 0$.

<u>Theorem 3.3</u>. $L^{\psi}_{2k-1}(Z_N) = 0$.

For k even, $L_{2k+1}(Z_{2N},\psi) = 0$ by [W2]. Hence 3.3 follows immediately from (3.1) and (3.2) in this case. So let k be odd. Then if N is also odd, $\exists\, s \in Z_{2N}$ with $\psi(s) = -1$ and $s^2 = 1$. Hence $L^{\psi}_{2k-1}(Z_N) = L_{2k-1}(Z_N) = 0$ by [L,B] or [W2]. So assume further that N is even. Our proof for this case will also yield calculations of even Browder-Livesay groups.

From (3.2) and the fact that i_* respects the Kervaire invariant, $\sigma_1\circ\alpha:L_{2k+1}(Z_{2N},\psi) \to L^{\psi}_{2k-1}(Z_N)$ is onto. From [W2], $L_3(Z_{2N},\psi) = Z_2 \oplus Z_2$. Therefore, it will suffice to show that $\operatorname{Im}\psi^! \supset Z_2\oplus Z_2$, since $\operatorname{Im}\partial$ is free abelian. Note first that the map $Z[Z_{2N}] \to Z_2$ that sends the generator T to 1 factors through $Z[Z_{2N}] \to Z$ where $T \to -1$. Hence the composite $c\circ j$, $c:L_2(Z_{2N}) \to Z_2$ the Arf-invariant, factors through the Arf-invariant map $L_0(e) \to Z_2$ and this is well-known to be trivial (e.g. see [W3]). Therefore $c\circ j = 0$, so $\psi^!(L_2(e)) = Z_2$.

To find the other Z_2, the Λ be the set of irreducible complex representations of Z_{2N} such that $\lambda(T^p) = \sqrt{-1}$, where p is the largest odd number dividing N. If $x \in L_2(Z_{2N})$, let $\bar{\varepsilon}(x) = \sum_{\lambda\in\Lambda} \chi_\lambda(x) \in 4Z$, and let $\varepsilon(x) = \bar{\varepsilon}(x) \pmod{8Z} \in 4Z/8Z$. The homomorphism $\bar{\varepsilon}$ can also be described as the composite

$$L_2(Z_{2N}) \xrightarrow{tr} L_2(Z_{2N/p}) \xrightarrow{\chi_\mu} 4Z,$$

where μ sends a generator to $\sqrt{-1}$, and tr is a transfer map.

It is not hard to see that we have the commutative diagram

$$\begin{array}{ccccc}
L_2^{\psi}(Z_N) & \xrightarrow{j} & L_2(Z_{2N}) & \searrow^{\bar{\varepsilon}} & \\
\downarrow t & & \downarrow & & \\
L_2^{\psi}(Z_{N/p}) & \xrightarrow{j} & L_2(Z_{2N/p}) & \xrightarrow{\chi_\mu} & Z \\
\downarrow & & \downarrow (\mu)_* & \nearrow_{\Sigma} & \\
L_0(e) & \xrightarrow{\times i} & L_2(Z[i]) & &
\end{array}$$

where Σ denotes signature and the unlabelled map is induced by $Z[Z_{2N/p}] \to Z$ that sends a generator to -1. (Recall N is even, p odd Since $L_0(e) = 8Z$, detected by the signature also, $\bar{\varepsilon}\circ j(L_2^{\psi}(Z_N)) \subseteq 8Z$. Thus $\varepsilon\circ j = 0$; i.e. ε factors through $\mathrm{Im}\,\psi^!$. Also, $\varepsilon(L_2(e)) = 0$ as $\bar{\varepsilon}(L_2(e)) = 0$.

Let λ be the irreducible representation with $\lambda(T) = \sqrt{-1}$. Let $x \in L_2(Z_{2N})$ with $\chi(x) = 4\lambda - 4\bar{\lambda}$. Then $\bar{\varepsilon}(x) \neq 0$; hence $\psi^! x \neq 0$. $\chi(t\,x) = (4\lambda-4\bar{\lambda})|Z_N = 0$. By 3.2, $c(t(x)) = 0$. Hence by the calculati of $L_2(Z_N)$ quoted above, $t(x) = 0$. Thus $\psi^! x \in \mathrm{Im}\,\alpha$, and $\psi^! x \notin \psi^!(L_2(e))$ as $\varepsilon(L_2(e)) = 0$. Hence $\mathrm{Im}\,\psi^! \supset \mathrm{Im}\,\alpha = Z_2 \oplus Z_2$, which completes the proof of 3.3.

The preceding argument, together with (3.2) for $i \equiv 0(4)$, and and the fact that $L_0(Z_{2N}, \psi) = 0$ for $N \neq 1$ even also proves the next result:

<u>Theorem 3.4</u>. <u>For</u> N <u>even the sequence</u>

$$0 \to L_2^{\psi}(Z_N) \to L_2(Z_{2N}) \xrightarrow{(t,c,\varepsilon)} L_2(Z_N) \oplus Z_2 \oplus Z_2 \xrightarrow{c\pi_1} Z_2 \to 0$$

is exact. (Here π_1 = projection on the 1st summand.)

For N odd, $L_3(Z_{2N};\psi) = 0$. Also, $L_0(Z_{2N},\psi) = Z_2$, but as $L_3^{\psi}(Z_N) = 0 = t_3(Z_N)$, Im α must equal Im $\psi^!$ = $Z_2 = L_3(Z_{2N})$. Thus

Theorem 3.5. For N odd the sequence

$$0 \to L_2(Z_N) \xrightarrow{j} L_2(Z_{2N}) \xrightarrow{t} L_2(Z_N) \xrightarrow{c} Z_2 \to 0$$

is exact.

Finally, $L_1(Z_{2N},\psi) = 0$, and $L_2(Z_N)$ maps surjectively to $L_2(Z_{2N},\psi) \cong Z_2$ (by the Arf-invariant). Therefore we have:

Theorem 3.6. The sequence

$$0 \to L_0^{\psi}(Z_N) \xrightarrow{j} L_0(Z_{2N}) \xrightarrow{t} L_0(Z_N) \to 0$$

is exact.

§4. Local representations, homotopy type, torsion and normal invariants for pseudo-free actions

Let $\alpha: Z_q \times S^m \to S^m$ be a pseudo free action; i.e. α restricts to free action on the complement of a finite set. The smallest such set will be called the set of singular points. By Smith theory, if α is not free, there must be at least two such points. It turns out [CS3] that the "general" pseudo-free action of a finite group has exactly two singular points, and that only a small list of exceptional groups can act with more than two singular points. In particular, if α is not free, there are two singular points and $m = 2k$. Using the Lefshetz fixed point theorem and Meyer-Vietoris sequence, it is not hard to show that α has no fixed point if and only if it has an orientation reversi element. In particular q will be even in this case.

From now on we study the case that α is piecewise linear (P.L.), pseudo free, without fixed points. Let $m = 2k$, and $q = 2N$, and let $k \geq 3$. (The smooth case is similar in theory, but the usual homotopy-theoretic difficulties make it hard to give a complete classification. The topological case is nearly identical, provided one assumes locally nice behavior about the singular points. The general topological case will be discussed in a further case.) Two P.L. actions will be called <u>equivalent</u> if there is an equivariant P.L. homeomorphism of th spaces on which the groups act. Let $T \in Z_{2N}$ be a fixed generator.

Let x and y be singular points of α. Let D_x and D_y be disjoint disks about x and y, so that $D_x \cup D_y$ is invariant and so that on $D_x \cup D_y$ α is given by radical extension from $\partial D_x \cup \partial D_y$; note that $T(D_x, x) = (D_y, y)$. (For example, $D_x \cup D_y$ may be obtained as the inverse image of a 2nd derived stellar neighborhood of the image of x i the quotient space. Further, $D_x \cup D_y$ is unique up to equivariant ambien sotopy.) Let $X_\alpha = S^{2k}\text{-Int}(D_x \cup D_y)$, and let $W_\alpha = X_\alpha/Z_{2N}$ be the quotient space. Then X_α is simply-connected and Z_{2N} acts freely on it. Hence we have an identification $\pi_1 W = Z_{2N}$.

Definition. The P.L. homeomorphism type of ∂W_α, respecting the given identification $\pi_1(\partial W_\alpha) = Z_N \subset Z_{2N}$, will be called the local type at the singular point (or just the "local type") of α. It is a P.L. invariant of α.

Note that $\partial W_\alpha = \partial D_x/Z_N$. Thus ∂W_α is an example of a fake lens space, i.e. a manifold with universal covering space a sphere. Let W_α have the local orientation that lifts to the restriction to X_α of a fixed orientation of S^{2k}.

Now the finite complex X_α has the homotopy type of S^{2k-1}. For definiteness we suppose that a homotopy equivalence, i.e. an orientation, has been chosen; it will not be hard for the reader to see that the invariants of our classification do not depend on this choice. The Reidemeister torsion $\Delta(X_\alpha)$ is defined as in [M1];

$$\Delta(X_\alpha) \in Q[Z_{2N}]/(\Sigma),$$

Σ the sum of the group elements, and is well-defined modulo $\{\pm T^i \mid 0 \le i \le 2N-1\}$. As in [W2,§14], it is not hard to see that

$$\Delta(X_\alpha) = (T-1)^k u_\alpha(T),$$

where $u_\alpha(T) \in (Z[Z_{2N}]/(\Sigma))^x$ is a unit in the indicated ring, defined modulo $\{\pm T^i\}$.

The notation is chosen to suggest that $u_\alpha(T)$ is a kind of polynomial. For example, if $u_\alpha(T)$ has the representative $\sum_{i=0}^{2N-1} a_i T^i$ then $u_\alpha(T)u_\alpha(-T) \in Z[Z_N]/(1+T^2+\ldots+T^{2N-2})$ is defined as the element represented by $(\sum_{i=0}^{2N-1} a_i T^i)(\sum_{i=0}^{2N-1} (-1)^i a_i T^i)$, a polynomial in T^2; this is easily seen to be well-defined in terms of $u_\alpha(T)$.

Proposition 4.1. The unit $u_\alpha(T)$ depends only on the equivalence class of α. Further

$$(T^2-1)^k u_\alpha(T)u_\alpha(-T) = \Delta(\partial W_\alpha) \quad (\text{modulo} \quad \{\pm T^{2i}\}).$$

The first part of 4.1 is standard. To prove the 2nd part one could proceed directly and first show that ∂W_α is simple homotopy equivalent to the 2-fold cover of W_α. Instead we apply the next result:

Proposition 4.2. There is a polarized fake lens space L^{2n-1} with $\pi_1 L = Z_{2N}$, and a degree one simple homotopy equivalence

$$h_\alpha:(W_\alpha,\partial W_\alpha) \to (E(\zeta),\partial E(\zeta)),$$

ζ the non-trivial line bundle over L. Further, h_α is unique up to homotopy and composition with a bundle map over a degree one homotopy equivalence of polarized fake lens spaces.

Assuming 4.2, $\Delta(\partial W_\alpha) = \Delta(\partial E(\zeta))$ and $\Delta(W_\alpha) = \Delta(L)$. But $\partial E(\zeta)$ is just the 2-fold cover of L_1. It follows (exercise left to the reader that if $p(T) = \Delta(L)$, then $\Delta(\partial E(\zeta)) = p(T)p(-T\ ) = (T-1)^k(-T-1)^k u_\alpha(T)u_\alpha$ $= (T^2-1)^k u_\alpha(T)u_\alpha(-T)$ (mod $\{\pm T^{2i}\}$).

Proof of 4.2. X_α has the homotopy type of S^{2k-1}. Hence, as in [W1 ,§14], $\exists$ a homotopy equivalence $h_1:L_1 \to W$, L_1 a (polarized) linear lens space, respecting the orientations and identifications of π_1. In particular, we have a normal map

$$\begin{array}{ccc} \nu_{L_1} & \xrightarrow{\ b\ } & \xi \\ \downarrow & & \downarrow \\ L_1 & \xrightarrow{\ h_1\ } & W_\alpha \end{array} .$$

But W_α is actually a simple Poincare complex of dimension $2k-1$.

For it is a Poincare complex by virtue of its homotopy equivalence to L_1. It is not hard to see by duality that, with respect to cellular bases for chains and cochains over $Z[\pi_1 W_\alpha]$, the torsion of the cap product with a fundamental cycle must have the form $x-x^*$, some $x \in Wh(Z_{2N})$. It is well known that such elements must be trivial (see [B] and compare [W1].)

Hence $\sigma(h_1,b) \in L_{2k-1}(Z_{2N}) = L^s_{2k-1}(Z_{2N})$ is defined. But [W2 ,5.4] the natural map

$$L_{2k-1}(Z_{2N}) \to L^h_{2k-1}(Z_{2N})$$

is monic, and $\sigma(h_1,b)$ obvious is in the kernel, since h_1 is a homotopy equivalence. Hence $\sigma(h_1,b) = 0$. Therefore there is a simple homotopy equivalence

$$h_2 : W_\alpha \to L\ ,$$

L a fake lens space, respecting the identifications of π_1.

Let ζ be the non-trivial line bundle over L. Then $\partial E(\zeta)$ is the double cover of L with $\pi_1(\partial E(\zeta)) = Z_N$. Hence $h_2|\partial W_\alpha$ lifts to a homotopy equivalence

$$h_3 : \partial W_\alpha \to \partial E(\zeta),$$

also inducing the identity on $\pi_1 = Z_N$. The obstructions to extending h_3 to

$$h_\alpha : (W_\alpha, \partial W_\alpha) \to (E(\zeta), \partial E(\zeta))$$

with $p_\zeta \circ h_2$, p_ζ the projection ζ, lie in $H^i(W_\alpha, \partial W_\alpha; \pi_{i-1}(D^1)) = 0$; thus h_α exists. Since p_ζ is a simple homotopy equivalence, h_α induces a simple homotopy equivalence $W_\alpha \to E(\zeta)$; by duality it is therefore a simple homotopy equivalence of pairs. We leave the uniqunes part of 4.2 as an exercise.

<u>Definition</u>. Let $\phi_L(\alpha) \in S(E(\zeta)) = S_{PL}(E(\zeta))$ be the class represented by h_α ($= h_\alpha(L)$ really). It is easy to see that $\phi_L(\alpha)$ depends only upon the equivalence class of α.

Let $g_L : L \to L^{2k-1}_{2N}(d_\alpha, 1, \ldots, 1)$ be a degree one homotopy equivalence of polarized lens spaces; g_L exists and is unique up to homotopy (see [W2].) Restriction to the common (2k-2)-skeleton induces an isomorphism

$$[L_{2N}(d_\alpha, 1, \ldots, 1); G/PL] \cong (L_{2N}(1, \ldots, 1); G/PL].$$

<u>Definition</u>. Let $t_{2r}(\alpha)$ and $T(\alpha)$ be $t_{2r}((g_L)_*(\eta(\phi_L(\alpha)))$ and $T((g_L)_*(\eta(\phi_L(\alpha)))$ respectively. Here

$$t_{2r} : [L_{2N-1}(1, \ldots, 1); G/PL] \to \begin{cases} Z_2 & r \text{ odd} \\ Z_{2^e} & n \text{ even} \end{cases},$$

(e-1) the largest power of 2 in N and

$$T : [L_{2N-1}(1, \ldots, 1); G/PL] \to Z_{2^{e+1}}$$

are as in [W2 ,§14], and $(g_L)_*$ is the map induced by <u>composition</u> with g_L. (Thus $(g_L)_*\eta(\phi_L(\alpha)) = \eta(g_L \circ h_\alpha) = \eta(g_L) + (g_L^{-1})^*\eta(h_\alpha)$.) It is easy to see that $t_{2r}(\alpha)$ and $T(\alpha)$ are also invariants of the equivalen class of α.

If $x \in Z_{2^a}$, let $\langle x \rangle$ denote the image in $Z_{2^{a+1}}$ of a representative between 0 and $2^a - 1$. Let $\tilde{g}_L : \partial E(\zeta) \to L_N(d_\alpha, 1, \ldots, 1)$ be a map of double covers lying over g_L.

<u>Definition</u>: Let $\tau_{4r}(\alpha) \in Z_2 = \{0,1\}$ be given by

$$\tau_{4r}(\alpha) = \frac{1}{2^{e-1}} (t_{4r}(\alpha) - \langle t_{4r}((\tilde{g}_L)_* \eta(h_\alpha | \partial W_\alpha)\rangle), \quad r > 1,$$

and $\tau_4(\alpha) = \frac{1}{2^e} (T(\alpha) - \langle T((\tilde{g}_L)_* \eta(h_\alpha | \partial W_\alpha)\rangle)$, N even. These are also invariants of the equivalence class of α. Note that for N odd, $\tau_{4r} = t_{4r}$, $r > 1$. For N odd, let $\tau_4(\alpha) = T_4(\alpha) \in Z_4$, and let $\tau_{4r-2}(\alpha) = t_{4r-2}(\alpha)$.

§5. The twisted desuspension theorem

Let γ be a free P.L. action of Z_{2N} on S^{2k-1}. The homomorphism $Z_{2N} \overset{\psi}{\rightarrow} \{\pm 1\} = S^0$ determines the action on S^0 given by $T(+1) = -1$. The join of the actions will be called the <u>twisted suspension of</u> γ <u>and denoted by</u> $\Sigma\gamma$. It is a pseudo-free P.L. action of Z_{2N} on S^{2k} with the suspension points as singular points.

<u>Theorem 5.1</u>. <u>Every pseudo-free P.L. action of</u> Z_{2N} <u>on</u> S^{2k} <u>with no fix point equivalent to a twisted suspension, for</u> $k \geq 3$.

However, the classification results below will show that a given pseudo-free action has infinitely many "desuspensions" ($N \neq 1$).

To prove Theorem 5.1, let $\alpha: Z_{2N} \times S^{2k} \rightarrow S^{2k}$ be a P.L. pseudo-free action. Let

$$h_\alpha : (W_\alpha, \partial W_\alpha) \rightarrow (E(\zeta), S(\zeta))$$

be the simple homotopy equivalence provided by 4.2. Then

$$\beta(h_\alpha) \in L^{\psi}_{2k-1}(Z_N)$$

is defined. By §3, $L^{\psi}_{2k-1}(Z_N) = 0$. Hence h_α is homotopic to h, where h is transverse to the zero-section $L \subset E(\zeta)$ and

$$h: (W_\alpha; h^{-1}L, W_\alpha - h^{-1}L) \rightarrow (E(\zeta), L, E(\zeta) - L)$$

is a simple homotopy equivalence of triples; note that

$$\pi_1(W_\alpha - h^{-1}L) = \pi_1(E(\zeta) - L) = Z_N.$$

Let $P = h^{-1}L$. Then the s-cobordism theorem implies that the inclusion $P \subset W_\alpha$ extends to a P.L. homeomorphism

$$f: E(\xi) \rightarrow W_\alpha,$$

ξ the non-trivial line bundle over P with associated [-1,1]-bundle $E(\xi)$. In the notation of §4, $E(\xi) = W_{\Sigma\gamma}$, where γ is the free action on S^{2k-1} with quotient space the fake lens space P. Hence f lifts to an equivariant P.L. homeomorphism

$$\tilde{f}: X_{\Sigma\gamma} \to X_{\alpha}$$

of spaces with free Z_{2N} actions. By radial extension, $\tilde{f}$ extends to an equivarient P.L. homeomorphism

$$\bar{f}: (S^{2k}, \Sigma\gamma) \to (S^{2k}, \alpha),$$

which proves 5.1.

The first case in which not every action desuspends is the quaternion group Q of order 8. In this case $L_3^{\psi}(Z_4) \neq 0$, and there are pseudo-free actions Q, of the simple homotopy type of a linear action, that do not desuspend at all.

§6. Signature invariants

For our first invariant of this type, let N be even and k odd, and let $\alpha: Z_{2N} \times S^{2k} \to S^{2k}$ be a pseudo free P.L. action. Let $h_\alpha:(W_\alpha, \partial W_\alpha) \to (E(\zeta), \partial E(\zeta))$, ζ a line bundle over L, be as in 4.2. Since $L^{\psi}_{2k+1}(Z_N) = 0$, it may be assumed that h_α is transverse regular to L and that if $P = h_\alpha^{-1}L$, then $h_\alpha|P:P \to L$ is a simple homotopy equivalence. Thus P is a (polarized) fake lens space.

Let ρ_P denote the Atiyah-Singer invariant of P [AS] [P] [W1]; i.e. the Atiyah-Singer invariant of the action of Z_{2N} on S^{2k-1} with quotient space P. Then ρ_P is a complex function defined on the non-trivial elements of Z_{2N}. In fact, let

$$\delta = \delta_{2N}: Z_{2N} \to C$$

be 1-dimensional representation $\delta(T) = e^{2\pi i/2N}$. Then $\rho_P \in Q[\delta]/(1+\delta+\ldots+\delta^{2}$ Q = rational numbers.

If N is a power of 2, let

$$a(P) = a_{N/2} - a_0,$$

where $a_0 + a_1\delta + \ldots + a_{2N-1}\delta^{2N-1} \in Q[\delta]$ projects to ρ_P. If $N = 2^{e-1}p$, p odd, let $a(P) = a(\tilde{P})$, $\tilde{P}$ the covering space of P with $\pi_1\tilde{P} = Z_{2^e}$. In general $a(P)$ can also be defined as follows: Let $A_1 = \{j \mid 0 \le j \le 2N-1 \text{ and } j \equiv 2^{e-2} \pmod{2^e}\}$ and let $A_2 = \{j \mid 0 \le j \le 2N-1 \text{ and } j \equiv 0 \pmod{2^e}\}$. Then

$$a(P) = \sum_{j \in A_1} a_j - \sum_{j \in A_2} a_j .$$

Note that the coefficients a_i will be unique if we require $a_0 = 0$. Hence we may sometimes write

$$\rho_P = a_1\delta+\ldots+a_{2N-1}\delta^{2N-1} \in Q[\delta].$$

Proposition 6.1. Modulo $8Z\subset Q$, $a(P)$ depends only upon α.

To discuss the values of this invariant, let $\xi = e^{\pi i/M}$ be a primitive 2M-th root of unity, $M = 2^{e-1}$. For any ℓ, one has

$$\left(\frac{1+\xi}{1-\xi}\right)^{\ell} = \sum_{i=0}^{M-1} c_i\xi^i \in Q(\xi)$$

$c_i = c_i(\ell,N)$ unique rational numbers.

Lemma. The c_i are all integers, and are odd if ℓ is odd.

Proof. $\frac{1+\xi}{1-\xi} = \xi+\xi^2+\ldots+\xi^{N-1} \in Z(\xi)$; hence the c_i are all (rational) integers. Further,

$$\left(\frac{1+\xi}{1-\xi}\right)^2 = -(M-1)-(M-2)\xi-(M-4)\xi^2-\ldots-(-M+2)\xi^{M-1};$$

i.e. $\left(\frac{1+\xi}{1-\xi}\right)^2 \equiv 1 \pmod{2Z(\xi)}$.

Thus, for ℓ odd, $\left(\frac{1+\xi}{1-\xi}\right)^{\ell} \equiv \left(\frac{1+\xi}{1-\xi}\right) \pmod{2Z(\xi)}$. Define $w(k,N) = c_{M/2}(k,N)$.

Proposition 6.2. Let $d_\alpha \equiv s^k$ (modulo 2M). Then

$$a(P) = \tfrac{1}{2}(-1)^{(s-1)/2}w(k,N) \pmod{4Z}.$$

Recall that d_α was the unique integer mod 2N so that W_α had the polarized[1] homotopy type of $L_{2N}(d_\alpha,1,\ldots,1)$. As in [W1], modulo 2N d_α is the image of $u_\alpha(T)$ under the map $Z[Z_{2N}]/(\Sigma) \to Z_{2N}$ sending T to 1. (Thus the notation for linear lens spaces is compatible with [M1].) Note also that s exists and is unique modulo 2M, as $(Z_{2M})^x$ has order $2^{e-1} = M$.

Definition[1]. For k odd and N even, let $\lambda(\alpha) \in 4Z/8Z$ be the element

[1]Changing the orientation on $X_\alpha \simeq S^{2k-1}$ will change $a(P)$, d_α, and s by a sign. Hence $\lambda(\alpha)$ remains unaltered.

represented by

$$a(P) - (-1)^{(s-1)/2}(\tfrac{1}{2}\, w(k,N)),$$

where $h_\alpha^{-1}L = P$ as above.

It remains to prove 6.1 and 6.2. Clearly it suffices to consider the case $N = M = 2^{e-1}$ $(e \geq 2)$, and this is assumed from now on.

Proof of 6.1. By the uniqueness part of 4.2, it suffices to see the effects of composition with a bundle map over a homotopy equivalence of fake lens space and of changing h_α by a homotopy. The former clearly has no effect, since the transverse inverse image P of the zero sectic will remain unaltered. Hence it suffices to consider a map

$$h:(W_\alpha,\partial W_\alpha) \to (E(\zeta),\partial E(\zeta)),$$

homotopic to h_α, transverse to L, so that

$$h|h^{-1}L:H^{-1}L_1 \to L$$

is a simple homotopy equivalence. Let $P' = h^{-1}L$. Then we wish to show that $a(P') \equiv a(P) \bmod 8Z$.

Let $H:W_\alpha \times [0,1] \to E(\zeta) \times [0,1]$ be a homotopy of h_α to h; i.e. $H(x,0) = (h_\alpha(x),0)$ and $H(x,1) = (h(x),1)$. Then

$$\beta(H) \in L_{2k}^{\psi}(Z_N)$$

is defined, and $j\beta(H) \in L_{2k}(Z_{2N})$ is just the usual surgery obstruction of the map

$$H|H^{-1}(L\times[0,1]):H^{-1}(L\times[0,1]) \to L\times[0,1];$$

here H is assumed transverse to $L \times [0,1]$. As in [P] or [W1], a traightforward argument from the definitions shows that* $\rho_{P'}-\rho_P = \chi(j\beta(H))$. Therefore

$$a(P')-a(P) = \chi_\mu(j\beta(H)),$$

here $\mu = \delta^{N/2}$. But by definition,

$$\chi_\mu(j\beta(H)) \equiv \varepsilon(j\beta(H)) \quad \text{in} \quad 4Z/8Z.$$

y 3.4, $\varepsilon\circ j = 0$. Thus $a(P') \equiv a(P) \bmod 8Z$. This proves 6.1.

Proof of 6.2. Let

$$g:P \to L_{2N}^{2k-1}(r_1,\ldots,r_k)$$

e a canonical homotopy equivalence to any linear lens space (e.g., the composite of $h_\alpha|P$ and the canonical map g_{L_1} of §4.) Then we first assert that, modulo $4Z$, $a(P)$ depends only upon the normal cobordism class of g. It is not assumed that g preserves polarizations.

To see this let $g':P' \to L_{2N}(r_1,\ldots,r_k)$ be normally cobordant to g_P. Then let

$$H:V^{2k} \to L_{2N}(r_1,\ldots,r_k) \times [0,1]$$

be the normal cobordism. By surgery it may be assumed that H is k-connected. View $Z[\sqrt{-1}] = Z[i]$ as a $Z[Z_{2N}]$ module via a map $Z[Z_{2N}] \to Z[i]$ sending a suitable generator of Z_{2N} to i, and let $M = H_{k+1}(F;Z[i]) = K_k(V;Z[i])$, the latter equality by definition. (Note that $M = K_R(V;Z[Z_{2N}]) \otimes_{Z[Z_{2N}]} Z[i]$.) Let b be the $(-1)^k$-skew-Hermitian form on M given by intersection numbers. Then by (Novikov) additivity,

*This equation is initially valid only in the ring of functions on $Z_{2N}-\{1\}$, which contains $Q[\delta]/(1+\delta+\ldots+\delta^{2N-1})$. But it holds in $R(Z_{2N})$ as well, if we view $\rho_P, \rho_{P'} \in Q[\delta] = R(Z_{2N})$ as above, because the $(-1)^k$-symmetry implies that $\chi(j\beta(H))$ has no constant term, k odd.

direct definition of χ_μ indicated in §3, and the definition of the ρ-invariant,

$$a(P')-a(P) = \Sigma$$

where Σ denotes the signature of the skew Hermitian form (M,b). But by the usual argument involving Poincare duality, b is non-singular; i.e. $\det b \in Z[i]$ is a unit. (In fact, a definition of b is as the form whose adjoint Ad_b is the composite

$$K_k(V;Z[i]) \to K^k(V, V, Z[i]) \to K^k(V;Z[i]),$$

where the first map is the Poincare duality map, an isomorphism, and the second is the natural map, an isomorphism because $K^*(\partial V;Z[i])$ is trivial. Further, $b = b_q$, q a quadratic form defined from the normal data. Hence $\Sigma \equiv 0 \pmod 4$, i.e. $a(P) \equiv a(P') \pmod 4$.

(It is well-known that the signature Σ of a unimodular skew-Hermitian form over $Z[i]$ associated to a quadratic form is trivial mod 4. One way to see this is to observe that the signature of the imaginary part of the form, viewed as a real form on the underlying Z-module, will have signature 2Σ and will be even and unimodular. Hence $2\Sigma \equiv 0 \pmod 8$ [O].)

Let $g_P : P \to L^{2n-1}_{2N}(d_\alpha,1,\ldots,1)$ be the canonical homotopy equivalence. Since k is odd, d_α is odd, and the group $(Z_{2N})^X$ of units has order 2^{e-1}, there is s with $d_\alpha \equiv s^k \bmod (2N)$. Hence (see [M1]) there is a homotopy equivalence $L^{2n-1}_{2N}(d_\alpha,1,\ldots,1) = L^{2n-1}_{2N}(s,s,\ldots,s)$; the latter and $L^{2n-1}_0 = L_{2N}(1,\ldots,1)$ are diffeomorphic, but not as polarized lens spaces. Hence there is a homotopy equivalence

$$g: P \to L^{2n-1}_0 .$$

Since $L_{2k-1}(Z_{2N}) = 0$ (or by [W1 ,14.E.4]), g is normally cobordant to a simple homotopy equivalence

$$g':P' \to L_0^{2n-1}.$$

Let us now change the polarizations of P' so that g' respects polarizations. Then the ρ-invariant with the original polarization is obtained from the invariant $\rho_{P'}$ with respect to the new one by replacing δ by δ^s; therefore, with the new polarization for P' we have

$$a(P) \equiv (-1)^{(s-1)/2} a(P') \pmod 4$$

According to [W1 ,14E.9], $\Sigma P'$ is normally cobordant to a fake lens space Q^{2k+1} which is the total space of a circle bundle over a fake complex projective space. The canonical maps $\Sigma P' \to L_0^{2n+1} = L_{2N}^{2n+1}(1,\ldots,1)$ and $Q \to L_0^{2n+1}$ will be simple homotopy equivalences; hence the normal cobordism between them will have an obstruction $x \in L_{2k+2}(Z_{2N})$, and $\chi(x) \in R_{k+1}(Z_{2N})$. As before,

$$\rho_Q - \rho_{\Sigma P'} = \chi(x) | (Z_{2N} - \{1\}).$$

Let $\rho_{P'} = \sum_{i=1}^{2N-1} a_i \delta^i$. Then $a_i = -a_{2N-i}$, by symmetry. Hence

$$\delta_P(T) = \sum_{i=1}^{N-1} (a_i + a_{N-i})\xi^i,$$

where $\xi = e^{\pi i/N}$ as above. Hence the coefficient of $i = \xi^{N/2}$ in $\delta_{P'}(T)$ is $2a(P')$. So we wish to show that this coefficient is $w(k,N)$, mod 8.

Let $Y = (1+\xi)/(1-\xi)$; then combining 14E.8 of [W1] with what we already have and multiplying by Y^{-1}, we obtain $(t = [k/2] + 1)$

$$\rho_{P'}(T) = Y^k + \sum_{r=1}^{t} 8s_{4r}(Y^{k-2r} - Y^{k-2r-2}) + Y^{-1}(\chi(x)(T)),$$

where s_{4r} are integers. Since $Y = \xi+\dots+\xi^{N-1}$ and $Y^{-1} = -\xi+\xi^2-\dots-\xi^{N-1}$ are in $Z(\xi)$, the coefficient of $\xi^{N/2}$ in the 2nd term is trivial mod 8. For the third term, it is not hard to see that since $\chi(x) \in R_{k+1}(Z_{2N})$,

$$\chi(x)T = 8\gamma_0 + 4\sum_{i=1}^{N/2-1} \gamma_i(\xi^i-\xi^{N-i}),$$

where $\gamma_0,\dots,\gamma_{(N/2)-1}$ are (rational) integers. Direct inspection then shows that in $Y^{-1}\chi(x)T$ the coefficient of $\sqrt{-1} = \xi^{N/2}$ is also divisible by 8. Hence $2a_{p'} \equiv w(k,N)$ (modulo 8), which proves 6.2.

Our final invariant is defined for k even, $k \geq 4$, and N arbitrary. Let

$$h_\alpha:(W_\alpha,\partial W_\alpha) \to (E(\zeta),\partial E(\zeta))$$

as before, and again suppose that h_α is transverse regular to L and that $h|h^{-1}L:h^{-L} \to L$ is a simple homotopy equivalence. Let $P = h^{-1}L$. If N is a power of two, let

$$a_0+a_1\delta+\dots+a_{2N-1}\delta^{2N-1} \in Q[\delta]$$

project to ρ_P, and define

$$b(P) = a_N-a_0.$$

If $N = 2^{e-1}p$, p odd, let $b(P) = b(\tilde{P})$, $\tilde{P}$ the covering space of P with $\pi_1\tilde{P} = Z_{2^e}$. In general $b(P)$ has a definition analogous to that of $a(P)$ above; we leave it to the reader to make this explicit. A priori, $b(P)$ is a rational number.

<u>Proposition 6.3</u>. <u>Modulo</u> $16Z$, $b(P)$ <u>depends only upon</u> α. <u>Further</u>, $b(P) \in 8Z$.

In view of Proposition 6.3 we define an invariant $\nu(\alpha) \in 8Z/16Z$ to be the element represented by $b(P)$. One can show that $b(P)$ has the following direct definition: Let $r\tilde{P} = \partial W$, as elements in the oriented cobordism of $K(Z_{2^e},1)$. Then

$$b(P) = \frac{1}{r}(2\sigma(W)-\sigma(\tilde{W})),$$

where $\tilde{W}$ is the 2-fold cover of W and σ denotes the usual (integral) signature.

To prove 6.3, assume $N = 2^{e-1}$ and suppose $P' \subset W_\alpha$ is obtained similarly to P, possibly using a different homotopy equivalence to a line bundle over a different fake lens space. Then the 2-fold cover of P' and P will be s-cobordant (compare §5) in that they are both s-cobordant to the local type of the singular points. So $\rho_P|(Z_N-\{1\}) = \rho_{P'}|(Z_N-\{1\})$.

On the other hand, P and P' are also normally cobordant. Hence, if $x \in L_{2k}(Z_{2N})$ is the obstruction of a normal cobordism,

$$\rho_{P'}-\rho_P = \chi(x)|(Z_{2N}-\{1\}).$$

It is not hard to see from this that $b(P')-b(P) = 8(a_0-a_N)$; e.g. evaluate at T and compare constant terms in an expansion with respect to the basis $\{1,\xi,\ldots,\xi^{N-1}\}$ of $Q(\xi)$ over Q. However $\chi(x)(T^2) = 0$; this implies $a_0 + a_N - 2a_{N/2} = 0$, similarly. So a_0-a_N is even; hence $b(P') = b(P) \bmod 16$.

In fact without using $\chi(x)|(Z_{2N}-\{1\}) = 0$, this argument remains valid to show that if P' is any fake lens space normally cobordant to P, then $b(P) \equiv b(P')$ modulo 8.

Let $g_P: P \to L^{2k-1}_{2N}(d,1,\ldots,1)$ be the canonical homotopy equivalence. Then $t_{2k-2}(g_P) = 0$, by [W1 ,14E.4]. Hence, since $L_{2k-3}(Z_{2N}) = 0$, it follows from the description of normal invariants in [W1] that P will be normally cobordant to a fake lens space P' obtained by taking a join of the standard action on S^1 with one, Q say, of dimension 2k-3; note that $k \geq 4$. Then

$$\rho_{P'} \equiv \rho_Q\left(\frac{1+\delta}{1-\delta}\right) \quad \text{and so}$$

$$\rho_{P'}(T) = \rho_Q(T)(\xi+\ldots+\xi^{N-1}).$$

Since $\rho_Q(T)$ has the form $\sum_{i=1}^{N-1} a_i\xi^i$ with $a_i = a_{N-i}$, it is clear that the constant term in the corresponding expression for $\rho_{P'}(T)$ will be trivial; i.e. $b(P') = 0$. Hence $b(P) \equiv 0 \pmod 8$. This proves 6.3

§7. Classification of pseudo-free actions of cyclic groups

To state the classification theorem with minimal relations among the invariants, a slight reformulation of the torsion invariant is needed. Let $V_N \subset (Z[Z_{2N}]/(\Sigma))^x$ be the subgroup of units of the form $u(-T)/u(T)$, $u(T)$ a unit in $Z[Z_{2N}]/(\Sigma)$. Let $\delta(\alpha)$ be the element in $V_N/\{\pm 1\}$ represented by $u_\alpha(-T)u_\alpha(T)^{-1}$, note that this is a well defined invariant of P.L. equivariant equivalence class even though $u_\alpha(T)$ is only defined modulo $\{\pm T^j\}$.

Theorem 7.1. Let α and α' be pseudo-free P.L. actions of Z_{2N} on S^{2k}, $k \geq 3$, without fixed points. Then there is an equivariant P.L. homeomorphism $(S^{2k},\alpha) \to (S^{2k},\alpha')$ if and only if the following hold:

(i) α and α' have the same local type at the singular points;

(ii) $\tau_{4r}(\alpha) = \tau_{4r}(\alpha')$ for $1 \leq r \leq (k-1)/2$;

(iii) For k even $\nu(\alpha) = \nu(\alpha')$;

(iv) For N odd, $\tau_{4r-2}(\alpha) = \tau_{4r-2}(\alpha')$ for $2 \leq r \leq (k-1)/2$

(v) For N even and k odd, $\lambda(\alpha) = \lambda(\alpha')$;

(vi) $\delta(\alpha) = \delta(\alpha')$.

Recall that the local type is a P.L. equivalence class of fake lens spaces with $\pi_1 = Z_N$ defined in §4. The elements $\tau_{4r}(\alpha) \in Z_2$, $\nu(\alpha) \in 8Z/16Z$, $t_{4r-2}(\alpha) \in Z_2$ and $\lambda(\alpha) \in 4Z/8Z$ were defined in §4 and §6 and shown to be P.L. invariants. Note also that the classification up to orientation preserving P.L. homeomorphism will be the same as the action of the generator of Z_{2N} gives an equivariant orientation reversing P.L. homeomorphism of every pseudo-free action without fixed points. In theorem 7.1, it is to be understood, for example, that if k is not even, (iii) is omitted, etc.

Theorem 7.2. Let M^{2k-1}, $k \geq 3$, be a fake lens space with $\pi_1 M = Z_N$. Let $\delta \in V_N/\{\pm 1\}$, $\tau_{4r} \in Z_2$, $2 \leq r \leq (k-1)/2$ be given. If k is even, let $\nu \in 8Z/16Z$ be given. If N is odd, let $\tau_1 \in Z_4$ and $\tau_{4r-2} \in Z_2$, $1 \leq r \leq k/2$ be given, but if N is even let $\tau_1 \in Z_2$ be given. If N is even and k odd, let $\lambda \in 4Z/8Z$. Then there is a pseudo-free action α of Z_{2N} on S^{2k}, without fixed points, with L as local type with $\delta(\alpha) = \delta$, with $\tau_{4r}(\alpha) = \tau_{4r}$, with $\nu(\alpha) = \nu$ for k even, with $\tau_{4r-2}(\alpha) = \tau_{4r-2}$ for N odd, and with $\lambda(\alpha) = \lambda$ for N even and k odd, if and only if there is $u(T)$ in $Z[Z_{2N}]/(\Sigma)$, so that $\Delta(M) = (T^2-1)^k u(T)u(-T)$ and $\delta = u(-T)u(T)^{-1}$.

These two theorems give a complete classification of pseudo-free actions without fixed points, at least if one assumes the classification of fake lens space to be known. In fact, fake lens spaces have been completely classified only for fundamental groups of odd order. However, case of even order can be handled using, among other things, some of the above arguments.

Let us say that α and α' have the same homotopy type if there is an equivariant homotopy equivalence of (S^{2k},α) and (S^{2k},α'). If the induced homotopy equivalence of W_α and $W_{\alpha'}$ is simple, α' and α will be said to have the same simple homotopy type. Theorem 7.2 says that once the local type L and δ are given completely, there exist actions with the other invariants taking arbitrary values. Clearly the simple homotopy type of α determines $\delta(\alpha)$, and will be seen in the pr of 7.1 to be determined it. Hence:

Theorem 7.3. With a given local type and simple homotopy type, there are exactly

$$\begin{cases} 2^{[\frac{1}{2}(k+1)]} \\ 2^{k-1} \end{cases}$$

distinct equivalence classes of pseudo-free actions of Z_{2N} on S^{2k}, $k \geq 3$, without fixed points, for

$$\begin{cases} N \text{ even} \\ N \text{ odd} \end{cases}$$

The simple homotopy types in with a given homotopy type and local type can be enumerated as follows: Let $V_N^{\#}$ be the units with $u(T)u(-T) \equiv 1$ (modulo $\{\pm T^j\}$). Then the indicated simple homotopy types are in 1-1 correspondence with (a coset of) the image of the map $V_N^{\#} \to V_N/\{\pm 1\}$ induced by $u(T) \to u(-T)u(T)^{-1}$. One can show that this image contains $(V_N/\{\pm 1\})^2$ and so has the same rank as $V_N/\{\pm 1\}$. Some transfer arguments show that at most $V_N/\{\pm 1\}$ has a single Z_2 as torsion, and that it is not in the image of the above map. Finally, transfer arguments and the Dirichlet unit theorem can be applied to calculate the rank of V_N. The result obtained gives:

Theorem 7.4. The simple homotopy types of pseudo-free actions of Z_{2N} on S^{2k} with given homotopy type and local type are in 1-1 correspondence with a free abelian group of rank

$$\frac{1}{2} \sum_r \phi(r),$$

where $\phi(r)$ is the Euler ϕ-function and r ranges over positive integers that divide $2N$ but don't divide N.

To prove theorem 7.1, note that the "only if" part has already been proven. So let α and α' have the same invariants. Let M be the common local type at the fixed points. Then by (4.1), and the definitio of $\delta(\alpha)$,

$$\Delta(M) \equiv (T^2-1)\delta(\alpha)u_\alpha(T)^2, \text{ modulo } \{\pm T^i\}$$

and similarly for α'. Hence $u_\alpha(T)^2 \equiv u_{\alpha'}(T)^2$. But $(Z[Z_{2N}]/(\Sigma))^x/\{\pm T$ has no torsion as a group. (This is well known; e.g. pass monomorphicall to a product of rings of integers in cyclic torsion fields and apply the Dirichlet unit theorem.) So $u_\alpha(T) \equiv u_{\alpha'}(T)$. Since W_α and $W_{\alpha'}$ have the homotopy type of fake lens spaces with torsions $(T-1)^k u_\alpha(T)$ and $(T-1)^k u_\alpha(T)$, it follows (see [M1]) that W_α and $W_{\alpha'}$ are of the same simple homotopy type.

Hence there is a fake lens space L with $\pi_1 L = Z_{2N}$ and with simpl homotopy equivalences

$$h_\alpha : (W_\alpha, \partial W_\alpha) \to (E(\zeta), \partial E(\zeta))$$
$$h_{\alpha'} : (W_{\alpha'}, \partial W_{\alpha'}) \to (E(\zeta), \partial E(\zeta)),$$

representing elements $\phi_L(\alpha)$, $\phi_L(\alpha') \in S_{PL}(E)$, $E = E(\zeta)$, ζ the non-trivial line bundle. It will be shown that $\phi_L(\alpha) = \phi_L(\alpha')$. This fact suffices, for it implies that $(W_\alpha, \partial W_\alpha)$ and $(W_{\alpha'}, \partial W_{\alpha'})$ are P.L. homeomorphic. Since the quotient spaces S^{2k}/α and S^{2k}/α' are obtaine from these spaces by attaching cones on the boundaries, S^{2k}/α and S^{2k}/α' will be P.L. homeomorphic by radial extension. It follows easily from this (or else simply lift to X_α and $X_{\alpha'}$ and extend radially) that α and α' are equivalent.

Now consider the commutative diagram

$$(7.5)\qquad \begin{array}{ccccccccc}
[\Sigma L; G/PL] \oplus L^\psi_{2k}(Z_N) & \xrightarrow{s-j} & L_{2k}(Z_{2N}) & \overset{\tau}{\to} & S(E) & \xrightarrow{i^*\eta} & [L; G/PL] & \overset{s}{\to} & L_{2k-1}(Z_{2N}) \\
\downarrow{\scriptstyle (p^*,0)} & & \downarrow{\scriptstyle t} & & \downarrow{\scriptstyle r} & & \downarrow{\scriptstyle p^*} & & \downarrow{\scriptstyle t} \\
[\Sigma(\partial E); G/PL] & \xrightarrow{s} & L_{2k}(Z_N) & \to & S(\partial E) & \overset{\eta}{\to} & [\partial E; G/PL] & \overset{s}{\to} & L_{2k-1}(Z_N).
\end{array}$$

Here the upper line is our structure sequence for line bundles (2.2), with account taken of the vanishing of $L^{\psi}_{2k-1}(Z_N)$. The lower line is the usual surgery sequence, r is a restriction map, and p is the projection of ζ.

Since α and α' have the same local type, $r\phi_L(\alpha) = r\phi_L(\alpha')$. Therefore $p^*i^*\eta(\phi(\alpha')) = p^*i^*\eta(\phi(\alpha'))$. From this, (ii) and (iv) in 7.1, and the description in [W1 ,§14] of the 2-primary part of [L;G/PL], it is a simple exercise to see that $i^*\eta(\phi(\alpha)) = i^*\eta(\phi(\alpha'))$. Hence there is $\gamma \in L_{2k}(Z_{2N})$ with

$$\tau(\gamma,\phi(\alpha)) = \phi(\alpha').$$

Further, $t(\gamma)\cdot(r\phi(\alpha)) = r\phi(\alpha')$. Hence

$$\chi(t(\gamma))|(Z_{2N}-\{1\}) = \rho(r\phi(\alpha))-\rho(r\phi(\alpha')) = 0,$$

so that $t(\gamma) \in L_{2k}(e) \subset L_{2k}(Z_N)$.

Suppose first that k is odd. By 3.4 or 3.5, $t(\gamma) = 0$, since $c(t(\gamma)) = 0$ and $c|L_2(e)$ is an isomorphism. Let N be odd. Then by 3.5, $\gamma \in$ Image j. Hence γ acts trivially; i.e. $\phi(\alpha) = \phi(\alpha')$. If N is even, then

$$\varepsilon(\gamma) = \lambda(\alpha')-\lambda(\alpha);$$

this follows as an exercise in the definitions and the usual additivity arguments. Thus $\varepsilon(\gamma) = 0$, so by 3.4, $\gamma \in \text{Im}\, j$ in this case also and hence acts trivially.

Suppose that k is even. Then $t(\gamma) \in L_{2k}(e) = 8Z$. From the direct definition indicated just after 6.3, it follows that $t(\gamma) \equiv \nu(\alpha')-\nu(\alpha)$ (modulo 16Z).. Thus $t(\gamma) \in 16Z$. But the image of $Z \cong \pi_{2k}(G/PL) \to [\Sigma L;G/PL] \xrightarrow{p^*} [\Sigma/\partial E);G/PL] \xrightarrow{\S} L_{2k}(Z_N)$ is 16Z; here the first map is induced by collapsing the complement of a cell in Int $(L\times I)$ to a point. Note that $p:\partial E \to L$ is just a double covering map. So there is $x \in [\Sigma L;G/PL]$, with $sp^*(x) = t(\gamma)$. Thus $t(\gamma - s(x)) = 0$; hence by 3.6, $\gamma - s(x) \in \mathrm{Im}\, j$. Therefore γ acts trivially in this case also; i.e. $\phi(\alpha) = \phi(\alpha')$.

To prove 7.2, we first observe the necessity of the conditions on torsion. Given α, let

$$h_\alpha:(W_\alpha,\partial W_\alpha) \to (E(\zeta),\partial E(\zeta))$$

be as in 4.2. By the vanishing of $L^\psi_{2k-1}(Z_N)$, we may suppose h_α transverse to $L \subset E(\zeta)$, the zero-section, and $h_\alpha:(W_\alpha,h_\alpha^{-1}L,W-h_\alpha^{-1}L) \to (E,L,$ a simple homotopy equivalence. It follows from the s-cobordism theorem that the local type M is the double cover of L. If $u(T) = u_\alpha(T)$, then $\delta = u(-T)u(T)^{-1}$ by definition. The usual type of transfer calculation shows that $\Delta(M) \equiv (T^2-1)^k u(T)u(-T)$.

Conversely, suppose given the prescribed values of the invariants in 7.2, with the torsion condition satisfied. We claim that there is a fake lens space L with $\pi_1 L = Z_{2N}$ and $\Delta(L) \equiv (T-1)^k u(T)$. In fact, by [W1 ,14.E.3] there is a simple Poincare complex X^{2k-1} with $\pi_1 X = Z_{2N}$, with the universal cover of X having the homotopy type of S^{2k-1}, and with $\Delta(X) = (T-1)^k u(T)$. Further, there is a homotopy equivalence

$$h:L^{2k-1}_{2N}(d,1,\ldots,1) \to X.$$

Hence the surgery obstruction $\sigma(h) \in L_{2k-1}(Z_{2N})$ is defined. But $L_{2k-1}(Z_{2N}) \to L^h_{2k-1}(Z_{2N})$ is monic [W2] and $\sigma(h)$ obviously vanishes in the latter group. Hence h is normally cobordant to a simple homotopy equivalence from a fake lens space L to X; clearly $\Delta(L) = \Delta(X)$ is as desired. In particular, the double cover of L, i.e. $\partial E(\zeta)$ for ζ

he non-orientable line bundle, will have the same simple homotopy type as M.

Now consider (7.5) again. Let $x \in S(\partial E)$ be represented by a simple homotopy equivalence of M and ∂E. Clearly $p_*:[L;G/PL] \to [\partial E;G/PL]$ induces an isomorphism on the odd primary parts of these finite abelian groups. If N is odd, $[\partial E;G/PL]$ has odd order and by the description of [W1] a complete set of invariants for the 2-primary part of $[L;G/PL]$ is given by the composites $t_{2r}\circ(g_L)_*:[L;G/PL] \to Z_2$, $2 \leq r \leq (k-1)$ and $T\circ(g_L)_*$ with the image in Z_4. Further, s is trivial for k odd, and can be identified with $t_{2k-2}\circ(g_L)_*$ for k even. Here g_L is the canonical map to a linear lens space, again. Hence in this case there is an element $y \in S(E)$ with $p^*i^*\eta(y) = \eta(x)$ and with $t_{2r}((g_L)_*i^*\eta(y)) = \tau_{2r}$ for the values of r indicated in 7.2.

In case N is even, a suitable $y \in S(E)$ also exists. In this case $t_{4r-2}\circ(g_L)_*$ factors through $[\partial E;G/PL]$ and so depends only on $\eta(x)$, and the diagram

$$\begin{array}{ccc} [L;G/L] & \xrightarrow{t_{4r}} & Z_{2^e} \\ \Big\downarrow p^* & & \Big\downarrow \\ [\partial E;G/PL] & \xrightarrow{t_{4r}} & Z_{2^{e-1}} \end{array} \qquad (N = 2^{e-1}\cdot(\text{odd})),$$

the unlabelled map the natural one, commutes. Hence in this case there is a normal invariant $\bar{y}$ in $[L;G/PL]$ with $\tau_{4r}(\bar{y}) = \tau_{4r}$ and $p^*\bar{y} = \eta(x)$. Since $t:L_{2k-1}(Z_{2N}) \to L_{2k-1}(Z_N)$ is an isomorphism in this case (see 3.2), $s(\bar{y}) = 0$. Hence there is y with $p^*i^*\eta(y) = \eta(x)$ and $\tau_{4r}(y) = \tau_{4r}$.

Let $\gamma \in L_{2k}(Z_N)$ with

$$\gamma\cdot r(y) = x.$$

Clearly, γ may be varied by elements of $L_{2k}(e) \subset L_{2k}(Z_N)$, as these act trivially. Hence by (3.2), it may be supposed that $\gamma = t(\beta)$ for $\beta \in L_{2k}(Z_{2N})$. Hence, if we replace y by $\tau(\beta,y)$, we have found y with $r(y) = x$, and with $\tau_{4r}(y) = \tau_{4r}$, and $\tau_{4r-2}(y) = \tau_{4r-2}$ if N is odd.

For k odd and N even, we claim that this completes the proof. For let

$$h:(W,\partial W) \to (E,\partial E)$$

represent y. Let $\tilde{W}$ be the universal cover of W; it has a free action of Z_{2N}. Let Σ be obtained from $\tilde{W}$ by attaching a cone to each of the two boundary components. Since $\partial\tilde{W}$ is the union of two homotopy spheres, the generalized Poincare conjecture says that they are P.L. spheres; hence Σ is a manifold. It is an excercise to see that Σ is a homotopy 2k-sphere; hence $\Sigma = S^{2k}$. A pseudo-free action α of Z_{2N} on S^{2k} is then obtained by radial extension of the free action of α on $\tilde{W}$. Clearly $W = W_\alpha$ and $h = h_\alpha$; i.e. $\phi(\alpha) = y$. Hence $\tau_{4r}(\alpha) = \tau_{4r}$, $\tau_{4r-2}(\alpha) = \tau_{4r-2}$, and $M = \partial\tilde{W}/Z_{2N} = \partial_+\tilde{W}/Z_N$, $\partial_+\tilde{W}$ one of the components of ∂W, is the local type. Also, $u_\alpha(T) = u(T)$, so that $\delta(\alpha) = \delta$, since $\Delta(W_\alpha) = \Delta(L) = (T-1)^k u(T)$.

Suppose k is odd and N even. As in the preceding paragraph, let α be an action with $\phi(\alpha) = y$. Then $\tau_{4r}(\alpha)$, the local type, and $\delta(\alpha)$ are prescribed. It remains to consider $\lambda(\alpha)$. Let $\gamma \in L_{2k}(Z_{2N})$, with $t(\gamma) = 0$ and $\varepsilon(\gamma) \neq 0$; such a γ exists by 3.4. Let $y' = \tau(\gamma,$ and let α' an action with $\phi_L(\alpha') = y'$. Then, as in the proof of 7.1,

$$\lambda(\alpha')-\lambda(\alpha) = \varepsilon(\gamma) \neq 0.$$

Clearly the local type of α' is still M, as $t(\gamma) = 0$, and the other invariants are also unchanged. Hence, since $\lambda(\alpha) \in 4Z/8Z$, either α or α' will have the given invariants.

The argument for the case k even is similar. If $\phi(\alpha) = y$, then all invariants are as prescribed except possibly $\nu(\alpha)$. Let $\gamma \in L_{2k}(Z_N)$, with $t(\gamma)$ a generator of $8Z = L_{2k}(e) \subset L_{2k}(Z_N)$. Such an element exists by (3.6) or (3.2). Let $g' = \tau(\gamma, y)$ and let α' have $\phi(\alpha') = y'$ again. Since $L_{2k}(e)$ acts trivially on $S(\partial E)$, α' still has M as local type. The other invariants are unchanged, and as in the proof of (7.1),

$$t(\gamma) \equiv \nu(\alpha') - \nu(\alpha) \pmod{16}.$$

Hence either α or α' will have the given invariants in this case also.

References

[AS] M.F. Atiyah and I.M. Singer. The index of elliptic operators III Ann. of Math. 87 (1968), 546-604.

[B1] H. Bass. L_3 of a finite groups, Ann. of Math. 99 (1974), 118-153

[B2] ________. Algebraic K-theory, Benjamin, 1968.

[B1] W. Browder. Surgery on simply connected manifolds, Springer-Verlag, 1972.

[BL] ________, and G.R. Livesay. Fixed point free involutions on homotopy spheres. Bull. A.M.S. 73 (1967), 242-5 (see also Tohoku Math. Journal 25 (1973), 69-88.)

[BP] ________, and T. Petrie. Diffeomorphisms of manifolds and semi-free actions on homotopy spheres. Bull. A.M.S. 1971 (77), 160-163.

[BPW] ________, and C.T.C. Wall. The classification of free actions of cyclic groups of odd order on homotopy spheres. Bull. A.M.S. 77 (1971), 455-459.

[CS1] S.E. Cappell and J.L. Shaneson. The codimension two placement problem and homology equivalent manifolds. Annals of Math. 99 (1974), 277-348.

[CS2] ________, these proceedings.

[CS3] Pseudo-free actions, II, to appear.

[H] F. Hegenbarth, preprint, Institute for Advanced Study.

[K] R. Kulkarni, to appear.

[L] R. Lee. Computation of Wall groups, Topology 10 (1971), 149-166.

[LM] S. Lopez de Medrano. Involutions on manifolds. Springer-Verlag, 1971.

[M1] J. Milnor. Whitehead torsion. Bull. A.M.S. 72 (1966), 358-426.

[MTW] I. Madsen, C.B. Thomas and C.T.C. Wall. The topological space form problem - II. Existence of free actions. Topology 15 (1976), 375-382.

[O] O.T. O'Meara. Introduction to Quadratic forms. Springer-Verlag, 1973.

[P] T. Petrie. The Atiyah-Singer invariant, the Wall groups $L_n(\pi,1)$ and the function te^x+1/te^x-1. Ann. of Math. 92 (1970), 174-187.

[R1] A.A. Ranicki. Algebraic L theory I. Proc. London Math. Soc. 27 (1973), 101-125.

[R2] _______________. The Algebraic theory of Surgery, to appear.

[R3] _______________. Algebraic L. Theory II, Proc. London Math. Soc. 27 (1973), 126-158.

[R] M.G. Rothenberg. Differentiable group actions on spheres. Proceedings of the Institute on Algebraic Topology, Aarhus University, 1970, Vol. II, 455-475.

[Sh1] J.L. Shaneson. Wall's surgery groups for Z×G. Annals of Math. 90 (1969), 296-334.

[W1] C.T.C. Wall. Surgery on compact manifolds. Academic Press, 1970.

[W2] ___________. Classification of Hermitian forms. VI, Group Rings. Ann. of Math. 103 (1976), 1-80.

[W3] ___________. Surgery on non-simply connected manifolds. Ann. of Math. 84 (1966), 217-276.

Semi-linear Group Actions on Spheres:
Dimension Functions.

Tammo tom Dieck

1. Introduction and results.

Let G be a finite group. Basic examples of G-actions on spheres are the linear G-spheres: the unit spheres S(V) in a (complex) representation V of G. A geometric understanding of G-spaces such as S(V) is desirable In particular one might ask: Can one describe the equivariant homotopy type of S(V) purely in terms of homotopy theory, without mentioning representation theory?

In order to deal with this question and related problems the homotopy representation group V(G) is introduced. This is a Grothendieck group of equivalence classes of actions of G on spheres with addition defined by join. The main result of this paper is a partial computation of V(G): a linearity theorem for G-actions on spheres for nilpotent groups G.

Definition 1. A semi-linear sphere is a finite G-CW-space X such that for each subgroup H of G the fixed point set X^H is an n(H)-dimensional space which is homotopy-equivalent to the sphere $S^{n(H)}$. Moreover we assume that n(H) is odd and that the NH (= normalizer) action on $H_{n(H)}(X^H;Z)$ is trivial. We orient X by choosing a generator for each $H_{n(H)}(X^H;Z)$.

With this definition we want to imitate complex linear spheres. Some of the following results are valid under weaker finiteness assumptions, as the reader will easily find out.

The join $X * Y$ of semi-linear spheres X and Y is again a semi-linear sphere. If X and Y are oriented, there is a canonical way to orient $X * Y$ such that forming the "oriented join" is associative.

Definition 2. Let X and Y be oriented semi-linear spheres. They are called oriented homotopy-equivalent, in symbols $X \sim Y$, if there exists a G-map $f : X \longrightarrow Y$ such that for all subgroups H of G the degree of f^H with respect to the given orientations is one.

Let $V^+(G)$ be the semi-group of oriented G-homotopy types of semi-linear spheres with join as composition law, and let V(G) be the associated Grothendieck group.

Definition 3. V(G) is called the homotopy representation group associated to G.

Let $\phi(G)$ be the set of conjugacy classes of subgroups of G and let C(G) be the ring of all functions $\phi(G) \longrightarrow Z$.

Definition 4. Let X be a semi-linear sphere. Its dimension function $\mathrm{Dim}\, X \in C(G)$ is given by

$$(\mathrm{Dim}\, X)(H) = \frac{1}{2}\,(\dim X^H + 1)\ .$$

The assignment $X \longmapsto \mathrm{Dim}\, X$ induces a homomorphism

$$\mathrm{Dim} : V(G) \longrightarrow C(G)\ .$$

We are now able to state the main result of this note.

Theorem. Let G be a nilpotent group. Let X be a semi-linear G-sphere. Then there exists a complex representation V such that

$$\mathrm{Dim}\, X = \mathrm{Dim}\, S(V).$$

This theorem gives in particular a description of the image of Dim for nilpotent groups G. A computation of V(G) for general groups G will be given in a joint paper with Ted Petrie (see also [3]). At this point I only mention that the kernel of Dim is always a finite group, a subgroup of the (oriented) Picard group of the Burnside ring (introduced in [2]). Moreover the class of nilpotent groups is best possible in the theorem: for every non-nilpotent group there exist semi-linear spheres with non-linear dimension function.

At the Aarhus conference I had several conversations with Mel Rothenberg and Ted Petrie which stimulated my interest in semi-linear spheres. Shortly afterwards I proved the theorem above. Later the group V(G) with its name homotopy representation group has been introduced in [3]. Semi-linear spheres have been studied earlier by Rothenberg (Torsion invariants and finite transformation groups. Proc.Symp.Pure Math.32, 267-311(1978).) We draw the readers attention to Theorem 6.4 of that paper. If one only considers free actions, then the analogous group V(G,free) has been determined by Swan (Periodic resolutions for finite groups. Ann.Math.72, 267-291(1960)), V(G,free) being the kernel of the finiteness obstruction, Generators $(\hat{H}^*(BG;\mathbb{Z})) \to \tilde{K}_0(\mathbb{Z}G)$. The use of the Schur index in the proof of Proposition 1 below was communicated to the author in a different context by Jørgen Tornehave.

2. Proof of the theorem.

We recall that a finite group G is nilpotent if and only if the Sylow subgroups are normal and G is the direct product of its Sylow subgroups. For the remainder of this note let G be nilpotent.

Let R(G) be the complex representation ring of G. The irreducible complex representations V of G are of three types: either V is not isomorphic to its conjugate V^* (complex type) or V is isomorphic to V^* and there exists a conjugate-linear G-map $J : V \longrightarrow V$ with $J^2 = \mathrm{id}$ (real type) or $J^2 = -\mathrm{id}$ (quaternionic type). If V is of quaternionic type then the dimensions $\dim_C V^H$ are all even. We define a homomorphism

$$d_1 : R(G) \longrightarrow C(G) \qquad (2.1)$$

by assigning to an irreducible representation V the function $(H) \longmapsto \frac{1}{2} \dim V^H$ if V is of quaternionic type and $(H) \longmapsto \dim V^H$ otherwise.

<u>Proposition 1.</u> <u>The image of d_1 is a direct summand.</u>

<u>Proof.</u> We recall that the kernel of d is generated by elements of the type $x - \psi^k x$ where ψ^k is the usual Adams operation and where k is prime to the order $|G|$ of G (see [5]). If V is irreducible then $\psi^k V$ is irreducible and we call V and $\psi^k V$ conjugate (k prime to $|G|$ always). Let $V_1, \ldots, V_n$ be a complete set of non-conjugate irreducible representations. Then $d_1(V_1), \ldots, d_1(V_n)$ is a Z-basis for the image of d_1. We have to show that a relation

$$mx = \sum a_i \, d(V_i) \qquad (2.2)$$

in C(G) with $m \in Z$, $a_i \in Z$ implies $a_i \equiv 0 \bmod m$. Put $e_i = \frac{1}{2}$ if V_i is quaternionic and $e_i = 1$ otherwise. Now (2.2) and Frobenius reciprocity ([4], V 16.5) implies

$$\sum a_i e_i \dim V_i^H = \langle \sum a_i e_i V_i, 1_H \rangle_H$$

$$= \langle \sum a_i e_i V_i, \mathrm{Ind}_H^G 1_H \rangle_G$$

$$\equiv 0 \bmod m .$$

The representations $\mathrm{Ind}_H^G 1_H$ generate the permutation representations. Since G is nilpotent the permutation representations generate the rational representation ring [7] . We therefore have for any rational representation W

(2.3) $$\langle \sum a_i e_i V_i, W \rangle_G \equiv 0 \bmod m .$$

For each i there exists precisely one irreducible rational representation W_i such that

(2.4) $$\langle V_j, W_i \rangle = m_i \delta_{ij}$$

where m_i is the rational Schur index of W_i (see [4] , V. 14). In our case $m_i = 2$ if V_i is quaternionic and $m_i = 1$ otherwise, [6] . Hence (2.3) and (2.4) give $a_i e_i m_i = a_i \equiv 0 \bmod m$ and this was to be shown.∎

The next step in the proof of the theorem will be proposition 2. Let

(2.5) $$d : R(G) \longrightarrow C(G)$$

be the homomorphism $d(V) : (H) \longmapsto \dim V^H$.

<u>Proposition 2.</u> <u>The homomorphisms</u> Dim : V(G) ⟶ C(G) <u>and</u> d : R(G) ⟶ C(G) <u>have the same image</u>.

<u>Proof</u>. Put d(G) = image d, D(G) = image Dim. Then we see from the definitions that d(G) ⊂ D(G). We begin by showing that d(G) and D(G) have the same rank.

It is well known that the rank of d(G) equals the number c(G) of conjugacy classes of cyclic subgroups of G. It is intructive to make this explicit as follows. Let V be a complex representation. Then we have the orthogonality relation

$$|G| \dim V^G = \sum_{g\in G} V(g). \tag{2.6}$$

Here V(g) is the value of the character of V at g. We can rewrite 2.6 as

$$|G| \dim V^G = \sum_C \sum_{D<C} \mu(|C/D|)\ |D| \dim V^D \tag{2.7}$$

where C runs through the cyclic subgroups of G and μ is the Möbius function. In particular we see that the fixed point dimensions $\dim V^H$, H < G, are determined by the $\dim V^C$ for cyclic C < G. On the other hand the rational representation ring R(G;Q) has rank c(G) (see [8] , Cor. 1 of Théorème 29) and the restriction of d to R(G;Q) ⊂ R(G) is injective ([8] , Cor. on p. 119). Hence rank d(G) = c(G).

We now prove that rank D(G) ≤ c(G). The basic geometric input is a result of Borel ([1] , XIII Theorem 2.3) which says that for G = Z/p x Z/p and a semi-linear sphere X formula 2.7 is still valid if we replace $\dim V^H$ with (Dim X)(H) throughout. Assuming this theorem of Borel we want to show that (Dim X)(G) can be computed from the (Dim X)(C), C < G cyclic. We use induction over the order of G. If G is cyclic there is

nothing to prove. If G is not cyclic we use the fact that G has a normal subgroup H such that $G/H \cong Z/p \times Z/p$ for some prime p (see [4], III. 7.1); and then we apply Borels theorem. (More precisely, this inductive proof shows the following:
(Dim X)(G) is a linear function of (Dim X)(C), $C < G$ cyclic, where the coefficients of this linear function are rational numbers, independent of X. Of course, as we finally see, this linear function is actually the analogue of 2.7.)

We remark that for G of odd order we have already proved proposition 2 because, by proposition 1, d(G) is a direct summand in C(G), so there is no larger subgroup of the same rank.

For general nilpotent groups we prove the equality $d(G) = D(G)$ by induction over $|G|$, using $d(G) \subset D(G)$ and rank $d(G)$ = rank $D(G)$. To begin with we note that these facts together with proposition 1 imply $D(G) \subset d_1(G)$, where $d_1(G)$ = image of d_1. Therefore given $x \in D(G)$ we can write $x = \sum a_j d(V_j)$ with $2a_j \in Z$ if V_j is quaternionic and $a_j \in Z$ otherwise. We have to show that $a_j \in Z$ for all j. So pick a quaternionic representation V_i. Let H_i be the kernel of V_i. If $H_i \neq \{1\}$ we consider H_i-fixed points and the resulting equality for the group G/H_i. By induction we see that $a_i \in Z$. Hence we need only consider the case $H_i = \{1\}$, i. e. the case of a faithful irreducible quaternionic representation $V = V_i$. We use the following

<u>Lemma.</u> If G is not a generalized quaternion group then there exists a subgroup H of index 2 in G and an irreducible quaternionic H-module W such that $V = \mathrm{ind}_H^G W$. Moreover the restriction $\mathrm{res}_H V$ splits as $W \oplus W_1$ where W and W_1 are not conjugate.
For a proof of this lemma see [6] .

Assuming this lemma and assuming that G is not generalized quaternion we consider $res_H x$ which must have the form $res_H x = a_i(d(W)+d(W_1))+\ldots$. So, again by introduction, we see that $a_i \in Z$. It remains to prove the proposition for generalized quaternion groups G.

For such a group the faithful representations are all conjugate; they are quaternionic and two-dimensional. Let X be a semi-linear sphere. We write

$$\mathrm{Dim}\, X = a_1\, d(V_1) + \ldots + a_n\, d(V_n)$$

and assume that V_1 is faithful . We have to show that $a_i \in Z$. By induction $a_i \in Z$ for $i > 1$. By taking suitable joins with linear spheres we can assume that all the a_i are positive. If we put $X_s = \bigcup_{H \neq 1} X^H$, then

$$\dim X - \dim X_s = 4a_1. \tag{2.8}$$

We have to show that this number lies in 4Z. Now note that G has periodic cohomology with a periodicity generator $y \in H^4(BG;Z)$. If we localize the equivariant cohomology $H^*(EG \times_G X)$ with respect to y we obtain an isomorphism

$$H^*(EG \times_G X)[y^{-1}] = H^*(EG \times_G X_s)[y^{-1}]$$

and this implies that 2.8 lies in 4Z. This finishes the proof of proposition 2. ■

We have now done enough work to finish the proof of the theorem. Let X be a semi-linear sphere. By proposition 2 be can write

$$\text{Dim } X = \sum a_j \, d(V_j) \tag{2.9}$$

where V_j runs through a complete set of non-conjugate irreducible representations of G and where the a_i are integers. We have to show that the a_i are non-negative. Again by induction over $|G|$ we only have to look at the faithful V_j. But either G is cyclic, in which case the theorem is easy to prove (each element in C(G) is a linear dimension function); or there exists a subgroup H of G of prime index $|G/H|$ such that the given faithful representation V is induced from H, $V = \mathrm{ind}_H^G W$. In the latter case V is the only irreducible representation such that its restriction to H contains W as a direct summand. Hence we can check the positivity of the coefficient of V in 2.9 by restricting to H. This finishes the proof of the theorem.

References

1. Borel, A.: Fixed point theorems for elementary commutative groups. In: Seminar on transformation groups. Princeton University Press, Princeton 1960.

2. tom Dieck, T., and T. Petrie: Geometric modules over the Burnside ring. Inventiones math. 47, 273 - 287 (1978).

3. tom Dieck, T., and T. Petrie: The homotopy structure of finite group actions on spheres. Proceedings of Waterloo topology Conference 1978.

4. Huppert, B.: Endliche Gruppen I. Springer Verlag, Berlin-Heidelberg-New York 1967.

5. Lee, Chung-Nim, and A. G. Wasserman: On the groups JO(G). Mem. Amer. Soc. 1959 (1975).

6. Roquette, P.: Realisierung von Darstellungen endlicher nilpotenter Gruppen. Arch. Math. 9, 241 - 250 (1958).

7. Segal, G. B.: Permutation representations of finite p-groups. Quart. J. Math. Oxford (2), 23, 375 - 381 (1972).

8. Serre, J.-P.: Représentations linéaires des groupes finis. 2. éd. Paris: Hermann 1971.

P-FREE LINEAR REPRESENTATIONS OF P-SOLVABLE FINITE GROUPS

Stefan Jackowski and Tomasz Zukowski
Institute of Mathematics, University of Warsaw, PL-00-901 Warszawa/Poland

§1. Introduction

This note is an attempt to investigate the orbit structure of linear representations of finite groups. We generalize the theorem of Vincent and Wolf /[7] Theorem 6.1.11/. According to this theorem a solvable group G has a free representation iff for arbitrary primes p, q every subgroup of order pq is cyclic /i.e. G satisfies all pq-conditions/.

Let P be a set of primes.

1.1 **Definition** A linear complex representation V of a finite group G is called P-free if for every $v \in V$, $v \neq 0$ the order of the isotropy subgroup G_v is not divisible by any prime belonging to P.

We prove that a P-solvable group admits a P-free complex representation iff it satisfies all pq-conditions for $p,q \in P$. The proof does not use the classification arguments of Vincent and Wolf.

We recall /cf. [1] Ex.XII.11/ the definition of a maximal P-generator in the cohomology ring of a finite group. Let Z_P denote the ring of integers localized at P. An element $g \in H^q(G:Z_P)$, $q > 0$ is called a maximal P-generator iff it is invertible in the Tate cohomology. We observe that if V is a P-free representation then its Euler class $eV \in H^{2d}(G:Z_P)$, $d = \dim_C V$ is a maximal P-generator. In the simplest case $P = \{p\}$ we show that a p-solvable group has a p-free representation of the minimal possible dimension. This result is strongly related to Swan's work /cf. [5], [6]/ on groups satisfying the p^2-condition.

Throughout the paper all representations are assumed to be complex however the results can be generalized to representations over other fields. We use the standard notation from group theory as in Gorenstein's book [3].

§2. p-solvable groups with Sylow p-subgroups either cyclic or generalized quaternionic.

For the proof of the main theorem we will need some facts about the structure of finite groups satisfying the p^2-condition for a given prime p. We start from a criterion for p-nilpotence of groups with cyclic Sylow p-subgroups.

2.1 Theorem Let G be a finite group with cyclic Sylow p-subgroup G_p and such that $G_p \cap Z(G) \neq 1$. Then G_p has a normal complement in G.

The proof is based on the Burnside criterion for existence of normal complements and the following elementary lemma.

2.2 Lemma Let G be a cyclic p-group and $f: G \to G$ be an automorphism such that $f^n = id$ and $(p,n) = 1$. If there exists $c \in G$, $c \neq 1$ such that $f(c) = c$ then $f = id$. □

Proof of Theorem 2.1. According to the Burnside theorem /[7] Theorem 5.2.9/ we have to prove that G_p is central in its normalizer. Let $g \in N_G(G_p)$ be decomposed into a commutative product of its p-regular and p-unipotent parts $g = g_r g_u$. Let $f_g: G_p \to G_p$ be the inner automorphism defined by g. Then it is clear that $f_g = f_{g_r}$. The order of the automorphism f_{g_r} is prime to p and because G_p contains a central element this automorphism has a nontrivial fixed point. Therefore $f_g = f_{g_r} = id$ according to the last lemma. Hence G_p is central in its normalizer. □

In case $p = 2$ we obtain a stronger result.

2.3 Theorem Let G be a finite group with cyclic Sylow 2-subgroup. Then G_2 has a normal complement in G.

Proof. The Sylow 2-subgroup G_2 contains the unique element of order 2 which is fixed under any automorphism. Lemma 2.2 and Burnside's theorem imply that G_2 has a normal complement in G. □

We will need also a classification theorem for groups with generalized quaternionic Sylow 2-subgroups. Recall that for a given set of primes P, $O_P(G)$ denotes a maximal normal P-subgroup of G. If P is a set of primes we denote its complement by P'.

2.4 Theorem Let G be a 2-solvable group such that its Sylow 2-subgroups are generalized quaternionic and $O_{2'}(G) = 1$. Then G is isomorphic to G_2, or to the binary tetrahedral group T^*, or to the binary octahedral group O^*.

The last theorem may be obtained combining the classification theorems of Gorenstein and Walter for groups with dihedral Sylow 2-subgroups /[3] p.462/ and the Brauer-Suzuki theorem /[3] p.373/. However in the case of 2-solvable groups the classification can be obtained using simpler arguments.

Sketch of the proof of Theorem 2.4. Let G be an arbitrary 2-solvabl group of order $2^s k$, k odd, $s>1$. Suppose that G has an element of order 2^{s-1}. Following Zassenhaus /cf.[7] Lemma 6.1.9/ we prove that if $O_{2'}(G) = 1$ then G has a normal cyclic 2-subgroup K such that G/K is isomorphic to Z_2, the tetrahedral group T $(\cong A_4)$ or the octahedral group O $(\cong S_4)$. The proof of this fact proceeds by induction on the order of Sylow 2-subgroups of G. For $s = 2,3$ the required fact follows easily from the classification of groups of orders 4 and 8. Now assume that Sylow 2-subgroups of G are generalized quaternionic and let us consider three possible cases.

1. If $G/K \cong Z_2$ then $G = G_2$.
2. If $G/K \cong A_4$ then G_2 is normal in G and $G/G_2 \cong Z_3$. The order of every automorphism of the generalized quaternionic group $Q2^n$, $n>3$ is a power of 2. The group $Q8$ has exactly 3 automorphis of odd orders and they permute canonical generators of $Q8$. Therefore $G \cong SL(2,3) \cong T^*$.
3. If $G/K = S_4$ then there is a normal subgroup G' of G such that $K \subset G'$ and $G'/K \cong A_4$. It is clear that G'_2 is normal in G' and $O_{2'}(G') = 1$. Hence it follows that $G' \cong T^*$. The group G is therefore an extension of Z_2 by T and it follows that $G \cong O^*$. □

§3. P-free representations of P-solvable groups.

Let P be a set of primes. Recall that a linear representation V of a group G is P-free iff for every nonzero vector $v \in V$ the order of its isotropy group G_v is not divisible by any prime $p \in P$. Observe that one can assume that all primes belonging to P divide the order of G. Finite groups which have free representations /i.e. P = all primes/ were investigated by Vincent and Wolf /cf. [7]/. From their results it follows that if a group G has a P-free representation then it satisfies all pq-conditions for all $p,q \in P$ /cf.[7] Theorem 5.3.1/. The aim of this section is to prove the converse theore for P-solvable groups.

3.1 Theorem Let G be a P-solvable group. Then G has a P-free representation iff it satisfies the pq-conditions for all primes $p,q \in P$.

For the proof we will use the following simple characterization of P-free representations.

3.2 Lemma A representation of a group G is P-free iff its restriction to every cyclic subgroup of order p , $p \in P$ is free. □

Recall that if G satisfies the p^2-condition for a prime p then its Sylow p-subgroups are cyclic or generalized quaternionic. Hence every Sylow p-subgroup G_p contains a unique cyclic subgroup C_p of order p . Thus all cyclic subgroups of order p are conjugate in G .

The proof of Theorem 3.1 is based on induction on the order of the group. The inductive step is provided by the following lemma.

3.3 Lemma Let G be a group satisfying the p^2-condition for every $p \in P$ and let H be a normal subgroup of G such that its order is divisible by all primes belonging to P . Then G has a P-free representation iff H has such representation.

Proof. It is clear that the restriction of P-free representation is P-free. We prove that if V is a P-free representation of H then $\mathrm{ind}_H^G(V)$ is a P-free representation of G . Let $p \in P$ and $C_p \subset G$ be a subgroup of order p . Then $C_p \subset H$ as $p \in H$ and all cyclic subgroups of order p are conjugate. According to Lemma 3.2 it is enough to prove that the representation $\mathrm{res}_{C_p} \mathrm{ind}_H^G(V)$ is free. This follows easily from the double cosets formula for restriction of an induced representation. □

Proof of Theorem 3.1. We proceed by induction on the order of the group. The theorem is true for the trivial group. Suppose it is true for groups of orders smaller then the order of G . If the maximal normal P'-subgroup $O_{P'}(G)$ is nontrivial then the quotient group satisfies the assumption of the theorem and has order smaller then G . Therefore it has a P-free representation $G/O_{P'}(G) \longrightarrow GL(V)$. The composition with the natural projection defines a P-free representation of G . Now let us investigate the case $O_{P'}(G) = 1$. The group G is P-solvable so there exists a minimal normal abelian p-subgroup for some $p \in P$. It is a cyclic subgroup C_p of order p . We may assume that p is odd. Otherwise $O_{2'}(G) = 1$ and Theorems 2.3 and 2.4 imply that $G = G_2$, T^* or O^*. All these groups have canonical free representations.

We consider the centralizer $C_G(C_p)$. It is easy to see that it is a normal subgroup. Let $q \in P$ and $q \neq p$. There exists a subgroup $C_q \subset G$ of order q. The order of the subgroup C_pC_q is clearly pq Therefore this subgroup is cyclic and hence $C_q \subset C_G(C_p)$. This implie that every prime $q \in P$ divides the order of $H := C_G(C_p)$.

If $H \neq G$ then from the inductive assumption and Lemma 3.3 we infer that G has a P-free representation.

Let us assume that $H = G$ i.e. C_p is a central subgroup. As we have assumed $p \neq 2$ a Sylow p-subgroup G_p is cyclic. Theorem 2.1 implies that G_p has a normal complement in G. Let $f: G \longrightarrow G_p$ be the resulting projection. The subgroup $f^{-1}(C_p)$ is normal in G and its order is divisible by every prime $p \in P$. Therefore again if $G_p \neq C_p$ then from the inductive assumption and Lemma 3.3 we obtain that G has a P-free representation. In case $G_p = C_p$ the group is a direct product $G = G_p \times N$. Let V be a free representation of G_p and W be a P-free representation of N which exists by the inductive assumption. It is easy to verify that the external tensor product $V \otimes W$ is a P-free representation of G. □

3.4 <u>Remark</u> The standard averaging process over the Galois group provides a construction of a P-free representation of G over the ring of rational integers.

In case of $P = \{\text{all primes}\}$ we obtain the theorem of Vincent and Wolf /cf.[7] Theorem 6.1.11/. The proof given above does not solve the classification problem for corresponding groups.

As one can expect the assumption of P-solvability is very important in Theorem 3.1. The alternating group A_5 satisfies the 3^2-condition and it does not have 3-free representation. It has a 5-free representation.

§4. <u>The Euler class of a P-free representation.</u>

Let G be a group admitting a P-free representation. As we observed in §3 it satisfies the p^2-conditions for $p \in P$. Cartan and Eilenberg /[1] Ex.XII.11/ proved that such group has a maximal P-generator in its cohomology /cf.§1/. Recall also that for a n-dimensional complex representaion V of a group G its Euler class is defined and $eV \in H^{2n}(G:Z)$. From [4] Corollary 2.2 we obtain the following result.

4.1 <u>Theorem</u> Let V be a P-free representation of a group G. Then its Euler class $eV \in H^{2n}(G:Z_p)$ is a maximal P-generator. □

Let us consider the simplest case $P = \{p\}$. The smallest dimension of a maximal p-generator is called the p-period of the group. The following theorem was proved by Swan [5].

4.2 <u>Theorem</u> Let G be a group satisfying the p^2-condition. Then its p-period $p(G)$ is given as follows:

$$p(G) = \begin{cases} 2\left|N_G(G_p):C_G(G_p)\right| & \text{if } p \text{ is odd} \\ 4 & \text{if } p = 2 \text{ and } G_2 = Q2^n \\ 2 & \text{if } p = 2 \text{ and } G_2 \text{ cyclic} \end{cases} \square$$

For a group G satisfying the p^2-condition Swan [6] constructed its free action on a mod p cohomological sphere. For p-solvable groups we have the following result.

4.3 <u>Theorem</u> Let G be a p-solvable group satisfying the p^2-condition. Then G has a p-free representation such that $2\dim_C V = p(G)$.

<u>Proof</u>. The existence of a p-free representation follows from Theorem 3.1. We repeat some arguments to check more precisely the dimension of the constructed representation.

Let $p \neq 2$ and let G be a group such that $O_{p'}(G) = 1$. Then Sylow p-subgroup G_p is normal and it coincides with its centralizer /cf.[3] Theorem 6.3.2/. If W is a free representation of G_p then $V = \mathrm{ind}^G W$ is a p-free representation of G of the required dimension /cf.Lemma 3.3/. Let G be an arbitrary p-solvable group satisfying the p^2-condition /p odd/. The projection $G \to G/O_{p'}(G)$ induces an isomorphism of the cohomology with Z_p coefficients. Therefore the Euler class of the representation $G \to G/O_{p'} \to GL(V)$ is a maximal P-generator in the cohomology of G. We have also $2\dim_C V = p(G/O_{p'}) = p(G)$.

Let $p = 2$. If G_2 is a cyclic group then by Theorem 2.3 it has a normal complement. We have 1-dimensional 2-free representation $G \to G_2 \to GL(1)$. If G_2 is a generalized quaternionic group then by Theorem 2.4 $G/O_{2'}$ is isomorphic to $Q2^n$, T^* or O^*. All these groups have 2-dimensional free representations. □

Theorem 4.1 implies that p-free representations of the group G constructed in the last theorem have minimal dimensions among p-free representations of G.

REFERENCES

1. H. Cartan, S. Eilenberg: Homological Algebra. Princeton University Press 1956.

2. T. tom Dieck: Lokalisierung aequivarianter Kohomologie-Theorien. Math. Z. 121 (1971) 253-262.

3. D. Gorenstein: Finite Groups. Harper and Row, New York, London 1968.

4. S. Jackowski: The Euler class in group cohomology and periodicity. Comment. Math. Helv. 53 (1978) 643-650.

5. R.G. Swan: The p-period of a finite group. Ill. J. Math. 4 (1960) 341-346.

6. R.G. Swan: Periodic resolutions for finite groups. Ann. Math. 72 (1960) 267-291.

7. J.A. Wolf: Spaces of constant curvature. Mc-Graw Hill New York 1967.

Orientation Preserving Involutions

Czes Kosniowski and Erich Ossa

University of Newcastle upon Tyne and Gesamthochschule Wuppertal

INTRODUCTION

Equivariant bordism theory was introduced in the middle sixties by P.E. Conner and E.E. Floyd [5, 6] as a means of studying group actions on oriented manifolds. During the last decade or so many workers have contributed to the program of determining various equivariant bordism groups. As a result, we now have a fairly good picture of the equivariant bordism theory of abelian group actions so long as we avoid complications connected with the prime 2. Thus, we can study abelian group actions on non-oriented manifolds or on unitary manifolds, or even on oriented manifolds so long as the group in question is finite of odd order.

There have been just a few results for actions of even order finite abelian groups on oriented manifolds. H.L. Rosenweig [10] and P.E. Conner [4] have studied the case of the group $\mathbb{Z}/2$. Their essential results were the determination of the bordism group modulo its torsion subgroup and the theorem that all torsion is of order two. Also, it was shown how the rank of the torsion subroup may be computed in each dimension. However, there were almost no results on the structure of the torsion as a module over the oriented bordism ring Ω_* or on the geometric construction of torsion manifolds.

The intention of our work is to fill this gap by determining completely the structure of the bordism module of orientation preserving involutions. At the same time we construct <u>most</u> of the geometric generators. The result (described below) is indeed complicated, and we have to admit that a good understanding of even order abelian group actions on oriented manifolds is still not quite in sight. We believe however that our methods, which will be outlined, will lend themselves to generalizations suited to attacking the equivariant bordism modules of a wider class of even order group actions.

Our basic method is the introduction of equivariant bordism with $\mathbb{Z}/2$ singularity in the sense of D. Sullivan, The resulting theory, which in our case of involutions we denote by $\sigma_*^{(2)}$ (all), may be thought of as the bordism theory of group actions on Wall manifolds. R.J. Rowlett in [11] attempted to study such objects but R.E. Stong in [12] pointed out an error in Rowlett's approach and gave the correct definitions, which turn out to be equivalent to ours.

The theory $\sigma_*^{(2)}$ (all) is in fact comparatively easy to handle. In some sense it is equivalent to the theory of group actions on non oriented manifolds with the restriction that the slice representations are those of oriented action.

In this paper we shall state the main results and give the main ingredients required for the proof. The complete proof is long and far exceeds the space that contributors are allowed for these proceedings.

We would like to express our thanks to the Gesamthochschule Wuppertal and the University of Newcastle upon Tyne whose support of our collaboration helped to make this work possible. We also profited from conversations with several people among whom we would like to mention B.J. Sanderson, R.E. Stong and W. Lellmann.

THE MAIN RESULT

To state the main result we need some notation: Let Γ be the set of all sequences $I = (i_1, i_2, \ldots, i_{2k})$ where $k > 0$ and $i_1 \geq i_2 \ldots \geq i_{2k} \geq 0$ are integers. Define subsets of Γ as follows:-

(a) $I \in \Gamma^o$ if there is some ℓ, $0 \leq \ell < k$ such that $i_{2j-1} = i_{2j} = 2n_j + 1$ for $1 \leq j \leq \ell$ and $i_{2\ell+1} = 2n_{\ell+1} + 1 > i_{2\ell+2}$.

(b) $\Gamma^o_R = \{(2n+1, 0);\ n \geq 0\}$

(c) $\Gamma^o_F = \{(2^{s+1}t-1, 2^s-2, 0, 0);\ s, t > 0\} \cup \{(2^k-3, 0, 0, 0);\ k > 2\}$

(d) $\Gamma^o_T = \Gamma^o - (\Gamma^o_R \cup \Gamma^o_F)$.

Furthermore, let Δ be the set of all sequences $I = (i_1, i_2, \ldots, i_k)$ where $k \geq 0$ and $i_1 > i_2 \cdots > i_k > 0$ are integers such that each i_j is even but not a power of 2. We define the degrees of such sequences as $|I| = \Sigma i_j$ and the length of the sequence is denoted by $\ell(I)$. Finally, denote the bordism module of all orientation preserving involutions by $\sigma_*(\text{all})$.

MAIN THEOREM As an Ω_* module

$$\sigma_*(\text{all}) = P_* \oplus T_* \oplus F_*$$

where

(i) P_* is the Ω_* polynomial algebra on $[\mathbb{C}P^{2n}, T_{2n}]$, $n \geq 1$, where $T_{2n} : \mathbb{C}P^{2n} \to \mathbb{C}P^{2n}$ is defined in terms of homogenous co-ordinates by $T_{2n}[z_o : z_1 : \ldots : z_{2n}] = [-z_o : z_1 : \ldots : z_{2n}]$.

(ii) T_* is a free $\Omega_* \otimes \mathbb{Z}/2$ module on generators $M(I, J)$ in $\sigma_{|I|+|J|+\ell(J)-1}(\text{all})$ for $I \in \Delta$, $J \in \Gamma_T^o$.

(iii) F_* is generated by $[\mathbb{Z}/2] \in \sigma_o(\text{all})$ and by certain elements

$$r_{4m} \in \sigma_{4m}(\text{all}), \; m > 1$$

$$t(m, n; I) \in \sigma_{m+n+|I|}(\text{all}), \; m > n > 4, \; I \in \Delta.$$

The $t(m, n; I)$ are torsion elements, and the submodule of F_* generated by $[\mathbb{Z}/2]$ and the r_{4m} has as defining relations

$$M[\mathbb{Z}/2] = 0 \text{ if } M \in \text{torsion } (\Omega_*),$$

$$2r_{4m} = W_{4m}[\mathbb{Z}/2] \text{ for } m > 1,$$

where the $W_{4m} \in \Omega_{4m}$, $m \geq 1$, are suitable polynomial generators for $\Omega_*/\text{Tors } \Omega_*$. A complete list of relations for F_* will not be given here, see [8] for these. We do not know whether F_* is indecomposable as an Ω_* module. We do know that F_* cannot be written as a direct sum of Ω_* submodules which at the same time are stable under a certain set of secondary operations.

Some of the $t(m;n;I)$ can be expressed in terms of the others; a minimal set of generaters for F_* will be given now.

PROPOSITION. $(\mathbb{Z}/2) \otimes_{\Omega_*} F_*$ is the $\mathbb{Z}/2$ vector space with basis the classes of:

(i) $[\mathbb{Z}/2]$;

(ii) r_{4m} <u>for</u> $m > 1$;

(iii) $t(2m-1, 2n; (2i_1, 2i_2 \ldots, 2i_k))$ <u>for</u> $k \geq 0$ <u>and</u> $m > n > i_1 > i_2 > \ldots > i_k > 0$ <u>with</u> m, n <u>and</u> $i_1, i_2, \ldots, i_k$ <u>not a power of</u> 2.

(iv) $t(2m, 2n; (2i_1, 2i_2, \ldots, 2i_k))$ <u>for</u> $k \geq 0$ <u>and</u> $m > n, m \geq i_1 > i_2 > \ldots > i_k > 0$ <u>with</u> $i_1, i_2, \ldots, i_k$ <u>not a power of</u> 2. <u>Furthermore if</u> $k = 0$ <u>then</u> m <u>and</u> n <u>are not both a power of</u> 2, <u>while if</u> $k = 1$ <u>then</u> $\{m, n\} \neq \{2^\ell, n\}$ <u>and if</u> $k = 2$ <u>then</u> $\{m, n\} \neq \{i_1, i_2\}$.

We can construct involutions r_{4m} and $t(m,n;I)$. (Essentially $r_{4m} = [\mathbb{C}P^{2m};$ conjugation$] + [\mathbb{C}P^2;$conjugation$]^m$.) Also, we can construct the $M(I,J)$ except if $J \in \Gamma$ is of the form $J = (j_1, j_2, j_3, j_4)$ with j_1 odd, j_2 even, $j_4 = 0$ and $|J| > 1$. The construction of the $M(\emptyset, J)$ for these missing sequences would in fact yield a construction of all generators of $\sigma_*(\text{all})$.

BACKGROUND AND DEFINITIONS

We first recall briefly the more well known parts of the bordism theory of oriente involutions as developed by H.L. Rosenweig in [10] or P.E. Conner in [4].

The basic technique of equivariant bordism, due to P.E. Conner and E.E. Floyd [6], gives an exact triangle of Ω_* module homomorphisms:

$$\begin{array}{ccc} \sigma_*(\text{free}) & \longrightarrow & \sigma_*(\text{all}) \\ {\scriptstyle \partial \text{ (degree } -1)} \nwarrow & & \swarrow {\scriptstyle j} \\ & \sigma_*(\text{rel}) & \end{array}$$

Here $\sigma_*(\text{free})$ is the bordism group of free involutions, $\sigma_*(\text{all})$ the -unknown- bordism group of all involutions, and $\sigma_*(\text{rel})$ is the relative bordism group formed from involutions on manifolds with boundary in which the involution is free on the boundary.

Taking the classifying map of a free involution defines an Ω_* module isomorphism:

$$\sigma_*(\text{free}) \cong \Omega_*(B\mathbb{Z}/2).$$

For the relative group, each connected component of the fixed point set has a tubular neighbourhood mapped into the universal R^{2j} bundle $\gamma_{2j} \to BO(2j)$ for some j. This gives rise to an Ω_* module isomorphism:

$$\sigma_*(\mathrm{rel}) \cong \bigoplus_{j\geq 0} \Omega_*(D\gamma_{2j}, S\gamma_{2j}).$$

In [10] H.L. Rosenweig used the interpretation of $\sigma_*(\mathrm{rel})$ resulting from the isomorphism

$$\Omega_n(D\eta, S\eta) \cong \Omega_{n-k+1}(D(\det\eta),\ S(\det\eta))$$

which is valid for any vector bundle η of fibre dimension k. Since $D(\det \gamma_{2j}) \cong BO(2j)$ and $S(\det \gamma_{2j}) = BSO(2j)$, this enabled him to prove that all torsion in $\sigma_*(\mathrm{all})$ is of order 2.

Based on [10] P.E. Conner in [4] essentially determined $\sigma_*(\mathrm{all})$ mod Torsion. This gives the structure of $\sigma_1(\mathrm{all}) \otimes \mathbb{Z}[\frac{1}{2}]$.

PROPOSITION $\sigma_*(\mathrm{all}) \otimes \mathbb{Z}[\frac{1}{2}]$ is the direct sum of an $\Omega_* \otimes \mathbb{Z}[\frac{1}{2}]$ polynomial algebra on generators $[\mathbb{C}P^{2n}, T_{2n}]$, $(n \geq 1)$, and a free $\Omega_* \otimes \mathbb{Z}[\frac{1}{2}]$ module generated by $\mathbb{Z}/2$.

Here $T_{2n}: \mathbb{C}P^{2n} \to \mathbb{C}P^{2n}$ is defined in homogeneous co-ordinates by $T_{2n}(z_0:z_1:z_2:\ldots:z_{2n}) = (-z_0:z_1:z_2:\ldots:z_{2n})$.

We intend to generalize the above to involutions with singularities. We start by saying a few words about the (non-equivariant) bordism theory of oriented manifolds with $\mathbb{Z}/2$ singularities - which is essentially the bordism theory of Wall-manifolds put into a form more suited to the equivariant situation.

DEFINITION. An n dimensional oriented manifold with $\mathbb{Z}/2$ singularity consists of

a) a compact oriented n manifold M,

b) a decomposition of the boundary of M into regular submanifolds

$\partial M = \partial_0 M \cup \partial_1 M \cup \partial_+ M$,

such that $\partial_0 M \cap \partial_1 M = \emptyset$ and

$(\partial_0 M \cup \partial_1 M) \cap \partial_+ M = \partial(\partial_0 M \cup \partial_1 M) = -\partial(\partial_+ M)$,

c) an orientation preserving diffeomorphism $\rho: \partial_0 M \cong \partial_1 M$.

We shall usually abbreviate such an object by the pair $(M, \partial_0 M)$. It is said to be closed if $\partial_+ M = \emptyset$.

With the objects above one defines in the usual way singular bordism groups, denoted $\Omega_n^{(2)}(X, A)$, for pairs of spaces (X, A). The relation to ordinary oriented bordism theory is given in a natural exact sequence

$$0 \to \Omega_n(X,A) \otimes \mathbb{Z}/2 \to \Omega_n^{(2)}(X,A) \xrightarrow{\beta'} \mathrm{Tor}(\Omega_{n-1}(X,A), \mathbb{Z}/2) \to 0$$

where β' maps $(M, \partial_o M)$ to $\partial_o M$. We refer the reader to the article [1] and the book [2] for a general discussion of bordism theories with singularities.

The theory $\Omega_*^{(2)}(X,A)$ is a multiplicative theory with a geometrically defined multiplication. The best description of this can be found in an article by C.T.C. Wal[l] [13]. There, the reader will also find a proof that the Bockstein homomorphism $\beta: \Omega_n^{(2)}(X,A) \to \Omega_{n-1}^{(2)}(X,A)$ (obtained from β' by composition with $\Omega_{n-1}(X,A) \to \Omega_{n-1}^{(2)}(X,A)$) is a derivation with respect to the multiplication.

For a manifold with singularity, $(M, \partial_o M)$ with structure map $\rho: \partial_o M \to \partial_1 M$, we can define a manifold $\tau(M, \partial_o M)$ by identifying $\partial_o M$ with $\partial_1 M$ in M by means of ρ. In general $\tau(M, \partial_o M)$ is non-orientable. This leads to a natural transformation of multiplicative homology theories:

$$\tau: \Omega_*^{(2)}(X,A) \to \mathfrak{N}_*(X,A).$$

It is easy to see from the bordism spectral sequence that τ is always a monomorphism so that calculations in $\Omega_*^{(2)}(X,A)$ can frequently be executed with advantage in $\mathfrak{N}_*(X,A)$. In particular we shall often use the following:

$$\Omega_n^{(2)}(D\gamma_j, S\gamma_j) \subset \mathfrak{N}_n(D\gamma_j, S\gamma_j) \cong \mathfrak{N}_{n-j}(BO_j).$$

We come now to the definition of involutions with singularities.

<u>DEFINITION.</u> An n dimensional <u>involution with $\mathbb{Z}/2$ singularity</u>, $(M, \partial_o M, T)$, consists of an n dimensional oriented manifold with singularity, $(M, \partial_o M)$ and an orientation preserving involution $T: M \to M$ such that $T(\partial_i M) = \partial_i M$ for $i \in \{0, 1, +\}$ and $T\rho = \rho T$ on $\partial_o M$.

Just as in the ordinary ("non-singular") case one defines the bordism groups $\sigma_n^{(2)}$(free), $\sigma_n^{(2)}$(all) and $\sigma_*^{(2)}$(rel) of involutions with singularities. For all of these there is a universal coefficient sequence:

$$0 \to \sigma_n(\mathcal{F}) \otimes \mathbb{Z}/2 \to \sigma_n^{(2)}(\mathcal{F}) \xrightarrow{\beta'} \mathrm{Tor}(\sigma_{n-1}(\mathcal{F}), \mathbb{Z}/2) \to 0$$

where $\mathcal{F}$ denotes free, all or rel. The sequences are derived as in the non-equivariant case. Moreover, the definition of the product as in [13] carries over to make $\sigma_*^{(2)}$(free), $\sigma_*^{(2)}$(all) and $\sigma_*^{(2)}$(rel) into algebras over $\Omega_*^{(2)}$, and again the Bockstein β acts as a derivation.

From the definition of the product, the forgetful homomorphisms

$$i: \sigma_*^{(2)}(\text{free}) \to \sigma_*^{(2)}(\text{all})$$

and, more important

$$j: \sigma_*^{(2)}(\text{all}) \to \sigma_*^{(2)}(\text{rel})$$

are homomorphisms of $\Omega_*^{(2)}$ algebras commuting with the Bockstein.

Just as in the "non-singular" situation we define

$$\partial: \sigma_n^{(2)}(\text{rel}) \to \sigma_{n-1}^{(2)}(\text{free})$$

by sending $(M, \partial_0 M, T)$ to the free involution $(\partial_+ M, -\partial(\partial_0 M), T|\partial_+ M)$, where $\partial M = \partial_0 M \cup \partial_1 M \cup \partial_+ M$. Observe that ∂ commutes with β. The familiar Conner + Floyd arguments go through to prove:

PROPOSITION. The following is an exact triangle of $\Omega_*^{(2)}$ modules.

$$\begin{array}{ccc}
\sigma_*^{(2)}(\text{free}) & \longrightarrow & \sigma_*^{(2)}(\text{all}) \\
{\scriptstyle \partial \,(\text{degree } -1)} \nwarrow & & \swarrow {\scriptstyle j} \\
& \sigma_*^{(2)}(\text{rel}) &
\end{array}$$

Similarly we get $\Omega_*^{(2)}$ algebra isomorphisms

$$\sigma_*^{(2)}(\text{free}) \cong \Omega_*^{(2)}(B\,\mathbb{Z}/2)$$

and

$$\sigma_*^{(2)}(\text{rel}) \cong \bigoplus_{j\geq 0} \Omega_*^{(2)}(D\gamma_{2j}, S\gamma_{2j})$$

commuting with the Bockstein.

To simplify our language we shall describe an $\Omega_*^{(2)}$ module with β action as an $\Omega_*^{(2)}\langle\beta\rangle$ module. Here $\Omega_*^{(2)}\langle\beta\rangle$ may be thought of as the non-commutative ring generated by $\Omega_*^{(2)}$ and an element β, subject to relations $\beta^2 = 0$ and $\beta.X = \beta(X) + X.\beta$ for $X \in \Omega_*^{(2)}$.

OUTLINE OF PROOF

The first step in the proof is to describe $\sigma_*^{(2)}(\text{rel}) = \oplus\, \Omega_*^{(2)}(D\gamma_{2k}, S\gamma_{2k})$. We start with some definitions.

For $n \geq 0$ let $\tilde{\xi}_{2n}$ be the normal bundle of RP^{2n} in RP^{2n+1} equipped with the orientation on its total space induced from the standard orientation on RP^{2n+1}. Then $D(\tilde{\xi}_{2n})$ defines an element of $\Omega_{2n+1}(D\gamma_1 S\gamma_1)$. The image of this element by the mod 2 reduction

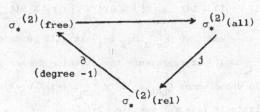

$$\Omega_{2n+1}(D\gamma_1, S\gamma_1) \to \Omega^{(2)}_{2n+1}(D\gamma_1, S\gamma_1)$$

will be denoted by $\xi_{2n} \in \Omega^{(2)}_{2n+1}(D\gamma_1, S\gamma_1)$.

Thinking of $\Omega^{(2)}_{2n+1}(D\gamma_1, S\gamma_1) \subset \mathfrak{N}_{2n}(BO(1))$ then of course, as an element of $\mathfrak{N}_{2n}(BO(1$ ξ_{2n} just represents $\tilde{\xi}_{2n}$.

Now $\tilde{\xi}_{2n}$ admits an involution $r_{2n} : \tilde{\xi}_{2n} \to \tilde{\xi}_{2n}$, the reflection in the fibres, which changes the orientation of its total space. Thus $D(\tilde{\xi}_{2n}) \in \Omega_{2n+1}(D\gamma_1, S\gamma_1)$ is an elemen of order 2 and must lie in the image of β'

$$\beta' : \Omega^{(2)}_{2n+2}(D\gamma_1, S\gamma_1) \to \mathrm{Tor}(\Omega_{2n+1}(D\gamma_1, S\gamma_1), \mathbb{Z}/2).$$

In fact, a counter image under β' can be constructed as the manifold with singularity $(D(\tilde{\xi}_{2n} \times I), D(\tilde{\xi}_{2n} \times \{0\}))$, where the (orientation preserving) structure map ρ

$$\rho: \partial_0 D(\tilde{\xi}_{2n} \times I) = D(\tilde{\xi}_{2n} \times \{0\}) \to D(\tilde{\xi}_{2n} \times \{1\}) = \partial_1 D(\tilde{\xi}_{2n} \times I)$$

is given by r_{2n}. As an element of $\Omega^{(2)}_{2n+2}(D\gamma_1, S\gamma_1)$ it will be denoted by ξ_{2n+1}. Regarde as an element in $\mathfrak{N}_{2n+1}(BO(1))$ it represents the exterior tensor product of ξ_{2n} with the "Mobius Band", (in other words $\tilde{\xi}_{2n+1} = (S^1 \times \xi_{2n})/(-1 \times r_{2n})$). The proof of the next result is not difficult (see for example [7]).

<u>LEMMA</u> $\Omega^{(2)}_*(D\gamma_1, S\gamma_1)$ <u>is the free</u> $\Omega^{(2)}_*$ <u>module on generators</u> ξ_{2n}, ξ_{2n+1}, $(n \geq 0)$.

More generally we shall describe $\Omega^{(2)}_*(D\gamma_{2k}, S\gamma_{2k})$, but first let Γ be the set of all sequences $I = (i_1, i_2, \dots, i_{2k})$ of even length $2k > 0$ with $i_1 \geq i_2 \geq \dots \geq i_{2k}$. For $I \in \Gamma$ let $\xi_I = \xi_{i_1}\xi_{i_2} \cdots \xi_{i_{2k}} \in \Omega^{(2)}_*(D\gamma_{2k}, S\gamma_{2k})$.

<u>PROPOSITION</u>

$\Omega^{(2)}_*(D\gamma_{2k}, S\gamma_{2k})$ <u>is a free</u> $\Omega^{(2)}_*$ <u>module on generators</u> ξ_I, <u>for</u> $I \in \Gamma$ <u>of length</u> $2k$. <u>Furthermore</u>

$$\beta(\xi_I) = \sum_{\substack{1 \leq j \leq k \\ i_j \text{ odd}}} \xi_{I-\delta_j}$$

<u>where</u> $I - \delta_j \in \Gamma$ is $(i_1, \dots, i_{j-1}, i_j - 1, i_{j+1}, \dots, i_{2k})$ - <u>reordered if necessary.</u>

Using the above results we can calculate the image of $\sigma^{(2)}_*(\mathrm{rel})$ in $\Omega^{(2)}_*(B\,\mathbb{Z}/2)$ under ∂. The module $\Omega^{(2)}_*(B\,\mathbb{Z}/2)$ splits as

$$\Omega^{(2)}_*(B\,\mathbb{Z}/2) \cong \Omega^{(2)}_* \oplus \tilde{\Omega}^{(2)}_*(B\,\mathbb{Z}/2).$$

The elements $\xi_n\xi_o$, $n \geq 0$, map via ∂ onto an $\Omega_*^{(2)}$ base of $\tilde{\Omega}_*^{(2)}(B\,\mathbb{Z}/2)$ while the image of $\sigma_*^{(2)}(\mathrm{rel})$ in $\Omega_*^{(2)}$ is E_* the submodule generated by $(M, \partial_o M) \in \Omega_*^{(2)}$ for which the euler characteristic $\chi(\tau(M, \partial_o M))$ is even. In fact the elements $\partial(\xi_I)$ for $I = (2^{s+1}t-1, 2^s-2, 0, 0)$, $(2^{s+1}t-2, 2^s-2, 0, 0)$ and $(2^k-3, 0, 0, 0)$ where $s, t > 0$, $k > 2$, give a basis for E_*. Furthermore, in $\Omega_*^{(2)}$, we have $\partial(\xi_I) = RP(\xi_I)$. Results of [3] and [9] are very useful in proving the above results.

Define $P_*^{(2)} \subseteq \sigma_*^{(2)}(\mathrm{all})$ as the subalgebra generated by $[CP^{2n+2}, T_{2n+2}]$, $n \geq 0$. Let $P_*^{(\mathrm{rel})} \subseteq \sigma_*^{(2)}(\mathrm{rel})$ be the image of $P_*^{(2)}$ under $j: \sigma_*^{(2)}(\mathrm{all}) \to \sigma_*^{(2)}(\mathrm{rel})$.

Define Γ^o to be the subset of Γ consisting of all sequences of the form $I = \alpha * I'$ where $*$ denotes juxtaposition,

$$\alpha = (2n_1+1, 2n_1+1) * (2n_2+1, 2n_2+1) * (2n_k+1, 2n_k+1)$$

with $k \geq 0$, and $I' = (i_1', i_2', \ldots, i_{2j}')$ with $i_1' > i_2' \geq \ldots \geq i_{2j}'$, i_1' odd and $j > 0$.

We then prove the following results.

<u>LEMMA</u> $\sigma_*^{(2)}(\mathrm{rel})/P_*^{(\mathrm{rel})}$ <u>is the free $\Omega_*^{(2)}\langle\beta\rangle$ module with basis</u> ξ_I, $I \in \Gamma^o$.

<u>PROPOSITION.</u> <u>There is a (non-canonical) splitting of $\Omega_*^{(2)}\langle\beta\rangle$ modules</u>

$$\sigma_*^{(2)}(\mathrm{rel}) = P_*^{(\mathrm{rel})} \oplus T_*^{(\mathrm{rel})} \oplus R_*^{(\mathrm{rel})} \oplus F_*^{(\mathrm{rel})}$$

<u>with the following properties</u>:

A) <u>Under the homomorphism</u> $\partial: \sigma_*^{(2)}(\mathrm{rel}) \to \sigma_*^{(2)}(\mathrm{free})$

(i) $P_*^{(\mathrm{rel})} \oplus T_*^{(\mathrm{rel})}$ <u>is sent to zero;</u>

(ii) $R_*^{(\mathrm{rel})}$ <u>is mapped isomorphically to</u> $\tilde{\Omega}_*^{(2)}(B\,\mathbb{Z}/2)$;

(iii) $F_*^{(\mathrm{rel})}$ <u>is mapped onto</u> $E_* \subset \Omega_*^{(2)} \subset \Omega_*^{(2)}(B\,\mathbb{Z}/2)$ <u>with the induced map</u> $\mathbb{Z}/2 \otimes_{\Omega_*^{(2)}\langle\beta\rangle} F_*^{(\mathrm{rel})} \to \mathbb{Z}/2 \otimes_{\Omega_*^{(2)}} E_*$ <u>being an isomorphism of degree -1.</u>

B) (i) $P_*^{(\mathrm{rel})}$ <u>is an $\Omega_*^{(2)}$ polynomial algebra on generators</u> $j[CP^{2n}, T_{2n}]$, $n \geq 1$;

(ii) $T_*^{(\mathrm{rel})}, R_*^{(\mathrm{rel})}$ <u>are free $\Omega_*^{(2)}\langle\beta\rangle$ modules.</u>

In fact $\Omega_*^{(2)}\langle\beta\rangle$ bases of $R_*^{(\mathrm{rel})}$ $F_*^{(\mathrm{rel})}$ and $T_*^{(\mathrm{rel})}$ respectively may be chosen in a one-one correspondence with the following sets:

$$\Gamma_R^o = \{(2n+1, 0);\ n \geq 0\}$$

$$\Gamma_F^o = \{(2^{s+1}t - 1, 2^s-2, 0, 0);\ s, t > 0\} \cup \{(2^k-3, 0, 0, 0);\ k > 2\}$$

$$\Gamma_T^o = \Gamma^o - (\Gamma_R^o \cup \Gamma_F^o)$$

Putting the information above into the Conner + Floyd sequence gives an exact sequence of $\Omega_*^{(2)}\langle\beta\rangle$ modules:

$$0 \to \Omega^{(2)}/E_* \xrightarrow{\bar{i}} \sigma_*^{(2)}(\text{all}) \to P_*^{(\text{rel})} \oplus T_*^{(\text{rel})} \oplus F_*^{(\text{rel})} \to E_* \to 0.$$

After some analysis this leads to the following result

THEOREM. As an $\Omega_*^{(2)}\langle\beta\rangle$ module

$$\sigma_*^{(2)}(\text{all}) = P_*^{(2)} \oplus T_*^{(2)} \oplus F_*^{(2)} \oplus \tilde{F}_*^{(2)}$$

where

(i) $P_*^{(2)}$ is an $\Omega_*^{(2)}$ polynomial algebra generated by $[CP^{2n}, T_{2n}]$, $n \geq 1$.

(ii) $T_*^{(2)}$ is a free $\Omega_*^{(2)}\langle\beta\rangle$ module with generators in a one-one correspondence with Γ_T^o.

(iii) $\tilde{F}_*^{(2)} \cong \Omega_*^{(2)}/E_*$ generated by $[\mathbb{Z}/2] \in \sigma_o^{(2)}(\text{all})$.

(iv) The map $j: \sigma_*^{(2)}(\text{all}) \to \sigma_*^{(2)}(\text{rel})$ induces an exact sequence of $\Omega_*^{(2)}\langle\beta\rangle$ modules:

$$0 \to F_*^{(2)} \xrightarrow[\text{degree } -1]{j'} \Omega_*^{(2)}\langle\beta\rangle \otimes ((\mathbb{Z}/2) \otimes_{\Omega_*^{(2)}\langle\beta\rangle} E_*) \to E_* \to 0$$

(v) The graded $\Omega_*^{(2)}\langle\beta\rangle$ module $F_*^{(2)}$ is indecomposable.

Finally, to get $\sigma_*(\text{all})$ define Ω_* submodules of $\sigma_*(\text{all})$ as follows: let P_* be the Ω_* subalgebra of $\sigma_*(\text{all})$ generated by the $[\mathbb{C}P^{2n}, T_{2n}]$, $n \geq 1$. Let T_* contain all torsion elements of $\sigma_*(\text{all})$ whose mod 2 reduction lies in $T_*^{(2)} \subset \sigma_*^{(2)}(\text{all})$. Finally, let F_* consist of all $x \in \sigma_*(\text{all})$ such that 2x is bordant to a free involution and such that the mod 2 reduction of x lies in $F_*^{(2)} \oplus \tilde{F}_*^{(2)}$. It is then not hard to prove.

PROPOSITION $\sigma_*(\text{all}) = P_* \oplus T_* \oplus F_*$

All that remains is an analysis (not too difficult) of T_* and F_* to get the description given in the Main Theorem.

We shall not describe the geometric generators of $\sigma_*(\text{all})$ and $\sigma_*^{(2)}(\text{all})$ here but refer the reader to a full account in [8]. Just to tempt the reader we mention the following:

$\sigma_o = \mathbb{Z} \oplus \mathbb{Z}$ generated by $[pt]$ and $[\mathbb{Z}/2]$

$\sigma_1 = 0$, $\sigma_2 = 0$, $\sigma_3 = 0$

$\sigma_4 = \mathbb{Z} \oplus \mathbb{Z} \oplus \mathbb{Z} \oplus \mathbb{Z}/2$ generated by $[\mathbb{C}P^2, T_2]$, $[\mathbb{C}P^2]$, $[\mathbb{C}P^2] \times [\mathbb{Z}/2]$ and

$[\mathbb{C}P^2, T_2] + [\mathbb{C}P^2, \text{conj}]$

$\sigma_5 = \mathbb{Z}/2 \oplus \mathbb{Z}/2$ generated by $[RP(\xi_2\xi_o^3)]$ and $[RP(\xi_2\xi_o \times \xi_o^2), 1 \times (-1)]$

Generators of $\sigma_6 = (\mathbb{Z}/2)^3$ and $\sigma_7 = (\mathbb{Z}/2)^2$ are described in [8]. The first place where at present we have a generator missing is in $\sigma_8 = (\mathbb{Z})^7 \oplus (\mathbb{Z}/2)^7$ - we cannot describe one of the torsion elements geometrically-$M(\emptyset, (3,2,0,0))$ in the notation of the Main Theorem.

REFERENCES

1. N.A. Baas. "On bordism theory of manifolds with singularities." Math. Scand. 33 (1973) 279-302.

2. S. Buoncristiano, C.P. Rourke and B.J. Sanderson. "A geometric approach to Homology Theory." L.M.S. Lecture Notes Vol. 18 C.U.P. 1976.

3. F.L. Capobianco, C. Kosniowski and R.E. Stong. "Free Involutions." To appear

4. P.E. Conner. "Lectures on the action of a finite group." Springer Lecture Notes Vol. 73. Springer 1968.

5. P.E. Conner and E.E. Floyd. "Differentiable Periodic Maps." Springer 1964.

6. P.E. Conner and E.E. Floyd. "Maps of odd period." Ann. of Math. 84 (1966) 132-156.

7. C. Kosniowski. "Actions of finite abelian groups." Pitman 1978.

8. C. Kosniowski and E. Ossa. "The bordism module of oriented involutions." To appear.

9. C. Kosniowski and R.E. Stong. "Involutions and characteristic numbers." Topology.

10. H.L. Rosenweig. "Bordism of involutions on manifolds." Ill. J. Math. 16 (1972) 1-10.

11. R.J. Rowlett. "Wall manifolds with involutions." Trans. A.M.S. 169 (1972) 153-162.

12. R.E. Stong. "Wall Manifolds." To appear.

13. C.T.C. Wall. "Formulae for surgery obstructions." Topology 15 (1976) 189-210.

UNIVERSITY OF NEWCASTLE UPON TYNE AND GESAMOCHSCHULE WUPPERTAL.

Obstructions to Equivariance

R. Lashof

0. Introduction

Let G be a compact Lie group and X and Y G-spaces. We study the problem of deforming an arbitrary map of X into Y to an equivariant map.

This problem arises in attempting to lift a G-action on X to a bundle E over X. In fact, if E is an A bundle, A a topological group, the equivalence classes of equivariant A bundles over X are in bijective correspondence with $[X,B_GA]_G$, where B_GA is the G-space defined in {5}. In particular, if we forget the G action, B_GA is a classifying space for A-bundles without G action. Thus if $f: X \to B_GA$ is a classifying map for the A bundle E, the G action in X lifts to E iff f may be deformed to an equivariant map. The specific applications to lifting group actions in bundles will be discussed in a later paper. Here we concentrate on the obstructions to equivariance, but many of the questions discussed are motivated by the lifting problem.

In section 1 we translate the problem to a question of cross-sections of spaces over X and over X/G. This is a direct generalization of the usual description in the case of free actions; i.e., principal G-bundle maps are given by cross-sections of an associated bundle. In the general case the spaces over X and X/G are not bundles globally, but only over each orbit type. This leads to an obstruction theory with local coefficients which we describe in section 2.

1. Equivariant maps as Cross-sections

If $f: X \to Y$ is a map, $(1,f): X \to X \times Y$, $(1,f)(x) = (x,f(x))$ is a section of $pr_1: X \times Y \to X$. If f is equivariant, then since for each $x \in X, G_x \in G_{f(x)}$ (G_x the isotropy subgroup), the image of $(1,f)$ is contained in the subspace $E = \{(x,y) \in X \times Y | G_y \supset G_x\}$

and defines a section of $\pi = pr_1|E: E \to X$. If we give $X \times Y$ the diagonal G-action, $E = E(X,Y)$ is a G-invariant subspace of $X \times Y$ and $\pi: E \to X$ is equivariant. Thus π induces $\bar{\pi}: \bar{E} \to \bar{X}$, $\bar{E} = E/G$, $\bar{X} = X/G$.

<u>Proposition 1.1</u>: The equivariant maps of X into Y are in bijective correspondence with the sections of $\bar{\pi}: \bar{E} \to \bar{X}$.

For the proof we need the

<u>Lemma 1.2</u>: E is G-equivalent to the pull back of X by $\bar{\pi}$.

Proof of the Lemma:

This result follows from the commutative diagram:

$$\begin{array}{ccc} E & \xrightarrow{\pi} & X \\ q_E \downarrow & & \downarrow q_X \\ \bar{E} & \xrightarrow[\bar{\pi}]{} & \bar{X} \end{array}$$

and the fact that, since $G_{(x,y)} = G_x$ if $G_y \supset G_x$, π is isovariant. (See Bredon {2}.)

<u>Proof of the Proposition</u>:

Since E is the pull back we have a bijective correspondence between sections $\bar{\sigma}: \bar{X} \to \bar{E}$ and equivariant sections $\sigma: X \to E$ as can be seen from the diagram:

$$\begin{array}{ccccc} X & \overset{\sigma}{\to} & E & \overset{\pi}{\to} & X \\ q_X \downarrow & & \downarrow q_E & & \downarrow q_X \\ \bar{X} & \overset{\bar{\sigma}}{\to} & \bar{E} & \overset{\bar{\pi}}{\to} & \bar{X} \end{array}$$

But if $\sigma: X \to E \subset X \times Y$ is an equivariant section $pr_2 \circ \sigma: X \to Y$ is an equivariant map. Conversely, if f is an equivariant map (1,f) is an equivariant section of E. Further $\sigma \to pr_2 \circ \sigma$ and $f \to (1,f)$ are inverse operations.

Restating this result we have:

<u>Proposition 1.1'</u>: A map $f: X \to Y$ is homotopic to an equivariant map iff the section s(f) of $pr_1: X \times Y \to X$ defined by f is homotopic to the pull back of a section of $\bar{\pi}: \bar{E} \to \bar{X}$ via $q_X: X \to \bar{X}$.

<u>Example 1</u>: If X is a free G-space, then $E = X \times Y$, and $\bar{E} = (X \times Y)/G$ is the associated bundle to $q_X: X \to \bar{X}$ with the fibre Y. If Y is also free, then Proposition 1 is just the usual theorem classifying G-bundle maps from X to Y.

<u>Example 2</u>: If we have a section $s: \bar{X} \to X$ and we let $C = s(\bar{X})$, $\pi^{-1}(C) \to C$ may be identified with $\bar{\pi}: \bar{E} \to \bar{X}$; i.e. since q_E is the pull back of q_X, $\pi^{-1}(C)$ is a section of q_E. Thus we get a

bijective correspondence between sections $\sigma: C \to \pi^{-1}(C)$ and equivariant maps $f: X \to Y$. Also note such a section is equivalent to a map $f_o: C \to Y$ such that $G_{f_o(x)} \supset G_x$, $x \in C$, and the corresponding equivariant map is an extension of f_o. (cf. Bredon {2}).

Remark: If $A \subset X$ is a G-invariant subspace of X, then $E(A,Y)$ is just $\pi^{-1}(A) \subset E = E(X,Y)$. We denote it by $E|A$. Similarly $E(A,Y)$ is $\bar{\pi}^{-1}(\bar{A}) \subset \bar{E}$ and is denoted $\bar{E}|\bar{A}$.

Neither π nor $\bar{\pi}$ is a bundle. However if we let $X_{(H)} = \{x \in X \mid (G_x) = (H)\}$ be the points of orbit type (H), we have

Proposition 1.3: $E|X_{(H)}$ and $\bar{E}|\bar{X}_{(H)}$ are bundles with fibre Y^H, and $q_E: E \to \bar{E}$ restricts to a bundle map of $E|X_{(H)}$ onto $\bar{E}|\bar{X}_{(H)}$.

Proof:

We may as well assume $X = X_{(H)}$ and $E = E|X_{(H)}$. Recall {2}, that in this case $\mu: G \underset{N(H)}{\times} X^H \to X$, $\mu[g,x] = gx$ is a G-equivalence (where $N(H)$ is the normalizer of H in G). Also since $X = X_{(H)}$, $X^H = \{x \in X \mid G_x = H\}$, and $\pi^{-1}(X^H) = X^H \times Y^H = (X \times Y)^H$. Likewise $\pi^{-1}(gX^H) = \pi^{-1}(X^{gHg^{-1}}) = (gX^H \times gY^H) = g(X \times Y)^H$. Thus $E = (X \times Y)_{(H)}$ and we have the commutative diagram.

$$\begin{array}{ccc} G \times_{N(H)} (X^H \times Y^H) & \xrightarrow{\mu_E} & E \\ \downarrow p & & \downarrow \pi \\ G \times_{N(H)} X^H & \xrightarrow[\mu_X]{} & X \end{array} ,$$

where $p[g,(x,y)] = [g,x]$, and μ_E and μ_X are G-equivalences.

Since p is a bundle with fibre Y^H (and group $\bar{N}(H) = N(H)/H$), so is π.

By passage to the quotients from the above diagram we get the commutative diagram:

$$\begin{array}{ccc} (X^H \times Y^H)/\bar{N}(H) & \xrightarrow{\bar{\mu}_E} & \bar{E} \\ \downarrow \bar{p} & & \downarrow \bar{\pi} \\ X^H/\bar{N}(H) & \xrightarrow{\bar{\mu}_X} & \bar{X} \end{array} ,$$

where $\bar{p}[x,y] = [x]$ and $\bar{\mu}_E$ and $\bar{\mu}_X$ are homeomorphisms. Since $\bar{p}$ is a bundle with fibre Y^H (and group $\bar{N}(H)$), so is $\bar{\pi}$.

Finally, the quotient map $q: G \times_{N(H)} (X^H \times Y^H) \to (X^H \times Y^H)/\bar{N}(H)$, $q\,[g,(x,y)] = [x,y]$ is a bundle map, and hence q_E is a bundle map.

Example 3: If H is normal in G, and $X = X_{(H)}$, then $G = N(H)$, $X = X^H$ and $E = X \times Y^H$.

Example 4: Suppose $X = G/H$. Then $X^H = \bar{N}(H) = N(H)/H$, and we have the commutative diagram:

$$\begin{array}{ccc} G \times_{N(H)} (X^H \times Y^H) & \xrightarrow{\mu_E} & E \\ & \phi \searrow \quad \nearrow \mu'_E & \\ & G/H \times Y^H & \end{array}$$

where ϕ is the G-equivalence $\phi[g,(\bar{n},y)] = (g\bar{n},n^{-1}y)$, and μ'_E is the G-equivalence $\mu'_E(\bar{g},y) = (\bar{g},gy)$. Also by passage to the quotients

$$\begin{array}{ccc} (X^H \times Y^H)/\bar{N}(H) & \xrightarrow{\mu_E} & \bar{E} \\ & \phi \searrow \quad \nearrow \mu'_E & \\ & Y^H & \end{array}$$

commutes,

and thus

$$\begin{array}{ccc} G/H \times Y^H & \xrightarrow{pr_2} & Y^H \\ \mu'_E \downarrow & & \downarrow \bar{\mu}'_E \\ E & \xrightarrow[q_E]{} & \bar{E} \end{array}$$

commutes,

where μ'_E is a G-equivalence and $\bar{\mu}'_E$ is a homeomorphism. Note however, that although q_E is equivalent to pr_2, it does <u>not</u> follow that an equivariant section of E corresponds to a constant map of X into Y. A section of $\bar{E}$ corresponds to a point $y_o \in Y^H$, which indeed pulls back to the section $\sigma(\bar{g}) = (\bar{g}, y_o)$ under pr_2. But $\mu'_E \sigma(\bar{g}) = \mu'_E(\bar{g}, y_o) = (\bar{g}, gy_o)$. I.e. an equivariant map $f: G/H \to Y$ is of the form $f(\bar{g}) = gy_o$, $y_o \in Y^H$, as one expects. In particular, when H is normal in G (see Ex.3), $\mu'_E : G/H \times Y^H \to G/H \times Y^H$ is not the identity.

<u>Proposition 1.4</u>: Suppose G acts freely on X and there exists a subspace $Z \subset Y^G$ which is a deformation retract of Y. Then

a) $[X,Y]_G = [X/G,Y]$

b) A map $f: X \to Y$ is homotopic to an equivariant map iff it is homotopic to a map which factors through X/G.

<u>Remark</u>: As we will see in part 2, if Y is a classifying space $B_G A$, such a $Z \subset Y^G$ always exists.

<u>Proof of the Proposition</u>:

Since X is free, $E = X \times Y$. We claim $\bar{E} = (X \times Y)/G$ is fibre homotopy equivalent to $X/G \times Y$, which will imply (a). In fact, the inclusion $(X \times Z)/G \to (X \times Y)/G$ is a homotopy equivalence on the fibre and hence a fibre homotopy evuivalence.
But $(X \times Z)/G = X/G \times Z$, since G acts trivially on Z. Thus we have the fibre homotopy equivalence $X/G \times Y \xrightarrow{1 \times r} X/G \times Z \xrightarrow{i} \bar{E}$,

where $r: Y \to Z$ is the retraction.

For (b) note the commutative diagram

$$\begin{array}{ccc} X \times Z & \xrightarrow{q_X \times 1} & X/G \times Z \\ \downarrow & & \downarrow \\ X \times Y & \xrightarrow[q_E]{} & (X \times Y)/G \end{array}$$

We see that a cross-section of E coming from one of $\bar{E}$ is homotopic to one coming from $X/G \times Z$. Thus if f is homotopic to an equivariant map, it is homotopic to a map $f': X \to Z \subset Y$ which factors through X/G. Conversely a map which factors through X/G deforms through such maps to a map into $Z \subset Y^G$, and hence equivariant.

If $\lambda: X' \to X$ is a G-map, then obviously a G-map f of X into Y induces the G-map $f \circ \lambda$ of X' into Y. We show how this is reflected in our construction:: Let $E = E(X,Y)$ and $E' = E(X',Y)$. Then the pull back $\lambda^* E = \{(x',(x,y)) \in X' \times E \mid \lambda(x') = x\}$ may be identified with the subspace of $E' = \{(x',y) \in X' \times Y \mid G_y \supset G_{x'}\}$ consisting of (x',y) with $G_y \supset G_{\lambda(x')}$, by $(x',(x,y)) \leftrightarrow (x',y)$, $\lambda(x') = x$. Then $\bar{\lambda}^* \bar{E}$ may be identified with the image $\overline{\lambda^* E}$ of $\lambda^* E$ in $\bar{E}'$. In fact, $\bar{\lambda}^* \bar{E} = \{(\bar{x}',\{x,y\}) \in \bar{X}' \times \bar{E} \mid \bar{\lambda}(\bar{x}') = \bar{x}\}$. We identify $\{x',y\} \in \overline{\lambda^* E}$ with $(x',\{\lambda(x'),y\}) \in \bar{\lambda}^* \bar{E}$ by passage to the quotients from the identification above. Thus by pulling back sections we get:

<u>Lemma 1.5</u>: If $\lambda: X' \to X$ is a G-map, $E = E(X,Y)$, $E' = E(X',Y)$, then a section $s: X \to E$ pulls back to a section $\lambda^* s: X' \to \lambda^* E \subset E'$ and a section $\bar{s}: \bar{X} \to \bar{E}$ pulls back to a section $\bar{\lambda}^* \bar{s}: \bar{X}' \to \bar{\lambda}^* \bar{E} \subset \bar{E}'$. Further, if s is induced from $\bar{s}$, $\lambda^* s$ is induced from $\bar{\lambda}^* \bar{s}$.

Now let $X_G = (UG) \times_G X$, UG the universal G-bundle. By choosing a point $p \in UG$ we have a map $j_p: X \xrightarrow[i_p]{} UG \times X \xrightarrow[q]{} X_G$, $i_p(x) = (p,x)$. The homotopy class j_p is independent of $p \in UG$ since UG is contractible.

<u>Proposition 1.6</u>: Suppose there exists a subspace $Z \subset Y^G$ which is a deformation retract of Y, then

1. If f is homotopic to a map which factors through X/G, f is homotopic to an equivariant map.

2. If f is homotopic to an equivariant map, f is homotopic to a map which factors through X_G.

<u>Proof of the Proposition</u>:

1. Since a map $f: X \xrightarrow{q} X/G \to Y^G \subset Y$ is equivariant, the result follows immediately.

2. The projection $\lambda: (UG) \times X \to X$ is a G-map, where G acts on $UG \times X$ by $g(z,x) = (zg^{-1}, gx)$. Hence a section $\bar{s}: \bar{X} \to \bar{E}$ pulls back to a section $\bar{\lambda}^* \bar{s}: X_G \to UG \underset{G}{\times} (X \quad Y)$. (Note that since $UG \times X$ is a free G-space, $E(UG \times X, Y) = UG \times X \times Y$.) If $s: X \to E$ is the pull back of $\bar{s}$, $\lambda^* s$ is the pull back of $\bar{\lambda}^* \bar{s}$ by Lemma 1.4.

Now $pr_2 \circ \lambda^* s \circ j_p: X \to UG \times X \to (UG \times X) \times Y \to Y$ is just $pr_2 \circ s: X \to Y$, since $\lambda \circ i_p = 1_X$. On the other hand by Proposition 1.4, $pr_2 \circ \lambda^* s: UG \times X \to Y$ factors through X_G up to homotopy. Thus $pr_2 \circ s$ is homotopic to a map which factors through X_G. But if f is homotopic to a G-map it is homotopic to $pr_2 \circ s$ for some $\bar{s}$.

<u>Proposition 1.7</u>: Suppose $X = X_{(H)}$.

1. If there is a subspace $Z \subset Y^{N(H)}$ which is a deformation retract of Y^H, then $[X,Y]_G \simeq [X/G, Y^H]$.

2. If there is a subspace $Z \subset Y^G$ which is a deformation retract of Y^H, then a map $f: X \to Y$ is a homotopic to an equivariant map iff it is homotopic to a map into Y^H which factors through X/G. The equivariant homotopy classes of equivariant maps homotopic to f are given by the preimage of [f] under the canonical map of $[X/G, Y^H]$ into $[X,Y]$.

3. If there exist $Z \subset Y^G$ which is a deformation retract of Y^H and $Z' \subset Y^H$ which is a deformation retract of Y, then a map $f: X \to Y$ is a homotopic to an equivariant map iff it is homotopic to a map which factors through X/G.

<u>Proof</u>:

1. $\pi\colon E \to \bar{E}$ is equivalent to $q\colon G \underset{N(H)}{\times} (X^H \times Y^H) \to (X^H \times Y^H)/N(H)$. Under the hypothesis on $Z \subset Y^{N(H)}$ we have the commutative diagram:

$$\begin{array}{ccc}
G \underset{N(H)}{\times} (X^H \times Y^H) & \overset{q}{\to} & (X^H \times Y^H)/N(H) \\
i \uparrow \text{f.h.e.} & & i \uparrow \text{f.h.e.} \\
G \underset{N(H)}{\times} (X^H \times E) & \overset{q}{\to} & (X^H \times Z)/N(H) \\
\| & & \| \\
(G \underset{N(H)}{\times} X^H) \times Z & \to & (X^H/N(H)) \times Z \\
\downarrow \mu_x \times 1 & & \downarrow \bar{\mu}_x \times 1 \\
X \times Z & \xrightarrow[q_x \times 1]{} & X/G \times Z \\
\text{f.h.e.} \uparrow 1 \times r & & \text{f.h.e.} \uparrow 1 \times r \\
X \times Y^H & \xrightarrow[q_x \times 1]{} & X/G \times Y^H
\end{array}$$

The result follows.

2. Under the hypothesis $Z \subset Y^G$ we get the commutative diagram

$$\begin{array}{ccc} (G \times_{N(H)} X^H) \times Z & = & G \times_{N(H)} (X^H \times Z) \\ \downarrow \mu_X \times 1 & & \downarrow \mu_E \\ X \times Z & \xrightarrow[i]{} & E \end{array},$$

since $\mu_E[g,(x,z)] = (gx,gz) = (gx,z)$. Similarly for $\bar{E}$ by passage to the quotients. So we have the commutative diagram

$$\begin{array}{ccc} E & \xrightarrow{\pi} & \bar{E} \\ i. \uparrow \text{f.h.e.} & & \uparrow \text{f.h.e.} \\ X \times Z & \xrightarrow[q_X \times 1]{} & X/G \times Z \end{array}$$

The result follows (cf. Proof of Prop. 1.4).

3. Under the hypothesis $Z \subset Y^G$ and $Z' \subset H^Y$, a map $X/G \to Y$ is homotopic to a map $X/G \to Y^H$ and hence to a map $X/G \to Z$. So a map $f: X \to Y$ which is homotopic to a map which factors through X/G is homotopic to $X \xrightarrow[q_X]{} X/G \to Z \subset Y^G \subset Y$, which is equivariant. The converse follows from (2).

Remark:

(a) Note that under hypothesis (1), it does not follow that the equivariant map corresponding to $f: X/G \to Y^H$ is homotopic to $f \circ q_X: X \to Y^H$. However, this is true under the stronger hypothesis (2). Under hypothesis (3) it is not true that $[X,Y]_G \approx [X/G,Y]$; we have only obtained a right inverse to $[X/G,Y^H] \to [X/G,Y]$.

(b) The hypothesis (1), (2), (3) will be fulfilled for $Y = B_G A$ with suitable choices of G and A.

Example 5: Let G act freely on X and suppose G acts on $Y = S^n$ with at least one fixed point. Then if $\dim X/G < n$, a map $f: X \to S^n$ is homotopic to an equivariant map iff it is homotopically trivial. (X/G a CW complex.)

In fact, by Proposition 1.1, $[X,Y]_G$ is given by the homotopy classes of cross-sections of $\bar{E} = (X \times S^n)/G \to X/G$. The fixed point gives us one cross-section and obstruction theory says any two cross-sections are homotopic. The equivariant map induced by the fixed point section is the constant map and all others are homotopic. The converse is obvious.

Note than $\dim X$ may be quite large. For example, $X = O(n)/O(1)$ and $G = O(n-1)$, so that $X/G = RP^{n-1}$. We can take $O(n-1) \subset O(n+1)$ acting on S^n as usual.

2. The obstruction theory: We will assume X is a G-CW complex. Thus X is built up by attaching G-n-cells, $G/H \times D^n$, to the n-1 G-skelton ${}_GX^{(n-1)}$, by an equivariant map $\partial\phi$: $G/H \times S^{n-1} \to {}_GX^{(n-1)}$. (Here G acts only on the first factor of $G/H \times D^n$.)

To understand the obstruction groups, consider the space of maps $f: G/H \times S^n \to Y$ such that $f(\bar{g},s_o) = gy_o$, for some fixed $y_o \in Y^H$ and s_o a base point of S^n. Let f correspond to $\hat{f}: G/H \to Y^{S^n}$ under the exponential law $Y^{G/H \times S^n} \simeq (Y^{S^n})^{G/H}$; i.e. $\hat{f}(\bar{g})(s) = f(\bar{g},s)$, $\bar{g} \in G/H$, $s \in S^n$. Then $\hat{f}(\bar{g})(s_o) = gy_o$. Given two such maps f_1, f_2, if $n \geq 1$, the usual loop composition defines a composition $\hat{f}_1 \circ \hat{f}_2$ by $(\hat{f}_1 \circ \hat{f}_2)(\bar{g}) = \hat{f}_1(\bar{g}) \circ \hat{f}_2(\bar{g})$: $(S^n,s_o) \to (Y,gy_o)$. We write $f_1 \circ f_2$ for the map corresponding to $\hat{f}_1 \circ \hat{f}_2$. This composition has a homotopy unit f_o, $f_o(\bar{g},s) = gy_o$ all $s \in S^n$, $\bar{g} \in G/H$, homotopy inverses, and is homotopy commutative if $n \geq 2$. The homotopy classes of such maps will be denoted $\Gamma_n^{G/H}(Y,Y_o)$. $\Gamma_n^{G/H}(Y,Y_o)$ is a group if $n \geq 1$, abelian if $n \geq 2$. $\Gamma_o^{G/H}$ is the set of homotopy classes of maps of G/H into Y with base point f_o, $f_o(\bar{g}) = gy_o$.

Now let $\Gamma_n^{G/H}(Y,Y^H,Y_o)$ be the homotopy classes of maps $f: G/H \times D^n \to Y$ such that $f(\bar{e},s) \in Y^H$ and $f(\bar{g},s) = gf(\bar{e},s)$, $s \in S^{n-1}$, and $f(\bar{e},s_o) = y_o \in Y^H$. Note that $\hat{f}(\bar{e})$: $(S^{n-1},s_o) \to (Y^H,y_o)$ and $\hat{f}(\bar{g})$: $(S^{n-1},s_o) \to (gY^H,gy_o)$. Again loop composition makes $\Gamma_n^{G/H}(Y,Y^H,Y_o)$ into a group for $n \geq 2$. $\Gamma_o^{G/H}(Y,Y^H,Y_o)$ is the set of homotopy classes of maps of G/H into Y.

Now assume $f: X \to Y$ is equivariant over the n-1 G-skelton ${}_GX^{(n-1)}$. If $G/H \times D^n$ is an n-cell attached by $\partial\phi: G/H \times S^{n-1}$, then by Lemma 1.5, the cross section $\bar{\sigma}_{n-1}$ of $\bar{E}|_G\bar{X}^{(n-1)}$ defined by $f|_GX^{n-1}$ pulls back to a cross section $\partial\bar{\phi}^*\bar{\sigma}_{n-1}$ of $\bar{E}(G/H \times D^n, Y)|(G/H \times S^{n-1})/G$ such that the equivariant map defined by $\partial\bar{\phi}^*\bar{\sigma}_{n-1}$ of $G/H \times S^{n-1} \to Y$ corresponds to $\partial\phi^*\sigma_{n-1}$, and hence is $f \circ \partial\phi: G/H \times S^{n-1} \to Y$. If $\phi: G/H \times D^n \to {}_GX^{(n)}$ is the defining map of the G-n-cell, $\phi|G/H \times S^{n-1} = \partial\phi$, then $f \circ \phi$ defines a cross-section of $G/H \times D^n \times Y \to G/H \times D^n$ which agrees with $\partial\phi^*\sigma_{n-1}$ over $G/H \times S^{n-1}$. Thus we have the following situation:

We are given a cross-section σ of $G/H \times D^n \times Y \to G/H \times D^n$ such that $\partial\sigma = \sigma|G/H \times S^{n-1}$ is equivariant, and we are required to deform σ rel $\partial\sigma$ to an equivariant cross-section.

Let $\partial\sigma(\bar{e}, s_o) = (\bar{e}, y_o)$, $y_o \in Y^H$. A homotopy class of cross-sections τ of $(G/H \times D^n) \times Y$ such that $\partial\tau$ is equivariant and $\tau(\bar{e}, s_o) = (\bar{e}, y_o)$ defines a class $[\tau]$ in $\Gamma_n^{G/H}(Y, Y^H, y_o)$. Further σ deforms rel $\partial\sigma$ to an equivariant section iff $[\sigma] = [f_o]$ (For $n = 0$, and hence $\partial\sigma$ not defined, this means that if $f: G/H \to Y$ is the map defined by σ, then f is homotopic to f_o when $f_o(\bar{g}) = gy_o$ some $y_o \in Y^H$.)

Thus we will have the usual obstruction theory situation once we see how the groups behave under change of base point in Y^H: If y_1 is in the same path component of Y^H as is y_o, then $\Gamma_n{}^{G/H}(Y,Y^H,y_1) \simeq \Gamma_n{}^{G/H}(Y,Y^H,y_o)$, but the isomorphism depends on the path. The argument is the same as for homotopy groups, and one gets an action of $\pi_1(Y^H,y_o)$ on $\Gamma_n{}^{G/H}(Y,Y^H,y_o)$. We say Y is <u>G-n-simple</u> if this action is trivial on $\Gamma_j{}^{G/H}$, $j \leq n$.

By the usual arguments of obstruction theory we have:

<u>Theorem 1.8</u>: Let X be a G-CW complex and Y G-n-simple. Suppose f: X → Y is equivariant over ${}_GX^{(n-1)}$. Then the obstruction to deforming f to be equivariant over ${}_GX^{(n)}$, keeping it fixed over ${}_GX^{(n-2)}$ is a class in $H^n(X/G;B_n)$; where B_n is the bundle of coefficients such that B_n over an n-cell of orbit type (H) is $\Gamma_n{}^{G/H}(Y,Y^H,y_o)$.

Similarly we have:

<u>Theorem 1.9</u>: Let X be the G-CW complex and Y G n+1 simple. Suppose f,f': X → Y are equivariant maps and we are given a homotopy of f to f' which is equivariant over ${}_GX^{(n-1)}$. Then the obstruction to deforming the homotopy to be equivariant over ${}_GX^{(n)}$, keeping it the same over ${}_GX^{(n-2)}$ is a class in $H^n(X/G;B_{n+1})$.

Note: One should beware that the dimensions of our coefficient groups are one higher than in the usual obstruction theory.

To understand the coefficient groups we will develop some exact sequences. First let us introduce the group $\mathring{\Gamma}_n^{G/H}(Y,y_o)$. This consists of homotopy classes of maps $f\colon G/H \times S^n \to Y$ such that $f(\bar{g},s_o) = gy_o$ and $f(\bar{e},s) = y_o$ for all $s \in S^n$. (Or equivalently, homotopy classes of maps $f\colon G/H \times S^n \cup e \times D^{n+1} \to Y$ with $f(\bar{g},s_o) = gy_o$.) Then it is easy to check that we get the following long exact braid:

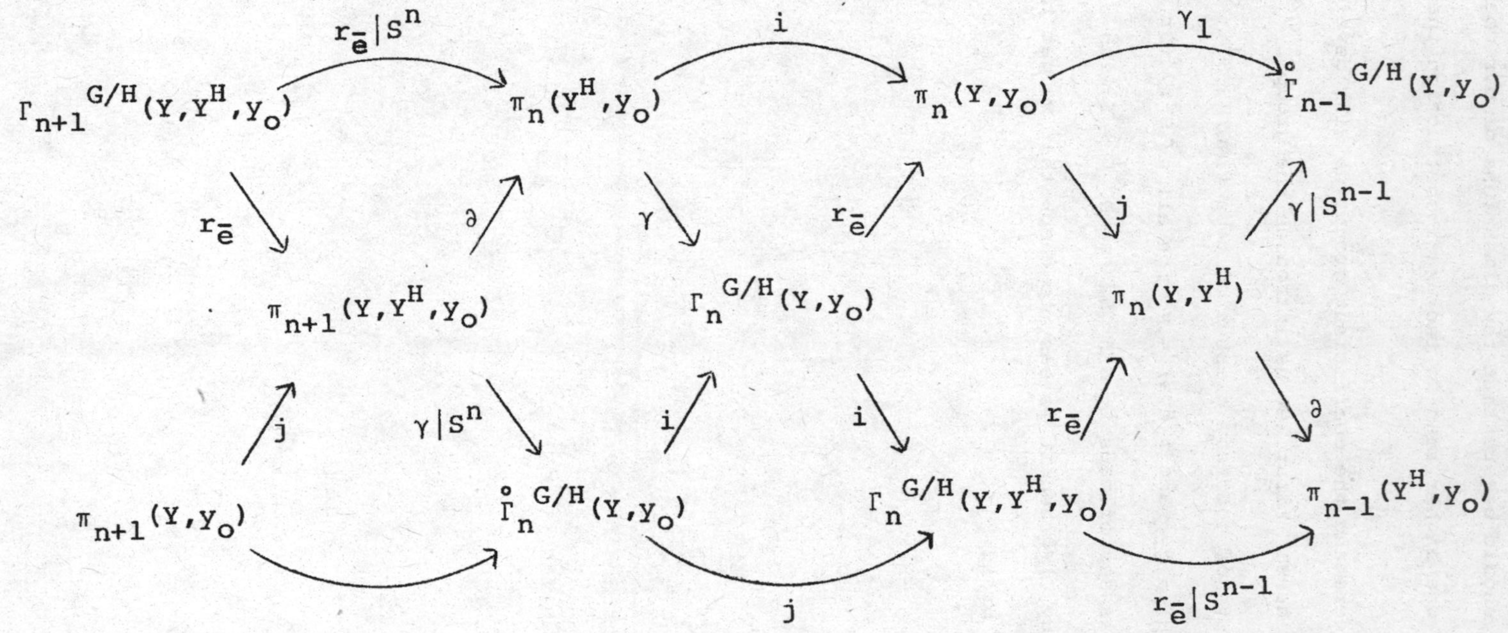

Γ_{n+1}^{G/H}(Y,Y^H,y_o)
π_n(Y^H,y_o)
π_n(Y,y_o)
Γ̊_{n-1}^{G/H}(Y,y_o)
r_ē|S^n
i
γ_1
r_ē
∂
γ
r_ē
j
γ|S^{n-1}
π_{n+1}(Y,Y^H,y_o)
Γ_n^{G/H}(Y,y_o)
π_n(Y,Y^H)
j
γ|S^n
i
i
r_ē
∂
π_{n+1}(Y,y_o)
Γ̊_n^{G/H}(Y,y_o)
Γ_n^{G/H}(Y,Y^H,y_o)
π_{n-1}(Y^H,y_o)
j
r_ē|S^{n-1}

where $r_{\bar{e}}[f] = [f|\bar{e} \times D^{n+1}]$ in $\Gamma_{n+1}{}^{G/H}(Y,Y^H,y_o)$; $\gamma[f] = [\tilde{f}]$, $\tilde{f}(\bar{g},s) = gf(s)$; and $\gamma_1[f] = [\tilde{f}_1]$, $\tilde{f}_1(\bar{g},s) = gy_o$, $\tilde{f}_1(\bar{e},d) = f(\bar{d})$ where $d \to \bar{d}$ is the map $D^n \to D^n/\partial D^n = S^n$.

<u>Lemma 1.10</u>:

1. If each path component of Y^H meets Y^G, then for $n > 0$, $\mathring{\Gamma}_n{}^{G/H}(Y,y_o) = [\Sigma^n G/H, Y]_o$ and $\Gamma_n{}^{G/H}(Y,y_o) = [\Sigma^n G/H, Y]_o \oplus \pi_n(Y,y_o)$.

2. If there is a subspace $Z \subset Y^G$ which is a deformation retract of Y^H, $\Gamma_n{}^{G/H}(Y,Y^H,y_o) = [\Sigma^n G/H, Y]_o \oplus \pi_n(Y,Y^H)$, $n > 0$.

<u>Proof</u>:

1. If y_1 Y^G is in the path component of Y^H containing y_o, $\Gamma_n{}^{G/H}(Y,y_o) \simeq \Gamma_n{}^{G/H}(Y,y_1)$ = homotopy classes of maps $G/H \times S^n \to Y$ such that $f(\bar{g},s_o) = gy_1 = y_1$. But $(G/H \times S^n)/G/H \times s_o \sim \Sigma^n G/H \vee S^n$. Similarly, for $\mathring{\Gamma}_n{}^{G/H}(Y,y_o) \simeq \mathring{\Gamma}_n{}^{G/H}(Y,y_1)$ = homotopy classes of map with $f(\bar{g},s_o) = f(\bar{e},s) = y_1$, and $(G/H \times S^n)/G/H \vee S^n = \Sigma^n(G/H)$.

2. With the hypothesis of (2) we get that $\Gamma_n{}^{G/H}(Y,Y^H,y_o) \simeq \Gamma_n{}^{G/H}(Y,Z,y_1)$ = homotopy classes of maps $f\colon G/H \times D^n \to Y$ such that $f(\bar{g},s_o) = y_1 \in Z$ and $f(\bar{g},s) = f(\bar{e},s) \in Z$, $s \in S^{n-1}$. Thus we have

$$\begin{array}{ccc} \Gamma_n{}^{G/H}(Y,Y^H,y_o) & \xrightarrow{r_{\bar{e}}} & \pi_n(Y,Y^H,y_o) \\ \updownarrow \simeq & & \updownarrow \simeq \\ \Gamma_n{}^{G/H}(Y,Z,y_1) & \xrightarrow{r_{\bar{e}}} & \pi_n(Y,Z,y_1) \end{array}$$

But $r_{\bar{e}}$ on $\Gamma_n^{G/H}(Y,Z,y_1)$ has a right inverse μ: $\pi_n(Y,Z,y_1)$ $\to \Gamma_n^{G/H}(Y,Z,y_1)$, $\mu[f] = [\tilde{f}]$, $\tilde{f}(\bar{g},d) = f(d)$, $\bar{g} \in G/H$, $d \in D^n$. Thus $\Gamma_n^{G/H}(Y,Y^H,y_o) \simeq \mathring{\Gamma}_n^{G/H}(Y,g_o) \oplus \pi_n(Y,Y_H,y_o)$, and the result follows from (1).

Remark: It is not difficult to show that for the conclusion of (1) it is sufficient that Ry_o: $G/H \to Y$, $Ry_o(\bar{g}) = gy_o$, is homotopically trivial; and for the conclusion of (2) that μ: $G/H \times Y^H \to Y$, $\mu(\bar{g},y) = gy$ is homotopic to a map which factors through pr_2: $G/H \times Y^H \to Y^H$.

Example 6: Let $G = S^1$ and $Y = CP^n = U(n+1)/U(1) \times U(n)$, and let S^1 act on CP^n via a representation of S^1 into $U(n-k) \subset U(n+1)$. Then for $H \subset S^1$ a closed subgroup, $Y^H = CP^q$, $q \geq k$. Let X be a G-CW complex with dim $X/G \leq 2k$, then $\pi_\ell(Y,Y^H) = 0$ all $\ell \leq 2k$, and $\Gamma_\ell^{G/H}(Y,Y^H,y_o) = \mathring{\Gamma}_\ell^{G/H}(Y,y_o) = [\Sigma^\ell G/H,Y]_o = 0$ for $\ell \neq 1$ and $\Gamma_1^{G/H}(Y,Y^H,y_o) = Z$. Thus there is a single obstruction in $H^1(X/G;B_1)$ to deforming a map f: $X \to Y$ into a G-map.*

If dim $X/G \leq 2k-1$, then two equivariant maps of X into Y which are homotopic are equivariantly homotopic.

Of course from a more pratical point of view one always has

*This class vanishes if for each $H \subset G$, $H^1(X^{(H)}/G, FrX^{(H)}/G;Z) = 0$.

Proposition 1.11: Let X be a G-CW complex and suppose $f: X \to Y$ is equivariant on the G-n-1 skeleton. If $\pi_n(Y,Y^H) = 0$ and $\bar{H}^i(G/H,\pi_{i+n}(Y)) = 0$ (reduced cohomology) all i and isotropy subgroups H; then f may be deformed to be equivariant on the n-skeleton.

Proof:

The obstruction is a class $[f] \in \Gamma_n^{G/H}(Y,Y^H,y_o)$ over a n-cell of type (H). If $\pi_n(Y,Y^H) = 0$, [f] comes from a class $[f_1] \in \mathring{\Gamma}_n^{G/H}(Y,y_o)$, $f_1: G/H \times S^n \to Y$, $f_1(\bar{g},s_o) = gf_1(\bar{e},s_o)$, $f_1(\bar{e},s) = y_o$. The obstructions to deforming f_1 to $f_o: G/H \times S^n \to Y$, $f_o(\bar{g},s) = gy_o$ lie in $H^i(G/H \times S^n, G/H \times s_o \cup \bar{e} \times S^n;\pi_i(Y)) = \bar{H}^i(\Sigma^n(G/H);\pi_i(Y))$
$= \bar{H}^j(G/H;\pi_{j+n}(Y))$.

Similarly, one has

Proposition 1.12: Let $f,f': X \to Y$ be equivariant maps and suppose we have a homotopy of f to f' which is equivariant on ${}_GX^{(n-1)}$. If $\pi_{n+1}(Y,Y^H) = 0$ and $\bar{H}^i(G/H,\pi_{i+n+1}(Y)) = 0$ all i and isotropy subgroups H, then the homotopy may be deformed to be equivariant on ${}_GX^{(n)}$.

Also, we note that by ordinary obstruction theory we have:

Theorem 1.13: Assume Y^H is n-simple all $H \subset G$. Let $f, f': X \to Y$ be equivariant maps such that $f|_G X^{(n-1)}$ is equivariantly homotopic to $f'|_G X^{(n-1)}$. Then the obstruction to determining $f|_G X^{(n)}$ to $f'|_G X^{(n)}$ through an equivariant homotopy which is the given homotopy on ${}_G X^{(n-2)}$ is a class in $H^n(X/G; C_n)$, where C_n is the bundle of coefficients such that C_n over an n-cell of type (H) is $\pi_n(Y^H)$.

Example 7: Let X be a G-CW comples. If Y^H is dim $\bar{X}_{(H)}$ connected for each closed $H \subset G$, Y is path connected and $Y^G \neq \emptyset$; then $f: X \to Y$ is homotopic to an equivariant map iff it is homotopically trivial.

Let $y_0 \in Y^G$. If f is homotopic to a constant map it is homotopic to $f_0^*: X \to y_0$. But f_0^* is equivariant. Conversely, since $X \times y_0 \subset E$, $X/G \times y_0 \subset \bar{E}$ and defines a section of $\bar{E}$ corresponding to the constant map f_0 into y_0. But any other section of $\bar{E}$ is homotopic to this section by 1.13.

Remark: Theorems 1.8, 1.9 and Propositions 1.11 and 1.12, as well as Theorem 1.13, all have relative versions - i.e. mod a sub G-CW complex.

Finally, we study the problem of getting a converse to Proposition 1.6 (2); namely: If $f: X \to Y$ is homotopic to a map which factors through X_G, when is f homotopic to an equivariant map?

Again we assume there is a space $Z \subset Y^G$ and a deformation retract $f: Y \to Z$. We let $E = E(X,Y)$ and $E_1 = E(UG \times X, Y)$. We have $(UG \times X) \times Z \subset (UG \times X) \times Y = E_1$. Then $X_G \times Z \subset \bar{E}_1$ and $X_G \times Y \xrightarrow{1 \times r} X_G \times Z \subset \bar{E}_1$ is a fibre homotopy equivalence (f.h.e.). Let $i = i_p: X \to UG \times X$ and $j = i \circ q_1: X \to X_G$, $q_1: UG \times X \to X_G$ the quotient map. The hypothesis implies f is homotopic to $f_1 \circ j$, $f_1: X_G \to Y$. Then by the f.h.e., the homotopy class of f_1 defines a homotopy class $[\bar{\sigma}_1]$ of sections of $\bar{E}_1$. Let $\lambda: UG \times X \to X$ be pr_2, and $\bar{\lambda}: X_G \to X/G$ the induced map of quotients. Suppose $[\bar{\sigma}_1] = \bar{\lambda}^*[\bar{\sigma}]$, $[\bar{\sigma}]$ a homotopy class of sections of $\bar{E}$. If $\phi: X \to Y$ is the equivariant map defined by $\bar{\sigma}$, then $\phi \circ \lambda: UG \times X \to Y$ is the equivariant map defined by $\bar{\lambda}^*\bar{\sigma}$. Now $\phi \circ \lambda \sim f_1 \circ q_1$, since $f_1 \sim f_1': X_G \to Z \subset Y$ and hence the equivariant map definec by $\bar{\sigma}$, is homotopic to $f_1' \circ q_1$. Thus $\phi = \phi \circ \lambda \circ i \sim f_1 \circ q_1 \circ i = f_1 \circ j \sim f$. We have proved:

<u>Lemma 1.14</u>: Let $f: X \to Y$ be homotopic to $f_1 \circ j$, $f_1: X_G \to Y$. The homotopy class of f_1 defines a homotopy class of sections $[\bar{\sigma}_1]$ of $\bar{E}_1$, via the f.h.e. $X_G \times Y \to \bar{E}_1$. If $[\bar{\sigma}_1] = \bar{\lambda}^*[\bar{\sigma}]$, f is homotopic to an equivariant map $\phi: X \to Y$, where the equivariant homotopy class of ϕ is that determined by $[\bar{\sigma}]$.

Corollary 1.15: Set $S\bar{E}$, $S\bar{E}_1$ be the homotopy classes of sections. If $\bar{\lambda}^*: S\bar{E} \to S\bar{E}_1$ is bijective, then $\bar{\lambda}^*$ induces a bijection $\bar{\lambda}^*: [X,Y]_G \approx [X_G,Y]$, and conversely.

Thus the problem comes down to studying the obstruction to deforming a section of $\bar{E}_1$ to one coming back from a section of $\bar{E}$, or equivalent ly of deforming an equivariant map $f: UG \times X \to Y$ equivariantly to one coming back from an equivariant map $\phi: X \to Y$.

By arguments completely analogous to those for Theorems 1.8, 1.9, we get:

Theorem 1.16: Let X be a G-CW complex and Y a G-n simple G space with $Z \subset Y^G$ a deformation retract of Y. Let $f: UG \times X \to Y$ be an equivariant map such that $f|UG \times ({}_GX^{(n-1)}) = \phi_{n-1} \circ \lambda$, where $\phi_{n-1}: {}_GX^{(n-1)} \to Y$ is equivariant. Then the obstruction to deforming f through equivariant maps, rel $UG \times ({}_GX^{(n-2)})$, to a map f' such that $f'|UG \times ({}_GX^{(n)}) = \phi_n \circ \lambda$, $\phi_n: {}_GX^{(n)} \to Y$ equivariant, is a class in $H^n(X/G; \mathcal{D}_n)$, where $\mathcal{D}_n$ over an n-cell of type (H) is $\Gamma_n^{BH}(Y, Y^H, Y_O)$.

Theorem 1.17: Let X and Y be as in 1.16. Let $\phi, \phi': X \to Y$ be equivariant maps such that $\phi \circ \lambda$, $\phi' \circ \lambda: UG \times X \to Y$ are equivariantly homotopic by a homotopy which pulls back over $UG \times ({}_GX^{(n-1)})$ from an equivariant homotopy of $\phi|_G X^{(n-1)}$ to $\phi'|_G X^{(n-1)}$. Then the obstruction to deforming the equivariant homotopy of $\phi\circ\lambda$ to $\phi\circ\lambda'$ to one coming back over $UG \times ({}_G X^n)$ from an equivariant homotopy of $\phi|_G X^{(n)}$ to $\phi'|_G X^{(n)}$, rel $UG \times ({}_G X^{(n-2)})$, is a class in $H^n(X/G;\mathcal{D}_{n+1})$.

Here,

$\Gamma_n{}^{BH}(Y,Y^H,y_o)$ = Equivariant homotopy classes of equivariant maps $f\colon UG \times G/H \times D^n \to Y$ such that $f|UG \times G/H \times S^{n-1} = \phi\circ\lambda$, $\phi\colon G/H \times S^{n-1} \to Y$ equivariant, $\phi(\bar{e},s_o) = y_o$.

If we let,

$\mathring{\Gamma}_n{}^{BH}(Y,y_o)$ = Equivariant homotopy classes of equivariant maps $f\colon UG \times G/H \times S^n \to Y$ such that $f(z,\bar{g},s_o) = gy_o$ and $f(z_o,\bar{e},s) = y_o$, $z \in U(G)$, $\bar{g} \in G/H$, $s \in S^n$,

we get a long exact sequence

$$\to \pi_{n+1}(Y,Y^H,y_o) \to \mathring{\Gamma}_n{}^{BH}(Y,y_o) \to \Gamma_n{}^{BH}(Y,Y^H,y_o) \to \pi_n(Y,Y^H,y_o) \to$$

Also note that looking at equivariant maps as sections of $\bar{E}_1$ we have the alternative description:

$\mathring{\Gamma}_n{}^{BH}(Y,y_o)$ = Homotopy classes of maps $f\colon BH \times S^n \to Y$ such that $f|BH \times s_o = f_o$ and $f(b_o,s) = f(b_o,s_o)$, where f_o is obtained as follows:

Let $\bar{\sigma}_o\colon BH \to \bar{E}_1|BH \times s_o$, $s_o \in S^{n-1}$, be a cross-section pulled back from a cross-section of $\bar{E}|s_o$ corresponding to an equivariant map $\phi_o: G/H \times S_o \to Y$, $\phi_o(\bar{g},s_o) = gy_o$. If $\psi: \bar{E}_1 \to X_o \times Y$ is the f.h.e., then $\bar{\sigma}_o$ defines a section $\psi\bar{\sigma}_o$ of $X_G \times Y$ and $f_o = pr_2\psi\bar{\sigma}_o$. Note that $\mathring{\Gamma}_n^{BH}(Y,y_o)$ depends, up to isomorphism, only on the homotopy class of f_o.

Just as for $\Gamma_n^{G/H}(Y,Y^H,y_o)$, $\Gamma_n^{BH}(Y,Y^H,y_o) = 0$ if $\pi_n(Y,Y^H,y_o) = 0$ and $\bar{H}^i(BH;\pi_{i+n}Y) = 0$ all i. Hence we have

<u>Proposition 1.18</u>: Let X be a G-CW complex and Y a G-space with $Z \subset Y^G$ a deformation retract of Y. Suppose $f: UG \times X \to Y$ is an equivariant map which comes back over $UG \times ({}_GX^{(n-1)})$ from an equivariant map of ${}_GX^{(n-1)}$ into Y. If $\pi_n(Y,Y^H,y_o) = 0$ and $\bar{H}^i(BH,\pi_{i+n}Y) = 0$ for all i and isotropy subgroups $H \subset G$, then f may be equivariantly deformed, rel $UG \times ({}_GX^{(n-1)})$, to f' where f' comes back over $UG \times ({}_GX^{(n)})$.

<u>Proposition 1.19</u>: Let X and Y be as in 1.18. Let $\phi,\phi': X \to Y$ be equivariant maps such that $\phi\circ\lambda$, $\phi'\circ\lambda: UG \times X \to Y$ are equivariantly homotopic by a homotopy coming back over $UG \times ({}_GX^{(n-1)})$ from an equivariant homotopy of $\phi|{}_GX^{(n-1)}$ to $\phi'|{}_GX^{(n-1)}$. If $\pi_{n+1}(Y,Y^H,y_o) = 0$ and $\bar{H}^i(BH,\pi_{i+n+1}(Y)) = 0$ all i and isotropy subgroups $H \subset G$, the homotopy of $\phi\circ\lambda$ to $\phi'\circ\lambda$ may be equivariantly deformed rel $UG \times ({}_GX^{(n-1)})$, to one coming back over $UG \times ({}_GX^{(n)})$.

References

1. Bierstone, E., The equivariant covering homotopy property for differentiable G-fibre bundles.

2. Bredon, G., Introduction to Compact Transformation Groups, Academic Press, New York, 1972.

3. Hattori, A., and Yoshida, T., Lifting compact group actions in fibre bundles, Japanese J. of Math., New Series 2 (1976), 13-26.

4. Lashof, R., Stable G-smoothing theory, in preparation.

5. Lashof, R., and Rothenberg, M., G-smoothing Theory, to appear in Proceedings AMS Topology Conference, Stanford.

SYMPLECTIC LIE GROUP ACTIONS

A. S. Mishchenko and A. T. Fomenko

During recent years great attention has been attracted to the problem of the full integrability (in Liouville sense) of different dynamical systems on symplectic manifolds. In this report, a new method of "noncommutative integration" worked out by the authors will be stated and some concrete applications of this method to the integrability problem of some important dynamical systems will be demontrated (for instance, the full integrability of the n-dimensional rigid body motion equations with or without a fixed point will be proved).

(1) As to the statement of the noncommutative integration method will be needed some preliminary remarks. Proofs will be omitted for want of place.

Let (M^{2n},w) be a symplectic manifold having the form w, and let V be a linear finite dimensional space of smooth functions on M^{2n}, closed with respect to Poisson's bracket $\{f,g\}$, i.e. V is a finite dimensional Lie algebra. It must be remembered that for every smooth function f, the term sgrad f denotes a vector field which is uniquely defined by the relation: $w(\text{sgrad } f, Y) = Y(f)$, where Y is an arbitrary vector field. Then Poisson's bracket of the pair of functions f and g is the following function:

$$\{f,g\} = w(\text{sgrad } f, \text{sgrad } g).$$

It must also be remembered that there is an identity

$$\operatorname{sgrad}\{f,g\} = [\operatorname{sgrad} f, \operatorname{sgrad} g].$$

So the Lie algebra V is represented as a vector field algebra via sgrad on M. We may consider the simply connected Lie group $\mathcal{G}$ as associated with the Lie algebra V. $\mathcal{G}$ symplectically acts on M (i.e. it preserves w) because the symplectic form w is invariant relative to the actions of any one-dimensional subgroup generated by vector fields $\operatorname{sgrad} f$, $f \in V$.

For every point $m \in M$ a linear functional $\bar{m} \in V$ is defined, where $\bar{m}(f) = f(m)$, $f \in V$. Let $P_\xi = \{m \in M: \xi(f) = f(m),\ \xi \in V^*\}$. Let the algebra V have a basis $f_1,\dots,f_k$ $(k = \dim V)$ such that the functions $f_1,\dots,f_k$ are functionally independent at the points of general position on M. Then P_ξ is a common level surface of the functions $f_1,\dots,f_k$:

$$P_\xi = \{m \in M: f_i(m) = \xi(f_i) = c_i\}.$$

If $\varphi: M \to V^*$, $\varphi(m) = \bar{m}$, then with the assumptions above, φ has inner points in V^* and the surface $P_\xi = \varphi^{-1}(\xi)$ is a smooth manifold, differentials df_i being linearly independent at every point $m \in P_\xi$, $i = 1,\dots,k$. Let us consider operators ad_f in the Lie algebra V; then in the conjugate space V^* there arise operators ad_f^*, where $\operatorname{ad}_f^*(\xi) = \xi(\{f,g\})$. Let $\{\xi,f\} = \operatorname{ad}_f^*(\xi)$ (not to be confused with Poisson's bracket) and define a subalgebra $H_\xi \subset V$ as the covector ξ's annulator, i.e. $H_\xi = \{h \in V: \{\xi,h\} = 0\}$. It may be found by direct calculations that if $h \in H_\xi$, $m \in P_\xi$, then $\operatorname{sgrad} h(m) \in T_m(P_\xi)$ where T_m is a tangent space of P_ξ at the point m. It implies that if $\mathcal{G}_\xi$ is a subgroup of $\mathcal{G}$, associated with the anullator $H_\xi \subset V$, then the surface P_ξ is invariant under the $\mathcal{G}_\xi$-action on M. Let us consider the restriction of the form w onto P_ξ: $\tilde{w} = w|_{P_\xi}$.

Then the kernel $\mathrm{Ker}(\tilde{w})$ of the form $\tilde{w}$ at the point $m \in P_\xi$ coincides with the subspace $H_m \subset T_m(P_\xi)$, generated by the subalgebra $H_\xi \subset V$, i.e. by vectors $\mathrm{sgrad}(H)$, $h \in H_\xi$.

These statements were discovered for the first time by J. Marsden and A. Weinstein (see [1]).

<u>Lemma 1</u>. Let the action of the group $\mathcal{G}_\xi$ on P_ξ be of unique closed orbit type, i.e. having a unique class of conjugate stationary (discrete-) subgroups. Then the factor manifold $N_\xi = P_\xi/\mathcal{G}_\xi$ has a nondegenerate symplectic form Ω with $p^*(\Omega) = w|P_\xi$ where $p: P_\xi \to N_\xi$ is the natural projection.

The proof is realized by immediate calculation (see [2]).

(2) Now let $\vec{v} = \mathrm{sgrad}\, F$ be a Hamilton current (vector field) on M, for which all the functions $f \in V$ are integrals, i.e. $\{F,f\} = 0$. Then the vector field $\vec{v}$ is tangent to the surface P_ξ and subgroup $\mathcal{G}_\xi$, acting on P_ξ, leaves it invariant because all the functions $f \in V$ commute with the Hamilton function F. Let $E(F)_\xi$ be a vector field on the factor manifold $N_\xi = P_\xi/\mathcal{G}_\xi$. It is easy to verify that the field $E(F)_\xi$ is a Hamilton one with respect to the symplectic form Ω on N_ξ for the Hamilton function $\tilde{F}$, which is equal to the projection of $F|_{P_\xi}$. Suppose now that the group $\mathcal{G}$, generated by the algebra V has only one stationary subgroup type when it acts in a neighbourhood of P_ξ. Let us consider the factor manifold $N = M/\mathcal{G}$; the function F is invariant with respect to $\mathcal{G}$ that is why the projection $p_N: M \to N$ transforms the field $\vec{v}$ into some field $E(F)$ on N. N will be referred to as an Euler manifold, and $E(F)$ - Euler's equation for the initial Hamiltonian system $\vec{v}$ on M. It is clear that $p_N(P_\xi) = N_\xi$.

In this way we can proceed from the system $\vec{v}$ on M to the new Hamiltinian system $E(F)_\xi$ on N, where $\dim N_\xi < \dim P_\xi$ if $\dim V > 0$. This reduction allows us to consider a new Hamiltonian system family (in this connection see also [1]) and, in particular, it permits us to obtain new integrals for the initial system v on M. Indeed, let q be an integral for the current $E(F)$ on N (i. e. the restriction $q_\xi = q|_{N_\xi}$ on every surface N_ξ of generel position is an integral of the Hamilton current $E(F)_\xi$). Then the function $q \circ p_N$ is invariant under the $\mathfrak{g}$-action on M; it follows that the function $q \circ p_N$ on M is an integral of the current $\vec{v}$ and is in involution with every function of the algebra V. Let us note that the Hamilton function F is considered as not belonging to the algebra V. Generally, for the sake of simplicity, we can consider not the whole manifold M but only some open neighbourhood $U(P_\xi)$ of the surface P_ξ in generel position, and then carry the constructions onto the whole manifold M.

<u>Proposition 1</u>. Let V be the finite dimensional Lie algebra of integrals of the Hamiltonian system $\vec{v} = \mathrm{sgrad}\, F$ on a symplectic manifold M, N an Euler manifold, $E(F)$ the Euler equation on N. Let V' be a finite dimensional Lie algebra of integrals for $E(F)_\xi$ on the factor manifolds $N_\xi \subset N$. Let $V'' = \{q \circ p_N : q \in V'\}$, $p_N : M \to N$ the projection. Then the space $V \oplus V''$ is a Lie algebra of integrals for $\vec{v} = \mathrm{sgrad}\, F$ with $[V, V''] = 0$, i.e. the subalgebras V and V'' commute.

For details and proof, see [2].

Note in particular that the Hamilton function F which is an integral of the current $E(F)_\xi$ on N_ξ is contained in the subalgebra V''. So the most interesting case is when the current $E(F)_\xi$

has some new integrals extending the subalgebra V'' besides the Hamilton function. If such additional integrals have been arrived at, then the process may be continued by turning from the symplectic manifold N_ξ to the new factor manifold of small dimension and so on.

(3) Let us consider an important special case, when $\dim V + \operatorname{rg} V = \dim M$, where $\operatorname{rg} V = \dim H_\xi$ (the covector ξ is in generel position). Since $\dim V + \dim P_\xi = \dim M$, we have $\dim P_\xi = \dim H_\xi$, i.e. $P_\xi = \mathcal{G}_\xi/\Gamma$ where Γ is a discrete subgroup in $\mathcal{G}_\xi$. The dependence of the subgroup $\mathcal{G}_\xi$ on the covector $\xi \in V^*$ is very simple. If ξ is in generel position, then ξ being little varied $\mathcal{G}_\xi$ is subject to an inner automorphism in $\mathfrak{g}$. The vector field $\operatorname{sgrad} F$ tangent to the surface P_ξ is a left-invariant vector field and consequently is defined by a vector of the Lie algebra H_ξ of the group $\mathcal{G}_\xi$.

<u>Proposition 2</u>. The transformation $\varphi: M \to V^*$ (see above) is equivariant under the coadjoint action of $\mathfrak{g}$ in V^*. If $\xi \in V^*$ is a covector in common position, then there is an open set $W \subset V^*$, $\xi \in W$, invariant under Ad^*, such that $U = \varphi^{-1}(W)$ may be represented as a bundle $\varphi: U \to W$ with a fiber diffeomorphic to P_ξ.

For details and proof see [2].

Let further $m_0 \in P_\xi$ and Γ be a stationary subgroup of m_0 in $\mathcal{G}_\xi$, let $W_0 \ni \xi$ be such a neighbourhood of $\xi \in V^*$ that $\varphi^{-1}(W_0) = W_0 \times P$. Since $\mathfrak{g}$ acts locally free on M, this neighbourhood may be represented as $\varphi^{-1}(W_0) = X_0 \times Y_0 \times P_\xi$, where $x \times (Y_0 \times P_\xi)$, $x \in X_0$ are orbits of the $\mathfrak{g}$-action. So there exists an equivariant diffeomorphism $\psi: X_0 \times Y_0 \times P_\xi \to X_0 \to \mathfrak{g}_0$, where $\mathfrak{g}_0$ is a unit neighbourhood in $\mathfrak{G}$. The mapping $\psi: Y_0 \times P_\xi \to \mathfrak{G}_0$ transforms fibers $y \times P_\xi$, $y \in Y_0$, inti right cosets by the subgroup $\mathfrak{H}_\xi$.

Proposition 3. The form w on M coincides on the submanifold $Y_0 \subset M$ with the Kirillov form w_{Y_0} on the orbit of the coadjoint representation of $\mathfrak{G}$ The manifold $X_0 \times P$ is symplectic with respect to the restriction of the form w on $X_0 \times P$.

For details and the proof see [2]. So, a typical neighbourhood of the surface P in M has a standard character. The neighbourhood is represented as a product: $X_0 \times P \times Y_0$, where Y_0 is a domain in the orbit $Ad^*\mathfrak{G}(\xi) \subset V^*$, $X_0 \times P_\xi$ is a symplectic manifold. The Hamilton function F, being invariant under $\mathfrak{G}$, depends only on the coordinates of the space X_0.

Theorem 4.

1. Let us consider a symplectic manifold M with a form w and a Hamilton current satisfying the previous conditions, i.e. $\dim V + rg\, V = \dim M$. Then the system motion realizes along tori P_ξ, provided this common surface is compact (i.e. every compact connected level surface P of general position is diffeomorphic to a torus), dimension is equal to the range of initial algebra of integrals (i.e. it is equal to the dimension of the annulator of the covector in generel position). In general case (i.e. when the algebra of integrals V is noncommutative) the dimension of these tori is less than a half of the dimension of manifold M. On every torus P_ξ we may choose such curvilinear coordinates that the vector field v will define a pseudoperiodical motion (i.e. it will have constant components defined by unique vector of algebra H). When the algebra of integrals V is commutative, we receive the classical Liouville

theorem about the integrability of systems, having a collection of integrals being in involution, the number of which is equal to half the dimension of manifold M.

For detail and the proof see [2]. We shall call this result the "noncommutative Liouville theorem". One of the fundamental observations which are the basis of the proof of theorem 1, is that (a fact discovered for the first time by M. Duflo and M. Vergne) if the covector $\xi \in V^*$ of the dual space to the Lie algebra V is in general position, then its annulator H_ξ is commutative (see [3], [4]). If the algebra V is noncommutative, then the level surface P_ξ and the orbit $\mathcal{G}(m)$ growing from the point $m \in M$ (of general position) are different, the surface P_ξ is not invariant with respect to $\mathcal{G}$ and is not contained in $\mathcal{G}(m)$. If V is commutative, then the surface P_ξ coincides with the orbit $\mathcal{G}(m)$, being a torus, on which $\mathcal{G}$ acts by left translations.

(5) So, if the conditions of theorem 1 are satisfied, then the system motion is realized along the tori, whose dimension is equal to the range of the algebra V and less than n, when this algebra is noncommutative. For example, if the algebra V is semi-simple, then $r = \mathrm{rg}\, V \simeq \sqrt{\dim V}$, i.e. by comparison with the "commutative Liouville theorem" situation, the tori P_ξ are strongly degenerate. But, it is possible that these "small-dimensional" tori may be organized in half dimensional tori and the motion along them will therefore be actually realized along the half dimensional tori to the tori of less dimensions.

Let us formulate this question in the following way: let a Hamiltonian system on the symplectic manifold M be fully integrable in the noncommutative Liouville sense, is the same system fully integrable in a usual commutative sense, i.e. does a commutative algebra V_0 of functionally independent functions with $2\dim V_0 = \dim M$ exist?

It is useful to know an answer because of the greater simplicity of the motion picture of integral trajectories, in spite of the fact that should the algebra V_0 exist, the invariant surfaces dimension is greater than $r = \dim P_\xi$. It appears that for the wide class of algebras V the answer is positive. Namely let us consider a Lie algebra V, satisfying the following simple condition (condition A): there exists a finite dimensional space of functionally independent functions F_0, defined on the dual space V^*, each pair of functions being in involution on the orbits of coadjoint representation $Ad^*_g: V^* \to V^*$, and $\dim F_0 = 1/2(\mathrm{rg}\, V + \dim V)$.

<u>Theorem 2</u>. Let M be a symplectic manifold, V a Lie algebra of functionally independent integrals of a Hamiltonian dynamical system, $\dim V + \mathrm{rg}\, V = \dim M$. If the algebra V satisfies the condition A, then there exists another commutative Lie algebra V_0 of functionally independent integrals with $2\dim V_0 = \dim M$. Hence, in this case "noncommutative Liouville integrability" implies classical integrability, i.e. a "commutative one". The algebra V_0^0 consists of functions which are functionally dependent on functions of algebra V.

For details and the proof see [2]. A question: do Lie algebras V satisfying the condition A exist?

<u>Theorem 3</u>. If V is a semisimple Lie algebra, then it satisfies the condition A.

For details and the proof see [2], [5], [6].

The proof is based on the Theorem about the full integrability of the Euler equations (of special kind) on semisimple Lie groups proved by the authors.

The condition A is also satisfied for reductive Lie algebras. So, if a Hamiltonian system permits a semisimple (or reductive) algebra of integrals V with condition: $\dim V + \mathrm{rg}\, V = \dim M$, then it is integrable both in the commutative and noncommutative senses. At the same time the existence of the noncommutative algebra V forces the system to move along the subtori of small dimension in the half-dimensional tori, defined by commutative algebra V_0 of integrals.

Hypothesis. Noncommutative full integrability in the Liouville sense on any symplectic manifold implies the commutative Liouville integrability.

The authors do not yet have any counter example for this hypothesis. Moreover, there are some observations confirming the hypothesis. Indeed, let us consider a mapping $\psi: M \to V^*$ and a covector in general position and a surface $P_\xi = \psi^{-1}(\xi)$, diffeomorphic to a torus (see theorem 1). Let us consider a rather small neighbourhood U of the point ξ and let $W = \psi^{-1}(U)$. As $\mathcal{G}$ acts on V^*, the point ξ generates an orbit $O(\xi)$ of the coadjoint representation $\mathrm{Ad}^*\mathcal{G}$. The dimension of $O(\xi)$ is equal to $k-r$, where $k = \dim V$, $r = \mathrm{rg}\, V$. Let us assume that we could construct in the neighbourhood U of the point ξ a collection of smooth functions $g_1, \dots, g'_{(k-2)/2}$; $\sigma_1, \dots, \sigma_r$ such that

(1) All these functions in pairs are in involution with respect to the standard symplectic structure on the orbit of the coadjoint representation (i.e. with respect to the Kirillov form),

(2) All these functions are functionally independent in U,

(3) The common level surface of functions $\sigma_1,\dots,\sigma_r$ (i.e. $\sigma_\alpha = c_\alpha = \text{const}$, $1 \leq \alpha \leq r$) coincides with the orbit $O(\xi)$ containing the point ξ, and the functions $\{\sigma_i\}$ are constant on the orbit $O(\eta)$, where $\eta \in U(\xi)$.

Remark: as the functions $\{\sigma_i\}$ are constant on orbits of the coadjoint representation, they are automatically in involution with all functions $g_1,\dots,g_{(k-r)/2}$.

Then we may consider the collection of functions $\tilde{g}_1,\dots,\tilde{g}_{(k-r)/2}$ $\tilde{\sigma}_1,\dots,\tilde{\sigma}_r$, where $\tilde{g}_\alpha = \tilde{g}_\alpha \circ \varphi$, $\tilde{\sigma}_i = \sigma_i \circ \varphi$, $1 \leq \alpha \leq (k-r)/2$, $1 \leq i \leq r$. These functions (numbering $(k+r)/2$) will be defined in the neighbourhood W on the manifold M and functionally independent, their common level surface will be a smooth submanifold of dimension $(k+r)/2$, i.e. it will be a half dimensional submanifold in M. Besides, this surface contains the torus P_ξ. The most important property of these functions is as follows: they are in pairs in involution with respect to the initial symplectic structure on M. It follows from proposition 3 because the submanifold Y_0 is turned exactly to the orbit $O(\xi)$. Let us now construct a collection of functions on V^* satisfying the properties (1) - (3). To do this, we shall introduce into V^* local coordinates $t_1,\dots,t_r$, orthogonal to the orbits $O(\xi)$ of coadjoint representation, $\{p_i,q_i\}$, $1 \leq i \leq \frac{k-r}{2}$ on the orbits $O(\xi)$ themselves. Here we make use of the fact that all these functions may be considered as euclidean coordinates in an open disk of euclidean space. We shall take $p_i^2 + q_i^2$ as functions $g_i,\dots,g_{(k-r)/2}$, $p_i^2 + q_i^2$ being obviously in involution to each other on the orbits $O(\xi)$ with respect to the Kirillov form. We shall take t_i, $1 \leq i \leq r$ as functions σ_i. It is clear that the functions so constructed satisfy properties (1) - (3). So we recieve a collection of $(k+r)/2$ integrals on the manifold M, being in pairs in involution and defining (in the neighbourhood W) a common level surface of dimension $(k+r)/2$, containing the torus P_ξ. It should be mentioned here that all the coordinates (p_i, q_i) of the covector ξ are nontrivial, then every equation $p_i^2 + q_i^2 = \text{const}_i$ defines a circle, here $\text{const}_i = (\xi_{p_i})^2 + (\xi_{q_i})^2$, where $\xi = (\xi_{p_i}, \xi_{q_i})$ - coordinates of covector ξ. The level surface: $p_i^2 + q_i^2 = \text{const}$, $1 \leq i \leq (k-r)/2$ in $O(\xi)$ is a torus of dimension $(k-r)/2$, hence, the level surface $g_i = \text{const}$ $1 \leq i \leq \frac{k-r}{2}$; $\tilde{\sigma}_j = \text{const}_j$, $1 \leq j \leq r$, is also a torus of dimension $\frac{k+r}{2}$, because all functions $(\tilde{g}_1, \tilde{\sigma}_j)$ are in involution on M. Perhaps, this torus may be obtained from the torus P_ξ by movements in M making use of the elements of $\mathfrak{g}$, which generate a torus $p_i^2 + q_i^2 = \text{const}_i$ in V^*; $1 \leq i \leq \frac{k-r}{2}$, being included into the orbit $O(\xi)$. So, in the neighbourhood W we have constructed a collection of commuting functions numbering half-dimension of M, turning W into a torus-bundle, where the tori contains surfaces $P_\eta, \eta \in U$. So, the collection $\tilde{g}_i$; $1 \leq i \leq \frac{k-r}{2}$; $\tilde{\sigma}_j$, $1 \leq j \leq r$, generate a commutative Lie algebra, satisfying (locally) the requirements of the commutative Liouville theorem; in particular all of them are functionally independent.

Since new tori include "previous" tori P_η, all functions $(\tilde{g}_i, \tilde{\sigma}_j)$, may be hence expressed (in some way) through the previous functions $f_1, \ldots, f_k$, i.e. the commutative algebra V_0 is found in an infinite dimensional Lie algebra, generated (functionally) by functions $f_1, \ldots, f_k$. Though the whole construction was done locally, it may, probably, be extended rather easily to the whole manifold M.

For example, we could try to make the functions (g_i, t_j) finite (preserving their functional independence and involutivity on their common support) and prolong them to the whole M. Then the commutative Liouville's integrability will occur in the open neighbourhood of the level surface P_ξ of general position.

We also mean that manifold M may be an algebraic one, so (if it is possible to choose polynomials as (g_i, t_j)) the construction may be extended into the whole M, it is possible with the exception of a set with a trivial measure, by algebraicity.

(6) It should not be supposed that the requirement on the functions $f_1, \ldots, f_k$ to be additive generators of a Lie algebra seems very important. The requirement that the linear space V of integrals generate a Lie algebra, may be changed by a slightly weaker condition. Later we shall state a result by A.V. Streltzov, who succeeded in eliminating the condition of the closeness of V relative to the Poisson's bracket. Let us consider a linear space V, generated by the linear combinations of functionally independent integrals $f_1, \ldots, f_{n+p}$; $0 \leq p \leq n-1$. Let A be a Lie algebra of functions, generated (in functional sense) by functions $f_1, \ldots, f_{n+p}$. Let us consider a mapping

$$\varphi: M^{2n} \to V \simeq R^{n+p}; \quad \varphi(x) = (f_1(x), \ldots, f_{n+p}(x)).$$

Let $\xi \in R^{n+p}$ be a regular point of the mapping φ and let submanifold $P_\xi = \varphi^{-1}(\xi)$ be compact and connected. Then there exists a neighbourhood $U(\xi)$ such that $\varphi: \varphi^{-1}(U(\xi)) \to U(\xi)$ is a locally trivial bundle which is diffeomorphic to the cartesian product $U(\xi) x \varphi^{-1}(\xi)$. Every element $f \in A$ is a function, functionally dependent on generators $f_1, \ldots, f_{n+p}$, that is why one can put every element $f \in A$ into a correspondence with a function $\tilde{f}$ on $U(\xi)$ such, that $f(x) = \tilde{f}(f_1(x), \ldots, f_{n+p}(x))$, $x \in \varphi^{-1}U(\xi)$. Let us put: $f_{ij} = \{f_i, f_j\}$; then we have functions f_{ij}, $1 \leq i, j \leq n+p$.

<u>Proposition 4</u>. Let ξ be a regular point of the mapping φ; $P_\xi = \varphi^{-1}(\xi)$ - a compact connected submanifold, and rang $\|f_{ij}(\xi)\| \leq 2p$. Then: (I) There exist n-p functions $g_i(x),\dots,g_{n-p}(x)$ with independent integrals: $\{g_i(x); f_j(x)\} = 0$, $i = 1,\dots,n-p$; $j = 1,\dots,n+p$. (2) The vector fields sgrad $g_i(x)$, $i = 1,\dots,n-p$, on the surface P_ξ generate a basis of the tangent space $T_x(P_\xi)$ in an arbitrary point $x \in P_\xi$ and commute on P_ξ, (3) The surface P_ξ is diffeomorphic to an (n-p)-dimensional torus.

One can take the functions $g_1(x),\dots,g_{n-p}(x)$ as belonging to the linear space V.

The requirement: rang$\|\tilde{f}_{ij}(\xi)\| \leq 2p$ - is equivalent to the requirement: rang$\|\tilde{f}_{ij}(\xi)\| = 2p$, for if a strong inequality rang$\|f_{ij}(\xi)\| < 2p$ holds then it could be shown that more than n-p functions $g_i(x)$ exist whose Poisson's bracket with the functions $f_j(x)$, $j = 1,\dots,n+p$ are trivial on P_ξ, which is impossible because of the linear independence of the differentials of the functions $g_i(x)$ and non-triviality of the form w.

<u>Theorem I (A)</u>. Let ξ be a regular point of the map φ, P_ξ - a compact manifold, a component of manifold $\varphi^{-1}(\xi)$. Let rang$\|\tilde{f}_{ij}(\xi)\| = 2p$ in any neighbourhood $U(\xi)$. Then there exists a neighbourhood of manifold P_ξ and functions $g_1(x),\dots,g_{n-p}(x)$, defined in this neighbourhood with linear independent differentials, such that $\{g_i(x), f_j(x)\} = 0$, $i = 1,\dots,n-p$; $j = 1,\dots,n+p$ for every point of this neighbourhood. There exists such a neighbourhood of surface P_ξ (which is a torus), in which we may fix I, p, q such that $w = d(\Sigma_{i=1}^{n-p} I_i \, d\varphi_i + \Sigma_{j=1}^{p} p_j dq_j)$, where w is a symplectic form. Here the coordinates I, p, q may by taken as

functions of $f_1, \ldots, f_{n+p}$.

Consequently, if (M,w,H) is a Hamiltonian dynamic system on the manifold $M, f_1(x), \ldots, f_{n+p}(x)$ a collection of integrals satisfying the conditions of the theorem 1(A), or see Proposition 4) and containing a Hamiltonian H, then the movement of the system is realized along the tori of dimension $n-p$ and is a pseudoperiodical one.

While functions $g_1(x), \ldots, g_{n-p}(x)$ may be taken to belong to V, we have the following picture of noncommutative integrability: the linear space V generates (a generally speaking infinite-dimensional) Lie algebra A closed relative to Poisson's bracket; while in every individual point there is only a finite number of linearly independent vector fields sgrad $f, f \in A$; other fields belong to the kernel of the representation: $f \to$ sgrad f (the kernel varies from point to point). Along P_ξ we can choose sgrad $f_1, \ldots,$ sgrad f_{n+p}, as linearly independent fields. These fields represent "the effective part" of the algebra A, acting nontrivially in a chosen point, in another point this "effective part" will vary. It now appears, that assuming rang $\|\tilde{f}_{ij}(x)\| = 2p$, among the functions of the space V there exist exactly $n-p$ commuting functions representing "an effective part" of, generally speaking, an infinite dimensional annulator of a covector in general position, whose skew gradients $\{\text{sgrad } g_i\}$ generate the tangent space to the surface P_ξ, which as a result of this becomes a torus of dimension $n-p$. Note, that generally speaking, the functions $g_1, \ldots, g_{n-p}$ commute only on the torus P_ξ. With ξ varying these functions will also be subjected to a variation. While $(n+p) + (n-p) = 2p$, and $n+p = \dim V$, $n-p = \dim(g_1, \ldots, g_{n-p})$, this condition is analogous to the condition: $\dim V + \mathrm{rg}\, V = \dim M$. So, the previous condition: V is a Lie algebra, - is changed to the condition: $\mathrm{rang}\, \|\tilde{f}_{ij}(\xi)\| = 2p$. This (A.V.Streltzov's) result, just stated by ourselves, is also connected with N.N. Nekhoroshev's results (see [11]). N.N. Nekhoroshev has shown some conditions, under which a symplectic manifold M^{2n} is diffeomorphic to the Cartesian product $T^s \times D^r$ ($r \geq s$, T^s-s-dimensional torus, D^r - a domain in an euclidian space) and permits a coordinate system $I, p\varphi (\mathrm{mod}\ 2\pi), q$, in which the symplectic form is:

$$w = d(\Sigma_{i=1}^{s} I_i d\varphi_i + \Sigma_{j=1}^{n-s} p_j dq_j)$$

These coordinates, generalizing variables: action-angle, are convinient in the investigation of a "degenerate" integrable system, i.e. such that its motion is realized along the tori of dimension less than n. To be more exact there is the following theorem (see [11]); let there be on a symplectic manifold (M^2, w) a collection of functions $f_1, \ldots, f_{n-p}; f_{n-p+1}, \ldots, f_{n+p}; p \geq 0$, whose differentials are linearly independent in every point $m \in M^{2n}$. Let every one of these functions be in involution on M with every one of the first $n-p$ functions. Let the components of manifolds: $\{m \in M : f_i(m) = c_i, i = 1, \ldots, n+p\}$, where c_i is a constant, - be compact. Then each such component has a neighbourhood U, where one may fix canonical coordinates $(I, p, \varphi (\mathrm{mod}\ 2\pi)$, such that $w = d(\Sigma_{i=1}^{n-p} I_i d\varphi_i + \Sigma_{i=1}^{p} p_j dq_j)$, where $I_i = I_i(f_1, \ldots, f_{n-p})$, and p_j, q_j are functions of $f_1, \ldots, f_{n-p}; f_{n-p+1}, \ldots, f_{n+p}$.

When $p = 0$, we obtain usual variables: action-angle $(I, \varphi (\mathrm{mod}\ 2\pi))$. So the linear space V, generated by $f_1, \ldots, f_{n+p}$, is not an algebra and contains in itself the centre generated by $f_1, \ldots, f_{n-p}$.

(7) Now let us apply the method obtained of "noncommutative Liouville integrability" to the problem of integrability of concrete dynamic systems. Let us first consider Hamiltonian currents on the cotangent fiber-bundles for Lie groups - geodesic currents. It appears that theorems 1, 2, 3 permit fully to integrate geodesic currents on the cotangent fiber-bundle of a semisimple Lie group for left invariant metrics, describing (in some exact sense) a rigid-body motion without fixed point.

Let $\mathfrak{G}$ be a Lie group, $T\mathfrak{G}$ - the tangent bundle, $T^*\mathfrak{G}$ - the cotangent fiberbundle; $\mathfrak{G}$ acts on $\mathfrak{G}$, $T\mathfrak{G}$, $T^*\mathfrak{G}$ in two ways - by using left and right translations L_g and R_g accordingly. The composition $L_g \cdot R_{g^{-1}} = R_{g^{-1}} L_g$ transform a fiber of the (co) tangent bundle $T_e(T_e^*)$ into itself. Let $\pi: T\mathfrak{G} \to \mathfrak{G}$, $\pi^*: T^*\mathfrak{G} \to \mathfrak{G}$ be projections of bundles; then $T^*\mathfrak{G}$ is supplied with symplectic structure w, equal to $w = d\Omega$, where Ω, a 1-form on $T^*\mathfrak{G}$ is defined as follows: let V be a tangent vector at a point $t \in T^*\mathfrak{G}$ $Y = d\pi(X)$ - tangent vector to $\mathfrak{G}$ at the point $g = \pi(t)$. Let us put $\Omega(X) = t(Y)$.

It is easy to check that w is a nondegenerate form. Every Hamilton's function $H(t)$, defined on $T^*\mathfrak{G}$, $t \in T^*\mathfrak{G}$ generates a Hamilton vector field $\vec{v}(H)_t : w(\vec{v}(H); X) = (dH, X), X \in T_t(T\mathfrak{G})$. We shall restrict ourselves to the left invariant functions; i.e. $H(L_g t) = H(t)$, $t \mp T^*\mathfrak{G}$, $g \in \mathfrak{G}$

Lemma 2. (see [5]). If a function $H(t), t \in T^*\mathfrak{G}$ is left invariant, then the field $\vec{v}(H)$ is also left invariant.

While $\vec{v}(H)$ is left invariant, the projection $l: T^*\mathfrak{G} \to T_e^*$, defined by the formula: $l(t) = L_{\pi(t)^{-1}}(t), t \in T^*\mathfrak{G}$ is in accordance with the field $\vec{v}(H)$, i.e. there exists a vector field $E(H)$ on

T_e^*, which is in accordance with the field $\vec{v}(H)$. It is clear that the equations $E(H)$ are precisely Euler's equations for the given Hamiltonian system $\vec{v}(H)$.

Let $\xi \in T_e^*, x \in T_e, \{\xi, x\} = ad_x^*(\xi)$. If f is a function on T_e^*, then $df \in T_e$.

Lemma 3. (see [5]). Let H be a leftinvariant function on $T^*\mathfrak{G}$. Then $E(H)_t = \{t, dH(t)\}, t \in T_e^*$. Here $Dh(t) \in T_e$ is the value of dH at the point $t \in T_e^*$.

It follows that if $f(t,g) = h(Ad^*_{g^{-1}}(t))$ (where h is a smooth function), $t \in T_e^*, g \in \mathfrak{G}$, then f is a first integral of the motion of system $E(H)$.

Before studying the integrability of the current $\vec{v}(H)$ on $T^*\mathfrak{G}$, let us investigate the integrability of the current $E(H)$ on T_e^*. Euler's equations $E(H)$ on T_e^* are:

$$\dot{t} = \{t, dH(t)\}, t \in T_e^*$$

and define a dynamical system on the dual space to the Lie algebra T_e.

Lemma 4. (see [5]). Let f be a function, constant on the orbits of the coadjoint representation on T_e^*. Then f is a first integral of the system $E(H)$.

So the vector field $E(H)$ is tangent to the orbits of coadjoint representation $O(t), t \in T_e^*$, where $O(t) = \{Ad_g^*(t), g \in \mathfrak{G}\}$. It is clear that $O(t) = \mathfrak{G}/\mathfrak{G}(t)$, where $\mathfrak{G}(t)$ is a stationary subgroup.

Lemma 5. (see [5]). A function f is constant on orbits of coadjoint representation $O(t)$, iff the identity: $\{f,df(t)\}=0$ holds. On orbits $O(t)$ there is a natural symplectic structure: if $X=\{t,x\}$, $Y=\{t,y\}$ - are two tangent vectors to the orbit $O(t)$, then $w(X,Y)=\langle t,[X,Y]\rangle$, where $\langle t,\rho\rangle$ denotes the value of covector t on the vector ρ.

Lemma 6. (see [5]). Euler's equations $E(H)$ generate a Hamiltonian vector field on every orbit $O(t)$, supplied by the above mentioned symplectic structure, with the Hamiltonian function being equal to the restriction of $H(t)$ into $O(t)$.

Let us now describe an important class of functions on the orbit $O(t)$ which are in pairs in involution with respect to the symplectic structure on $O(t)$. Each of them will be automatically an integral for the hamiltonian vector field generated by any other function from this class.

Theorem 4. (see [5]). Let f, g be two functions on T_e^*, constant on orbits $O(t)$, $a\in T_e^*$ a fixed covector, $\lambda,\mu\in\mathbb{R}$ fixed numbers. Then functions $f(t+\lambda a)$ and $g(t+\mu a)$ are in involution on every orbit $O(t)$ with respect to the above mentioned symplectic structure.

Up to the present we considered arbitrary Hamiltonian functions H. Euler equations define a dynamical system on T_e^*, the dual space to T_e, which is a Lie algebra. Denote it by G. So, on $T_e^*=G^*$ there is considered an Euler equation $\dot{t}=\{t,dH(t)\}$,

$t \in G^*$. The case of the geodesic current for the leftinvariant metrics on the group $\mathfrak{G}$ is of a special interest. In this case H is a nondegenerate quadratic form on G^*, dH is a linear transformation $dH: G^* \to G$. The self-conjugate condition: $\langle t_1, dH(t_2)\rangle = \langle t_2, dH(t_1)\rangle$ - holds. Vice versa, if a linear self-conjugate operator $\varphi: G^* \to G$ is given, then there exists a quadratic function $H(f)$, such that $\varphi(t) = dH(t)$. So, in this case the Euler's equations are: $\dot{t} = \{t, \varphi(t)\}$.

Let us apply Theorem 4 to find such quadratic Hamiltonian functions, which have sufficiently many first integrals. In this case the function H must functionally depend on functions $f(t+\lambda a)$.

<u>Theorem 5</u>. (see [5]). Let F be a set of functions, functionally generated by functions $f(t+\lambda a)$, where f are functions constant on orbits of coadjoint representation, $a \in G^*$ is a fixed covector, λ is a real number. The quadratic polynomial $H(t)$ belongs to F, iff there exists a vector $b \in G$, s.b. $\{a, dH(t)\} = \{t, b\}, t \in G^*, \{a, b\} = 0$.

So, in a natural way, some special class of quadratic Hamiltonians is arrived at which have sufficiently many first integrals. As we shall see, this class contains, for example, Hamiltonians corresponding to the motion of n-dimensional rigid body with a fixed point.

Let us study this question for the case of semisimple Lie algebras.

(8). Semisimple complex Lie algebras are characterized by the existence of a nontrivial complexvalued symmetric Cartan-Killing form $B(x,y) = (x,y), x, y \in G; (x,[y,z]) = (y,[z,x])$. Since the homomorphism $B:G \to G^*$, generated by the form B, is an isomorphism, the Euler's equation $\dot{t} = \{t, dH(t)\}$ may be rewritten in variables of algebra G, taking $x = B^{-1}(t)$.

Let $H(B(x)) = \tilde{H}(x)$; then it is easy to see that $\dot{x} = [x, \operatorname{grad} \tilde{H}(x)]$. This formula has the classical form of the Euler equation for the semisimple group . If $f(t)$ is a function, constant on the orbits of the coadjoint action, then the function $f(B^{-1}(t))$ is constant on the orbits of the adjoint action.

<u>Corollary 1</u>. (see [5]). Let G be the semisimple complex Lie algebra, $a \in G$ - a vector in general position, F - the collection of the functions on G, functionally generated by the functions $f(x+\lambda a)$, where f is a function constant on the orbits of the adjoint action. The quadratic function $H(x)$ belongs to F if and only if there exists a vector $b \in G$, such that $[\operatorname{grad} H(x), a] = [x,b], [a,b] = 0$. (The condition B).

The proposition follows from the Theorem 5. Hence, the linear self-conjugate (relative to the Cartan-Killing form) operators $\varphi: G \to G$, satisfying the condition B (see the Corollary 1), may be bescribed as follows: let us consider the Cartan subalgebra $H(a)$, containing $a \in G$ (where a is in general position): let Σ be the root system of G; $\{X_\alpha\}_{\alpha \in \Sigma}$ - the root vectors; $V \subset G$ - the subspace generated by the vectors $X_\alpha, \alpha \in \Sigma$, $G = H(a) \oplus V$. Then the operator $\operatorname{ad}_a: G \to G$ is identically equal to zero on $H(a)$ and leaves V invariant and is an isomorphism on V, if a belongs to the Weil chamber, that is, when $\alpha(a) \neq 0, \alpha \in \Sigma$. Hence, any operator $\varphi: G \to G$, $b \in H(a)$, satisfying the condition B, has the form: $\varphi = \begin{pmatrix} \varphi_1 & 0 \\ 0 & \varphi_2 \end{pmatrix}$, in the decomposition $G = H(a) \oplus V$; $\varphi_1: H(a) \to H(a)$; $\varphi_2: V \to V$. Here φ_1

is any self-conjugate operator on $H(a)$, φ_2 is defined as follows: $\varphi_2(x) = ad_a^{-1} ad_b(x)$. Changing the operator φ_1 and the vector b, we receive the family of operators φ. Hence, we receive a series of self-conjugate operators whose Euler equations have enough integrals in involution on the orbits of the coadjoint action.

Theorem 6. Let G be a complex semisimple Lie algebra, and let the quadratic function $H(x)$ satisfy the condition B. Then Euler equations are fully integrable in the classical commutative Liouville sense on the orbits of the adjoint action which are in general position.

The proof of this theorem is rather nontrivial and was obtained by us in [5]. Let us note that the full integrability of the system occurs not only on the orbits in general position, but on the singular orbits of the adjoint action also; this result was proved by Dao Chong Tchio. He analyzed our method obtained for the investigation of the orbits in general position.

It turned out more deeply that there exists other rich classes of fully integrable systems, connected with other forms of the semisimple groups. There exists two types of natural subalgebras in the semisimple Lie algebras connected with the root systems. They are the compact and normal forms of the Lie algebra. The compact form G_μ of the Lie algebra G is defined as the set of the fixed points of the anticomplex involution σ of the algebra G, where the involution σ is defined as follows: $\sigma(x_\alpha) = x_{-\alpha}$; $\sigma(h_\alpha) = -h_\alpha$, where $h_\alpha \in H$ are the vectors from the Cartan subalgebra conjugated to the roots $\alpha \in \Sigma$. It is easy to see that if a self-conjugated operator $\varphi: G \to G$ leaves the subalgebra G_μ invariant and has the form:

$\varphi = \begin{pmatrix} \varphi_1 & 0 \\ 0 & \varphi_2 \end{pmatrix}$; $\varphi_2 = ad_a^{-1} ad_b$, then the vectors a,b must be taken from the subalgebra $H = G_\mu \cap H$.

Theorem 7. (see [5]). Let G_μ be a compact form of the semisimple complex Lie algebra G; $\dot{x} = [x, \varphi(x)]$ - the Euler equation with the operator φ satisfying the condition B. Then this system is fully integrable in the classical, commutative Liouville sense on the orbits of adjoint action in algebra G_μ of the group $\mathfrak{G}_\mu$ which are in general position.

This result may be obtained from the Theorem 6, because the algebra G may be represented as the complexification of G_μ. It is interesting to note that the full integrability of Euler system also occurs on singular orbits (proved by Dao Chong Tchio).

In the algebra G there exists another anticomplex involution τ, defined as follows: $\tau(X_\alpha) = X_\alpha$; $\tau(h_\alpha) = h_\alpha$. Then the subalgebra $G_n = (G_\mu)^\tau$ is called a normal form and its complexification coincides with the fixed points of the complex involution $\sigma\tau : \mathbb{C} \oplus G_n \simeq G^{\tau\sigma}$. It is easy to see that, if $a,b \in Hu$, then the operator φ leaves subalgebras $G_n, G^{\tau\sigma}$ invariant. Hence we may speak about Euler equations on the subalgebra G_n. It is interesting to see that the vectors a,b do not belong to the normal form G_n.

Theorem 8. (see [5]). Let $G_n \subset G_u$ be the normal subalgebra of the semisimple Lie algebra G; $\dot{x} = [x, \varphi(x)]$ the Euler equation on the subalgebra G_n satisfying the condition B (that is $\varphi_2 = ad_a^{-1} ad_b$). Then this system is fully integrable in the classical, commutative Liouville sense, on the orbits of algebra G_n which are in general position.

The full integrability also takes place on singular orbits (Dao Chong Tchio). Let us note that the collection of the functions $f(x+\lambda a)|G_n$, where f are the functions which are constant on the orbits ofthe algebra G_n, is the commutative family and all these functions are integrals of Euler equations. It appears (see [5]) that from the above mentioned collection of integrals we may choose the functionally independent integrals and their number is equal to half the dimension of the algebra G_n.

(9). Some special examples of the integrals were received earlier by several authors. For example, if $G = gl(n,C)$, then $G_u = u(n)$, $G_n = O(n)$, and the operators φ correspond to the Hamilton functions considered in [8]. If a is a diagonal matrix from $gl(n,C)$, and $b = -a^2$, then the operator φ defines the Euler equation of the motion of "n-dimensional rigid body" considered by S.V. Manakov in [7]. In these papers were contained some examples of the integrals of the Euler equations, but the completeness of this system of the integrals was not proved, i.e. that is it was not proved that this set of integrals is sufficient for the full integrability of the corresponding Euler systems. The integrals of "normal" series for $\varphi: SO(n) \to SO(n)$ contain the integrals which were discovered for the first time by A.S. Mischenko in [8]; the involutivity of this special series of the integrals was proved by L.A. Dikii in [9]. Let $SO(n)$ be the normal form in $su(n)$; $a = \begin{pmatrix} i\lambda_1 & 0 \\ 0 & i\lambda_n \end{pmatrix}$; $b = \begin{pmatrix} i\lambda_1^2 & 0 \\ 0 & i\lambda_n^2 \end{pmatrix}$, $a, b \in u(n)$; $\lambda_i \neq \lambda_j$ if $i \neq j$. Then $\varphi: so(n) \to so(n)$ and the operator φ multiplies the vector $E_{pq} = \begin{pmatrix} 0 & 1 \\ -1 & 0 \end{pmatrix}$ by $\lambda_p + \lambda_q$, that is, $\varphi(X) = XI + IX$, where $I = -ia^2$. The first series of the integrals from [8] coincide with the functions $Spur(X^k)$, $1 \leq k \leq n$, $X \in so(n)$, where $so(n)$ is considered in the standard representation of the minimal dimension as the space of the skew-symmetric matrices. The second series of the integrals in [8] coincides with the functions $P_{k-2}(X,a)$ in the decomposition of $\mathrm{Spur}(X+\lambda a)^k$ (We mean the coefficients $P_\alpha(X,a)$, where α is the degree of variable $\lambda : P_\alpha(X,a)\lambda^\alpha$).

Let us consider, as an example, the operators of the normal series for $G_n = so(4) \subset G_u = su(4) \subset G = sl(4,C)$. The algebra $so(4)$ is realized in G as the skew-symmetric matrices and is generated by the vectors $E_{ij} = E_\alpha + E_{-\alpha}$ of the standard basis. Let us represent $X \in so(4)$ as: $X = \alpha E_{12} + \beta E_{13} + \gamma E_{14} + \delta E_{23} + \rho E_{24} + \varepsilon E_{34}$. The orbits in general position in $so(4)$ are four-dimensional manifolds $S^2 \times S^2$. Let $a, b \in H_u \subset su(4)$, then

$$\psi(x) = \alpha\frac{b_1-b_2}{a_1-a_2}E_{12} + \beta\frac{b_1-b_3}{a_1-a_3}E_{13} + \gamma\frac{b_1-b_4}{a_1-a_4}E_{14} + \delta\frac{b_2-b_3}{a_2-a_3}E_{23} + \rho\frac{b_2-b_4}{a_2-a_4}E_{24} + \varepsilon\frac{b_3-b_4}{a_3-a_4}E_{34}$$

The direct calculations show that four integrals, which give the full integrability of Euler equations, and describe the motion of the 4-dimensional rigid body, are:

$$h_1(X) = \alpha^2 + \beta^2 + \gamma^2 + \delta^2 + \rho^2 + \varepsilon^2, \quad h_2(X) = \alpha\varepsilon - \beta\rho + \gamma\delta,$$

$$h_3(X) = \alpha^2(a_1+a_2) + \beta^2(a_1+a_3) + \gamma^2(a_1+a_4) + \delta^2(a_2+a_3) + \rho^2(a_2+a_4) + \varepsilon^2(a_3+a_4),$$

$$h_4(X) = \alpha^2(a_1^2+a_1a_2+a_2^2) + \beta^2(a_1^2+a_1a_3+a_3^2) + \gamma^2(a_1^2+a_1a_4+a_4^2) + \delta^2(a_2^2+a_2a_3+a_3^2) + \rho^2(a_2^2+a_2a_4+a_4^2) + \varepsilon^2(a_3^2+a_3a_4+a_4^2).$$

These four integrals are functionally independent and the integrals h_3, h_4 commute on the orbits $S^2 \times S^2$. The $SO(4)$ group case was also considered in [8], [10].

(10). Let G be the semisimple Lie algebra, φ - the self-jugate (with respect to Cartan-Killing from) operator. Let us consider the dynamical system $x = [x, \varphi(x)]$ and also investigate the question of the existence of analytic one-parameter λ family of systems (which are equivalent one to another) $\frac{d}{dt} L(x,\lambda) = [L(x,a); A(x,\lambda)]$, including the system $\dot{x} = [x,\varphi(x)]$.

Proposition 5. (see [5]). Any analytic family of equivalent systems of the above mentioned kind is reduced to the linear family (on parameter λ).

(11). The special interest is concentrated on the problem of full integrability of the Euler equations on the Lie algebras, which are not semisimple, first of all, on the solvable and nilpotent Lie algebras. Direct (although somewhat cumbersome) calculations prove the following proposition.

Proposition 6. Let G be a Lie algebra of triangular matrices of the order 2 or 3; $a \in G^*$ - a covector in a general position, H - a Hamilton function on the space G belonging to the class F, which is generated in the functional sense by the functions $f(t+\lambda a)$, where f - are the functions which are constant on the orbits of the coadjoint representation. Then the Euler equation $\dot{t} = \{t,dH(t)\}, t \in G^*$ is fully integrable in the classical, commutative Liouville sense on the orbits of general position.

Full integrability appears to take place on the singular orbits too. As direct calculations have shown, the further progress, namely the proof of the full integrability of Euler equations on the Lie algebra of triangular matrices (of order $n, n \geq 4$) is impossible using the functions which are constant on the orbits of coadjoint representation. It appears that the class F is too poor for generating the full collection of the integrals. In this sense the solvable case is very different from the semisimple one. Nevertheless, recently, the full integrability of Euler equations on the triangular matrices algebra (of arbitrary order) by A.A. Archangelsky was discovered. He discovered a new algebraic situation which generates new, supplementary integrals of Euler equations. Let us give an account the Archangelsky's result. It appears that it is useful to consider the functions f which are defined on G^*, and are invariant for the coadjoint (group) representation, acting on the orbits of this representation; to be more precise we must consider such functions f, for which the following identity realized: $f(Ad_g^* t) = \chi(g) f(t), g \in \mathfrak{G}$, where χ is some one-dimensional representation of $\mathfrak{G}$ (that is - a character). It is obvious that the functions which are constant on the orbit of the coadjoint representation are the invariant functions corresponding to the character $\chi(g) = 1$. Consequently, the investigation of the invariant functions for the non-trivial characters is the very natural expansion of the old class of functions guaranteeing the full integrability of Euler equations for the semisimple algebras.

__Proposition 7__. Let $f_1, \ldots, f_q$ be invariant functions on G which are in involution. Then for every $a \in G^*$ the following functions: $f_i(t+\lambda a), \lambda \in R, 1 \le i \le q$, are in involution also.

Let V_n be the upper-triangular matrices algebra of the order n, $\mathfrak{G}_n$ - the corresponding solvable Lie group. The dual space V_n^* may be identified with the space of lower-triangular matrices. Let $X \in V_n^*$; and let us denote the minor of the matrix X in rows heaving the numbers $i_1, \ldots, i_k$ and in columns having the numbers

$j_1,\ldots,j_k$, ad $M^{i1\ldots ik}_{j1\ldots jk}(X)$. Let $n=2k-1, 2k, (k\geq 1)$ and let us consider the following functions on V_n^* (lower-triangular matrices):

$$S_i(X) = \Sigma_{j=i}^{n-i+1} M^{j,n-i+2,\ldots,n}_{1,2,\ldots,i-1,j}(X), \quad i=1,2,\ldots,k,$$

$$S_i(X) = M^{i,\ldots,n}_{1,\ldots,n-i+1}(X), \qquad i=k+1,\ldots,n,$$

and the following homomorphism of the group $\mathfrak{g}_n$ in the group of the real numbers:

$$\Phi_i(g) = M^{1,\ldots,n-i+1}_{1,\ldots,n-i+1}(g) M^{i,\ldots,n}_{i,\ldots,n}(g^{-1}) = \frac{g_{11}g_{22}\cdots g_{n-i+1,n-i+1}}{g_{ii}g_{i+1,i+1}\cdots g_{nn}},$$

$$g = \|g_{ij}\| \in \mathfrak{g}_n.$$

<u>Proposition 8</u>. The functions $S_i (i=1,\ldots,n)$ on V_n^* are invariant with the characters Φ_i, and are in involution. Hence for any $i,j=1,\ldots,n$ and $\lambda,\mu \in R$ there is an identity: $\{S_i(t+\lambda a); S_j(t+\mu a)\} = 0$.

A direct calculation shows that the codimension of the orbit (in general position) in V_n^* is equal to $[\frac{n+1}{2}]$. Let us remind you that to every Hamilton function H on V_n^* there corresponds the Euler equation: $\dot{t} = \{t, dH(t)\}$.

<u>Theorem 9</u>. Let the covector $a \in V_n^*(n=2k-1,2k)$ satisfy the condition: $S_i(a) \neq 0$; $i=n-k+1,\ldots,n$ (that is, the covector a is in general position); let the Hamilton function H on V_n^* functionally depend on the functions $S_i(t+\lambda a)$, $i=1,\ldots,n$. Then the system of Euler equations $\dot{t} = \{t, dH(t)\}$ is fully integrable in the classical, commutative Liouville sense on the orbits in general position.

Hence, as A.A. Arkhangelsky has shown, the Euler equations on the simplest solvable Lie algebra of upper-triangular matrices (for the above mentioned Hamilton function) are fully integrable.

Our hypothesis is as follows: for every finite-dimensional Lie algebra there exists a family of functions on the space, dual to the Lie algebra, such that all these functions form a commutative family and the Euler equations corresponding to Hamilton functions belonging to this family are fully integrable on the orbits (in general position) of the coadjoint representation. This hypothesis is valid for the semisimple Lie algebras, for the normal and compact forms, and for the upper-triangular matrices solvable Lie algebra also. Let us stress that for the construction of the commutative families of functions on the orbits of coadjoint representation, perhaps not only the functions invariant for coadjoint representation will be needed, but also the invariant subspaces in the space of all functions on the orbits in general position; which subspaces appear under the coadjoint action on the space of functions. The functions which are constant on the orbits, form a very poor class; this was demonstrated by A.V. Strelzov who proved that for many series of solvable Lie algebras which continuously depend on some parameter (for different values of which non-isomorphic algebras correspond) no functions which are constant on the orbits in general position (besides the constants) exist.

However, it is not clear that there are any examples of leftinvariant matrices (on the semisimple Lie groups), such that the corresponding Euler equations do not admit the full Liouville (in the commutative sense) integration.

<u>Hypothesis</u>: for every quadratic Hamilton function $H(t)$ the corresponding Euler equations $\dot{t} = \{t, dH(t)\}$ are fully integrable on the semisimple Lie algebra.

(12) There exists one more discription of the class of symplectic manifolds admitting fully integrable Hamiltonian systems of the type "dynamic of n-dimensional rigid body". Let us consider the simply connected compact homogeneous space (manifold) $\mathfrak{G}/\mathfrak{H}$, where $\mathfrak{G}$ is the simply connected Lie group. Let w be the symplectic structure on $\mathfrak{G}/\mathfrak{H}$ and let w be invariant under the natural $\mathfrak{G}$-action on $\mathfrak{G}/\mathfrak{H}$. It is known that $\mathfrak{G}/\mathfrak{H}$ may be embedded in the Lie algebra G of the group $\mathfrak{G}$ such that: (1) the natural action $\mathfrak{G}$ on $\mathfrak{G}/\mathfrak{H}$ coincides with the coadjoint action of $\mathfrak{G}$ on G; (2) $\mathfrak{G}/\mathfrak{H}$ will be one of the orbits (possible singular) of $Ad_{\mathfrak{G}}$-action on G; (3) the form w coincides on $\mathfrak{G}/\mathfrak{H}$ with Kirillov's form on the orbit $\mathfrak{G}/\mathfrak{H}$. Hence, on every compact $\mathfrak{G}/\mathfrak{H}$ with the invariant symplectic form w there exists a natural dynamic system - the restriction of the current $x = [x, \varphi(x)]$ from the algebra G on the orbit $\mathfrak{G}/\mathfrak{H}$. Will this system by integrable? The answer is - yes. This result follows from our theorem about the full integrability of the systems on the orbits in general position, and from Dao Chong Tchi's theorem about the integrability of the systems on the singular orbits. In this way, on every compact homogeneous symplectic manifold $\mathfrak{G}/\mathfrak{H}$ with $\mathfrak{G}$-invariant symplectic form w there exists natural dynamical systems (the analogs of the dynamics of a rigid body with a fixed point), which are fully integrable in the Liouville sense.

(13) Let us return to the non-commutative integration in the Liouville sense. We may aply the same ideas as in (12), to the

case, when the finite-dimensional Lie algebra V on the symplectic manifild M is not functionally independent, but has a unique orbit-type of the annulator $\mathfrak{f}$-action in the coadjoint representation. As an example, let us consider the classical geodesic current in the phase-spase of the linear elements on the sphere S^n with the standard riemannian metric. Applying our method, we receive the full integrability (in the commutative sense) of the geodesic current on T^*S^n. Let us note that the non-commutative Lie algebra V of integrals on T^*S^N is isomorphic to the Lie algebra of the group SO(n+1). It appears (applying theorem 2) that this algebra may be replaced by a commutative algebra of integrals of the dimension n.

(14) Let us make use of the above mentioned results concerning the non-commutative integration in the Liouville sense to the important problem of integrating the Hamiltonian system on the phase-space (a co-tangent bundle) of a semi-simple Lie group with the left-invariant metric. Let us investigate the question about the integration of the geodesic current corresponding to these metrics. Let $\mathfrak{G}$ be the semisimple Lie group, G - its Lie algebra, H - the left-invariant Hamilton function defined on the co-tangent bundle $T^*\mathfrak{G}$. As far as the group $\mathfrak{G}$ acts on the symplectic manifold $T^*\mathfrak{G}$, then, according to the Noether (?) theorem, there exists a finite-fimensional Lie algebra V og the integrals on $T^*\mathfrak{G}$ and the corresponding Euler equation is defined on the orbits of the coadjoint representation of the Lie group $\mathfrak{G}$ in the following way: $\xi = \{\xi, dH(\xi)\}$. Let us establish the relation with the previous indications. Let us set up $M^{2n} = T^*\mathfrak{G}$ - a symplectic manifold having the natural form w (for its definition see the

previous item); $V \cong G$; $\vec{v}$ = sgrad H - the geodesic current on $T^*\mathfrak{G}$ with the Hamilton function H. Then V_1 is the space of the functions on $M^{2n} = T^*\mathfrak{G}$, $f_X(\eta,g) = \langle R^*_{g^{-1}}(\eta);X\rangle$, where R_g is the operator of the right translation, $x \in G = T_e$; $\eta \in T^*_g$, - the value of the linear functional on the vector. The space V_1 is the Lie algebra relative to the Poisson's brackets; the correspondence $x \dashrightarrow f_X$ is an isomorphism of the Lie algebra G on the algebra V_1. Let us represent the co-tangent bundle $T^*\mathfrak{G}$ as a cartesian product: $G^* \times \mathfrak{G}$, where the operators of the left translations L_g act in the following way: $L^*_g(\xi,g') = (\xi,gg')$. Then the operators of the right translations R_g act like this: $R^*_g(\xi,g') = (Ad^*_g\xi,gg')$. The tangent vector X in the point $(\xi,g) \in G^* \times \mathfrak{G}$ has two coordinates $X = (x_1,x_2)$, $x_1 \in G^*$, $x_2 \in G$. The value of the symplectic form w on the tangent vectors $x = (x_1,x_2)$, $Y = (Y_1,Y_2)$ is: $w(x,Y) = \langle x_1,Y_2\rangle - \langle x_2,Y_1\rangle - \langle \xi,[x_2,Y_2]\rangle$. The manifold P^1_ξ (the level surface), corresponding to the covector ξ, may be described in the manifold $M^{2n} = T^*\mathfrak{G}$ as the point set: $\{(Ad^*_g(\xi),g);g \in \mathfrak{G}\}$. If $m = (Ad^*_g\xi,g)$ is the point on the manifold P^1_ξ, then the tangent space $T_m(P^1_\xi)$ consists of vectors $X = (ad^*_x\xi,x)$. In this case the projection P'_N maps $T^*\mathfrak{G}$ on the coalgebra G^* which coincides with the Euler manifold $N' = M/\mathfrak{G}$. The projection P'_N maps the level surface P'_ξ on the manifold N'_ξ, which is the orbit $O(\xi)$ of the covector ξ. The formula of this projection is: $P_N(Ad^*_g\xi,g) = Ad^*_g\xi$; $P_N: P'_\xi \dashrightarrow O(\xi)$. It is evident that $P^{-1}_N(\xi)$ consists of the points $(Ad^*_g\xi,g)$ satisfying the condition: $Ad^*_g\xi = \xi$. If $X = (ad^*_x\xi,x)$, $Y = (ad^*_y\xi,y)$ - two tangent vectors in the point $m = (Ad^*_g\xi,g) \in P_\xi$, then $w(X,Y) = \langle\xi,[x,y]\rangle$ that is, Kirillov's form on the orbit $O(\xi)$ actually coincides with the form w. The Euler manifold $N' = G^*$ is the collection of all manifolds N'_ξ, that is the orbits $O(\xi)$. On every orbit $O(\xi)$ the Euler equation coincides with the Euler equation $\xi = \{\xi, dH(\xi)\}$, which was considered in the previous items. As has been proved these Euler equations are fully integrable in the Liouville commutative sense, that is, there exist $k = \frac{1}{2}\dim O(\xi)$ functionally independent integrals on the orbits $O(\xi)$, which are in involution. This series F of the left-invariant Hamilton functions consists of the functions which are (in functional sense) generated by the function: $f(t + \lambda a)$, where f - a function which is constant on the orbits of the coadjoint representation, λ - the arbitrary number, a - a fixed (for the whole series) covector, $a \in G^*$. Let $n = \dim G$, $r = \text{rang } G$. Then we can choose k functionally independent integrals, generating the commutative Lie algebra V_o of the integrals, $\dim V_o = k$ on every orbit $O(\xi) \subset G^* = T_e^*$.

Caution: Don't confuse the level surfaces P_ξ' and P_ξ; the surfaces P_ξ' are common level surfaces for the algebra V_o, and P_ξ - for the algebra V.

<u>Theorem 10</u>. Let $\mathfrak{G}$ be the semisimple Lie group, H - the left invariant Hamilton function of the series F. Then: (1) the Hamilton system on the phase-space $T^*\mathfrak{G}$ is integrable in the non-commutative Liouville sense, that is, there exists finite-dimensional algebra V of the integrals, such that $\dim V + \text{rang } V = \dim T^*$.

(2). If the common level surface is compact, then this surface is diffeomorphic to the torus of the dimension $\dim P_\xi = \frac{1}{2}(n+r)$; the Hamilton vector field is invariant under the parallel translation and the integral trajectories, in general, realize the pseudo-periodic motion along these tori.

For proof see [2]. The scheme of the proof is as follows. Let V_o be a commutative algebra of the functions on G^* which represent the full collection of integrals of the Euler equation. Let us extend the functions from V_o to the left-invariant functions on $T^*\mathfrak{G}$. Then we obtain the commutative algebra V_o of the functions on $T^*\mathfrak{G}$. Moreover, if V_1 is the algebra of integrals, corresponding to the left $\mathfrak{G}$-action on $T^*\mathfrak{G}$, then all the functions of V_o commute with the functions from V_1. The algebra V_1 is isomorphic to Lie algebra G. Hence, we obtain a new Lie algebra $V = V_o \oplus V_1$ of the integrals on the phase-space $T^*\mathfrak{G}$. If the group $\mathfrak{G}$ is semisimple, then the algebra has a locally constant orbit type of the coadjoint action, and the annulator $H_\xi \subset V$ of the covector $\xi \in V^*$ (which is in general position) is commutative and equal to the sum $H_\xi = H \oplus V_o$, where $H \subset G$ is the Cartan subalgebra. Hence, $\dim V = n + \frac{1}{2}(n-r)$; $\operatorname{rang} V = r + \frac{1}{2}(n-r)$. Then $\dim V + \operatorname{rang} V = 2n = \dim T^*$.

So it remains us to learn about the periodic coordinates on the level surface P_ξ and about the invariant vector field on it. Let us represent $T^*\mathfrak{G}$ as a cartesian product $T^*\mathfrak{G} = T_e^* \times \mathfrak{G} = G^* \times \mathfrak{G}$ where the left translation of the group $\mathfrak{G}$ does not change the coordinate in G^*. Then every function $f \in V_1$ is right-invariant and has the following values: $f(\xi, g) = \langle Ad^*_{g^{-1}}(\xi), g\rangle$, $f \in V_1 = G$, $\xi \in G^*$, $g \in \mathfrak{G}$ Hence, as was demonstrated previously, P'_ξ is the bundle with the base $o(\xi)$ and the fiber $\mathfrak{H}_\xi$, where $O(\xi)$ is the orbit of the coadjoint action; and $\mathfrak{H}_\xi$ is the Cartan subalgebra, leaving the covector ξ fixed. Since all the functions from algebra V_o are left-invariant, these functions are constant on the fibers $\mathfrak{H}_\xi$ of the fibration $P' \longrightarrow O(\xi)$. The Hamilton function H belongs to the class F, that is functionally expressed in terms of the functions of the algebra V_o. Hence, the differential dH(ξ), the corresponding choice of the function H being made, may have any value from the subspace $T_\xi P_\xi$ which is equal to $H_\xi \oplus d\,V_o$. In particular, on some level surfaces of the integrals, the trajectories of the dynamical system realize the pseudo-periodic motion.

Hence, we have succeeded in proving the full integrability of the geodesic current corresponding to the left-invariant metrics (of the special kind) which we have introduced on the Lie groups $\mathfrak{G}$ (that is the metrics, which describe "the motion of a rigid body with group $\mathfrak{G}$"). The classical metric of the n-dimensional rigid body on the group SO(n) certainly belongs to this class. The geodesic current of this metric describes the motion of a rigid body having no fixed point.

It would be useful to obtain other examples of non-commutative integration (in the example above from the non-commutative integrability there follows commutative integrability) where the reduction $P_N: M \dashrightarrow N$ would be realized in several steps, but not in one step.

REFERENCES

1. J. Marsden, A.Weinstein, Reports of Math.Ph. 5:1 (1974), 121-130.

2. A.S. Mischenko, A.T. Fomenko, A generalized Liouville method for the integration of the Hamilton systems. Funk.Anal. and its appl., 1978, v. N.2.

3. M. Duflo et M. Vergne, C.R. Acad.Sc.Paris, 268 (1969), 583-585.

4. P. Bernat, N. Conze, M. Vergne, Representation de groupes de Lie résolubles. Paris, 1972.

5. A.S. Mischenko, A.T. Fomenko, The Euler equations on the finite-dimensional Lie groups. Izv.Akad.Nauk SSSR. Ser. math., 1978, N 3.

6. A.S. Mischenko, A.T. Fomenko. On the integration of the Euler equations on the semisimple Lie algebras. Dokl. Akad.Nauk. SSSR, 231:3, 1976, pp.536-538.

7. S.V. Manakov, A note about the integration of the Euler equation of the dynamic of an n-dimensional rigid body. Funk.Anal. and its appl. v.10, N. 4 (1976), pp.93-94.

8. A.S. Mischenko, Funk.Anal. v. 4 N 3 (1970), 75-78.

9. L.A. Dikii, A note about the Hamilton system connected with the rotation group. Funk.Anal. and its appl. v 6, N 4, 1972.

10. M. Langlois, Contribution a l'étude du mouvement du corps rigide a N dimensions autour d'un point fixe. Thése présenté a la faculté des sciences de l'univ. de Besançon, 1971.

11. N.N. Nekhoroschev, The variables action-angle and their generalizations. Trudi MMO, v.26, 1972, pp.181-198.

Free compact group actions on products of spheres

Robert Oliver*

This paper has two main results about free actions on products of spheres. The first (Theorem 2) is that the alternating group A_4 has no free action on $S^k \times S^k$ for any k; in fact, no free action on any finite CW-complex X with $H^*(X;\mathbb{Z}) \cong H^*(S^k \times S^k;\mathbb{Z})$. The second result (Theorem 5) is that a compact Lie group has a free action on some product of spheres $\prod^n S^k$ if and only if it has no subgroup isomorphic to $SO(3)$.

Theorem 2 was motivated by Swan's result [5] which (together with Theorem XII.11.6 in [2]) says that a finite group has a free action on a finite complex with the homotopy type of a sphere if and only if it has "maximal rank" one (by the maximal rank of a group is meant the maximum of ranks of abelian subgroups). Conner [3] showed that a group having a free action on a finite dimensional space with the cohomology of $S^k \times S^k$ has maximal rank at most two (his restriction that the induced action on homology be trivial is easily removed). Theorem 2 shows that Swan's result does not generalize to a converse of Conner's theorem: A_4 has maximal rank two but cannot act freely on any finite CW-complex with the cohomology of $S^k \times S^k$.

The fact that $SO(3)$ cannot act freely on any product of spheres of the same dimension is a simple corollary of the results for A_4-actions. The converse is proven by explicitely constructing free actions for groups not containing $SO(3)$; this is simplified by the fact that the only compact simple Lie groups not containing $SO(3)$ are $SU(2)$ and $Sp(2)$.

*This work was supported, partly by an NSF summer grant, and partly by a Sloan Fellowship.

We first consider free A_4-actions. The subgroup $\mathbb{Z}_2^2 \triangleleft A_4$ of index 3 induces an inclusion

$$H^*(A_4;\mathbb{Z}_2) \subseteq H^*(\mathbb{Z}_2^2;\mathbb{Z}_2) \cong \mathbb{Z}_2[x,y];$$

and $H^*(A_4;\mathbb{Z}_2)$ can be identified with the ring $\mathbb{Z}_2[x,y]^\alpha$ of polynomials invariant under the automorphism α given by: $\alpha(x) = y$, $\alpha(y) = x + y$. Fix an element

$$\varepsilon = xy(x+y) \in H^3(A_4;\mathbb{Z}_2).$$

Lemma 1. Let $M \subseteq H^k(A_4;\mathbb{Z}_2)$ be a non-zero subgroup (any $k \geq 0$) such that the ideal $H^*(A_4;\mathbb{Z}_2) \cdot M$ generated by M is stable under the action of the Steenrod algebra. Then $3|k$, and $M = \{\varepsilon^{k/3}, 0\}$.

Proof. All cohomology is assumed to be with $\mathbb{Z}_2$-coefficients. The lemma will be proven by induction on k; it is clear when $k = 0$. First assume k is even (and positive).

Any homogeneous polynomial $a \in \mathbb{Z}_2[x,y]$ of even degree can be written in a unique fashion in the form $a = r^2 + xys^2$ (just separating the terms with even or odd exponents of x and y.) We can thus define a homomorphism

$$\varphi : \mathbb{Z}_2[x,y]^{2i} \to \mathbb{Z}_2[x,y]^i$$

by setting $\varphi(r^2+xys^2) = r + (x+y)s$. This commutes with the automorphism α, and thus restricts to a homomorphism

$$\varphi : H^{2i}(A_4) \to H^i(A_4).$$

We clearly have $\varphi(a^2b) = a \cdot \varphi(b)$; in particular $\varphi(a^2) = a$.

Since $H^1(A_4) = 0$, we must have $Sq^1(m) = 0$. Thus, for any

$a = r^2 + xys^2 \in M$,

$$0 = Sq^1(a) = (x^2y+xy^2)s^2;$$

$s = 0$, and a is a square. So all elements of M are squares, and setting $M' = \varphi(M)$, we have

$$M = \{a^2 | a \in M'\}.$$

Now choose any $a \in M'$. Since $H^*(A_4) \cdot M$ is stable under the action of the Steenrod algebra, we can write, for any $i > 0$,

$$Sq^i(a))^2 = Sq^{2i}(a^2) = \Sigma_j r_j b_j^2$$

for some $r_j \in H^{2i}(A_4)$ and $b_j \in M'$. Applying φ gives

$$Sq^i(a) = \Sigma_j \varphi(r_j) b_j$$

with $\varphi(r_j) \in H^i(A_4)$; so $H^*(A_4) \cdot M'$ is stable under the action of the Steenrod algebra. Applying the induction hypothesis to M' gives that $6|k$, and

$$M' = \{\varepsilon^{k/6}, 0\}; \quad M = \{\varepsilon^{k/3}, 0\}.$$

Now assume k is odd; $Sq^1(M) = 0$ as before. This time, any $a \in M$ can be written in the form $a = xr^2 + ys^2$, giving

$$0 = Sq^1(a) = x^2r^2 + y^2s^2 = (xr+ys)^2; \quad r = yt \text{ and } s = xt \text{ (some } t).$$

Thus, $a = (xy^2+yx^2)t^2 = \varepsilon t^2$ for some t. In particular, $M = \varepsilon \bar{M}$ for some $M \subseteq H^{k-3}(A_4)$, and the Cartan formula easily shows that $H^*(A_4) \cdot \bar{M}$ is stable under the action of the Steenrod algebra. Applying the induction hypothesis to $\bar{M}$ gives $M = \{\varepsilon^{k/3}, 0\}$. $\square$

This now applies directly to show:

Theorem 1. There is no free action of A_4 on any finite dimensional space X such that $H^*(X;\mathbb{Z}_2) \cong H^*(\prod^n S^k;\mathbb{Z}_2)$, with A_4 acting trivially on cohomology.

Proof. Assume that A_4 does have a free action on such an X inducing the trivial action on cohomology. As usual, the fiber bundle

$$X \to EA_4 \times_{A_4} X \to BA_4 \qquad (EA_4 \times_{A_4} X \simeq X/A_4)$$

induces a spectral sequence

$$E_2 \cong H^*(\prod^n S^k) \otimes H^*(A_4) \Rightarrow H^*(X/A_4)$$

(again, all cohomology is with $\mathbb{Z}_2$-coefficients). Since X/A_4 must have finite cohomological dimension ([4], Proposition A.11), we can get a contradiction by showing that the E_∞-term must be infinite dimensional.

The first non-zero differential is d^{k+1}; let $M \subseteq H^{k+1}(A_4)$ be the image of the transgression. Since transgressions commute with Steenrod powers, and Sq^i is zero on $H^*(X)$ for all $i > 0$, $H^*(A_4) \cdot M$ is stable under the action of the Steenrod algebra. So $M = 0$, or $3 | k+1$ and $M = \{\varepsilon^{(k+1)/3}, 0\}$.

Thus, either $E_{k+2} = E_2$, or

$$E_{k+2} \cong H^*(\prod^{n-1} S^k) \otimes (H^*(A_4)/H^*(A_4) \cdot \varepsilon^{(k+1)/3}).$$

In either case, E_{k+2} is generated (as an algebra) by elements of vertical dimension zero and k; all later differentials vanish, and so $E_\infty = E_{k+2}$. Since

$$H^*(A_4) \quad \text{and} \quad H^*(A_4)/H^*(A_4) \cdot \varepsilon^{(k+1)/3}$$

are infinite dimensional ($H^*(A_4)$ has Krull dimension two by [4], Corollary 7.8), E_∞ is infinite dimensional in both cases. □

Theorem 2. A_4 cannot act freely on any finite CW-complex with $H^*(X, \mathbb{Z}) \cong H^*(S^k \times S^k; \mathbb{Z})$.

Proof. If A_4 does have such an action, the induced action on $H^*(X;\mathbb{Z})$, and thus on $H^*(X;\mathbb{Q})$, must be non-trivial (Theorem 1). Let $g \in A_4$ be any element of order 3; then the only possibility is that g act trivially on $H^0(X)$ and $H^{2k}(X)$, and with trace -1 on $H^k(X)$. So the action of g has Lefshetz number (2 ± 1), g has a fixed point, and A_4 does not act freely. □

Since $A_4 \subseteq SO(3)$, another immediate corollary to Theorem 1 is:

Theorem 3. SO(3) has no free action on any finite dimensional space X with $H^*(X;\mathbb{Z}_2) \cong H^*(\prod^n S^k;\mathbb{Z}_2)$. □

Remark. SO(3) acts freely on the twisted product $SO(3) \times_{S^1} S^3$, where S^1 acts freely on S^3, and acts on SO(3) as a subgroup. This space can, as in the proof of Lemma 3 below, be shown to be diffeomorphic to $S^2 \times S^3$. So Theorems 1, 2, and 3 are all false if one allows products of spheres of different dimensions.

The rest of this paper now deals with proving a converse to Theorem 3, by explicitely constructing actions of arbitrary compact Lie groups not containing SO(3) on products of copies of S^7.

Lemma 2. Let G be a compact connected simple Lie group, not containing any subgroup isomorphic to SO(3). Then G is isomorphic to SU(2) or Sp(2).

Proof. Both SU(2) and Sp(2) are simply connected with center $\mathbb{Z}_2$. So the only other groups locally isomorphic to them are

$$SU(2)/\mathbb{Z}_2 \cong SO(3) \text{ and } Sp(2)/\mathbb{Z}_2 \cong SO(5),$$

both of which contain SO(3).

It remains to show that for any other local isomorphism class of simple groups, some representative contains either SU(3) or $SU(3)/\mathbb{Z}_3$. This then implies that any representative contains SU(3) or $SU(3)/\mathbb{Z}_3$ (SU(3) being simply connected with center $\mathbb{Z}_3$), both of which in turn contain SO(3). For the remaining classical groups, such inclusions are obvious:

$$SU(3) \subseteq SO(n) \qquad (n \geq 6)$$

$$SU(3) \subseteq SU(n) \subseteq Sp(n) \qquad (n \geq 3).$$

The inclusion for the exceptional groups can be checked in the table on page 219 of [1]. □

It remains to construct free actions of SU(2) and Sp(2) on products of spheres. Since we want to deal with arbitrary compact Lie groups, not just the semisimple ones, actions must actually be constructed for the twisted products $SU(2) \times_{\mathbb{Z}_2} S^1$ ($\cong U(2)$) and $Sp(2) \times_{\mathbb{Z}_2} S^1$ (SU(2) and Sp(2) both have center $\mathbb{Z}_2$).

<u>Lemma</u> 3. $SU(2) \times_{\mathbb{Z}_2} S^1$ has a smooth action on S^7 whose restriction to SU(2) is free. $Sp(2) \times_{\mathbb{Z}_2} S^1$ has a smooth action on $S^7 \times S^7$ whose restriction to Sp(2) is free.

<u>Proof</u>. $SU(2) \times_{\mathbb{Z}_2} S^1 \cong U(2)$ has a unitary representation on $\mathbb{C}^4$ (twice the standard representation), whose restriction to SU(2) is free on the unit sphere $S(\mathbb{C}^4)$.

To construct an action for Sp(2), first regard S^7 as the set

$$\{x = (x_1, x_2) \in \mathbb{H}^2 \mid |x_1|^2 + |x_2|^2 = 1\}$$

($\mathbb{H}$ denoting the quaternions). Then Sp(1), the group of unit quaternions, acts on S^7 on both the left and right:

$$a(x_1, x_2)b = (ax_1b, ax_2b) \in S^7 \quad \text{for } a, b \in Sp(1) \text{ and } (x_1, x_2) \in S^7.$$

Define a manifold M to be the twisted product

$$M = Sp(2) \times_{Sp(1)} S^7 = (Sp(2) \times S^7)/\sim$$

with the relation given by

$$(A\begin{bmatrix} a & 0 \\ 0 & 1 \end{bmatrix}, x) \sim (A, ax) \text{ for } A \in Sp(2),\ a \in Sp(1),\ x \in S^7.$$

The equivalence class of (A, x) will be denoted [A, x].

As usual, M is a smooth fiber bundle over

$$Sp(2)/Sp(1) \cong S^7$$

with fiber S^7 and structure group Sp(1). The Sp(1)-action on S^7 is unitary, thus contained in the U(4)-action; and so M may be regarded as a U(4)-bundle over S^7. Since

$$\pi_7(B\,U(4)) \cong \pi_7(B\,U) \cong 0,$$

the bundle is trivial, and so $M \cong S^7 \times S^7$.

It remains to define the action of $Sp(2) \times_{\mathbb{Z}_2} S^1$ on M. Regard S^1 as the group of unit complex numbers. For all $B \in Sp(2)$, $z \in S^1$, and $[A, x] \in M$, set

$$(B, z)([A, x]) = [BA\begin{bmatrix} z & 0 \\ 0 & z \end{bmatrix}, z^{-1}xz] .$$

This is easily checked to be well defined as an action of $Sp(2) \times_{\mathbb{Z}_2} S^1$ (the element $(-I, -1)$ acts trivially).

Restricted to $Sp(2)$, the action is just the standard one: $B([A, x]) = [BA, x]$. Since $Sp(1)$ acts freely on S^7, $Sp(2)$ acts freely on M. □

Theorem 4. Let G be a compact connected Lie group with no subgroup isomorphic to $SO(3)$. Then G has a smooth free action on $\prod^n S^7$, where $n = rk(G)$.

Proof. Let $H \triangleleft G$ be the maximal semisimple subgroup; so $G/H \cong T^m$ (some m). By Lemma 2, the universal covering group $\widetilde{H}$ of H is a product:

$$\widetilde{H} = H_1 \times \ldots \times H_t \quad \text{and} \quad H = \widetilde{H}/\Gamma$$

with $H_i \cong SU(2)$ or $Sp(2)$, and $\Gamma \subseteq \widetilde{H}$ some finite central subgroup. The standard $SU(2) \subseteq Sp(2)$ contains the center $Z(Sp(2))$, so there is a subgroup $\prod^t SU(2) \subseteq \widetilde{H}$ containing Γ. Since $SO(3) \cong SU(2)/\mathbb{Z}_2$ cannot be a subgroup of H, Γ must be trivial, and thus

$$H = H_1 \times \ldots \times H_t .$$

Now, for $1 \leq i \leq t$, fix manifolds M_i:

$$M_i \cong \begin{cases} S^7 & \text{if } H_i \cong SU(2) \\ S^7 \times S^7 & \text{if } H_i \cong Sp(2) . \end{cases}$$

Set $G_i = G/H_1 \times \ldots \times H_{i-1} \times H_{i+1} \times \ldots \times H_t$. There are two possibilities:

(1) $G_i \cong H_i \times T^m$, and G_i surjects onto H_i

(2) $G_i \cong H_i \times_{\mathbb{Z}_2} T^m$ (some $\mathbb{Z}_2 \subseteq T^m$), and G_i surjects onto $G_i \times_{\mathbb{Z}_2} S^1$.

In either case, Lemma 3 applies to show the existence of a smooth action of G_i (and thus G) on M_i whose restriction to H_i is free. Also, $G/H \cong T^m$ has a free action on $\prod^m S^7$, and so the product action of G on

$$(\prod^m S^7) \times (\prod_{i=1}^t M_i) \cong \prod^n S^7 \quad (n=rk(G))$$

is free. □

Let G be any compact Lie group, $G' \subseteq G$ a subgroup of index $k < \infty$, and X a space with G'-action. By the "induced" action of G will be meant the obvious G-action on the space $\mathrm{Map}_{G'}(G, X) \cong X^k$ (the space of all G'-equivariant maps). Clearly, the induced action is smooth if the original one is, and any element of G' acting freely on X still acts freely on X^k.

This can now be used to extend Theorem 4, proving:

Theorem 5. Let G be a compact Lie group. Then G has a smooth free action on some (finite) product of spheres of the same dimension, if and only if G contains no copy of SO(3).

Proof. If $SO(3) \subseteq G$, then G has no such action by Theorem 3. Conversely, if G contains no copy of SO(3), then its identity component G_0 has a free action on some product $\prod^n S^7$. Let $k = |G/G_0|$; then G_0 acts freely under the induced G-action on $\prod^{kn} S^7$.

For any $1 \neq g \in G/G_0$, the cyclic group generated by g has a free action on S^7; let M_g denote the induced action of G/G_0. Then G acts freely on

$$(\prod_{1 \neq g \in G/G_0} M_g) \times (\prod^{kn} S^7),$$

and this is again a product of S^7's. (My thanks to Jørgen Tornehave for pointing out this last construction to me.) □

References

1. A. Borel and J. de Siebenthal, Sur les sous-groupes fermés de rang maximum des groupes de Lie compacts connexes, Comm. Math. Helv. 23 (1949), 200-221.

2. H. Cartan and S. Eilenberg, Homological Algebra, Princeton Univ. Press (1956).

3. P. Conner, On the action of a finite group on $S^n \times S^n$, Ann. of Math. 66 (1957), 586-588.

4. D. Quillen, The spectrum of an equivariant cohomology ring I, Ann. of Math. 94 (1971), 549-572.

5. R. Swan, Periodic resolutions for finite groups, Ann. of Math. 72 (1960), 267-291.

THREE THEOREMS IN TRANSFORMATION GROUPS

Ted Petrie, Rutgers University U.S.A.

§1 Introduction

Let G be a finite group of order $|G|$ and $\mathcal{C}$ a category of G spaces. E. g.: $\mathcal{L}$ the category consisting of unit spheres S(V) and unit disks D(V) of complex representations V of G, $\mathcal{D}$ the category of smooth G manifolds and $\mathcal{H}$ the category of finite G C.W. complexes. Some typical invariants I(Y) for $Y \in \mathcal{C}$ are: Iso(Y) - the set of isotropy groups G_p for $p \in Y$, T_pY the isotropy representation of G_p on the tangent space at p, Dim Y - the function which assigns to the subgroup H of G Dim Y(H) the dimension of the H fixed point set Y^H and $\{Y\}$ - the G homotopy type of Y. Let $\underline{Y}$ denote the homotopy type of Y without G action. Much of the subject of transformation groups is encompassed in the following

1.0. Basic Project: Describe $\{ I(Y) \mid Y \in \mathcal{C} , \underline{Y} = M \}$. Here M is some fixed homotopy type. Typically $M = S^n$ or D^n. In other words, describe the set of values of an invariant I(Y) as Y ranges over a fixed homotopy type.

Three contributions to this project are discussed here. These deal with the invariants I(Y) : Dim Y, T_pY and $\{Y\}$ when Y is a homotopy sphere. We present some of the relevant history to motivate theorems A - C which treat these invariants.

Concerning the invariant Dim Y when $Y \in \mathcal{C}$ is a homotopy sphere, we have this result of Artin which treats the case $\mathcal{C} = \mathcal{L}$:

<u>1.1. Theorem [9] : Let</u> V <u>be a complex representation of</u> G. <u>Then</u> Dim V(G) <u>is a function of</u> $\{\text{Dim } V(H) \mid H \text{ cyclic}\}$.

In fact

$$\operatorname{Dim} V(G) = |G|^{-1} \sum_C \sum_{D<C} u(|C/D|)\,|D|\ \operatorname{Dim} V(D)$$

where C runs through the cyclic subgroups of G and μ is the Möbius function.

When G is an elementary abelian p group, Borel gives the formula

$$\operatorname{Dim} Y(1) - \operatorname{Dim} Y(G) = \sum_{H \in \mathcal{F}} (\operatorname{Dim} Y(H) - \operatorname{Dim} Y(G))$$

for any homotopy sphere $Y \in \mathcal{H}$. Here $\mathcal{F}$ is the family of subgroups of index p. Again Dim Y(G) is expressed in terms of Dim Y(H) for H in a family of subgroups of G.

tom Dieck has generalized 1.1 to the category $\mathcal{H}$ provided G is a p group. In fact if Y is a homotopy G sphere such that Dim Y(H) is odd for all H, then Dim Y = Dim S(V) for some complex representation V of G.

Motivated by these results we might ask: For which G does there exist a function f_G such that $\operatorname{Dim} Y(G) = f_G(\operatorname{Dim} Y(H) \mid H \neq G)$ whenever $Y \in \mathcal{C}$ is a homotopy sphere? In view of the above results this is now completely answered by

<u>Theorem A [8] : If there is a function f_G such that</u>

$$\operatorname{Dim} Y(G) = f_G(\operatorname{Dim} Y(H) \mid H \neq G)$$

<u>whenever $Y \in \mathcal{D}$ and $Y = S^n$ for some n, then G is a p group.</u>

Concerning the invariant T_pY, we mention this theorem of Atiyah-Bott and Milnor:

1.2. Theorem [1] : Let G be a compact Lie group acting smoothly on a homotopy sphere Y preserving orientation with Y^G consisting of 2 points p and q. If G acts freely on $Y - Y^G$, then the isotropy representations T_pY and T_qY are equal.

The hypothesis that G act freely on $Y - Y^G$ is very restrictive (in particular G must have periodic cohomology). What is the situation when this hypothesis is removed?

1.3. Theorem (Sanchez) 1.2 remains true without the assumption that G act freely on $Y - Y^G$ provided G is a p group.
Sanchez, a student of Bredon, has published this result from his thesis but unfortunately I haven't a reference. The proof is similar to that in [1] .)

Let R(G) be the complex representation ring and $\mathcal{P}$ the family of groups of prime power order. If H is a subgroup of G, restriction defines a homomorphism of R(G) to R(H) called Res_H. Set

1.4. $$I = Ker(R(G) \xrightarrow{Res} \prod_{P \in \mathcal{P}} R(P))$$

Combining 1.3 with the elementary fact that T_xY is constant on components of Y^H $H = G_x$ and that Y^P is a mod p homology sphere if Y is and P is a p group, we see that

1.5. $$T_pY - T_qY \in I$$

whenever Y is a smooth G homotopy sphere with $Y^G = p \amalg q$. Conversely

Theorem B [17] : Every odd order abelian group G with at least 4 non cyclic Sylow subgroups acts smoothly on a homotopy sphere Y with

$Y^G = p \amalg q$ and $T_pY - T_qY$ equal to any given element of I.

The study of the invariant $\{Y\}$ is interesting even for $\mathcal{C} = \mathcal{L}$. If V_1 and V_2 are two complex representations of G, write

$$1.6. \qquad S(V_1) \sim S(V_2)$$

if there is a G map $f : S(V_1) \longrightarrow S(V_2)$ such that degree $f^H = 1$ for all H. (Degree is defined with respect to the orientation defined by complex structure. We say $S(V_1)$ and $S(V_2)$ are G oriented homotopy equivalent. See [3], [4] and [6].

Actually the notion in 1.4 goes back to the study of the homotopy type of Len's spaces [12] and more generally the equivariant homotopy type of free actions on homotopy spheres. Even this classification is not complete and forms a good show case to motivate Theorem C below. If G acts freely, on a homotopy sphere Y preserving orientation, then for each conjugacy class of Sylow subgroups $P \subset G$, $\mathrm{Res}_P Y$ is P oriented homotopy equivalent to $S(V_P)$ for some complex representation V_P of P. These representations all have the same dimension n independent of the Sylow subgroup. Then $\lambda^n(V_P)$ is a complex one dimensional representation of P. Since G acts freely on a homotopy sphere, it has periodic cohomology so the Sylow subgroups are cyclic (P odd) and generalized quaternion. In the cyclic case, the one dimensional representation $\lambda^n(V_P)$ is completely determined by an integer mod $|G|$. This integer mod$|P|$ again called $\lambda^n(V_P)$ is prime to $|P|$ because P acts freely on $S(V_P)$. Carrying this discussion to its conclusion we have

Proposition 1.7. Let G act freely on a homotopy sphere Y of dimension 2n-1. Then $\{Y\}$ is determined by n and $\{\lambda^n(V_P) \mid P \text{ Sylow}\}$.

Thus the value group for the invariant $\{Y\}$ for free actions on spheres is $Z \times Z^{*}_{|G|}$ where $Z_{|G|}$ is cyclic of order $|G|$ and $*$ means the units in this ring. The question which naturally arises is:

Which values occur?

The study of this question has a long history. Fundamental contributions occur in the papers [12], [14], [10] and [19]. Aside from the case of cyclic and generalized quaternion groups and the metacyclic groups $Z_{p,q}$ p,q prime, it is not solved.

This history should motivate the depth and the difficulty of the characterization of the invariant $\{Y\}$ as Y ranges over the homotopy spheres in $\mathcal{C}$. The following treatment of this question was motivated by the above discussion.

Let $\mathcal{S}$ denote the set of Sylow subgroups of G up to conjugacy. Let $P \in \mathcal{S}$ be a Sylow subgroup and V_P a complex representation of P. We require $\dim_C V_P$ to be independent of $P \in \mathcal{S}$ and set

$$\mathcal{V} = \{V_P \mid P \in \mathcal{S}\}$$

$\mathrm{Dim}(\mathcal{V}) = \dim_C V_P$ for any P.

If Y is a G space, $\mathrm{Res}_H Y$ denotes the H space obtained by restricting the action to H. We say $\mathcal{V}$ is <u>free</u> if P acts freely on $S(V_P)$ for each $P \in \mathcal{S}$. We say $\mathcal{V}$ <u>is G invariant</u> if for each $P \in \mathcal{S}$ and $H \subset P \cap gPg^{-1}$, we have $\mathrm{Res}_H V_P = \mathrm{Res}_{gHg^{-1}} V_P$ as representations of H. We say $\mathcal{V}$ is G <u>invariant up to homotopy</u> if

$$S(\mathrm{Res}_H V_P) \sim S(\mathrm{Res}_{gHg^{-1}} V_P) \qquad (1.4)$$

$Iso(\mathcal{V})$ denotes the set of subgroups of G which are conjugate to a subgroup of $Iso(S(V_P))$ for some $P \in \mathcal{S}$. Associated to $\mathcal{V}$ is a subgroup $B(\mathcal{V})$ of the reduced projective class group $K_o(Z(G))$ which only depends on $Iso(\mathcal{V})$ and where $B(\mathcal{V})$ contains $B(\mathcal{W})$ whenever $Iso(\mathcal{V})$ contains $Iso(\mathcal{W})$. Moreover $B(\mathcal{V}) = 0$ if $\mathcal{V}$ is free.

1.8. Definition: $f : X \longrightarrow S(\mathcal{V})$: This means X is a G space and a P map $f_P : Res_P X \longrightarrow S(V_P)$ is given for all $P \in \mathcal{S}$. We say $\mathcal{V}$ is realized by X if each f_P is a homotopy equivalence.

1.9. Realization Problem: Given $\mathcal{V}$ when does there exist a G homotopy sphere $X \in \mathcal{C}$ which realizes $\mathcal{V}$?

The answer to 1.9 (and 1.0 more generally) depends on $\mathcal{C}$ both in treatment and result.

We have a complete answer to 1.9 for $\mathcal{C} = \mathcal{H}$ and a quite general result for $\mathcal{C} = \mathcal{D}$ which is mentioned in §4.

Theorem C $\mathcal{V}$ is realized by $X \in \mathcal{H}$ iff $\mathcal{V}$ is G invariant up to homotopy and an invariant $\chi(\mathcal{V}) \in K_o(Z(G))/B(\mathcal{V})$ vanishes; moreover, $\chi(\mathcal{V} + \mathcal{W}) = \chi(\mathcal{V}) + \chi(\mathcal{W})$ whenever $Iso(\mathcal{V}) = Iso(\mathcal{W})$. There is an integer $n = n(G)$ such that $n\mathcal{V}$ is realized.

Theorem C is a generalization of Swan's paper [19] on free actions on spheres. To give a feeling for Theorem C, let $G = Z_{p,q}$ be the metacyclic group with presentation

$$\left\{ x,y \mid x^p = y^q = 1,\ y^{-1}\, x\, y = x^a \quad a^q \equiv 1\ (p) \right\}$$

where a has order q in Z_p^* and p and q are prime. Then $2q = \text{period}(G)$

and if $\mathcal{V}$ is free, it is G invariant up to homotopy iff Dim $\mathcal{V} \equiv 0(q)$. Let m = Dim $\mathcal{V}$.

Theorem 1.10 [15] and [20] : *Let* $G = Z_{p,q}$, Dim $\mathcal{V} \equiv 0(q)$ *and* $\mathcal{V}$ *be free*. *Then* $\chi(\mathcal{V}) = 0$ *iff* $\lambda^m(V_{Z_p})$ *is a* q th *power* mod p.

Corollary 1.11. *Let* $G = Z_{p,q}$. *Then* G *acts freely and smoothly on a homotopy sphere* Y *of dimension* 2m-1 *with equivariant homotopy type* $\{Y\}$ *equal to* $(m, \{\lambda^m(V_P) \mid P \text{ Sylow}\}) \in Z \times Z^*_{|G|}$ *iff* $m \equiv 0(q)$ *and* $\lambda^m(V_{Z_p})$ *is a* q-th *power* mod p.

We remark that there is an analog of Theorem C in $\mathcal{D}$. The techniques involved go back to the author's construction of free metacyclic actions on spheres. As we see here the methods are applicable much more generally.

The method for treating 1.0 and in particular Theorems A - C is G surgery (§2) which is summarized by the sequence 2.6. The basic project 1.0 and the exact sequence 2.6 are intimately related. Among those invariants of a G-map $f : X \longrightarrow Y$ in $\mathcal{D}$, we must find those which are relevant to pseudo-equivalence abreviated by p. e. A p. e. is a G map f which is a homotopy equivalence. We associate to f a set $\pi(f)$ of invariants of X,Y and f which are relevant to the process of G surgery and pseudo equivalence. In particular $\pi(f)$ specifies these invariants: Iso(Z), T_pZ $p \in Z$, Dim Z for Z equal to X,Y and Deg f where Deg f is the function which assigns to $H \subset G$ degree f^H. Here f^H is the mapping on the H fixed set. A fixed value set for $\pi(f)$ will be denoted by λ. See §2 and [17] for precise details. The point is that the invariants specified by λ play a distinguished role among all possible invariants.

The aims of this paper are: To give an impression of the subject of transformation groups and in particular 1.0 through Theorem A - C; To illustrate the versality of the sequence 2.6 as a tool for treating 1.0 and in particular Theorems A - C.

The author thanks tom Dieck for many useful conversations resulting from our joint work [6] in treating the invariant $\{Y\}$.

§2. The G Surgery Sequence

A set of subgroups of G invariant under conjugation is called a family and typically denoted by $\mathcal{F}$.

2.1. G Surgery Problem: Let $\mathcal{F}_0$ be given and $f : X \longrightarrow Y$ be a map in $\mathcal{C}$. When can we alter (X,f) (rel. a G neighborhood of $\bigcup_{H \in \mathcal{F}_0} X^H$) to (X',f') where $f' : X' \longrightarrow Y$ is a p. e.

The answer as well as the definition of "alter" depends on $\mathcal{C}$. When $\mathcal{C} = \mathcal{H}$, we require $X \subset X'$ and f' extends f. When $\mathcal{C} = \mathcal{D}$, we require (X',f') to be G cobordant to (X,f). We summarize the answer to 2.1 in both categories; however, to simplify this discussion we restrict to the case:

2.2. Y^P is connected $\forall P \in \mathcal{P}$.

The treatment of 2.1 for $\mathcal{C} = \mathcal{H}$ is joint work with Oliver. A family $\mathcal{F}$ is called connected if for any $P \in \mathcal{P}$, there is a unique minimal subgroup $P \in \mathcal{F}$ containing P; when this holds we write $\hat{\mathcal{F}} = \{\hat{P} | P \in \mathcal{P}\}$. The motivation for this property is this: If Y is a smooth G manifold satisfying 2.2, then Iso(Y) is connected. The answer to 2.1 under assumption 2.2 is [13].

Theorem 2.3. Let $f : X \longrightarrow Y$ be a map in $\mathcal{H}$. Let $\mathcal{F}$ be any connected family containing $\mathrm{Iso}(X \amalg Y)$, and let $\mathcal{F}' \subset \mathcal{F}$ be any subfamily containing $\hat{\mathcal{F}}$. Assume also that $X^G \neq \emptyset$ or $\hat{\mathcal{F}} \subset \mathcal{P}$. Then there is an $X' \supset X$ such that $\mathrm{Iso}(X'-X) \subset \mathcal{F}'$ and a p. e. $f' : X' \longrightarrow Y$ extending f, iff

$$[Y] - [X] \in \Delta(G,\mathcal{F}) + \Omega(G,\mathcal{F}') \text{ in } \Omega(G).$$

This is a special case of Theorem 3.2 of [13] .

Here $\Omega(G)$ denotes the Burnside ring of G (under tom Dieck's definition [5]) and [Y] denotes the class of Y in $\Omega(G)$. $\Omega(G, \mathcal{F}')$ is the subgroup generated by [G/H] for $H \in \mathcal{F}'$; while $\Delta(G, \mathcal{F})$ is characterized by

$$\Delta(G, \mathcal{F}) = \left\{ [X] - 1 \in \Omega(G) \;\middle|\; \begin{array}{l} (X,x) \text{ is a finite} \\ \text{contractible based } G \\ \text{complex, } \mathrm{Iso}(X-x) \end{array} \right\}$$

Remark: 2.1 is solved by 2.3 whenever $\mathcal{F}_0 \cap \mathcal{F}' = \emptyset$.

To illustrate 2.3 in a situation which is relevant to the classification of equivariant homotopy types of actions on spheres consider:

Corollary 2.4: *Let V be a complex representation of G. Then there is a G homotopy sphere $\Sigma \in \mathcal{H}$ with $\mathrm{Iso}(\Sigma) \subset \mathcal{P}$ and a p. e. $f : \Sigma \to S(V)$.*

Proof: Apply 2.3 with Y = S(V), X = $\emptyset$, $\mathcal{F} = \mathrm{Iso}(Y) \cup \mathcal{P}$, $\mathcal{F}' = \mathcal{P}$. Since Y^H is an odd dimensional manifold for all H, $[Y] - [X] = 0$ in $\Omega(G)$ so 2.3 applies.

Remark: 2.4 is false in general for a real representation of G. 2.4 remains true in the smooth G category with some additional hypothesis. Note that $Y \longmapsto \mathrm{Iso}(Y)$ is an invariant of the G homotopy type of Y in the category of smooth closed G manifolds; however, it is not an invariant on the category $\mathcal{H}$.

The treatment of 2.1 in $\mathcal{D}$ is considerably more difficult as it involves much from the theory of surgery and the theory of transformation groups. We summarize the answer in a form applicable to the ideas here. First some generalities: <u>All G manifolds are given with an orientation for the fixed set of every subgroup</u>. <u>Such a manifold is said to be oriented</u>. Let E denote a contractible complex on which G acts freely. Let $\hat{X} = E \times_G X$ for any G space X. Let $\hat{f} : \hat{X} \longrightarrow \hat{Y}$ be the map induced by $f : X \longrightarrow Y$. Note that if η is a G vector bundle over X, $\hat{\eta}$ is a vector bundle over $\hat{X}$.

We require this added data for treating 2.1 in $\mathcal{D}$ under assumption 2.2:

2.5.) i) degree $f = 1$

ii) $\hat{f}^* \xi = \hat{TX}$ for some stable vector bundle ξ over $\hat{Y}$.

iii) $f^{H*} \eta(H) = \nu(X^H, X) \quad \forall H \in \mathcal{P}$

iv) $[Y] - [X] \in \Delta(G, \mathcal{F}_f) + \Omega(G, \mathcal{F}'(X))$ in $\Omega(G)$

v) Connectivity condition: $f_o : \pi_o(X^H) \longrightarrow \pi_o(Y^H)$ is a bijection $\forall H \in \mathcal{P}$.

vi) $\pi_1(Y) = 0$

vii) Gap Hypothesis: $\dim X^H < \frac{1}{2} \dim X^K$ whenever $K > H$ and $X^H \neq X^K$.

The meaning of 2.5 iii) is this: $\nu(X^H, X)$ is the H normal bundle of X^H in X and 2. iii) asserts the existence of an H vector bundle $\eta(H)$ over Y^H so that the stated equality of H vector bundles holds. The families $\mathcal{F}_f$ and $\mathcal{F}'$ occuring in iv) are $\mathcal{F}_f = \mathrm{Iso}(X \amalg Y)$ and $\mathcal{F}'(X) = \mathrm{Iso}(X) - \mathcal{F}_o$.

The solution to 2.1 in $\mathcal{D}$ is provided by the G surgery sequence:

2.6.) $hS_G(Y,\lambda) \xrightarrow{d} N_G(Y,\lambda) \xrightarrow{\sigma} I(G,\lambda)$.

This we briefly describe. Denote the terms of 2.6 from left to right by A_i. $i = 0,1,2$. Elements of A_i are equivalence classes of pairs $z = (X,f)$ where $f : X \longrightarrow Y'$ is a map in $\mathcal{D}$ satisfying 2.5 and $\pi(F) = \lambda$. In particular $z \in A_0$ iff $Y' = Y$ and f is a p. e., $z \in A_1$ iff $Y' = Y$ while $z \in A_2$ has no further requirements other than 2.5 and $\pi(f) = \lambda$.

<u>Theorem 2.7.</u> [17] <u>The G Surgery Sequence is an exact sequence of sets.</u>

This is actually a solution to 2.1 in $\mathcal{D}$.

<u>Theorem 2.7'. Let</u> $f : X \longrightarrow Y$ <u>be a map in</u> $\mathcal{D}$ <u>satisfying</u> 2.5. <u>Let</u> $\lambda = \pi(f)$. <u>Then</u> (X,f) <u>is</u> G <u>cobordant rel a</u> G <u>neighborhood of</u> $\bigcup_{H \in \mathfrak{F}_0} X^H$ <u>to</u> (X',f') <u>with</u> f' <u>a p. e. iff</u> $\sigma(X,f) = 0$ <u>in</u> $I(G,\lambda)$.

Of course the G cobordism in 2.7' is required to preserve the conditions of 2.5. The process used in 2.6 - 2.7' is G surgery [17]. Relatively simple restrictions on λ insure that $I(G,\lambda)$ is a group [7]. When G acts freely on Y, $I(G,\lambda)$ is the Wall group $L_n(G)$ $n = \dim Y$.

In order to be able to use this theorem, we must be able to recognize the zero object in $I(G,\lambda)$. Instead of defining zero, we give a sufficient geometric property which implies that $z = (X,f)$ represents zero in $I(G,\lambda)$. Let $f : X \longrightarrow Y$, $\pi(f) = \lambda$. Suppose $X = \partial W$, $Y = \partial Z$, $\pi_0(X^H) \longrightarrow \pi_0(W^H)$ and $\pi_0(Y^H) \longrightarrow \pi_0(Z^H)$ are surjective for all H, and suppose there is a map $F : W \longrightarrow Z$ with $\pi(F) = \lambda$ which satisfies 2.5 and the conditions of 2.5 for F extend those for f. Then we write $(X,f) = \partial(W,F)$.

2.8. If $z = (X,f) = \partial(W,F)$, then z represents zero in $I(G,\lambda)$.

2.9. Properties of the G surgery sequence: Let $f : X \longrightarrow Y$ $\pi(f) = \lambda$.

i) If $\dim Y^H$ is odd for all H, then $I(G,\lambda)$ is finite [8].

ii) If $\dim Y^H \geq G$ whenever $Y^H \neq 0$, then $I(G,\lambda)$ is an abelian group.

iii) If $Y = S(V)$ where V is a representation of G with $\dim_{\mathbb{R}} V \geq 6$, then $N_G(Y;\lambda)$ has an additive structure and σ is additive [7].

iv) If G is an odd order abelian group and $[Y] - [X] \in \Delta(G, \mathfrak{F}_f) + 2\Omega(G, \mathfrak{F}'(X))$ (compare 2.5 iv), then $\sigma(X,f) = 0$ iff $\mathrm{Res}_H \sigma(X,f) = 0$ in $I(H,\lambda_H)$ for all hyperelementary subgroups $H \subseteq G$. Here $\mathrm{Res}_H : I(G,\lambda) \longrightarrow I(H,\lambda_H)$ is defined by restriction of G data to H data.

§ 3. Outline of Theorems A and B

Outline of Theorem A: There is a virtual G set X with Euler characteristics $\chi(X) = 1$ and $\chi(X^G) = 0$ iff $G \notin \mathcal{P}$ for such a G there is a complex representation A and a proper G map $a : A \longrightarrow A$ such that degree $a = 1$ and degree $a^G = 0$. The point is that equivariant homotopy classes of proper equivariant maps $a : A \longrightarrow A$ correspond bijectively to elements of $\Omega(G)$ [18] . The correspondence sends a virtual G set X to a map a such that degree $a^H = \chi(X^H)$ for all $H \subseteq G$.

Let $Y = S(V)$ where V is a complex representation of G such that Iso(Y) contains all subgroups of G, the Gap Hypothesis 5 vii) holds for Y and $\dim Y^G \geq 6$. Let h' be the self map of $Y \times A$ defined by $1 \times a$. Then h' is properly G homotopic to a map h transverse to $Y \times 0$ written $h \pitchfork Y$. Set $X = h^{-1}(Y)$ and $f = h_{|X} : X \longrightarrow Y$. Then $h \pitchfork Y$ implies $h^K \pitchfork Y^K$ $K \subseteq G$ and this implies $\dim X^K = \dim Y^K$ whenever $X^K \neq 0$ and degree f^K = degree a^K. In particular if $X^G = \emptyset$, degree $f^G = 0$ = degree a^G. Conversely if degree $a^G = 0$, we can arrange $X^G = \emptyset$ and $X^H \neq \emptyset$ for $H \neq G$; so $\dim X^H = \dim Y^H$ iff $H \neq G$.

It is not a difficult matter to check the properties 2.5 for $z = (X,f)$. Only 2.5 iii and v are not immediate and these can be arranged. Let $\lambda = \pi(f)$. Then $z \in N_G(Y, \lambda)$. Using 2.9 i - iii, there is an integer n such that $\sigma(nz) = 0$; so by 2.6, $nz = d(Y',f')$ $f' : Y' \longrightarrow Y$ a p. e.

Since $\pi(f') = \lambda = \pi(f)$ and since λ specifies the dimension functions of both source and target of these maps, it follows that $\dim Y'^H = \dim X^H$ for all H. This means that the dimension functions Dim Y and Dim Y' agree on all subgroups of G except G itself. Since both Y and Y' are smooth homotopy spheres; Theorem A is established.

Outline of Theorem B: This theorem is much deeper. Though some important details must be suppressed, enough remains to give a good illustration of the ideas.

Let R(G) denote the complex representation ring of G and K($\cup$) the complex K theory of $\cup$. Let x be the G space consisting of one point and observe that $\hat{x} = B_G$ is the classifying of G. If we view a complex representation A of G as a G vector bundle over x, then $\hat{A}$ is a vector bundle over $\hat{x}$. This defines a homomorphism

$$\wedge : R(G) \longrightarrow K(\hat{x}) .$$

3.1. Theorem [O] : I = ker($\wedge$) (See 1.4)

Let $\xi \in$ I. Then there exist complex representations V and V' of G such that V - V' = ξ and [17] there exist smooth G spheres $\cup$(A) for A = V, V' such that if $\cup = \cup(A)$ and $h_{\cup(A)} = h_{\cup} : \cup \longrightarrow x$ is the point map:

i) $\cup^G = p_A$

ii) $T_x\cup = A$ for $x = p_A$

iii) $T\cup = h_{\cup}^{*}(A)$ as a stable G vector bundle

iv) $\nu(\cup^K, \cup) = h_{\cup}^{K*}(\mathrm{Res}_K A/A^K)$ whenever $K \in \mathcal{P}$.

For each hyperelementary group H in G, there is a smooth H manifold D(A,H) = D such that if $h_D : D \longrightarrow x$ is the point map:

i') $\partial D = \mathrm{Res}_H \cup(A)$

ii') $\pi_0(\cup^K) \longrightarrow \pi_0(D^K)$ is surjective for all $K \subset H$

iii') $TD = h_D^{*}(\mathrm{Res}_H A)$

iv') $\nu(D^K, D) = h_D^K(\mathrm{Res}_K A/A^K)$ whenever $K \in \mathcal{P}$ $K \subset H$.

Let X = $\cup(V) \amalg \cup(V')$ Y = S(V $\oplus$ $\mathbb{R}$), Z = D(V $\oplus$ $\mathbb{R}$) and for each hyperelementary subgroup H of G let $W(H) = D(V,H) \amalg D(V',H)$. Above $\mathbb{R}$ is the trivial one dimensional real representation of G and Z is the unit disk of $V \oplus \mathbb{R}$.

Define a degree one G map $f : X \longrightarrow Y$ by the Thom map which collapse the exterior of an invariant ball about p_V to one point. (Note the Thom space of the G vector bundle $T_pX = V$ over $p = p_V$ is Y).

Observe that $X^G = p_V \amalg p_{V'}$ and $T_pX = A$ for $p = p_A$. If X were a homotopy sphere, we'd be done. The aim is to use 2.6 and 2.7' to produce a homotopy sphere Y' with these same properties. In particular we don't want to alter X in a G neighborhood of X^G. This dictates the choice $\mathcal{F}_o = \{G\}$.

We need to verify the conditions of 2.5. We note that v) is violated since X is not connected; however, this can be rectified by zero dimensional G surgery in the presence of iv), so we suppose this achieved. A judicious choice of V and V' will guarantee iv), vi) and vii) though iv) is non trivial.

We show how 2.5 ii) and iii) are consequences of the assumption $\xi \in I$. Let $h : X \longrightarrow x$ be the point map. It clearly suffices to verify these conditions for h rather than f. Since $V - V' \in I$, $V = V'$ by 3.1) and since $T \cup (A) = h^*_{\cup} A$ (iii) for $A = V,V'$, $\hat{TX} = \hat{h}^*(\hat{V})$. Since $\mathrm{Res}_K V = \mathrm{Res}_K V'$ for $K \in \mathcal{P}$ and since $\nu(\cup^K, \cup) = h^{K*}_{\cup}(\mathrm{Res}_K A/A^K)$ for $A = V,V'$ (iv), it follows that $\nu(X^K,X) = h^{K*}(\mathrm{Res}_K V/V^K)$. This verifies 2.5 ii) and iii).

Let $\lambda = \pi(f)$. Then $\mathfrak{z} = (X,f) \in N_G(Y,\lambda)$. We apply 2.9 iii) to show that $\sigma(\mathfrak{z}) = 0$. The map $\mathrm{Res}_H f : \mathrm{Res}_H X \longrightarrow \mathrm{Res}_H Y$ extends to an H map $F(H) : W(H) \longrightarrow \mathrm{Res}_H Z$ for all hyperelementary subgroups H of G. The degree of F(H) is one. The properties i' - iv' are used to show that $\mathrm{Res}_H(X,f) = \partial(W(H),F(H))$; so by 2.8, $\mathrm{Res}_H \sigma(\mathfrak{Z}) = 0$. Now apply 2.9 iii) to conclude $\sigma(\mathfrak{Z}) = 0$.

Now apply 2.7') with $\mathfrak{F}_o = \{G\}$. Then (X,f) is G cobordant rel. an invariant neighborhood of X^G to (Y',f') where f' is a p. e. In particular Y' is a smooth homotopy sphere with $Y'^G = X^G = p_V \amalg p_{V'}$ and $T_p Y' = T_p X$ for $p \in Y'^G$. Since $T_p X = \Lambda$ for $p = p_A$, subtracting the isotropy representations of Y' at p_V and $p_{V'}$ gives ξ. This establishes Theorem B.

§4. Outline of Theorem C

One of the essential points for treating Theorem C, is the relation between the degree of an equivariant map and homological dimension over the group ring. The notion of degree requires a definition of orientation. In $\mathcal{D}$ this was defined in §2. In $\mathcal{H}$ we define it this way: $X \in \mathcal{H}$ is oriented if for each $H \subset G$ with $X^H \neq \emptyset$ $H_{n(H)}(X^H) = Z$ for $n(H) = \dim X^H$ and in addition a specific generator for this homology group is given.

If X is oriented and $f : X \longrightarrow S(\mathcal{V})$, then degree f^H is defined for $H \in \mathcal{P}$ by degree f^H = degree f_P^H where $H \subset P \in \mathcal{S}$. Each $S(V_P)$ is oriented by the complex structure of V_P. Note in particular that degree f_P must be constant for $P \in \mathcal{S}$. The fundamental lemma connecting degree and homological dimension is

Lemma 4.1. Let $f : X \longrightarrow S(\mathcal{V})$ be a map with degree f = 1. Suppose $H_i(X) = 0$ for $i \neq k$ and dim X. Then hom $\dim_{Z(G)} H_k(X) \leq 1$.

If there exists an X with the properties of Lemma 4.1 and if Iso(X) Iso($\mathcal{V}$), we define

4.2. $$\chi(\mathcal{V}) = (-1)^k \, [H_k(X)] \in \tilde{K}_o(Z(G))/B(\mathcal{V}) .$$

Here $\tilde{K}_o(Z(G))$ is regarded as the Grothendieck group of Z(G) modules of finite homological dimension modulo free modules. If X_i $i = 1,2$ are two elements of $\mathcal{C}$ satisfying the hypothesis of Lemma 4.1 and if $Iso(X_i) = Iso(\mathcal{V})$, then the two elements in $\tilde{K}_o(Z(G))$ associated to X_1 and X_2 differ by an element of $B(\mathcal{V})$; so $\chi(\mathcal{V})$ is well defined.

There are two steps in satisfying the hypothesis of

Lemma 4.1: i) Produce $f : X \longrightarrow S(\mathcal{V})$ where degree $f = 1$ and X is oriented and X^H is connected for all H. ii) Arrange $H_i(X) = 0$ for $i \neq k$, dim X. The first step is achieved via

<u>Lemma 4.3.</u> <u>Suppose that</u> $\mathcal{V}$ <u>is</u> G <u>invariant</u> (G <u>invariant up to homotopy</u>). <u>Then there exists</u> $f : X \longrightarrow S(\mathcal{V})$ <u>with degree</u> $f = 1$ <u>where</u> $X \in \mathcal{D}$ <u>is oriented and</u> X^H <u>is connected for all</u> H($X \in \mathcal{H}$ <u>is oriented and</u> X^H <u>is a connected Poincaré duality space</u> $\forall$ H).

Lemma 4.3 goes as follows. Choose integers a_P $P \in \mathcal{S}$ such that

4.4. $$\sum_{P \in \mathcal{S}} a_P \, |G|/|P| = 1.$$ Set

4.5. $$X_o = \coprod_{P \in \mathcal{S}} a_P \, G \times_P S(V_P)$$

Here the signs of the integers a_P are incorporated into the orientation of the smooth G manifold X_o. Observe that

4.6. $$\mathrm{Iso}(X_o) = \mathrm{Iso}(\mathcal{V})$$

Since $\mathcal{V}$ is G invariant (G invariant up to homotopy) there is a map $f_o : X_o \longrightarrow S(\mathcal{V})$ of degree 1.

Now $X_o \in \mathcal{D}$ is oriented but X_o^H is not connected for $H \subset G$. To connect the various components of X_o^H by zero dimensional G surgery on X_o and remain in $\mathcal{D}$, we need the isotropy representations $T_P X$ for $p \in X^H$ to be independent of p. In order for this to hold for all H, $\mathcal{V}$ must be G invariant. If however we are willing to settle for an X such that X^H is a P.D. space for all H, we need only require $\mathcal{V}$ be G invariant up to homotopy.

When either of these requirements is fullfilled, we do the zero dimensional G surgery producing the required X and f.

We remark that if Lemma 4.1 holds, then

4.7. $f_P^H : X^H \longrightarrow S(V_P)^H$ is a mod p equivalence wherever $H \subset P$ $|P| = p^k$.

Conversely if 4.7 holds for all $H \in \mathcal{P}-1$, then Lemma 4.1 can be established. The conditions of 4.7 are achieved inductively on the partial order on the groups in $\mathcal{P}$. Again degree plays a role; namely, we need degree f_P^H to be a unit mod p. This is the reason for requiring X^H to be oriented for all H; so degree is defined. Since degree f_P is one for all P, degree f_P^H is a unit mod $|P|$ by [2] . This uses the fact that initially X is a P. D. space (Lemma 4.3) and the subsequent alterations do not disturb degree.

Now the treatment of Theorem C in $\mathcal{D}$ and $\mathcal{H}$ diverge in the same way the treatment of 2.1 diverged in section 2. First Lemma 4.1 has to be established. For $\mathcal{C} = \mathcal{H}$ no further hypothesis are required. By adding handles $G/H \times D^i$ $0 < i < n(H)-1$ for $H \in Iso(\mathcal{V})$ to the X of Lemma 4.3, we produce an extension $X' \supseteq X$ with $Iso(X') = Iso(X)$ and an extension $f' : X' \longrightarrow S(\mathcal{V})$ of f such that Lemma 4.3 is achieved with $k = \dim X - 1$. If $\chi(\mathcal{V}) = 0$, there is a further extension $X'' \supseteq X$ and $f'' : X'' \longrightarrow S(\mathcal{V})$ which realizes $\mathcal{V}$.

The process for treating Theorem C and Lemma 4.1 for $\mathcal{C} = \mathcal{D}$ is G surgery (G cobordism). We need these modifications of the requirements 2.5: Let $h : X \longrightarrow x$ be the point map

4.8. i) $\hat{TX} = \hat{h}^* \xi$ for some stable vector bundle ξ over $\hat{x}$.

ii) $\nu(X^H, X) = h^{H*}\eta(H)$ for some H vector bundle $\eta(H)$ over x. This must hold for all H .

iii) The Gap Hypothesis holds X.

There are apriori conditions on V which guarantee these for the manifold X_o of 4.5. Let N(P) be the Normalizer of P. Then

$$V \in \prod_{P \in \mathcal{S}} R(P)^{N(P)} \quad \text{so} \quad \hat{V} \in \prod_{P \in \mathcal{S}} \hat{R(P)}^{N(P)}$$

4.9. i) $\hat{V} \in \text{Image } (R(G) \xrightarrow{\text{Res}} \prod_{P \in \mathcal{S}} R(P)^{N(P)})$

ii) V is G invariant

iii) $\text{Dim}(V)$ satisfies the Gap Hypothesis: $\dim V_P^H < \frac{1}{2} \dim V_P^K$ whenever $P \supset H > K$ and $V_P^H \neq V_P^K$.

If 4.9 holds, then 4.8 holds for $X = X_o$. Since the conditions in 4.8 are invariant by G surgery, they remain valid for any manifold and map produced by G surgery.

Suppose then that 4.9 holds. Apply, G surgery to (X_o, f_o) to construct a G cobordism to (X,f) where $\text{Iso}(X) = \text{Iso}(X_o) = \text{Iso}(V)$ and $f: X \longrightarrow S(V)$ satisfies Lemma 4.1 with $2k + 1 = \dim X$. Then $\chi(V)$ is defined.

<u>Proposition 4.10.</u> <u>If</u> $\chi(V) = 0$, <u>then</u> (X,f) <u>defines an element</u> $\sigma(V)$ <u>in the Wall group</u> $L_n(G)$ $n = \dim X$. <u>If</u> $\sigma(V) = 0$, (X,f) <u>is</u> G <u>cobordant to</u> (Y', f') <u>where</u> Y' <u>realizes</u> V.

As an application, suppose that V is free. Then G has periodic cohomology. If $|G|$ is odd, Res is onto in 4.9 i); thus the conditions of

4.9 are always achieved whenever $\mathcal{V}$ is G invariant and this is the case iff V_P is left invariant by the normalizer of P. This uses the fact that P is cyclic for $P \in \mathcal{S}$. At any rate it is easy to produce a G invariant free $\mathcal{V}$ for any odd order G having periodic cohomology. Since $\chi(\mathcal{V})$ is additive in $\mathcal{V}$, we can even suppose $\chi(\mathcal{V}) = 0$ as $\tilde{K}_0(Z(G))$ is finite. Then $\sigma(\mathcal{V}) \in L_n(G)$ is defined. Since $n = 2\dim \mathcal{V}-1$ is odd and $|G|$ is odd, $L_n(G) = 0$; so $\sigma(\mathcal{V}) = 0$.

<u>Corollary 4.11.</u> <u>Let</u> G <u>be an odd order group having periodic cohomology</u>. <u>Then</u> G <u>acts freely and smoothly on a homotopy sphere</u>.

<u>Remarks 4.12.</u> Note that 4.11 is achieved by performing free G surgery on the manifold X_0 of 4.5. It seems clear that with a little more effort this method will yield the result of [10] characterizing those groups which act freely and smoothly on a homotopy sphere. It should be noted that this is essentially the method used by the author in 1969 [14] and [15] to construct free actions of odd order periodic groups on homotopy spheres. There the essential geometric tool was the construction of free actions on Brieskorn varieties. These varieties can be used to show $\chi(\mathcal{V}) = 0$ for appropriate $\mathcal{V}$; moreover, they are freely G cobordant to an X_0 of 4.5 for an appropriate $\mathcal{V}$.

References

0. Atiyah, M. F.: Characters and cohomology of finite groups, Pub. I.H.E.S. No. 9 (1961)
1. Atiyah, M. F. and Bott, R.: The Lefschetz fixed point theorem for elliptic complexes II, Ann. of Math. 86 (1967) 451 - 491.
2. Bredon, G.: Fixed point sets of actions on Poincaré Duality Spaces, Topology 12 (1973) 159 - 175.
3. tom Dieck, T.: Homotopy equivalent group representations, J. reine angew. Math. 298 (1978) 182 - 195.
4. tom Dieck, T.: Homotopy equivalent group representations and Picard groups of the Burnside ring and the character ring, Manuscripta math., to appear.
5. tom Dieck, T.: The Burnside ring of a compact Lie group I, Math. Ann. 215 (1975), 235 - 250.
6. tom Dieck, T. and Petrie, T.: The homotopy structure of finite group actions on spheres, Proceedings of Waterloo topology Conference 1978, to appear.
7. Dovermann, H.: Thesis, Rutgers University 1978.
8. Dovermann, H. and Petrie, T.: G surgery III, to appear.
9. Feit, W.: Characters of Finite Groups, Benjamin (1967).
10. Madsen, I, Thomas, C. and Wall, C. T. C.: The spherical space form problem II, Topology (1977).

11. Milnor, J.: Whitehead Torsion, Bull. AMS 72 (1966), 358 - 426.

12. Milnor, J.: Groups which act on S^n without fixed ppoints, Am. J. Math. 79 (1967) 86 - 110.

13. Oliver, R. and Petrie, T.: G-CW surgery and $K_o(Z(G))$, to appear.

14. Petrie, T.: Free metacyclic group actions on homotopy spheres, Ann. of Math. (1972).

15. Petrie, T.: Representation theory surgery and free actions of finite groups on varieties and homotopy spheres. Springer Verlag 168 (1970).

16. Petrie, T.: G surgery I - Proceedings of Santa Barbara topology Conference 1977, to appear.

17. Petrie, T.: G surgery II, to appear.

18. Segal, G.: Equivariant stable homotopy, Proc. I.C.M. (1970) 59-63.

19. Swan, R.: Periodic resolutions for finite groups, Ann. of Math. 72 (1960) 267 - 291.

20. Wall, C. T. C.: Periodic projective resolutions, preprint.

HOMOTOPY TYPE OF G SPHERES

Mel Rothenberg

Let G be a finite group. We consider CW - actions of G on finite CW-complexes. A semilinear action of G, or equivalently a semilinear G sphere, is a piecewise linear action of G on a sphere S such that each fixed set S^H is p.l. homeomorphic to a sphere $S^{n(H)}$ for all subgroups H of G. The action is called orientable if N(H), the normalizer of H in G, acts trivially on $H_*(S^H)$, the homology of the fixed set. We will deal only with orientable semilinear actions in this paper. In a previous paper [R], we introduced the Reidemeister torsion invariant for such actions. We proved that this invariant determines the action when the action is actually linear. Further, we showed how to characterize the G homotopy type of linear G actions in terms of this invariant when G is Abelian. In this paper, we will extend that result to linear actions of more general groups G, more precisely, to all groups such that the Sylow 2-subgroup $G_{(2)}$ is very nice in a sense to be defined (Theorem 1.7). It remains an open question whether or not all groups are very nice. We show that all Abelian groups, as well as dihedral and generalized quaternion groups are very nice (Theorem 1.6 and remarks preceding). We also prove that for those groups all of whose Sylow subgroups are very nice, our characterization of G homotopy type of linear actions extends to the G homotopy types of semilinear actions (Theorem 1.5). Finally, as an application of these results, we prove a desuspension theorem for semilinear actions of groups whose Sylow 2-subgroup is very nice (Theorem 1.8).

The study of the G homotopy type of linear actions has a fairly long history, beginning with the work of Atiyah-Tall [A, T] and continuing up to recent results of tom Dieck [tD]. Recently, Petrie and tom Dieck [P, D] have obtained substantial results on semilinear actions. Their methods are aimed at calculating a homotopy version of the representation ring and are centered on the concept of oriented stable G homotopy equivalence. We, on the other hand, are concerned with actual not stable equivalences, and while we need the notion of <u>orientable</u> we do not require the assumption of <u>oriented</u> – a subtle but nontrivial distinction in the equivariant category. In any case, the relations between the Petrie-tom Dieck invariants and ours are interesting problems for future research.

1. Notation, Background, and Statement of Results

To state our results precisely we need some machinery from [R]. In that paper we defined the notion of a Reidemeister torsion functor τ and to it we associate an abelian group in which it lives $\overline{K}_1(B(G), \tau)$. The notation was intended to emphasize the connection with classical algebraic K-theory. In this paper, we consider only the specific torsion given by oriented torsion over the rationals. Then we can simplify the original cumbersome notation and write $A(G) = \overline{K}_1(B(G))$. We recall the following definitions and results.

Let R be a commutative ring with unit. For H any subgroup of G we define $R(G/H) = R(G) \otimes_{R(H)} R$, where $R(H)$ acts on R via the augmentation $\varepsilon: R(H) \to R$. $R(G/H)$ is a left $R(G)$-module. We define $\overset{o}{R}(G/H) = \ker \hat{\varepsilon}$: $R(G) \otimes_{R(H)} R \to R \otimes_{R(H)} R = R$, a left $R(G)$-module and hence a $\overset{o}{R}(G)$-module, where $\overset{o}{R}(G)$ is a ring with unit when R contains $1/o(G)$. By Burnside(G) we mean the semiring of finite G sets under disjoint union and Cartesian product. For $x \in$ Burnside(G), R(x) is the free abelian R-module generated by x. R(x) is a left R(G)-module and is isomorphic as such to $\sum_{H_i \subset G} R(G/H_i)$, the isomorphism being determined up to the order of the factors and the conjugacy class (H_i) of H_i. We have an augmentation $\varepsilon: R(x) \to R(x/G)$ whose kernel is by definition $\overset{o}{R}(x)$. (Here G acts trivially on x/G.)

A <u>Burnside complex</u> over G is a finite R(G)-complex C: $0 \to C_n \to C_{n-1} \cdots \to C_0 \to 0$, with each $C_i = R(x_i)$, $x_i \in$ Burnside(G). Associated to each Burnside complex C over G and each $H \subset G$, is the Burnside complex over N(H) $C^H: 0 \to C_n^H \to C_{n-1}^H \to \cdots \to C_0^H \to 0$, where $C_i^H = R(x_i^H)$.

A Burnside complex C is called <u>special</u> if N(H) acts trivially on $H_*(C^H; R)$ for all $H \subset G$. For C special, the restriction of the action to a subgroup K makes C a special K-complex. Further, C^H is a special N(H) and a special N(H)/H-complex.

Here is a useful property of special complexes:

<u>1.1 Proposition.</u> Leg G be a solvable group, C a special G-complex. Then $\chi(C) = \chi(C^G)$. (Here χ stands for the usual Euler characteristic, calculated using any field F for which there is a nontrivial homomorphism of R into F.)

<u>Proof.</u> We argue by induction on $o(G)$ and thus can assume G acts effectively on C. Then $0 \to H \to G \to Z/pZ \to 0$ for some normal subgroup H. Then C is a special H-complex, and thus by induction $\chi(C) = \chi(C^H)$. Now C^H is again a special

G/H complex, so once more by induction, $\chi(C^H) = \chi((C^H)^{G/H}) = \chi(C^G)$. It follows that it suffices to prove it for $G = Z/pZ$, p a prime. We now apply the Smith formula $p\chi(C/G) = \chi(C) + (p-1)\chi(C^G)$. Since G acts trivially on $H_*(C)$, $\chi(C/G) = \chi(C)$. Thus $\chi(C) = \chi(C^G)$. Q.E.D.[1]

<u>1.2 Conjecture</u>. Proposition 1.1 is true for any group G.

Now assume R(G) is semisimple (i.e., R is a field containing $1/o(G)$). Then R(G) splits, $R(G) = \overset{o}{R}(G) + \overset{1}{R}(G)$, with $\overset{1}{R}(G) \cong R$. Thus every R(G)-module splits, and in particular, the Burnside complex C splits, $C = \overset{o}{C} + \overset{1}{C}$. Similarly, $C^H = \overset{o}{C}{}^H + \overset{1}{C}{}^H$ as a $Q(N(H))$ complex. The homology of C^H splits then as $H(C^H) = \overset{o}{H}(C^H) + \overset{1}{H}(C^H)$.

The condition that C is special is equivalent to the condition that $\overset{o}{H}(C^H) = 0$ for all $H \subset G$, which is equivalent to the condition that $\overset{o}{Q}(N(H)/H)C^H = \overset{o}{Q}(C^H)$ is acyclic.

A <u>strict</u> exact sequence of Burnside complexes $0 \to C^1 \xrightarrow{\alpha^1} C^2 \xrightarrow{\alpha^2} C^3 \to 0$ is an exact sequence such that $C^i_j = R(x^i_j)$, $x^2_j = x^1_j \cup x^3_j$, $x^1_j \cap x^2_j = \emptyset$, and the sequence

$$0 \to R(x^1_j) \xrightarrow{\alpha^1_j} R(x^2_j) \xrightarrow{\alpha^2_j} R(x^3_j) \to 0,$$

are the obvious split maps.

We now take our coefficient ring R to be Q. An element of the abelian group A(G) is represented by a special Burnside complex. These satisfy the following relations in A(G) [see R, sect. 1].

(I) If $0 \to C^1 \to C^2 \to C^3 \to 0$ is a strict exact sequence of special complexes, then $C^2 = C^1 + C^3$ in A(G).

(II) If C is isomorphic to C' by an isomorphism which comes from a family of G isomorphisms $\psi_i : x_i \cong x_i'$, where $C_i = Q(x_i)$, $C_i' = Q(x_i')$, then $C = C'$ in A(G).

(III) If C is a special complex of the form $0 \to 0 \to 0 \to Q(x) \xrightarrow{f} Q(x) \to 0 \to 0$, and where $\overset{o}{Q}(f): \overset{o}{Q}(x) = \overset{o}{Q}(G) \otimes_{Q(G)} Q(x) \to \overset{o}{Q}(G) \otimes_{Q(G)} Q(x) = \overset{o}{Q}(x)$ is $\pm$ id, then $C = 0$ in A(G). If the group G is abelian, then A(G) is known [R, Thms. 1.23, 1.24].

We define $Wh(G; Z) = \sum_{(H) \in conj(G)} Wh(N(H)/H)$. A generator of Wh(G; Z) is represented by a free, acyclic, based Z(N(H)/H) complex C, i.e., a Z Burnside complex over N(H)/H which is acyclic. Letting $\overline{C} = Q(G) \otimes_{Z(H)} C$, yields a special Burnside complex and thus defines a homomorphism $\gamma: Wh(G; Z) \to A(G)$. For t_1, t_2 in $A(G)$, we write $t_1 \equiv t_2$ if there exists an $a \in Wh(G; Z)$ with $t_1 - t_2 = \gamma(a)$, i.e., t_1, t_2 represent the same element in $A(G)/\gamma(Wh(G; Z))$.

Let X be a finite G CW-complex. Then the n-cells of X are an element of Burnside(G). The rational cellular chain complex $C(X; Q)$ is a Burnside complex and even a special complex if $N(H)$ acts trivially on $H_*(X; Q)$ for all $H \subset G$. In this case, $C(X; Q)$ determines an element in $A(G)$ which we denote by $\tau(X)$ (or $\tau_G(X)$ when we wish to explicitly note G). If X and X' are two special G CW-complexes and $f: X \to X'$ a G-homotopy equivalence then f defines, in an obvious way, an element $Wh(f)$ in $Wh(G; Z)$ and $\tau(X) - \tau(X') = \gamma(Wh(f))$. Hence, a necessary condition that two special G-complexes X and X' are G-homotopy equivalent is that $\tau(X) \equiv \tau(X')$. We wish to investigate circumstances under which this condition is sufficient.

If X is a special G CW-complex, $\tau_G(X)$ is defined. If H is a subgroup, we can restrict the action and get $\tau_H(X)$ and, considering the action of $N(H)$ on X^H, we get $\tau_{N(H)}(X^H)$ and $\tau_{N(H)/H}(X^H)$ as well. The power of the invariant τ is that $\tau(X)$ determines all these other invariants. In particular, if $\tau_G(X) \equiv \tau_G(X')$, then $\tau_H(X) \equiv \tau_H(X')$ and $\tau_{N(H)}(X^H) \equiv \tau_{N(H)}(X'^H)$.

We call a family F of special G CW-complexes <u>simple</u> if for $X, X' \in F$, $\tau(X) \equiv \tau(X')$ implies $X \underset{G}{\sim} X'$ ($\underset{G}{\sim}$ means G-homotopy equivalent). In [R], we showed that the family of G-linear n-spheres is simple for G abelian.

In a simple family, the torsion must determine $\dim X^H$ for every $H \subset G$. That this is not an unreasonable requirement is shown by the following.

<u>1.3 Theorem:</u> Let G be a solvable group, X and X' semilinear spheres. If $\dim X^G = \dim X'^G$ and $\tau(X) \equiv \tau(X')$, then $\dim X^H = \dim X'^H$ for all $H \subset G$.

<u>Proof.</u> We first prove that the hypothesis implies $\dim X = \dim X'$. By the proof of 1.2, we can reduce to the case where $G = Z/pZ$. If $\dim X^G = \dim X'^G \neq 0$, let $x \in X^G$ and $x' \in X'^G$. Then $X \underset{G}{\sim} S(Lk(x))$ and $X' \underset{G}{\sim} S(Lk(x'))$, where S denotes suspension, Lk denotes link. By [4], $Lk(x)$ and $Lk(x')$ also satisfy the hypothesis; hence we can reduce to the case where $G = Z/pZ$ acts freely on the spheres X, X'. But then it is known that $X \underset{G}{\sim} \Sigma$ and $X' \underset{G}{\sim} \Sigma'$, where Σ and Σ' are linear spheres. For linear Z/pZ spheres, we know the theorem is true, [R, 4.5]. Hence we have shown $\dim X = \dim X'$. If for some $H \subset G$, $\dim X^H = \dim X'^H + n$, then n must be even since $\chi(X^H)$ is determined by the torsion. Then $S^n(X')$ and X are H semi-

linear spheres which satisfy the hypothesis of the theorem [R]. Hence, as we have already proven, $\dim S^n(X') = \dim X$. Thus $n = 0$.

Corollary. If G is solvable, $\dim X = \dim X'$ and $\tau(X) \equiv \tau(X')$, then $\dim X^H = \dim X'^H$ for all $H \subset G$. (For by the argument above, $\dim S^n(X') = \dim X$, where $n = \dim X^H - \dim X'^H$, for any H G.)

1.4 Conjecture: For any finite G the family of n-dimensional semilinear G-spheres is simple.

We now state some theorems which can be adduced as evidence for the above conjecture. We need one more definition. For any finite group G, let $N = \sum_{g\in G} g$, an element of $Z(G)$. Let $\overline{Z}(G) = Z(G)/(N)$. We have the injection $\overline{Z}(G) \hookrightarrow Q(G)/(N) = \overset{o}{Q}(G)$. We then have the two homomorphisms

$$\iota: K_1(\overline{Z}(G)) \to K_1(\overset{o}{Q}(G)) \quad \text{and} \quad \bar{\varepsilon}: K_1(\overline{Z}(G)) \to K_1(Z/o(G)Z)) = U^*(Z/o(G)).$$

We say G is nice if $\ker(\iota) \subset \ker(\bar{\varepsilon})$. We say G is very nice if every quotient group of a subgroup is nice. Observe that an abelian group is nice and hence very nice since if $a \in \ker(\iota)$, then $\det(a)$ is 1 in $U^*(\overline{Z}(G))$ and thus $\det(\bar{\varepsilon}(a)) = 1$. Since $\det: K_1(Z/o(G)) \to U^*(Z/o(G))$ is an isomorphism, G is very nice.

1.4' Conjecture: All finite groups are nice, hence very nice.

The connection between the two conjectures is the following theorem.

1.5 Theorem. If G is a finite group all of whose Sylow subgroups are very nice and if X and X' are two semilinear G-spheres such that $\dim X^H = \dim X'^H$ for all $H \subset G$, and $\tau(X) \equiv \tau(X')$, then $X \underset{G}{\sim} X'$.

Corollary: If G is a solvable group all of whose Sylow subgroups are very nice, then the family of n-dimensional semilinear G-spheres is simple.

Evidence beyond abelian groups for Conjecture 1.4' is given by

1.6 Theorem: If G is an extension of an abelian p-group by Z/pZ, p a prime, then G is nice, thus very nice.

This has some interest since it shows that all the groups with periodic cohomology satisfy the hypothesis of 1.5. Further, it allows us to prove the following about linear G-spheres.

1.7 Theorem: If G is a group whose Sylow 2-subgroups are very nice, then the family of oriented n-dimensional linear spheres is simple.

Finally, as an application of our results we prove the following destabilization theorem.

1.8 Theorem: Let G be a finite group whose Sylow 2-subgroups are very nice. Let X, Y, and Z be semilinear G-spheres and suppose $X \# Z \underset{G}{\simeq} X \# Z$. Then $X \underset{G}{\simeq} Y$ (# denotes join).

Remark. Much of our results goes through for the weaker notion of semilinear sphere studied by Petrie and tom Dieck [P, tD]. To them a semilinear sphere is a finite G CW-complex X with X^H a subcomplex of dimension n_H homotopically equivalent to S^{n_H}. In particular, Theorem 1.5 is true for such G-complexes. However, our proof of 1.3 and 1.8 does not go through to this broader category.

2. Proof of Theorem 1.5

Step I: Algebraic Preliminaries.

Let A be a ring, $C: 0 \to 0 \to C_n \to C_{n-1} \to \cdots \to C_1 \to 0$ a finite R-complex i.e., each C_k is finitely generated over A. We let $C^e = \sum_k C_{2k}$ and $C^o = \sum_k C_{2k+1}$. If C is acyclic and each C_k is A-projective, then such a complex determines an isomorphism $\psi: C^e \cong C^o$. If further, A is semisimple and each C_k is represented as the dirct sum of simple R-modules, then the number of factors of a given simple type in C^e and C^o must be the same and we have an isomorphism $\lambda: C^o \cong C^e$ defined uniquely up to a permutation of the simple factors of C^o. Then $\lambda\psi: C^e \cong C^e$ defines a unique element k(C) in $K_1(R)$.

Now let C be a special Burnside complex. Then $\overset{o}{Q}(G)C = \overset{o}{C}$ satisfies the above condition for $A = \overset{o}{Q}(G)$. Since $\overset{o}{C}_k = \sum_{H_i} \overset{o}{Q}(G/H_i)$ one must fix a factorization of the $\overset{o}{Q}(G/H_{i_o})$ into simple $\overset{o}{Q}(G)$-modules. The construction above then induces a $\mu: A(G) \to K_1(\overset{o}{Q}(G))/(\pm g)$ given by $C \to k(C)$. In general, of course, μ depends on the factorization of the $\overset{o}{Q}(G/H)$ chosen.

Now let $\chi_G: A(G) \to K_0(\text{Burnside } G)$ be the G Euler characteristic [R, 1.27] and $\overset{o}{A}(G) = \ker \chi_G$. Then $\gamma(Wh(G; Z)) \subset \overset{o}{A}(G)$. Then $\mu | \overset{o}{A}(G)$ is independent of the factorizations of $\overset{o}{Q}(G/H)$ chosen. Further, if $C \subset \overset{o}{A}(G)$, then the restricted action of $H \subset G$ on C yields an element of $\overset{o}{A}(H)$, and $C^H \in \overset{o}{A}(N(H))$.

We can identify $K_1(\overset{o}{Q}(H))$ with $\ker \varepsilon$, $\varepsilon: K_1(Q(H)) \to K_1(Q) = Q^*$. Since ε factors into $Q(H) \to Q(G) \to Q$, for $H \subset G$ we have an induced homomorphism $\lambda_{H,G}: K_1(\overset{o}{Q}(H)) \to K_1(\overset{o}{Q}(G))$ and this defines $\mu^H: \overset{o}{A}(G) \to K_1(\overset{o}{Q}(G))/(\pm g)$ by $\mu^H(C) = \lambda_{N(H),G}(\mu(C^H))$. Note that $\mu = \mu^e$. Let $a \in Wh(N(H)/H) \subset Wh(G; Z)$. We can represent a by a two-term $Z(G)$ acyclic complex $D: D_1 \to D_0$, where $D_1 = D_0 = kZ(G/H) = Z(G/H) + \dots + Z(G/H)$, and where $D^H \in K_1(Z(N(H)/H))$ is in the kernel $K_1(Z(N(H)/H)) \to K_1(Z) = (\pm 1)$. A simple calculation shows $\mu^K(\gamma(a)) = \mu(\gamma(a)$ if $(K) \subset (H)$ and $= 0$ if $(K) \not\subset (H)$. Since for $C \in \overset{o}{A}(G)$ and $C^H = 0$, $\mu^H(C) = 0$, by an easy induction over the subgroups of G, applying μ^K above we have: Let $C \in \overset{o}{A}(G)$ with $C^H = 0$ for $H \neq (e)$, and suppose $C = \sum_{H \in \text{cong}(G)} \gamma(a_H)$ in $A(G)$. Then $\mu(C) = \mu\gamma(a_e)$.

Step II: Geometric Preliminaries.

For any G CW-complex X let $X^s = \{x \in X \mid G_x \neq (e)\}$. The following is true from elementary equivariant obstruction theory, [C, p. 91-97].

2.1 Proposition: Let X, X' be two semilinear G-spheres with $\dim(X^H) = \dim(X'^H)$ for all $H \subset G$.

(a) If $Y \subset X$ is any G subcomplex and $f: Y \to X'$ a G-map, there exists an equivariant extension to X.

(b) If $f_i: X \to X'$ are G-maps with $f_1|X^s \underset{G}{\sim} f_2|X^s$, then degree f_1 = degree f_2 mod $o(G)$.

(c) If G acts effectively on X and if $f: X \to X'$ is a G-map and $Y \subset X$ a proper G subcomplex and if n = degree f mod $o(G)$, then there exists a G-map $f': X \to X'$ with $f|Y = f'|Y$ and degree $f' = n$.

Step III.

2.2 Proposition: Let G be a group all of whose Sylow subgroups are very nice; let X, X' be semilinear spheres with $\dim X^H = \dim X'^H$ for all $H \subset G$ and such that $\tau(X) \equiv \tau(X')$; let $\bar{f}: X^s \to X'^s$ be a G-homotopy equivalence. Then $\bar{f}$ extends to a G-homotopy equivalence $f^{\#}: X \to X'$.

Proof. If G fails to act effectively on X and thus X', there is nothing to prove since then $X = X^s$ and $X' = X'^s$. Assuming otherwise, let $\bar{g}$ be a G-homotopy inverse to $\overline{f}$. By 2.1 there exists extension f of $\overline{f}$ and g of $\overline{g}$ and degree $gf = 1$ mod $o(G)$. Thus degree f is a unit mod $o(G)$. By 2.1(c), it suffices to show that degree $f = \pm 1$ mod $o(G)$.

Construct the mapping cone $T(f)$ of f and observe that $\overline{H}_n(T(f)) \cong Z/$ degree$(f)Z$, where degree(f) is defined only up to sign and $\overline{H}_i(T(f)) = 0$ for $i \neq n$. Since $T(f)^s = T(f^s) = T(\overline{f})$ and $\overline{f}$ is a G-homotopy equivalence, $T(f)^s$ is a contractible G complex. We have $\tau(T(f),*) = \tau(X') - \tau(X) = \gamma(a)$, $a \in Wh(G; Z)$. Also $\tau(T(f),*) = \tau(T(f), T(f)^s) \mp \tau(T(f)^s,*)$. Since $T(f)^s$ is contractible, $(T(f)^s,*) = \gamma(b)$, and thus $\tau(T(f), T(f)^s) = \gamma(d)$, for $d \in Wh(G; Z)$, and $H_*(T(f), T(f)^s) = H_n(T(f), T(f)^s) \cong Z/\deg(f)Z$. Now let $C = C(T(f), T(f)^s)$ be the complex of cellular chains, a free $Z(G)$ complex. Then $C \otimes Q$ represents an element of $A(G)$, and by the above $C \otimes Q = \gamma(d)$ in $A(G)$. By the conclusion of Step I, we can find a free, acyclic $Z(G)$ complex D with $\mu((C \oplus D) \otimes Q) = 0$. Since replacing C by $C \oplus D$ does not change the homology, we have produced a free $Z(G)$ complex A with $H_*(A) = H_n(A) \cong Z/\text{degree}(f)Z$. Further, the construction permits us to assume $A_i = 0$ for $i > n+1$. Then the usual splitting argument allows us to assume $A_i = 0$ for $i \neq n+1, n$. The cell structures of X, X' determines a basis for A up to multiplication by $\pm g$, $g \in G$. Choosing a basis for A gives a lifting of $\mu(A \otimes Q)$ from $K_1(\overset{o}{Q}(G))/(\mp g)$ to $\psi(A) \in K_1(\overset{o}{Q}(G))$. Since $\mu(A \otimes Q) = 0$ we can fix a basis for A such that $\psi(A) = 0$. Then $\det(d) \in Z$, where $d: A_{n+1} \to A_n$ is well-defined and $H_n(A) = Z/\det(d)Z$. Thus, specifying such a basis of A determines the degree of $f = \det(d)$, removing the sign ambiguity.

The complex A is such that $A \otimes Q$ and $A = \overline{Z}(G) \otimes_{Z(G)} A$ are acyclic. Hence $\overset{o}{A}$ defines elements in $K_1(Q(G))$ and $K_1(\overline{Z}(G))$ and since $\mu(A \otimes Q) = 0$ in in $K_1(\overset{o}{Q}(G))$ and ker $K_1(Q(G)) \to K_1(\overset{o}{Q}(G))$ is represented by units of the form $1+qN$, $q \in Q$, we must have $A \otimes Q = 1+qN$ in $K_1(Q(G))$. On the other hand, $\varepsilon: Z(G) \to Z$ induces $\hat{\varepsilon}: \overline{Z}(G) \to Z/o(G)Z$. It is not difficult to show that $\hat{\varepsilon}(A) \in K_1(Z/o(G)Z) = U^*(Z/o(G)Z)$ is exactly $\det(d) = \text{degree}(f)$.

Consider a p Sylow subgroup $G_{(p)}$ of G. The complex A is also a free, based $Z(G_{(p)})$ complex. Since under restriction $f: K_1(Q(G)) \to K_1(Q(G_{(p)}))$, the unit $1+qN$ goes to the unit $1+q(o(G/G_{(p)}))N_{(p)}$, it follows that the complex $A \otimes Q$ also represents 0 in $K_1(\overset{o}{Q}(G_{(p)}))$. A represents an element of $K_1(\overline{Z}(G_{(p)}))$ and once again $\hat{\varepsilon}(A) \in U^*(Z/o(G_{(p)})Z) = \det(d) = \text{degree}(f)$ mod $o(G_{(p)})$. Since $G_{(p)}$ is nice we must have degree$(f) = 1$ mod $o(G_{(p)})$. Since this holds for all Sylow subgroups of G we must have degree$(f) = 1$ mod $o(G)$ and 2.2 is proved.

Theorem 1.5 follows immediately from 2.2 through a step by step construction of f over GX^H once we have constructed f over $GX^{H'}$ for $H' > H$.

3. Proof of Theorem 1.6

Since we already know the theorem is true for abelian groups, we assume G nonabelian. By hypothesis we have an exact sequence $(e) \to H \to G \to Z/pZ \to (e)$ with H abelian and $o(G) = p^n$. We now consider coefficients in the integers localized at p, $Z_{(p)}$. We have $\overline{Z}(G) \to \overset{o}{Q}(G)$ factors through $\overline{Z}(G) \to Z_{(p)}(G)/(N) \to \overset{o}{Q}(G)$, and $\overline{Z}(G) \to Z/o(G)Z$ factors through $\overline{Z}(G) \to Z_{(p)}(G)/(N) \to Z/o(G)Z$. Further, since $Z_{(p)}(G)$ is local, $K_1(Z_{(p)}(G)) \to K_1(Z_{(p)}(G)/(N))$ is onto. Thus it suffices to prove the following. If $x \in K_1(Z_{(p)}(G))$ and $1(x) = 0$ in $K_1(\overset{o}{Q}(G))$ then $\varepsilon(x) = 1 \bmod (p^n)$, where $\varepsilon: K_1(Z_{(p)}(G)) \to K_1(Z_{(p)}) = U^*(Z_{(p)})$ where $U^*(R)$ denotes multiplicative group of units. Finally, since $Z_{(p)}(G)$ is local we have $U^*(Z_{(p)}(G)) \to K_1(Z_{(p)}(G))$ is onto and since $1(x) = 0$ if and only if $x = 1+kN$ in $K_1(Q(G))$ we have reduced the problem to the following:

3.1 Proposition. Let $x \in U^*(Z_{(p)}(G))$, $c \in Z_{(p)}$ with $c \neq 0 \bmod (p)$, then $x \neq 1 + (c/p)N$ in $K_1(Q(G))$.

Proof. Suppose $x \equiv 1+(c/p)N$ (where $\equiv$ means $=$ in $K_1(Q(G))$. We have the diagram of homomorphisms defining Det:

$$\text{Det:}\quad \begin{array}{ccccccc} U^*(Z_{(p)}(G)) & \longrightarrow & K_1(Z_{(p)}(G)) & \xrightarrow{r} & K_1(Z_{(p)}(H)) & \overset{\det}{\cong} & U^*(Z_{(p)}(H)) \\ \cap & & \downarrow & & \cap & & \cap \\ U^*(Q(G)) & \longrightarrow & K_1(Q(G)) & \xrightarrow{r} & K_1(Q(H)) & \overset{\det}{\cong} & U^*(Q(H)) \end{array}$$

where r is the restriction map and det is the usual determinant defined and an isomorphism because H is abelian. Write t for a lift of a generator of Z/pZ back to G. Then every element of $Z_{(p)}(G)$ can be written uniquely in the form $\sum_{i=0}^{p-1} a_i t^i$, $a_i \in Z_{(p)}(H)$. Write $N_0 = \sum_{h \in H} h$. Then $N = \sum_{i=0}^{p-1} N_0 t^i = N_0 \cdot N_1$, where $N_1 = \sum_{i=0}^{p-1} t^i$. Since $x \equiv 1+(c/p)N$, $\text{Det}(x) = \text{Det}(1+c/p)N) = 1+cN_0$ by an easy calculation. According to [W, 3.5], Det extends (uniquely) to an additive homomorphism $\widehat{\text{Det}}: Z_{(p)}(G) \to Z_{(p)}(H)/T$, where T is the trace subgroup defined as follows: Since G is an extension of H by Z/pZ, Z/pZ acts on H and thus on

$Z_{(p)}(H)$. This action is given by the conjugate action of t^i, $a \to t^i a t^{-i}$. The trace subgroup $T = \{\sum_{i=0}^{p-1} t^i b t^{-i} \mid b \in Z_{(p)}(H)\}$.

We then compute $\widehat{\mathrm{Det}}$ on additive generators at^i of $Z_{(p)}(G)$, $a \in H$ and this will determine $\widehat{\mathrm{Det}}$ fully. Since Z/pZ acts on H, this determines H as a $Z(Z/pZ)$-module, and we introduce $N_1 = \sum_{i=0}^{p-1} t^i$ as an element of $Z(Z/pZ)$. For $a \in H$, $\mathrm{Det}(at^j) = \mathrm{Det}(a)\mathrm{Det}(t^j)$. $\mathrm{Det}(a) = N_1(a)$ and $\mathrm{Det}(t^j) = \mathrm{Det}(t)^j = a_0^j$, where $a_0 = t^p$ is in (center of G) $\cap$ H. Thus $\mathrm{Det}(at^j) = N_1(a)a_0^j \in H \subset U^*(Z_{(p)}(H))$. These formulae will allow us to factor $\widehat{\mathrm{Det}}$.

Let $\overline{H} = N_1(H) \subset H$, and let $H' = \{x \in H \mid N_1(x) = 0\}$. Then we have the exact sequence

$$0 \to H' \to H \xrightarrow{N_1} \overline{H} \to 0 .$$

Further, H' is a normal subgroup of G. Let $\overline{G} = G/H'$. We have the following commutative diagram of exact sequences

$$\begin{array}{ccccccccc}
 & & 0 & & 0 & & & & \\
 & & \downarrow & & \downarrow & & & & \\
 & & H' & \xrightarrow{\mathrm{Id}} & H' & & & & \\
 & & \downarrow & & \downarrow & & & & \\
0 & \longrightarrow & H & \longrightarrow & G & \longrightarrow & Z/pZ & \longrightarrow & 0 \\
 & & \downarrow & & \downarrow{\scriptstyle \pi} & & \| {\scriptstyle \mathrm{id}} & & \\
0 & \longrightarrow & \overline{H} & \longrightarrow & \overline{G} & \longrightarrow & Z/pZ & \longrightarrow & 0 \\
 & & \downarrow & & \downarrow & & & & \\
 & & 0 & & 0 & & & &
\end{array}$$

Further, since $\overline{H} \subset H^{Z/pZ}$, $\overline{G}$ is abelian.

Each element of $Z_{(p)}(\overline{G})$ can be written uniquely as $\sum_{i=0}^{p-1} b_i \bar{t}^i$, with $b_i \in Z_{(p)}(\overline{H})$ and $\bar{t}^i = \pi(t^i)$, and the map $\pi_*: Z_{(p)}(G) \to Z_{(p)}(\overline{G})$ is given on additive generators by $\pi_*(at^j) = N_1(a)\bar{t}^j$, $a \in H$. Thus we can factor $\widehat{\mathrm{Det}}$ as follows:

$$\begin{array}{ccc}
Z_{(p)}(G) & \xrightarrow{\widehat{\mathrm{Det}}} & Z_{(p)}(H)/T \\
 & {\scriptstyle \overline{\pi}} \searrow \quad \nearrow {\scriptstyle \gamma} & \\
 & Z_p(\overline{G}) &
\end{array}$$

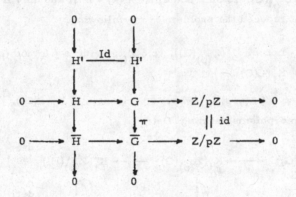

where $\gamma(\sum_{i=0}^{p-1} b_i t^i) = \sum_{i=0}^{p-1} b_i a_0^i$. We also have the following commutative diagram of morphisms some of which are isomorphisms, and some of which are injections (remember $\overline{G}$ is abelian):

$$\begin{array}{ccccccc}
 & & U^*Z_{(p)}(G) & \longrightarrow & K_1(Z_{(p)}(G)) & & \\
 & \swarrow & \big\downarrow \overline{\pi} & & \big\downarrow & & \\
U^*(Q(G)) & \cong & U^*Z_{(p)}(\overline{G}) & \cong & K_1(Z_{(p)}(\overline{G})) & \cong & U^*Z_{(p)}(\overline{G}) \\
 & \searrow^{\overline{\overline{\pi}}} & \cap & & \cap & & \cap \\
 & & U^*Q(\overline{G}) & \cong & K_1(Q(\overline{G})) & \cong & U^*Q(G)
\end{array}$$

Thus if $x \equiv 1 + (c/p)N$, we can compute $\overline{\pi}(x)$ by computing $\overline{\overline{\pi}}(1 + (c/p)N)$. Let $N_{00} = \sum_{\overline{h} \in \overline{H}} \overline{h}$. Then $\overline{\pi}(x) = \overline{\overline{\pi}}(1 + (c/p)(\sum_i N_1 t^i) = 1 + (c\, o(H')/p)(\sum_i N_{00} t^i)$. It follows that $H' \neq (e)$. Further $\widehat{\mathrm{Det}}(x) = 1 + (c\, o(H')/p)N_{00} + (c\, o(H')/p)(\sum_{i>0} N_{00} a_0^i)$ in $Z_{(p)}(H)/T$.

We are now ready to deduce the contradiction.

Case 1. $o(H') = p^k$, $k > 1$. Since N_{00} and a_0 are invariant under the action of Z/pZ on $Z_{(p)}(H)$, pa_0 and $pN_{00}a_0^i$ are in T. Thus the above formula says $\mathrm{Det}(x) = 1$ in $Z_{(p)}(H)/T$. However, our earlier formula $\mathrm{Det}(x) = 1 + cN_0$ then implies that $cN_0 \in T$. But then $c\overline{N} \in T$, where $\overline{N} = \sum_{b \in H^{Z/pZ}} b$. But the trace subgroup of $Z_{(p)}(H^{Z/pZ})$ is $pZ_{(p)}(H^{Z/pZ})$, and $c \not\equiv 0 \mod (p)$. This yields a contradiction.

Case 2. $o(H') = p$. Consider the operator $T = (t-1) \in Z(Z/pZ)$ on H. Since $0 \to H' \to H \xrightarrow{N_1} H \to 0$ and $N_1 T = 0$, $T: H \to H'$. Since Z/pZ acts nontrivially on H (G is not abelian), $T(H) \neq (e)$. Since $H' = Z/pZ$, $T(H) = H'$. Thus H splits as the direct sum of $Z(Z/pZ)$-modules, $H = H' + H$. However Z/pZ acts trivially on H and since $H' = Z/pZ$, Z/pZ must act trivially on H', hence on H - a contradiction. This proves Proposition 3.1 and thus Theorem 1.6.

4. The Stable Homotopy of Representations

Theorem 1.7 will follow from 1.8 once we have some stable homotopy theory for the representations of p-groups. Here the methods of tom Dieck yield the results we need, and since the arguments are just step by step repitition of those of [tD] we indicate them very briefly. We consider the real representation ring, RO(G). An element of RO(G) is given by a difference a - b, where a, b are real representations of G. There is no harm in assuming a, b orthogonal, and of course a-b and $(a\oplus c) - (b\oplus c)$ represent the same element. For $H \subset G$ we have the homomorphism, $\dim(H): RO(G) \to Z$, given by $\dim(H)(a-b) = \dim(a^H) - \dim(b^H)$. We define $RO_0(G) = \bigcap_{H \subset G} \ker \dim(H)$. tom Dieck defines $RO_1(G)$, $RO_1(G) \subset RO_0(G)$, by purely algebraic conditions. He then considers $RO_h(G) = \{(a-b) \in RO_0(G) \mid S(a\oplus c) \underset{G}{\simeq} S(b\oplus c)\}$, where $S(x)$ is the unit sphere of x. He proves that $RO_1(G) \subset RO_h(G)$ and considers the surjection $i(G) = RO_0(G)/RO_1(G) \sim RO_0(G)/RO_h(G) = J(G)$. He then proves:

4.1 Theorem (tom Dieck): If G is a p-group, then $i: i(G) \cong J(G)$.

Emulating tom Dieck, let us define $RO_s(G) = \{(a-b) \in RO_0(G) \mid$ there exists c such that $\tau S(a\oplus c) \equiv \tau S(b\oplus c))\}$. Note, for $a - b \in RO_s(G)$, we can always find c with $S(a\oplus c)$, $S(b\oplus c)$ G-orientable. It is not difficult to check that $RO_s(G)$ is a subgroup of $RO_0(G)$, with $RO_h(G) \subset RO_s(G)$ and there exists a natural surjection $\mu: J(G) = RO_0(G)/RO_h(G) \to RO_0(G)/RO_s(G) = K(G)$.

4.2 Theorem: Let G be a p-group, then $\mu: J(G) \cong K(G)$.

Proof. In view of 4.1, it suffices to show $\mu i: i(G) \cong K(G)$. By the induction theorem of tom Dieck [tD, 3.2], it suffices to consider representations induced up from cyclic, dihedral or generalized quaternion groups. We have shown these groups are very nice, and thus the theorem is true for these groups. The proof now follows exactly as in tom Dieck's proof of 4.1.

4.3 Corollary: If G is a p-group and a,b, and c oriented linear n-dimensional G-spheres such that $\tau(a \# c) \equiv \tau(b \# c)$, then there exists an oriented linear G-sphere c' such that $a \# c' \underset{G}{\simeq} b \# c'$. (Note the dimension condition is taken care of by the corollary to 1.3.)

5. Stable versus Unstable G-Homotopy Equivalences of Semilinear Spheres

If a and b are complex linear G-spheres with $\dim a^H \leq \dim b^H$ for every $H \subset G$, then the G-homotopy class of any G-map $f: a \to b$ is determined by $\mathrm{degree}(f^H): a^H \to b^H$, where the degree is unambiguously determined as an integer since a^H and b^H have natural orientations. For complex representations, this allows us to reduce questions of G-homotopy type to questions of stable G-homotopy type. For real representations, not to speak of semilinear spheres, there is no natural orientation, and the general problem of reducing the study of G-homotopy type to stable G-homotopy type is mathematically formidable. In this section we will prove some partial results which will allow us to deduce 1.8 and 1.7.

For a, b semilinear spheres, we say $a \overset{s}{\underset{G}{\simeq}} b$ if there exists a semilinear c such that $a \# c \underset{G}{\simeq} b \# c$. We study the question: When does $a \overset{s}{\underset{G}{\simeq}} b$ imply $a \underset{G}{\simeq} b$?

For X a G-complex and $H \subset G$, let $X_+^H = \bigcup_{H < H'} X^{H'}$. The crucial algebraic invariant of a semilinear sphere a for homotopy purposes is $a(H) = H_{n(H)}(a^H, a_+^H)$, where $n(H) = \dim a^H$. We observe

1. a has an orbit of type H if and only if $a(H) \neq 0$;
2. if a has an orbit of type H, then $a(H) \cong kZ$, $k > 0$, and $k = 1$ if and only if there does not exist H' such that $H < H'$ and $\dim(a^H) - \dim(a^{H'}) = 1$.

<u>For the remainder of this section we assume</u> $\dim a^H \leq \dim b^H$, <u>for all $H \subset G$.</u>

Given a G-map $f: a \to b$, $f^H: (a^H, a_+^H) \to (b^H, b_+^H)$ induces homomorphisms and a commutative diagram

$$\begin{array}{ccc} \bar{H}_{n(H)}(a^H) & \longrightarrow & a(H) \\ \Big\downarrow \deg_H(f) & & \Big\downarrow \widehat{\deg_H}(f) \\ \bar{H}_{n(H)}(b^H) & \longrightarrow & b(H) \end{array}$$

where $\deg_H(f)$ and $\widehat{\deg_H}(f)$ are taken to be 0 when $\dim a^H < \dim b^H$. Since a, b are G-orientable, there is a well-defined identification $H_{n(H)}(a^H) \cong H_{n(H')}(b^{H'})$ and $a(H) \cong a(H')$ when H and H' are conjugate and thus $\deg_H(f)$ and $\widehat{\deg_H}(f)$ depend only on the conjugacy class (H) of H. By elementary obstruction theory $\widehat{\deg_H}(f)$ contains all the homotopy information about f, precisely:

<u>5.1 Proposition:</u> If $f, f': a \to b$ are G-maps, then $f \underset{G}{\simeq} f'$ if and only if $\widehat{\deg_H}(f) = \widehat{\deg_H}(f')$ for all $H \subset G$. Further, f is a G-homotopy equivalence if and only if $\deg_H(f)$ is invertible for all $H \subset G$.

Now suppose we are given a G-map $f: a \# c \to b \# c$, where we assume $c^H \neq 0$ and $\dim c^H = k_H$. We then have maps

$$\overline{H}_{n_H}(a^H) \overset{\lambda_1}{\cong} \overline{H}_{n_H+k_H}(a^H \# c^H) = \overline{H}_{n_H+k_H}((a\# c)^H) \xrightarrow{f_*} \overline{H}_{n_H+k_H}((b\# c)^H)$$

$$= \overline{H}_{n_H+k_H}(b^H \# c^H) \overset{\lambda_2}{\cong} \overline{H}_{n_H}(b^H).$$

The isomorphisms λ_1, λ_2 depend on choosing a generator of $\overline{H}_{k_H}(C)$, but the composite does not if we choose the same generator at both ends, hence the composite is a well-defined map $\deg^s_H(f): \overline{H}_{n_H}(a^H) \to \overline{H}_{n_H}(b^H)$ which again is 0 if $\dim a^H < \dim b^H$. Further, $\deg^s_H(f)$ is a stable invariant. That is, $\deg^s_H(f) = \deg^s_H(f \# \mathrm{Id})$, $f \# \mathrm{Id}: a \# c \# d \to b \# c \# d$.

Suppose $f: a \# c \to b \# c$ is a G-map. Now suppose $C^G \neq 0$. We want to know when there exist an $f': a \to b$ with $\deg_H(f') = \deg^s_H(f)$ for all $H \subset G$. Such an f' will be a G-homotopy equivalence if f is. One difficulty that may arise is the following: Suppose $H < H'$, $a^H = a^{H'}$, and $\deg^s_H(f) \neq \deg^s_{H'}(f)$. Then no such f' can exist. In fact, this is the only obstruction.

5.2 Theorem: Let $C^G \neq 0$ and $f: a \# c \to b \# c$ be a G-map such that for $H < H'$ and $a^H = a^{H'}$, $\deg^s_H(f) = \deg^s_{H'}(f)$. Then there exists $f': a \to b$ with $\deg_H(f') = \deg^s_H(f)$ for all $H \subset G$.

Proof. By replacing c by $c \# d$ if necessary, where d is an appropriate complex linear G-sphere, we can assume $c(H) = Z$ for all $H \subset G$. Then $(a \# c)(H) = 0$ or Z for all H and similarly for $b \# c$. Then $\widehat{\deg}_H(f)$ is determined by $\deg_H(f)$. To prove 5.2 it will suffice to prove 5.3 below, for then we can construct f' step by step over all strata of a.

5.3. Let $H \subset G$ and suppose we have constructed a N(H)-map $g: a^H_+ \to b^H_+$ with the following property: For all H', $H < H' \subseteq N(H)$, $(g^{H'})^* = \deg^s_{H'}(f)$, where $(g^{H'})^*: \overline{H}_{n_{H'}}(a^{H'}) \to \overline{H}_{n_{H'}}(b^{H'})$ is the map induced from $g^{H'}: a^{H'} = (a^H_+)^{H'} \to (b^H_+)^{H'} = b^{H'}$. Then g can be extended to a N(H)-map $g: a^H \to b^H$ with $\deg(g) = \deg^s_H(f)$.

Proof of 5.3. If $a^H = a^{H'}$ for $H < H'$ then $a^H = a^H_+$ and the hypothesis on $\deg^s_H(f) = \deg^s_{H'}(f)$ allows us to take $g = \hat{g}$. Otherwise, $a^H - a^H_+ \neq 0$ and by elementary obstruction theory we can extend g to $g': a^H \to b^H$. Consider $g' \# \mathrm{Id}: (a' \# c)^H \to (b \# c)^H$ and observe that $\deg_{H'}(g') = \deg_{H'}(g' \# \mathrm{Id}) = \deg^s_{H'}(f)$ for all H', $H < H' \subset N(H)$. Since $(a \# c)^H$, $(b \# c)^H$ are N(H)/H spaces, it follows (from an easy extension of 5.2) that $g' \# Id | (a\# c)_+^H \underset{G}{\simeq} f^H | (b\# c)_+^H$. It follows (see 2.1) that $\deg_H(g' \# Id) = \deg_H(g') = \deg^s_H(f) \mod o(N(H)/H)$. Therefore (2.1 again), one can rechoose g' to get $\hat{g}$ with $\deg_H(\hat{g}) = \deg^s_H(f)$. Q.E.D.

If a, b, c are complex linear G-spheres and $f: a\# c \to b\# c$ is an oriented homotopy equivalence, the hypothesis of 5.2 holds automatically, and thus a and b are G-oriented homotopically equivalent. Thus we recover Theorem 5 of [tD]. For certain groups we can say more:

5.4 Theorem: Let G be an odd order group such that for every pair of subgroups H, H', $H < H'$, $H \neq$ normalizer of H in H', (i.e., G is a p-group or G is abelian). If $f: a\# c \to b\# c$ is a G-homotopy equivalence, then $a^H = a^{H'}$ and $H < H'$ implies $\deg^s_H(f) = \deg^s_{H'}(f)$. Thus $a \underset{G}{\simeq} b$.

Proof. As above we can assume that $c(H) = Z$ for all $H \subset G$. Let H be a maximal subgroup of G such that there exists an H', $H < H'$, $a^H = a^{H'}$, but $\deg^s_H(f) \neq \deg^s_{H'}(f)$. Following the proof of 5.3, we can construct $\bar{f}: A_+^H \to B_+^H$ with $\deg_{H'}(\bar{f}) = \deg^s_{H'}(f)$ for all H', $H < H'$. Since $a^H = a^{H'}$ and f is a G-homotopy equivalence, dimension restrictions imply $b^H = b^{H'}$. Thus $a^H = a_+^H$ and $b^H = b_+^H$.

We have $(a\# c)_+^H = (\bigcup_{H<H'} a^{H'} \# c^{H'})_+^H$. Similarly for $(b\# c)_+^H$. Then $\bar{f} \# Id: (a\# c)^H = a^H \# c^H \to b^H \# c^H = (b\# c)^H$ is a N(H)/H homotopy equivalence, which by the assumption on degrees is N(H)/H homotopy equivalent to f^H on $(a\# c)_+^H$. Since both are homotopy equivalences, $\deg_H(\bar{f} \# Id) = \pm \deg_H(f)$. On the other hand, since they are N(H)/H homotopic on $(a\# c)_+^H$, $\deg_H(\bar{f} \# Id) = \deg_H(f) \mod o(H(H)/H)$. Since $o(N(H)/H)$ is odd and $\neq 1$, we must have $\deg_H(\bar{f} \# Id) = \deg_H(f)$. Then $\deg^s_H(f) = \deg_H(f) = \deg_H(\bar{f}) = \deg_{H'}(\bar{f}) = \deg^s_{H'}(f)$ by hypothesis. Hence we have a contradiction. Q.E.D.

Combining 5.4 and 4.3 yields

5.5 Theorem: If G is an odd p-group and a and b are real oriented linear spheres of the same dimension, then $a \underset{G}{\simeq} b$ if and only if $\tau(a) \equiv \tau(b)$ if and only if there exists a real linear oriented c with $\tau(a \# c) \equiv \tau(b \# c)$.

For semilinear spheres a, b we introduce the notation for stable algebraic G-equivalence $\tau(a) \overset{s}{\equiv} \tau(b)$ if there exists a semilinear c with $\tau(a \# c) = \tau(b \# c)$. We should remark that $\tau(a) \equiv \tau(b)$ implies $\tau(a \# c) \equiv \tau(b \# c)$ for any c, but as far as we know $\tau(a \# c) \equiv \tau(b \# c)$ does not imply $\tau(a) \equiv \tau(b)$ unless G is abelian. Before proving our main theorem we need a preliminary observation.

5.6 Lemma: If a is a semilinear sphere and $a^G \neq \emptyset$, then there exists a G-homotopy equivalence $f: a \to a$ with $\deg_G(f) = -1$.

Proof. The conclusion depends only on the G-homotopy type of G. However, an easy induction shows $a \underset{G}{\simeq} a^G \# a'$, where $a' = Lk(x) \cap Lk(a^G)$, for any $x \in a^G$. Hence we can choose $f = (-Id) \# Id$. Theorems 1.7 and 1.8 both follow from the following theorem.

5.7 Destabilization Theorem: Let G be any group with $G_{(2)}$, the Sylow 2-subgroup of G, very nice. Let a, b be any semilinear G-spheres such that

1. $\dim a^H = \dim b^H$ for all $H \subset G$,
2. $a \underset{G_{(p)}}{\overset{s}{\sim}} b$, for the odd Sylow subgroups $G_{(p)}$,
3. $\tau(a) \overset{s}{\equiv} \tau(b)$.

Then $a \underset{G}{\simeq} b$.

Proof. Since the conditions are all inherited by a^H, b^H as $N(H)/H$ spheres, we can argue by induction on strata. Thus it suffices to prove that any G-homotopy equivalence $\overset{o}{f}: a_+^{(e)} \to b_+^{(e)}$ extends to a G-homotopy equivalence $f: a \to b$. For p odd, select $\psi_{(p)}: a \to b$, a $G_{(p)}$-homotopy equivalence which exists by 5.4. Let

$$a_{(p)} = \bigcup_{\substack{H \subset G_{(p)} \\ H \neq (e)}} a^H .$$

$a_{(p)} \subset a_+^{(e)}$ is a $G_{(p)}$-invariant subspace. Let $f_{(p)} = \overset{o}{f} | a_{(p)}$ and let $\psi'_{(p)} = \psi_{(p)} | a_{(p)}$. Both $f_{(p)}$ and $\psi'_{(p)}$ are $G_{(p)}$-homotopy equivalences. Since $G_{(p)}$ is an odd p-group, $a(H) \cong b(H) = Z$ or 0 for $H \subset G_{(p)}$. Then $f_{(p)} \underset{G_{(p)}}{\sim} \psi'_{(p)}$ if and only if $\deg_H(f) = \deg_H(f_{(p)}) = \deg_H(\psi'_{(p)})$ for all $H \neq (e)$, $H \subset G_{(p)}$. Since for all $H \subset G_{(p)}$, $N(H)/H \neq \emptyset$ if $H \neq G_{(p)}$, it follows by a simple induction and by 2.1 that $\deg_H(f_{(p)}) = \deg_H(\psi'_{(p)})$ for all $H \subset G_{(p)}$, $H \neq (e)$, if and only if $\deg_{G_{(p)}}(f_{(p)}) = \deg_{G_{(p)}}(\psi'_{(p)})$. By 5.6 we can alter $\psi_{(p)}$ by composing with a $G_{(p)}$-homotopy equivalence of b to assure this condition, and thus we can choose $\psi_{(p)}$ with $\psi'_{(p)} \underset{G_{(p)}}{\sim} f_{(p)}$ as maps $a_{(p)} \to b_{(p)}$. We now extend $\overset{o}{f}$ to a G-map $\overline{f}: a \to b$. Then $\deg(\overline{f}) = \deg_{(e)}(\overline{f}) = \deg(\psi_{(p)}) \bmod o(G_{(p)})$. In calculating $\deg(\overline{f})$ we can replace a, b by $a \# c$, $b \# c$, $\overline{f}$ by $\overline{f} \# \mathrm{Id}$, $\psi_{(p)}$ by $\psi_{(p)} \# \mathrm{Id}$, etc. Hence we may as well assume $\tau(a) \equiv \tau(b)$. We must show $\deg(\overline{f}) = \mp 1 \bmod o(G)$.

We now perform the construction of A as in the proof of 2.2. This selects generators of $H_{n_e}(a)$, $H_{n_e}(b)$ and allows us to consider $\deg(\overline{f})$, $\deg(\psi_{(p)})$ as actual integers. Since $\psi_{(p)}$ is a homotopy equivalence, $\deg \psi_{(p)} = \mp 1$. However, by restricting to a nontrivial abelian subgroup of G, say Z/pZ which is nice, we must have $\deg(\overline{f}) = 1 \bmod p$. Since $\deg(\overline{f}) = \deg(\psi_{(p)}) \bmod p^n$, we must have $\deg(\psi_{(p)}) = 1$. Thus $\deg(\overline{f}) = 1 \bmod o(G_{(p)})$. Since G is nice, the argument of the proof of 2.2 implies $\deg(\overline{f}) = 1 \bmod o(G_{(Z)})$. Thus $\deg(\overline{f}) = 1 \bmod o(G)$ and we can replace $\overline{f}$ by f, with $\deg(f) = 1$. Q.E.D.

Bibliography

[A, T]. Atiyah, M.F. and D.O. Tall, Group representations, λ-rings, and the $\mathfrak{J}$-homomorphism, Topology 8, (1969), 253-297.

[L, W] Lee, Chung-Nim and A.G. Wasserman, On the groups JO(G), Memoirs American Math. Soc. 159 (1975).

[tD] tom Dieck, T, Homotopy-equivalent group representations, J. reine angew. Math. 298 (1978), 182-195.

[P, D] Petrie, T and tom Dieck, T., The homotopy structure of finite group actions on spheres, to appear in Proceedings of 1978 Waterloo conference on topology.

[C] Cohen, M., A Course in Simple Homotopy Theory, Springer-Verlag, 1973.

[R] Rothenberg, M., Torsion invariants and finite transformation groups, Proceedings of Symposia in Pure Mathematics, Vol. XXXII, Part I, AMS (1978), 267-313.

[W] Wall, C.T.C., Norms of units in group rings, Proceedings of L.M.S. Vol. XXIX, Dec. 1974.

Notes

[(1)] The use of the Smith formula in this proof was suggested to me by R. S. Kulkani I previously had a less elegant argument.

The author was partially supported by the National Science Foundation under grant NSF MCS 77-01623.

FINDING FRAMED $\mathbb{Z}_p$ ACTIONS ON EXOTIC SPHERES

Reinhard Schultz
Purdue University

A smooth action of the group G on a manifold M^n is called <u>stably frameable</u> if its equivariant tangent bundle is stably trivial as a G-vector bundle; if an explicit stable trivialization is specified, we shall say M^n has been (equivariantly stably) <u>framed</u>. If M^n is a homotopy sphere, then M^n is nonequivariantly frameable by a result of Kervaire and Milnor [7]; this is an important preliminary step towards their classification of all homotopy spheres in dimension ≥ 5. Consequently, the question of equivariant frameability of G-action on homotopy spheres arises naturally if one thinks about classifying such objects. In fact, if $G = \mathbb{Z}_p$ (p an odd prime) it is possible to obtain considerable information about equivariantly framed actions by homotopy-theoretic methods (compare [14,16]), and thus the frameability question is the basic obstacle to generalizing this information to all $\mathbb{Z}_p$ actions.

If a $\mathbb{Z}_p$ action on a homotopy sphere M^n resembles a linear action on S^n in some sense, then M^n is stably frameable by results of P. Löffler (see [8]; this is also true if p = 2 [9]). However, the examples of [15] show that in general M^n is not stably frameable. In this paper we shall prove a substitute for stable frameability that allows us to bypass this question in some cases:

DECOMPOSITION THEOREM. <u>Let</u> $M^n (n \geq 7)$ <u>be a homotopy sphere with smooth</u> $\mathbb{Z}_p$ <u>action and nonempty fixed point set. Then there is an integer</u> $u \equiv 1 \bmod p$ <u>such that the u-fold connected sum</u> uM^n <u>(along the fixed point set) splits equivariantly as</u> $N^n \# P^n$, <u>where</u> P^n <u>is nonequivariantly diffeomorphic to</u> S^n <u>and</u> N^n <u>is stably frameable.</u>

Of course, M^n and N^n are nonequivariantly diffeomorphic. Therefore we have the following consequence:

COROLLARY. <u>The homotopy sphere</u> M^n <u>admits a smooth</u> $\mathbb{Z}_p$ <u>action with k-dimensional fixed point set if and only if it admits a stably frameable action of this sort.</u>

Partially supported by NSF Grants MPS74-03609, MCS76-08794, and MCS78-02913.

In fact, the construction yields a stably frameable action with fixed point set a homotopy k-sphere, provided $n-k > 2$.

We shall say that the $\mathbb{Z}_p$ action on M^n is <u>exceptional</u> if the tangent bundle of M^n determines a nontrivial element of $\widetilde{KO}_{\mathbb{Z}/p}(M^n)\otimes Q$ and <u>unexceptional</u> otherwise. If the action is unexceptional, the proof of the theorem is fairly simple (see Section 1). Results of J. Ewing imply the exceptional case occurs only if the fixed point set is S^2 and 2 has odd multiplicative order mod p [3]. Thus the balance of our paper is devoted to looking at this special case in detail. By the results of [14], we are led to evaluate the p-localized normal invariants of certain homotopy equivalences. The latter are best understood by relationships of the form h=gf where f is the original homotopy equivalence and h,g are very nicely behaved $\mathbb{Z}_{(p)}$-homology equivalences — namely, they are compositions of cyclic branched coverings having degrees prime to p. Therefore, we first develop machinery for computing the p-local normal invariants of such maps in Section 3, write out the desired relationships h=gf in Section 4, and put everything together in Section 5 to finish the Decomposition Theorem's proof.

I would like to thank John Ewing for bringing 2.1 and 2.2 to my attention.

1. The unexceptional case

It will be helpful to have some basic examples of stably framed $\mathbb{Z}_p$ actions with which to work:

PROPOSITION 1.1. (Löffler [8,9]) <u>Suppose $\mathbb{Z}_p$ acts smoothly on the homotopy sphere Σ^n with fixed point set a homotopy sphere F^k. Then the action is equivariantly stably frameable if and only if it is unexceptional</u>∎

PROPOSITION 1.2. <u>Assume that the Pontrjagin Thom construction of Σ^n in π_n may be chosen to lie in the image of $\{S^{k+1}(S(V)/\mathbb{Z}_p),S^0\}$ under the suspended projection $S^n = S^{k+1}S(V)\longrightarrow S^{k+1}(S(V)/\mathbb{Z}_p)$. Then Σ^n admits a stably frameable action with fixed point set S^k, where S^k is embedded with trivial normal bundle</u>.

This follows from an exact sequence due to Rothenberg [10]. Here is an important special case, slightly strengthened using an idea of Bredon [compare [11]):

COROLLARY 1.3. <u>If Σ^n is divisible by p in Θ_n, then Σ^n admits a $\mathbb{Z}_p$ action as in Proposition</u> 1.2∎

Finally, we dispose of homotopy spheres that bound parallelizable manifolds.

PROPOSITION 1.4. <u>Suppose $\sum^n$ bounds a parallelizable</u> (n+1)-<u>manifold</u>.
(i) <u>If</u> n-k > 2, <u>then the conclusion of Proposition</u> 1.2 <u>is valid for</u> $\sum$.
(ii) <u>If</u> n-k=2, <u>the conclusion is valid except that the fixed point set might not be</u> S^k.

METHOD OF PROOF. One uses the Rothenberg exact sequence again to prove (i); the key extra piece of information required is that the transfer map of Wall groups $L_*(\mathbb{Z}_p) \longrightarrow L_*(1)$ is onto for p odd [19]. The Brieskorn examples discussed in [14,§4] yield (ii); by construction, they are stably framed■

Given a smooth $\mathbb{Z}_p$ action on the homotopy sphere $\sum$, a <u>knot invariant</u> ω of the action is defined in the localized homotopy group $\pi_k(F_{\mathbb{Z}_p}(V)/C_{\mathbb{Z}_p}(V))_{(p)}$, where

(a) k = dimension of fixed point set,

(b) V = space of normal vectors to the fixed point set at a given fixed point,

(c) $F_{\mathbb{Z}_p}(V)$ = space of equivariant self maps of the unit sphere in V, and $C_{\mathbb{Z}_p}(V)$ = orthogonal centralizer.

(see [14]). The action is stably frameable if and only if the image of ω under the map induced by

$$\sigma : F_{\mathbb{Z}_p}(V)/C_{\mathbb{Z}_p}(V) \xrightarrow{\text{stabilize}} F_{\mathbb{Z}_p} \longrightarrow BC_{\mathbb{Z}_p} = (BU)^{(p-1)/2}$$

is zero (since the codomain has torsion free homotopy, this is equivalent to saying the image in $\pi_k(BC_{\mathbb{Z}_p}) \otimes Q$ is zero).

We can now dispose of unexceptional actions:

THEOREM 1.5. <u>The Decomposition Theorem is valid for unexceptional actions</u>.

<u>Remark</u>. Actions that are unexceptional but not stably frameable have been constructed in the Appendix to [15].

PROOF. We denote the lens space $S(V)/\mathbb{Z}_p$ by L(V) henceforth. Consider the following commutative diagram, whose lower row is exact sequence [13,(1.1)] and whose right hand column is a surgery exact sequence:

$$
\begin{array}{ccccc}
 & & L_{k+1}(1) & \xrightarrow{\times L(V)} & L_n(\mathbb{Z}_p) \\
 & \text{onto} \swarrow & \downarrow & & \downarrow \Delta \\
 & bP_{k+1} \xrightarrow{\subseteq} & \Theta_k & & \\
CS^n(\mathbb{Z}_p,V) & \xrightarrow{\Lambda} & \oplus & \xrightarrow{E} & hS_k(L(V)) \\
 & & \pi_k(F_{\mathbb{Z}_p}(V)/C_{\mathbb{Z}_p}(V)) & & \downarrow q \\
 & & & & [S^k(L(V)^+),F/O]
\end{array}
$$

Since $L(V)$ is odd dimensional and the odd Wall groups of 1 and $\mathbb{Z}_p$ vanish, the top map is zero.

Given our action on Σ^n with knot invariant ω, let $\theta \in \pi_k(F/O)_{(p)}$ denote the Pontrjagin-Thom construction of the fixed point set (an oriented $\mathbb{Z}_{(p)}$ homology k-sphere by P. A. Smith theory). Choose a positive integer $a \equiv 1 \bmod p$ so that $a\omega$ is the image of an unlocalized class $\omega' \in \pi_k(F_{\mathbb{Z}_p}(V)/C_{\mathbb{Z}_p}(V))$, and choose $\theta' \in \Theta_k$ such that $a\theta$ is the Pontrjagin-Thom invariant of θ'. We would like to show that (θ',ω') or something similar lies in the image of Λ.

By the results of [14,§§2-3], the classes θ and ω combine to determine an element Q in $[S^k(L(V)^+),F/O]_{(p)}$ which is trivial since (θ,ω) comes from a $\mathbb{Z}_p$ action on Σ[14,Prop. 3.1]. On the other hand, by construction Q is equal to the p-localization of $aqE(\theta',\omega')$; thus $0 = baqE(\theta',\omega')$ for some large $b \equiv 1 \bmod p$. It follows that $E(b\theta',b\omega') \in$ Image Δ. But modulo the image of $\Delta | L_n(1)$, which is finite, everything in the image of Δ is detected by an Atiyah-Singer invariant as presented in [15,§2]. The latter only depends on $\sigma_*\omega'$, however, and therefore it is zero by our assumption that Σ^n was unexceptional (hence $\sigma_*\omega' = 0$). Finally, we claim that $E(b\theta',b\omega')$ is 2-primary; this is clear if $n \not\equiv 0 \bmod 4$ because the image of $\Delta | L_n(1)$ is at most $\mathbb{Z}_2$, while if $n \equiv 0 \bmod 4$ this follows from [17,Thm.3.1]. Choose $c \equiv 1 \bmod p$ so that $0 = E(cf\theta',cb\omega')$. Then there is a $\mathbb{Z}_p$ action on some Σ' with $\Lambda(\Sigma') = (cb\theta',cb\omega')$ by exactness in (1.6).

It follows from [14,Thm.3.4] that Σ' is (nonequivariantly) diffeomorphic to $abc\Sigma \# \Sigma''$, where Σ'' is a sum of exotic spheres satisfying the conditions of 1.2, 1.3, or 1.4; in fact $abc \equiv 1 \bmod p$ means we may similarly write $\Sigma' = \Sigma \# \Sigma'''$ by 1.3. Take the $\mathbb{Z}_p$ action on Σ''' given by adding the actions of 1.2-1.4. Then the desired decomposition is given with $N^n = \Sigma' \# -\Sigma'''$ and $P^n = \Sigma \# -\Sigma' \# \Sigma'''$ (connected sums along the k-dimensional fixed point set) ∎

Acknowledgment. Several years ago Wu-Chung Hsiang outlined to me a different approach for proving such a result based on the work of L. Jones[5,6].

2. Exceptional equivariant normal bundles

Suppose now that $\mathbb{Z}_p$ acts smoothly on Σ^{2n} with fixed point set S^2 (the necessary conditions for exceptionality). Express the tangent bundle $\tau(\Sigma)|S^2$ as a sum

$$\tau(S^2) \oplus \sum \xi_j \otimes_C W_j,$$

where ξ_j is a complex vector bundle over S^2 and W_j is the realification of $t^j \in R(\mathbb{Z}_p)$, $1 \leq j \leq (p-1)/2$. For $(p+1)/2 \leq j \leq p-1$, define W_j as likewise and set $\xi_j = -\xi_{p-j}$; extend both definitions to arbitrary $j \not\equiv 0$ mod p in the obvious way. The bundles ξ_j are completely determined by the Chern class numbers $x_j = c_1(\xi_j) \cap [S^2] \in \mathbb{Z}$. It is necessary for us to know exactly which sequences $\{x_j\}$ are allowed by the G-signature theorem. In other words, if we let $\zeta^j = \exp(2\pi ij/p)$ and we set $\phi_j = 2/(\zeta^j - \zeta^{-j})$, we want all solutions of the equation

$$\sum_{j=1}^{(p-1)/2} x_j \phi_j = 0$$

(see [3] for the derivation of these equations).

I am indebted to J. Ewing for the following two formulas:

(2.1) <u>Let</u> $\alpha_j = (2 + \zeta^j + \zeta^{-j})/(\zeta^j - \zeta^{-j})$. <u>Then</u> $\phi_j = \alpha_j - \alpha_j^{-1}$.

(2.2) $\alpha_j + \alpha_j^{-1} = 2\alpha_{2j}$.

Both verifications are very routine algebra.

If we add up the relations (2.2) for $j, 2j, 4j, \ldots, 2^M j$ (where $0 < M$ is minimal so that $2^M \equiv 1$ mod p), we get the following result:

THEOREM 2.3. <u>Assume M is odd. Then the set of sequences</u> $\{x_j\}$ <u>satisfying</u> $\sum x_j \phi_j = 0$ <u>is free abelion on</u> $(p-1)/2M$ <u>generators. Specifically, let</u> ε_j <u>be the sequence with</u> $(\varepsilon_j)_k = \pm 1$ <u>if</u> $k \equiv \pm j$ mod p <u>and</u> 0 <u>elsewhere. Then a free basis of solutions is given by</u>

$$\sigma_m = \sum_{k=0}^{M-1} \varepsilon_{2^k m},$$

<u>where</u> m <u>runs over a set of representitives for the cosets of</u> Units $(\mathbb{Z}_p)$/Subgp. gen by 2∎

<u>Remark</u>. If M is even, adding up the relations 2.2 merely yields the trivial identity 0=0.

The following consequence of Theorem 2.3 was used previously in [17]:

COROLLARY 2.4. <u>Suppose</u> $\mathbb{Z}_{p^r}$ <u>acts exceptionally on a homotopy sphere</u> Σ^{2n}. <u>Then</u> $n \geq M+1$, <u>where</u> M <u>is the order of</u> 2 mod p.

PROOF. Suppose first that r=1. Then Theorem 2.3 implies that the number of nonzero x_j's is divisible by M. Hence dim V $\geq$ 2M, so that dim $\Sigma \geq$ 2M+2.

If $r > 1$, things proceed similarly, but one now needs Ewing's extended calculations for composite numbers [4] and some formalism to deal with the $\mathbb{Z}_{p^r}$ equivariant tangent bundle in this more complicated setting (e.g., the machinery of [15])■

3. Normal invariants of cyclic branched coverings

Suppose now that $\pi: E \to X$ is a smooth r-fold cyclic branched covering along a codimension 2 smooth submanifold B; throughout our discussion we assume everything is oriented. In the general study of branched coverings, it is unusual to assume that π is a $\mathbb{Z}[r^{-1}]$-homology equivalence, but for our purposes we must make this assumption. The results of [14,§1] show that a normal invariant $q(\pi) \in [X, F/O] \otimes \mathbb{Z}[r^{-1}]$ may be defined. We must compute $q(\pi)$ for the following class of examples:

EXAMPLES 3.1. (i) Consider the free complex representation $t^a + \sum t^{b_i}$ of $\mathbb{Z}_p$, and let

$$F_r: E(t^a + \sum t^{b_i}) \to E(t^{ra} + \sum t^{b_i})$$

(E = representation space) be the equivariant map defined by $\tilde{f}_r(z,w) = (z^r, w)$. Denote the induced map of lens spaces by $f_r: L(t^a + \sum t^{b_i}) \to L(t^{ra} + \sum t^{b_i})$; then f_r is a cyclic r-fold branched covering with branch set $L(\sum t^{b_i})$.

(ii) Let Y be a smooth manifold with trivial $\mathbb{Z}_p$ action, and let $\xi_o \otimes t^a + \sum \xi_i \otimes t^{b_i}$ be a sum of complex $\mathbb{Z}_p$ line bundles over Y. Then one has an analogous map $F_r: L(\xi_o \otimes t^a + \sum \xi_i \otimes t^{b_i})$ whose restriction to each fiber is the previous map f_r. The branch set is $L(\sum \xi_i \otimes t^{b_i})$.

The first step toward computing the normal invariant is to calculate its restriction to the branch set in the general situation. To do this, we need a piece of Adams' original work on the $J(X)$-groups; namely, the construction of a fiber homotopy trivialization for $r^e(\eta - \psi_C^r \eta)$, where η is a complex line bundle. In principle, this corresponds to constructing a map

$$A_r: BU_1 \to F/O[r^{-1}]$$

such that the composite $BU_1 \to F/O[r^{-1}] \to BO[r^{-1}]$ is $\eta - \psi_C^r$.

THEOREM 3.2. Let $\eta = \eta(B,E)$ be the oriented normal bundle of B in E with its unique complex structure, and let $\chi \in H^2(B;\mathbb{Z}) \cong [B, BU_1]$ classify η (using $BSO_2 = BU_1$). Then $q(\pi)|B = A_{r*}(\chi)$.

PROOF. We may as well confine attention to a neighborhood of the branch set, assuming E is the disk bundle $D(\eta)$ and X is the disk bundle of $\psi_C^r \eta = \eta \otimes \ldots \otimes \eta$ (r terms) $= \eta^r$. Then the branched covering simply takes a vector $v \in D(\eta)$ and sends it to $v^r = v \otimes \ldots \otimes v$.

The first step in forming $q(\pi)$ is to construct an embedding $D(\eta) \to D(\eta^r) \times \mathbb{R}^M$ approximating the map $(\pi,0)$. A simple explicit choice is given by writing $B \times \mathbb{R}^M = E(\eta \oplus \eta^\perp)$ to η and sending $v \in D(\eta)$ to $(v^r, v, 0) \in D(\eta^r) \times \mathbb{R}^M \subseteq E(\eta^r \oplus \eta \oplus \eta^\perp)$. Next, we must construct the umkehr map for this embedding. Since our embedding is constructed canonically on each fiber, the umkehr map will behave uniformly on each fiber, and thus it suffices to study a single fiber provided we keep everything equivariant with respect to the structure group of η, which is S^1. But if $t, t^r \in R(S^1) = \mathbb{Z}[t,t^{-1}]$ have their usual meaning, then the induced fiber map $\pi: D(t) \to D(t^r)$ is S^1 equivariant, so this poses no problem. It follows that the fiberwise model for the umkehr map is given by taking an S^1 equivariant map

$$g_o: E(t)^\circ \to E(t^r)^\circ \ (^\circ = 1 \text{ pt. compactification})$$

and smashing it with the compactified fiber of $\eta^\perp$. By construction, g_o sends the zero and infinity points to their counterparts, and therefore by obstruction theory g_o is S^1 homotopic to the one point compactification of the map $z \mapsto z^r$. Thus the umkehr map $S^M(B^+) \to B^{\eta^r \oplus \eta^\perp}$ is homotopic to the one point compactification of the proper map

$$B \times \mathbb{R}^M = E(\eta \oplus \eta^\perp) \to E(\eta^r \oplus \eta^\perp)$$

sending (x,y) to (x^r,y). Finally, a direct check of the definitions from [14, §1] now shows that the normal invariant of π is essentially given by a similar map with an inverse to η^r replacing $\eta^\perp$. Explicitly, one takes the stabilized one point compactification $B^{\eta - \eta^r} \to B^+$, projects from B^+ to the local sphere $S^o[r^{-1}]$, and divides by r. But this is equivalent to the construction of A_r as given in [1]∎

We now specialize to the examples of 3.1(ii), which we call <u>special</u> r-fold cyclic branched coverings. In three examples the normal bundle η is simply the bundle $\xi_o \otimes t^a / \mathbb{Z}_p$ over $S(\sum \xi_i \otimes t^{b_i}) / \mathbb{Z}_p = L(\sum \xi_i \otimes t^{b_i})$.

THEOREM 3.3. <u>Let</u> F_r <u>be a special</u> r-<u>fold cyclic branched covering, and let</u> $\eta\#$ <u>be the bundle</u> $\xi_o \otimes t^a / \mathbb{Z}_p$ <u>over</u> $L(\xi_o^r \otimes t^{ra} \oplus \sum \xi_i \otimes t^{b_i})$. <u>Then</u> $q(F_r) = A_r \, \chi(\eta\#)$.

PROOF. Consider the special r-fold cyclic branched coverings $F_r(\xi_o \otimes t^a, \sum \xi_i \otimes t^{b_i} \oplus \xi_o^r \otimes t^{ra} \oplus \sum_{\eta_j} \otimes t^{c_j})$, where enough line bundles $\eta_j \otimes t^{c_j}$ are added to ensure that all equivariant linear embeddings of $\xi_o^r \otimes t^{ra} \oplus \sum \xi_i \otimes t^{b_i}$ in $k\xi_o^r \otimes t^{ra} \oplus \sum \xi_i \otimes t^{b_i} \oplus \sum_{\eta_j} \otimes t^{c_j}$ are equivariantly linearly isotopic with $k=1$ or 2. Since F_r maps $L(\xi_o^r \otimes t^{ra} + \sum \xi i \otimes t^{b_i})$ to itself and is linear on a tubular neighborhood, it follows that

$$q(F_r(\xi_o \otimes t^a, \sum \xi_i \otimes t^{b_i} \oplus \xi_o^r \otimes t^{ra} \oplus \sum_{\eta_j} \otimes t^{c_j}) | L(\xi_o^r \otimes t^{ra} \oplus \sum \xi_i \otimes t^{b_i} \oplus 0) =$$

$$q(F_r(\xi_o \otimes t^a, \sum \xi_i \otimes t^{b_i}).$$

On the other hand, if we restrict to $L(0 \oplus \sum \xi_i \otimes t^{b_i} \oplus \xi_o^r \otimes t^{ra} \oplus 0)$, then we get the normal invariant restricted to the branch set of $F_r(\xi_o \otimes t^a, \sum \xi_i \otimes t^{b_i} \oplus \xi_o^r \otimes t^{ra})$. By 3.2 this is $A_r\ \chi(\eta\#)$. But the two linear embeddings of $L(\xi_o^r \otimes t^{ra} \oplus \sum \xi_i \otimes t^{b_i})$ in the big lens space bundle are isotopic, and therefore $q(F_r)$ must equal $A_r\ \chi(\eta\#)$■

4. Special branched coverings and fiber homotopy trivializations

In this section we shall show that certain homotopy equivalences $f:X \to M$ of closed manifolds satisfy identities of the form $h \simeq gf$, where both $g:M \to N$ and $h:Y \to N$ are r-fold special cyclic branched coverings. Since the r-local normal invariants satisfy the equation $g(h) = q(g) + (g^*)^{-1}(q(f))$[14,(1.1)], this allows one to calculate $q(f)$ fairly directly.

Throughout this section p is an odd prime, the multiplicative order M of 2 mod p is assumed to be odd, and r is some fixed primitive root of unity modulo p^2.

THEOREM 4.1. *Let V be a free $\mathbb{Z}_p$ module of real dimension $2m \geq 4$. Given another 2m-dimensional free $\mathbb{Z}_p$ module W, let $F_{\mathbb{Z}_p}(V,W;k)$ be the space of all equivariant maps $S(V) \to S(W)$ having degree $k>0$. If $g:S(W) \to S(W')$ is equivariant of degree $\ell>0$, let*

$$\alpha_{k,k\ell} \circ F_{\mathbb{Z}_p}(V,W;k) \to F_{\mathbb{Z}_p}(V,W';k\ell)$$

be induced by composition. Then the space $F'_{\mathbb{Z}_p}(V) = \operatorname{colim} \alpha_{k,k\ell}$ is naturally homotopy equivalent to the localization $F_{\mathbb{Z}_p}(V)_{(p)}$.

The first step in proving this is to note that the subsystem with $W = W' = V$ is cofinal. Given this, the result may be checked in a variety of fairly standard ways (e.g., using [12]). In fact, using the results of [12] we get the following specific information:

COROLLARY 4.2. *The following sequence is exact:*

$$\mathbb{Z}_2 = \pi_2(F_{\mathbb{Z}_p}(V)) \to \pi_2(F_{\mathbb{Z}_p}(V)/C_{\mathbb{Z}_p}(V)) \to \pi_2(F'_{\mathbb{Z}_p}(V)/C_{\mathbb{Z}_p}(V)) \to \mathbb{Z}_2 \to 0,$$

where p is the composite

$$\pi_2(F'_{\mathbb{Z}_p}(V)/C_{\mathbb{Z}_p}(V) \overset{\cong}{\to} \pi_1(C_{\mathbb{Z}_p}(V)) = \pi_1(U^m) \overset{\oplus}{\to} \pi_1(U) \to \pi_1(O) = \mathbb{Z}_2,$$

m being the number of inequivalent irreducible real representations contained in V■

We can now start finding the desired relations $h=gf$.

PROPOSITION 4.3. Let r be a primitive root of unity mod p^2, let $\varepsilon_r = 2$ or 1 as r is even or odd, let $\xi \downarrow S^2$ be the canonical complex line bundle, and let α be a free $\mathbb{Z}_p$ module containing at least ε_s copies of t^s, where $s \not\equiv 0$ mod p. Then for every K>0 there exist maps

$$h_{k,\alpha,s}: L(K_{\varepsilon_r}(r^{p-1}-1)\xi \otimes t^s \oplus \alpha) \to S^2 x L(K_{\varepsilon_r}(r^{p-1}-1)(t^s \oplus \alpha)$$

with the following properties:

(i) $h_{K,\alpha,s}$ is a composite of special r-fold cyclic branched coverings.

(ii) If g_o is the restriction of $h_{K,\alpha,s}$ to a fiber, then $h_{K,\alpha,s}$ is homotopic to $(id(S^2)xg_o) \circ f$, where f is the orbit map of an equivariant fiber homotopy equivalence (hence a fiber homotopy equivalence itself).

PROOF (Sketch). Let $q = r^{p-1}$. Since the complex bundles $(q-1)\xi$ and $\xi^q \oplus \xi^{-1}$ over S^2 are isomorphic up to trivial factors needed to equate dimensions, we may use the latter instead of $(q-1)\xi$. In fact, we need $\xi^{K\varepsilon q} \oplus \xi^{-K\varepsilon}$ to treat the general situation. Now take a succession of (p-1) special cyclic branched coverings:

$$F_r(\xi^{-K\varepsilon} \otimes t^s, --), F_r(\xi^{-rK\varepsilon} \otimes t^{rs}, --), F_r(\xi^{-r^2K\varepsilon} \otimes t^{r^2s}, --), \text{etc.}$$

The composition of these maps gives the desired $h_{K,\alpha,\varepsilon}$, and the existence of f and the factorization follow from 4.1 and 4.2∎

Although these examples are interesting in their own right, the vector bundles discussed in Section 2 are far more important to us. In this case we have a similar result:

THEOREM 4.4. Let α be a free $\mathbb{Z}_p$ module as in 4.3, and let M be the multiplicative order of 2, which we assume is odd. Then for every K>0 divisible by $\varepsilon_r(q-1)/p$ there exist maps

$$h: L(K\xi \otimes \{\sum_o^{M-1} t^{s2^j}\} \oplus \alpha) \to S^2 x L(\sum K t^{s2^j} \oplus \alpha)$$

satisfying the conditions of (i) and (ii) in 4.3.

PROOF (Sketch).One forms a sequence of special r-fold cyclic branched coverings from the alleged domain of h_o to $L(\{\sum K\xi^{2^{M-j}}\} \otimes t^s)\alpha)$. But the latter bundle is isomorphic to $L(K(2^M-1)\xi \otimes t^s \oplus \alpha')$, and since $K(2^M-1)$ is divisible by $\varepsilon_r(q-1)$ there is another composite of special cyclic branched coverings h_1 from the latter bundle to $S^2 x L\left(K(2^M-1)t^s \oplus \alpha\right)$. Finally, there is a sequence of special cyclic branched coverings from $L(K(2^M-1)t^s \oplus \alpha')$ to $L(Kt^s \oplus \alpha)$; cross this with $id(S^2)$ to get h_3. Then $h_3h_2h_1$ is the desired map h, and everything now proceeds as in 4.3∎

5. Proof of Decomposition Theorem

In order to work effectively with the preceding machinery, we need some control over sums of the form $\sum A_r(\alpha_i)$, where the α_i are complex line bundles. Specifically, we need an additivity statement about solutions to the Adams Conjecture. The following result, which is a consequence of V. Snaith's further study of the Becker-Gottlieb solution [2,18], gives this to us:

THEOREM 5.1. There is a solution $A: BU_{(p)} \to F/U_{(p)}$ of the complex Adams conjecture at p such that A homotopy commutes with Whitney sum and A restricted to BU_1 is A_r, where r is a primitive root of unity mod p^2 ∎

To obtain the conclusion we want from this, it will be helpful to use some notation from earlier papers. Let $\mathcal{L}: \pi_2(BC_{\mathbb{Z}_p}(V)) \to \mathbb{R}^{(p-1)/2}$ be defined as in [15, §2] using the Atiyah-Singer invariant, and let $\partial: \pi_2(F_{\mathbb{Z}_p}(V)/C_{\mathbb{Z}_p}(V)) \to \pi_2(BC_{\mathbb{Z}_p}(V))$ be given by the connecting homomorphism for the principal bundle $C_{\mathbb{Z}_p}(V) \to F_{\mathbb{Z}_p}(V) \to F_{\mathbb{Z}_p}(V)/C_{\mathbb{Z}_p}(V)$. Also, let E_o be the homomorphism from $\pi_2(F_{\mathbb{Z}_p}(V)/C_{\mathbb{Z}_p}(V))$ to $hS_2(L(V))$ that lies in exact sequence [13,(1.1)], and let qE_o be the normal invariant homomorphism into $[S^2(L(V)^+), F/O]$.

THEOREM 5.2. Let $\omega \in \pi_2(F_{\mathbb{Z}_p}(V)/C_{\mathbb{Z}_p}(V))$ satisfy $\mathcal{L}\partial(\omega) = 0$. Then the p-localization of $qE_o(\omega)$ lies in the image of

$$p_* A_*: \tilde{K}(S^2(L(V)^+))_{(p)} \to [S^2(L(V)^+), F/O]_{(p)},$$

where $p: F/U \to F/O$ is induced by realification.

PROOF. The group $\pi_2(F_{\mathbb{Z}_p}(V)/C_{\mathbb{Z}_p}(V))$ is isomorphic to $\pi_2(F_{\mathbb{Z}_p}(V\oplus W)/C_{\mathbb{Z}_p}(V\oplus W))$ by suspension, where W is a sum of subrepresentations of V. Furthermore, the results of [14, §2] imply that nothing is lost if we consider $qE_o(\omega\oplus W)$ instead of $qE_o(\omega)$. Therefore, without loss of information we may assume that V has an arbitrarily large number of copies of each irreducible representation type it contains.

Recall that Theorem 2.3 and Corollary 4.2 specify completely the kernel of $\mathcal{L}\partial$. Suppose that ω lies in the kernel of $\mathcal{L}\partial$, and suppose further that ω is divisible by $\varepsilon_r(r^{p-1}-1)/p$ mod torsion (notice this number is prime to p since $r^{p-1} \not\equiv 1 \bmod p^2$). Then the homotopy smoothing f of $S^2 \times L(V)$ determined by $E_o(\omega+I)$— where I is an indeterminacy from $\pi_2(F_{\mathbb{Z}_p}(V)) = \mathbb{Z}_2$— satisfies a relation $gf \simeq h$, where g and h are both composites of special r-fold cyclic branched coverings.

By the composition laws for normal invariants (i.e., $q(ab) = q(b) \oplus b^{*-1} q(a)$ [14,(1.1)]), the results of Sections 3 and 4, and Theorem 5.1, it follows that $q(g)$, $q(h)$, and hence the p-localized $q(f) = qE_o(\omega)$ all lie in the image of A_*(notice that $2I=0$ implies that $qE_o(I)$ disappears upon localization). Although we have assumed ω is divisible by $d = \varepsilon_r(r^{p-1}-1)/p$ up till now, it is immediate that this restriction can be lifted at this point because d is a unit mod p and we are looking at the p-localization of the homomorphism qE_o■

When combined with [14,§3], Theorem 5.2 yields the final step in proving the Decomposition Theorem in the introduction.

THEOREM 5.3. Let $\mathbb{Z}_p$ act smoothly on the homotopy sphere Σ^{2n} with fixed point set S^2; assume 2 has odd multiplicative order mod p. Then M^{2n} admits a stably frameable action with S^2 as the fixed point set.

The Decomposition Theorem follows immediately from this and Theorem 1.5. In the nonexceptional case, let N^{2n} be the associated stably frameable action. Then $P = \Sigma^{2n} \#\text{-}N$ is nonequivariantly a standard sphere, and it has a natural action if connected sums are taken along the fixed point set.

PROOF OF THEOREM 5.3. Let $\omega \in \pi_2(F_{\mathbb{Z}_2}(V)/C_{\mathbb{Z}_2}(V))$ be the action's knot invariant. Then by [14, Thm. 3.4] we have that

$$q\Theta[\omega\oplus M] \in [S^2(L(V\oplus M)^+), F/O]_{(p)}$$

comes from $-q(M\Sigma) \oplus ? \in \pi_{2n}(F/O)_{(p)} \oplus \pi_{2n+1}(F/O)_{(p)}$ under the collapse map

$$c: S^2L(V\oplus M) \to S^2[L(V\oplus M)/L(V)] \simeq S^{2n} \vee S^{2n+1}.$$

Now ρA is the first factor inclusion of a splitting $F/O_{(p)} \simeq BSO_{(p)} \times CokJ_{(p)}$, and thus Theorem 5.3 says that $q\Theta[\omega\oplus M] \in \text{Image } \rho_* A_*$. On the other hand, $-q(\Sigma)$ comes from the $CokJ_{(p)}$ factor, and therefore $c^*q(\Sigma) = 0$. By the exactness of the Puppe sequence, this implies that $q(\Sigma)^{2n}$ comes from $\{S^3L(V), S^o\}$. Therefore the exotic sphere Σ^{2n} also admits a stably frameable action with fixed point set S^2 by Proposition 1.2■

PURDUE UNIVERSITY
WEST LAFAYETTE, INDIANA 47907

REFERENCES

1. J. F. Adams, On the groups J(X)-I, Topology 2(1963), 181-195.

2. J. C. Becker and D. H. Gottlieb, The transfer map and fiber bundles, Topology 14(1975), 1-12.

3. J. Ewing, Spheres as fixed point sets, Quart. J. Math. Oxford (2) 27(1976), 445-455.

4. ________, Semifree actions of finite groups on homotopy spheres, Trans. Amer. Math. Soc., to appear.

5. L. Jones, The converse to the fixed point theorem of P. A. Smith: I, Ann. of Math. 94(1971), 52-68.

6. ________, Ibid: II, Indiana Univ. Math. J. 22(1972), 309-325; correction 24(1975), 1001-1003.

7. M. Kervaire and J. Milnor, Groups of homotopy spheres, Ann. of Math. 78(1963), 514-537.

8. P. Löffler, Über die G-Rahmbarigkeit von G-Homotopiesphären, Arch. Math. (Basel) 29(1977), 628-634.

9. __________, Equivariant frameability of involutions on homotopy spheres, Manuscripta Math. 23(1978), 161-171.

10. M. Rothenberg, Differentiable group actions on spheres, Proc. Adv. Study Inst. on Algebraic Topology (Aarhus 1970), 455-475. Matematisk Institut, Aarhus Universitet, 1970.

11. R. Schultz, $\mathbb{Z}_2$-torus actions on homotopy spheres, Proc. Second Conf. on Compact Transformation Groups (Amherst, Mass., 1971), Lecture Notes in Math. Vol. 298, 117-118. Springer, New York, 1972.

12. __________, Homotopy decompositions of equivariant function spaces I, Math. Z. 131(1973), 49-75.

13. __________, Homotopy sphere pairs admitting semifree differentiable actions, Amer. J. Math. 96(1974), 308-323.

14. __________, Differentiable group actions on homotopy spheres: I, Invent. Math. 31(1975), 105-128.

15. __________, Spherelike G-manifolds with exotic equivariant tangent bundles, Studies in Algebraic Topology (Adv. in Math. Suppl. Studies, Vol.5), 1-38. Academic Press, New York (to appear in 1979).

16. __________, Smooth actions of small groups on exotic spheres, Proc. A.M.S. Sympos. Pure Math. 32, Pt.1(1978), 155-160.

17. __________, Isotopy classes of periodic diffeomorphisms on spheres, Proc. Waterloo Alg. Top. Conference (June, 1978), to appear.

18. V. P. Snaith, Algebraic Cobordism and K-Theory, Memoirs. Amer. Math. Soc., to appear.

19. C. T. C. Wall, Classification of Hemitian forms-VI. Group rings, Ann. of Math. 103(1976), 1-80.

The rational homotopy groups of Diff (M) and Homeo (M^n) in the stability range

by D. Burghelea

Ch. I: Introduction

For a differentiable (topological) manifold M^n, let Diff (M^n) (Homeo (M^n)) be the group of C^∞-diffeomorphisms (homeomorphisms) of M^n which restrict to the identity on ∂M endowed with the C^∞-topology (compact-open topology).

The study of the homotopy type of Diff (M^n) (Homeo M^n) seems to be a fascinating problem because of its implications and significance inside and outside topology as well as because of new connections between various fields of mathematics and new ideas it has stimulated.

Although the problem is far from being solved, a good amount of information about the homotopy type of Diff (M^n) (Homeo M^n) has been obtained due to the combined work of Cerf, Morlet, Hatcher, Quinn, Antonelli-Burghelea-Kahn, Lashof, Rothenberg, Burghelea, Hsiang, etc. and more recently to Waldhausen whose algebraic K-theory of topological spaces provides the possibility of rational computations and clarifies the relationship with the algebraic K-theory.

The key geometric result in this area is, in my opinion the "stability range" which is an increasing function $\omega^D(n)$ for differential manifolds respectively $\omega^T(n)$ for topological manifolds, which tends to ∞ when $n \to \infty$. Its interest consists in the fact that the homotopy type of the $(\omega(n)-1)$-Postnikov term of Diff (M^n) (Homeo (M^n)) away from the prime "2" can be described as a twisted product of $T_1(M)$ and $T_2(M)$

i.e. the total space of a principal fibration with fibre $T_1(M)$ and base $T_2(M)$,

$$*(D) \qquad T_1^D(M) \to T_1^D(M) \times_t T_2^D(M) \to T_2^D(M)$$

$$*(T) \qquad T_1^T(M) \to T_1^T(M) \times_t T_2^T(M) \to T_2^T(M)$$

where $T_1(M)$ is a homotopy type invariant, $T_2(M)$ is a geometric invariant and the twisting t, the classifying map for the fibration (*), $t: T_2(M) \to BT_1(M)$[+] is essentially a homotopy invariant in a sense which will be explained below. For the moment the best value of the stability range is not yet established but we do have the following estimates: $\omega^D(n) \geq \frac{n}{12} - 6$, $\omega^T(n) \geq \frac{n}{6} - 4$.

$T_2(M)$ can be entirely understood by means of surgery theory, in particular its homotopy groups can be at least theoretically computed and tensored by the rational numbers explicitly computed at least in the case of 1-connected manifolds. $T_1(M)$ is a homotopy type invariant of the manifold with boundary M^n which means it depends only on the Poincaré Duality structure of M (it depends on the homotopy type of the space M^n and of the element in the Adams group $J(M)$ defined by its tangent bundle) and rationally only on the rational homotopy type of M. Its homotopy groups tensored by rationals can be upper bound estimated at least for M 1-connected. Exact estimates have been obtained in some particular cases, as for instance $X = D^n$, $X = K(Z,2r)$, $X = K(Z \oplus Z,2r)$, $X = K(G,1)$ for G satisfying some additional properties, see [16],[8],[3],[4], [10].

The computation (estimation) of these homotopy groups tensored by the rationals reduces to the computation of the algebraic K-theory of topological spaces as described by Waldhausen, which brings back to

---

[+] BX denotes the classifying space (or the deloop) of the loop space X.

our attention the classical invariant theory and exploits recent results about the homology of arithmetic groups.

The twisting $t: T_2(M) \to BT_1(M)$ factors through a map β (defined below) which depends only on the Poincaré Duality space M, and rationally only on the rational homotopy type of M.

The decomposition as a twisted product of $T_1(M)$ and $T_2(M)$ is a remarkable fact; in some particular cases, when the twisting is trivial this decomposition is the homotopy theoretic analogue of the decomposition of a vector space as a sum of eigenspaces produced by a linear involution.

As an application of this description and computations, we give two theorems

Theorem 1[+]. If k is either the field C of complex numbers or H of quaternions and $P_n(k)$ denotes the corresponding n-th projective space, then for $i \leq$ minimum $(\omega(2un+r), 2un-2)$

$$\pi_i(\mathrm{Diff}\,(P_n(k)\times D^r)) \otimes Q = \pi_{i+r}(PGL_{n+1}(k)) \otimes Q + \widetilde{KO}(\Sigma^{r+i+1}(P_{n-1}(k)\cup p)) \otimes Q + \pi_{i+2}(Wh^D_{\varepsilon(r)}(K(Z,2u)) \otimes Q$$

where $u = 1$ if $k = C$ and $u = 2$ if $k = H$, $\varepsilon(s) = +$ if s is even and $\varepsilon(s) = -$ if s is odd and $PGL_n(k)$ denotes the projective group over k.

$$\pi_i(Wh^D_+(K(Z,2u)) \otimes Q = 0$$

$$\pi_i(Wh^D_-(K(Z,2u)) \otimes Q = \begin{cases} Q & \text{if } i = 4k+1, \ k = 1,2,3\ldots \\ 0 & \text{elsewhere} \end{cases}$$

[+]This theorem is based on a statement announced as Theorem by Hsiang and Jahren; although very likely true the proof is not yet available. Without this statement Theorem 1 gives the right value for the direct sum $\pi_i(\mathrm{Diff}(P_n(k)\times D^r)) \otimes Q + \pi_i(\mathrm{Diff}(P_n(k)\times D^{r+1})) \otimes Q$.

Analogously for $i \leq \inf(\omega^T(2nu+r)\ 2un-2)$

$\pi_i(\mathrm{Homeo}(P_n(k)\times D^r)) \otimes Q = \pi_{i+r}(PGL_{n+1}(k) \otimes Q + \widetilde{KO}(\Sigma^{r+i+1}(P_{n-1}(k)\cup p)) \otimes Q +$
$\pi_{i+2}(Wh^T_{\varepsilon(r)}(K(Z,2u)) \otimes Q$ where

$$\pi_i(Wh^T_+(K(Z,2u))) \otimes Q = 0$$

$$\pi_i(Wh^T_-(K(Z,2)) \otimes Q = \begin{cases} 0 \text{ if } i \text{ is odd} \\ Q^{j-1} \text{ if } i = 4j \\ Q^{j-1} \text{ if } i = 4j+2 \end{cases}$$

$$\pi_i(Wh^T_-(K(Z,4)) \otimes Q = \begin{cases} 0 \text{ if } i = \text{odd} \\ 0 \text{ if } i = 4j \\ Q^{j-1} \text{ if } i = 4j+2 \end{cases}$$

Theorem 2. If M is a 1-connected compact differentiable manifold, then $\dim \pi_i(\mathrm{Diff}\ (M^n)) \otimes Q < \infty$ if $i \leq \omega^D(n)-1$ and $\dim (\pi_i(\mathrm{Homeo}\ (M^n)) \otimes Q < \infty$ if $i \leq \omega^T(n)-1$.

The computations of $\pi_i(\mathrm{Diff}\ (M^n) \otimes Q$ in the stability range have been previously done(* for $M^n = D^n, S^n$, $K(\pi,1)$ if π satisfies some additional algebraic properties(for instance $\pi = Z\oplus\ldots\oplus Z$) and for $M^n = S^n/\Sigma$(if Σ is finite group which acts freely on S^n in terms of the algebraic K-theory of Σ (see [17],[8], [9]); from the point of view of this paper they reduce to the computation of $\pi_i(T_1(M)) \otimes Q$ hence of the algebraic K-theory. Both the triviality of the twisting and the computation of the homotopy groups of $T_2(M)$ being immediate.

In Chapter 2 of this paper we give the precise description of the homotopy type of Diff (M^n) (Homeo M^n) away from the "prime 2" in the stability range, reviewing our joint work with R. Lashof [6]. In Chapter 3 the

---

(*They faced the same problem as we have remarked with our Theorem 1.

Waldhausen algebraic K-theory of topological rings and its connection with our problem and in particular with the computation of the rational homotopy groups of $T_1(M)$ is briefly presented: we survey in this chapter only in part the Waldhausen theory.

In Chapter 4 we sketch part of our results [3],[4] about the computation of the algebraic K-theory tensored by the rationals for 1-connected spaces (similar results are claimed by Hsiang and Staffeldt [11]).

In Chapter 5 we sketch the proofsof Theorems 1 and 2.

Chapter II: The structure of Diff (M^n) and Homeo M^n in the stability range and away from "2"

Among the groups of diffeomorphisms and homeomorphisms, $C^D(M)$ ($C^T(M)$) the group of C^∞-diffeomorphisms (homeomorphisms) of $M\times I$ which restrict to the identity on $M\times\{0\}\cup\partial M\times I$ endowed with the C^∞-topology (compact open topology) often called differentiable (topological) concordances or pseudoisotopies[+] enjoy the following remarkable properties:

(1) Transfer and stability: For any differentiable (topological) locally trivial bundle $E^{n+k}\to M^n$ with E^{n+k}, M^n compact manifolds, there exists a well defined and natural (up to homotopy) map $C(M^n)\to C(E^{n+k})$ which is a ω (n)-homotopy equivalence if the fibre is contractable.

(2) The canonical involution: There exists a well defined and natural (up to homotopy) homotopy involution on $C(M)$. To describe this involution in the differentiable case (the description is entirely analogous in topological case) we denote by Diff $(M\times I/\partial M\times I)$ the group of C^∞-diffeomorphisms on $M\times I$ which restrict to the identity on $\partial M\times I$ (endowed with the C^∞-topology) and by $\mathrm{Diff}_s(M\times I/\partial M\times I)$ its subgroup consisting of those diffeomorphisms which commute with the projection on I. The composition $C^D(M)\to \mathrm{Diff}\,(M\times I;\partial M\times I)\to \mathrm{Diff}\,(M\times I;\partial M\times I)/\mathrm{Diff}_s(M\times I;\partial M\times I)$ is a homotopy equivalence and therefore a natural involution on Diff $(M\times I;\partial M\times I)/\mathrm{Diff}_s(M\times I;\partial M\times I)$ will produce an involution up to homotopy on $C^D(M)$. The involution on Diff $(M\times I;\partial M\times I)/\mathrm{Diff}_s(M\times I;\partial M\times I)$ is given by the conjugacy with $id_M\times\nu_I$ where ν denotes the reflection through middle point of I.

[+] Cerf was the first to notice the interest of these groups.

(3) Loop structure: C (M) has a (k+1)-loop structure if $M = N\times I^k$, in a very precise way.

These geometric properties combine and lead to Theorem 2.1 below. To state the theorem let us denote by CW the category of the topological spaces homeomorphic with finite CW-complexes, by $\overset{\infty}{\Omega}$ the category of ∞-loop spaces and by $\overset{\infty}{\Omega}{}^h$ the homotopy category of $\overset{\infty}{\Omega}$.(+

Theorem 1.A: There exists two homotopy functors ${}^D S$ and ${}^T S : CW \longrightarrow \overset{\infty}{\Omega}{}^h$ so that:

(1) For any M^n differentiable (topological) compact manifold, there exists a $\omega^D(n)+1$ $(\omega^T(n)+1)$-homotopy equivalence $i^D(M) : BC^D(M^n) \to {}^D S(M^n)$ $(i^T(M) : BC^T(M^n) \to {}^T S(M^n))$.

(2) There exists a natural transformation $\gamma(X) : {}^D S(X) \to {}^T S(X)$ whose fibre as functor from CW with values in $\overset{\infty}{\Omega}{}^h$ is the nonreduced homology theory associated to the ∞-loop space ${}^D S(pt)$.

(3) There exists two natural maps (up to homotopy) $\theta^D(X) : \underline{H}(X) \to {}^D S(X)$, $\theta^T(X) : \underline{H}(X) \to {}^T S(X)$ where $\underline{H}(X)$ denotes the topological associative monoid of simple homotopy equivalences so that the diagram below is homotopy commutative

$$\begin{array}{ccc} & \xrightarrow{\theta^D} & {}^D S(X) \\ \underline{H}(X) & & \downarrow \gamma(X) \\ & \xrightarrow{\theta^T} & {}^T S(X) \end{array}$$

(4) For any stable vector bundle (microbundle) $\xi \in \widetilde{KO}(X)$ $(\xi \in \widetilde{KT}(X))$(++ there exists an involution $\tau(\xi) : S(X) \to S(X)$ in $\overset{\infty}{\Omega}{}^h$ which decomposes

(+Technical difficulties oblige us to take as morphisms in $\overset{\infty}{\Omega}{}^h$ "visible" homotopy classes of ∞-loop space maps instead of ordinary homotopy classes.

(++One denotes by $\widetilde{KT}(X)$ the reduced K-theory based on topological microbundles.

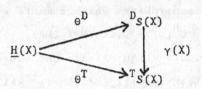

$S(X)_{odd}$ as $S^{\xi}(X)_{+} \times S_{-}^{\xi}(X)$; moreover

i) $\gamma(X)$ commutes with these involutions

ii) The "0" localization of $\tau(\xi)$, $\tau(\xi)_{(0)}: S(X)_{(0)} \to S(X)_{(0)}$ is independent of ξ; $\tau^{T}(\xi)$ depends only on the stable spherical class of ξ.

iii) $\Omega i(M): C(M^n) \to \Omega S(M^n)$ is homotopy equivariant if we consider $C(M^n)$ endowed with the natural involution and $S(M^n)$ with the involution $(-1)^n \tau(T(M^n))$.

5) The functors S satisfy a long list of properties like: (a) - strong connectivity, (b) - transfer, (c) - external products $(S(X) \times S(X') \to S(X \times X'))$, (d) - weak Kuneth property (see [6] and [17]).

The construction of the functor S can be obtained in two essentially different ways:

1) <u>the geometric construction</u>: One obtains the space $S(X)$ from the classifying spaces of concordances $BC(M^n)$ of all M^n with the homotopy type of X. This construction is done [2], [6] or [14]. The geometric construction makes clear the relationship between ${}^{D}S(M^n)$ and Diff (M^n) respectively of ${}^{T}S(M^n)$ with Homeo (M^n).

2) <u>the homotopy theoretical way</u>: (invented by Waldhausen [16]) produces the construction of these functors without using manifolds; moreover the Waldhausen construction can be connected with computable functors.

The equivalence between these definitions is also due to Waldhausen. Part of the Waldhausen results will be stated in Chapter 3.

Let us come back to the topological groups Diff (M^n) and Homeo (M^n) which from now on will be regarded as simplicial groups (replacing them with their singular simplicial groups[+]). The reason to regard Diff (M^n) and Homeo (M^n) as simplicial groups comes from our need to compare them with some bigger groups, $\widetilde{\text{Diff}}\,(M^n)$ and $(\widetilde{\text{Homeo}}\,(M^n))$ which can be defined only as simplicial groups. Their k-simplexes, see [7], are self-diffeomorphisms respectively self-homeomorphisms of $\Delta[k] \times M$ which restrict to the identity on $\Delta[k] \times \partial M$ and are face preserving; clearly Diff $(M^n) \subseteq \widetilde{\text{Diff}}\,(M^n)$ and Homeo $(M_n) \subseteq \widetilde{\text{Homeo}}\,(M^n)$ and therefore we have the principal fibrations:

$$\Omega(\widetilde{\text{Diff}}(M^n)/\text{Diff}(M^n)) \to \text{Diff}(M^n) \to \widetilde{\text{Diff}}(M^n) \dashrightarrow \widetilde{\text{Diff}}(M^n)/\text{Diff}(M^n)$$

$$(\widetilde{\text{Homeo}}(M^n)/\text{Homeo}(M^n)) \to \text{Homeo}(M^n) \to \widetilde{\text{Homeo}}(M^n) \dashrightarrow \widetilde{\text{Homeo}}(M^n)/\text{Homeo}(M$$

<u>Theorem 2.B</u> ([6],[2]). There exists a $\omega^D(n)$-homotopy equivalence $(\widetilde{\text{Diff}}(M^n)/\text{Diff}(M^n))_{odd} \to ({}^DS^\xi(M^n))_{\varepsilon(n)}$, $\xi = T(M^n)$, respectively, $(\widetilde{\text{Homeo}}(M^n)/\text{Homeo}(M^n))_{odd} \to ({}^TS^\xi(M^n))_{\varepsilon(n)}$, $\xi = T(M^n)$ so that the following diagrams are commutative

$$\begin{array}{ccc} \widetilde{\text{Diff}}(M^n) & \to & \widetilde{\text{Diff}}(M^n)/\text{Diff}(M^n) \\ \downarrow & & \downarrow \Theta^D_{\varepsilon(n)} \\ \underline{H}(M^n) & \to & ({}^DS^\xi(M^n))_{\varepsilon(n)} \end{array} \quad \text{and} \quad \begin{array}{ccc} \widetilde{\text{Homeo}}(M^n) & \to & \widetilde{\text{Homeo}}(M^n)/\text{Homeo}(M^n) \\ \downarrow & & \downarrow \Theta^T_{\varepsilon(n)} \\ \underline{H}(M^n) & \to & {}^TS^\xi(M^n) \end{array}$$

where $\varepsilon(n) = \begin{cases} + \text{ if } n \text{ is even} \\ - \text{ if } n \text{ is odd} \end{cases}$, $\Theta_{\varepsilon(n)}$ is the composite of

$\Theta : \underline{H}(M^n) \to S(M^n)$ with the projection $S(M) \to S^\xi(M^n)_{\varepsilon(n)}$ and $\xi = T(M^n)$.

[+] For Diff (M^n) we will consider as singular simplexes only the C^∞-differentiable singular simplexes.

If we take ${}^{D}T_2(M) = \widetilde{\mathrm{Diff}}\,(M^n)_{odd}$ (respectively ${}^{T}T_2(M) = \widetilde{\mathrm{Homeo}}\,(M^n)_{odd}$) and ${}^{D}T_1(M^n) = \Omega^{D}S^{\xi}(M^n)$ (respectively ${}^{T}T_1(M^n) = \Omega^{T}S^{\xi}(M^n)$) for $\xi = T(M^n)$, and t the composition $\widetilde{\mathrm{Diff}}(M^n)_{odd} \to \underline{H}(M^n)_{odd} \to {}^{D}S^{\xi}(M^n)_{\varepsilon(n)}$ (respectively $\widetilde{\mathrm{Homeo}}(M)_{odd} \to \underline{H}(M)_{odd} \to S^{\xi}(M^n)_{\varepsilon(n)}$), Theorem 2.B says that in stability range $\mathrm{Diff}\,(M)_{odd}$ (respectively $\mathrm{Homeo}\,(M)_{odd}$) occurs as a twisted product of $T_1(M)$ and $T_2(M)$ with t a twisting function. As we can see, our twisting function factors through $\Theta_{\varepsilon(n)}$, the composite of $\Theta:\underline{H}(X) \to S(X)$ and the projection $S(X) \to S^{\xi}(X)_{\varepsilon(n)}$; Θ is a homotopy invariant by Theorem 1.A, while the projection $S(X) \to S^{\xi}(X)_{\varepsilon(n)}$ $\xi = T(M^n)$ depends on the homotopy type of X as a space, the element $\xi \in \widetilde{KO}(X)$ or $\widetilde{KT}(X)$ up to a stable spherical eiquivalency and the parity of n. If X is a compact manifold with boundary, it is well known that n and ξ up to a stable spherical equivalence class are "compact manifold" homotopy invariants in the sense that they depend only on the Poincaré Duality structure of M. If we localize to "0" then $S(X) \to (S(X)_{\varepsilon(n)})_0$ depends only on the homotopy type of X and on the parity of n.

If $E(X,\xi) \xrightarrow{I_{\varepsilon(n)}} \underline{H}(X)$ is the homotopy theoretic fibre of $\underline{H}(X) \to (S^{\xi}(X))_*$ and M^n is a compact manifold then $\mathrm{Diff}\,(M^n)_{odd}$ (respectively $\mathrm{Homeo}\,(M^n)_{odd}$) in stability range are the fibre products of $I_{\varepsilon(n)}$ and $II_{(odd)}$, $II:\mathrm{Diff}(M^n) \to \underline{H}(M^n)$ (respectively $II:\mathrm{Homeo}(M^n) \to \underline{H}(M^n)$) which are basically understood by the means of surgery [1] and [15].

<u>Theorem</u> ([1],[16]). There exists the fibrations

$$H(M^n)/\widetilde{\mathrm{Diff}}(M^n) \to \mathrm{Maps}(M^n, G/O) \to \Omega^n L(\pi_1(M), \omega_1(M))$$

$$H(M^n)/\widetilde{\mathrm{Homeo}}(M^n) \to \mathrm{Maps}(M^n, G/Top) \to \Omega^n L(\pi_1(M), \omega_1(M))$$

where $H(M^n)$ denotes the space of simple homotopy equivalences of M which restrict to the identity on ∂M and $L(\pi_1(M),\omega_1)$ denotes the Quinn space whose homotopy groups $\pi_i(L(G,\omega_1))$ are the i-th surgery groups, $L_i^s(G,\omega_1)$. Here ω_1, the first Stiefel-Whitney class, is regarded as a group homomorphism $\pi_1(M) \to \mathbb{Z}_2$.

Corollary: If M is 1-connected manifold then

$$\pi_i(H(M^n)/\widetilde{\mathrm{Homeo}}(M^n)) \otimes Q = \widetilde{KO}(\Sigma^i(M \cup p)) \otimes Q$$

$$\pi_i(H(M^n)/\widetilde{\mathrm{Diff}}(M^n)) \otimes Q = KO(\Sigma^i(M \cup p)) \otimes Q.$$

Corollary 2.C. If $M^n = N^k \times D^{n-k}$ and $\xi = \tau(M^n)$, then

$$\mathrm{Diff}(M^n)_{odd} \underset{\sim}{\overset{D_{\omega(n)-1}}{\longrightarrow}} \widetilde{\mathrm{Diff}}(M) \times \Omega\ ({}^{D}S^{\xi}(M^n)_{\varepsilon(n)}) \qquad \xi = T(M^n)$$

$$\mathrm{Homeo}(M^n)_{odd} \underset{\sim}{\overset{T_{\omega(n)-1}}{\longrightarrow}} \widetilde{\mathrm{Homeo}}(M) \times \Omega\ ({}^{T}S^{\xi}(M^n)_{\varepsilon(n)}) \qquad \xi = T(M^n).$$

Proposition 2D: The composition $PGL_{n+1}(k) \to \mathrm{Diff}\ (P_n(k)) \to \underline{H}(P_n(k))$ is a rational homotopy equivalence.

Corollary 2.E.

$$\pi_i(\widetilde{\mathrm{Diff}}(P_n(k) \times D^r)) \otimes Q = \pi_i(\widetilde{\mathrm{Homeo}}(P_n(k) \times D^r)) \otimes Q =$$

$$\pi_{i+r}(PGL_{n+1}(k)) \otimes Q + \widetilde{KO}(\Sigma^{r+i+1}(P_{n-1}(k) \cup p)) \otimes Q$$

for $n \geq 3$ if $k = C$ and $n \geq 2$ if $k = H$.

Chapter III. Some results of Waldhausen

Let $R.$ be a topological (semisimplicial) ring with unit and $\pi: R. \to \pi_0(R.)$ be the projection of $R.$ on the ring of its connected components. In [17] Waldhausen associated to $R.$ an ∞-loop space $\underline{K}(R.)$ by imitating Quillen definition of the algebraic K-theory of rings; in this way he produces a functor $\underline{K}$ from the category of topological (semisimplicial) rings with values in the category $\overset{\infty}{\Omega}$. The definition of $\underline{K}(R.)$ goes as follows.

Denote by $\widetilde{GL}_n(R.)$ the associative monoid of $n\times n$ matrices $\{a_{ij}\}$ with $\{\pi a_{ij}\}$ an invertible matrice, and define $\widetilde{GL}_\infty(R.) = \lim_{n\to\infty} \widetilde{GL}_n(R.)$ (with respect to the obvious inclusion $\widetilde{GL}_n(R.) \to \widetilde{GL}_{n+1}(R.)$ given by $\{a_{ij}\} \to \{\begin{array}{c|c} a_{ij} & 0 \\ \hline 0 & 1 \end{array}\}$). Since $\widetilde{GL}_\infty(R.)$ is an associative H-space one can consider $B\widetilde{GL}_\infty(R.)$ and remark that $\pi_1(B\widetilde{GL}_\infty(R.)) = GL_\infty(\pi_0(R.))$ is a group whose commutator is perfect, therefore one can apply the Kervaire-Quillen "+"-construction; one defines $\underline{K}(R.)$ as $B\widetilde{GL}_\infty(R)_+ \times K_0(\pi_0(R.))$; here $K_0(A)$ denotes the Grothendieck K_0-group of the discrete ring A. It turns out that $\underline{K}(R.)$ is an ∞-loop space and the correspondence $R. \rightsquigarrow \underline{K}(R.)$ is a functor. If $R.$ comes equipped with a continuous (semisimplicial) antiinvolution, one obtains an ∞-loop space involution on $\underline{K}(R.)$. If G_r is the category of semi-simplicial groups and S Ring the category of semi-simplicial rings, the functor $Z(\ ): G_r \rightsquigarrow$ S Ring, associates with G the semi-simplicial group ring $Z(G)$; $Z(G)$ has a natural antiinvolution produced by the anti-involution $g \rightsquigarrow g^{-1}$ in G. If Top_* is the category of the based point connected semi-simplicial complexes, the free group version of the loop space functor Ω, invented by D. Kan, defines a functor $G: Top_* \rightsquigarrow G_r$ and with the previous construction one obtains the functor $\underline{K}: Top_* \rightsquigarrow$ {The category of ∞ loop spaces with involution} given by $\underline{\underline{K}}(X) = \underline{K}(Z(GX))$.

The functor $\underline{\underline{K}}$ is a homotopy functor which is strongly connected in the sense that it transforms a k-cartesian (m,n)-connected diagram in a k-cartesian (m,n)-connected diagram. Recall that by a k-cartesian (m,n)-connected diagram we understand a commutative diagram

$$\begin{array}{ccc} X_{11} & \xrightarrow{\beta_1} & X_{12} \\ \downarrow{\scriptstyle \alpha_1} & & \downarrow{\scriptstyle \alpha_2} \\ X_{21} & \xrightarrow{\beta_2} & X_{22} \end{array}$$

with horizontal arrows m-connected, vertical arrows n-connected and the induced map (Homotopy fibre α_1) $\to$ (Homotopy fibre α_2) (k+1)-connected. Clearly $\underline{\underline{K}}(X)$ decomposes naturally as $\underline{\underline{K}}(X) = \widetilde{\underline{\underline{K}}}(X) \times \underline{\underline{K}}(*)$ where $\widetilde{\underline{\underline{K}}}(X) = \{$homotopy fibre of $\underline{\underline{K}}(X) \to \underline{\underline{K}}(*)\}$ with $\underline{\underline{K}}(X) \to \underline{\underline{K}}(*)$ induced by the projection of X on its base point.

There exists a natural unreduced homology theory(+ $\underline{\underline{K}}^s$, the "completion" of $\underline{\underline{K}}$, and a natural transformation $\alpha : \underline{\underline{K}}(X) \to \underline{\underline{K}}^s(X)$; the pair $(\underline{\underline{K}}^s, \alpha)$ is "universal" in an obvious sense. Actually $\alpha(X) : \underline{\underline{K}}(X) = \widetilde{\underline{\underline{K}}}(X) \times \underline{\underline{K}}(*) \to \widetilde{K}^s(X) \times K^s(*)$ is $\alpha(X) = \widetilde{\alpha}(X) \times \alpha(*)$, $\widetilde{\alpha}(X) : \widetilde{\underline{\underline{K}}}(X) \to \widetilde{\underline{\underline{K}}}^s(X)$ respectively $\alpha(*) : \underline{\underline{K}}(*) \to \underline{\underline{K}}^s(*)$ being defined as follows: The diagram $\begin{array}{ccc} X & \to & CX \\ \downarrow & & \downarrow \\ CX & \to & \Sigma X \end{array}$ induces $\widetilde{\underline{\underline{K}}}(X) \xrightarrow{s(X)} \Omega\widetilde{\underline{\underline{K}}}(\Sigma X)$; we take $\widetilde{K}^s(X) = \underrightarrow{\lim}_n (\widetilde{\underline{\underline{K}}}(X) \xrightarrow{s(X)} \Omega\widetilde{\underline{\underline{K}}}(\Sigma X) \xrightarrow{\Omega s(\Sigma X)} \Omega^2\widetilde{\underline{\underline{K}}}(\Sigma^2 X) \to \dots)$ and $\widetilde{\alpha}(X)$ the canonical map $\widetilde{\underline{\underline{K}}}(X) \to \underrightarrow{\lim}_r \Omega^r \widetilde{K}(\Sigma^r X)$

$\underline{\underline{K}}^s(*) = \underrightarrow{\lim} \{\underline{\underline{K}}(*) \xrightarrow{\alpha_0} \Omega\widetilde{\underline{\underline{K}}}(S^1) \xrightarrow{\Omega s(S^1)} \Omega^2\widetilde{K}(S^2) \to \dots\}$ where α_0 is a familiar map to algebraic K-theorists

(+Here we regard a homology theory as a homotopy functor $L : Top_* \rightsquigarrow \Omega^h$ wh satisfies the Meyer-Vietoris condition; $X \rightsquigarrow \pi_i(L(X))$ defines what is usually called a generalized homology theory.

$\alpha_0 = \underline{\underline{K}}(*) = \underline{\underline{K}}(Z) \to \widetilde{\Omega K}(Z[t,t^{-1}]) = \widetilde{\Omega K}(Z(Z)) = \widetilde{\Omega \underline{\underline{K}}}(S^1)$ defined for instance in [14] and $\alpha(*)$ is the canonical map $\underline{\underline{K}}(*) \to \varinjlim_r \Omega K(S^r)$.

The involution we have started with produces an involution on $\underline{\underline{K}}^s(X)$ so that the natural transformation is equivariant.

If we denote by $\mathcal{W}h^D(X)$ the homotopy theoretic fibre of $\alpha(X):\underline{\underline{K}}(X) \to \underline{\underline{K}}^s(X)$ one obtains a new functor endowed with involution.

<u>Theorem 3.1</u>. 1) (Waldhausen [17]) There exists a natural transformation $S^D \rightsquigarrow \Omega\, \mathcal{W}h^D$ which for any X in Top_* is a rational homotopy equivalence.

2) (Hsiang-Jahren)(+ This natural transformation commutes up to homotopy with the involution defined on $\mathcal{W}h^D(X)$ and the involution associated with any $\xi \in \widetilde{KO}(X)$ on $S^D(X)$ if the first Stiefel Whitney class of ξ is trivial.

There exists a second way to approximate the functor $\underline{\underline{K}}$ with an unreduced homology theory, more precisely, there exists a natural transformation $\beta:h(\ldots;\underline{\underline{K}}(*)) \rightsquigarrow \underline{\underline{K}}(\ldots)$ from the unreduced homology theory given by the ∞-loop space $\underline{\underline{K}}(*)$ regarded as an Ω-spectrum with 0-component $\underline{\underline{K}}(*)$ to $\underline{\underline{K}}$. This natural transformation is again compatible with the involution where the involution on $h(X:\underline{\underline{K}}(*))$ is the one induced by the involution on $\underline{\underline{K}}(*)$. This natural transformation is well known in algebraic K-theory and for the case of $X = K(G,1)$, it is described in [14].

$\beta(X)$ is an infinite loop space map so if we define $\Omega \mathcal{W}h^T(X)$ as the homotopy fibre of $\beta(X)$ one can deloop once and produce the functor $\mathcal{W}h^T$.

(+ The proof of this statement is not yet available.

Theorem 3.2 (1) (Waldhausen) There exists a natural transformation $S^T \rightsquigarrow \Omega Wh^T$ which for any X in Top_* is a rational homotopy equivalence.

(2) (Hsiang-Jahren)[+] This natural transformation commut up to homotopy with the involution defined on $Wh^T(X)$ and any involutio associated with $\xi \in \widetilde{KT}(X)$ on $S^T(X)$ if the first Stiefel Whitney class of ξ is trivial.

[+] The proof of this statement is not yet available.

Chapter IV: Rational computations of the Waldhausen algebraic K-theory

The main future of $\underline{\underline{K}}(X)$ comes from the possibility of computing its homotopy groups tensored by the rationals and therefore the homotopy groups tensored by the rations of $Wh^D(X)$ and $Wh^T(X)$, at least for the class of simple spaces whose fundamental group is finite abelian. Our statements will consider only the case of 1-connected spaces for which we have worked out some computations. Of course our attention will be focused on the computations which are necessary for proving Theorems 1 and 2.

It is convenient to state the results in terms of topological (semisimplicial) rings $R.$ which satisfy the following conditions:

a) $\pi_0(R.) = Z$ and the ring has a unit, i.e. there exists $i:Z \to R.$ a ring homomorphism with $\pi i = id$

b) $r_i = \dim \pi_i(R.) \otimes Q < \infty$ for all i

c) this is a quite complicated requirement (may be not necessary) which is fulfilled by the semisimplicial rings which are free in any degree, for instance by $Z(GX)$.

If X is a 1-connected base pointed semisimplicial complex with $\dim \pi_i(X) \otimes Q < \infty$, clearly $Z(GX)$ satisfies the above conditions since $\pi_i(ZGX) \otimes Q = H_i(GZ:Q)$ whose dimension can be expressed in terms of $\dim \pi_k(X) \otimes Q$, $2 \leq k \leq i+1$.

Let us consider the following commutative diagram whose horizontal lines are fibrations.

$$\begin{array}{llcccc}
(*) & T(R.) & \longrightarrow & B\widetilde{GL}_+(R.) & \underset{B\tilde{i}_+}{\overset{B\tilde{\pi}_+}{\rightleftarrows}} & BGL_+(Z) \\
 & \uparrow & & \updownarrow & & \uparrow \\
(**) & BM_\infty(\mathring{R}.) & \xrightarrow{Bi_\infty} & B\widetilde{GL}(R.) & \underset{B\tilde{i}}{\overset{B\tilde{\pi}}{\rightleftarrows}} & BGL(Z)
\end{array}$$

where $\mathring{R}.$ denotes the 0-connected component of $R.$, $M_\infty(\mathring{R}.) = \varinjlim_n M_n(\mathring{R}.)$; $M_n(\mathring{R}.)$ is the associative H-space of $(n\times n)$ matrices with coefficients in $\mathring{R}.$ endowed with the composition $M*N = M + N + M\cdot N$ and $M_n(\mathring{R}.)$ is embedded in $M_{n+1}(\mathring{R}.)$ by $|M| \longmapsto \left|\begin{array}{c|c} M & 0 \\ \hline 0 & 0 \end{array}\right|$. We define for $n = 1,\ldots,\infty$, $i_n : M_n(\mathring{R}.) \to \widetilde{GL}_n(R.)$ by $i_n(M) = M + I$. $\tilde{i}$ is the crossection induced by $i: Z \to R.$, and $T(R.)$ is the homotopy fibre of $B\tilde{\pi}_+$.

The fibrations $BM_n(\mathring{R}.) \xrightarrow{Bi_n} B\widetilde{GL}_n(R.) \xrightarrow{B\tilde{\pi}_n} BGL_n(Z)$ are the fibrations associated with the action $\bar{\rho}_n : GL_n(Z) \times BM_n(\mathring{R}.) \to BM_n(\mathring{R}.)$ obtained by delooping the adjoint action $\rho_n : GL_n(Z) \times M_n(\mathring{R}.) \to M_n(\mathring{R}.)$ with $\rho_n(A,M) = \tilde{i}(A)\cdot M\cdot \tilde{i}(A^{-1})$. The canonical base point of $BM_n(\mathring{R}.)$ is left fixed by $\bar{\rho}_n$, consequently $\pi_i(BM_n(\mathring{R}.)) \otimes Q$ are linear representations of $GL_n(Z)$, precisely $\rho_n \oplus \rho_n \oplus \ldots \oplus \rho_n$ where ρ_n is the adjoint representation of $GL_n(Z)$ on $M_n(Q)$, the vector space of $n\times n$ matrices with coefficients in Q. For a linear representation ξ of the group G, we denote by $\operatorname{Inv}\xi$ the biggest trivial subrepresentation of ξ and by $\operatorname{Cov}\xi$ the quotion representation ξ/η where η is the smallest subrepresentation of ξ so that ξ/η is trivial. Clearly $\operatorname{Inv}\xi \to \xi \to \operatorname{Cov}\xi$ are morphisms of representations and if ξ is semisimple (direct sum of finite dimensional irreducible representations, the composition $\operatorname{Inv}\xi \to \operatorname{Cov}\xi$ is an isomorphism.

<u>Theorem 4.1</u>. If $R.$ satisfies the conditions a), b) and c) then:

1) $T(R.)$ is rationally homotopy equivalent to a

product of Eilenberg-MacLane spaces

2) $\phi_*: H_*(B\overset{\circ}{M}_\infty(\overset{\circ}{R}):Q) \to H_*(T(R.):Q)$ and $\phi^*: H^*(T(R.):Q) \to H^*(BM_\infty(\overset{\circ}{R}.):Q)$ identify to the canonical morphisms $H_*(BM_\infty(\overset{\circ}{R}):Q) \to \mathrm{Cov}(H_*(B\widetilde{M}_\infty(R.):Q))$ and $\mathrm{Inv}\,(H^*(BM_\infty(\overset{\circ}{R}.):Q)) \to H^*(BM_\infty(\overset{\circ}{R}.):Q)$.

3) $GL_n(Z)$ acts semisimply on the rational homotopy type of $BM_n(\overset{\circ}{R})$ (i.e. one can produce an action on the Sullivan's minimal model which describes the rational homotopy type of $BM_n(\overset{\circ}{R}.)$ which regarded as a representation is semisimple; moreover these actions are compatible with n)

4) there exist the polynomials $R_i(x_1,\dots,x_{i-1})$ so that $\dim \pi_i(T(\overset{\circ}{R})) \otimes Q \le R_i(r_1,\dots,r_{i-1})$ where $r_i = \dim \pi_i(R.) \otimes Q$.

b) The inequalities given in d) are equalities for $R. = ZG(K(Z,2n))$ in which case $\pi_i(T(R.)) \otimes Q = Q$ if $i = 2ns$ and 0 elsewhere.

There are many other cases in which inequalities stated in d) are equalities as also cases in which theyare not and in fact we can improve d) to rather complicated formulas which involve Whitehead products of $BM_1(\overset{\circ}{R})$.

The proof of Theorem 4.1(+ is long and it will be described in the forthcoming paper [4]. However for the purpose of Theorems 1 and 2 we need only (S) and the finiteness of $\dim \pi_i(B\widetilde{GL}_+(R.) \otimes Q$ which is contained in [3] or indicated below.

The finiteness of $\dim \pi_i(B\widetilde{GL}_+(R.)) \otimes Q$ is a consequence of the finiteness of $\dim (H_i(B\widetilde{GL}_+(R.):Q)$ which is a consequence of the finiteness of $\dim H_r(BGL(Z);\{H_i(BM_\infty(\overset{\circ}{R}.):Q)\})$ through Serre-spectral sequence.

(+Similar results have been announced by Hsiang and Staffeld following a slightly different approach from mine. (It turns out that in order to feel a gap in my arguments for 3) I must use Quillen's rational homotopy theory in a way which was mentioned to me by Hsiang).

of the fibration (**). To verify the last finiteness, one uses the Milnor-Rothenberg-Steenrod spectral sequence ${}^{(n)}E^{i}_{p,q}$ which computes the homology of $H_*(BM_n(\mathring{R}.);Q)$ and have

$${}^{(n)}E^{1}_{p,q} = \sum_{i_1+i_2+\dots i_q=q} \tilde{H}_{i_1}(M_n(\mathring{R}.);Q) \otimes \tilde{H}_{i_2}(M_n(\mathring{R}.);Q) \otimes \dots \tilde{H}_{i_q}(M_n(\mathring{R}.);Q)$$

$GL_n(Z)$ acts on the n-th spectral sequence and all representations ${}^{n}E^{i}_{p,q}$ are semisimple. The classical H. Weyl invariant theory implies $\dim {}^{\infty}E^{1}_{p,q} < \infty$ and the vanishing theorem [8] implies $\dim H_r(BGL_\infty(Z):\{H_i(BM_\infty(\mathring{R}.);Q)\}) \leq \sum_{p+q=i} \dim H_r(BGL_\infty(Z);Q) \otimes$ $\mathrm{Inv}\, {}^{\infty}E^{\infty}_{p,q} \leq \sum_{p+q=i} \dim (H_r(BGL_\infty(Z);Q) \otimes \mathrm{Inv}\, {}^{\infty}E^{1}_{p,q})$. The dimension of $H_r(BGL_\infty(Z);Q)$ was computed by A. Borel, $\mathrm{Inv}\, {}^{\infty}E^{1}_{p,q} = \mathrm{Inv}\, {}^{n}E^{1}_{p,q}$ for $n >> p,q$ and $\dim (\mathrm{Inv}\, {}^{n}E^{1}_{p,q})$ is finite and can be calculated using the invariant theory in terms of r_i.

In [3], $\underline{\underline{K}}(X)$ for $X = K(Z,2r)$ is explicitly computed after localization at zero and as a consequence of [3] and the theorem above we have:

<u>Corollary 4.2</u>: If $X = K(Z,2r)$, then

$$\pi_i(\mathcal{W}h^D(X)) \otimes Q = \pi_i(\mathcal{W}h^D(*)) \otimes Q = \begin{cases} Q & 4k+1,\ k \geq 1 \\ 0 & \text{elsewhere} \end{cases}$$

$$\pi_i(\mathcal{W}h^D_{\varepsilon}(X)) \otimes Q = \pi_i(\mathcal{W}h^D_{\varepsilon}(*)) \otimes Q = \begin{cases} Q & i = 4k+1,\ k \geq 1,\ \varepsilon = - \\ 0 & \text{elsewhere} \end{cases}$$

$$\pi_i(\mathcal{W}h^T(X)) \otimes Q = \tilde{h}_{i+1}(X:\underline{\underline{K}}(*)_{(0)}) = \tilde{H}_{i+1}(X;Q) \oplus \sum_{k=1}^{\infty} \tilde{H}_{i-4k}(X;Q)$$

$$\pi_i(\mathcal{W}h^T_+(X)) \otimes Q = \tilde{h}_{i+1}(X;\underline{\underline{K}}_+(*)_{(0)}) = \tilde{H}_{i+1}(X;Q)$$

$$\pi_i(\mathcal{W}h^T_-(X)) \otimes Q = \tilde{h}_{i+1}(X;\underline{\underline{K}}_-(*)_{(0)}) = \sum_{k=1}^{\infty} \tilde{H}_{i-4k}(X;Q)$$

2) If X is a 1-connected CW-complex with $\dim \pi_i(X) \otimes Q < \infty$

then $\dim \pi_i(Wh^D(X)) \otimes Q < \infty$ and $\dim \pi_i(Wh^T(X)) \otimes Q < \infty$.

Recall that the involution we have mentioned decomposes $Wh_{(odd)}(X)$ as $Wh_+(X) \times Wh_-(X)$ the symmetric respectively antisymmetric parts (up to homotopy).

Recall also from [16] that $\pi_i(\underline{\underline{K}}(*)_{(0)}) = \begin{cases} Q & i=0 \text{ and } 4k+1,\ k > 1 \\ 0 & \text{elsewhere} \end{cases}$

and from [8] that $\pi_i(Wh_+^D(*)) \otimes Q = 0$, $\pi_i(Wh_-^D(*)) \otimes Q = Q$ for $i = 4k+1$, $k \geq 1$ and 0 elsewhere.

Chapter V. Proof of Theorems 1 and 2

Theorems 1 and 2 are immediate consequences of the results discussed in the previous sections.

Proof of Theorem 1. If $M^{2un+r} = P_n(k) \times D^r$ and $\xi = T(P_n(k))$ th $\mathrm{Diff}\,(M^{2un+r})_{(0)} \overset{\omega(2un+r)}{\sim} \widetilde{\mathrm{Diff}}\,(M^{2un+r})_{(0)} \times \Omega^D S^{\xi}_{\varepsilon(2un+r)}(P_n(k))_{(0)}$. For $r \geq 1$ this follows from Corollary 2.6. For $r = 0$, Proposition 2.D implies that $\mathrm{Diff}\,(P_n(k))_{(0)} \to \underline{H}(P_n(k)) = \underline{H}(P_n(k))_{(0)}$ is surjective in the homotopy category and since the composition $\mathrm{Diff}\,(P_n(k)) \to \widetilde{\mathrm{Diff}}\,(P_n(k)) \to \underline{H}(P_n(k)) \to \Omega^D S^{\xi}_{\varepsilon(2un)}(P_n(k))$, $\varepsilon(2un) = +$ is $(\omega(2un)-1)$-homotopy trivial then $\underline{H}(P_n(k))_{(0)} \to \Omega(^D S^{\xi}_{+}(P_n(k))_{(0)}$ is. This implies the composition $\widetilde{\mathrm{Diff}}\,(P_n(k))_{(0)} \to \underline{H}(P_n(k))_{(0)} \to \Omega^D S^{\xi}_{+}(P_n(k))_{(0)}$ is $(\omega(2un)-1)$-homotopy trivial which by Theorem 2.B implies the statement.

$\widetilde{\mathrm{Diff}}\,(M^{2un+r})_{(0)}$ is calculated in Corollary 2.E and $\Omega^D S^{\xi}_{+}(P_n(k))_{(0)}$ is homotopy equivalent to $\Omega^2 \mathcal{W}h^D_{\pm}(P_n(k))_{(0)}$ (by Theorem (3.1)) which is (2un-2)-homotopy equivalent to $\Omega^2 \mathcal{W}h^D_{\pm}(K(Z,2u)_{(0)}$ which is calculated in Corollary 4.2. The proof is similar for Homeo (M^n).

Proof of Theorem 2. We consider the fibration

$\Omega\,\widetilde{\mathrm{Diff}}\,(M)/\mathrm{Diff}\,(M) \to \mathrm{Diff}\,(M) \to \widetilde{\mathrm{Diff}}\,(M)$ and

$\Omega\,\widetilde{\mathrm{Homeo}}\,(M)/\mathrm{Homeo}(M) \to \mathrm{Homeo}\,(M) \to \widetilde{\mathrm{Homeo}}\,(M)$.

Since for a 1-connected compact manifold M, $\dim \pi_i(\mathrm{Diff}\,(M^n) \otimes Q$ respectively $\dim \pi_i(\widetilde{\mathrm{Homeo}}\,(M^n)) \otimes Q < \infty$, by Theorem (ch 2), and in stability range $\dim \pi_i(\Omega\,\mathrm{Diff}\,(M)/\mathrm{Dim}\,(M)) \otimes Q$ or $\dim \pi_i(\Omega\,\widetilde{\mathrm{Homeo}}\,(M)/\mathrm{Homeo}\,(M)) \otimes Q$ are finite as consequence (by Theorem 3.1) of the finiteness of $\dim \pi_i(\underline{\underline{K}}(M)) \otimes Q$ stated in Corollary 4.2, the result follows.

References

[1] P. Antonelli, D. Burghelea, P. Kahn, Concordance homotopy groups of geometric automorphism groups, Lecture Notes in Math. Vol. 215, Springer-Verlag.

[2] D. Burghelea, Automorphisms of manifolds, Proceedings of Symposia in Pure Mathematics, Vol. 32, pp. 347-371.

[3] ____________, Some rational computations of the Waldhausen algebraic K-theory, Comment. Math. Helv., 1978 (to appear).

[4] ____________, Algebraic K-theory of topological spaces, rational computations (in preparation).

[5] D. Burghelea, R. Lashof, Stability of concordances and suspension homeomorphism, Ann. of Math., Vol. 105 (1977), pp. 449-472.

[6] ____________________, The homotopy type of the groups of automorphisms of compact manifolds in stability range and away of "2" (to be published).

[7] D. Burghelea, R. Lashof, M. Rothenberg, Groups of automorphisms of compact manifolds, Lecture Notes in Math., Vol. 473, Springer-Verlag, Berlin-New York, Vol. 473.

[8] T. Farrell, W.C. Hsiang, On the rational homotopy groups of the diffeomorphism groups of discs, spheres and aspherical manifolds, Proceedings Symposia in Pure Mathematics, Vol. 32, pp. 325-337.

[9] W.C. Hsiang, B. Jahren, On the homotopy groups of the diffeomorphism groups of spherical space forms (preprint).

[10] W.C. Hsiang, R. Staffeldt, to be published.

[11] W.C. Hsiang, R.W. Sharpe, Parametrized surgery and isotopy, Pacific Journal of Mathematics, 67 (1976), pp. 401-459.

[12] A. Hatcher, Concordance spaces, Proceedings Symposia in Pure Mathematics, Vol. 32, 1978.

[13] __________, Higher simple homotopy theory, Ann. of Math. 102 (1975), pp. 101-137.

[14] J.L. Loday, K-theorie algebrique et representations des groupes, Ann. Sci. Ecole Normale Sup. 9 (1976), pp. 309-377.

[15] F. Quinn, Thesis, Princeton Univ., Princeton, N.J., 1969.

[16] F. Waldhausen, Algebraic K-theory of topological spaces, I, Proceedings Symposia in Pure Math, Vol. 32, pp. 35-60.

[17] F. Waldhausen, Mathematical Arbeitstagung 1978, Bonn (Summary of the lecture).

INCREST - Bucharest, Romania
Rutgers University, New Brunswick, N.J., USA

A COUNTEREXAMPLE ON THE OOZING PROBLEM FOR CLOSED MANIFOLDS

by Sylvain E. Cappell and Julius L. Shaneson

In classifying manifolds in a fixed homotopy type the procedure is conveniently broken into two stages :

1) Tangential information given by the "normal invariant".

2) algebraic K-theoretic information given by the surgery obstruction.

It is then all important to see how the non-simply connected surgery obstruction arises from the normal invariant. The normal invariant can be read off from the simply connected surgery obstructions, e.g. signatures and Kervaire-Arf invariants, along (generalized) submanifolds [9] ; we may then ask for a description of how these contribute "upwards" to the non-simply connected surgery obstruction. This problem has been aptly called by John Morgan the oozing problem. It is particularly important for closed manifolds where the possible surgery obstructions are more limited. This has special consequences for the classification of closed manifolds (see [5]) .

For infinite groups, the oozing problem is related to the Novikov higher signature conjecture and an implicit solution for many classes of groups is given in [2] [3] [4] .

For finite groups, the surgery obstructions of closed manifolds are quite special. They can be viewed as the image of a bordism group and are all, apart from the classical signature invariant, 2-torsion. Indeed, as according to Wall they can be detected by transfer to the 2-Sylow subgroup, the oozing problem is reduced to the case of 2-groups. When these are <u>abelian</u> , a complete description has been given by Morgan and Pardon [6] ;

for closed manifolds, the surgery obstruction can be read off from the simply-connected obstructions along (generalized) submanifolds of codimension 0 and 1, and sometimes 2. It was hoped that a similar situation would pertain for non-abelian 2-groups.

In this note, we check that the surgery obstructions that arise from closed manifolds with fundamental group a generalized quaternion group are more interesting. The oozing problem for them reaches down to codimension three.

It is well-known, and indeed obvious from cobordism-theoretic considerations, that the critical case is the surgery obstruction [1] $\sigma(f) \in L_5(\pi_1 M)$ of the map :

$$f : M^3 \times T^2 \longrightarrow M^3 \times S^2$$

where M is a 3-dimensional space form with $\pi_1 \cong G$ and G contains Q, the quaternion group of order 8. Here T^2 is the torus $S^1 \times S^1$ equipped with the Lie group invariant framing. This map f has several interesting features :

1) The non-triviality of the normal invariant of f is not detectable by simply connected surgery obstructions along (generalized) submanifolds of codimension 0, 1, or 2. It is, of course detected by the codimension 3 Kervaire-Arf invariant of the map $p \times T^2 \longrightarrow p \times S^2$, $p \in M$.

2) Besides performing surgery on f to replace it by a map inducing an isomorphism on π_1, it is also possible to furthermore make f_*, the induced map on homology with local coefficients, which is onto, have kernel just a finite group of odd order.

However, we check here that :

<u>Proposition</u> : $\sigma(f) \neq 0$

---

[1] All the results of this paper apply equally to the L_n^S or L_n^h theories. In fact the computations will be done over Q and $Wh(\pi_1 Q) = 0$.

We use a decomposition of M and peel it, layer by layer, until we reach codimension 3. Of course, using the transfer map associated to $Q \subset G$, to check that $\sigma(f) \neq 0$ it suffices to consider only the case when M is the 3-dimensional space form with

$$\pi_1 M \cong Q \cong \{x,y | x^2 = y^2 \ , \ yxy^{-1} = x^{-1}\} \ ,$$

the quaternion group of order 8.

We use the following description of this M. Let K be a Klein bottle with $\pi_1(K) = \{X,Y | YXY^{-1} = X^{-1}\}$ and with the orientation character φ of K given by $\varphi(X) = -1$, $\varphi(Y) = +1$. There is a unique $I = [-1,+1]$ bundle over K with oriented total space E; the boundary of E is a torus with fundamental group generated by Y^2 and X. Then $M \cong E \cup_{S^1 \times S^1} D^2 \times S^1$, where the $D^2 \times S^1$ is attached to kill $Y^2 X^{-2} \in \pi_1(S^1 \times S^1)$. To see this recall that M is the quotient of $S^3 = \{(z_1,z_2) \in \mathbb{C}^2 \mid |z_1|^2 + |z_2|^2 = 1\}$ by the action of the quaternion group $\{\pm 1, \pm i, \pm j, \pm k\}$; the inclusion $i : K \longrightarrow M$ is the quotient of the torus $T^2 \subset S^3$ where $T^2 = \{(z_1,z_2) \in \mathbb{C}^2 \mid |z_1| = \frac{1}{2} \ , \ |z_2| = \frac{1}{2}\}$.

Remark on $\mathrm{Diff}(M)$: This decomposition of M, and the analogous decomposition into E^3 and $D^2 \times S^1$ for any generalized quaternion space form in dimension three, can be used to show that in many ways these spaces behave like the "sufficiently large" 3-manifolds. For example, by keeping track of the image of the Klein bottle K, one can see, as was also observed by H. Rubinstein [1] [8] that for any M possessing such a decomposition into E and a solid torus $\pi_0(\mathrm{Aut}(M)) \cong \pi_0(\mathrm{Diff}(M))$, $\mathrm{Aut}(M)$ = the space of auto-homotopy equivalences of M. However, unlike the sufficiently large 3-manifolds, one can check that $\mathrm{Aut}(M)$ and $\mathrm{Diff}(M)$ are not homotopy equivalent.

---

1 He also has results on yet larger classes of 3-manifolds.

Remarks : 1) The methods of [5] can be used to often show that many elements of surgery groups do not arise from closed manifolds.

2) Recall [5] that elements that do not arise from closed manifolds act freely on sets of manifolds in a fixed homotopy type.

3) The present result gives some interesting examples of product formulas and provides, for example, a non-trivial element in Ranicki's [7] group $L^3(\pi_1 M)$.

4) The result on diffeomorphism groups referred to above, combined with some analysis of the desuspension invariants of [5] , lead to computable invariants which are obstructions to 4-dimensional s-cobordisms with fundamental group a generalized quaterniom group being a product. Realization for these invariants is not known, but is "predicted" by analogous high-dimensional surgery theory.

5) The proof of the proposition, which geometrically is by peeling away the manifold layer by layer, is of independent interest and should be usable in many contexts. It suggests that they might be an algebraic version of this, involving tracing through something like a spectral sequence for L-groups of G where $1 \longrightarrow H \longrightarrow G \longrightarrow \mathbb{Z}_2 \longrightarrow 1$ is exact, begining with suitable L groups over H of the kind discussed in [5] .

Problem : What is this "spectral sequence" ?

The proof uses results and methods of § 1 of [5] ; references and more details of the results quoted may be found there. When writing surgery groups, we omit reference to orientation characters when these are trivial.

Proof of the Proposition : The element $\sigma(f)$ determine an element , which is also denoted $\sigma(f)$ in $L_5(\mathbb{Q},\mathbb{Z}_4)$, where $\mathbb{Z}_4$ is the subgroup of Q generated by $x = i_*(X)$; that is, we have a short exact sequence

$$1 \longrightarrow \mathbb{Z}_4 \longrightarrow Q \xrightarrow{\psi} \{\pm 1\} \longrightarrow 1 \ ,$$

where $\psi(x) = +1$ and $\psi(y) = -1$. From the above description of M , $\sigma(f)$ is readily seen to be just $\psi!(\sigma(g))$, where :

$$g : K \times T^2 \longrightarrow K \times S^2$$

is the surgery problem with $K \subset M$ and T^2 framed as before. (Recall from section 1 of [5] that $\psi!$ is represented by the appropriate surgery problem produced by an I-bundle over g). Here $\sigma(g)$ is the surgery obstruction of g evaluated not in $\pi_1 K$ but, under i_*, in $\pi_1 Q$, $i:K \subset M$. We proceed to show :

1) $\psi! \ L_4(Q;\psi) \longrightarrow L_5(Q,\mathbb{Z}_4)$ is injective, and

2) $\sigma(g) \neq 0$ in $L_4(\pi_1 Q;\psi)$.

<u>Note</u> : It is easy to see, using for example the methods of [2], [3], [4], that g represents a non-zero element in $L_4(\pi_1 K;\psi i_*)$ but this is not as direct in $L_4(\pi_1 Q;\psi)$.

From [5] , the kernel of $\psi!$ is the image of :

$$j : L_4^{\psi}(\mathbb{Z}_4) \longrightarrow L_4(Q;\psi) \ .$$

Here $L_4^{\psi}(\mathbb{Z}_4)$ is the L_4 group of the theory associated to the ring with antistructure triple $(\mathbb{Z}[\mathbb{Z}_4], I, -x^2)$, x the generator of $\mathbb{Z}_4$ and I = Identity conjugation. That $L_0^{\psi}(\mathbb{Z}_4) = 0$ can be seen in a number of maps ; for example this follows from the long exact sequence of the pull-back diagram [11].

$$\begin{array}{ccc}
(\mathbb{Z}[\mathbb{Z}_4], I, -x^2) & \longrightarrow & (\mathbb{Z}(i), I, 1) \\
\Big\downarrow & & \Big\downarrow \\
(\mathbb{Z}[\mathbb{Z}_2], I, -1) & \longrightarrow & (\mathbb{Z}_2[\mathbb{Z}_2], I, 1)
\end{array}$$

as $L_4(\mathbb{Z}(i), I, 1) = 0$, and the L_4 of the lower 2 terms are both $\mathbb{Z}_2$, given by the Arf invariant, and $L_1(\mathbb{Z}_2[\mathbb{Z}_2], I, 1) = 0$, see [12].

To show that $\sigma(g) \neq 0$, we repeat the procedure. In fact, we show $\sigma(g) \neq 0$ in $L_4(Q,\mathbb{Z}_4;\psi)$; here $\mathbb{Z}_4$ is the subgroup of Q generated by $y = i_*(Y)$ so that $\psi(\mathbb{Z}_4) = \{\pm 1\}$. Now, letting C be an embedded curve on K representing $XY \in \pi_1(K)$ and letting $h : C \times T^2 \longrightarrow C \times S^2$, with associated surgery obstruction <u>evaluated in</u> $L_3(Q;\rho\cdot\psi)$, with $\rho : Q \longrightarrow \{\pm 1\}$ with kernel $\mathbb{Z}_4$, we have,

$$\sigma(g) = \rho!(\sigma(h)) ,$$

where $\rho! : L_3(Q;\rho\cdot\psi) \longrightarrow L_4(Q,\mathbb{Z}_4;\psi)$.

Using again the exact sequence of [5], to see that $\sigma(g) \neq 0$ we show that :

1) $\rho!$ is injective

2) $\sigma(h) \neq 0$.

To see that $\rho!$ is injective, identify as above $Ker(\rho!) = Im\, j$,

$$j : L_3(\mathbb{Z}[\mathbb{Z}_4], D, \pm y^2) \longrightarrow L_3(Q,\rho\cdot\psi) \text{ where } \mathbb{Z}_4 \text{ is}$$

generated by y and D is the conjugation with $D(y) = -y$. Now a quadrant of rings, similar to the above and again the simplest results of [11] show again that $L_3(\mathbb{Z}[\mathbb{Z}_4], D, \pm y^2)$ is 0. The checking is left as an exercise.

Lastly, to check that $\sigma(h) \neq 0$, recall that

$$\mathbb{Z}_2 \cong L_2(e) \longrightarrow L_3(\mathbb{Z}) \longrightarrow L_3(\mathbb{Z}_4)$$

are isomorphisms with the first map being multiplication by S^1 and the second the reduction mod 4. Thus $\sigma(h)$ is the image of the non-trivial element of $L_3(\mathbb{Z}_4)$ and we finish by showing :

<u>Lemma</u> : $L_3(\mathbb{Z}_4) \rightarrow L_3(Q;\nu)$ <u>is injective</u>.

Here, $\nu : Q \longrightarrow \{\pm 1\}$ is non-trivial and $\mathbb{Z}_4 = Ker(\nu)$. To check the lemma, consider the diagram :

$$\begin{array}{ccccc}
L_3(Q) & \cdots\cdots\searrow & & & \\
\downarrow{\scriptstyle \nu !} & & & & \\
L_4(Q,\mathbb{Z}_4;\nu) & \longrightarrow & L_3(\mathbb{Z}_4) & \longrightarrow & L_3(\mathbb{Q};\nu) \\
\downarrow & & & & \\
L_2(\mathbb{Z}[\mathbb{Z}_4],\ I\ ,\ g^2) & & & &
\end{array}$$

The generator of $\mathbb{Z}_4$ is here denoted g and the horizontal row is the usual exact sequence of the pair while the vertical exact sequence is that of [5] . Again by an easy exercise $L_2(\mathbb{Z}[\mathbb{Z}_4],\ I\ ,\ g^2) = 0$ and so it is sufficient to prove that the dotted line, which is geometrically the transfer homomorphism, is zero. But as [12] $L_3(Q) \cong \mathbb{Z}_2 \oplus \mathbb{Z}_2$, and as $L_3(\mathbb{Z}_4) \xrightarrow{\cong} L_3(\mathbb{Z}_2) \cong \mathbb{Z}_2$, one sees easily that $L_3(Q)$ is generated by the images of $L_3(\mathbb{Z}_4) \longrightarrow L_3(Q)$ under two maps $\mathbb{Z}_4 \longrightarrow \mathbb{Q}$. On these images, it is obvious geometrically that this transfer is 0 .

Note : It is not too difficult to see geometrically that $\sigma(g)$ is represented by a hyperbolic form on a projective module. Using an analogue of the methods of [5] for L-theory of projective modules, it follows that $\sigma(f)$ does represent zero in that theory. Thus, the corresponding surgery problem $M \times T^2 \times R \rightarrow M \times S^2 \times R$ is solvable, but the solution has non-trivial Siebenmann obstruction to being "boundarizable".

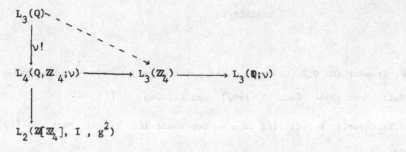

REFERENCES

[1] W. Browder and G.R. Livesay, Fixed point free involutions on homotopy spheres, Bull. Amer. Math. Soc. 73 (1967), pp. 242-245 .

[2] S.E. Cappell, A splitting theorem for manifolds, Invent. Math. 33 (1976) pp. 69-170 .

[3] S.E. Cappell, On homotopy invariance of higher signatures, Invent. Math. 33 (1976) 171-179 .

[4] S.E. Cappell, Mayer-Victoris sequences in Hermitian K-theory, Proc. Battelle K-theory Conf., Springer Lecture Notes in Math. 343 (1973) pp. 478-512.

[5] S.E. Cappell and J.L. Shaneson, Group actions with isolated fixed points, in Proceedings of the 1978 Aarhus Topology Conference, Springer lecture notes, (to appear).

[6] J. Morgan and W. Pardon , (to appear).

[7] A. Ranicki, Algebraic L-theory n , (to appear).

[8] H. Rubinstein, Homotopy implies isotopy for a class of M^3 , (to appear)

[9] J.L. Shaneson, Wall's surgery obstruction group for $Z \times G$, Ann. of Math. 90 (1969) pp. 296-334.

[10] D. Sullivan, Triangulating and smoothing homotopy equivalences, seminar notes, Princeton University (1967).

[11] C.T.C. Wall, On the classification of Hermitian forms, V, Global rings, Inventiones Math. 23 (1974), pp. 261-288.

[12] C.T.C. Wall, Classification of Hermitian forms. VI Group rings, Ann. of Math. 103 (1976), pp. 1-80 .

[13] C.T.C. Wall, Surgery on compact manifolds, Academic Press 1970.

Remarks on Novikov's Conjecture and the Topological-Euclidean Space Form Problem

by

F. T. Farrell[1] and W. C. Hsiang[2]

1. Let M^n be an oriented manifold and $x \in H^i(\pi , \mathbb{Q})$ a rational cohomology class of $K(\pi , 1)$. Given a homomorphism $\pi_1 M^n \longrightarrow \pi$, we have a natural map

(1) $$f : M^n \longrightarrow K(\pi , 1) .$$

Let $L_*(M^n) \in H^{4*}(M^n , \mathbb{Q})$ be the total L-genus of M^n. Consider the value

(2) $$L(x)(M^n) = \langle L_*(M^n) \cup f^*(x) , [M^n] \rangle \in \mathbb{Q} .$$

It is called the higher signature of M^n associated to x.

In 1965, Novikov made the following conjecture.

Novikov's Conjecture. *If M^n is a closed manifold, then $L(x)$ is a homotopy invariant.*

More precisely, if M_1^n, M_2^n are closed manifolds and if $g : M_1^n \longrightarrow M_2^n$ is a homotopy equivalence such that we have the following homotopically commutative diagram

$$(3) \qquad \begin{array}{ccc} M_1^n & \xrightarrow{f_1} & \\ \downarrow g & & K(\pi, 1) \\ M_2^n & \xrightarrow{f_2} & \end{array}$$

where f_1, f_2 are the maps for M_1^n, M_2^n as defined in (1), then $L(x)(M_1^n) = L(x)(M_2^n)$. Since then, the conjecture has been verified for various π : Novikov [15] ; Rohlin [16] ; Farrell and Hsiang [4], [8], [9] ; Kasparov [10] [11] ; Lusztig [13] ; Miščenko [14] and Cappell [3]. The most interesting case for this conjecture is when M^n itself is a compact $K(\pi, 1)$. In this special case, the conjecture has the following form.

Novikov's Conjecture for aspherical manifolds. Let M^n *be a closed manifold which is a* $K(\pi, 1)$, *i.e.,* M^n *is an aspherical manifold. Then, the surgery map of* [17]

$$\theta : [M^n ; G/Top] \otimes \mathbb{Q} \longrightarrow L_n(\pi, w_1(M^n)) \otimes \mathbb{Q}$$

is a monomorphism.

In fact, there is no counterexample to the following much stronger conjecture.

Conjecture 1. *Let* M^n *be a closed manifold which is a* $K(\pi, 1)$, *i.e.,* M^n *is an aspherical manifold ; then*

$$\theta : [M^n \times D^k, \partial ; G/Top, *] \longrightarrow L_{n+k}(\pi, w_1(M^n))$$

is an isomorphism of groups if G/Top *is endowed with the* H-*space structure*

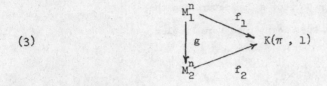

<u>from the identification to the Quinn's surgery space</u> $\underline{\underline{L}}(1)$.

This conjecture is equivalent to $S^{Top}(M^n \times D^k, \partial) = 0$ for $n + k > 4$. In particular, it implies that homotopic asperical manifolds of dimension greater than 4 are homeomorphic.

The algebraic K-theory analogue of Conjecture 1 can be formulated as follows. Let $\underline{\underline{K}}(Z)$ be the spectrum for the integers, $h_*(\ ;\underline{\underline{K}}(Z))$ the generalized homology theory with respect to $\underline{\underline{K}}(Z)$. Recall Loday's homomorphism [12]

(4) $$\lambda_* : h_*(B\pi ; \underline{\underline{K}}(Z)) \longrightarrow K_*(Z[\pi]) .$$

<u>Conjecture</u> 2 (Cf. [7]) . <u>If</u> M^n <u>is a closed manifold which is a</u> $K(\pi, 1)$, <u>then</u>

$$\lambda_* : h_*(M^n ; \underline{\underline{K}}(Z)) \longrightarrow K_*(Z[\pi])$$

<u>is an isomorphism. In particular, we should have</u> $Wh(\pi) = \tilde{K}_0(\pi) = 0$ <u>for aspherical manifolds.</u>

2. We now consider a very special class of aspherical manifolds. Let $E(n)$ be the group of rigid motions of n-dimensional Euclidean space and let Γ be a torsion-free uniform discrete subgroup of $E(n)$. The Riemannian manifold $M^n = \mathbb{R}^n / \Gamma$ is the usual (Riemannian) flat manifold. Γ is called a Bieberbach group.

Bieberbach showed that the group Γ is fitted into a short exact sequence

(5) $$1 \longrightarrow A \longrightarrow \Gamma \longrightarrow G \longrightarrow 1$$

where A is the unique normal abelian group such that the factor group

$G = \Gamma / A$ is a finite group which acts faithfully on A. G is called the holonomy group of Γ. A is isomorphic to the free abelian group of rank n (denoted by T^n). We shall refer to A as the holonomy representation of Γ, and we define the rank of Γ to be the rank of A. (I.e., Γ is of rank n if A is isomorphic to T^n the free abelian group of rank n.) For any positive integer s, define $\Gamma_s = \Gamma / sA$ and $A_s = A / sA$ where sA is the subgroup of A consisting of all elements divisible by s; Γ_s is an extension of A_s by G and in fact a semidirect product if $(s, |G|) = 1$. In [6], we proved that $Wh(\Gamma) = \tilde{K}_0(\Gamma) = 0$ for Γ a Bieberbach group, and we also verified Conjecture 1 for flat Riemannian manifolds with odd holonomy group and of dimension greater than 4.

In this note, we announce that Conjecture 1 is true for <u>all</u> flat Riemannian manifolds of dimension greater than 4. Since the proof is somewhat complicated, we shall only give a brief sketch of the steps and publish the details somewhere else.

Here is the precise statement of the result.

<u>Theorem</u>. <u>Let</u> M^n $(n > 4)$ <u>be a closed connected Riemannian flat manifold</u>. <u>Let</u> N^n <u>be a topological manifold and</u> $f : N^n \longrightarrow M^n$ <u>be a homotopy equivalence</u>. <u>Then</u>, f <u>is homotopic to a homeomorphism</u>.

We follow the same general line of philosophy of [6], but we have to generalize the argument to crystallographic group in order to treat the even holonomy case. By crystallographic group Γ, we mean that Γ is a uniform subgroup of $E(n)$ (possibly containing torsion). We again have a short exact sequence (5) for Γ. Let Γ_s, A_s etc. be defined as before. We observe that we have the following structure theorem for Γ. (Cf. Theorem 1.1 of [6].)

(i) $\Gamma = \pi \times_\alpha T$; i.e., Γ is a semidirect product of a crystallographic subgroup of rank $n - 1$ and the infinite cyclic group T ; or

(ii) there is an epimorphism from Γ to a non-trivial crystallographic group $\bar{\Gamma}$ where $1 \longrightarrow \bar{A} \longrightarrow \bar{\Gamma} \longrightarrow \bar{G} \longrightarrow 1$ is the short exact sequence of (5) for $\bar{\Gamma}$, and an infinite sequence of positive integers $s \equiv 1 \mod |\bar{G}|$ such that any hyperelementary subgroup of $\bar{\Gamma}_s$ which projects onto $\bar{G}$ is isomorphic to $\bar{G}$; or

(iii) G is an elementary abelian 2-group and either

(a) $\Gamma = A \times_\alpha T_2$; i.e., a semidirect product of the cyclic group of order 2 and A with T_2 acting on A via the multiplication by -1, or

(b) Γ maps epimorphically onto a crystallographic group $\bar{\Gamma}$ such that if we write $\bar{\Gamma}$ as $1 \longrightarrow \bar{A} \longrightarrow \bar{\Gamma} \longrightarrow \bar{G} \longrightarrow 1$, then $\bar{A} = T^2$ and $\bar{G} = (T_2)^2$, and the holonomy representation is either $\begin{pmatrix} \pm1 & 0 \\ 0 & \pm1 \end{pmatrix}$, or $\pm\begin{pmatrix} 1 & 0 \\ 0 & 1 \end{pmatrix}$ and $\pm\begin{pmatrix} 0 & 1 \\ 1 & 0 \end{pmatrix}$.

We next observe that the epimorphism given in (ii) of a Bieberbach group Γ (i.e., Γ torsion free) of rank n onto a crystallographic group $\bar{\Gamma}$ of rank m allows us to build a singular Siefert fibration

$$f : B\Gamma = \mathbb{R}^n / \Gamma \longrightarrow B\bar{\Gamma} = \mathbb{R}^m / \bar{\Gamma} . \tag{6}$$

Even though $\mathbb{R}^m / \bar{\Gamma}$ is not a manifold in general, it has the expansive maps of Epstein-Shub type. Moreover, the recent result of Ferry-Chapmann-Quinn spplies to singular fibrations of the above situation as the substitute for the local contractibility of Černovskii-Edwards-Kirby (as it was used in [6]). Using these facts, we prove the theorem by induction on the rank of Γ (Γ torsion free) and on the order of G just as we did in [6]. If the structure of Γ satisfies (i), then the theorem follows from the argument of [5], [6]. If Γ has an epimorphism onto $\bar{\Gamma}$ as given in (ii), then we construct a singular fibration $B\Gamma = \mathbb{R}^n/\Gamma \longrightarrow B\bar{\Gamma} = \mathbb{R}^m/\bar{\Gamma}$. (Note that even though Γ is torsion free, $\bar{\Gamma}$ may still have torsion!) Using the expansive maps of $B\bar{\Gamma}$ and the result of Quinn, we give a more complicated version of an analogous argument of [6] to finish the proof for this case. Finally, we come to case (iii). We first observe that (iii a) does not occur since Γ is torsion free. Using the specific structure of (iii b), we see that Γ can be expressed as $B *_D C$ in two essentially different ways where B, C, D are Bieberbach groups of rank $n - 1$ and $(B : D) = (C : D) = 2$. (Note that $B\Gamma = \mathbb{R}^n/\Gamma$ is possibly non-orientable in the present situation.) If we can apply Cappell's splitting theorem [2] to $B\Gamma$, we can finish off our theorem by an induction argument. A priori, we should have some UNil obstruction of [1]. (This is the place where we were unable to do anything about it in [6].) Since we now only have to worry about the specific case (iii b), we can interplay the two essentially different decomposition of Γ as $B *_D C$ to show that the UNil obstruction actually vanishes. So, we can finish off the proof of the theorem.

Footnotes. (1) The first-named author was partially supported by NSF grant number MCS-7701124.

(2) The second-named author was partially supported by NSF grant number GP 34324X1.

References

[1] S. E. Cappell, Unitary nilpotent groups and Hermitian K-theory, Bull. AMS 80 (1974), 1117-1122.

[2] ___________, A splitting theorem for manifolds, Invent. Math. 33 (1976), 69-170.

[3] ___________, On homotopy invariance of higher signatures, Invent. Math. 33 (1976), 171-179.

[4] F. T. Farrell and W. C. Hsiang, Manifolds with $\pi_1 = G \times_\alpha T$, Amer. J. Math. 95 (1973), 813-848.

[5] ______ ______________, Rational L-groups of Bieberbach groups, Comment. Math. Helv. 52 (1977), 89-109.

[6] ____________________, The topological Euclidean space form problem, Invent. Math. 45 (1978), 181-192.

[7] ____________________, On the rational homotopy groups of the diffeomorphism groups of discs, spheres and aspherical manifolds, Proc. Symp. Pure Math. 32 part 1, Algebraic and geometric topology, AMS 1978, 325-337.

[8] W. C. Hsiang, A splitting theorem and Künneth formula in algebraic K-theory, Algebraic K-theory and its geometric applications, Lecture Notes in Math., Vol. 108, Springer-Verlag, Berlin and New York, 1969, 72-77.

[9] ___________, Manifolds with $\pi_1 = Z^k$, Manifolds-Amsterdam 1970, Lecture Notes in Math. Vol. 197, Springer-Verlag, Berlin and New York, 1971, 36-43.

[10] G. G. Kasparov, The homotopy invariance of the rational Pontrjagin numbers, Dokl. Akad. Nauk. SSSR 190 (1970), 1022-1025.

[11] ____________, Topological invariance of elliptic operator I , K-homology, Izv. Akad. Nauk SSSR, Ser. Mat. 39 (1975), 796-838.

[12] J.-L. Loday, K-théorie algebrique et représentations de groupes, Ann. Sci. E'cole Norm. Sup. 9 (1976), 309-377.

[13] G. Lusztig, Novikov's higher signature and families of elliptic operators, J. Differential Geometry 7 (1972), 229-256.

[14] A. S. Miščenko, Infinite-dimensional representations of discrete groups and higher signatures, Izv. Akad. Nauk. SSSR Ser. Mat. 38 (1974), 81-106.

[15] S. P. Novikov, Homotopic and topological invariance of certain rational classes of Pontrjagin, Dokl. Akad. Nauk SSSR 162 (1965), 1248-1251.

[16] V. A. Rohlin, Pontrjagin-Hirzebruch class of codimension 2 , Izv. Akad. Nauk SSSR Ser. Math. 30 (1966) , 705-718.

[17] C. T. C. Wall, Surgery on compact manifolds, Academic Press, New York, 1970.

Isotopy classes of diffeomorphisms of

(k-1)-connected almost-parallelizable 2k-manifolds

M. Kreck[+]

§ 1 Results

The group of isotopy classes of orientation preserving diffeomorphisms on a closed oriented differentiable manifold M is denoted by $\pi_0 \mathrm{Diff}(M)$; the group of pseudo isotopy classes is denoted by $\tilde{\pi}_0 \mathrm{Diff}(M)$. In this paper we will compute $\pi_0 \mathrm{Diff}(M)$ for M a closed differentiable (k-1)-connected almost-parallelizable 2k-manifold in terms of exact sequences for $k \geq 3$, and classify elements in $\tilde{\pi}_0 \mathrm{Diff}(M)$ for any simply-connected closed differentiable 4-manifold.

In the following M stands for a closed differentiable (k-1)-connected almost-parallelizable 2k-manifold if $k \geq 3$ and a simply-connected manifold if k=2. To describe our results we need some invariants. We denote by Aut $H_k(M)$ the group of automorphisms of $H_k(M) := H_k(M;\mathbb{Z})$ preserving the intersection form on M and (for $k \geq 3$) commuting with the function $\alpha : H_k(M) \longrightarrow \pi_{k-1}(SO(k))$, which assigns to $x \in H_k(M)$ the classifying map of the normal bundle of an embedded sphere representing x. As the induced map in homology of any orientation preserving diffeomorphism lies in Aut $H_k(M)$, we obtain a homomorphism

$$\pi_0 \mathrm{Diff}(M) \longrightarrow \mathrm{Aut}\, H_k(M),\ [f] \longmapsto f_*.$$

We denote the kernel of this map by $\pi_0 S\,\mathrm{Diff}(M)$.

Our next invariant is defined for elements $[f]$ in $\pi_0 S\,\mathrm{Diff}(M)$. It assigns to $[f]$ a homomorphism $H_k(M) \longrightarrow S\,\pi_k(SO(k))$, where S is the map $\pi_k(SO(k)) \longrightarrow \pi_k(SO(k+1))$ induced by inclusion. If $x \in H_k(M)$ is represented by an embedded sphere $S^k \subset M$ we can assume that $f|_{S^k} = \mathrm{Id}$. As the stable normal bundle of S^k in M is trivial the operation of f on $\nu(S^k) \oplus 1$ given by the differential of f corresponds to an element of $\pi_k SO(k+1)$. It is obvious that this element lies in the image of $\pi_k(SO(k)) \longrightarrow \pi_k(SO(k+1))$.

[+]This work was begun in Bonn in 1976 and was partially supported by the Sonderforschungsbereich (SFB 4o). It was finished during a stay in Aarhus (Denmark) in 1978. I would like to thank the University of Aarhus for the invitation and the stimulating atmosphere there.

Lemma 1: The construction above leads to a well defined homomorphism

$$\chi : \pi_0 S\,\mathrm{Diff}(M) \longrightarrow \mathrm{Hom}(H_k(M), S\pi_k(SO(k))).$$

The proof of this Lemma for $k>3$ is contained in the papers of Wall ([19]; [20], Lemma 2 3), the case k=3 follows from Lemma 2 below. I want to repeat here the warning of Wall that it is not obvious that χ and similar invariants are well defined. The difficult point is to show that $\chi(f)$ depends only on the isotopy class of f.

From the work of Kervaire ([5]) one can easily deduce the following list for $S\pi_k(SO(k))$ for $k>2$ and $k\neq 6$:

k mod 8	0	1	2	3	4	5	6	7
$S\pi_k(SO(k))$	$\mathbb{Z}_2\oplus\mathbb{Z}_2$	$\mathbb{Z}_2$	$\mathbb{Z}_2$	$\mathbb{Z}$	$\mathbb{Z}_2$	0	$\mathbb{Z}_2$	$\mathbb{Z}$

For k=6 we have $S\pi_6(SO(6)) = 0$.

Thus, for k=3 mod 4 we can identify $\mathrm{Hom}(H_k(M), S\pi_k(SO(k))) = \mathrm{Hom}(H_k(M),\mathbb{Z})$ with $H^k(M)$. In this case we can describe $\chi(f)$ by an invariant defined by Browder using the Pontrjagin class of the mapping torus $M_f = I\times M_{(o,x)\sim(1,f(x))}$ ([2]). The definition is as follows. We consider the map $c : M_f \longrightarrow M_{f/\{0\}\times M} = \Sigma M^+$ From the Wang sequence we know that $i^* : H^k(M_f) \longrightarrow H^k(M)$ is surjective, if $f_* = \mathrm{Id}$. Thus we obtain an isomorphism $c^* : H^{k+1}(\Sigma M^+) \longrightarrow H^{k+1}(M_f)$. The invariant $p'(f)\in H^k(M)$ is defined as the image of the inverse suspension isomorphism applied to $c^{*-1}(p_{(k+1)/4}(M(f))$. It is not difficult to see that $f \longmapsto p'(f)$ is a homomorphism. It is related to $\chi(f)$ in the following way.

If $x\in H_k(M)$ is represented by an embedded sphere S^k and $f|_{S^k} = \mathrm{Id}$ then $S^1\times S^k/_{\{1\}\times S^k}$ represents the image of x in $H_{k+1}(\Sigma M^+)$ under the suspension isomorphism. We denote it by y. Now we consider the stable vector bundle E over $S^1\times S^k/_{\{1\}\times S^k}$ classified by $\chi(f)(x)$. By the classification of vector bundles over spheres we know that the Kronecker product $\langle p(E), [S^1\times S^k/_{\{1\}}\times S^k]\rangle = \pm a_{(k+1)/4}((k-1)/2)!\,\chi(f)(x)$, where $a_m = 2$ for m odd and 1 for m even. But it is obvious that $(c|_{S^1\times S^k})^*(E)$ is equal to the restriction of the stable tangent bundle of M_f to $S^1\times S^k$. Thus $\langle p'(f), y\rangle = \pm a_{(k+1)/4}((k-1)/2)!\,\chi(f)(x)$. This implies:

Lemma 2: If k=3 mod 4

$$p'(f) = \pm a_{(k+1)/4}((k-1)/2)!\,\chi(f),$$

where $a_m = 2$ for m odd and 1 for m even.

Now we are ready to formulate our results.

Theorem 1: For a simply-connected closed differentiable 4-manifold the homomorphism

$$\tilde{\pi}_0 \mathrm{Diff}(M) \longrightarrow \mathrm{Aut}\, H_2(M),\ [f] \longmapsto f_*$$

is injective.

Remarks:

1) This result is completely analogous to the classification of isotopy classes of diffeomorphisms of an oriented connected surface by the operation of the diffeomorphism on the fundamental group.

2) It seems very hard to determine the image of the map $\tilde{\pi}_0 \mathrm{Diff}(M) \longrightarrow \mathrm{Aut}\, H_2(M)$. Wall has shown that it is surjective if M is of the form $M = N \# S^2 \times S^2$ and the intersection form is indefinite or has rank ≤ 8 ([18]).

In the following examples it is obvious that the map is surjective and we obtain the following results:

$$\tilde{\pi}_0 \mathrm{Diff}(S^2 \times S^2) = \mathbb{Z}_4$$

$$\tilde{\pi}_0 \mathrm{Diff}(P_2\mathbb{C} \# \overline{P_2\mathbb{C}}) = \mathbb{Z}_2 \oplus \mathbb{Z}_2$$

$$\tilde{\pi}_0 \mathrm{Diff}(\underbrace{P_2\mathbb{C} \# \ldots \# P_2\mathbb{C}}_{k}) = O(k; \mathbb{Z}).$$

$O(k; \mathbb{Z})$ is the group of matrices with exactly one coefficient ± 1 in each row and column and zero otherwise. It is the group of automorphisms of a k-dimensional cube. For instance: $\tilde{\pi}_0 \mathrm{Diff}(P_2\mathbb{C}) = \mathbb{Z}_2$; $\tilde{\pi}_0 \mathrm{Diff}(P_2\mathbb{C} \# P_2\mathbb{C}) = D_8$, the dihedral group.

$$\tilde{\pi}_0 \mathrm{Diff}(\underbrace{P_2\mathbb{C} \# \ldots \# P_2\mathbb{C}}_{k} \# \underbrace{\overline{P_2\mathbb{C}} \# \ldots \# \overline{P_2\mathbb{C}}}_{l}) \subset O(k,l; \mathbb{Z})$$

the group of integer matrices preserving the form $\begin{pmatrix} I_k & 0 \\ 0 & -I_l \end{pmatrix}$.

Theorem 2: $k \geq 3$. The following sequences are exact:

$$0 \longrightarrow \pi_0 S\,\mathrm{Diff}(M) \longrightarrow \pi_0 \mathrm{Diff}(M) \longrightarrow \mathrm{Aut}\, H_k(M) \longrightarrow 0$$

$$0 \longrightarrow \Theta_{2k+1}/\Sigma_M \longrightarrow \pi_0 S\,\mathrm{Diff}(M) \xrightarrow{\chi} \mathrm{Hom}(H_k(M), S\pi_k(SO(k))) \longrightarrow 0$$

Σ_M is an element in the group of (2k+1)-dimensional homotopy spheres Θ_{2k+1} of order 2 and depending only on M.

Σ_M can be described as follows. We consider the embedding of $S^1 \times D^{2k}$ into $S^1 \times M$ obtained from a product embedding twisted by the nontrivial element of $\pi_1 SO(2k) = \mathbb{Z}_2$. Then we replace $S^1 \times M$ by a homotopy sphere Σ_M by a sequence of surgeries first killing the fundamental group with this embedding and then killing the k-th homotopy by arbitrary surgeries.

The map $\Theta_{2k+1}/\Sigma_M \longrightarrow \pi_0 S\,\mathrm{Diff}(M)$ is induced by the following map $\Theta_{2k+1} \to \pi_0 S\mathrm{Diff}(M)$. Consider $\Sigma \in \Theta_{2k+1}$ as $D^{2k+1} \cup_f D^{2k+1}$ and assume that f is the identity on a neighbourhood of the lower hemisphere $D^{2k}_- \subset S^{2k}$. Then we get an element of $\pi_0 S\,\mathrm{Diff}(M)$ by the diffeomorphism on M which is the identity outside an embedded disk in M and is equal to $f_{|D^{2k}_+}$ on this disk.

Part of this result is contained in the work of Wall ([19], [2o]) where he computes the group of pseudo isotopy classes of diffeomorphisms of M minus an open embedded disk. Complete results were known for the case $M = S^k \times S^k$ and $k \geq 4$ (compare [17]) and for homotopy spheres. Then Σ_M coincides with the $\gamma(M)$ of [21].

To complete our computation we have to determine Σ_M. First, I state some properties of Σ_M.

Lemma 3:

a) $\Sigma_{M \# N} = \Sigma_M + \Sigma_N$

b) If M bounds a framed manifold then $\Sigma_M = 0$

c) If M is a homotpy sphere we get Σ_M from the Milnor-Munkres-Novikov pairing $\Theta_{2k} \times \pi_1(SO) \to \Theta_{2k+1}$ as the image of (M, η) where $\eta \in \pi_1(SO)$ is the nontrivial element.

Σ_M is closely related to the following diffeomorphism on M. We consider an embedding of $2\cdot D^{2k}$ into M and a differentiable map $\alpha : [1,2] \longrightarrow SO(2k)$ which maps a small neighbourhood of the boundary to the identity matrix and represents the nontrivial element in $\pi_1(SO(2k))$. Then we get a diffeomorphism f_α of M by taking the identity on D^{2k} and outside $2\cdot D^{2k}$ and by mapping $x \in 2\cdot D^{2k} \setminus D^{2k}$ to $\alpha(|x|)\cdot x$.

Lemma 4: $\Sigma_M = 0 \Longleftrightarrow f_\alpha$ is isotopic to the identity rel. D^{2k}.

To formulate our main result about Σ_M we have to distinguish between the case where M can be framed and the case where it cannot. Under our assumptions M can automatically be framed if $k \neq 0 \bmod 4$ and in the case $k = 0 \bmod 4$ it can be framed if and only if the signature $\tau(M)$ vanishes.

We identify a framed manifold $[M,\beta] \in \Omega^{fr}_{2k}$ by the Pontrjagin-Thom construction with the corresponding element in π^s_{2k}. We denote the map $\Theta_n \longrightarrow \operatorname{cok} J_n$ by T ([6]) and the projection map $\pi^s_n \longrightarrow \operatorname{cok} J_n$ by P.

Theorem 3:

a) If M is an s-parallelizable manifold then
$T(\Sigma_M) = P(\eta \circ [M,\beta])$
where β is any framing on M. $\circ$ denotes the composition map in the stable homotopy groups.

b) If $\tau(M^{4n}) = s\,\sigma_n, s \neq 0$, where $\sigma_n/8$ is the order of bP_{4n}, then
$T(\Sigma_M) \subset P([s\varrho, 2, \eta])$
where ϱ is the element of order 2 in $\operatorname{im} J_{4n-1}$ and $[s\varrho, 2, \eta]$ denotes the Toda bracket.

c) If M is an s-parallelizable manifold then

$$\Sigma_M = 0 \Longleftrightarrow \begin{cases} \text{there exists a framing } \beta \text{ on M such that} & \\ \eta \circ [M,\beta] = 0 & k \text{ odd or } k \text{ even an } bP_{2k+2} = 0; \\ \text{there exists a framing } \beta \text{ on M such that } \eta\circ[M,\beta] \in \operatorname{im} J & \\ \text{and an invariant } \alpha(M) \in \mathbb{Z}_2 \text{ vanishes} & k \text{ even and } bP_{2k+2} \neq 0 \end{cases}$$

$\alpha(M)$ is defined as the Arf invariant of Σ_M. It is only defined if the first condition is fulfilled. For then we will show that $\Sigma_M \in bP_{2k+2}$.
Especially it follows that for k odd $\Sigma_M = 0 \Leftrightarrow \Sigma_M \in bP_{2K+2}$. This extends a result of Levine ([21], Prop.8).

Remark: I have no example where $\mathfrak{a}(M) \neq 0$. Thus it may be that the condition $\mathfrak{a}(M) = 0$ can be omitted.

Now, we will discuss some consequences of our theorems. First we will give some examples where Σ_M is nonzero. In the case of stably parallelizable manifolds we can use Toda's tables ([15]) to get complete information about Σ_M in low dimensions. As $\eta\mu_{8k+2} \neq 0$ ([1]) we get, furthermore, a series of examples in higher dimensions. This completes the computations of ([21], 16).

Corollary 1: Notations as in Toda's tables ([15]). If M is a framed manifold which represents one of the following elements in π_*^s then Σ_M is nonzero

$$\bar{\nu}, \varepsilon ; \eta\circ\mu, \ \eta\circ\mu + \beta_1 ; \kappa, \kappa + \sigma^2 ; \eta^*, \ \eta^* + \eta\circ\xi ; \eta\circ\bar{\mu}, \eta\circ\bar{\mu} + \nu^*$$

For all other framed manifolds of dim≤ 18 Σ_M is zero.
If M^{8k+2} is a framed manifold representing μ_{8k+2} then Σ_M is nonzero.

In the case of non s-parallelizable manifolds we get examples of M with nonzero Σ_M in dim 8k. For Adams has proved that $e_c[\xi, 2, \eta]$ is nonzero for all elements of this Toda bracket, where $\xi \in J_{8k-1}$ is the element of order 2 ([1], 11.1). But ([1], 7.19) implies that no element of $[\xi, 2, \eta]$ is contained in im J_{8k+1}. Thus $\Sigma_M \neq 0$, if M^{8k} has signature $(2r+1)\sigma_{2k}$.

Corollary 2: If the signature of M^{8k} is an odd multiple of σ_{2k} then

$$\Sigma_M \neq 0.$$

From these examples we can see that in most cases π_0Diff(M) depends on the differentiable structure on M. This was known in some dimensions for a sphere ([11]). But our examples show that this is the case for all highly connected s-parallelizable 8k+2-dim manifolds. For if M is such a manifold with $\Sigma_M = 0$ then we can change the differentiable structure on M by replacing M by the connected sum of M with a framed homotopy sphere representing μ_{8k+2}. By Lemma 3 and Corollary 1 we know that for M with this differentiable structure Σ_M is nonzero. Thus π_0Diff(M) has changed. On the other hand on every M there exists a differentiable structure such that $\Sigma_M = 0$. For if Σ_M is nonzero we know that M is framed bordant to a homotopy sphere N. By Lemma 3 we know that $\Sigma_M = \Sigma_N$ and that $\Sigma_{M \# (-N)} = 0$.

<u>Corollary 3:</u> For every highly connected s-parallelizable 8k+2-manifold M the group $\pi_0 \mathrm{Diff}(M)$ depends on the differentiable structure on M.

The proofs of our results are very much in the spirit of Kervaire-Milnor's work on homotopy spheres and are based on direct surgery arguments. They make no use of the general machinery of surgery as developed by Browder, Novikov, Sullivan, Wall. This machinery leads to very interesting informations about the rational homotopy type of Diff(M) ([14]; [16]; compare the report of Burghelea at this conference). But it seems hard to get complete information from it. I want to indicate this very briefly.

For a 1-connected manifold M^n of dim ≥ 5 the general surgery theory gives the following information ([17]). There are exact sequences:

$$\begin{array}{ccccccc}
0 \longrightarrow bP_{n+2} & \longrightarrow & \mathcal{S}(M \times I,\ M \times \dot{I}) & \longrightarrow & [\Sigma M,\ G/O] \\
& & \downarrow \emptyset & & \\
& & \pi_0 \mathrm{Diff}(M)^{\pi} = \text{isotopy classes of diffeomorphisms homotopic to Id} & & \\
& & \downarrow & & \\
& & 0 & &
\end{array}$$

It seems that for highly connected almost parallelizable manifolds $\pi_0 \mathrm{Diff}(M)^{\pi} = \pi_0 S\,\mathrm{Diff}(M)$. The difficulties in applying these sequences to the computation of $\pi_0 \mathrm{Diff}(M)$ are 1) the computation of $[\Sigma M,\ G/O]$ and with it of $\mathcal{S}(M \times I,\ M \times \dot{I})$ and 2) the computation of Ker $\emptyset$. I have no idea how to solve especially the last problem. Perhaps the knowledge of the results for sufficiently many examples would suggest the solution. The present paper could be understood as a first step into this direction.

§ 2 Proofs

Before we give the proof of theorem 1 we formulate a general criterion for the problem, which diffeomorphism on the boundary of a 1-connected manifold can be extended for the interior and spezialize it to the problem of existence of pseudo-isotopies.

Proposition 1: (compare [3], 2.3 ;[9], Lemma 7) Let N be a 1-connected manifold of diemsnion$\geq$5 and f an orientation preserving diffeomorphism of ∂N. f can be extended to a diffeomorphism on N if and only if the twisted double $N \cup_f -N$ bounds a 1-connected manifold W such that all relative homotopy groups $\pi_k(W,N)$ and $\pi_k(W,-N)$ are zero, where N and -N mean the two embeddings of N into $N \cup_f -N$.

Proof: If we introduce corners along the boundary of a tubular neighbourhood of ∂N into $N \cup_f -N$ we see that W is a relative h-cobordism between $(N, \partial N)$ and $(-N, \partial N)$. Then the proposition is a standard application of the relative h-cobordism theorem ([12]).

If we spezialize this proposition to the case where N is equal to $M \times I$, M a 1-connected manifold of dimension$\geq$4, and consider diffeomorphisms of $\partial N = M + (-M)$ of the form f + Id we obtain the following criterion for the existence of pseudo-isotopies between f and Id. For dim M$\geq$5 we get the existence of isotopies using the deep result of Cerf ([4]).

Proposition 2: Let M be a 1-connected manifold of dimension $\geq$4. An orientation preserving diffeomorphism of M is pseudo-isotopic (isotopic, if dim M$\geq$5) to Id if and only if the mapping torus $M_f = I \times M/_{(0,x)\sim(1,f(x))}$ bounds a 1-connected manifold W with $\pi_k(W,M) = \{0\}$ for all k.

Remark: The conditions of Proposition 1 can be reformulated as: M_f is h-cobordant to $M \times S^1$.

Proof of Theorem 1: We consider an orientation preserving diffeomorphism $f : M \longrightarrow M$ of a simply-connected closed differentiable 4-manifold with $f_* : H_2(M) \longrightarrow H_2(M)$ the identity. All we have to do is to construct a 6-manifold W with the conditions of Proposition 2. The idea is to start with an arbitrary manifold W bounding M_f and to modify this manifold by surgery in the interior of W until the properties are fulfilled. But in general this does not work, for we can only do surgery if we can represent homology classes by embedded spheres with trivial normal bundle. As we are in the oriented case each embedded 1-sphere has trivial normal bundle and each bundle over S^3 is trivial, so the only problem arises at embedded 2-spheres. But the normal bundle of an embedded 2-sphere is trivial if and only if the Stiefel Whitney class w_2 is zero. So there is no problem if W is a spin-manifold. We will see that we can choose W as a spin-manifold if M is a spin-manifold and that we don't need any condition for W if M is not a spin-manifold.

Using this idea we first have to check that for a diffeomorphism $f : M \longrightarrow M$ with $f_* = \mathrm{Id}$ the mapping torus M_f bounds an oriented 6-manifold W which can be chosen as a spin-manifold if M admits a spin structure. As $f_* = \mathrm{Id}$ the Wang sequence shows that the inclusion induces an isomorphism $H^2(M_f) \longrightarrow H^2(M)$. Thus if M admits a spin structure, which means $w_2(M) = 0$, then M_f admits one. But the bordism group of 5-dimensional spin-manifolds is zero ([13]), so M_f bounds a spin-manifold W.

If M admits no spin structure, we want to show that M_f bounds an oriented 6-manifold W (without any additional condition). The only obstruction for this is the Stiefel Whitney number $w_2(M_f)w_3(M_f)$. But by a formula of Lusztig, Milnor and Peterson:
$w_2(M_f)w_3(M_f) = \dim H_2(M_f;\mathbb{Q}) + \dim H_4(M_f;\mathbb{Q}) - \dim H_2(M_f;\mathbb{Z}_2) - \dim H_4(M_f;\mathbb{Z}_2) \bmod 2$,
the mod 2 difference of the semicharacteristics with coefficients in $\mathbb{Q}$ and $\mathbb{Z}_2$ resp. ([10]). But as M is simply-connected and $f_* = \mathrm{Id}$ the Wang sequence shows that $H_*(M_f)$ is torsion free. Thus the semicharacteristics with coefficients in $\mathbb{Q}$ and $\mathbb{Z}_2$ are the same and $w_2(M_f)w_3(M_f) = 0$.
Now we want to do surgery on W to kill $\pi_1(W)$ and $\pi_i(W,M)$ for $i \geq 2$, which is equivalent to killing $\pi_1(W)$ and $H_i(W,M)$ for all i. It is well known that we can kill $\pi_1(W)$ by a sequence of surgeries and can do this in such a manner that the resulting simply-connected manifold is a spin-manifold if W was. We denote this simply connected manifold again by W.

The next step is to kill $H_2(W,M)$. As $H_2(W) \longrightarrow H_2(W,M)$ is surjective we can represent an element x of $H_2(W,M)$ by $\bar{x} \in H_2(W)$. As $\pi_2(W) \cong H_2(W)$ we can represent $\bar{x}$ by an embedded $S^2 \hookrightarrow W$. This sphere has trivial normal bundle if and only if the Kronecker product $\langle w_2(W),\bar{x}\rangle$ is zero. If M admits a spin structure we have supposed that W has one and so $w_2(W) = 0$. If M admits no spin structure there exists $z \in H_2(M)$ with $\langle w_2(M),z\rangle \neq 0$. If $\langle w_2(W),\bar{x}\rangle \neq 0$ we replace $\bar{x}$ by $\bar{x} + i_* z$, i the inclusion $M \longrightarrow W$. In $H_2(W,M)$ the element $\bar{x} + i_* z$ again represents x, but $\langle w_2(W),\bar{x} + i_* z\rangle = 0$.

So we can represent each element x of $H_2(W,M)$ by an embedded sphere $S^2 \hookrightarrow W$ with trivial normal bundle. Surgery with this S^2 kills x and so we can kill $H_2(W,M)$ by a sequence of surgeries giving a simply-connected manifold, again denoted by W with $H_2(W,M) = \{0\}$.

Now we come to the final step namely killing $H_3(W,M)$. If we can do this we are finished for by Poincaré duality

$$H_k(W,M) \cong H^{6-k}(W, \partial W-M) \cong H^{6-k}(W,M) .$$

Again from Poincaré duality and the universal coefficient theorem it follows that $H_3(W,M)$ is torsion free.

To see how to kill $H_3(W,M)$ we consider the following situation. Let $x \in H_3(W,M)$ be a primitive element representable by an embedded sphere $S^3 \hookrightarrow W$. This sphere has trivial normal bundle. Now an easy generalization of a standard argument of surgery theory (compare [6]) shows that if we do surgery with this embedded sphere the resulting manifold W' is again simply-connected, $H_2(W',M) = \{0\}$ and $H_3(W',M) = H_3(W,M)/\mathbb{Z}x+\mathbb{Z}y$ where y is an element of $H_3(W,M)$ such that the intersection number of the embedded sphere S^3 with y is 1.

This shows that we can kill $H_3(W,M)$ by a sequence of surgeries if there exists a direct summand U in $H_3(W,M)$ with the following properties:

1.) $\dim U = \frac{1}{2} \dim H_3(W,M)$

2.) each $x \in U$ can be represented by an embedded sphere $S^3 \hookrightarrow W$

3.) for $x,y \in U$ the intersection number $x \circ y$ vanishes.

Then we choose a basis of $H_3(W,M)$ of the from $x_1,\dots,x_k$, $y_1,\dots,y_k$ such that $x_1,\dots,x_k$ is a basis of U and $x_i \circ y_i = 1$ for all i. But by condition 2.) we can represent each x_i by an embedded sphere $S^3_i \hookrightarrow W$ and condition 3.) allows us to choose these embeddings disjointly. According to the considerations above it follows that we can kill $H_3(W,M)$ by a sequence of surgeries with S_i^3.

To show that such a subspace $U \subset H_3(W,M)$ exists we first compute the dimension of $H_3(W,M)$. We consider the following exact sequences:

$$\begin{array}{ccccccccc}
 & & 0 & & & & & & \\
 & & \downarrow & & & & & & \\
 & & H_4(W,\partial W) & & & & & & \\
 & & \downarrow & & & & & & \\
 & & H_3(\partial W) & & & & & & \\
 & & \downarrow{\scriptstyle i_*} & & & & & & \\
0 & \longrightarrow & H_3(W) & \xrightarrow{k_*} & H_3(W,M) & \longrightarrow & H_2(M) & \longrightarrow & H_2(W) \longrightarrow 0 \\
 & & \downarrow{\scriptstyle j_*} & & & & & & \\
 & & H_3(W,\partial W) & & & & & &
\end{array}$$

The zero at the top results from the fact that the map $H_4(W) \longrightarrow H_4(W, \partial W)$ is the Poincaré dual of $H^2(W, \partial W) \longrightarrow H^2(W)$ which factorizes through $H^2(W,M) = \{0\}$.
From these exact sequences it follows:

$$\begin{aligned}\dim H_3(W,M) &= \dim H_3(W) + \dim H_2(M) - \dim H_2(W)\\ &= \operatorname{rank} j_* + \operatorname{rank} i_* + \dim H_2(M) - \dim H_2(W).\end{aligned}$$

But $\operatorname{rank} i_* = \dim H_3(\partial W) - \dim H_4(W, \partial W)$ and $\dim H_3(\partial W) = \dim H_2(M)$ by the Wang sequence and $\dim H_4(W, \partial W) = \dim H_2(W)$ by Poincaré duality. So $\dim H_2(M) - \dim H_2(W) = \operatorname{rank} i_*$ and we have:

$$\dim H_3(W,M) = \operatorname{rank} j_* + 2 \operatorname{rank} i_*.$$

As $H_3(W,M)$ is torsion free, the same holds for $H_3(W)$. We decompose $H_3(W)$ into subspaces $S \oplus V$ such that $\operatorname{im} i_* \subset S$ and $\dim S = \operatorname{rank} i_*$. From this it follows that for $x \in S$ and $y \in H_3(W)$ the intersection number $x \circ y$ vanishes. Furthermore it follows that $\dim V = \operatorname{rank} j_* =$ rank of the intersection form on W. The restriction of the intersection form to V is non-degenerate and as this form is antisymmetric there exists a direct summand T of V such that $\dim T = \frac{1}{2} \dim V$ and the intersection form vanishes on T. Thus $U = k_*(S \oplus T)$ is a direct summand in $H_3(W,M)$, of dimension $\frac{1}{2} \dim H_3(W,M)$, on which the intersection form vanishes.

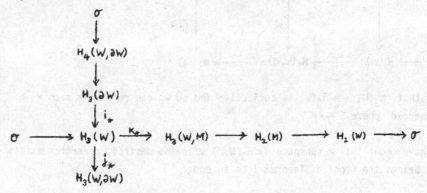

To show that U fulfils condition 2.) we consider the following commutative diagramm:

$$\begin{array}{ccccccc}
 & & \pi_3(W) & \longrightarrow & \pi_3(W,M) & \longrightarrow & \pi_2(M) \\
 & & \downarrow & & \downarrow \wr\!\!\mid\!\mid & & \downarrow \wr\!\!\mid\!\mid \\
0 & \longrightarrow & H_3(W) & \xrightarrow{k_*} & H_3(W,M) & \longrightarrow & H_2(M)
\end{array}$$

It shows that $\pi_3(W) \longrightarrow H_3(W)$ is surjective and so we can represent each $x \in U$ by an embedded sphere $S^3 \hookrightarrow W$.

Thus we have shown that a subspace $U \subset H_3(W,M)$ with the desired properties exists and this brings the proof of Theorem 1 to an end.

The proof of Theorem 2 splits into two parts. First, we compute $\tilde{\pi}_0\mathrm{Diff}(M \text{ rel } D^{2k})$, the group of pseudo-isotopy classes of diffeomorphisms leaving an embedded disk D^{2k} fixed. This is easier than the computation of $\pi_0\mathrm{Diff}(M)$. But $\pi_0\mathrm{Diff}(M)$ can be expressed as a quotient of $\tilde{\pi}_0\mathrm{Diff}(M \text{ rel } D^{2k})$ and this leads to the proof of theorem 2.

<u>Proposition 3:</u> $k \geq 3$. The following sequences are exact:

$$0 \longrightarrow \tilde{\pi}_0\, S\,\mathrm{Diff}(M \text{ rel } D^{2k}) \longrightarrow \tilde{\pi}_0\mathrm{Diff}(M \text{ rel } D^{2k}) \longrightarrow \mathrm{Aut}\, H_k(M) \longrightarrow 0$$

$$0 \longrightarrow \Theta_{2k+1} \longrightarrow \tilde{\pi}_0\, S\,\mathrm{Diff}(M \text{ rel } D^{2k}) \longrightarrow \mathrm{Hom}(H_k(M), S\pi_k(SO(k))) \longrightarrow 0$$

The maps are defined as in Theorem 1.

<u>Proof:</u> We denote the manifold obtained from M by removing a disk disjoint from D^{2k} by N. Wall has shown that every element of Aut $H_k(N)$ = Aut $H_k(M)$ can be realized by a diffeomorphism on N rel D^{2k} ([19], Lemma 1o). This follws rather easily using a handle decomposition of N. A similar argument shows that every element of Hom $(H_k(N), S\pi_k(SO(k)))$ = Hom $(H_k(M), S\pi_k(SO(k)))$ can be realized by an element

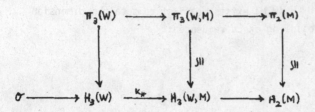

of S Diff(N rel D^{2k}). Thus the sequences would be exact on the right-hand side if every diffeomorphism on N could be extended to a diffeomorphism on M and this is equivalent to the fact that the restriction of any diffeomorphism of N to $\partial N=S^{2k-1}$ beeing isotopic to Id. But if we identify the restriction of diffeomorphisms of N to ∂N with the inertia group of M we see from the work of Kosinski that all diffeomorphisms of N can be extended to M ([7]).

To finish our proof we have to show that the homomorphism $\theta_{2k+1} \longrightarrow \tilde{\pi}_0 S$ Diff(M rel D^{2k}) is injective and that its image is equal to the kernel of $\tilde{\pi}_0 S$ Diff(M rel D^{2k}) $\longrightarrow$ Hom($H_k(M)$, $S\pi_k(SO(k))$). We show this by constructing an inverse σ from this kernel to θ_{2k+1}.

The map σ is defined as follows. We fix embeddings $(S^k \times D^{k+1})_i \subset M \times (0,1)$, disjoint from D^{2k}, representing a basis of $H_k(M)$. Now, for a diffeomorphism $f \in$ ker $\tilde{\pi}_0$ S Diff(M rel D^{2k}) $\longrightarrow$ Hom($H_k(M)$, $S\pi_k(SO(k))$) we take its mapping torus M_f. We want to kill $\pi_1(M_f)$ by a sequence of surgeries. We do this using the embedding $S^1 \times D^{2k} \subset M_f$, which exists since $f|_{D^{2k}} = Id$, and the embeddings $(S^k \times D^{k+1})_i \subset M \times (0,1) \subset M_f$. From the work of Kervaire-Milnor ([6]) together with the fact that $H_k(M_f) = H_k(M)$ is torsion free it follows that the resulting manifold is a homotopy sphere which depends only on the pseudo-isotopy class of f rel D^{2k} and is denoted by $\sigma(f)$.

We get a bordism between M_f and $\sigma(f)$ by adding handles to $M_f \times I$ using the embeddings above. This bordism W is a k-connected manifold and its k+1-homology is isomorphic to $H_{k+1}(M_f)$ by inclusion. For our proof we need an additional property of this bordism, namely that all elements of $H_{k+1}(W)$ can be represented by embedded spheres with trivial normal bundle. I don't know whether this is already true for this bordism. But in any case we can get such a manifold by two surgeries on this bordism. First we do surgery with $S^1 \times D^{2k+1} \subset M_f \times (0,1)$ which is contained in our original bordism. The resulting manifold already has the desired property for H_{k+1}. For this we use that $\chi(f) = 0$. But its second homology is now equal to $\mathbb{Z}$ which can be killed by a second surgery.

We summarize the properties of the bordism W

1) W is k-connected

2) the inclusion $H_{k+1}(M_f) \longrightarrow H_{k+1}(W)$ induces an isomorphism and all elements of $H_{k+1}(W)$ can be represented by embedded spheres with trivial normal bundle. This implies that the signature of W is zero.

3) The embedding of $S^1 \times D^{2k}$ into M_f coming from the fact that $f|_{D^{2k}} = \mathrm{Id}$ can be extended to an embedding of $D^2 \times D^{2k}$ into W meeting M_f transversally.

Remark: It's an easy exercise in elementary surgery to show that if W is any manifold with these properties and ∂W is equal to M_f and a homotopy sphere then this homotopy sphere is equal to $\sigma(f)$.

Remark: If M_f is a framed manifold and the embeddings above are compatible with the framing we get W as a framed manifold and in particular we get a framing on $\sigma(f)$ from the framing on M_f. We need this for the proof of theorem 3.

Now, we show that σ is a homomorphism. For diffeomorphisms f and f' in $\ker \pi_0 S\,\mathrm{Diff}(M \text{ rel } D^{2k}) \longrightarrow \mathrm{Hom}\,(H_k(M), S\pi_k(SO(k)))$ we consider manifolds W and W' as above. Let S denote the bordism between $M_f + M_{f'}$ and $M_{ff'}$ given by the fibration with fibre M over the twice punctured disk D^2_2 classified by f and f' as indicated in the following picture.

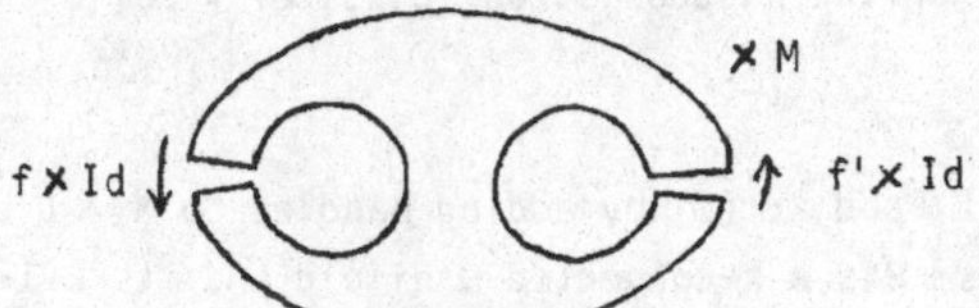

Now, we consider the manifold $S \cup W \cup W'$ with boundary consisting of $M_{ff'}$ and $\sigma(f) + \sigma(f')$. It follows again from a standard surgery argument that we can by a sequence of surgeries replace this manifold by one which fulfils the conditions above. Together with the first remark above this implies that $\sigma(ff') = \sigma(f) + \sigma(f')$.

σ is surjective. This follows from the fact that for a diffeomorphism f which is the image of a homotopy sphere Σ under the homomorphism $\theta_{2k+1} \rightarrow \tilde{\pi}_0 S\,\mathrm{Diff}(M \text{ rel } D^{2k})$ it is known that $M_f = M \times S^1 \# \Sigma$ ([2], Lemma 1). This implies $\sigma(f) = \Sigma$.

We finish the proof by showing that σ is injective. If $\sigma(f) = S^{2k+1}$ we consider

$W \cup D^{2k+2}$, where W is as above a bordism between M_f and $\sigma(f)$. Then we attach to this manifold a handle along $S^1 \times D^{2k} \subset M_f$. The resulting 1-connected manifold $\tilde{W}$ has the following properties, which can be verified rather easily.

1) $\partial\tilde{W} = N \times I \underset{f \cup Id}{\cup} N \times I$, where $N = M - \mathring{D}^{2k}$ and $f \cup Id$ is the diffeomorphism on $\partial(N \times I) = N \cup N$ given by f and Id.

2) $H_2(\tilde{W}) \cong \mathbb{Z}$, generated by an embedded sphere with trivial normal bundle. $H_i(\tilde{W}) = \{0\}$ for $2 < i \leq k$.

3) $H_{k+1}(W) \xrightarrow{\cong} H_{k+1}(\tilde{W})$ and we have an exact sequence
$$0 \longrightarrow H_{k+1}(\tilde{W}) \longrightarrow H_{k+1}(\tilde{W},N) \longrightarrow H_k(N) \longrightarrow 0$$

As $H_{k+1}(\tilde{W})$ is a subspace of half dimension in $H_{k+1}(\tilde{W},N)$ in which all elements can be represented by embedded spheres with trivial normal bundle we can kill $H_*(\tilde{W},N)$ by a sequence of surgeries. Now, Proposition 1 implies that the diffeomorphism $f \cup Id$ on $\partial(N \times I)$ can be extended to $N \times I$. But this implies that $f|_N$ is pseudo-isotopic to Id rel $\partial N = S^{2k-1}$. Thus f is pseudo-isotopic to Id in Diff(M rel D^{2k}).

q.e.d.

To complete the computation of π_0Diff(M) we use the following exact sequence for a 1-connected manifold ([19], p.265):
$$\mathbb{Z}_2 = \pi_1(SO(2k)) \longrightarrow \tilde{\pi}_0\text{Diff}(M \text{ rel } D^{2k}) \longrightarrow \pi_0\text{Diff}(M) \longrightarrow 0$$
The homomorphism $\pi_1(SO(2k)) \longrightarrow \tilde{\pi}_0\text{Diff}(M \text{ rel } D^{2k})$ is defined as follows. We extend the embedding of D^{2k} into M to an embedding of $2 \cdot D^{2k}$ into M. For $\gamma: (I, \partial I) \longrightarrow (SO(2k),e)$ we define a diffeomorphism on M by the identity on D^{2k} and outside $2D^{2k}$ and by $x \mapsto \gamma(|x|-1) \cdot x$ for $x \in 2D^{2k} - D^{2k}$.

It is obvious that this diffeomorphism is contained in ker $\tilde{\pi}_0 S\,\text{Diff}(M \text{ rel } D^{2k}) \longrightarrow \text{Hom}(H_k(M), S\pi_k(SO(k)))$. Thus we can apply σ to it. If γ is the nontrivial element in $\pi_1 SO(2k)$ we denote the image under σ of the corresponding diffeomorphism by Σ_M. Now, it is clear that Theorem 2 follows from Propositon 3 and the exact sequence above. Then the definition of Σ_M gives Lemma 4.

Remark: It is useful to have the following description of Σ_M. Let f be the diffeomorphism corresponding to the nontrivial element in $\pi_1(SO(2k))$. There is a diffeomorphism $M_f \longrightarrow S^1 \times M$ which is the identity outside $S^1 \times 2D^{2k}$ and whose restriction to $S^1 \times D^{2k}$ corresponds to the twisting by the nontrivial element $\gamma \in \pi_1 SO(2k)$. Thus Σ_M can be obtained from $S^1 \times M$ by a sequence of surgeries starting with the embedding of $S^1 \times D^{2k}$ into $S^1 \times M$, which maps $(x,y) \mapsto (x, \gamma(x)\cdot y)$ and then killing $H_k(S^1 \times M)$ by arbitrary surgeries.

Now we come to the proof of Theorem 3.

Proof of Theorem 3: If M is a framed manifold with framing α we can obtain Σ_M by framed surgeries on $S^1 \times M$ with the product of the nontrivial framing on S^1 and the framing α on M. Then we obtain Σ_M as a framed manifold which is framed bordant to $\eta \circ [M,\alpha]$.
This gives the proof of theorem 3, a.

For the proof of part b) and c) we need the δ-invariant of a framed manifold ([8]). For a framed manifold (V^{4n-1}, α) there exists an $r>0$ such that $r(V,\alpha)$ bounds a framed manifold (W, β). $\delta(V,\alpha) := \frac{1}{r}\cdot \tau(W) \in \mathbb{Q}$. It can be considered as the defect of the signature theorem for any manifold bounding V, where we have to use relative characteristic classes with respect to α in the L-polynomial. We need the following properties of this invariant. If we fix a framing β on V then - with respect to this framing - the set of all homotopy classes of framings on V is euqal to $[V,SO]$. The following formula is true.

$$\delta(V, \gamma_1\cdot\gamma_2) = \delta(V,\gamma_1) + \delta(V,\gamma_2) - \delta(V,\beta)$$

where $\gamma_1, \gamma_2 \in [V,SO]$ and (V,γ_i) denotes the framed manifold corresponding to γ_i with respect to β. If we fix the restriction to S^{4n-1} of the framing of D^{4n} then $\delta: \pi_{4k-1}(SO) = \mathbb{Z} \longrightarrow \mathbb{Q}$ is an injective homomorphism. The framings on S^{4k-1} are classified by δ. The δ-invariant mod 1 is a framed bordism invariant and is equal to $\pm a_n \cdot 2^{2n+1}(2^{2n-1}-1)\cdot e_R$, the real Adams invariant, where $a_n = \begin{cases} 1 \text{ for } n \text{ even} \\ 2 \text{ for } n \text{ odd.} \end{cases}$

For the proof of b) we consider a manifold M^{4n} with $\tau(M) = s\cdot\sigma_n$, $s \neq 0$. We consider a framing β on $M - \mathring{D}^{4n}$. The restriction of β to S^{4n-1} is a nontrivial element in $\pi_{4n-1}(SO)$, as the δ-invariant is equal to $\tau(M)$. Since im J has even order this element has even order. Thus there exists a framing $\tilde{\beta}$ on S^{4n-1} such that $2\tilde{\beta} = \beta|_{S^{4n-1}}$ regarded as elements in $\pi_{4n-1}(SO)$. From the correspondence between the δ-invariant and the real e-invariant it follows that the framed bordism class $[S^{4n-1},\tilde{\beta}]$ is equal to $s\cdot\varsigma$, where ς is the element of order 2 in im J_{4n-1}.

Now we construct an element in the Toda bracket $[s\cdot\varsigma, 2, \eta]$ as follows. We consider the standard framed bordism between $2(S^{4n-1},\tilde{\beta})$ and $(S^{4n-1}, \beta|_{S^{4n-1}})$ and glue the product of this manifold with (S^1,γ) to $(S^1\times I\times S^{4n-1}, \gamma\times \mathrm{Id}\times\tilde{\beta})$ along $2(S^1\times S^{4n-1})$ with an appropriate orientation preserving diffeomorphism to obtain a framed manifold (V,ζ) with boundary $(S^{4n-1},\beta|_{S^{4n-1}})$. The union of (V,ζ) with $(S^1\times(M-\mathring{D}^{4n}), \gamma\times\beta)$ along $S^1\times S^{4n-1}$ is contained in $[s\cdot\varsigma, 2, \eta]$. To finish the proof we have to show that this manifold is framed bordant to Σ_M with a suitable framing.

We will show that (V,ζ) is framed bordant modulo boundary to a manifold which is diffeomorphic to $D^2\times S^{4n-1}$ by a diffeomorphism which is equal to $(x,y)\overset{g}{\longmapsto}(x,\gamma(x)\cdot y)$ on the boundary. γ is the nontrivial element in $\pi_1(SO(4n))$. But this implies that our manifold above is framed bordant to $D^2\times S^{4n-1}\cup_g S^1\times(M-\mathring{D}^{4n})$ with some framing. Now, this manifold is obtained from $S^1\times M$ by surgery with the embedding $(x,y)\longmapsto(x,\gamma(x)\cdot y)$ and by the remark on p. 16 we can obtain Σ_M from it by a sequence of framed surgeries.

To show that (V,ζ) is framed bordant modulo boundary to a manifold diffeomorphic to $D^2\times S^{4n-1}$ we do surgery on it. V has the following homology:

$$H_1(V)\cong\mathbb{Z}\oplus\mathbb{Z}\ ;\ H_2(V)\cong\mathbb{Z}\ ;\ H_3(V)=\cdots=H_{4n-2}(V)=\{0\}\ ;\ H_{4n-1}(V)\cong H_{4n}(V)\cong\mathbb{Z}.$$

Now, we kill $H_1(V)$ and $H_2(V)$ by framed surgery and obtain a framed manifold S with the desired properties. This can be seen as follows. We consider $\tilde{S} := S-\mathring{D}^2\times S^{4n-1}$, where $D^2\times S^{4n-1}$ is a tubular neighbourhood of an embedded $S^{4n-1}\subset\mathring{S}$ which is isotopic to $\{1\}\times S^{4n-1}\subset\partial(S)=S^1\times S^{4n-1}$. $\tilde{S}$ fulfils the condition of the Browder-Levine fibration theorem ([3]). Thus the fibration $\partial\tilde{S}=S^1\times S^{4n-1}+S^1\times S^{4n-1}\longrightarrow S^1$ can be extended to a fibration $\tilde{S}\longrightarrow S^1$. From the homology of V it is easy to see that the fibre is a h-cobordism between S^{4n-1} and S^{4n-1}. Thus it is diffeomorphic to $S^{4n-1}\times I$. This implies that S is diffeomorphic to $D^2\times S^{4n-1}$

by a diffeomorphism whose restriction to the boundary is given by an element of $\pi_1(SO(4n-1))$. But as the framing on ∂S given by $\gamma \times \beta|_{S^{4n-1}}$ can be extended to S, this must be the nontrivial element. This ends the proof of part b.

For the proof of part c we begin with the case k odd. Suppose $\Sigma_M = S^{2k+1}$. For a framing β on M we have shown in the second remark on p.14 that we can extend the framing $\gamma \times \beta$ on $S^1 \times M$ to a framing on W. We denote the restriction of this framing to $\Sigma_M = S^{2k+1}$ by ζ. We are done if ζ extends to D^{2k+2} and this is equivalent to $\delta(S^{2k+1}, \zeta) = 0$. But $\delta(S^{2k+1}, \zeta) = \delta(S^1 \times M, \gamma \times \beta)$ as $\tau(W) = 0$. Since γ considered as an element of $\pi_1(SO)$ has order 2 the formula for the δ-invariant above implies:

$$\delta(S^1 \times M, \tau \times \beta) = 2 \cdot \delta(S^1 \times M, \gamma \times \beta) - \delta(S^1 \times M, \tau \times \beta)$$

where τ is the trivial framing on S^1. On the other hand $\delta(S^1 \times M, \tau \times \beta) = 0$, as $(S^1 \times M, \tau \times \beta)$ bounds the framed manifold $D^2 \times M$ with signature 0. Thus $\Sigma_M = 0$ implies $\eta \circ [M, \beta] = 0$.

If $\eta \circ [M, \beta] = 0$ then (Σ_M, ζ) bounds a framed manifold (V, ϱ). Thus $\Sigma_M \in bP_{2k+2}$ and is determined by the signature of V ([6]). But $\tau(V) = \delta(\Sigma_M, \zeta)$ and this is zero as shown above.

The case k even and $bP_{2k+2} = \{0\}$ can be seen in a similar but even simpler way. For the case k even and $bP_{2k+2} \neq \{0\}$ we first have to show that if $\eta \circ [M, \beta] = 0$ then $\Sigma_M \in bP_{2k+2}$. If $\eta \circ [M, \beta] = 0$ it follows that (Σ_M, γ) is framed bordant to zero. Thus Σ_M bounds a framed manifold. Now, the case k even and $bP_{2k+2} \neq 0$ follows as the cases above using in addition the fact that bP_{2k+2} is classified by the Arf invariant ([6]).

q.e.d.

Proof of Lemma 3: a) Let V be the standard bordism between M + N and M#N. We consider the manifold $S := W_M + W_N \cup S^1 \times V \cup W_{M\#N}$ where W_M is the bordism between $S^1 \times M$ and Σ_M as in the definition of Σ_M. We want by a sequence of surgeries to replace S by an h-cobordism between $\Sigma_M + \Sigma_N$ and $\Sigma_{M\#N}$.

S is 1-connected and has the following homology. $H_i(S) = \{0\}$ for $0 < i \leq k$ and $i \neq 2$. $H_2(S) \cong \mathbb{Z} \oplus \mathbb{Z}$. The second Stiefel-Whitney class $w_2(S)$ is zero. This follows from the fact that the product of the non-trivial spin-structure of S^1 with the spin-structure on M,N and M#N can be extended to W_M, W_N, $W_{M\#N}$ and $S^1 \times V$. This gives a spin-structure on S. Thus all elements in $H_2(S)$ can be represented by embedded spheres with trivial normal bundle.

For $H_{k+1}(S)$ one obtains the following information from a Mayer-Vietoris sequence. There is an exact sequence

$$0 \longrightarrow H_{k+1}(S^1 \times M) \oplus H_{k+1}(S^1 \times N) \longrightarrow H_{k+1}(S) \longrightarrow H_k(S^1 \times M) \oplus H_k(S^1 \times N) \longrightarrow 0$$

As the map on the left side factorizes through W_M and W_N and all elements in $H_{k+1}(W_M)$ and $H_{k+1}(W_N)$ can be represented by embedded spheres with trivial normal bundle we get a subspace of half the dimension in $H_{k+1}(S)$ with the same property.

It is well known that these properties imply that we can replace S by a sequence of surgeries by a h-cobordism between $\Sigma_M + \Sigma_N$ and $\Sigma_{M\#N}$.

b) If M bounds a framed manifold V than Σ_M bounds the s-parallelizable manifold $S := W \cup S^1 \times V$. Thus $\Sigma_M \in bP_{2k+2}$. If k is odd the vanishing of the signature of W and the Novikov-additivity imply that $\tau(S) = 0$. Thus $\Sigma_M = 0$ in this case.

If k is even we have to show that the Arf-invariant of Σ_M is zero. First we can assume that V is k-1-connected and that $H_k(V,M) = \{0\}$. This implies that $H_i(S) = \{0\}$ for $0 < i \leq k$ and $i \neq 2$ and that $H_2(S) \cong \mathbb{Z}$. A Mayer-Vietoris argument similar to that in a) shows that there is a direct summand in $H_{k+1}(S)$ of half the dimension in which all elements can be represented by spheres with trivial normal bundle. So the Arf-invariant of Σ_M vanishes.

c) This follows immediately from the definition of Σ_M and the geometric description of the Milnor-Munkres-Novikov pairing.

q.e.d.

Fachbereich Mathematik
Universität Mainz
Saarstr. 21
D 65oo Mainz
West Germany

References

[1] J.F. Adams: On the group J(x) - IV, Topology 5, 21-71 (1966)

[2] W. Browder: Diffeomorphisms of 1-connected manifolds, Trans. A.M.S. 128, 155-163 (1967).

[3] W. Browder and J. Levine: Fibering manifolds over a circle, Comm. Math. Helv. 4o, 153-16o (1965/66)

[4] J. Cerf: The pseudo-isotopy theorem for simply connected differentiable manifolds, Manifolds Amsterdam, Springer Lecture notes 197, 76-82 (197o)

[5] M. Kervaire: Some nonstable homotopy groups of Lie groups, Illinois J. Math 4, 161-169 (196o)

[6] M. Kervaire and J. Milnor: Groups of homotopy spheres, Ann. of Math. 77, 5o4-537 (1963)

[7] A. Kosinski: On the inertia group of π-manifolds, Am. J. Math. 89 227-248 (1967)

[8] M. Kreck: Eine Invariante für stabil parallelisierte Manigfaltigkeiten, Bonner Math. Schriften Nr. 66 (1973)

[9] M. Kreck: Bordismusgruppen von Diffeomorphismen, Habilitationsschrift, Bonn (1976)

[1o] G. Lusztig, J. Milnor and F.P. Peterson: Semi-characteristics and cobordism, Topology 8, 357-36o (1969)

[11] H. Saito: Diffeomorphism groups of $S^p \times S^q$ and exotic spheres . Quart. J. Math. Oxford, 2o, 255-276 (1969)

[12] S. Smale: On the structure of manifolds, Amer. J. Math. 84, 387-399 (1962)

[13] R.E. Stong: Notes on cobordism theory. Math. notes, Princeton University Press (1968)

[14] D. Sullivan: Infinitesimal computations in topology, Publ. I.H.E.S. 47 Paris, 1978

[15] H. Toda: Composition methods in homotopy groups of spheres, Ann. of Math. Study 49, Princeton University Press, 1962

[16] E.C. Turner: Diffeomorphisms homotopic to the Identity, Trans. A.M.S 186, 489-498 (1973).

[17] E.C. Turner: A survey of diffeomorphism groups. Algebraic and geometrical methods in topology. Springer lecture notes 428, 2oo-219 (1974)

[18] C.T.C.Wall: Diffeomorphisms of 4-manifolds, J. London Math. Soc. 39, 131-14o (1964)

[19] C.T.C. Wall: Classification problems in differential topology II: Diffeomorphisms of handlebodies, Topology 2, 263-272 (1963)

[2o] C.T.C. Wall: Classification problems in differential topology III: Applications to special cases, Topology 3, 291-3o4 (1964)

[21] J. Levine: Inerta groups of manifolds and diffeomorphisms of spheres, Ann.J.Math.92, 243-258 (1970).

Inefficiently embedded surfaces in 4-manifolds

by Steven H. Weintraub

In this paper we are interested in surfaces embedded in 4-manifolds. In particular, we are interested in when two such embeddings must be "inefficient" with respect to each other, i.e. when their geometric intersection number must be greater than their algebraic intersection number. In §1, we find a lower bound for the inefficiency in some cases, and apply it to the geometry of surfaces in $S^2 \times S^2$, showing that if some homology classes are represented by embedded spheres the embeddings must be rather complicated. In §2 we record a folk construction of embeddings, and apply it to homology classes in CP^2. Also, we obtain a lower bound for the unknotting number of certain torus knots which is approximately half of the conjectured value.

§1. A bound on the inefficiency

Definition 1. If M and M' are two connected surfaces embedded smoothly and transversely in a 4-manifold N, define their inefficiency to be

$$i(M,M') = \frac{1}{2}(\#\text{points in } M \cap M' - |[M][M']|)$$

where $[M][M']$ denotes the intersection number of M and M'.

If $i(M,M') > 0$, we will say that M is embedded inefficiently with respect to M', or vice versa.

In the interest of simplicity, we assume henceforth that $[M][M'] \geq 0$; this can be arranged by a suitable choice of orientation. Also, the genus of a surface M will be denoted by g_M. All embeddings will be smooth.

Our first results will all be applications of the following theorem:

Theorem 2. Suppose $\pi_1(N^4) = 0$ and the sum of the homology classes $[M] + [M']$ is divisible by d in the free abelian group $H_2(N)$. Let $m = \max(0,[M][M'] - 1)$.

Then

$$2(g_M + g_{M'} + i(M,M')) \geq \left(\frac{([M]+[M'])^2}{d^2}\right)(d^2-1)/2$$
$$-2m - \operatorname{rank}(H_2(N)) - \operatorname{Index}(N)$$

for d odd, and similarly for d even, with $(d^2-1)/2$ replaced by $d^2/2$.

Proof. We shall identify M and M' with their images in N.

Let $M \cap M' = \{p_1,\dots,p_{k+i},q_1,\dots,q_i\}$ where the intersection number is $+1$ at $p_1,\dots,p_{k+i}$ and -1 at $q_1,\dots,q_i$. Then $k = [M][M']$ and $i = i(M,M')$.

Now we perform i surgeries on M in order to eliminate the pairs of intersections $(p_1,q_1),\dots,(p_i,q_i)$, as follows:

Identify the normal bundle $\nu(M')$ with a tubular neighborhood of M', chosen sufficiently small so as to be embedded in N, and having $\nu(M') \cap M = \{D_j^2\}$, with each D_j^2 containing one point of $M \cap M'$.

Connect p_1 to q_1 by a path I_1 on M'. Then we may identify $\partial(\nu(M') \mid I_1)$ with $I_1 \times S^1$, and now set

$\overline{M} = (M - D_1^2) \cup I_1 \times S^1$ with corners smoothed.

Do this for each pair (p_j, q_j), choosing the paths I_j disjoint. Then we obtain a new oriented surface $\overline{M}$ with $g_{\overline{M}} = g_M + i(M,M')$, and $\overline{M} \cap M' = \{p_{i+1}, \ldots, p_{i+k}\}$. If this set is empty, let S be the connected sum $\overline{M} \# M'$ in N_1 and so $g_S = g_{\overline{M}} + g_{M'}$.

Otherwise, in a neighborhood of each p_j, $\overline{M} \cap M'$ is diffeomorphic to two transverse 2-disks in $\mathbb{R}^4$, and we may delete their interiors and join their two boundaries by an embedded $S^1 \times I$. For p_{i+1}, this has the effect of taking the connected sum $\overline{M} \# M'$; afterwards, it has the effect of adding a handle, so if S is the surface obtain by performing this process at all of the points p_j, $g_S = g_{\overline{M}} + g_{M'} + ([M][M'] - 1)$.

In any case, $g_S = g_M + g_{M'} + i(M,M') + \max(0,[M][M'] - 1)$ (*).

Now S is an embedded surface representing $[M] + [M']$, a class divisible by d in $H_2(N)$, and so we have a lower bound on g_S ([2,3,4]), namely

$$2g_S \geq \frac{[S]^2}{d^2}\left(\frac{d^2-1}{2}\right) - \operatorname{rank}(H_2(N)) - \operatorname{Index}(N) \qquad (**)$$

for d odd, and similarly for d even (with $d^2/2$ instead of $(d^2-1)/2$). Substituting for g_S from (*) then yields the theorem.

Similar results to Theorem 2 have also been obtained by Patrick Gilmer in his thesis (Berkeley, 1978).

Let $N = S^2 \times S^2$, and let $H_2(N)$ be the free group on the two generators a and b, the homology classes represented by the first and second factors respectively. If $c \in H_2(N)$, S_c will denote an embedded sphere representing c.

It is known that $pa+qb$ can be represented by a smoothly embedded sphere if $|p| \leq 1$ or $|q| \leq 1$, and cannot be if $(p,q) > 1$; all other cases are unknown. We shall henceforth assume that $q > p > 1$ and $(p,q) = 1$.

Corollary 3. Suppose $p > q/2$. If M is a surface representing $pa+qb$, and M is efficiently embedded with respect to S_a, then $g_M \geq 1$. If $p > 4$, and M is efficiently embedded with respect to S_b, then $g_M \geq 1$.

Proof. Suppose M, of genus g, has inefficiency i with respect to S_a. Then M intersects q-p copies of S_a in $(q-p)(q+2i)$ points. Performing the construction at the end of the proof of Theorem 2, we obtain an embedded surface of genus $g + (q-p)(q+i-1)$ representing $qa+qb$ in $H_2(S^2 \times S^2)$. Then by (**), with $d = q$,

$$2(g+(q-p)(q+i-1)) \geq q^2-3 \quad (\text{or } q^2-2) \text{ if } q \text{ is odd (or even)}$$

and elementary algebra yields

$$2(g+(q-p)i) \geq (q-1)[2p+1-q] - [(3+(-1)^{q+1})/2] > 0 \quad \text{for } p > q/2.$$

Thus if $i = 0$, $g > 0$ (and vice versa).

For the second part of the theorem, adding $(q-p)$ copies of S_{-b} to M, a similar analysis shows that i and g cannot both be zero for $p \geq 2q/3$, except in the cases $(p,q) = (2,3)$ or $(3,4)$. Adding $(2p-q)$ copies of S_b to M then excludes all the cases where $p < 2q/3$.

We clearly get the strongest bound when $p = q-1$. The above calculation shows that an efficiently embedded surface must have genus at least $(q^2-2q-1)/2$ for q odd, and $(q^2-2q)/2$ for q even. These numbers are also a lower bound for the inefficiency of an embedded 2-sphere representing $(q-1)a + qb$, with respect to S_a.

By proper choice of c, one may derive bounds for $i(M,S_c)$ in numerous situations. For example:

Corollary 4. Suppose M is a surface representing $pa+qb$, with $(p+1,q+1) = k > 1$. Then

$$2(g_M + i(M,S_{a+b})) \geq k((p+1)^2 - 1) - 2(p+q) > 0.$$

Proof. This is a direct application of Theorem 2.

The known examples of classes representing $pa+qb$ of lowest genus are given by algebraic surfaces of genus $(p-1)(q-1)$. Since algebraic surfaces have a canonical orientation, any two of these must be efficiently embedded with respect to each other. In contrast, if any primitive class $pa+qb$, $p \geq 2$, $q \geq 2$ is represented by an embedded sphere, this embedding must be inefficient with respect to some other embedded sphere. To be precise,

Corollary 5. Suppose $p \geq 2$, $q \geq 2$, $(p,q) = 1$. If $p \equiv 0(3)$, let $c = b$. If $p \equiv q \equiv 1$ or $2(3)$, let $c = a+b$. If $p \equiv 1$ and $q \equiv 2(3)$, let $c = 2a+b$. Then if M is a smoothly embedded 2-sphere representing $pa+qb$, $i(M,S_c) > 0$.

Proof. This is a direct application of Theorem 2 with $d = 3$.

§2. Constructing embeddings

We now prove a lemma originally due to Boardman (see Proc. Comb. Soc. vol. 60).

<u>Lemma 6</u>. Let N be an oriented 4-manifold. Suppose a class $\alpha \in H_2(N^4)$ is represented by a smoothly immersed surface M of genus g, with M having k self-intersections of sign $+1$ and ℓ self-intersections of sign -1. Then $\alpha \in H_2(N \# (k\mathbb{C}P^2) \# (\ell(-\mathbb{C}P^2)))$ can be represented by a smoothly embedded surface $\overline{M}$ of genus g.

<u>Proof</u>. Let p be a self-intersection point of M of sign $+1$, and let D^4 be a small disk in N around p. Then $M \cap \partial D^4 \subset \partial D^4 = S^3$ consists of two unknotted linked circles with linking number $+1$. Now in $\mathbb{C}P^2$ the generator α and its negative $-\alpha$ are represented by two embedded 2-spheres S_1 and S_2 which intersect at a single point p', and if D^4 is a disk around p', $(S_1 \cup S_2) \cap \partial D^4 \subset \partial D^4$ consists of two unknotted circles with linking number -1.

Since in taking the connected sum we reverse the orientation of the boundary, we may arrange to take $N \# \mathbb{C}P^2$ "around" p and p' so that $M \cap \partial D^4$ and $(S_1 \cup S_2) \cap \partial D^4$ match up. The result will be a new surface $\overline{M}$, still of genus g, which represents $\alpha + \gamma + (-\gamma) = \alpha \in H_2(N \# \mathbb{C}P^2)$.

Perform this construction at every self-intersection, using $-\mathbb{C}P^2$ instead of $\mathbb{C}P^2$ at self-intersections with sign -1.

<u>Theorem 7</u>. Let N be an oriented 4 manifold, and let $N_0 = N - D^4$, D^4 a smoothly embedded 4-disk. Suppose $\alpha \in H_2(N_0, \partial)$ is represented by a smooth embedding $\Phi : (D^2, S^1) \to (N_0, \partial)$ and let K denote the knot given by $\Phi : S^1 \to \partial N_0 = S^3$. If n is the unknotting number of K, then α is represented by a smoothly embedded 2-sphere in $\overline{N} = N \# (n(\pm\mathbb{C}P^2))$.

Here α is identified with its image $H_2(N_0,\partial) \cong H_2(N_0) \to H_2(\bar{N})$, and the choice of signs is explained below.

Proof. Let $N_1 = N_0 \cup S^3 \times I \cup D^4$, so N_0 is diffeomorphic to N. Extend Φ to an immersion of $D^2 \cup S^1 \times I$ (i.e. a larger D^2) which switches the n crossings of K between $\Phi|S^1 \times \{0\}$ and $\Phi|S^1 \times \{1\}$ (and $\Phi(S^1 \times t) \subset (S^3 \times t)$). Now $\Phi(S^1 \times 1)$ is unknotted so extends to an embedding of D^2 in D^4. This gives an immersion of S^2 in N representing α with n self-intersection points, one for each crossing that was switched. Now apply Lemma 7.

The sign of a self-intersection point is determined as follows: Look at a presentation of the oriented knot. If, in a crossing, the overpass has to be rotated counter-clockwise to agree with the underpass, the sign is +1, if clockwise -1 (see [1]).

+1 -1

(Observe that the embeddings constructed above all have inefficiency 1 with respect to 2-spheres representing the generators of the second homology group of each $\pm CP^2$ added in. Indeed, if the surface did not intersect such a 2-sphere, that copy of $\pm CP^2$ could be dispensed with.)

Corollary 8. If γ denotes a generator of $H_2(CP^2)$, $n\gamma$ can be represented by an embedded sphere in $CP^2 \# (n-1)(n-2)/2\ CP^2$.

Proof. Represent $n\gamma$ by n embedded 2-spheres in general position; these will have n(n-1)/2 self-intersections. Use (n-1) of these intersections to connect them, and then apply the above lemma.

It is conjectured that the unknotting number of a torus knot of type (p,q) is equal to $(p-1)(q-1)/2$. Our method yields the following result:

Corollary 9. The unknotting number of the torus knot of type $(n-1,n)$ is at least $(n^2-5)/4$ if n is odd, and $(n^2-4)/4$ if n is even.

Proof. By [1], $n\gamma$ can be represented by an embedded disk with boundary the torus knot of type $(n-1,-n)$, where γ is the generator of $H_2(CP^2 - D^4,\partial)$. If the unknotting number of this knot is k, Theorem 7 shows that $n\gamma$ is represented by an embedded 2-sphere in $\#(k+1)CP^2$.

Then by (**),

$$0 \geq \left(\frac{n^2-1}{2}\right) - 2(k+1) \qquad \text{for } n \text{ odd}$$

and

$$0 \geq \left(\frac{n^2}{2}\right) - 2(k+1) \qquad \text{for } n \text{ even}$$

and elementary algebra yields the corollary.

Finally, note that the unknotting number in Theorem 7 and Corollary 9 may be replaced by the "slicing number," where the slicing number of a knot is the minimum number of crossings which must be switched in order to make the knot slice.

References

1. M. Kervaire and J. Milnor, On 2-spheres in 4-manifolds, Proc. National Academy of Sciences 47 (1961), 1651-1657.

2. V. A. Rokhlin, Two-Dimensional Submanifolds of Four-Dimensional Manifolds, Functional Anal. Appl. 5 (1971), 39-48.

3. P. E. Thomas and J. Wood, On Manifolds Representing Homology Classes in Codimension 2, Invent. Math. 25 (1974), 63-89.

4. S. H. Weintraub, $\mathbb{Z}_p$-actions and the Rank of $H_n(N^{2n})$, J. London Math. Soc. (2), 13 (1976), 565-572.

Local Surgery: Foundations and Applications

Laurence Taylor and Bruce Williams*

In sections 1 through 7 of this paper we collect the basic results of local surgery theory. Sections 1 through 6 merely collect results found in Quinn [16]. We incorporate a twist motivated by Barge's work [3], and rearrange the material to suit our needs in sections 7, 8, and 9. The theory parallels the integral theory until one goes to calculate the normal map set. Here Quinn found an extra obstruction (see section 6).

Section 7 is a general section in which we try to handle Quinn's extra obstruction and the surgery obstruction simultaneously. We give two applications of the general theory to embedding theory in sections 8 and 9. Hopefully more applications will be forthcoming.

We must apologize to the many people who have worked in this area but are not mentioned here. A combination of ignorance and lack of space prevents a detailed look at the historical foundations of local surgery. Our thanks go to Frank Quinn for helpful conversations on the material in [16].

§1. Basics.

We begin by fixing some notation. We let P denote an arbitrary subset of primes in Z, and we let P' denote the complementary set. We let R denote the subring of Q consisting of all rationals with

*Both authors were partially supported by NSF Grant MCS76-07158.

denominators relatively prime to the primes in P, and we use R' to denote the complementary subring.

We use a localization process which preserves the geometry coming from π_1. If X is a CW complex, consider the map

$$1.1) \quad u: X \to K(\pi_1, 1) = B\pi$$

which classifies the universal cover. We convert u to a fibration and apply the fibrewise localization functor of Bousfield - Kan [4] p. 40. We get a commutative diagram

$$\begin{array}{ccccc} \tilde{X} & \to & X & \to & B\pi \\ \downarrow & & \downarrow & & \downarrow \\ \tilde{X}_{(P)} & \to & X_{(P)} & \to & B\pi \end{array}$$

where $\tilde{X}_{(P)}$ is the usual localization of the simply connected space $\tilde{X}$.

A map $f: X \to Y$ is a P-equivalence if the induced map $f_P: X_{(P)} \to Y_{(P)}$ is a homotopy equivalence. A space is P-local if the map $X \to X_{(P)}$ is a homotopy equivalence.

§2. Local Poincaré spaces.

We say that a P-local space, denoted X, is a simple P-local P. D. space if there exists a finite complex, K, and a P-equivalence $\rho: K \to X$, together with

i) a homomorphism $w_1: \pi_1 X \to Z/2$ and

ii) a class $[X] \in H_m(X; R^t)$ such that

$$\xi\cap : \mathrm{Hom}_\Lambda (C_{m-*}(X); R\pi) \to C_*(X) \otimes_\Lambda R\pi$$

is a simple equivalence, where $\Lambda = Z\pi$ and ξ is a chain representative for [X]. For more details, see Anderson [1] p. 39 and Wall [24] p. 21. In particular, the notion of a simple P-local Poincaré n-ad should be clear.

Remark: The choice of K and ρ determines the P-local simple homotopy type of X.

Definition 2.1: An oriented P-local Poincaré space consists of a simple P-local P. D. space X; a specific choice of [X]; and a fixed P-local simple homotopy type for X. We denote such a gadget by (X;[X]), suppressing the simple type.

§3. Normal maps.

We agree to let C stand for O, PL, or TOP: then $BSC_{(P)}$ denotes the localization of the classifying space BSC. Given a C-manifold, M, we have the map u: $M \to B\pi$ (1.1). The homomorphism w_1: $\pi_1 M = \pi \to Z/2$ gives rise to a line bundle λ over $B\pi$. If ν_M denotes the normal bundle of M, $\nu_M \oplus u^*(-\lambda)$ is orientable. Hence we get a map

$$3.1)\quad \eta_M: M \to BSC \times B\pi$$

from which we can recover both u and ν_M. In fact, w_1 can be used to get a map μ: $BSC \times B\pi \to BC$ such that $\mu \circ \eta_M = \nu_M$ and $(2^{nd}\text{ projection}) \circ \eta_M = u$.

Definition 3.2: An "oriented" manifold is a manifold M together with a choice of class $[M] \varepsilon H_m(M, \partial M; Z^t)$.

Remark: The bundle $\nu_M \oplus u^*(-\lambda)$ is now oriented.

Definition 3.3: A degree 1, P-normal map is a map $f\colon M \to X$ and a map $\zeta_P\colon X \to BSC_{(P)}$ such that

i)
$$\begin{array}{ccc} M & \xrightarrow{\quad f \quad} & X \\ \downarrow \eta_M & & \downarrow \zeta_P \times u \\ BSC \times B\pi_1 M & \to & BSC_{(P)} \times B\pi_1 X \end{array}$$
commutes

ii) $f^* w_1$ is the first Stiefel-Whitney class of M, and

iii) $f_*[M] = [X]$.

There is an obvious generalization to n-ads. This permits us to define the set of bordism classes of degree 1, P-normal maps over the oriented Poincaré complex $(X;[X])$.

We denote this set by $N(X;[X])$.

§4. Surgery.

Our goal is to define and interpret a surgery obstruction map

$$4.1) \quad \sigma_* : N(X;[X]) \to L_m^s(R\pi_1 X; w_1).$$

To begin, we form the pullback
$$\begin{array}{ccc} E(\zeta_P) & \to & BSC \\ | & & | \\ X & \xrightarrow{\zeta_P} & BSC_{(P)} \end{array}$$

Given a degree 1, P-normal map $f\colon M \to X$ and ζ_P, we get a map $\hat{f}\colon M \to E(\zeta_P)$. We will need

Lemma 4.2: Let K be a finite complex and let $g\colon K \to F$ be a map. Suppose there exists a finite complex, L, and a P-equivalence $L \to F_{(P)}$. Then there exists a finite complex L_∞ such that

i) g factors as $K \xrightarrow{g_\infty} L_\infty \xrightarrow{r_\infty} F$

ii) r_∞ is a P-equivalence.

Proof: We shall define a series of spaces L_i and maps g_i, r_i such that $g = r_i \circ g_i$. Let $L_0 = K$; $g_0 = 1_K$; $r_0 = g$.

Since $\pi_1 F$ is finitely presented, we can attach a finite number of cells to L_0 to get a complex L_1 and a map $r_1 \colon L_1 \to F$ which is an isomorphism on π_1. The map g_1 is the obvious inclusion $K = L_0 \subset L_1$.

Suppose we have constructed L_i, g_i, and r_i so that $(r_i)_P \colon (L_i)_{(P)} \to F_{(P)}$ is an i-equivalence. Then $\pi_{i+1}(F_{(P)}, (L_i)_{(P)})$ is a finitely generated $R\pi$-module (e.g. [24] Lemma 2.3 (b)). We can choose a finite set of elements in $\pi_{i+1}(F, L_i)$, attach cells to get L_{i+1}, and extend the maps. As usual, $(r_{i+1})_P$ is now an (i+1)-equivalence.

Construct L_i, g_i, r_i for $i = \max(\dim L, 2)$. Then Lemma 2.3 of Wall [24] shows that $\pi_{i+1}(F_{(P)}, (L_i)_{(P)})$ is s-free over $R\pi$. By adding more (i+1)-cells to L_i, we can assume it free and choose elements in $\pi_{i+1}(F, L_i)$ to give a basis for $\pi_{i+1}(F_{(P)}, (L_i)_{(P)})$. Then $L_\infty = L_{i+1}$; $g_\infty = g_{i+1}$; $r_\infty = r_{i+1}$ satisfy all the requirements. //

Once upon a time we had a map $\hat{f} \colon M \to E(\zeta_P)$. Use Lemma 4.2 to find a finite complex K and a factorization of $\hat{f} \colon M \xrightarrow{g} K \xrightarrow{r} E(\zeta_P)$. Over $E(\zeta_P)$ we have a C-bundle, $\zeta \oplus \lambda$, where $\zeta \colon E(\zeta_P) \to BSC$ and λ is the line bundle given by

$\lambda\colon E(\zeta_P) \to X \xrightarrow{u} B\pi \xrightarrow{Bw_1} RP^{\infty}$. The bundle $\zeta \oplus \lambda$ restricts to a bundle $r^*(\zeta \oplus \lambda)$. With this bundle over K, the map $g\colon M \to K$ becomes a normal map in the sense of Anderson [1] and so has a well-defined surgery obstruction. Using an n-ad version of Lemma 4.2, we see that the obstruction in $L^s_m(R\pi; w_1)$ depends only on the degree 1, P-normal map. We get

<u>Theorem 4.3</u>: The map σ_* (4.1) has the property that $\sigma_*(f, \zeta_P) = 0$ iff $f\colon M \to X$ is normally bordant to a simple P-equivalence (provided, as usual, dimension $M \geq 5$).

Even more is true. Let

$$\begin{array}{ccc} M & \xrightarrow{f} & F \\ & \searrow^{\nu_M} \quad \swarrow_{\zeta} & \\ & BC & \end{array}$$

commute, and suppose there is a P-equivalence $\beta\colon F \to X$ such that $\beta \circ f$, and $\zeta_P \circ \beta_P^{-1}$ give a degree 1, P-normal map $M \to X$. Then, if $\sigma_*(\beta \cdot f, \zeta_P \cdot \beta_P^{-1}) = 0$, $f\colon M \to F$ is normally bordant over F to a map $f_1\colon M_1 \to F$ which is a simple P-equivalence. Furthermore, if F is a finite complex, then f_1 can be chosen to be $\left[\frac{m-2}{2}\right]$-connected.

<u>Proof</u>: One uses Lemma 4.2 and the material in Anderson [1] to prove all but the last sentence. This follows as in Cappell - Shaneson [6] Addendum to 1.7, p. 293. //

Remark: Theorem 4.3 has a straightforward n-ad version. The experts can amuse themselves by considering non-simple, P-local,

P. D. spaces; doing surgery to get P-equivalences with exotic torsions; introducing Γ-groups [6]; etc.

Remark: If we define $s_C(X;[X])$ to be the set of degree 1, simple P-equivalences $f: M \to X$ (M a C-manifold) modulo the relation of P-local s-cobordism, then the usual long exact sequence (e.g. Wall [24] 10.3 and 10.8) is valid.

§5. The local Spivak normal fibration and local lifts.

As usual X is a P-local Poincaré space. Let $\rho: K \to X$ be a P-equivalence from a finite complex K. We can embed K in some large sphere and take a regular neighborhood $(N^{m+k}, \partial N)$. Make the inclusion map $\partial N \to N$ into a fibration, and let F denote the fibre.

We can localize the entire fibration and it is easy to redo Spivak [21] to prove that $F_{(P)}$ is a local sphere and that the associated stable spherical fibration

$$\nu_X: X \to N_{(P)} \to BSG_{(P)} \times K(R^*,1)$$

is unique (R^*= units of R). (Recall that $BSG_{(P)} \times K(R^*,1)$ is the classifying space for P-local spherical fibrations, Sullivan [22] p. 4.14 and May [13].)

More is available from our geometry. Instead of considering $F_{(P)}$ we can use Serre class theory and compute $H_*(F;Z)$ modulo the class of P'-torsion groups. One easily discovers that $H_*(F;Z)$ is P'-torsion, $* \neq k-1$, and $H_{k-1}(F;Z)$/Torsion is a rank 1 abelian group. The cohomology groups have a similar description. The universal coefficients theorem and Fuks [8], p.111, Prop. 85.4, then show that $H_{k-1}(F;Z)/P'\text{-torsion} = Z$. Hence the map ν_X factors through $BSG_{(P)} \times K(Z^*,1)$, and the map $X \to K(Z^*,1) = RP^{\infty}$ is just w_1. Hence, just as for manifolds, we can define a map

$$5.1) \quad \eta_X \colon X \to BSG_{(P)} \times B\pi$$

Over $BSG_{(P)} \times B\pi$ we have the universal fibration $\mu_P \times \lambda$. If we pull this fibration back over X, we get ν_X and we can form the Thom spectrum $\mathfrak{J}(\nu_X)$.

Note: All Thom spectra are indexed so that the Thom class has dimension 0.

In $\pi_m(\mathfrak{J}(\nu_X))$ there are elements c_X, which, once we orient X, map to [X] under the Hurewicz and Thom maps. We choose one of these once and for all and refer to it as the local reduction of the Thom spectrum for the Spivak normal fibration of X.

Definition 5.2: We define $\mathrm{Lift}(\eta_X)$ to be the set of lifts of η_X to $BSC_{(P)} \times B\pi$. We suppress which C as it is either clear from context or irrelevant.

We have the usual map

$$5.3) \quad \ell \colon N(X;[X]) \to \mathrm{Lift}(\eta_X)$$

The map ℓ is defined as follows. The map $X \to BSC_{(P)} \times B\pi$ is given by $\zeta_P \times u$ and the specific equivalence of the underlying local spherical fibration with η_X is specified by choosing the equivalence which takes the reduction of $\mathfrak{J}(\nu_M)$ to c_X using the map $\mathfrak{J}(\nu_M) \to \mathfrak{J}(\zeta_P \times \lambda)$ induced by our normal map. Kahn [11] and May [13] may be profitably consulted here.

Remark: If $\mathrm{Lift}(\eta_X) \neq \phi$ it is in one to one correspondence with $\left[X, (G/C)_{(P)}\right]$.

Remark: If $P \neq \phi$, there is no reason to suppose that ℓ is an isomorphism. Anderson [1] considers a less natural definition of degree 1, P-normal map and gets a map similar to ℓ but taking values in the set of lifts of ν_X to BC. He claims, Thm. 3 p. 51, that his map is an isomorphism, but we are unable to follow his proof (in particular, the first two lines).

§6. Normal maps again.

We need to calculate $N(X;[X])$ since the map ℓ (5.3) is no longer an isomorphism. This was done by Quinn [16] and we display the result following Barge [3]. Rather than interrupt the presentation later, we pause to prove

Lemma 6.1: Consider the square of connected CW complexes

$$6.2) \qquad \begin{array}{ccc} A & \longrightarrow & B \\ \downarrow & & \downarrow f \\ C & \xrightarrow{g} & D \end{array}$$

Suppose that g induces an isomorphism on π_1. Further suppose that f is a P-equivalence and that C and D are P-local spaces.

Then, if 6.2 is a fibre square, it is a cofibre square. If $\pi_1 A = 0$, then the converse holds.

Proof: Define F to be the fibre of f. Show that $H_*(F;Z)$ is P'-torsion. As in [16], the spectral sequence $H_*(D,C;H_*(F;Z))$ $H_*(B,A)$ shows $H_*(D,C) = H_*(B,A)$. The converse is easy. //

To fix notation, let $\mathfrak{J}\pi$ denote the Thom spectrum of the line bundle λ over $B\pi$. (We should probably call it $\mathfrak{J}(\pi_1, w_1)$, but we won't.) Given a lift $X \to BSC_{(P)} \times B\pi$ the composite

$$S^m \xrightarrow{c} X \to \mathfrak{J}(\nu_X) \to MSC_{(P)} \wedge \mathfrak{J}\pi$$

defines a homomorphism

$$6.3)\quad \beta_P \colon \mathrm{Lift}(\eta_X) \to MSC_m(\mathfrak{J}\pi; R) .$$

We also have a map

$$6.4)\quad \beta \colon N(X;[X]) \to MSC_m(\mathfrak{J}\pi)$$

defined by sending $\begin{array}{ccc} M & \xrightarrow{f} & X \\ \downarrow \eta_M & & \\ BSC \times B\pi_1 M & & \end{array}$ to the composite

$$S^m \xrightarrow{c} M \to \mathfrak{J}(\nu_M) \to MSC \wedge \mathfrak{J}\pi_1 M \to MSC \wedge \mathfrak{J}\pi .$$

Clearly $\begin{array}{ccc} N(X;[X]) & \xrightarrow{\ell} & \mathrm{Lift}(\nu_X) \\ \downarrow \beta & & \downarrow \beta_P \\ MSC_m(\mathfrak{J}\pi) & \to & MSC_m(\mathfrak{J}\pi; R) \end{array}$

commutes. Hence a necessary condition for a lift to be in the image of ℓ is that β_P of it must correspond to an honest manifold. This is also sufficient as 6.5 below shows.

To fix notation, let $\beta' \colon N(X;[X]) \to MSC_m(\mathfrak{J}\pi; R')$ denote β followed by the obvious coefficient homomorphism. If $\alpha \in N(X;[X])$ is given, $\ell(\alpha)$ determines a map $X \to BSC_{(P)} \times B\pi$. If λ denotes the line bundle over X induced from the fixed one on $B\pi$, we get a homomorphism

$$\theta_\alpha \colon \pi_{m+1}(\mathfrak{J}\lambda; R) \oplus MSC_{m+1}(\mathfrak{J}\pi; R') \to MSC_{m+1}(\mathfrak{J}\pi; Q) .$$

Quinn's Theorem 2.3, as reformulated by Barge, now reads

Theorem 6.5: There is an exact sequence of sets

$$N(X;[X]) \xrightarrow{\ell \times \beta'} \mathrm{Lift}(\nu_X) \times MSC_m(J\pi;R') \to MSC_m(J\pi;Q) .$$

The group $MSC_{m+1}(J\pi;Q)$ acts on $N(X;[X])$ so that two elements α_1 and $\alpha_2 \in N(X;[X])$ satisfy $(\ell \times \beta')(\alpha_1) = (\ell \times \beta')(\alpha_2)$ iff α_1 and α_2 lie in the same orbit under this action.

The isotropy subgroup of an element α is just the image of θ_α.

Proof: The proof is clear from studying Quinn [16] and Barge [3]. Lemma 6.1 is used extensively. //

Remark: Quinn [16] has also proved an n-ad version of 6.5.

§7. Surgery again.

Ranicki [17] has defined a symmetrization map

$$1+T : L_m^s(R\pi;w_1) \to L^m(R\pi;w_1) .$$

The goal of this section is to understand $1+T$ composed with the surgery obstruction map 4.1. We shall do this in terms of a homomorphism $\sigma^*: MSC_m(J\pi) \to L^m(R\pi;w_1)$ and an element $\sigma^*(X;[X]) \in L^m(R\pi;w_1)$, both defined by Ranicki [17] (or Mischenko [14] if $2 \in P'$). The formula is

$$7.1) \quad (1+T)\, \sigma_*(\ \ ) = \sigma^*\, \beta(\ \ ) - \sigma^*(X;[X])$$

This gives a solution to our problem, but we wish more. We want to define maps

$$\sigma'_*: MSC_m(\mathfrak{I}\pi;R') \to L^m(R\pi;w_1) \otimes R'$$

$$\tau^*: Lift(\nu_X) \to L^m(R\pi;w_1) \otimes R$$

such that

Theorem 7.2: The diagram

$$\begin{array}{ccc} N(X;[X]) & \longrightarrow & Lift(\nu_X) \times MSC_m(\mathfrak{I}\pi;R') \\ \downarrow (1+T)\,\sigma_* & & \downarrow \tau^* \times \sigma'_* \\ 0 \to L^m(R\pi;w_1) & \longrightarrow & L^m(R\pi;w_1)\otimes R \oplus L^m(R\pi;w_1)\otimes R' \end{array}$$

commutes.

Remark: If we think of $Lift(\nu_X)$ as the P-part of the set of normal maps, and of $MSC_m(\mathfrak{I}\pi;R')$ as the P'-part of the set of normal maps, then Theorem 7.2 says that the P-local part of the symmetrized surgery obstruction is determined by the P-local part of the normal map set, with a similar statement for P'.

The map σ'_* is easily defined: one just takes the map $\sigma^*(\ ) - \sigma^*(X;[X]): MSC_m(\mathfrak{I}\pi) \to L^m(R\pi;w_1)$ and localizes it with respect to P'. The map τ^* is almost as easy. Take the map $\Psi: Lift(\nu_X) \xrightarrow{\beta_P} MSC_m(\mathfrak{I}\pi;R) \xrightarrow{(\sigma^*)_P} L^m(R\pi;w_1) \otimes R$ and let $\tau^*(\ ) = \Psi(\ ) - \sigma^*(X;[X])_{(P)}$. The proof of Theorem 7.2 is easy.

Remark 7.3: The map $L^s_m(R\pi;w_1) \otimes R' \to L^m(R\pi;w_1) \otimes R'$ is an isomorphism by Ranicki [17], so we have determined the P'-local part of the surgery obstruction from the P'-local part of the normal map set.

Remark 7.4: If $2 \in P'$, the map $L^s_m(R\pi;w_1) \rightarrow L^m(R\pi;w_1)$ is an isomorphism. Hence we can determine each part of the surgery obstruction from the corresponding part of the normal map set.

Remark: If $2 \in P$, there is a very involved construction of a map τ_*: $\mathrm{Lift}(\nu_X) \rightarrow L^s_m(R\pi;w_1)$ so that we can compute σ_* from σ'_* and τ_*. We neither need nor pursue this refinement here.

§8. A metastable embedding theorem.

Dax [7], Laramore [12], Salomonsen [20], Rigdon [18], Rigdon-Williams [19], etc. have shown that the best metastable embedding codimension is a 2-local phenomenon. This suggests the following "converse"

Theorem 8.1: Given a smooth manifold, M^m, whose Novikov higher signature (defined below) vanishes, there exists a smooth manifold, N^m, and a map $f: N \rightarrow M$ such that

i) N embeds in S^{m+k} if $m+3 \leq 2k$

ii) f is a $(\frac{1}{2})$-local equivalence

iii) f is $\left\lceil \frac{m-2}{2} \right\rceil$-connected

iv) $f^*\nu_M = \nu_N$.

Definition 8.2: The Novikov higher signature of a manifold M is defined to be

$$\mathcal{L} \setminus (\eta_M)_*([M]) \in H_*(B\pi;Z^t_{(2)})$$

where η_M: M $\to$ BSC $\times$ Bπ is the map 3.1 and $\mathfrak{L}$ is the Morgan-Sullivan L-class in $H^{4*}(BSC;Z_{(2)})$ [15].

Remark: In the proof of 8.1 we assume only that $\sigma^*(M;[M])$ is an odd torsion element in $L^m(R\pi;w_1)$, where R denotes $Z[\frac{1}{2}]$ for the rest of sections 8 and 9. The Novikov higher signature is more easily calculated than $\sigma^*(M;[M])$. The relation between them is supplied by

Lemma 8.3: There is a homomorphism

$$A: H_*(B\pi;Z^t_{(2)}) \to L_*(R\pi;w_1) \otimes Z_{(2)}$$

such that $A(\mathfrak{L} \setminus (\eta_M)_*([M])) = \sigma^*(M;[M]) \otimes 1$.

Proof: Ranicki's methods define an assembly map $\mathbb{L}^\circ(R) \wedge \mathfrak{J}\pi \to \mathbb{L}^\circ(R\pi;w_1)$ and a map MSTOP $\wedge\, \mathfrak{J}\pi \to \mathbb{L}^\circ(R) \wedge \mathfrak{J}\pi$ so that the composite $S^m \xrightarrow{c} M \to \mathfrak{J}(\nu_M) \to \text{MSTOP} \wedge \mathfrak{J}\pi \to \mathbb{L}^\circ(R) \wedge \mathfrak{J}\pi \to \mathbb{L}^\circ(R\pi;w_1)$ is just $\sigma^*(M;[M])$. See [17] for more details.

In [23] we showed that $\mathbb{L}^\circ(R)$ is a product of Eilenberg-MacLane spectra. Anderson [2] has shown that

$$\pi_*(\mathbb{L}^\circ(R)) = \begin{cases} Z \oplus Z/2 & * \equiv 0 \pmod 4 \\ 0 & * \not\equiv 0 \pmod 4 \end{cases} \qquad (R = Z[\tfrac{1}{2}])$$

Classical quadratic form theory and the methods of [23] provide classes $L_i \in H^{4i}(\mathbb{L}^\circ(R);Z_{(2)})$ and $h_i \in H^{4i}(\mathbb{L}^\circ(R);Z/2)$ which give the decomposition. The map MSTOP $\to \mathbb{L}^\circ(R)$ is described at 2 by the fact that the h_i restrict to 0 and the L_i restrict to the Morgan-Sullivan L-class. This proves 8.3. //

Remark: This proof was our original motivation for [23].

We need one more lemma.

Lemma 8.4: If $(X;[X])$ is an oriented P-local Poincaré space $(2 \in P')$ then

$$\sigma^*(X;[X]) = \sigma^*(X;4^i[X]) .$$

Proof: Miscenko's version of symmetric L-theory with 2 invertible, [14], shows that $\sigma^*(X;[X])$ is determined by $C_*(X)$ and the map $\xi\cap : C^*(X) \to C_{m-*}(X)$. Multiplication by 2^i gives a chain map $C_*(X) \to C_*(X)$ which induces an equivalence from $C_*(X)$ and $4^i[X]$ to $C_*(X)$ and $[X]$. Hence they have the same σ^*. //

We can now prove 8.1. Our first goal is to produce a finite complex having the homotopy properties N is to enjoy.

Let V denote the pullback
$$\begin{array}{ccc} V & \to & BO(r) \\ \downarrow & & \downarrow \\ M & \to & BO \end{array}$$
r fixed below.

We wish to find a finite complex X and a map $g: X \to V$ such that the composite $X \to V \to M$ is an r-connected, $\frac{1}{2}$- equivalence. If k is odd, set $k = r$. If k is even, set $k-1 = r$. Note r is odd.

To begin, let X_r be an r-skeleton for M. It is easy to map $X_r \to V$ so that $X_r \to V \to M$ is the inclusion, hence r-connected. Define X_i and g_i: $X_i \to V$ inductively by adding i-cells to X_{i-1} so that g_i is $\frac{1}{2}$-locally, i-connected. Since the map $V \to M$ is $\frac{1}{2}$-locally, $(2r+1)$-connected, this is easy to do up to X_m since $m \leq 2k-2$.

Now $\pi_{m+1}(V,X_m) \to \pi_{m+1}(V_{(\frac{1}{2})},(X_m)_{(\frac{1}{2})}) \to \pi_{m+1}(M_{(\frac{1}{2})},(X_m)_{(\frac{1}{2})}) \to 0$ and, clearly, $\pi_{m+1}(M_{(\frac{1}{2})},(X_m)_{(\frac{1}{2})})$ is s-free over $R\pi$. As usual, we may assume that it is free. One can then choose elements in $\pi_{m+1}(V,X_m)$ to give a basis in $\pi_{m+1}(M_{(\frac{1}{2})},(X_m)_{(\frac{1}{2})})$ and attach cells to get X and $g: X \to V$ as required.

Over X there is a k-plane bundle, ν^k, and a stable bundle equivalence $h^*\nu_M = \nu^k$, where h is $X \to V \to M$. Hence we get a stable map $\mathfrak{T}(\nu^k) \to \mathfrak{T}(\nu_M)$ which is easily seen to be a $\frac{1}{2}$-equivalence. Hence there exists an element $c \in \pi_m(\mathfrak{T}(\nu^k))$ such that c goes to $4^e c_M$ for some positive integer e. We also have the stabilization map $\pi_{m+k}(T(\nu^k)) \to \pi_m(\mathfrak{T}(\nu^k))$, where $T(\nu^k)$ is the Thom space.

Since $m \leq 2k-2$, Theorem 0.2 of [25] assures us that we can find an integer, d, such that, for all $i \geq d$ we have an element $\gamma_i \in \pi_{m+k}(T(\nu^k))$ which goes to $4^i c_M$ under stabilization and the map $\mathfrak{T}(\nu^k) \to \mathfrak{T}(\nu_M)$.

Associated to each γ_i we get a normal map $\alpha_i\colon N_i \to M$ which is degree 1 if we consider the $\frac{1}{2}$-local oriented Poincaré space $(M;4^i[M])$. If $\sigma_*(\alpha_i) = 0$, and if $m \geq 5$, then, since $m \leq 2k-3$, Levine's work (see [19], Embedded Surgery Lemma) shows that we can do the surgery inside S^{m+k}, proving 8.1.

But we can calculate $(1+T)\ \sigma_*(\alpha_i)$. It is $\sigma^*(N_i;[N_i]) - \sigma^*(M;4^i[M])$. Lemmas 8.3 and 8.4 show that this is $4^i\sigma^*(M;[M]) - \sigma^*(M;[M])$. Since the order of $\sigma^*(M;[M])$ is odd, we can choose $i \geq d$ so that $\sigma_*(\alpha_i) = 0$ (see 7.4).

If $m < 5$, then M itself embeds in S^{m+k}, $m \leq 2k-3$. //

§9. The False - Hirsch conjecture is half true.

We begin with a local analogue of a theorem of Browder [5].

Theorem 9.1: Let M^m be a manifold such that

i) ν_M desuspends to a bundle η^k with $k \geq 2$;

ii) the image of c_M in $\pi_m(T(\nu_M)_{(P)})$ comes from an element in $\pi_{m+k}(T(\eta^k)_{(P)})$.

Then M^m embeds in Σ^{m+k+1} with normal bundle $\eta \oplus \epsilon^1$, where Σ^{m+k+1} is a framed P-local homotopy sphere which bounds a framed P-local homotopy disc.

We postpone the proof to discuss a corollary. Hirsch [9] has conjectured that every framed manifold M^m embeds in S^{m+k} with $m \leq 2k$. The statement that they embed with trivial normal bundle is known to be false (e.g. [10]) and is usually labeled the False-Hirsch conjecture (even though it was never conjectured by Hirsch.) We can prove

Theorem 9.2: Every framed manifold of dimension m embeds in a framed $\frac{1}{2}$-homotopy sphere Σ^{m+k} with trivial normal bundle if $m \leq 2k-1$. We may choose Σ to bound a framed $\frac{1}{2}$-disc.

Proof of 9.2: Clearly ν_M desuspends to ϵ^{k-1} for any k we wish, so the problem is to desuspend the normal invariant when $P' = \{2\}$.

The space $T(\epsilon^{k-1})$ is $S^{k-1} \vee X$ for some space X. Theorem 0.2 of [25] shows that, if $m \leq 2(k-1)+1$, the image of c_M in $\pi^S_{m+k-1}(X_{(P)})$ desuspends to $\pi_{m+k-1}(X_{(P)})$. The image of c_M in $\pi^S_{m+k-1}(S^{k-1}_{(P)})$ desuspends to $\pi_{m+k-1}(S^{k-1}_{(P)})$ by classical EHP sequence arguments.

Hence 9.1 can be applied to yield the result. //

To prove 9.1 we will need

<u>Theorem 9.3</u>: Assume (X,A) is a CW pair with X and A simply connected. Let $\beta \in \pi^S_r(X,A)$ and $\alpha \in \pi_r(X_{(P)},A_{(P)})$ be elements whose images agree in $\pi^S_r(X_{(P)},A_{(P)})$. Then, if $r \geq 6$, there is a framed manifold $(W^r, \partial W)$ and maps $f: (W,\partial W) \to (X,A)$ and $g: (W,\partial W) \to (D^r_{(P)}, S^{r-1}_{(P)})$ such that

i) the framed bordism class represented by W and f is just β;

ii) g is a P-local equivalence such that $f_P \circ g_P^{-1}$ is α.

<u>Proof of 9.3</u>: Lemma 6.1 implies that the following fibre square is also a cofibre square.

$$\begin{array}{ccc} (E,E_1) & \xrightarrow{j} & (X,A) \\ \downarrow \pi & & \downarrow \\ (D^r_{(P)}, S^{r-1}_{(P)}) & \xrightarrow{\alpha_P} & (X_{(P)},A_{(P)}) \end{array}$$

Hence there is an element $\mathfrak{b} \in \pi_r^S(E,E_1)$ such that $j(\mathfrak{b}) = \beta$ and $\pi(\mathfrak{b}) = \iota_P$ where ι_P denotes the stable homotopy class of the localization map $(D^r,S^{r-1}) \to (D^r_{(P)},S^{r-1}_{(P)})$.

The element $\mathfrak{b}$ corresponds to a framed bordism class. Let it be represented by a manifold $(W_1^r,\partial W_1)$ and a map $f_1\colon (W_1,\partial W_1) \to (E,E_1)$. Since π is a P-equivalence, f_1 corresponds to a degree 1 normal map. The 2-ad version of Theorem 4.2 gives a local π-π theorem, so we can do the surgery (provided $r \geq 6$) to get a P-equivalence $f\colon (W,\partial W) \to (E,E_1)$ still representing $\mathfrak{b}$. //

Proof of 9.1: If $m+k+1 < 6$, it is easy to prove 9.1 case by case, so assume that $m+k+1 \geq 6$. The proof which follows is essentially Browder's [5] (also see [24] Thm. 11.3).

Let $c \in \pi_{m+k}(T(\eta^k)_{(P)})$ be the element promised in 9.1 ii). Let $\alpha \in \pi_{m+k+1}(\mathrm{Cone}(T(\eta^k)_{(P)}),T(\eta^k)_{(P)})$ be the unique element whose boundary is c. Define $\beta \in \pi_{m+k+1}(\mathrm{Cone}(T(\nu_M)),T(\nu_M))$ to be the element whose boundary is c_M. Theorem 9.3 applies so we can find $(W,\partial W)$ and a map $g\colon \partial W \to T(\eta^k)$.

$$\begin{array}{ccccc} S(\eta^k \oplus \epsilon^1) & \to & T(\eta^k) & \to & T(\eta^k)\cup_g W \\ \downarrow & & & & \downarrow \\ D(\eta^k) & & \longrightarrow & & A \end{array}$$

is defined to be a pushout. Browder's proof shows that A is a P-local S^{m+k+1} and that there is an element $\gamma \in \pi^S_{m+k+1}(A)$ going to c_M under the map $A \to T(\eta^k \oplus \epsilon^1)$. Let $\beta \in \pi^S_{m+k+2}(\mathrm{Cone}\ A,A)$

be the element whose boundary is γ. Let $\alpha \in \pi^S_{m+k+2}(\text{Cone } A_{(P)}, A_{(P)})$ be the stabilization of $(D^{m+k+2}_{(P)}, S^{m+k+1}_{(P)}) \rightarrow (\text{Cone } A_{(P)}, A_{(P)})$.

Use 9.3 again to get Σ_1^{m+k+1} and a P-equivalence g: $\Sigma_1 \rightarrow A$ so that g represents γ. Following Browder, make the composite

$$\Sigma_1 \rightarrow A \rightarrow T(\eta \oplus \epsilon^1)$$

transverse to the zero section. The result is normally bordant to M, so we finish just as Browder does using the local π-π theorem in place of the integral π-π theorem. //

§10: References.

[1] G. A. Anderson, Surgery with coefficients, Lecture Notes in Math., Vol. 591, Springer - Verlag, 1977.

[2] G. A. Anderson, Computation of the surgery obstruction groups $L_{4k}(1;Z_p)$, Pacific J. Math. 74 (1978), 1 - 4.

[3] J. Barge, Structures différentiables sur les types d'homotopie rationelle simplement connexes, Ann. Sci. École Norm. Sup. 9 (1976), 469 - 501.

[4] A. K. Bousfield and D. M. Kan, Homotopy limits, completions and localizations, Lecture Notes in Math., Vol. 304, Springer - Verlag, 1972.

[5] W. Browder, Embedding 1-connected manifolds, Bull. Amer. Math. Soc. 72 (1966), 225 - 231 and 736.

[6] S. Cappell and J. Shaneson, The codimension two placement problem and homology equivalent manifolds, Ann. of Math. 99 (1974), 277 - 348.

[7] J. P. Dax, Étude homotopique des espàces de plongements, Ann. Sci. École Norm. Sup. 5 (1972), 303 - 377.

[8] L. Fuks, Infinite Abelian Groups II, Academic Press, 1973.

[9] M. W. Hirsch, Problems in differential and algebraic topology, Seattle Conference, 1963 (ed. R. Lashof), Ann. of Math. 81(1965), 563 - 591.

[10] W. C. Hsiang, J. Levine, and R. Szczarba, On the normal bundle of a homotopy sphere embedded in Euclidean space, Topology 3(1965), 173 - 181.

[11] P. J. Kahn, Mixing homotopy types of manifolds, Topology 14(1975), 203 - 216.

[12] L. L. Laramore, Obstructions to embedding and isotopy in the metastable range, Rocky Mountain J. Math. 3(1973), 355 - 375.

[13] J. P. May, Fibrewise localization and completion, to appear.

[14] A. S. Miscenko, Homotopy invariants of non-simply connected manifolds, I. Rational invariants, Izv. Akad. Nauk Ser. Mat. Tom 34(1970), 501 - 514. English trans. Math. USSR - Izv 4(1970), 506 - 519.

[15] J. Morgan and D. Sullivan, The transversality characteristic class and linking cycles in surgery theory, Ann. of Math. 99(1974), 463 - 544.

[16] F. Quinn, Semifree group actions and surgery on PL homology manifolds, in Geometric Topology, Lecture Notes in Math. Vol. 438, Springer - Verlag, 1975, 395 - 414.

[17] A. Ranicki, The algebraic theory of surgery, preprint, Princeton University, 1978.

[18] R. Rigdon, P-equivalences and embeddings of manifolds, Proc. London Math. Soc. 11(1975), 233 - 244.

[19] R. Rigdon and B. Williams, Embeddings and immersions of manifolds, in Geometric Applications of Homotopy Theory I, Lecture Notes in Math., Vol. 657, Springer-Verlag, 1978, 423 - 454.

[20] H. Salomonsen, On the existence and classification of differentiable embeddings in the metastable range, Aarhus notes.

[21] M. Spivak, Spaces satisfying Poincaré duality, Topology 6 (1967), 77 - 102.

[22] D. Sullivan, Geometric topology, Part I. Localization, Periodicity, and Galois symmetry, Lecture notes, M. I. T. 1970.

[23] L. Taylor and B. Williams, Surgery spaces: formulae and structure, in Proceedings of the Waterloo Conference, Lecture Notes in Math., Springer - Verlag, to appear.

[24] C. T. C. Wall, Surgery on compact manifolds, Academic Press 1970.

[25] B. Williams, Hopf invariants, localization, and embeddings of Poincaré complexes, Pacific J. Math., to appear.

University of Notre Dame
Notre Dame, In. 46556
U. S. A.

Vol. 609: General Topology and Its Relations to Modern Analysis and Algebra IV. Proceedings 1976. Edited by J. Novák. XVIII, 225 pages. 1977.

Vol. 610: G. Jensen, Higher Order Contact of Submanifolds of Homogeneous Spaces. XII, 154 pages. 1977.

Vol. 611: M. Makkai and G. E. Reyes, First Order Categorical Logic. VIII, 301 pages. 1977.

Vol. 612: E. M. Kleinberg, Infinitary Combinatorics and the Axiom of Determinateness. VIII, 150 pages. 1977.

Vol. 613: E. Behrends et al., L^P-Structure in Real Banach Spaces. X, 108 pages. 1977.

Vol. 614: H. Yanagihara, Theory of Hopf Algebras Attached to Group Schemes. VIII, 308 pages. 1977.

Vol. 615: Turbulence Seminar, Proceedings 1976/77. Edited by P. Bernard and T. Ratiu. VI, 155 pages. 1977.

Vol. 616: Abelian Group Theory, 2nd New Mexico State University Conference, 1976. Proceedings. Edited by D. Arnold, R. Hunter and E. Walker. X, 423 pages. 1977.

Vol. 617: K. J. Devlin, The Axiom of Constructibility: A Guide for the Mathematician. VIII, 96 pages. 1977.

Vol. 618: I. I. Hirschman, Jr. and D. E. Hughes, Extreme Eigen Values of Toeplitz Operators. VI, 145 pages. 1977.

Vol. 619: Set Theory and Hierarchy Theory V, Bierutowice 1976. Edited by A. Lachlan, M. Srebrny, and A. Zarach. VIII, 358 pages. 1977.

Vol. 620: H. Popp, Moduli Theory and Classification Theory of Algebraic Varieties. VIII, 189 pages. 1977.

Vol. 621: Kauffman et al., The Deficiency Index Problem. VI, 112 pages. 1977.

Vol. 622: Combinatorial Mathematics V, Melbourne 1976. Proceedings. Edited by C. Little. VIII, 213 pages. 1977.

Vol. 623: I. Erdelyi and R. Lange, Spectral Decompositions on Banach Spaces. VIII, 122 pages. 1977.

Vol. 624: Y. Guivarc'h et al., Marches Aléatoires sur les Groupes de Lie. VIII, 292 pages. 1977.

Vol. 625: J. P. Alexander et al., Odd Order Group Actions and Witt Classification of Innerproducts. IV, 202 pages. 1977.

Vol. 626: Number Theory Day, New York 1976. Proceedings. Edited by M. B. Nathanson. VI, 241 pages. 1977.

Vol. 627: Modular Functions of One Variable VI, Bonn 1976. Proceedings. Edited by J.-P. Serre and D. B. Zagier. VI, 339 pages. 1977.

Vol. 628: H. J. Baues, Obstruction Theory on the Homotopy Classification of Maps. XII, 387 pages. 1977.

Vol. 629: W. A. Coppel, Dichotomies in Stability Theory. VI, 98 pages. 1978.

Vol. 630: Numerical Analysis, Proceedings, Biennial Conference, Dundee 1977. Edited by G. A. Watson. XII, 199 pages. 1978.

Vol. 631: Numerical Treatment of Differential Equations. Proceedings 1976. Edited by R. Bulirsch, R. D. Grigorieff, and J. Schröder. X, 219 pages. 1978.

Vol. 632: J.-F. Boutot, Schéma de Picard Local. X, 165 pages. 1978.

Vol. 633: N. R. Coleff and M. E. Herrera, Les Courants Résiduels Associés à une Forme Méromorphe. X, 211 pages. 1978.

Vol. 634: H. Kurke et al., Die Approximationseigenschaft lokaler Ringe. IV, 204 Seiten. 1978.

Vol. 635: T. Y. Lam, Serre's Conjecture. XVI, 227 pages. 1978.

Vol. 636: Journées de Statistique des Processus Stochastiques, Grenoble 1977, Proceedings. Edité par Didier Dacunha-Castelle et Bernard Van Cutsem. VII, 202 pages. 1978.

Vol. 637: W. B. Jurkat, Meromorphe Differentialgleichungen. VII, 194 Seiten. 1978.

Vol. 638: P. Shanahan, The Atiyah-Singer Index Theorem, An Introduction. V, 224 pages. 1978.

Vol. 639: N. Adasch et al., Topological Vector Spaces. V, 125 pages. 1978.

Vol. 640: J. L. Dupont, Curvature and Characteristic Classes. X, 175 pages. 1978.

Vol. 641: Séminaire d'Algèbre Paul Dubreil, Proceedings Paris 1976–1977. Edité par M. P. Malliavin. IV, 367 pages. 1978.

Vol. 642: Theory and Applications of Graphs, Proceedings, Michigan 1976. Edited by Y. Alavi and D. R. Lick. XIV, 635 pages. 1978.

Vol. 643: M. Davis, Multiaxial Actions on Manifolds. VI, 141 pages. 1978.

Vol. 644: Vector Space Measures and Applications I, Proceedings 1977. Edited by R. M. Aron and S. Dineen. VIII, 451 pages. 1978.

Vol. 645: Vector Space Measures and Applications II, Proceedings 1977. Edited by R. M. Aron and S. Dineen. VIII, 218 pages. 1978.

Vol. 646: O. Tammi, Extremum Problems for Bounded Univalent Functions. VIII, 313 pages. 1978.

Vol. 647: L. J. Ratliff, Jr., Chain Conjectures in Ring Theory. VIII, 133 pages. 1978.

Vol. 648: Nonlinear Partial Differential Equations and Applications, Proceedings, Indiana 1976–1977. Edited by J. M. Chadam. VI, 206 pages. 1978.

Vol. 649: Séminaire de Probabilités XII, Proceedings, Strasbourg, 1976–1977. Edité par C. Dellacherie, P. A. Meyer et M. Weil. VIII, 805 pages. 1978.

Vol. 650: C*-Algebras and Applications to Physics. Proceedings 1977. Edited by H. Araki and R. V. Kadison. V, 192 pages. 1978.

Vol. 651: P. W. Michor, Functors and Categories of Banach Spaces. VI, 99 pages. 1978.

Vol. 652: Differential Topology, Foliations and Gelfand-Fuks-Cohomology, Proceedings 1976. Edited by P. A. Schweitzer. XIV, 252 pages. 1978.

Vol. 653: Locally Interacting Systems and Their Application in Biology. Proceedings, 1976. Edited by R. L. Dobrushin, V. I. Kryukov and A. L. Toom. XI, 202 pages. 1978.

Vol. 654: J. P. Buhler, Icosahedral Golois Representations. III, 143 pages. 1978.

Vol. 655: R. Baeza, Quadratic Forms Over Semilocal Rings. VI, 199 pages. 1978.

Vol. 656: Probability Theory on Vector Spaces. Proceedings, 1977. Edited by A. Weron. VIII, 274 pages. 1978.

Vol. 657: Geometric Applications of Homotopy Theory I, Proceedings 1977. Edited by M. G. Barratt and M. E. Mahowald. VIII, 459 pages. 1978.

Vol. 658: Geometric Applications of Homotopy Theory II, Proceedings 1977. Edited by M. G. Barratt and M. E. Mahowald. VIII, 487 pages. 1978.

Vol. 659: Bruckner, Differentiation of Real Functions. X, 247 pages. 1978.

Vol. 660: Equations aux Dérivée Partielles. Proceedings, 1977. Edité par Pham The Lai. VI, 216 pages. 1978.

Vol. 661: P. T. Johnstone, R. Paré, R. D. Rosebrugh, D. Schumacher, R. J. Wood, and G. C. Wraith, Indexed Categories and Their Applications. VII, 260 pages. 1978.

Vol. 662: Akin, The Metric Theory of Banach Manifolds. XIX, 306 pages. 1978.

Vol. 663: J. F. Berglund, H. D. Junghenn, P. Milnes, Compact Right Topological Semigroups and Generalizations of Almost Periodicity. X, 243 pages. 1978.

Vol. 664: Algebraic and Geometric Topology, Proceedings, 1977. Edited by K. C. Millett. XI, 240 pages. 1978.

Vol. 665: Journées d'Analyse Non Linéaire. Proceedings, 1977. Edité par P. Bénilan et J. Robert. VIII, 256 pages. 1978.

Vol. 666: B. Beauzamy, Espaces d'Interpolation Réels: Topologie et Géometrie. X, 104 pages. 1978.

Vol. 667: J. Gilewicz, Approximants de Padé. XIV, 511 pages. 1978.

Vol. 668: The Structure of Attractors in Dynamical Systems. Proceedings, 1977. Edited by J. C. Martin, N. G. Markley and W. Perrizo. VI, 264 pages. 1978.

Vol. 669: Higher Set Theory. Proceedings, 1977. Edited by G. H. Müller and D. S. Scott. XII, 476 pages. 1978.

Vol. 670: Fonctions de Plusieurs Variables Complexes III, Proceedings, 1977. Edité par F. Norguet. XII, 394 pages. 1978.

Vol. 671: R. T. Smythe and J. C. Wierman, First-Passage Perculation on the Square Lattice. VIII, 196 pages. 1978.

Vol. 672: R. L. Taylor, Stochastic Convergence of Weighted Sums of Random Elements in Linear Spaces. VII, 216 pages. 1978.

Vol. 673: Algebraic Topology, Proceedings 1977. Edited by P. Hoffman, R. Piccinini and D. Sjerve. VI, 278 pages. 1978.

Vol. 674: Z. Fiedorowicz and S. Priddy, Homology of Classical Groups Over Finite Fields and Their Associated Infinite Loop Spaces. VI, 434 pages. 1978.

Vol. 675: J. Galambos and S. Kotz, Characterizations of Probability Distributions. VIII, 169 pages. 1978.

Vol. 676: Differential Geometrical Methods in Mathematical Physics II, Proceedings, 1977. Edited by K. Bleuler, H. R. Petry and A. Reetz. VI, 626 pages. 1978.

Vol. 677: Séminaire Bourbaki, vol. 1976/77, Exposés 489–506. IV, 264 pages. 1978.

Vol. 678: D. Dacunha-Castelle, H. Heyer et B. Roynette. Ecole d'Eté de Probabilités de Saint-Flour. VII-1977. Edité par P. L. Hennequin. IX, 379 pages. 1978.

Vol. 679: Numerical Treatment of Differential Equations in Applications, Proceedings, 1977. Edited by R. Ansorge and W. Törnig. IX, 163 pages. 1978.

Vol. 680: Mathematical Control Theory, Proceedings, 1977. Edited by W. A. Coppel. IX, 257 pages. 1978.

Vol. 681: Séminaire de Théorie du Potentiel Paris, No. 3, Directeurs: M. Brelot, G. Choquet et J. Deny. Rédacteurs: F. Hirsch et G. Mokobodzki. VII, 294 pages. 1978.

Vol. 682: G. D. James, The Representation Theory of the Symmetric Groups. V, 156 pages. 1978.

Vol. 683: Variétés Analytiques Compactes, Proceedings, 1977. Edité par Y. Hervier et A. Hirschowitz. V, 248 pages. 1978.

Vol. 684: E. E. Rosinger, Distributions and Nonlinear Partial Differential Equations. XI, 146 pages. 1978.

Vol. 685: Knot Theory, Proceedings, 1977. Edited by J. C. Hausmann. VII, 311 pages. 1978.

Vol. 686: Combinatorial Mathematics, Proceedings, 1977. Edited by D. A. Holton and J. Seberry. IX, 353 pages. 1978.

Vol. 687: Algebraic Geometry, Proceedings, 1977. Edited by L. D. Olson. V, 244 pages. 1978.

Vol. 688: J. Dydak and J. Segal, Shape Theory. VI, 150 pages. 1978.

Vol. 689: Cabal Seminar 76–77, Proceedings, 1976–77. Edited by A.S. Kechris and Y. N. Moschovakis. V, 282 pages. 1978.

Vol. 690: W. J. J. Rey, Robust Statistical Methods. VI, 128 pages. 1978.

Vol. 691: G. Viennot, Algèbres de Lie Libres et Monoïdes Libres. III, 124 pages. 1978.

Vol. 692: T. Husain and S. M. Khaleelulla, Barrelledness in Topological and Ordered Vector Spaces. IX, 258 pages. 1978.

Vol. 693: Hilbert Space Operators, Proceedings, 1977. Edited by J. M. Bachar Jr. and D. W. Hadwin. VIII, 184 pages. 1978.

Vol. 694: Séminaire Pierre Lelong – Henri Skoda (Analyse) Année 1976/77. VII, 334 pages. 1978.

Vol. 695: Measure Theory Applications to Stochastic Analysis, Proceedings, 1977. Edited by G. Kallianpur and D. Kölzow. XII, 261 pages. 1978.

Vol. 696: P. J. Feinsilver, Special Functions, Probability Semigroups, and Hamiltonian Flows. VI, 112 pages. 1978.

Vol. 697: Topics in Algebra, Proceedings, 1978. Edited by M. F. Newman. XI, 229 pages. 1978.

Vol. 698: E. Grosswald, Bessel Polynomials. XIV, 182 pages. 1978.

Vol. 699: R. E. Greene and H.-H. Wu, Function Theory on Manifolds Which Possess a Pole. III, 215 pages. 1979.

Vol. 700: Module Theory, Proceedings, 1977. Edited by C. Faith and S. Wiegand. X, 239 pages. 1979.

Vol. 701: Functional Analysis Methods in Numerical Analysis, Proceedings, 1977. Edited by M. Zuhair Nashed. VII, 333 pages. 1979.

Vol. 702: Yuri N. Bibikov, Local Theory of Nonlinear Analytic Ordinary Differential Equations. IX, 147 pages. 1979.

Vol. 703: Equadiff IV, Proceedings, 1977. Edited by J. Fábera. XIX, 441 pages. 1979.

Vol. 704: Computing Methods in Applied Sciences and Engineering, 1977, I. Proceedings, 1977. Edited by R. Glowinski and J. L. Lions. VI, 391 pages. 1979.

Vol. 705: O. Forster und K. Knorr, Konstruktion verseller Familien kompakter komplexer Räume. VII, 141 Seiten. 1979.

Vol. 706: Probability Measures on Groups, Proceedings, 1978. Edited by H. Heyer. XIII, 348 pages. 1979.

Vol. 707: R. Zielke, Discontinuous Čebyšev Systems. VI, 111 pages. 1979.

Vol. 708: J. P. Jouanolou, Equations de Pfaff algébriques. V, 255 pages. 1979.

Vol. 709: Probability in Banach Spaces II. Proceedings, 1978. Edited by A. Beck. V, 205 pages. 1979.

Vol. 710: Séminaire Bourbaki vol. 1977/78, Exposés 507–524. IV, 328 pages. 1979.

Vol. 711: Asymptotic Analysis. Edited by F. Verhulst. V, 240 pages. 1979.

Vol. 712: Equations Différentielles et Systèmes de Pfaff dans le Champ Complexe. Edité par R. Gérard et J.-P. Ramis. V, 364 pages. 1979.

Vol. 713: Séminaire de Théorie du Potentiel, Paris No. 4. Edité par F. Hirsch et G. Mokobodzki. VII, 281 pages. 1979.

Vol. 714: J. Jacod, Calcul Stochastique et Problèmes de Martingales. X, 539 pages. 1979.

Vol. 715: Inder Bir S. Passi, Group Rings and Their Augmentation Ideals. VI, 137 pages. 1979.

Vol. 716: M. A. Scheunert, The Theory of Lie Superalgebras. X, 271 pages. 1979.

Vol. 717: Grosser, Bidualräume und Vervollständigungen von Banachmoduln. III, 209 pages. 1979.

Vol. 718: J. Ferrante and C. W. Rackoff, The Computational Complexity of Logical Theories. X, 243 pages. 1979.

Vol. 719: Categorial Topology, Proceedings, 1978. Edited by H. Herrlich and G. Preuß. XII, 420 pages. 1979.

Vol. 720: E. Dubinsky, The Structure of Nuclear Fréchet Spaces. V, 187 pages. 1979.

Vol. 721: Séminaire de Probabilités XIII. Proceedings, Strasbourg, 1977/78. Edité par C. Dellacherie, P. A. Meyer et M. Weil. VII, 647 pages. 1979.

Vol. 722: Topology of Low-Dimensional Manifolds. Proceedings, 1977. Edited by R. Fenn. VI, 154 pages. 1979.

Vol. 723: W. Brandal, Commutative Rings whose Finitely Generated Modules Decompose. II, 116 pages. 1979.

Vol. 724: D. Griffeath, Additive and Cancellative Interacting Particle Systems. V, 108 pages. 1979.

Vol. 725: Algèbres d'Opérateurs. Proceedings, 1978. Edité par P. de la Harpe. VII, 309 pages. 1979.

Vol. 726: Y.-C. Wong, Schwartz Spaces, Nuclear Spaces and Tensor Products. VI, 418 pages. 1979.

Vol. 727: Y. Saito, Spectral Representations for Schrödinger Operators With Long-Range Potentials. V, 149 pages. 1979.

Vol. 728: Non-Commutative Harmonic Analysis. Proceedings, 1978. Edited by J. Carmona and M. Vergne. V, 244 pages. 1979.